ROUTE SURVEYING AND DESIGN

SERIES IN CIVIL ENGINEERING

Brinker and Wolf, *Elementary Surveying, Sixth Edition*
Clark, Viessman, and Hammer, *Water Supply and Pollution Control, Third Edition*
McCormac, *Design of Reinforced Concrete*
McCormac, *Structural Analysis, Third Edition*
McCormac, *Structural Steel Design, Second Edition*
Meyer and Gibson, *Route Surveying and Design, Fifth Edition*
Mikhail, *Observations and Least Squares*
Moffitt and Mikhail, *Photogrammetry, Third Edition*
Moffitt and Bouchard, *Surveying, Sixth Edition*
Salmon and Johnson, *Steel Structures: Design and Behavior, Second Edition*
Viessman, Harbaugh, Knapp, and Lewis, *Introduction to Hydrology, Second Edition*
Wang and Salmon, *Reinforced Concrete Design, Third Edition*

ROUTE SURVEYING AND DESIGN

FIFTH EDITION

CARL F. MEYER
Professor Emeritus, Worcester Polytechnic Institute

DAVID W. GIBSON
University of Florida, Gainesville

HarperCollins*Publishers*

Sponsoring Editor: Charlie Dresser
Production Manager: Marion A. Palen
Compositor: Syntax International Pte. Ltd.

Art Studio: J & R Technical Services, Inc.

ROUTE SURVEYING AND DESIGN, Fifth Edition

Library of Congress Cataloging in Publication Data

Meyer, Carl F

Route surveying and design.

(Series in civil engineering)
Includes index.
1. Roads—Surveying. 2. Railroads—Surveying.
3. Route surveying. I. Gibson, David W., 1944–
joint author. II. Title.
TE209.M48 1980 625 79-15267
ISBN 0-700-22524-2

Contents

PART III TABLES

APPENDIXES

Preface

The fifth edition of *Route Surveying and Design* is characterized by a reframing and updating of the chapters on surveying and design procedures. Each of the last three editions contained one new chapter written to cover evolving developments in field and office methods. As a result, there was a tendency to consider the material as frozen in a time frame, being either "traditional," "recent," or "new." In truth, however, all the methods described had their ideal place of application, depending on the nature and scope of the project.

The approach of a fifth edition gave the opportunity to revise the foregoing chapters by eliminating the "time frame" chapter titles. Instead, all the material, whether formerly thought of as "old" or "new," is now treated under method or job titles. At the same time, new material has been added and older material updated.

It is a privilege to introduce Professor David W. Gibson as coauthor of the fifth edition. He is solely responsible for the writing in Chapters 10 through 13. In addition, he has lent valuable assistance in reviewing and suggesting improvements in the other chapters.

Chapters 1 through 9 contain numerous small changes intended to clarify and improve on the former treatment. Worthy of special mention is a new treatment of compound curves, which takes advantage of frequently used surveying procedures that are "hard wired" into some desk-top calculators or are available as prerecorded program cards that can be inserted into some pocket calculators.

Users of the end-of-chapter problems will be glad to find all new problems with answers given to most of them.

CARL F. MEYER
DAVID W. GIBSON

ROUTE SURVEYING AND DESIGN

East Los Angeles Interchange. (Courtesy of California Division of Highways.)

Part I
BASIC PRINCIPLES

Chapter 1
Route Location

1-1. Introduction

This chapter comprises an outline of the basic considerations affecting the general problem of route location. The material is nonmathematical, but it is necessary for a clear understanding of the purposes served by the technical matters in the remaining chapters of Part I. Specific practical applications of these basic considerations to the location of highways, railroads, and other routes of transportation and communication are given in Part II.

1-2. Route Surveying and Systems Engineering

Route surveying includes all field work and requisite calculations (together with maps, profiles, and other drawings) involved in the planning and construction of any route of transportation. If the word *transportation* be taken to refer not only to the transportation of persons but also to the movement of liquids and gases and to the transmission of power and messages, then route surveying covers a broad field. Among the important engineering structures thus included are: highways and railroads; aqueducts, canals, and flumes; pipe lines for water, sewage, oil, and gas; cableways and belt conveyors; and power, telephone, and telegraph transmission lines.

Though this definition of route surveying distinguishes the subject from other branches of surveying, it is assumed that projects involving route surveys have considerable magnitude. There will usually be definite termini a fairly long distance apart. In such a situation the surveys serve two purposes: (1) to determine the best general route between the termini; (2) to find the optimum combination of alignment, grades, and other details of the selected route. To accomplish these purposes requires not only expert survey technique but also experience in the *art of engineering*. This combination of creative planning, surveying, and design is a good example of what is called *systems engineering*.

1-3. Relation of Project to Economics

Every route-surveying project involves economic problems both large and small. *By far the most important question is whether or not to construct the project*. Essentially, this decision is based on a comparison of the cost of the enterprise with the probable financial returns or social advantages to be expected. In some cases the question can be answered after a careful preliminary study without field work; in others, extensive surveys and cost estimates must first be made.

However simple or complex the project may be, it is rarely possible for the engineer alone to answer this basic economic question. To his studies must be added those of the persons responsible for the financial and managerial policies of the organization. In the case of a public project the broad social, environmental, and political objectives also carry weight.

The engineer responsible for conducting route surveys is not solely a technician. In addition to his indispensable aid in solving the larger economic problems, he is continually confronted with smaller ones in the field and office. For example, the relatively simple matter of deciding which of several methods to be used in developing a topographic map of a strip of territory is, basically, an economic problem that involves survey, terrain, and equipment and personnel available.

1-4. Relation of Project to Design

Design problems in route location are closely related to route surveying. Some matters of design must precede the field work; others are dependent on it. For example, in order that field work for a proposed new highway may be done efficiently, the designers must have chosen—at least tentatively—not only the termini and possible intermediate connections but also such design details as the number of traffic lanes, width of right-of-way, maximum grade, minimum radius of curve, and minimum sight distance. On the other hand, considerable field work must be done before the designers can fix the exact alignment, grade elevations, shoulder widths, and culvert locations to fit the selected standards safely and with the greatest overall economy. The relationships between modern route surveying and design are described in detail in Chapter 11.

1-5. Basic Factors of Alignment and Grades

In route location it is usually found that the termini and possible intermediate controlling points are at different elevations. Moreover, the topography and existing physical features rarely permit a straight location between the points. These circumstances invariably require the introduction of vertical and horizontal changes in direction; therefore, grades, vertical curves, and horizontal curves are important features of route surveying and design.

Curvature is not inherently objectionable. Though a straight line is the shortest distance between two points, it is also the most monotonous—a consideration of some aesthetic importance in the location of scenic highways. The device of curvature gives the designer limitless opportunities to fit a location to the natural swing of the topography in such a way as to be both pleasing and economical. Excessive or poorly designed curvature, however, may introduce serious operating hazards, or may add greatly to the costs of constructing, maintaining, or operating over the route.

Steep grades are likely to have the same effects on safety and costs as excessive curvature. It should be emphasized, nevertheless, that problems of curves and grades are ordinarily interrelated. Thus, on highway and railroad location it is often the practice to increase the distance between two fixed points in order to reduce the grade. This process, known as "development," necessarily adds to the total curvature. It is not always a feasible solution, for added curvature may be more objectionable than the original steep grade.

The aim of good location should be the attainment of consistent conditions with a proper balance between curvature and grade. This is especially true in highway location, owing to the fact that each vehicle is individually operated and the driver often is unfamiliar with a particular highway. Many highway accidents occur at a place where there is a sudden and misleading variation from the condition of curvature, grade, or sight distance found on an adjacent section of the same highway. To produce a harmonious balance between curvature and grade, and to do it economically, requires the engineer to possess broad experience, mature judgment, and a thorough knowledge of the objectives of the project.

1-6. Influence of Type of Project

The type of route to be built between given termini has a decided influence on its location. As an example, the best location for a railroad would not necessarily be the most suitable one for a power-transmission line. A railroad requires a location having fairly flat grades and curves. Moreover, there are usually intermediate controlling points such as major stream and highway crossings, mountain passes, and revenue-producing markets. In contrast, power is transmitted as readily up a vertical cable as along a horizontal one. Grades, therefore, have no significance, and river and highway crossings present no unusual problems. Where changes in direction are needed, they are made at angle towers. Consequently, the alignment is as straight as possible from generating station to substation.

1-7. Influence of Terrain

Character of the terrain between termini or major controlling points is apt to impress a characteristic pattern upon a route location, particularly in the case of a highway or a railroad. Terrain may be generally classified as *level*, *rolling*, or *mountainous*.

In comparatively level regions the line may be straight for long distances, minor deviations being introduced merely to skirt watercourses, avoid poor foundations, or possibly to reduce land damages. On an important project, however, the artificial control imposed by following section lines or other boundaries should not be permitted to govern.

In rolling country the location pattern depends on orientation of the ridges and valleys with respect to the general direction of the route. Parallel orientation may result in a *valley line* having flat grades, much curvature, frequent culverts and bridges, and fill in excess of cut; or it may permit a *ridge line* having simpler alignment and drainage problems. To connect two such situations, and also when ridges are oblique to the general direction of the route, there may be a *side-hill line*. This has the characteristics of uniformly rising grades, curvature fitted to the hillsides, and relatively light, balanced grading.

Where ridges and valleys are approximately at right angles to the general direction of the route, the typical pattern that results may be called a *cross-drainage line*. There the location of passes through the ridges and of crossings over major streams constitute important controlling points between which the line may be the side-hill type. Generally, a cross-drainage line involves steep grades, heavy grading with alternate cuts and fills, expensive bridges, and curvature considerably less than that on a valley line.

Mountainous terrain imposes the severest burden upon the ingenuity of a locating engineer. No simple pattern or set of rules fits all situations. Short sections of each type of lines previously described must be inserted as conditions require. "Development," even to the extent of switchbacks and loops, may be the only alternative to expensive tunnel construction.

1-8. The Basic Route Survey and Design System

Figure 1-1 indicates the basic route survey and design system that has been successfully used in this country for most projects. Although these operations vary with different organizations, and particularly with the nature and scope of the project, a typical outline of the field and office work is represented.

1-9. Importance of the Reconnaissance

Second in importance to the primary question—whether or not to build the project—is selection of the *general route* between the termini. This is usually determined by *reconnaissance.*

The statement by Wellington,* "*The reconnaissance must not be of a line, but of an area,*" is a most apt one. Extent of the area depends, of course, on the type of project and nature of the terrain, but the area must be broad enough to cover all practicable routes joining the termini. Of particular importance is the need to guard against the natural tendency to favor an obviously feasible location. It is possible that country which is covered with tangled undergrowth, or otherwise rough for foot travel on reconnaissance, may hide a much better location than available in more settled or open territory.

With regard to the importance of the "art of reconnaissance" and attitude of an engineer toward it, nowhere will more effective comments be found than in Wellington's classic treatise. Though written by that author in 1887 for the instruction of engineers on railroad location, the following statements are timeless in their application to all types of route location:

> . . . there is nothing against which a locating engineer will find it necessary to be more constantly on his guard than the drawing of hasty and unfounded conclusions, especially of an unfavorable character, from apparent evidence wrongly interpreted. If his conclusions on reconnaissance are unduly favorable, there is no great harm done—nothing more at the worst will ensue than an unnecessary amount of surveying; but a hasty conclusion that some line is not feasible, or that further improvements in it cannot be made, or even sometimes—often very absurdly—that no other line of any kind exists than the one which has chanced to be discovered—these are errors which may have disastrous consequences.
>
> On this account, if for no other, the locating engineer should cultivate . . . what may be called an optimistic habit of mind. He should not allow himself to enter upon his work with the feeling that any country is seriously difficult, but rather

* Reprinted by permission from *Economic Theory of the Location of Railways* by A. M. Wellington, published by John Wiley & Sons, Inc., 1915.

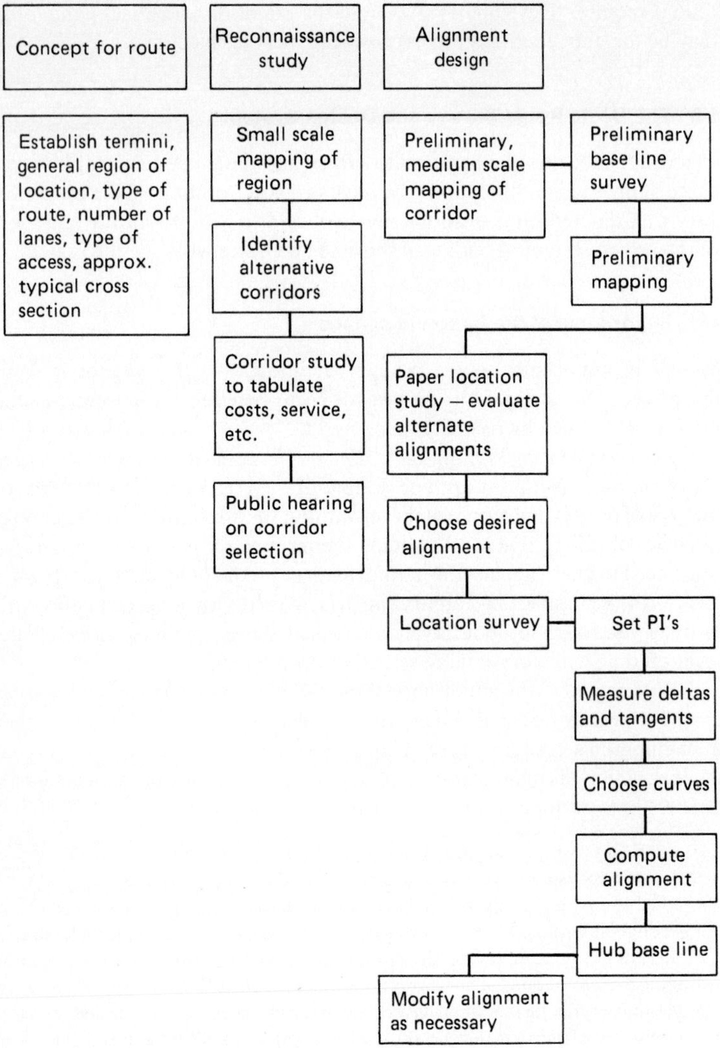

Figure 1-1 The basic route survey and design system.

that the problem before him is simply to find the line, which undoubtedly exists, and that he can only fail to do so from some blindness or oversight of his own, which it will be his business to guard against.

For the reason that there is so much danger of radical error *in the selection of the lines to be surveyed* (or, rather, of the lines not to be examined), it results that THE WORST ERRORS OF LOCATION GENERALLY ORIGINATE

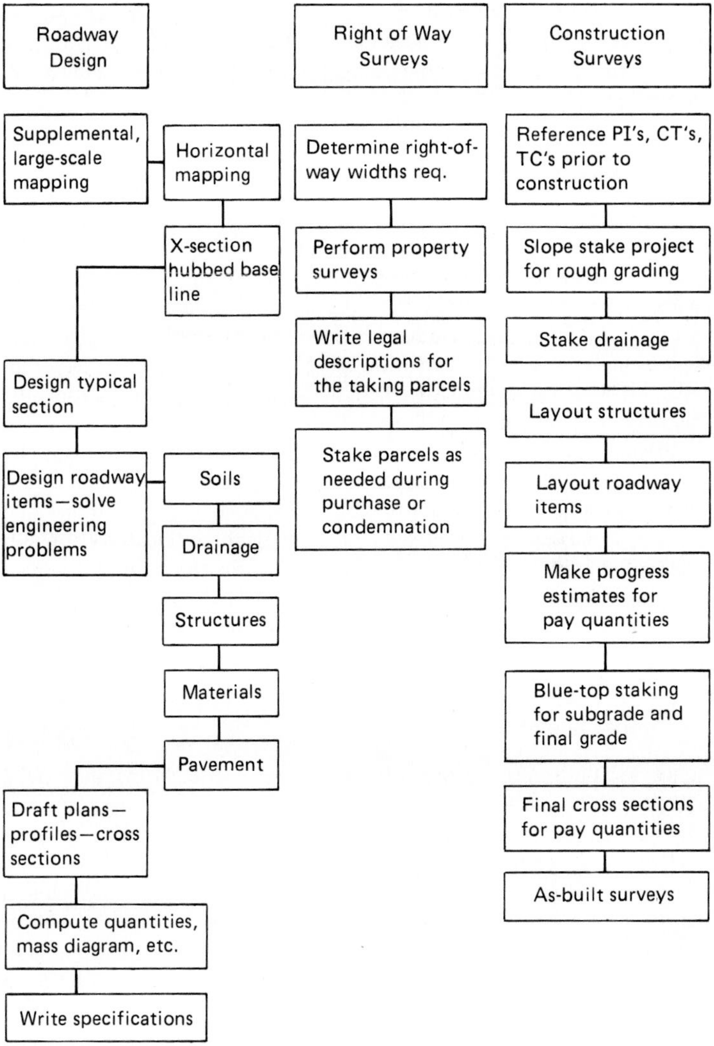

Figure 1-1 (*continued*)

IN THE RECONNAISSANCE. This truth once grasped, the greatest of all dangers, over-confidence in one's own infallibility, is removed.

If, as often happens, the reconnaissance is entrusted to one engineer, he should have mature experience in the promotional, financial, and engineering aspects of similar projects. It is not enough that he be an experienced locating engineer, for such a man is likely to concentrate upon the purely physical possibilities of a route. Furthermore, he should be able to sense the significance of present trends and their probable effect upon the future utility of the project and realize when to seek a specialist's advice in such matters.

1-10. Purposes of Preliminary Surveys

A preliminary survey follows the general route recommended in the reconnaissance report. The most important purpose of such a survey is to obtain data for plotting an accurate map of a strip of territory along one or more promising routes. This map serves as the basis for projecting the final alignment and profile, at least tentatively. Enough data are also obtained from which to make an estimate of earthwork quantities, drainage-structure and needed right-of-way. Taken together, these data permit compilation of a fairly close cost estimate.

Preliminary surveys differ greatly in method and precision. Invariably, however, there is at least one traverse (compass, stadia, or transit-and-tape) which serves as a framework for the topographical details. Elevations along the traverse line and tie measurements to existing physical features are essential. Accurate contours may or may not be needed, the requirement depending on the type of project.

Detailed methods of running preliminary surveys including photogrammetric methods will be found in Chapter 11.

1-11. Proper Use of Topography

On new locations of routes over which grades are particularly important, an accurate contour map is indispensable. Relocation of an existing route, such as a highway, may sometimes be made by revising the preliminary survey directly on the ground. The method, termed "field location" or "direct location," is not recommended for a new line. It is true that some engineers seem to have uncanny ability for locating a satisfactory line—though not necessarily the best one—by direct field methods. Such a natural gift is not to be belittled, but it should be subordinated in difficult terrain to careful office studies aided by a contour map.

The primary purpose of a contour map is to serve as a basis for making a "paper location" of the final center line. On such a map the locating engineer is able to scan a large area at once. By graphical methods he can study various locations in a small fraction of the time required for a field party to survey lines on the ground. Furthermore, he is not subject to the natural optical illusions that often mislead even the most experienced engineer in the field. An added advantage of the contour map, provided it is extensive enough, is to supply visible evidence that no better line has been overlooked.

It is possible, however, to put too much reliance upon map topography. Particularly to be avoided is temptation to control the work from the office by making such a meticulous paper location, even to the extent of complete notes for staking all curves, that the field work of final location becomes a mere routine of carrying out "instructions from headquarters." A contour map, no matter how accurate it is, cannot impress upon the mind as forcibly as field examination such details as the true significance of length and depth of cuts and fills; nature of the materials and foundations; susceptibility to slides, snow drifting, and other maintenance difficulties; or the aesthetic values of a projected location. At best, the map facilitates making what might be termed a "semifinal location," which is to be further revised in minor details during the location survey.

1-12. Function of Location Survey

The purpose of the location survey is to transfer a paper location, complete with curves, to the ground. It is too much to expect that this ground line will conform to the paper location in every detail. Almost certainly there will be minor deviations, resulting usually from errors in the preliminary traverse or in the taking or plotting of the topography. An exact agreement does not assure excellence of the location; it merely proves the geometric accuracy of the field and office work. Consequently, regardless of the "fit" with the paper location, the engineer should be constantly on watch for opportunities to make those minor adjustments in alignment or grades that only close observation of the field conditions will reveal.

When staked, the final location is usually cross sectioned for closer determination of earthwork quantities. In addition, tie measurements to property lines are made to serve for preparing right-of-way descriptions, and all necessary field data are obtained to permit detailed design of miscellaneous structures.

1-13. Relation of Surveying to Engineering

Before we leave these basic considerations to study the technical aspects of route surveying, it should be pointed out that surveying and mapping, as ordinarily practiced, are not engineering; they are merely methods of obtaining and portraying data needed as a prelude to the design and construction of engineering works.

During study of the chapters that follow, it will be natural for the students to concentrate on geometrical and instrumental techniques. However, the course in Route Surveying will not reach its potential value unless it is more than drill in field and office practice. Students stimulated by the instructor's examples and illustrations, should attempt to look beyond technical details and gain some insight into the factors that lead to the conception of a particular project. Knowledge of those factors will give them a better appreciation of the engineering surveys—their planning, the controlling specifications, and the usefulness of the data to the designers.

To be of the greatest usefulness, without being unduly costly, surveys, maps, and computations should be only as complete and accurate as needed for the ultimate job. For some purposes utmost accuracy is required. For others, extreme niceties are too costly and time-consuming; they may be replaced by approximate methods and short cuts. As an example, much time is often wasted in "exact" calculation of yardage estimates prior to construction, only to find that shrinkage, compaction, overbreak, or stripping allowances change the estimated quantities by large amounts. This is not to imply that grading calculations may always be done by approximate methods. Accuracy is invariably required, for example, in determination of yardage for payment to contractors. One trait of a good engineer is his judgment of the degree of precision required in obtaining data and computing values for use.

Surveying and mapping are essential prerequisites to engineering design for mass transportation. In designing a large transportation project, extensive surveying operations are involved in the early reconnaissance, detailed preliminary and location surveys, and all the work leading to the preparation of topographic maps, profiles, cross sections, and other working drawings. If to these there are added the construction layout and "as-built" record surveys, it is apparent that a large portion of the total engineering costs is absorbed by surveying and mapping.

In contrast to the leisurely pace of highway construction in the early part of this century, wherein the ordinary piecemeal survey served the purpose, we now have vast and costly projects. Noteworthy among these are the heavily-traveled California Freeways and the toll highways in Massachusetts, Connecticut, New York, New Jersey, Pennsylvania, Ohio, Indiana, Florida, and several other states. Even these are dwarfed by the Interstate Highway System. Already conceived is the proposed Mississippi River Parkway, for which map reconnaissance of 50,000 square miles has yielded over 8000 miles of alternate routes. Survey and mapping methods must be selected to keep pace with the advanced design and construction techniques used on such vast projects.

Applications of photogrammetry and new automated devices are given in Chapters 10, 11, and 12. These represent advances in route surveying and design that save time and reduce interest charges during construction of a major project.

In fitting surveying and mapping into the plans for a transportation project, saving in time by use of shortcut methods should not be achieved at the expense of reduction in ultimate required accuracy. Time saved by short cuts in control surveys, for example, may be lost many times over in transferring the paper location to the ground or in monumenting the right-of-way.

During construction, one field change caused by poor original surveys may delay the work longer than the time saved earlier by shortcut survey methods.

By observing the ultimate accuracies needed in the various phases of a project and adopting well-planned methods of surveying and mapping, even at the expense of some extra time, the whole project will have a firmer base around which design and construction operations can be planned. This kind of surveying and mapping is necessary for route location in systems engineering.

Chapter 2
Simple Curves

2-1. Foreword

Every route has a calculable geometric alignment. The form may be a simple series of straight lines and angles, as on a power transmission line, or it may be an intricate combination of straight lines and curves.

The survey location of every staked point along such a route is identified by its *stationing*, that is its total horizontal distance from the point of beginning. Stationing is expressed as the number of 100-ft horizontal distances, or *stations*, plus any remaining horizontal distance in feet. Thus, a stake located 1452.65 ft from the point of beginning is designated as station (or sta.) 14+52.65.

The mathematical computations required for staking horizontal curves are not difficult. All calculations in this chapter, as well as those on *compound curves* (Chapter 3) and most of the *special curve problems* (Chapter 7), are done with the aid of elementary algebra, geometry, and trigonometry. The trigonometry used is almost entirely restricted to the relations between sides and angles of right triangles.

The primary purpose of any curve is to provide the required change in direction in the form best suited to the operating characteristics. Secondary considerations are reasonable economy in construction cost and ease of staking the curve in the field.

Horizontal curves are usually arcs of circles or of spirals. Generally, a circular arc makes up the greater portion of a curve. The arcs of varying radii, or spirals, provide a gradual transition between the circular arc and tangents.

Vertical curves joining straight sections of grade line are invariably parabolic arcs. When vertical and horizontal curves overlap, their study is simplified by considering them separately.

2-2. Definitions and Notation

A *simple curve* consists of a circular arc tangent to two straight sections of a route. Though spiral transitions are commonly used at the ends of circular arcs on modern highways and railroads, a thorough knowledge of the simple curve—its basic geometry, calculation, and method of staking—is necessary to understand more complex curve problems.

There is no universally accepted notation. That shown in Fig. 2-1, however, is commonly used in recent practice. (In this diagram, parts of the curve layout commonly surveyed in the field are drawn as solid lines, whereas

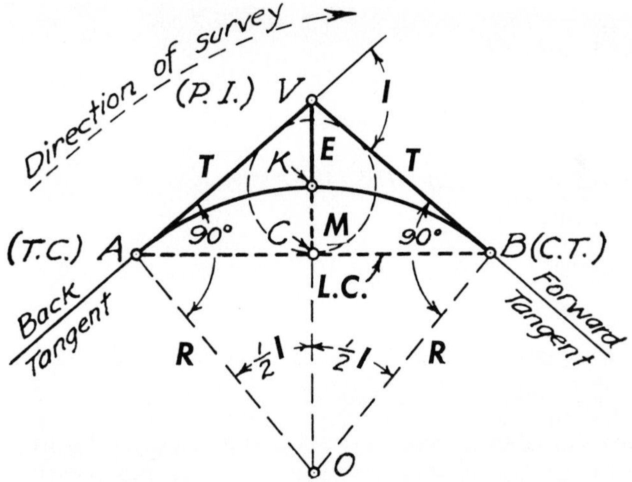

Figure 2-1 Simple-curve layout.

geometric construction lines and curve parts infrequently surveyed are shown dotted.) The intersection of tangents at V is called the *vertex*, or *point of intersection*, abbreviated P.I. The deflection angle between tangents is denoted by I; it is equal to the *central angle* of the curve. For a survey progressing in the direction indicated, the tangent up to the P.I. is called the *initial tangent*, or *back tangent;* that beyond the P.I., the *forward tangent*. The beginning of the circular arc at A is known as the T.C. (tangent to curve); the end at B, the C.T. (curve to tangent). In a simple curve the T.C. and the C.T. are equidistant from the P.I. The T.C. is sometimes designated as the P.C. (point of curve) or B.C. (beginning of curve). Corresponding terms for the C.T. are P.T. (point of tangent) and E.C. (end of curve).

Certain lines on the curve layout are very useful in calculations and field work. Those shown in Fig. 2-1 are: the distance from the P.I. to the T.C. (or C.T.), known as the *tangent distance*, T; the distance from the P.I. to the midpoint K *of* the *curve*, called the *external distance*, E; and the *radius* of the circular arc, designated by R. Also shown, though of lesser importance, are the *long chord*, L.C., which is the distance between the T.C. and the C.T.; and the *middle ordinate*, M, the distance from the midpoint C of the long chord to the midpoint K of the curve.

2-3. General Formulas

The basic formulas may be derived with the aid of Fig. 2-1.

In the right triangle VAO, $\tan \frac{1}{2}I = T \div R$ whence

$$T = R \tan \tfrac{1}{2}I \tag{2-1}$$

In right triangle VAO, $\cos \frac{1}{2}I = R/(R + E)$ whence $R \cos \frac{1}{2}I + E \cos \frac{1}{2}I = R$. Solving for E gives

$$E = R\left(\frac{1}{\cos \frac{1}{2}I} - 1\right) \tag{2-2}$$

In right triangle ACO, $\sin \frac{1}{2}I = \frac{1}{2}\text{L.C.} \div R$ whence

$$\text{L.C.} = 2\,R \sin \tfrac{1}{2}I \tag{2-3}$$

In right triangle ACO, $\cos \frac{1}{2}I = (R - M)/R$ whence

$$M = R(1 - \cos \tfrac{1}{2}I) \tag{2-4}$$

The foregoing formulas are expressed in a form convenient for use with electronic pocket calculators that can deliver values of sin, cos, and tan directly.

Some other useful expressions may be derived by combining the basic formulas and using trigonometric conversions. However, their derivation directly from sketches will illustrate further interesting properties of the simple curve.

A portion of Fig. 2-1 is reproduced in Fig. 2-2. If arc VA' is drawn with radius OV (A' being on OA produced), the triangle $A'OV$ will be isosceles and angle A' must equal $90° - \frac{1}{4}I$. Therefore the angle at V in the right triangle $A'AV$ must equal $\frac{1}{4}I$, from which

$$E = T \tan \tfrac{1}{4}I \tag{2-5}$$

In Fig. 2-1 the circle inscribed within triangle ABV must have its radius equal to M and its center located at point K. From this fact it follows that

$$M = E \cos \tfrac{1}{2}I \tag{2-6}$$

Another useful relation is

$$E = R \tan \tfrac{1}{2}I \tan \tfrac{1}{4}I \tag{2-7}$$

If computations are to be made without an electronic calculator that has trigonometric functions, then the trigonometric functions known as the external secant (*exsec*) and versed sine (*vers*) are convenient. By definition, for any angle A, exsec $A = \sec A - 1$.

From the definition, the parenthetical term in formula 2-2 equals exsec $\frac{1}{2}I$, whence

$$E = R \operatorname{exsec} \tfrac{1}{2}I \tag{2-8}$$

Also, by definition, the parenthetical term in formula 2-4 equals vers $\frac{1}{2}I$, whence

$$M = R \operatorname{vers} \tfrac{1}{2}I \tag{2-9}$$

Numerical values of sin, cos, tan, vers, and exsec of angles are listed in Table XX, thereby facilitating machine computation using these formulas.

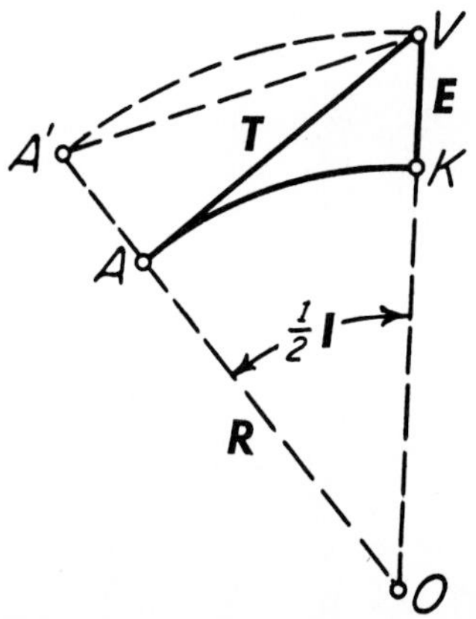

Figure 2-2

2.4. Degree of Curve, Arc Definition

The curvature of a circular arc is perfectly defined by its radius. However, where the radius is long, as on modern highway and railroad alignment, the center of the curve is inacessible or remote. In this case the radius is valueless for surveying operations, though it is still needed in certain computations; it must be replaced by a different characteristic of the curve which is directly useful in the field. The characteristic commonly used is known as the *degree of curve, D*. Though several definitions of degree of curve may be found, all are based upon the fact that a circle is a curve having a constant angular change in direction per unit of distance. The two most widely used are the chord definition and the arc definition of D.

The chord definition of D was formerly used by most railroads and a few highway departments; but the arc definition is employed almost exclusively in modern practice. Therefore the arc definition is given precedence in this book. To avoid confusion, formulas for the chord definition are developed separately in Section 2-13.

For the *arc definition*, the degree of curve is the central angle subtended by a 100-*ft arc*. It is denoted by D_a, as shown in Fig. 2-3.

By proportion, $D_a:100 = 360^\circ:2\pi R$ from which

$$D_a = \frac{18{,}000}{\pi R} \tag{2-10}$$

Values of R for given values of D_a are listed in Table I.

Some formulas involving the degree of curve are valid for both the arc and the chord definitions of D. In such cases the subscript is omitted from the

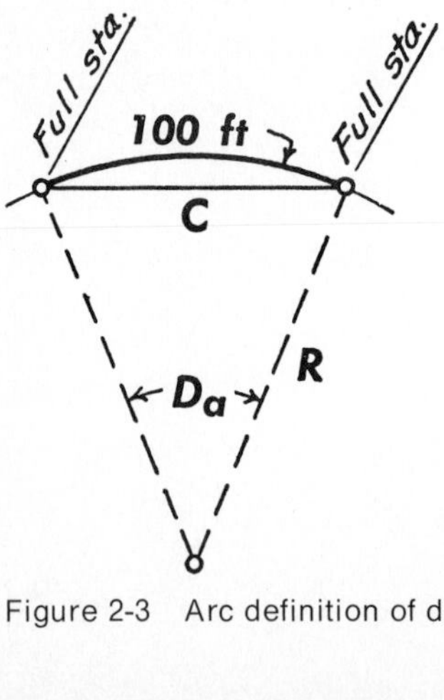

Figure 2-3 Arc definition of degree of curve.

term *D*. (*See* formula 2-11 for example.) If a formula is valid only for the arc or the chord definition, the degree of curve in the formula is expressed as D_a or D_c. Formula 2-10 is an example of the former; formula 2-17 of the latter.

In a practical problem the numerical values of *I* and D_a will usually be given. Section 2-7 shows how the required curve parts may be obtained either by the electric pocket calculator or by computing machine and tables.

2-5. Measurements on Curves

Linear measurements along a curve must be made by taping a series of chords. An isolated curve may be staked conveniently by dividing the central angle into an equal number of parts. The resulting chords with their directions determined by any appropriate method then form an inscribed polygon of equal sides. However, on most route surveys the curves are parts of the continuous alignment over which it is convenient to carry the regular survey stationing without a break. In consequence, a curve rarely begins or ends at a whole station; accordingly, chords less than 100 ft long, called *subchords*, will adjoin the T.C. and the C.T.

Figure 2-4 represents a portion of a circular arc having certain stakes located thereon. For the arc definition of *D*, stations 16 and 17 will be exactly 100 ft apart as measured around the arc. The chord distance between them must, by definition, be slightly less than 100 ft.

It is frequently necessary to set stakes on a curve at points between full stations, such as at sta. 16+50, for example. The only logical position for this stake is exactly midway between stations 16 and 17. Thus, sta. 16+50

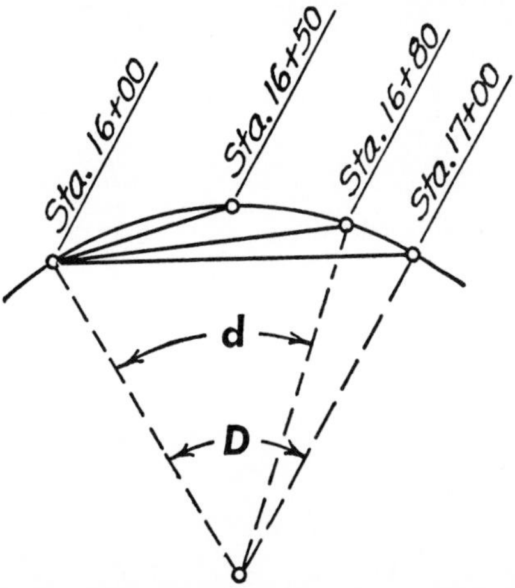

Figure 2-4 Subchords.

would be separated from the adjacent full stations by chords equal in length and shorter than 50 ft. Such chords are loosely referred to as "50-ft subchords."

On precise work, however, it is advisable to use a more definite terminology. All uncertainty is eliminated by calling the actual length of a chord its *true length*, and designating as the *nominal length* the value found by taking the difference between the stationing at the ends of the chord. The nominal length may be further illustrated by imagining the arc in Fig. 2-4 between stations 16 and 17 to be divided into exactly 100 equal parts. Theoretically, sta. 16+80 is at the 80th division, and it is joined to sta. 16 by an 80-ft *nominal subchord;* but it is located in the field by taping a *true subchord* slightly less than 80 ft long.

If d, Fig. 2-4, is the central angle subtended by any nominal subchord c_n, then $d:D = c_n:100$, or

$$d = \frac{c_n D}{100} \tag{2-11}$$

The true subchord is

$$c_t = 2\,R \sin \tfrac{1}{2}d \tag{2-12}$$

Note the analogy of formula 2-12 to formula 2-3. True subchords are given in Table II.

2-6. Length of Curve

The *length of curve*, L, is the nominal distance around the curve. It equals the difference between the stationing of the C.T. and the T.C.

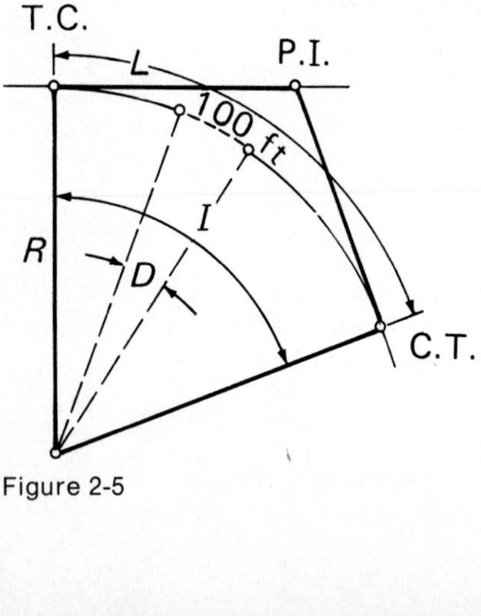

Figure 2-5

From Fig. 2-5 by proportion

$$L:100 = I:D$$

or

$$L = \frac{100\,I}{D} \tag{2-13}$$

For the arc definition of D, formula 2-13 obviously gives the true length of the total circular arc. This length is also given by $L = RI$ where I is in radians.

2-7. Computation of Curve Parts

Given: a simple curve with $I = 36°43'30''$ and $D_a = 8°00'$.

● *Method (a): Computations by Electronic Pocket Calculator. Note:* One or two extra digits beyond those considered to be significant may be carried in intermediate computations; these extra figures are indicated by superimposed bar lines. *See also* Section 2-18.

Find R from formula 2-10, or $R = 18{,}000 \div 8\pi = 716.19\overline{72}$; and store R for future use if calculator permits.
Convert I to degrees and decimals by a method adapted to the calculator used. Result: $I = 36.725°$. Store I if possible.
Find T from formula 2-1. Result: $T = 237.72\overline{66}$.
Find E from formula 2-2. Result: $E = 38.423\overline{5}$. The value of E may be checked by formula 2-5.
Find L.C. from formula 2-3. Result: L.C. $= 451.24\overline{43}$.
Find M from formula 2-4. Result: $M = 36.46\overline{70}$. The value of M may be checked by formula 2-6.
Find L from formula 2-13. Result: $L = 459.06\overline{25}$.
Find the true length of 1 station (100-ft arc) from analogy to formula 2-12. $c_{100} = 2\,R \sin \frac{1}{2}D_a = 2\,R \sin 4° = 99.91\overline{88}$. Also, $c_{1/2\text{ sta.}} = 2\,R \sin 2° = 49.98\overline{98}$. (These values are verified by Table II.)

● *Method (b): Computations by Machines-and-Tables*

Take R from Table I and needed trigonometric functions from Table XX.
Find T from formula 2-1; $T = 716.197 \times 0.331929 = 237.73$
Find E from formula 2-8: $E = 716.197 \times 0.053649 = 38.42$
Find M from formula 2-9: $M = 716{,}197 \times 0.050917 = 36.47$
Find L.C. from formula 2-3: L.C. $= 716.197 \times 0.315028 = 451.24$

These results agree with those found by method (a).

2-8. Locating the T.C. and C.T.

In locating a curve on a projected alignment, the tangents are run to an intersection. at the P.I., the angle I is measured, and the stationing is carried forward along the back tangent as far as the P.I. (see section 7-3 for a description of the process when the P.I. is inaccessible). The degree of curve will usually have been selected from the paper-location study. If it has not been, a suitable value may be determined by measuring the approximate E (or T) needed to give a good fit with the topography. The tabulated E (or T) for a 1° curve (Table VIII) divided by the approximate E (or T) gives a value of D which will fit the field conditions. Usually this D is rounded off to a figure convenient for calculation, and of course it must be within the limiting specifications for the project.

The values of T and L are next computed, and from them the stationing of the T.C. and C.T. are determined as follows:

$$\left.\begin{array}{l}\text{Sta. P.I.} - T = \text{Sta. T.C.} \\ \\ \text{Sta. T.C.} + L = \text{Sta. C.T.}\end{array}\right\} \begin{array}{l}\text{Arrange} \\ \text{computations} \\ \text{as on page 24.}\end{array}$$

For staking and checking the curve it is necessary to set hubs at the T.C. and the C.T. This is done by taping the calculated T backward and forward from the P.I. Sometimes it is more convenient to set the T.C. by taping from a tacked hub (P.O.T.) on the back tangent. In case the curve is long or the terrain is difficult for taping, it is also advisable to set a check hub at the midpoint of the curve by taping the *exact value* of E from the P.I. along the bisector of the angle.

2-9. Deflection-Angle Method

The convenient deflection-angle method of locating points on a simple curve is based on the fact that in geometry both *an inscribed angle and an angle formed by a tangent and a chord are measured by one-half the intercepted arc.*

In Fig. 2-6, *a*, *b*, and *c* represent 100-ft stations on a portion of a simple curve, *a* being less than one station beyond the T.C. By definition, each central angle between full stations is equal to D; the angle subtended by the first subchord is d_1. The angles *VAa*, *VAb*, and *VAc* are known as the *deflection angles*, or *total deflections*, to the stations on the curve. From the foregoing proposition they are equal to one-half the corresponding central angles. Thus, the deflection angle to locate *c* from a setup at the T.C. is $\frac{1}{2}(d_1 + 2D)$.

The *subdeflection* for the fractional station is $\frac{1}{2}d_1$. Once this first subdeflection has been computed, the total deflections to succeeding points are found by adding the inscribed angles *aAb*, *bAc*, ... Since an inscribed angle equals one-half its central angle, the total deflections to full stations are found by adding successive increments of $\frac{1}{2}D$. A second subchord c_2 falls between

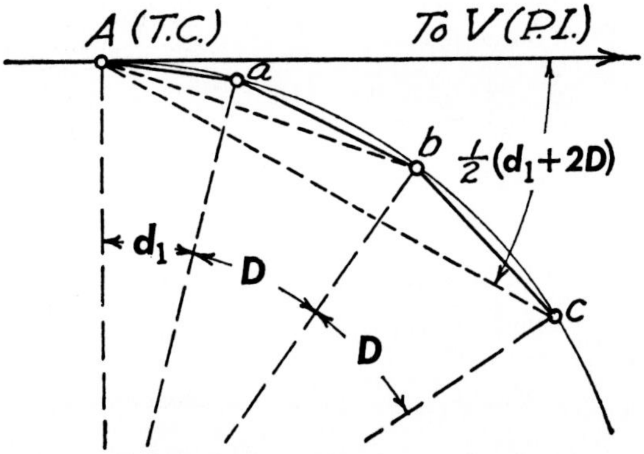

Figure 2-6 Deflection-angle method.

the C.T. and the preceding full station. Its subdeflection $\frac{1}{2}d_2$ should be added to the total deflection for the last full station on the curve. If the sum equals $\frac{1}{2}I$, the arithmetic is checked.

In practice, subdeflections are usually small and best computed in minutes from either of the following formulas:

$$\tfrac{1}{2}d \text{ (in minutes)} = 0.3\ c_n\ D^\circ \tag{2-14}$$

$$\tfrac{1}{2}d \text{ (in minutes)} = c_n \times \text{defl. per ft*} \tag{2-15}$$

The distances taped in connection with the deflection angles *are not* the rays from the T.C. to the various points on the curve. Theoretically they are the successive true chords from point to point, starting at the T.C. Therefore, once the deflection angles have been calculated, there is practically no added computation. However, the noncoincidence of tape and line of sight may be slightly confusing in the field work.

Example

Given: Sta. P.I. = 25+06.81; $I = 36°43'30''$; $D_a = 8°$. Stakes are to be set at full stations and half stations; this degree of precision is needed on final location. Deflections are needed to 1/100th minute to ensure that the final deflection equals $\frac{1}{2}I$ without being affected by rounding off. In the field a transitman will set the proper estimation of the tabulated deflection on the

* In formula 2-16, *defl. per ft* means "deflection in minutes per foot of nominal chord" (or per foot of station). Thus, if $D = 8°00' = 480'$, $\frac{1}{2}D = 240'$ and defl. per ft = 2.4'. Values of defl. per ft are listed in Table I.

vernier. (This is the curve in section 2-7 from which needed computed values are taken.)

Sta. P.I. = 25+06.81		
$-T =$ 2+37.73		
Sta. T.C. = 22+69.08	$c_1 =$ 30.92	$\frac{1}{2}d_1 = 2.4 \times 30.92 =$ 1°14.21′
$+L =$ 4+59.06	23 to 27 = 400.00	$\frac{1}{2}D \times 4$ = 16°00′
Sta. C.T. = 27+28.14	$c_2 =$ 28.14	$\frac{1}{2}d_2 = 2.4 \times 28.14 =$ 1°07.54′
	Sum = 459.06	Sum = 18°21.75′
	$= L$	$= \frac{1}{2}I$
	(check)	(check)

A typical form for setting up the notes on the left-hand page of the field book follows. Notes run upward in order to simplify orienting sketches on the right-hand page while facing along the forward line. The center line is used to represent the survey line.

FORM OF NOTES FOR SIMPLE CURVE

STA.	POINT	TOTAL DEFL.	CURVE DATA
28			
+50			
+28.14	⊙ C.T.	18°21.75′	$I = 36°43'30''$
27		17°14.21′	$D_a = 8°00'$
+50		15°14.21′	P.I. = 25 + 06.81
26	⊙	13°14.21′	$T = 237.73$
+50		11°14.21′	$L = 459.06$
25		9°14.21′	$E = 38.42$
+50		7°14.21′	Defl. per ft
24		5°14.21′	= 2.4′
+50	⊙	3°14.21′	
23		1°14.21′	
+69.08	⊙ T.C.	0°00.00′	
22			

2-10. Transit Setups on Curve

In the preceding description the implication is that the entire curve is staked from a setup at the T.C. This is often true. However, there are circumstances (see section 2-11) which make this procedure impracticable or even impossible. Should such be the case, *deflections are computed as though for a setup at the T.C.* By proper manipulation of the transit it can be set up at any staked point on the curve, and the staking then continues from that setup without altering the previously computed deflections.

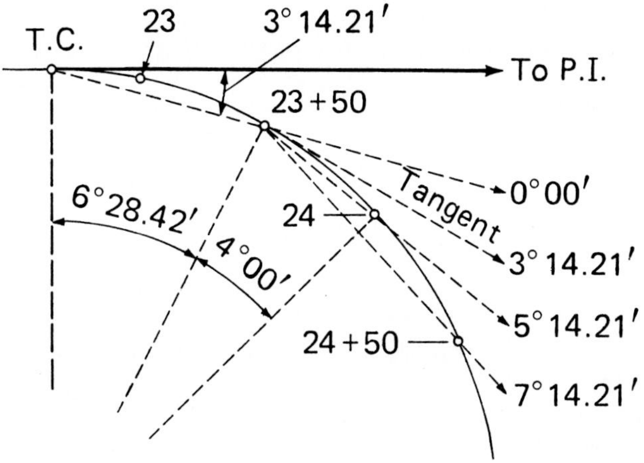

Figure 2-7

Figure 2-7 shows the first portion of the curve computed in section 2-9. It is assumed that the curve has been staked as far as sta. 23 + 50 from a setup at the T.C., but conditions make it difficult to sight beyond from that setup. The total deflection turned thus far is 3°14.21′, which equals one-half the intercepted arc. If the transit were moved to sta. 23 + 50 and then backsighted to the T.C. with 0°00′ on the vernier, the telescope (after plunging) would be sighted along the dotted line marked 0°00′. By turning clockwise through 3°14.21′, the telescope would then be sighted along the local tangent; this condition is shown by the dotted line marked 3°14.21′. To set the stake at the next point (sta. 24) the telescope must be turned further clockwise by one-half the intercepted arc. This is a 50-ft arc which, by definition, intercepts $\frac{1}{2}D_a$ at the center of the curve; therefore the added deflection for locating sta. 24 is $\frac{1}{4}D_a$ or 2°00′. The condition would then be as shown by the dotted line marked 5°14.21′.

Without moving the transit from sta. 23 + 50, the stake at sta. 24 + 50 would be sighted by turning clockwise another increment of one-half the intercepted arc, or 2°00′. This result is illustrated by the dotted line marked 7°14.21′. Successive increments of 2°00′ would position the telescope for sighting additional stakes at 50-ft intervals.

If another setup is required at sta. 26, the telescope can be oriented by backsighting to sta. 23 + 50 with the vernier set to 3°14.21′ *on the correct side of 0°00′* (clockwise in Fig. 2-7). As before, the telescope should then be plunged and turned clockwise through successive increments of 2°00′ to sight further stakes at 50-ft intervals. By following this procedure the vernier reading when sighting at any station is the total deflection opposite that station in the notes. No added calculations are required if this method is followed; the only record of the setups is the symbol ⊙ in the Point column of the notes.

The preceding description may be summarized in a rule-of-thumb:

> To move up on curve and retain original notes, occupy any station and backsight to any other station with the *vernier* set to the total deflection of the *station sighted.*

Occasionally it may be desired to have the vernier read 0°00′ when the line of sight is on the *local tangent* to the curve at a new setup, as at the C.C. of a compound curve (see section 3-13). To do this the vernier must be set at the *difference between the tabulated deflections for station sighted and station occupied.** As in applying the rule-of-thumb, the surveyor must be careful to set the *verneir reading* for the backsight *on the correct side of 0°00′*.

2-11. Comments on Field Work

Transit setups for staking the curve must be started at the T.C. or C.T. For short curves that are entirely visible from one setup, it is preferable to occupy the C.T. and to tape the chords toward it from the T.C. By so doing, the longer sights are taken before possible settlement of the transit occurs. Moreover, one setup is eliminated, for the transit is then in position for lining in the stakes along the forward tangent. For a setup at the C.T., the transit is oriented by backsighting to the T.C. with the vernier reading 0°00′ or to the P.I. with the vernier set at $\frac{1}{2}I$. In either case it is wise to check the angle to the other point, in order to verify the equality of tangent distances.

In case the field conditions require one or more setups on the curve, it is good practice to occupy the T.C. first and tape in a forward direction, moving up the transit according to the rule-of-thumb stated in section 2-10. The final portion of the curve is best located by setting up at the C.T. and taping *backward* to the previous setup. This practice ensures good tangencies at the T.C. and C.T., and throws any slight error into the curve where it is more easily adjusted.

A long curve, or one having a large central angle, may justify intermediate setups even though the entire curve is visible from either end. The consideration here is the required degree of accuracy in setting points on the curve. When the tape and line of sight intersect at a large angle, it is possible for the chainman to swing one end of the tape through a certain distance without

* When this book is used as a textbook for a course in Route Surveying, it is recommended that students be given the opportunity to compute and stake at least one simple curve at this point in their study. Doing so will help to fix the principles more firmly in mind and contribute toward a better appreciation of the practical suggestions which follow.

detection by the transitman, thus throwing off the position of the point and introducing an accumulative taping error. The amount of this swing—and its relative importance on the particular job—is best determined by trial in the field. As a general guide, the deflection angle from the transit to the taping point should be kept under 25° or 30°.

Field checks should be made at every convenient opportunity. In locating the T.C. and the C.T., it is good insurance to double-chain the tangent distances, at least until the chainmen have demonstrated the reliability of their first measurements. Less time will be lost if the check chaining is done while curve deflections are being computed. Setting a check point at the middle of the curve may also be advisable (see section 2-8).

Before definitely setting each stake on the curve, the head chainman should sight to the second stake back and verify the middle ordinate m to the intervening stake. (See Fig. 2-8.) A sudden change in middle ordinate (for equidistant points on the curve) will reveal a mistake in setting off the deflection angle. Otherwise, such an error might go undetected at the time or even fail to show up in the final closing check if the curve is flat.

The final check occurs at the extremity of the last chord taped. This may be at the T.C., at the C.T., or at any intermediate stake previously set. The best method is to mark an independent point by measuring the required distance and deflection from the final setup. The distance between this point and its theoretical equivalent at the stake previously set is a direct measure of the error of closure in the traverse around the tangent distances and curve chords. The error limit should be consistent with the survey methods, difficulty of the terrain, and requirements of the project.

Chord lengths used in running curves depend principally on the degree of curve and purpose of the survey. On *flat curves* in easy terrain, full stations suffice for the field work and most construction purposes. Ordinarily the subchords adjoining the T.C. and C.T. may be taken equal to their nominal lengths, but the chord for each of a long series of full stations may require a slight correction in order to reduce the systematic error in taping (see Tables II and IV).

On *sharp curves*, or in difficult terrain requiring frequent cross sections, additional points between full stations are needed. Whether it is necessary to use true subchords instead of nominal subchords depends on the accuracy required. Table II shows at a glance the effect of the sharpness of the curve

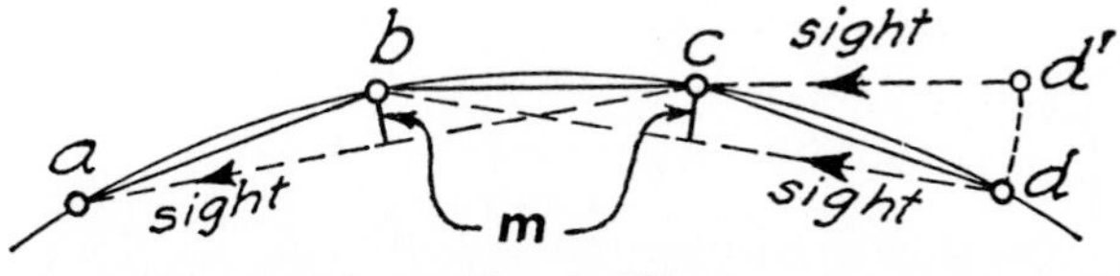

Figure 2-8 Check of stake positions.

on the true lengths of nominal full chords and subchords. When Table II is used in conjunction with Table IV any degree of accuracy is readily obtainable. Study of Table II shows in general that chord corrections are unnecessary on curves flatter than about 5°. On sharper curves, shorter subchords can be used to avoid chord corrections.

Stakes are set on curves where needed for subsequent field work or for controlling construction. A tacked *hub*, with a *guard stake* designating the point and stationing, is set at each transit station. Important points, such as the P.I., T.C., and C.T., are carefully referenced. Points not occupied by the transit are identified by *station stakes*, which are driven and marked as in profile leveling.

Station stakes used for cross sectioning prior to construction need not be individually centered by the transit. Doing this is of no practical value, and it wastes time if the stakes are hidden by grass or brush. A rapid method is shown in Fig. 2-9. The head chainman first obtains line from the transitman and then sets a range pole (flag) off line about the width of a stake and a few tenths of a foot beyond the station. He next verifies the middle ordinate (see Fig. 2-8). Then he stretches the tape tangent to the range pole and sets a chaining pin at the exact distance. Finally, he uses the point of the range pole to make a hole beside and slightly behind the point where the pin enters the ground, and proceeds to the next station with the pole and tape. The stakeman immediately drives a marked station stake into the hole at the pin, at the same time keeping the moving tape to one side with his foot. The pin should not be disturbed in the process; it remains in place until pulled by the rear chainman after the next forward pin is set.

By following the foregoing method on tangents and flat curves, station stakes will always be close enough in both line and distance. On sharp curves where the angle between tape and line of sight is large, it may be slightly better to set the range pole practically on the line of sight.

In setting points on curves the head chainman can expedite the work by placing himself quickly on line at the point to be set. One method of doing this, especially suited to open country, is for the head chainman to range forward the proper chord distance along a line through the last two stakes, placing

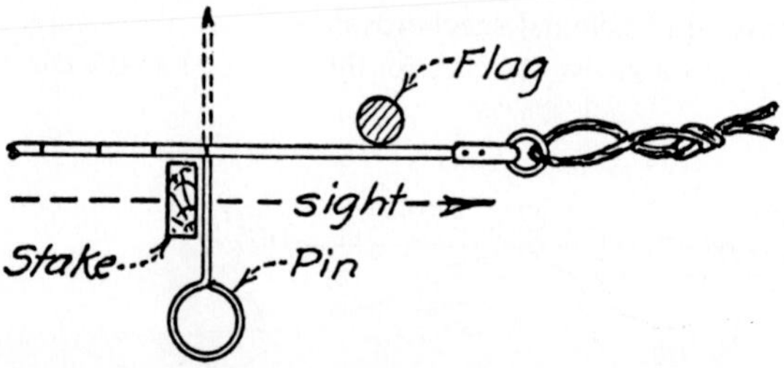

Figure 2-9 Setting station stakes.

himself in position as at *d′* in Fig. 2-8. By then pacing inward the known chord-offset distance (section 2-15) he will be in approximately the correct position for setting the required point at *d*. With experience the head chainman will invariably be in the field of view by the time the transitman sets off the deflection angle and sights.

Many variations are possible in the technique described. The range pole may be dispensed with for short sights in open country. Another device used where frequent *plus points* are required is to set those station stakes directly at the proper division on the tape and to move the tape forward at each full station. The important factors to consider are speed, accuracy consistent with the needs, and avoidance of taping errors. In all cases, carrying the taping forward by *chaining from pin to pin* is advisable.

After grading, all points on curves which are set to control construction details must be more carefully centered than in the methods just described.

On pavements or in other locations where stakes cannot be driven, the points are chiseled or marked with keel, or a nail may be driven flush through a small square of red cloth. Guard stakes marked with the stationing are set at a convenient offset distance to one side.

Minor obstacles to setting calculated points on curves are often met in running location prior to clearing or grading. These should not be permitted to delay the field work.

Figure 2-10 shows another sketch of the first portion of the curve calculated in section 2-9. If the line of sight to sta. 24 were to hit a large tree, for example, it could be backed off to the left of the tree for setting a stake at any convenient plus point. Thus, if sta. 23+90 were in the clear, its deflection would be 40 × 2.4′ greater than that used to set sta. 23+50. The calculations are $40 \times 2.4' = 96' = 1°36'$, and $3°14.21' + 1°36' = 4°50.21'$ for the deflection to 23+90. (To check, subtract $10 \times 2.4'$ from the deflection at sta. 24.) After the station stake at 23+90 is set, another new plus point to the right of the tree (such as 24+20) could be staked by following a similar process; or else the taping could proceed from sta. 23+90 directly to the next calculated point at sta. 24+50.

A minor obstacle on the curve itself (such as a large boulder at sta. 24) could be by-passed by the same method.

Major obstacles are treated in Chapter 7.

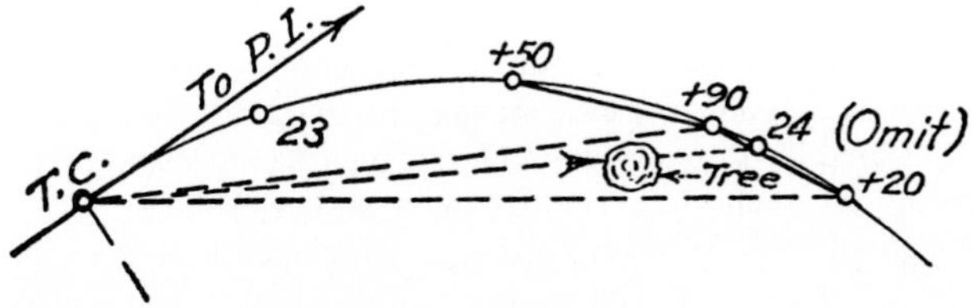

Figure 2-10 Passing minor obstacle.

2-12. Even-Radius Curves

Some organizations, including a few highway departments, do not use the degree of curve designation either in calculations or in field work. Instead, a curve is chosen on the basic of its radius, which usually is a round number. Deflections are calculated theoretically from the following proportion:

$$\text{deflection}:\tfrac{1}{2}\text{ arc length} = 360^\circ : 2\,\pi\,R$$

which reduces to

$$\text{defl. (in minutes)} = \frac{5400}{\pi\,R} \times \text{arc length} \tag{2-16}$$

This method has the advantage that radii may be selected which are convenient for calculating curve parts. The *radius* would also mean more to the average person than the *degree*, should it become the practice to post warning values of R at approaches to highway curves. Calculation of deflections, however, is much less simple than with a rounded value of D, and their fractional values are inconvenient in the field.

The process of calculating and staking these "even-radius curves" is expedited by use of Tables V and VI. The latter table is used to compute L when I and R are given. Other curve parts are found from the standard formulas (see Section 2-3). Table V gives directly most of the data concerning deflections and chord lengths needed in the field. Subchord deflections are readily computed from the tabulated deflections per foot.

2-13. Degree of Curve, Chord Definition

For the *chord definition*, the degree of a curve is the central angle subtended by a *100-ft chord*. It is denoted by D_c, as shown in Fig. 2-11.

From the right triangle

$$\sin \tfrac{1}{2}D_c = \frac{50}{R} \tag{2-17}$$

In a practical problem the numerical values of I and D_c will usually be given. Formula 2-17 is used to find R, and the required curve parts then calculated as shown in section 2-7.

Formulas 2-11 and 2-13 are valid for both definitions of D. However, the term L in formula 2-13 does not have the same meaning with both definitions. When D_a is used for D in formula 2-13, L is the true arc length of the curve between the T.C. and C.T.; whereas when D_c is substituted for D, L is the total length of the 100-ft chords and nominal subchords between the T.C. and C.T.

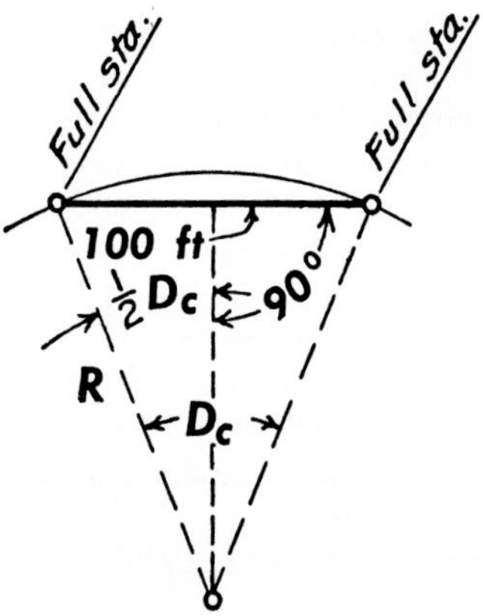

Figure 2-11 Chord definition of degree of curve.

If needed, the true length of a chord definition curve can be computed from $L = IR$, where I is in radians; or it may be obtained with the aid of Tables II, III, and IV.

2-14. Metric Curves

In countries having the metric system, distances are usually measured with a 20-m tape, and a "station" is nominally a distance of 10 m. On route surveys either the radius in meters or the degree of curve may be used in selecting and staking curves. However, the customary metric "degree of curve" D_m differs from those already defined not only in unit of measurement but in angle and length of chord (or arc) as well. Specifically, D_m is the *deflection angle* for a chord (or arc) of *20 m*. Thus, by analogy from equations 2-17 and 2-10,

$$\sin D_m \text{ (chord def.)} = \frac{10}{R_m} \tag{2-18}$$

and

$$D_m \text{ (arc def.)} = \frac{1800}{\pi R_m} \tag{2-19}$$

Also, by analogy from equation 2-13

$$L_m = \frac{10\,I}{D_m} \tag{2-20}$$

Problem 2-10 at the end of this chapter is worked according to these equations.

It is not necessary to use special metric-system tables to compute and stake metric curves. The foregoing simple relations used in connection with the tables in Part III of this book will enable an engineer to handle any metric-curve problem. If Table VIII is used, for example, the tangents and externals for a 1° metric curve are *one-tenth* the tabulated values (for the same central angle). This fact follows from equations 2-10 and 2-19.

For a quick comparison between *metric* and *foot* curves, it is well to remember that for curves of the same radius the ratio between D and D_m is about 3 to 1 (actually 3.048 to 1). Thus, a curve of given D_m is about 3 times as sharp as one having the same numerical value of D.

With the probable adoption of the metric system in the United States in (or before) the 1980s, changes may be made in the definitions of metric "station" and metric "degree of curve." As an example, it has been proposed by representatives of 12 states attending a metric workshop in 1975 that two standard lengths of tapes be used: 30 m and 50 m; and further, that a "station" should be 100 m long. In any event the surveyor will have to adhere to the practice used in the country in which he is working. The basic simple-curve theory in this chapter will enable him to do this.

2-15. Staking by Offsets

There is no good substitute for the deflection-angle method when points must be set quickly and accurately on long-radius curves. It is mathematically possible, however, to devise several other methods based either upon *angles from the P.I.* or upon *offsets*. The former basis is rarely practicable, but there are occasions where an offset method is very useful.

The offset methods* are known as:

a. Chord offsets
b. Middle ordinates
c. Tangent offsets
d. Ordinates from a long chord

The *chord offset*, *tangent offset*, and *middle ordinate* for a *full station* are designated by C.O., T.O., and M.O. respectively; corresponding terms for any other distance are *c.o.*, t, and m. The *true chord* for a *full station* is *100 ft* or C, the length depending on the definition of D.

* Following the derivation of some needed mathematical relations, only the chord-offset method is described in detail. Abbreviated descriptions of the remaining methods are enough to serve as a basis for applying them.

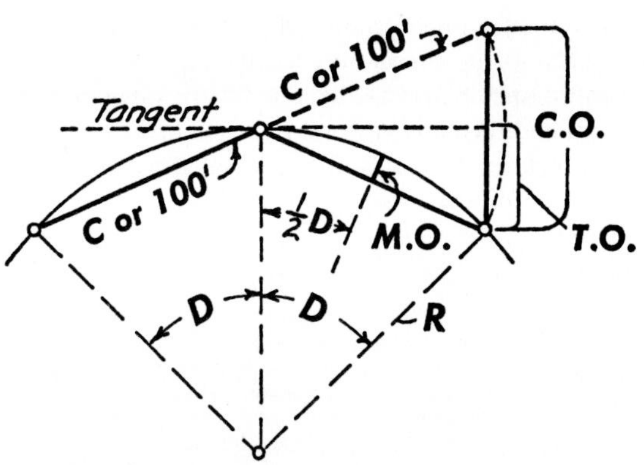

Figure 2-12 Offsets for full station.

From Fig. 2-12 the following *exact* relations may be written practically by inspection:

$$C = 2\,R \sin \tfrac{1}{2}D_a \qquad (2\text{-}21)$$

$$\text{C.O. (chord def.)} = 200 \sin \tfrac{1}{2}D_c \qquad (2\text{-}22)$$

$$\text{C.O. (arc def.)} = 2\,C \sin \tfrac{1}{2}D_a \qquad (2\text{-}23)$$

$$\text{T.O.} = \tfrac{1}{2}\text{C.O.} \qquad (2\text{-}24)$$

$$\text{M.O.} = R \text{ vers } \tfrac{1}{2}D \qquad (2\text{-}25)$$

Also, by similar triangles,

$$\text{C.O.} : 100 \text{ (or } C) = 100 \text{ (or } C) : R$$

Hence,

$$\text{C.O. (chord def.)} = \frac{100^2}{R} \qquad (2\text{-}26)$$

and

$$\text{C.O. (arc def.)} = \frac{C^2}{R} \qquad (2\text{-}27)$$

Four useful *approximate* relations are:

$$\text{M.O.} = \tfrac{1}{8}\text{C.O.} \qquad (2\text{-}28) \text{ Approx.}$$

$$\text{C.O.} = 1\tfrac{3}{4}D \qquad (2\text{-}29) \text{ Approx.}$$

$$t = \tfrac{7}{8}s^2\,D \text{ (} s \text{ is chord in } \textit{stations}) \qquad (2\text{-}30) \text{ Approx.}$$

$$m \text{ (in inches for a 62-ft chord)} = D \qquad (2\text{-}31) \text{ Approx.}$$

The approximate 8:4:1 ratio of the chord offset, tangent offset, and middle ordinate is worth memorizing (mnemonic: "eight for one").

Offsets for any subchord (or subarc) may be found by first determining the values for a full station and then applying the principle that *c.o.*, t, and m are proportional to the *squares* of the subchords (or subarcs). They may also be found from the *exact* relations:

$$c.o. = \frac{c^2}{R} \tag{2-32}$$

$$t = \frac{c^2}{2R} \tag{2-33}$$

$$m = R \text{ vers } \tfrac{1}{2}d \tag{2-34}$$

Table I contains values of certain of the full-station offsets. The explanatory notes preceding that table show its usefulness in obtaining any offsets accurately.

Chord offsets sometimes called "deflection distances," provide a means of setting station stakes rapidly when a transit is not available. If the tapeman sights past his plumbbob cord, surprising accuracy is possible. It is invariably adequate for surveys made prior to clearing and grading.

In Fig. 2-13 (curvature greatly exaggerated) the P.I. and the tangent points are assumed to have been set by the usual methods. If the transitman and some of the field party now proceed along the forward tangent for purposes of setting stakes or measuring the angle I for the next curve, three men (or even two) can set the station stakes on the curve left behind.

The procedure follows: Knowing the stationing of the T.C., calculate the subchords c_1 and c_2 to the adjacent full stations at A and B (sta. A is on the curve produced backward). Then calculate t_1 and t_2, the tangent offsets for the subchords. Field work may now be started after range poles have been set at convenient points on the tangent. Establish a temporary point at A

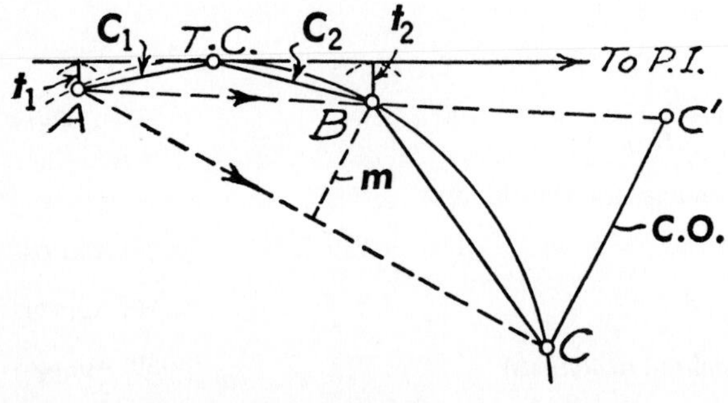

Figure 2-13 Staking by offsets.

by the "swing offset" method. Locate the station stake at B similarly. Range the line AB ahead a full-station chord to C', and locate the next station stake at C by the C.O. Continue around the curve by full-chord offsets (obviously, half-stations or quarter-stations could be used instead) until the full station preceding the C.T. has been set. Make a final check by measuring the last subchord and swing offset to the forward tangent; check them against their theoretical values.

Middle ordinates are also useful in staking a curve by tape alone. Their most valuable use, however, is in checking and realigning existing curves on railroads. (See section 9-4.)

In staking problems, first locate full stations either side of the T.C. by swing tangent offsets, as in the chord-offset method just described. Then set the next full station by sighting from the rear station along a line m ft inside the forward station, and measure the full chord distance from the forward station (see Fig. 2-13). Continue this process to the stake before the C.T., and make the final check as in the chord-offset method. (Fractional stations could be used instead of full stations.)

It should be noted that, when stakes are set at full stations, the distance m is the middle ordinate for two stations; consequently, m = T.O. = $\frac{1}{2}$C.O.

Topographical conditions sometimes make this method preferable to chord offsets. It is probably better on sharp curves, regardless of topography, because the offsets are only half as long.

Tangent offsets provide a handy means of setting station stakes by tape alone. The method is essentially one of rectangular coordinates, the T.C. being used as the origin and the tangent being used as an axis.

The deflection $\frac{1}{2}d$ to the first regular station is computed. The tangent distance to this station is $c \cos \frac{1}{2}d$, and its offset $c \sin \frac{1}{2}d$. Coordinates of remaining full stations are obtained by adding increments of $C \cos (d + \frac{1}{2}D)$ and $C \sin (d + \frac{1}{2}D)$, $C \cos (d + \frac{3}{2}D)$ and $C \sin (d + \frac{3}{2}D)$, and so forth, where C is the chord for a full station. Since in most cases $C = 100$ ft, the computation is very simply performed by using natural sines and cosines. As a check on the computations, the coordinates of the C.T. should equal $R \sin I$ and $R \text{ vers } I$. The long tangent offsets to the second half of a sharp curve can be avoided by computing that half with reference to the C.T. and the forward tangent.

In the field, set temporary stakes at the calculated tangent distances. Then locate station stakes at the intersections of tangent offsets and chords from the preceding stations.

Ordinates from a long chord are also based upon rectangular coordinates. As in the tangent-offset method, the origin is at the T.C., but the axis is along a convenient long chord. If the chord is the L.C. of the curve, simple relations for the coordinates of the station stakes may be expressed (these will not be derived). However, it is difficult to imagine many practical situations in which one of the foregoing methods would not be better than this.

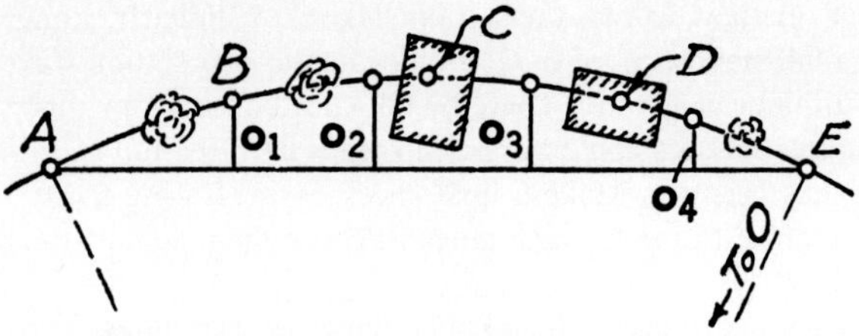

Figure 2-14 Ordinates from a long chord.

Perhaps the most useful application of this method is in setting stakes at odd plus points between a series of minor obstacles on a curve. Figure 2-14 illustrates such a situation. Station E is set by turning a deflection angle with the transit at A and taping the chord AE, which is $2\,R \sin \frac{1}{2}AOE$. Stakes set sufficiently close for cross sectioning are then located by taping the offsets $o_1, o_2, \ldots$ at convenient points along AE. If the segments of AE at any ordinate are designated by s_1 and s_2 (units are stations), then the length of the ordinate in feet is found from the relation

$$o = \tfrac{7}{8} s_1 s_2 D \qquad \text{(2-35) Approx.}$$

Note the resemblance between formulas 2-30 and 2-35.

2-16. Parallel Curves

It is often necessary to determine the length of a curve parallel to a curved center line. This problem occurs in connection with finding lengths of curb, guardrail, or right-of-way along curved alignment. Moreover, stakes are often set along a parallel "offset" line in order to avoid traffic or construction operations. Figure 2-15 shows offset curves outside and inside a center-line curve of length L.

Since the true length of a circular arc, where I is in radians, equals IR, $L_o = I(R + p_o)$ and $L_i = I(R - p_i)$ from which

$$L_o = L + Ip_o \qquad \text{(2-36)}$$

and

$$L_i = L - Ip_i \qquad \text{(2-37)}$$

A parallel offset curve is usually staked by setting stakes at points radially opposite the center-line stakes. This is done either by the deflection-angle method or by one of the offset methods described in section 2-15. Values

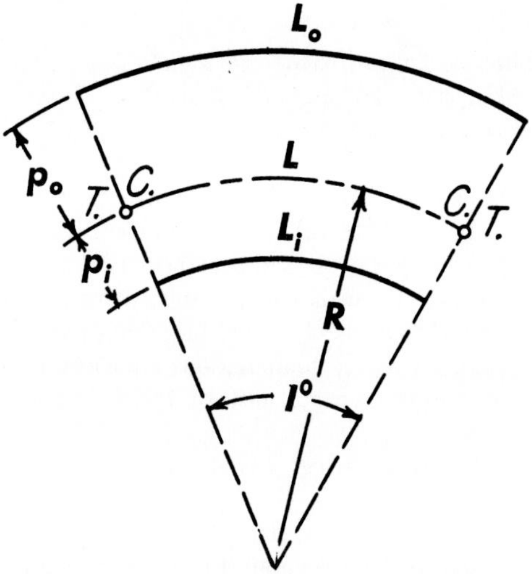

Figure 2-15 Parallel curves.

of T, L, and chord lengths for the offset curves are proportional to those for the center-line curve. True chord lengths can be found from this proportionality or from formula 2-12; they can also be taken from Table II with sufficient accuracy.

As an illustration, refer to the example in section 2-7 and assume that it is required to stake one offset curve 60 ft outside and another 25 ft inside the given curve. From formulas 2-36 and 2-37,

$$L_o = 459.06\overline{25} + 36.725\left(\frac{\pi}{180}\right)60 = 497.52\overline{1}; \quad \text{and}$$

$$L_i = 459.06\overline{25} - 36.725\left(\frac{\pi}{180}\right)25 = 443.03\overline{8}.$$

The true lengths of 50-ft nominal chords are equal to

$$\text{center line} = 2 \times 716.19\overline{72} \sin 2^\circ = 49.990$$

$$\text{60 ft outside} = 2(716.19\overline{72} + 60) \sin 2^\circ = 54.177$$

$$\text{25 ft inside} = 2(716.19\overline{72} - 25) \sin 2^\circ = \underline{4}8.245$$

Deflection angles to stakes radially opposite the center-line stakes are the same as for the center line.

2-17. Methods of Computation

Proper methods of computation and use of significant figures are so important that this and the following section are inserted here so the suggestions can be weighed and then used, wherever applicable, in problems throughout the book.

Route surveying computations require relatively simple mathematics—arithmetic, geometry, and trigonometry. The solution of a problem is often tedious in that it involves many repetitions of the same mathematical operations. As a result, students (particularly) find the manipulation of surveying instruments much more interesting. They are likely to conclude that surveying field work is of primary importance, whereas computations are a chore to complete as quickly as possible. This is a serious mistake. Field work and calculations are equally important.

Surveying field work alone is of little value; it is always put to use in some practical form that involves computations. Unless the calculations are made correctly, they negate the field work and may cause serious errors when used as a basis for the design or construction of engineering works. A student can take as much pride in obtaining the answer to a tedious surveying problem—the correct answer, with the proper number of significant figures and with the decimal point in the right place—as he does in his ability to "run a transit" rapidly and accurately. There is no better place to develop the habits of computation which ensure this result than a course in surveying.

Surveying computations are made algebraically, graphically, and mechanically, or by a combination of these methods. Useful tables reduce much of the drudgery (see Part III for examples of tables).

A surveying problem can usually be solved by more than one exact method. This should always be done on important work. Moreover, all computations should be checked, preferably by a different computer. If an exact algebraic solution cannot be found, the unknowns may be scaled from an accurate drawing on which the measured data have been carefully plotted. This method is reliable to three significant figures if the scale of the drawing is fairly large.

Among the mechanical aids are the *slide rule*, *planimeter*, *calculating machine*, and *electronic computer*. The slide rule is an excellent device for checking the first three significant figures. The planimeter will measure irregular areas (see section 6-8) to an accuracy within 1%.

Important surveying computations require accuracy to at least five significant figures. This degree of refinement is easily obtained by an electronic pocket calculator and by machine-and-tables (Table XX). Some organizations prefer to use desk calculators which give a permanent printout for office use. Logarithms have been made practically obsolete by the convenience and portability of pocket calculators.

The electronic computer is employed by large organizations where the number of repetetive operations is great enough to justify the high cost of computer purchase or rental. Such a machine is not a "brain." It does no thinking, but only follows coded commands fed into it. Instead of downgrading the ability of the surveyor, the electronic calculator requires him to be even more expert in order to "program" the calculator to operate with the highest efficiency.

Some practical suggestions for making surveying computations are:

1. Do all computations on "squared paper" to simplify making sketches and arranging work in columns.
2. After transferring data from field book to computation sheet, *check the copying.* Mistakes made at this point are common and are not normally disclosed by the computing operations.
3. Draw a small sketch, approximately to scale, showing the known data and unknowns sought. Aided by the sketch, make an estimate of answers to the problem and record them.
4. Show the formula at the head of the block of computations to which it applies.
5. Adopt a clear and logical arrangement of computations; a tabular form fits many problems (see examples in section 4-4). Avoid an arrangement that requires numbers to be added or subtracted horizontally. Line up digits and decimal points in a column of numbers.
6. Wherever possible, perform computations in a systematic order, for example, addition, subtraction, multiplication, and division. Do no work on "scratch paper." Incidental computations may be written in a margin on the computation sheet.
7. Use a straight edge in order to avoid taking a value from the wrong line of a table. Check additions, subtractions, extractions of tabular values, and so forth as they are made and indicate by a check mark.
8. Follow good practice in the use of significant figures and rounding off. In general, carry one extra significant figure in intermediate calculations. (See section 2-18 for detailed suggestions.)
9. Take advantage of all automatic checks on the accuracy of calculations (see example on page 24 for illustrations). Make a slide-rule check at convenient points; this will verify the arithmetic to three significant figures though it will not disclose errors resulting from the use of the wrong formula. Any such blunder, or a mistake in arithmetic, will usually be caught by applying a graphical check.
10. Label the final answers and all intermediate results clearly. Compare the answers with estimated values obtained from the sketch and do not accept the work unless the agreement is reasonable.

2-18. Significant Figures

Both *exact numbers* and *round numbers* are used in computations. Exact numbers come from tallies and theoretical considerations; round numbers emerge from measurements. For example, the number 2 in the formula for L.C. (equation 2-3) is an exact number that comes from geometric theory. On the contrary, the measured length of a line is a round number; it can never be exact, for it would have to be correct to an infinite number of digits.

In computations it is often desirable to change a round number to an equivalent one having fewer digits. This process is called "rounding off." Common practice in rounding off is as follows:

> When dropping one or more digits, round off to the nearer of the two round numbers between which the given round number lies. If the digit to be dropped is 5, use the nearest even number for the preceding digit. (Consider zero to be an even number.)

According to this practice, the round number 18.1827 is rounded off to 18.183, 18.18, or 18.2, the result depending on the number of digits desired. Also, the round numbers 85.155 and 85.165 are rounded off to 85.16; the numbers 85.095 and 85.105 are rounded off to 85.10. When properly rounded off, a measured quantity contains the number of *significant figures* consistent with the data.

The significant figures in any round number are those digits (including zeros) that have real meaning. The maximum number of such figures includes those digits that are certain, plus one that is estimated. In making measurements the number of significant figures *sought* is governed by economic considerations; whereas the number *obtained* is limited by the precision with which the measurements are made.

Confusion as to which digits are significant is avoided by observing the following principles (in each example there are five significant figures and all significant digits and zeros are printed in *italics*):

1. The digits 1 through 9 are always significant.
2. Zeros that lie between any two digits are significant, regardless of the position of the decimal point. Examples are 21006 and 21.006.
3. Zeros that lie to the left of the leftmost digit *are not* significant, regardless of the position of the decimal point. Examples are 0.71700 and 0.00071700.
4. Zeros that lie to the right of any digit *and also* to the *right* of the decimal point are significant. An example is 717.00.
5. Zeros that lie to the right of any digit *and also* to the *left* of the decimal point may or may not be significant. For example, in 71700. the zeros are significant if they were actually measured. If they were not measured, doubt as to the number of figures which are significant is resolved by writing the number as 717×10^2.

6. The number of significant figures in any round number is independent of the units in which it is expressed. An example is 586.08 ft = 0.11100 miles.

It is a waste of time, as well as misleading, to carry computed results beyond the precision inherent in the measurements on which the computations are based. A few examples will clarify this statement.

Example of Addition or Subtraction

Add the round numbers: 9.47, 241.3, 7.6625, and 0.891.

SOLUTION AND COMMENTS

GIVEN NUMBER	NO. OF SIG. FIGS.	WRONG METHOD	COMMON METHOD	PREFERRED METHOD
9.47	3	9.47	9.5	9.47
241.3	4	241	241.3	241.3
7.6625	5	7.66	7.7	7.66
0.891	3	0.891	0.9	0.89
	Sum =	259.021	259.4	259.32
	Rounded answer =	259	259.4	259.3

The "wrong method" comes from a mistaken interpretation of the adage that a chain is no stronger than its weakest link. The fault lies in assuming that each number, before adding, must be rounded off to the least number of significant figures possessed by any of the numbers (three significant figures in this example), and also that the sum must be similarly rounded off.

In the "common method" the numbers are first rounded off to the limit of accuracy of the least accurate number (tenths in this example). Their sum is accepted without further rounding.

In the **preferred method** the numbers are first rounded off to one decimal place beyond that of the least accurate number. Their sum is then rounded off to the nearest digit in the preceding decimal place (nearest tenth in this example). This method recognizes the fact that the principal sources of error lie in the numbers that are *not* rounded off. In this example, the final 6 and the final 9 in the last two numbers are certain. Consequently, the final 2 in the sum may not be far in error, and the rounded answer 259.3 has a somewhat higher probability than the answer 259.4.

Example of Multiplication

Multiply the round numbers 362.56 and 2.13.

SOLUTION AND COMMENTS

When two or more round numbers are to be multiplied, the common method is to round off each number to the same number of significant figures as appears in the least accurate number. Then multiply, and round off each intermediate product and the final answer to this same number of significant figures. As applied to this example, the common method is $363 \times 2.13 = 773.19$, after which 773 is used as the intermediate product or the final answer.

A **preferred method** is to round off each number to one more significant figure than appears in the least accurate number. Then multiply, and round off each intermediate product to this number of significant figures. Finally round off the answer to the number of significant figures in the least accurate number. As applied to this example, the preferred method is $362.6 \times 2.13 = 772.3$ as the intermediate product or 772 as the final answer.

Example of Division

Divide the round number 675.41 by the round number 87.5.

SOLUTION AND COMMENTS

As in multiplication, the common method is $675 \div 87.5 = 7.71$ as the final answer. A **preferred method** is to round off the divisor to the same number of significant figures as appears in the dividend, or round off the dividend to one more significant figure than appears in the divisor (whichever case applies). Then divide, and round off the final answer to the number of significant figures in the rounded divisor. One more significant figure is temporarily carried in the answer if the quotient is to be used in subsequent computation involving other round numbers. As applied to this example, the preferred method is $675.4 \div 87.5 = 7.719$ as the intermediate quotient or 7.72 as the final answer.

The difference between the results of the common and preferred methods will never be large. Both methods recognize that final computed results can be no more accurate than the least accurate factor used to obtain them. In more intricate computations involving intermediate arithmetical operations, it is good practice to carry one (not more than two) extra figures so as to avoid the buildup of errors caused by premature rounding off. Such extra figures may be indicated by bar lines, as $625.42\overline{83}$.

PROBLEMS

2-1. Given the following values of I and D_a. Compute T, E, L.C., M, and L from basic formulas. Verify answers for T, E, and L by Tables VIII and VI.

(a) $I = 20°34'$; $D_a = 2°05'$. *Answers:* $T = 498.97$; $E = 44.90$; L.C. = 981.91; $M = 44.18$; $L = 987.20$.

(b) $I = 31°18'30''$; $D_a = 5°$. *Answers:* $T = 321.11$; $E = 44.14$; L.C. = 618.41; $M = 42.50$; $L = 626.17$.

(c) $I = 43°52'15''$; $D_a = 8°30'$. *Answers:* $T = 271.46$; $E = 52.61$; L.C. = 503.61; $M = 48.80$; $L = 516.13$.

(d) $I = 67°20'18''$; $D_a = 9°18'$. *Answers:* $T = 410.40$; $E = 124.18$; L.C. = 683.11; $M = 103.35$; $L = 724.07$.

(e) $I = 70°44'30''$; $D_a = 12°$. *Answers:* $T = 338.95$; $E = 108.08$; L.C. = 552.78; $M = 88.13$; $L = 589.51$.

2-2. Given the following values of I and D_c. Compute T, E, L.C., M, and L from basic formulas. Find true length of arc by Table VI. Verify answers for T, E, and true length of arc by Tables VIII, IX, X, and II where applicable.

(a) $I = 16°22'$; $D_c = 1°40'$. *Answers:* $T = 494.38$; $E = 35.37$; L.C. = 978.70; $M = 35.01$; $L = 982.00$; arc = 982.04.

(b) $I = 32°44'30''$; $D_c = 5°00'$. *Answers:* $T = 336.73$; $E = 48.44$; L.C. = 646.16; $M = 46.47$; $L = 654.83$; arc = 655.04.

(c) $I = 44°13'15''$; $D_c = 8°20'$. *Answers;* $T = 279.58$; $E = 54.62$; L.C. = 518.03; $M = 50.61$; $L = 530.65$; arc = 531.12.

(d) $I = 68°34'12''$; $D_c = 9°30'$. *Answers:* $T = 411.81$; $E = 127.02$; L.C. = 680.43; $M = 104.97$; $L = 721.79$; arc = 722.62.

(e) $I = 71°30'40''$; $D_c = 12°00'$. *Answers:* $T = 344.43$; $E = 111.10$; L.C. = 559.01; $M = 90.16$; $L = 595.93$; arc = 597.02.

2-3. Prepare deflection angles for staking the curves in problem 2-1. In every example assume that Sta. P.I. = 64 + 72.38. Carry the results to the precision used in section 2-9.

(a) Set full stations. *Partial answers:* Sta. T.C. = 59 + 73.41; C.T. = 69 + 60.61; Deflection T.C. to 60 + 00 = 0°16.62′.

(b) Set half-stations. *Partial answers:* Sta. T.C. = 61 + 51.27; C.T. = 67 + 77.44; Deflection T.C. to 62 + 00 = 1°13.10′.

(c) Set half-stations. *Partial answers:* Sta. T.C. = 62 + 00.92; C.T. = 67 + 17.05; Deflection T.C. to 62 + 50 = 2°05.15′.

(d) Set half-stations. *Partial answers:* Sta. T.C. = 60 + 61.98; C.T. = 67 + 86.05; Deflection T.C. to 61 + 00 = 1°46.07′.

(e) Set quarter-stations. *Partial answers:* Sta. T.C. = 61 + 33.43; C.T. = 67 + 22.94; Deflection T.C. to 61 + 50 = 0°59.65′.

2-4. Prepare deflection angles for staking the curves in problem 2-2. In every example assume that Sta. P.I. = 37 + 53.21. Carry the results to the precision used in section 2-9.

(a) Set full stations. *Partial answers:* Sta. T.C. = 32 + 58.83; C.T. = 42 + 40.83; Deflection T.C. to 33 + 00 = 0°20.58′.

(b) Set half-stations. *Partial answers:* Sta. T.C. = 34 + 16.48; C.T. = 40 + 71.31; Deflection T.C. to 34 + 50 = 0°50.28′.

(c) Set half-stations. *Partial answers:* Sta. T.C. = 34 + 73.63; C.T. = 40 + 04.28; Deflection T.C. to 35 + 00 = 1°05.92′.

(d) Set half-stations. *Partial answers:* Sta. T.C. = 33+41.40; C.T. = 40+63.19; Deflection T.C. to 33+50 = 0°24.51′.

(e) Set quarter-stations. *Partial answers:* Sta. T.C. = 34+08.78; C.T. = 40+04.71; Deflection T.C. to 34+25 = 0°58.39′.

2-5. List the nominal and true chord lengths needed for staking regular stations on the curves in problems 2-3 and 2-4.

Answers:

PROBLEM	c_n	c_t	PROBLEM	c_n	c_t
2-3(a)	100	99.99	2-4(a)	100	100
(b)	50	50	(b)	50	50.01
(c)	50	49.99	(c)	50	50.03
(d)	50	49.99	(d)	50	50.04
(e)	25	25	(e)	25	25.04

2-6. Refer to the field notes for problem 2-3 (c):

(a) By the simplest method, compute the total deflection to sta. 64 + 27. *Answer:* 9°36.50′.

(b) With transit at sta. 66, what vernier reading must be set before backsighting to sta. 63 for the vernier to read 0°00′ when the telescope is along the local tangent? *Answer:* 12°45′.

(c) With transit at sta. 66, what vernier reading must be set before backsighting to sta. 63 for the vernier to read 21°12.65′ when the telescope is plunged and sighted at sta. 67? *Answer:* 4°12.65′.

(d) With transit at the C.T., what vernier reading must be set before backsighting to sta. 66 for the vernier to read 0°00′ when the telescope is sighted to the P.I.? *Answer:* 4°58.47′.

2-7. Refer to the field notes for problem 2-4(d):

(a) Compute the total deflection to sta. 35+81. *Answer:* 11°22.86′.

(b) With transit at sta. 38, what vernier reading must be set before backsighting to sta. 35 for the vernier to read 0°00′ when the telescope is along the local tangent? *Answer:* 14°15′.

(c) With transit at sta. 38, what vernier reading must be set before backsighting to sta. 35 for the vernier to read 31°17′00″ when the telescope is plunged and sighted to sta. 40? *Answer:* 7°32.01′.

(d) With transit at the C.T., what vernier reading must be set before backsighting to sta. 38 for the vernier to read 0°00′ when the telescope is sighted at the P.I.? *Answer:* 12°30.09′.

2-8. Find D to the nearest $\frac{1}{2}°$ for each of the following curves, given I and the approximate T or E:

(a) $I = 30°18'$; T = approx. 440 ft. *Answer:* $3\frac{1}{2}°$.

(b) $I = 35°50'$; E = approx. 50 ft. *Answer:* 6°.

(c) $I = 54°30'$; T = approx. 360 ft. *Answer:* 8°.

(d) $I = 71°52'$; E = approx. 142 ft. *Answer:* $9\frac{1}{2}°$.

2-9. Prepare deflection angles for staking the following even-radius curves, given I and R. In every example assume that Sta. P.I. = 53+18.45. Carry the results to the precision used in section 2-9.

(a) $I = 20°34'$; $R = 2500$ ft. Set full stations. *Partial answers:* Sta. T.C. = 48+64.87; C.T. = 57+62.26; Deflection T.C. to 49+00 = 0°24.15′.

(b) $I = 31°18'30''$; $R = 1200$ ft. Set half-stations. *Partial answers:* Sta. T.C. = 49+82.18; C.T. = 56+37.90; Deflection T.C. to 50+00 = 0°25.52′.

(c) $I = 43°52'15''$; $R = 675$ ft. Set half-stations. *Partial answers:* Sta, T.C. = 50+46.62; C.T. = 55+63.46; Deflection T.C. to 50+50 = 0°08.61′.

(d) $I = 70°44'30''$; $R = 500$ ft. Set quarter-stations. *Partial answers:* Sta. T.C. = 49+85.38; C.T. = 56+02.72; Deflection T.C. to 50+00 = 0°50.26′.

2-10. Given the following data for a metric curve: $I = 70°30'$; D_m (arc def.) = 4°00′; sta. P.I. = 31+06.285. Prepare notes for setting stakes at every 10-m station with deflections to the nearest 10″. *Partial answers:* Sta. T.C. = 21+05.054; C.T. = 39+01.304; Deflection T.C. to 22+00 = 0°59.35′.

2-11. Compute the tangent offset t_2 (Fig. 2-10) to sta. 62 for the curve in problem 2-3(*e*). Use four different methods: (1) Equation 2-29 for C.O., then $t_2 = \frac{1}{2}(\text{C.O.})(0.6657)^2$; (2) Equation 2-30; (3) Equation 2-33; and (4) $t_2 = c_2 \sin 3°59'40''$. *Answer:* $t_2 = 4.65$ (or 4.64).

2-12. Compute the tangent offset t_1 to sta. 61 for the curve in problem 2-3(e). Verify answer by at least two different methods. *Answer:* $t_1 = 1.17$.

2-13. Find the indicated offsets to hundredths for the following curves (refer to the notes preceding Table I):

(a) $D_c = 6°00'$. Find T.O., M.O., and t for a 50-ft chord. *Answer:* T.O. = 5.23 (or 0.24); M.O. = 1.31; $t_{50} = 1.31$.

(b) $D_a = 6°00'$. Find T.O., M.O., and m for a 50-ft chord. *Answer:* T.O. = 5.23 (or 0.24); M.O. = 1.31; $m_{50} = 0.33$.

(c) $D_c = 20°00'$. Find T.O., M.O., and t for a 20-ft chord. *Answer:* T.O. = 17.36 (or 0.37); M.O. = 4.37; $t_{20} = 0.69$.

(d) $D_a = 20°00'$. Find T.O., M.O., and m for a 20-ft chord. *Answer:* T.O. = 17.27 (or 0.28); M.O. = 4.35; $m_{20} = 0.17$.

2-14. Compute tangent distances and offsets for staking full stations on the curve in problem 2-3(a) by tangent offsets. Use a systematic tabular form and make an independent check of the coordinates of the C.T.

2-15. Find the offsets o_1, o_2, and so forth, from the long chord (see Fig. 2-11) to the regular full stations on the curve in problem 2-3(a). Partial answers: $o_{60} = 6.3$; $O_{64} = 43.4$; $o_{68} = 22.2$.

2-16. Find D_a (to the nearest 10′) for a curve to pass through a point 35 ft inside the initial tangent and at a chord distance 240 ft beyond a selected

T.C. Make an approximate check of the answer by equation 2-30. *Answer:* $D_a = 6°50'$.

2-17. Compute L, L_o, L_i, and the true half-station chords needed for staking:

(a) Curves 120 ft outside and 120 ft inside the curve in problem 2-1(b). *Answers:* $L_o = 691.74$; $L_i = 560.60$; c_t on center line = 50.00; $c_o = 55.23$; $c_i = 44.76$.

(b) Curves 175 ft outside and 175 ft inside the curve in problem 2-1(c). *Answers:* $L_o = 650.13$; $L_i = 382.13$; c_t on center line = 49.99; $c_o = 62.97$; $c_i = 37.01$.

2-18. Derive: (a) formula 2-30; (b) formula 2-31; and (c) formula 2-35.

Chapter 3
Compound and Reverse Curves

3-1. Definitions

Stated generally, a compound curve consists of two or more consecutive curves that are tangential. In the terminology of route surveying, however, a *compound curve* is a two-arc simple curve having its centers on the same side of the common tangent at the junction; whereas a *reverse curve* is one having its centers on opposite sides. A *multicompound curve* has three or more centers on the same side of the curve.

COMPOUND CURVES

3-2. Use

Owing to the inequality of their tangent distances, compound curves permit the fitting of a location to the topography with much greater refinement than do simple curves. Conditions often occur in railroad and highway location where the changes in direction between established tangents can only be accomplished economically by compound curves. This is true in mountainous terrain or along a large river winding close to a rock bluff.

The flexibility of compound curves may tempt the locating engineer to use them merely to reduce grading quantities or to expedite the field work (as in section 3-14). This is not good practice, as it complicates design details related to superelevation and introduces certain permanent operating disadvantages. A compound curve should not be used where a simple curve is practicable. If a large difference in radii cannot be avoided, use a combining spiral (see section 5-16).

3-3. Requirements for a Rigid Solution

Figure 3-1 shows a compound curve, the notation for which is self-explanatory. The layout has *seven* important *parts* T_S, T_L, R_S, R_L, I_S, I_L, and I. However, since $I = I_S + I_L$, there are only *six independent variables*, namely, the four lengths and any two angles. Trial with compass and ruler will show that:

> For a rigid solution four parts must be known, including at least one angle and at least two lengths.

Of the many possible combinations of four known parts giving a rigid solution, some are more readily solved than others. In practice the more difficult problems may often be converted to simpler cases by measuring one more angle or distance. Certain combinations rarely occur in the field.

The treatment which follows is not exhaustive; yet it is complete enough to serve as a basis for solving any compound-curve problem.

3-4. Solution Through Vertex Triangle

The most obvious method of solving a compound-curve problem is by means of the triangle VV_SV_L, Fig. 3-1, formed by lines joining the P.I. and the vertices of the two simple curves. (The base of the triangle is the common tangent at the C.C.) If this vertex triangle can be solved, all unknown parts of the layout are easily determined. For example, if I_S, I_L, R_S, and R_L are known, solve first for the tangent distances of the individual simple curves. Their sum equals the base of the vertex triangle. Solve that triangle for the other sides by two applications of the sine law. Then find the missing compound-curve tangents, T_S and T_L, by adding those sides to the proper individual tangent distances. Of course, $I = I_S + I_L$.

It should be emphasized that the vertex-triangle method of solution is possible only if at least two of the known compound-curve parts are angles (fixing its shape), and then only if one side of the triangle can be found (fixing its size). The method is *not* applicable if, along with any two angles, the parts T_S and T_L, R_S and T_L, or R_L and T_S are known.

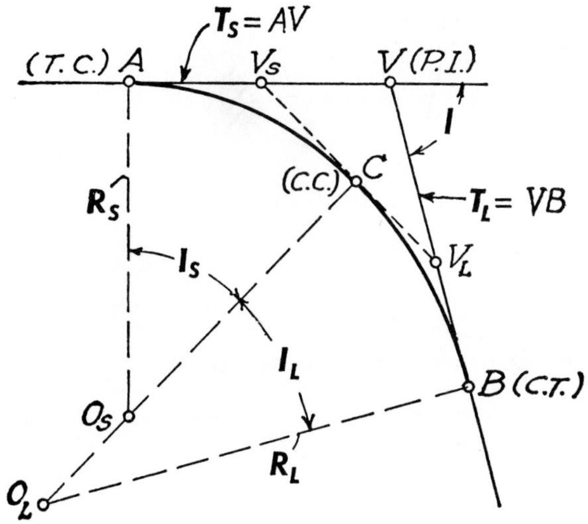

Figure 3-1 Compound-curve layout.

3-5. Solution by Traverse

One of the most useful cases occurs when I has been measured, and topographical conditions fix the T.C. and C.T. within narrow limits at unequal distances from the P.I. If one more variable is assumed, the layout is fixed, but the problem may not be solved easily. Should the assumed variable be the degree of the sharper curve (a practical situation, since D is often limited by specifications), then R_S is the fourth known part, and solution through the vertex triangle is impossible.

This problem may be solved by using the fact that the algebraic sums of the latitudes and departures of a closed traverse must each equal zero.

Figure 3-2 is the same as Fig. 3-1 except for the omission of the common tangent at C. The seven parts of the layout are interrelated by the five-sided traverse $AVBO_LO_S$. The general rules of procedure are:

1. Draw traverse to include all seven parts.
2. Take 0°00′ azimuth parallel or perpendicular to an unknown length and proceed clockwise around the traverse.
3. Set Σ latitudes and Σ departures equal to zero, obtaining two equations.
4. Solve the equation containing one unknown. If both equations have two unknowns, divide one by the other in such a way as to eliminate one unknown.

In expressing the latitudes and departures, consider that $I < 90°$ and recall that *latitude = length × cosine of azimuth* and *departure = length ×*

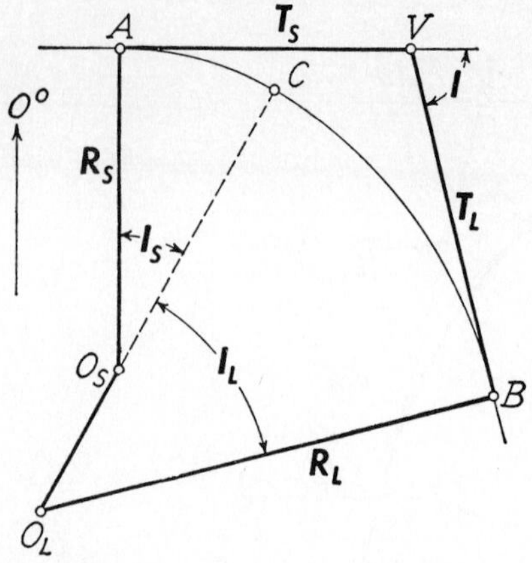

Figure 3-2

sine of azimuth. The following tabulation is convenient:

SIDE	LENGTH	AZIMUTH	LATITUDE	DEPARTURE
O_SA	R_S	0°	R_S	0
AV	T_S	90°	0	T_S
VB	T_L	$90° + I$	$-T_L \sin I$	$T_L \cos I$
BO_L	R_L	$180° + I$	$-R_L \cos I$	$-R_L \sin I$
O_LO_S	$R_L - R_S$	I_S	$(R_L - R_S) \cos I_S$	$(R_L - R_S) \sin I_S$

Then, from Σ *latitudes*,

$$R_S - T_L \sin I - R_L \cos I + (R_L - R_S) \cos I_S = 0$$

or

$$\cos I_S = \frac{T_L \sin I + R_L \cos I - R_S}{R_L - R_S} \tag{3-1}$$

The term R_S in the numerator may be eliminated by subtracting both sides from 1, canceling R_S, and converting cosines to versines. The result is

$$\text{vers } I_S = \frac{R_L \text{ vers } I - T_L \sin I}{R_L - R_S} \tag{3-2}$$

From Σ *departures*,

$$T_S + T_L \cos I - R_L \sin I + (R_L - R_S) \sin I_S = 0$$

or

$$\sin I_S = \frac{R_L \sin I - T_L \cos I - T_S}{R_L - R_S} \tag{3-3}$$

But vers A/sin A = tan $\frac{1}{2}A$ (for proof see page 517).

Thus, if formula 3-2 is divided by 3-3, the result is

$$\tan \tfrac{1}{2}I_S = \frac{R_L \operatorname{vers} I - T_L \sin I}{R_L \sin I - T_L \cos I - T_S} \tag{3-4}$$

The foregoing formulas fit the layout of Fig. 3-2 in which the curve starts with the shorter-radius arc. For the reverse layout starting with the longer-radius arc, the corresponding formulas have the subscripts reversed. Thus,

$$\cos I_L = \frac{R_L - T_S \sin I - R_S \cos I}{R_L - R_S} \tag{3-1R}$$

$$\operatorname{vers} I_L = \frac{T_S \sin I - R_S \operatorname{vers} I}{R_L - R_S} \tag{3-2R}$$

$$\sin I_L = \frac{T_L + T_S \cos I - R_S \sin I}{R_L - R_S} \tag{3-3R}$$

$$\tan \tfrac{1}{2}I_L = \frac{T_S \sin I - R_S \operatorname{vers} I}{T_L + T_S \cos I - R_S \sin I} \tag{3-4R}$$

The traverse method is valuable not only in the "standard" compound-curve problems just described but also in many special-curve problems. (Chapter 7 contains some practical examples.)

3-6. Missing Data Method

Surveyors will recognize that the five-sided closed traverse in Fig. 3-2 has two parts missing. The missing data can be found by an established method commonly used in plane surveying.

The general procedure is to draw a sketch of the closed traverse. If the defective courses are not adjacent, one such course is shifted parallel to itself into a position adjacent to the other. The closing side of the triangle thus formed with the two defective courses becomes the unknown course of a closed traverse containing all known courses. After solving for the bearing and length of the closing side all parts of the triangle can be found, plus the missing data.

From a trigonometric standpoint the foregoing method is simpler than using one of the four basic formulas developed in section 3-5 but, until the development of electronic calculators, not as rapid. However, the convenience of preprogrammed calculators justifies a description of their use as an alternative to the basic formulas.

Advantage may be taken of frequently-used surveying procedures that are "hardwired" into some desk top calculators or available as prerecorded program cards that can be inserted in some pocket calculators. For solving route surveying problems the following computer programs are convenient:

1. Bearing-bearing intersect.
2. Bearing-distance intersect.
3. Distance-distance intersect.
4. Inverse from coordinates.
5. Bearing traverse.
6. Triangle traverse.

The ten practical cases in which I is given are solved in outline form as follows:

• *Case No. 1. Given:* I_L, I_S, T_S, R_S. In Fig. 3-3 the missing lengths shown by heavy dotted lines are already adjacent. (Lines AC and CB are used only in Case 4.)

1. Assume direction O_SA as 0°00′ azimuth and point V as the origin of coordinates.
2. Find I from $I = I_L + I_S$.
3. By traverse program ascertain the coordinates of C from the traverse VAO_SC.

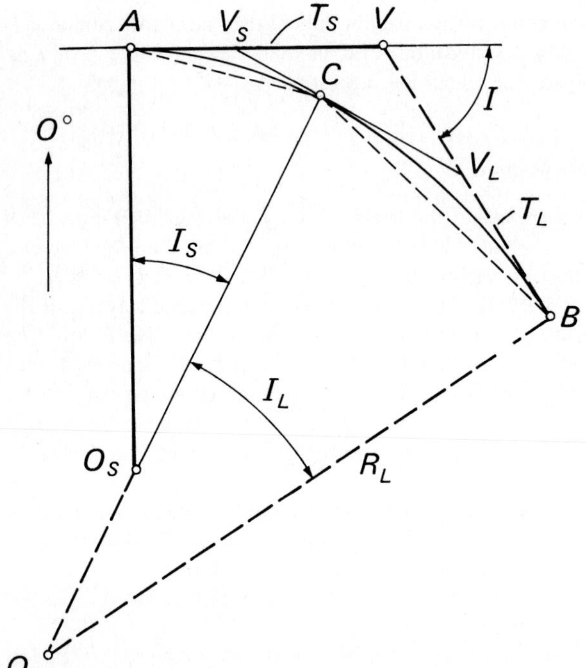

Figure 3-3

4. By bearing-bearing intersect program from V and C procure the coordinates of V_L.
5. By inverse program find the length V_LC ($V_LC = V_LB$).
6. By inverse program compute length VV_L. (VV_L may also be found from traverse VAO_SCV_L.)
7. Compute $VB = T_L$ from $VB = VV_L + V_LB$.
8. Compute R_L from $R_L = V_LB \div \tan \frac{1}{2}I_L$.

• Case No. 2. *Given:* I_L, I_S, T_L, R_L. This case is the reverse of Case No. 1. Draw a sketch in the reverse direction and follow a procedure similar to Case No. 1.

• Case No. 3. *Given:* I_L, I_S, R_L, R_S. In this case the missing lengths VA and VB are adjacent; Fig. 3-3 may be used by taking care to visualize these lengths as heavy dotted lines.

1. Assume direction O_SA as 0°00′ azimuth and point A as the origin of coordinates.
2. Find I from $I = I_L + I_S$.
3. By traverse program determine the coordinates of B from the traverse AO_SO_LB.
4. By bearing-bearing intersect program from A and B define the coordinates of V.
5. By inverse programs secure the lengths T_S and T_L from the known coordinates of A, V, and B.

• Case No. 4. *Given:* I_L, I_S, T_L, T_S. In this case the missing lengths R_L and R_S are adjacent; Fig. 3-3 may be used by taking care to visualize these lengths as heavy dotted lines.

1. Assume direction O_SA as 0°00′ azimuth and point A as the origin of coordinates.
2. Find I from $I = I_L + I_S$.
3. By traverse program find coordinates of B from the traverse AVB.
4. By bearing-bearing intersect program from A and B find the coordinates of C.
5. By inverse programs determine the lengths of the chords AC and CB from known coordinates of A, C, and B.
6. Find R_S from $R_S = AC \div 2 \sin \frac{1}{2}I_S$.
7. Find R_L from $R_L = CB \div 2 \sin \frac{1}{2}I_L$.

• Case No. 5. *Given:* I_L, I_S, T_S, R_L. See Fig. 3-4. In this case the missing lengths are not adjacent. To make them so, move VB parallel to itself to position AB'. Draw the light dotted lines.

1. Assume direction O_SA as 0° azimuth and point B' as the origin of coordinates.

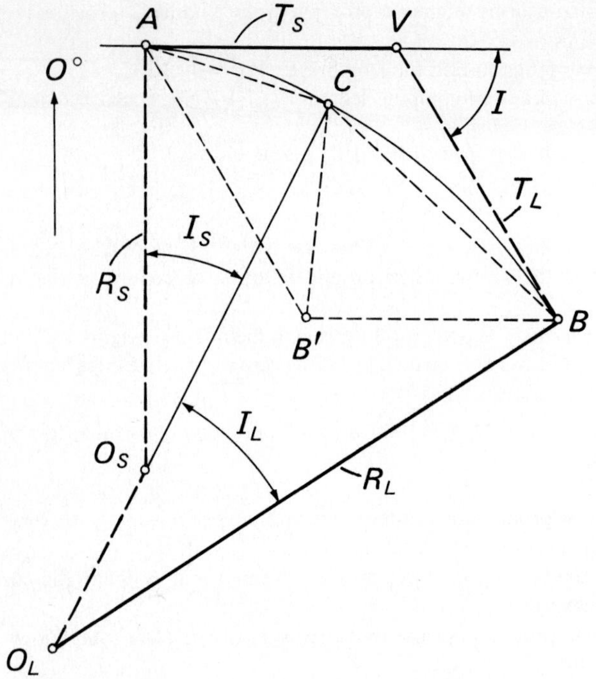

Figure 3-4

2. Find I from $I = I_L + I_S$.
3. By traverse program find the coordinates of C from the traverse $B'BO_LC$.
4. By inverse program find the length and bearing of $B'C$.
5. In triangle $AB'C$ one side ($B'C$) and all the angles are known; therefore, calculate the unknown sides AC and AB' from the triangle computer program.
6. Finally, $T_L = AB'$ and $R_S = AC \div 2 \sin \frac{1}{2} I_S$.

● Case No. 6. *Given:* I_L, I_S, T_L, R_S. This case is the reverse of Case No. 5. Draw a sketch in the reverse direction and follow a procedure similar to Case No. 5.

● Case No. 7. *Given:* I, T_L, T_S, R_S. See Fig. 3-5 in which solid lines are used in the primary traverse.

1. Assume direction O_SA as 0°00′ azimuth and point O_S as the origin of coordinates.
2. By traverse program find the coordinates of B from the traverse O_SAVB.

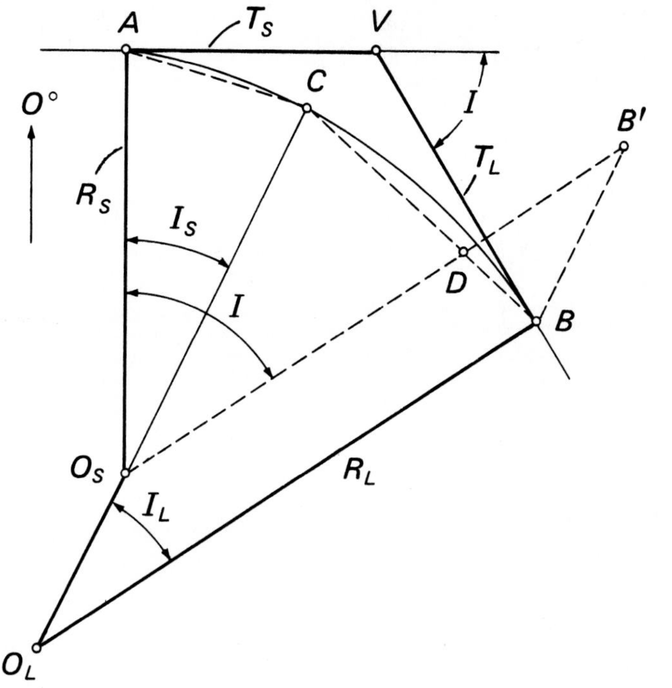

Figure 3-5

3. Draw the construction lines shown dotted, O_SB' and BB' being drawn parallel to O_LB and O_LO_S. Note that point D on arc CD must fall at the intersection of CB and O_SB'.
4. By bearing-distance intersect program ascertain the coordinates of point D at the intersection of bearing O_SB' and an arc drawn with center at O_S and radius equal to O_SC $(=R_S)$.
5. By inverse program procure the length and bearing of BD.
6. From the known bearings of BD and BV secure angle DBV; double it to obtain I_L.
7. Find I_S from $I_S = I - I_L$, from which obtain the bearing of O_SC.
8. By traverse program from O_SC find the coordinates of C.
9. By inverse program find length BC.
10. Finally, $R_L = BC \div 2 \sin \frac{1}{2}I_L$.

● *Case No. 8. Given: I*, T_L, T_S, R_L. This case is the reverse of Case No. 7. Draw a sketch in the reverse direction and follow a procedure similar to Case No. 7.

● *Case No. 9. Given: I*, T_S, R_L, R_S. See Fig. 3-6 in which solid lines are used in the primary traverse.

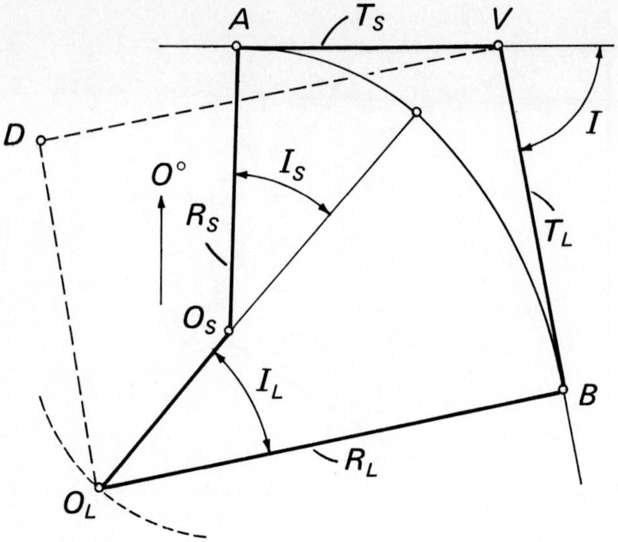

Figure 3-6

1. Assume direction O_S as 0°00′ azimuth and O_S as the origin of coordinates.
2. Draw VD parallel and equal to BO_L, and use traverse program O_SAVD to find the coordinates of point D.
3. By bearing-distance intersect program ascertain the coordinates of point O_L at the intersection of bearing DO_L and an arc drawn with center at O_S and radius equal to O_SO_L.
4. By inverse program secure the bearing of O_SO_L.
5. Determine angle I_L from the bearings of O_LO_S and O_LB.
6. Obtain I_S from $I_S = I - I_L$.
7. By inverse program procure length and bearing of DO_L which, by construction, equals T_L.

● *Case No. 10. Given:* I, T_L, R_L, R_S. This case is the reverse of Case No. 9. Draw a sketch in the reverse direction and follow a procedure similar to Case No. 9.

3-7. Practical Cases of Compound Curves

In a practical problem in which only one of the four known parts is an angle, it will be angle I. If any two angles are known, the first step is to compute the remaining angle from $I = I_S + I_L$. After this the remaining unknowns can be solved either by the electronic calculator method (section 3-6) or by use of the formulas developed in section 3-5. In certain cases the vertex triangle method is applicable.

Assuming nonuse of the electronic calculator method, the following tabulation shows how to approach the problem in the 10 practical cases enumerated in section 3-6.

CASE NO.	GIVEN	SOLUTION BY SECTIONS 3-4 OR 3-5
1	2 angles, T_S, R_S	Vertex triangle, or 3-2R for R_L
2	2 angles, T_L, R_L	Vertex triangle, or 3-2 for R_S
3	2 angles, R_L, R_S	Vertex triangle, or 3-2 for T_L
4	2 angles, T_L, T_S	3-4R for R_S, or 3-4 for R_L
5	2 angles, T_S, R_L	3-2R or 3-1R for R_S
6	2 angles, T_L, R_S	3-2 or 3-1 for R_L
7	I, T_L, T_S, R_S	3-4R for $\frac{1}{2}I_L$
8	I, T_L, T_S, R_L	3-4 for $\frac{1}{2}I_S$
9	I, T_S, R_L, R_S	3-2R or 3-1R for I_L
10	I, T_L, R_L, R_S	3-2 or 3-1 for I_S

3-8. Example of Calculation

Given: $I = 67°21'50''$; $T_S = 981.63$ ft; $T_L = 1{,}401.56$ ft; $D_L = 2°00'$ (arc def.). The curve begins with the shorter-radius arc (Fig. 3-2). Determine* I_S, I_L, and D_S (arc def.), each to the nearest 1″.

Use formula 3-4, which is

$$\tan \tfrac{1}{2}I_S = \frac{R_L \operatorname{vers} I - T_L \sin I}{R_L \sin I - T_L \cos I - T_S}$$

$$\begin{aligned} R_L \operatorname{vers} I &= 2864.79 \times 0.615123 = 1762.198 \\ -T_L \sin I &= -1401.56 \times 0.922967 = -1293.594 \\ &\qquad\qquad\text{Alg. sum} = 468.604 = \text{Num.} \end{aligned}$$

$$\begin{aligned} R_L \sin I &= 2864.79 \times 0.922967 = 2644.107 \\ -T_L \cos I &= -1401.56 \times 0.384877 = -539.428 \\ -T_S &= -981.63 \\ &\qquad\qquad\text{Alg. sum} = 1123.049 = \text{Den.} \end{aligned}$$

$$\tan \tfrac{1}{2}I_S = \frac{\text{Num.}}{\text{Den.}} = \frac{468.604}{1123.049} = 0.417261^{-}$$

whence

$$\tfrac{1}{2}I_S = 22°38'56'' \quad \text{and} \quad I_S = 45°17'52''$$

* *Note.* This problem illustrates Case No. 8 in section 3-7. It also shows the utility of Table XX, which lists all needed natural functions in one place.

Use formula 3-2, which is

$$R_L - R_S = \frac{R_L \text{ vers } I - T_L \sin I}{\text{vers } I_S}$$

$$R_L - R_S = \frac{\text{Num.}}{0.296578} = 1580.036$$

$$R_L = \underline{2864.79}$$

$$\text{whence } R_S = 1284.754 \quad \text{and} \quad D_S = 4°27'34''$$

$$I_L = I - I_S = 67°21'50'' - 45°17'52'' = 22°03'58''$$

3-9. Multicompound Curves

In especially difficult terrain a compound curve may be chosen to fit the situation better by using more than two circular arcs. Such a multicompound curve may be located by the trial field method described in section 3-14. A detailed example of the replacement of substandard alignment by a multicompound curve is given in section 7-14.

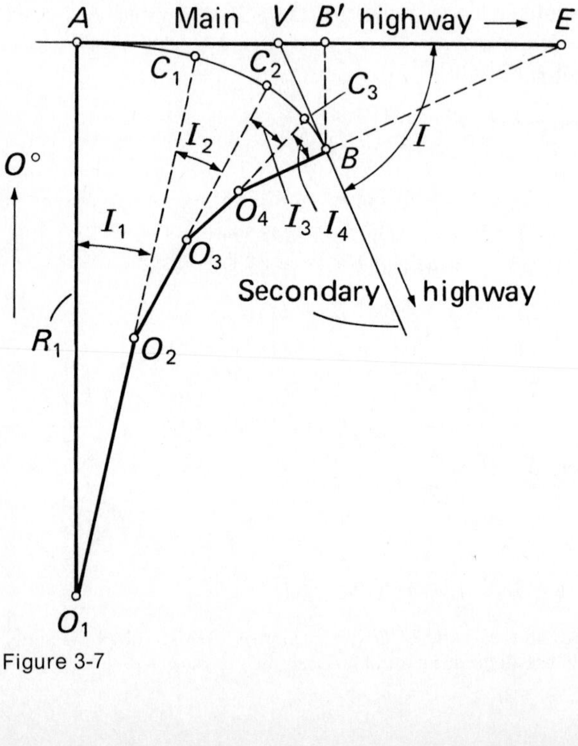

Figure 3-7

At highway interchanges, restricted physical conditions often prevent the use of simple curves on ramps. Spirals (Chapter 5) are sometimes chosen in such situations. Another common solution is to use multicompound curves consisting of three or more consecutive simple curves having their centers on the same side of the ramp.

Figure 3-7 shows a multicompound curve $AC_1C_2C_3B$ such as might be used for an interchange exit ramp. In practice angle I between the main and secondary highways is known. Convenient integral values of the four radii and four central angles would be chosen by trial to fit the terrain, leaving the tangent distances T_L (or AV) and T_S (or VB) as the unknowns to be determined.

One method of solution is that of section 3-6 as follows:

1. Assume direction O_1A as 0°00′ azimuth and point A as the origin of coordinates.
2. From the traverse $AO_1O_2O_3O_4B$ find the coordinates of point B (these are AB' and $B'B$).
3. Compute T_S from $T_S = B'B \div \sin I$.
4. Determine T_L from $T_L = AB' - T_S \cos I$. Symmetrical multicompound curves have many uses outside of route surveying, especially in architecture. The "three-centered" oval, or "basket-handle" arch, is a common form.

REVERSE CURVES

3-10. Limitations and Uses

When conditions do not permit a simple curve AB inside the P.I. of established tangents (see Fig. 3-8), the change in direction may be accomplished by locating a reverse curve $A'CB'$ in the area beyond the P.I. Points A' and B' may lie on either side of the P.I. and point B, the positions depending on the radii.

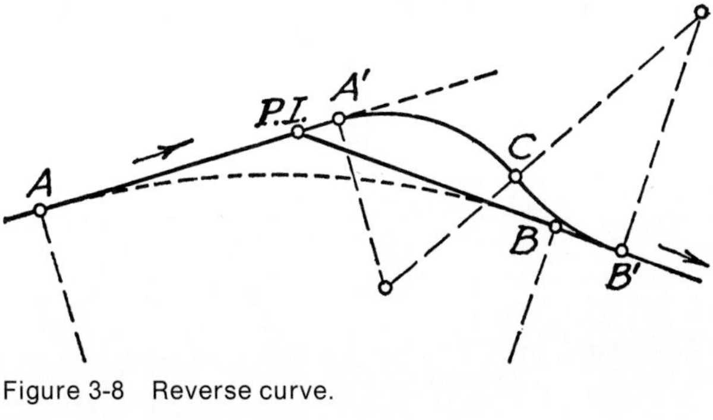

Figure 3-8 Reverse curve.

On high-speed routes, reverse curves are inadmissible. This is particularly true on highways and railroads because opposite superelevation at the point of reversal cannot be provided. If the area beyond the P.I. must be used for the location, the two arcs of the reverse curve should be separated by a tangent long enough to permit proper operating conditions. This should be at least as long as the length of the longest vehicle using the road.

Reverse curves may be used to advantage on closed conduits such as aqueducts and pipe lines; on flumes and canals where erosion is no problem; and on local roads, in railroad yards, or in any similar location where speeds are low.

3-11. Case of Parallel Tangents

The simplest case of a reverse curve occurs when the tangents are parallel (see Fig. 3-9).

With the aid of the perpendicular dropped from C to the radii, it is clear that $p = AE + FB$ and $AD = EC + CF$, or

$$p = (R_L + R_S) \text{ vers } I \tag{3-5}$$

and

$$AD = (R_L + R_S) \sin I \tag{3-6}$$

Usually, p is known and two more variables must be assumed. More commonly, $R_L = R_S$, which reduces the number of variables to four, of which two must be known or assumed.

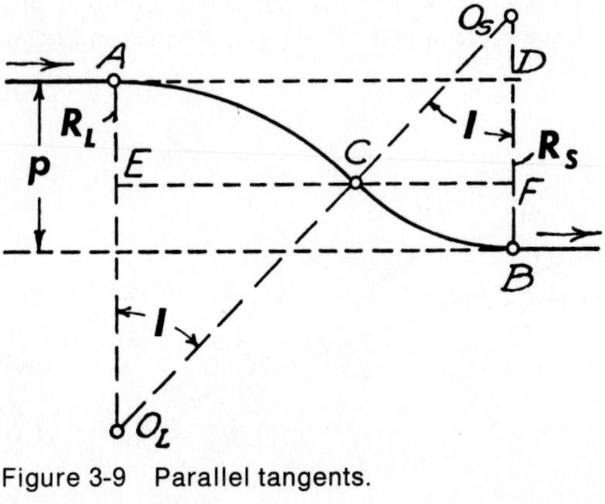

Figure 3-9 Parallel tangents.

3-12. Case of Nonparallel Tangents

In essential theory, the general case of a reverse curve between nonparallel tangents is no different from a compound curve. The same seven parts are present, and the identical requirement for a rigid solution must be met (see section 3-3). The problem can be treated as two separate simple curves, provided sufficient data are available. More complicated cases may be solved by the computer method of section 3-6 or by the traverse method of section 3-7. (See problems 3-8 and 3-9.)

COMMENTS ON NOTES AND FIELD WORK

3-13. Notes

Notes for staking compound and reverse curves by deflection angles are computed by treating the branches as separate simple curves, though they are set up in continuous stationing according to the form given in section 2-9. Owing to the change in curvature at the C.C. (or at the point of reversal in curvature, or C.R.C., on a reverse curve), that point must be occupied in moving up the transit from the T.C. The deflection angle would not be correct for a sight spanning the C.C. (or C.R.C.). Consequently, the safest form of notes is that in which the deflection at the C.C. (or C.R.C.) is $0°00'$ for starting the second arc. Some other method may be used, but it is likely to confuse the transitman.

3-14. Field Work

Setups and field checks follow the general scheme outlined in section 2-11. If a setup is made at the C.C. (or C.R.C.) and the deflections are recorded as just described, the transit is oriented so that the vernier will read $0°00'$ when the line of sight is along the common tangent. Generally, setups are best made at the T.C. and C.T., with the final check occurring at the C.C. (or C.R.C.).

Location by trial in the field often expedites fitting compound curves to particular situations. An example of an obstacle on the tangent is given in section 7-6.

Another use of the trial method is in running semifinal location in mountainous or otherwise difficult terrain not accurately mapped. Figure 3-10 represents a situation in which the direction of the back tangent and the location of the T.C. at *A* have been fixed by the topography. The direction of the forward tangent is indefinite, though it should pass near a distant prominent point. The back tangent plunges into inaccessible territory toward the P.I.

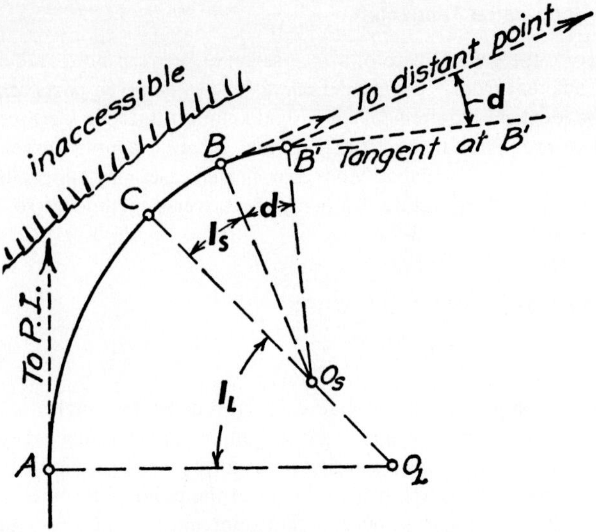

Figure 3-10 Location by trial.

The procedure is as follows: From the relation $t = \frac{7}{8} s^2 D$ (see equation 2-30) select the degree of an initial curve fitting the conditions. Continue this curve until it requires compounding (as at C) to fit the topography. Select the degree of the second arc by the same method, and repeat. Do this as many times as necessary. When the distant point is visible from a set-up on the last arc (as at B'), measure the angle d. To find the stationing of the C.T. at B, consider that the directions to the distant point from B and B' are parallel. Then, sta. B = sta. $B' - 100d/Ds$, where Ds is the degree of the final arc. Check the field work, if desired, by locating B and closing back to A by a convenient random traverse.

PROBLEMS

3-1. Given four parts of each of the following compound curves. Compute the three unknown parts. Each curve may be paired with the corresponding case in problem 3-2 if desired.

(a) Case No. 1. $I = 69°52'20''$; $I_S = 33°16'30''$; $T_S = 521.62$; D_S (arc def.) $= 10°00'$. *Answers:* $I_L = 36°35'50''$; $T_L = 703.11$; $R_L = 1151.029$ ($D_a = 4°58'40''$).

(b) Case No. 2. $I_S = 37°18'30''$; $I_L = 23°40'10''$; $T_L = 518.22$; D_L (arc def.) $= 5°30'$. *Answers:* $I = 60°58'40''$; $T_S = 413.08$; $R_S = 635.156$ ($D_a = 9°01'15''$).

(c) Case No. 3. $I = 85°31'30''$; $I_L = 40°54'00''$; $R_L = 1000$; $R_S = 600$. *Answers:* $I_S = 44°37'30''$; $T_L = 809.13$; $T_S = 652.83$.

(d) Case No. 4. $I = 54°11'40''$; $I_S = 32°32'10''$; $T_S = 812.25$; D_L (chord def.) $= 2°00'$. *Answers:* $I_L = 21°39'30''$; $T_L = 1167.98$; $R_S = 1325.57$ ($D_c = 4°19'24''$).

(e) Case No. 5. $I_S = 25°29'50''$; $I_L = 23°18'20''$; $T_L = 1239.64$; D_S (chord def.) $= 6°00'$. *Answers:* $I = 48°48'10''$; $T_S = 703.06$; $R_L = 3442.12$ ($D_c = 1°39'53''$).

(f) Case No. 6. $I = 77°32'10''$; $I_L = 30°43'20''$; $T_L = 861.47$; $T_S = 660.26$. *Answers:* $I_S = 46°48'50''$; $R_L = 1313.621$ ($D_a = 4°21'42''$); $R_S = 714.999$ ($D_a = 8°00'48''$).

(g) Case No. 7. $I = 68°29'45''$; $T_L = 1039.17$; $T_S = 716.83$; D_S (arc def.) $= 7°30'$. *Answers:* $I_L = 34°24'18''$; $I_S = 34°05'27''$; $R_L = 1810.192$ ($D_a = 3°09'55''$).

(h) Case No. 8. $I = 105°58'12''$; $T_L = 2641.08$; $T_S = 2110.17$; D_L (chord def.) $= 2°30'$. *Answers:* $I_L = 55°50'45''$; $I_S = 50°07'27''$; $R_S = 1223.47$ ($D_c = 4°41'04''$).

(i) Case No. 9. $I = 107°21'42''$; $T_S = 800.76$; D_L (arc def.) $= 6°$; D_S (arc def.) $= 12°00'$. *Answers:* $I_L = 45°45'25''$; $I_S = 61°36'17''$; $T_L = 1036.71$.

(j) Case No. 10. $I = 112°43'18''$; $T_L = 571.75$; $R_L = 500$; $R_S = 200$. *Answers:* $I_L = 49°18'18''$; $I_S = 63°25'00''$; $T_S = 413.75$.

3-2. Prepare deflection angles for staking the curves in problem 3-1. In every example assume that Sta. P.I. = 86+21.56 and setups are made at the T.C. and C.T. (see section 3-14).

(a) Curve begins with D_L; set half-stations. *Partial answers:* Sta. T.C. = 79+18.45; C.C. = 86+53.66; C.T. = 89+86.41. Defl. T.C. to 79+50 = 0°47.11'; defl. C.T. to 89+50 = 1°49.23'.

(b) Curve begins with D_S; set half-stations. *Partial answers:* Sta. T.C. = 82+08.48; C.C. = 86+22.06; C.T. = 90+52.41. Defl. T.C. to 82+50 = 1°52.36'; defl. C.T. to 90+50 = 0°03.98'.

(c) Curve begins with R_S; set full stations. *Partial answers:* Sta. T.C. = 79+68.73; C.C. = 84+36.04; C.T. = 91+49.88. Defl. T.C. to 80+00 = 1°29.58'; defl. C.T. to 91+00 = 1°25.74'.

(d) Curve begins with D_L; set full stations. *Partial answers:* Sta. T.C. = 74+53.58; C.C. = 85+36.50; C.T. = 92+89.06. Defl. T.C. to 75+00 = 0°27.85'; defl. C.T. to 92+00 = 1°55.52'.

(e) Curve begins with D_S; set full stations. *Partial answers:* Sta. T.C. = 79+18.51; C.C. = 83+43.46; C.T. = 97+43.52. Defl. T.C. to 80+00 = 2°26.7'; defl. C.T. to 97+00 = 0°21.7'.

(f) Curve begins with D_L; set half-stations. *Partial answers:* Sta. T.C. = 77+60.09; C.C. = 84+64.46; C.T. = 90+48.66. Defl. T.C. to 78+00 = 0°52.22'; defl. C.T. to 90+00 = 1°56.98'.

(g) Curve begins with D_L; set full stations. *Partial answers:* Sta. T.C. = 75+82.39; C.C. = 86+69.38; C.T. = 91+23.92. Defl. T.C. to 76+00 = 0°16.72'; defl. C.T. to 91+00 = 0°53.82'.

(h) Curve begins with D_S; set full stations. *Partial answers:* Sta. T.C. = 65+11.39; C.C. = 75+81.42; C.T. = 98+15.25. Defl. T.C. to 66+00 = 2°04.5′; defl. C.T. to 98+00 = 0°11.44′.

(i) Curve begins with D_L; set full stations. *Partial answers:* Sta. T.C. = 75+84.85; C.C. = 83+47.47; C.T. = 88+60.84. Defl. T.C. to 76+00 = 0°27.27′; defl. C.T. to 88+00 = 3°39.02′.

(j) Curve begins with R_S; set quarter-stations. *Partial answers:* Sta. T.C. = 82+07.81; C.C. = 84+29.18; C.T. = 88+59.44. Defl. T.C. to 82+25 = 2°27.7′; defl. C.T. to 88+50 = 0°32.4′.

3-3. Given the first portion of a highway interchange ramp in the form of a multicompound curve *AB* (see Fig. 3-7).

$$D_1 = 3^\circ \qquad D_2 = 6^\circ \qquad D_3 = 12^\circ \qquad D_4 = 24^\circ$$
$$I_1 = 10^\circ \qquad I_2 = 15^\circ \qquad I_3 = 20^\circ \qquad I_4 = 15^\circ$$

Compute the tangent distances T_S and T_L of the multicompound curve. *Answers:* $T_S = 286.98$; $T_L = 546.49$.

3-4. Compute I and R_S for a reverse curve between parallel tangents 500 ft apart (Fig. 3-9), given $R_L = 1800$ ft and $AD = 1500$ ft. *Answers:* $I = 36^\circ 52' 12''$; $R_S = 700$ ft.

3-5. A detour to a parallel highway must be inserted in the form of an 800-ft common-radius reverse curve. If the offset p is 150 ft (Fig. 3-9), how much space (distance *AD*) is required? *Answer:* $AD = 676.39$ ft.

3-6. Find the flattest common-radius reverse curve (R to the nearest foot) that can be inserted between parallel tangents 100 ft apart without the distance *AD* (Fig. 3-9) exceeding 800 ft. *Answer:* $R = 1625$ ft.

3-7. Given a reverse curve between nonparallel tangents with I, T_L, T_S, and R_S known. The layout starts with R_L. Derive the formula for finding I_L using the traverse method. *Answer:*

$$\tan \tfrac{1}{2} I_L = \frac{T_S \sin I + R_S \operatorname{vers} I}{T_S \cos I + R_S \sin I \pm T_L}$$

Compare with (3-4R). *Note:* When V is to left of A use $-T_L$; to right, use $+T_L$.

3-8. Numerical data for problem 3-7 are: $I = 14^\circ 00'$; $T_L = 450$; $T_S = 1500$; $R_S = 800$. Compute the three unknown parts. *Answers:* $I_L = 35^\circ 44' 49''$; $I_S = 21^\circ 44' 49''$; $R_L = 1252.33$.

3-9. Given a reverse curve between nonparallel tangents with I_L, I_S, R_L, and R_S known. The layout starts with R_S. Derive the formulas for finding T_L and T_S using the traverse method. *Note:* Answers are omitted purposely; after their derivation, compare with formulas 3-1 and 3-3.

3-10. Numerical data for problem 3-9 are: $I_L = 31^\circ 00'$; $I_S = 62^\circ 00'$; $R_L = 1400$; $R_S = 1000$. Compute the three unknown parts. *Answers:* $I = 31^\circ 00'$; $T_L = 2083.93$; $T_S = 388.25$.

Chapter 4
Parabolic Curves

4-1. Uses

Parabolic, instead of circular, arcs may be used for horizontal curves. Where the curves are flat, there is no discernible difference between the two types. However, a parabola cannot be staked out readily by the deflection-angle method. Moreover, determination of the radius of curvature at any point requires higher mathematics, thus complicating superelevation and related calculations. Also, parabolic alignment does not permit making simple right-of-way descriptions. For these reasons parabolic arcs on horizontal curves are restricted to such locations as park drives and walks, where they may be easily located by tape alone.

For curves in a vertical plane the situation is the reverse. Here, parabolic arcs are almost always used because elevations can be computed much more easily than on circular arcs. The vertical curve used is a portion of a vertical-axis parabola, the particular parabola and portion being chosen with regard to certain practical considerations.

4-2. Equal-Tangent Vertical Curve

Figure 4-1 portrays a vertical-axis parabola, PP', of indefinite length. The general equation of such a parabola with respect to random rectangular axes is $y = ax^2 + bx + c$. The magnitude of a controls sharpness of the parabola; its sign controls the orientation. When a is positive the parabola is turned upward; when negative, downward.

The equal-tangent vertical curve used in route alignment is simply a portion of a vertical-axis parabola selected so its length and tangent gradients are best suited to the design requirements. Such a selection is represented in Fig. 4-1 by curve AB which has the numerically unequal tangent gradients G_1 and G_2. Grades rising in the direction of stationing are positive; falling grades are negative.

It is convenient to use the beginning of a vertical curve (point A) as the origin of x coordinates. Term c then becomes the elevation of point A measured above a chosen datum for elevations. The slope dy/dx of any tangent is $2ax + b$. But at point A, $x = 0$; consequently term b in the general parabola equation equals the slope at A, the tangent gradient G_1.

The second derivative d^2y/dx^2 of the general function $y = ax^2 + bx + c$ equals $2a$, a constant. This means that a tangent to a vertical-axis parabola changes a constant amount of *grade* for each increment of distance. (In contrast, a tangent to a circular arc changes direction a constant amount of *angle* for equal increments of distance along the arc.) The useful consequence is that *the rate of change of grade on a vertical curve is constant, and equals $2a\%$ per station.* On a vertical curve the total change in direction between the profile grades is $G_2 - G_1$, termed $A\%$. If this change is accomplished on a curve L ft long, the constant rate of change must be

$$2a = \frac{100(G_2 - G_1)}{L} = \frac{100\,A}{L} \qquad (4\text{-}1)$$

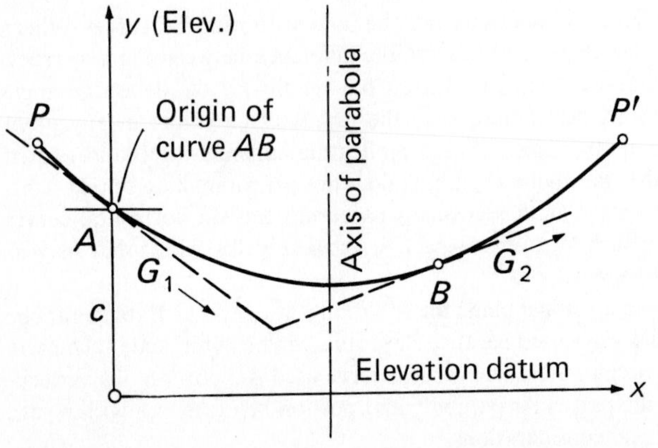

Figure 4-1

with G and A in percent, L in feet. It is convenient to think of $2a$ as a measure of the sharpness of a vertical curve. In this respect it is analogous to D for a circular curve. Another measure of curvature that is used in sight-distance calculations (see Chapter 8) is the term K, which equals $L \div A$. This is the horizontal distance in feet required to effect a 1% change in gradient on a vertical curve. Therefore

$$K = \frac{L}{A} = \frac{50}{a} \tag{4-2}$$

The practical formula for an equal tangent vertical curve is therefore

$$y = aX^2 + G_1X + c \tag{4-3}$$

where y is elevation in feet, X is the horizontal distance in feet from point A to the point being computed, G_1 is the incoming grade in percent, and a is one-half the rate of change of grade in percent.

The terms in equations 4-1 and 4-3 have the graphical significance shown in Fig. 4-2. *Tangents drawn from any two points on a vertical-axis parabola always intersect midway between the points of tangency.* This is the reason why the portion of a vertical-axis parabola used in route alignment is called an "equal-tangent vertical curve." For purpose of computation, the length L of a vertical curve *is not* the distance along the curve; it is the difference in stationing (foot units) between the ends of the curve regardless of the signs or numerical values of the tangent gradients. Thus in Fig. 4-2 the length L is the horizontal projection of AB (600 ft), and the projection of V lies $L/2$ (300 ft) from both A and B.

The first term of equation 4-3, aX^2, is the offset in feet from the incoming grade line AV, leading to the *rule of offsets.*

Vertical offsets from a tangent to a parabola are proportional to the squares of the distances from the point of tangency.

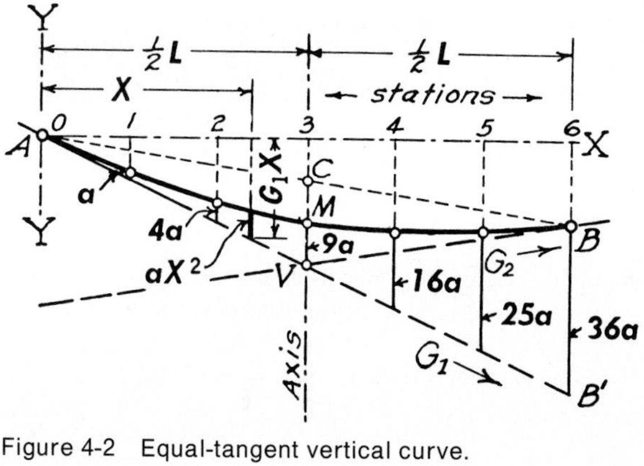

Figure 4-2 Equal-tangent vertical curve.

For the 600-ft curve in Fig. 4-2, the term aX^2 gives offsets from AB' of a for the first station, $4a$ for the second station, $9a$, $16a$, $25a$, and $36a$; those from BV(extended) are the same. In fact, offsets depend only on a and X and are identical for every tangent drawn to the curve.

The second and third term in equation 4-3, $G_1X + c$, yields elevations on the incoming tangent (extended) above or below the curve point at distance X. The second term by itself gives the elevation rise or fall from the curve's beginning point A to the required tangent point.

The parabola also has the important characteristic that *the external distance E and the middle ordinate M are equal.* (This is not quite true for a circular arc. See Fig. 2-1.) In Fig. 4-2, $VC = \frac{1}{2}B'B$ (similar triangles). From the rule of offsets, $VM = \frac{1}{4}B'B$. Thus, $VM = MC$.

The term $G_2 - G_1$ is the amount, in percent, by which the tangent grades diverge. It is analogous to the angle I of a simple curve. Formulas are simplified by replacing $G_2 - G_1$ by the term A, defined previously as the algebraic difference in grades. In Fig. 4-2, $B'B$ equals the amount the grades diverge in the distance $\frac{1}{2}L$, or $B'B = (L/2)(G_2 - G_1)/100$, *L being in foot units.* Since $VM = \frac{1}{4}B'B$, the external distance, designated as E, is

$$E = \left(\frac{L}{800}\right)(G_2 - G_1) = \frac{AL}{800} \qquad (4\text{-}4)$$

4-3. Methods of Calculation of Vertical Curves

The object of vertical-curve calculations is to determine the elevations at specified stations on the designed grade line. These elevations are needed for cross sectioning prior to grading and for setting construction grade stakes.

Before starting the calculations, a simple sketch should be drawn which shows G_1 and G_2 in their correct relations with regard to sign and magnitude. This will show to which of the six possible types the vertical curve conforms, and will help in making a common-sense check of the results.

The calculation is simpler than for horizontal circular curves; no trigonometric formulas or special curve tables are needed. Either the *algebraic method* or the *method of chord gradients* may be used.

4-4. Algebraic Method

The algebraic method (also called the tangent-offset method) uses the basic formula 4-3. It is accomplished very quickly by an electronic pocket calculator, as in the following example.

Given: Sta. $V = 73+40$; Elev. $V = 254.16$; $G_1 = +5.2\%$; $G_2 = +0.8\%$; $L = 550$ ft.

STATION	COMPUTATIONS	CURVE ELEV.
(A) 70 + 65	$254.16 - 2.75 \times 5.2 = c$	= 239.86
71	$+ 5.2 \times 0.35 - 0.4 \times 0.35^2$	= 241.63
72	$c + 5.2 \times 1.35 - 0.4 \times 1.35^2$	= 246.15
73	$c + 5.2 \times 2.35 - 0.4 \times 2.35^2$	= 249.87
74	$c + 5.2 \times 3.35 - 0.4 \times 3.35^2$	= 252.79
75	$c + 5.2 \times 4.35 - 0.4 \times 4.35^2$	= 254.91
76	$c + 5.2 \times 5.35 - 0.4 \times 5.35^2$	= 256.23
(B) 76 + 15	$c + 5.2 \times 5.5 - 0.4 \times 5.5^2$	= 256.36

Find $2a$ from formula 4-1. $2a = (+0.8 - 5.2) \times 100 \div 550 = -0.8\%$ per sta., or $a = -0.4\%$ per sta.

Find Elev. A from $254.16 - 2.75 \times 5.2 = 239.86$, and store this value in computer memory. Then find the curve elevations with the aid of formula 4-1 in the expression: Curve Elev. = Elev. $A + G_1X + aX^2$. As each curve elevation is computed record it in the calculation form; the display is automatically cleared when Elev. A is recalled from memory before computing the next curve elevation.

It is advisable to check the elevation of B from Elev. B = Elev. $V + G_2L \div 2$. Thus, Elev. $B = 254.16 + 0.8 \times 5.5 \div 2 = 256.36$ (check).

4-5. Chord-Gradient Method

Chords between full stations on a curve such as that in Fig. 4-2 must have the successive gradients $G_1 + a$, $G_1 + 3a$, $G_1 + 5a$, . . . , differing (algebraically) by $2a$. Since the offsets to the curve from BV are the same as those from AV, the calculated chord gradients are correct if the last chord gradient plus a equals G_2. For the 600-ft curve in Fig. 4-2, for example, $G_1 + 12a$ should equal G_2; it does when $L = 600$ in formula 4-1. If the calculations are started with the known elevation of A, successive addition of chord gradients gives the elevations of full stations on the curve. All elevations are checked if the elevation at B equals its value as computed around the tangent grades AVB. This method has two pronounced advantages: simple theory and automatic checks.

The characteristics of an equal-tangent vertical curve give rise to two useful principles:

1. The change in gradient between a tangent and chord equals "a" times the station length of the chord.
2. The change in gradient between two adjacent chords equals "a" times the sum of their station lengths.

An application of these principles to the solution of the example in section 4-4 is shown below.

Example of Solution by Chord Gradients

Given: Sta. $V = 73 + 40$; Elev. $V = 254.16$; $G_1 = +5.2\%$; $G_2 = +0.8\%$; $L = 550$ ft.

STATION	CHORD GRADIENT	CHORD IN STATIONS	CURVE ELEV.	PRELIMINARY CALCULATIONS
On tangent	$G_1 = +5.20$			
(*A*) 70+65	$0.35a = -0.14$		239.86	$2a = \dfrac{0.8 - 5.2}{5.5}$
	$G_1 + 0.35a = +5.06$	0.35	+1.77	$= -0.8\%$ per sta.
71	$1.35a = -0.54$		241.63	
	$G_1 + 1.7a = +4.52$	1	+4.52	$G_2 = G_1 + \dfrac{2aL}{100}$
72	$2a = -0.80$		246.15	
	$G_1 + 3.7a = +3.72$	1	+3.72	$= G_1 + 11a$
73	$2a = -0.80$		249.87	Elev. $A = 254.16$
	$G_1 + 5.7a = +2.92$	1	+2.92	$-2.75 \times 5.2 = 239.86$
74	$2a = -0.80$		252.79	
	$G_1 + 7.7a = +2.12$	1	+2.12	Elev. $B = 254.16$
75	$2a = -0.80$		254.91	$+2.75 \times 0.8 = 256.36$
	$G_1 + 9.7a = +1.32$	1	+1.32	
76	$1.15a = -0.46$		256.23	
	$G_1 + 10.85a = +0.86$	0.15	+0.13	
(*B*) 76+15	$0.15a = -0.06$		$256.36 = B$ (check)	
On tangent	$G_1 + 11a = +0.80 = G_2$ (check)			

The curve elevation at any plus point may also be found by the chord-gradient method. The procedure is to find the gradient of the subchord to the plus point (by principle 2), and then to calculate the elevation to the required plus point. As an illustration, calculations to find the curve elevation at sta. 74+30 in the foregoing example are as follows:

Chord gradient sta. 73 to 74	$= +2.92\%$
Change in gradient at 74 = $1.30a$	$= -0.52$
Subchord gradient 74 to 74 + 30	$= +2.40\%$
Elevation at sta. 74	$= 252.79$
Change in elevation = $+2.40 \times 0.30$	$= +0.72$
Required elevation at sta. 74+30	$= 253.51$

It is not essential to compute the complete vertical curve in order to find the elevation at any plus point. *Principle 1* can be used in this same example as follows:

Gradient at 70+65 = G_1	= +5.20%
Change in gradient 70+65 to 74+30 = 3.65a	= −1.46
Chord gradient 70+65 to 74+30	= +3.74%
Elevation at 70+65	= 239.86
Change in elevation = +3.74 × 3.65	= +13.65
Elevation at 74+30	= 253.51

4-6. Unequal-Tangent Vertical Curve

A vertical parabola having tangent grades of unequal station lengths is analogous to a compound curve (Chapter 3). It consists of two (or more) equal-tangent vertical curves having a common tangent where they join; and it is used where a single equal-tangent vertical curve cannot be made to fit imposed conditions so well.

Figure 4-3 shows an unequal-tangent vertical curve. It approximates a parabola having an inclined axis. However, for ease in calculating elevations, it is best treated as two consecutive equal-tangent vertical curves AM and MB. Since the vertices V_1 and V_2 of the separate parabolas are at the mid-points of AV and VB, it follows that $VM = MC$.

The gradients of AB and the parallel tangent V_1V_2 are both equal to $(G_1L_1 + G_2L_2)/(L_1 + L_2)$. Consequently, V_1M and V_1V diverge at the rate of $(G_1L_1 + G_2L_2)/(L_1 + L_2) - G_1$; and VM, the divergence in the distance $\frac{1}{2}L_1$, equals $(L_1/200)[(G_1L_1 + G_2L_2)/(L_1 + L_2) - G_1]$, which reduces to

$$E = \frac{L_1L_2A}{200L} \tag{4-5}$$

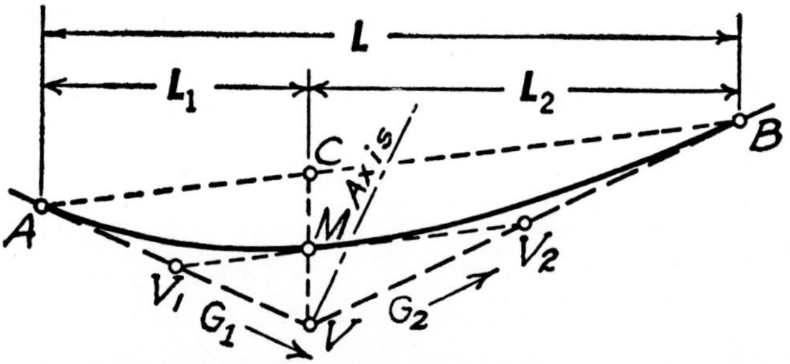

Figure 4-3 Unequal-tangent vertical curve.

In formula 4-5 the lengths are in feet; gradients, in percent. Obviously the two consecutive parabolas will have different values of $2a$. Either chord gradients or tangent offsets may be used to calculate elevations along each parabola.

The less-practical, unequal-tangent parabola, having an *inclined axis*, is treated by R. P. Vreeland in Paper No. 5739, *Journal of the Surveying and Mapping Div., ASCE*, Vol. 94, No. SU 1, January, 1968.

4-7. Lowest Point on Vertical Curve

The lowest or highest point on a vertical curve is sometimes needed. (This point is sometimes called the turning point on the curve.) For installing a culvert at the low point, the approximate stationing may be determined quickly by interpolating between calculated elevations. The stationing of a high point or low point may be computed more accurately either by applying the chord-gradient principle or from specific formulas.

From the definition of K, the distance X_t in feet from the beginning of a vertical curve to the turning point (where the gradient is zero) must be $-KG_1$. This may be derived mathematically by setting dy/dX equal to zero in the expression $y = aX^2 + G_1X$. Thus, $2aX + G_1 = 0$ at the turning point. The result of solving this equation and expressing X in foot units is $X_t = -100G_1/2a$, which reduces to

$$X_t = \frac{LG_1}{G_1 - G_2} = -KG_1 \tag{4-6}$$

After X_t has been computed, the elevation of the turning point may be found from the fact that *the gradients from the ends of the curve to the turning point are exactly one-half the gradients on the tangents.* This comes from *Principle 1*, which requires the gradient from A to the turning point to be $G_1 + aX_t/100$ (algebraically). This is equivalent to

$$G_1 + \left[\frac{100(G_2 - G_1)}{2L}\right]\left[\frac{LG_1}{100(G_1 - G_2)}\right]$$

which reduces to $\frac{1}{2}G_1$. From the turning point to B the gradient is $\frac{1}{2}G_2$.

On an unequal-tangent vertical curve the turning point may occur on either of the two parabolas, its position depending on the tangent grades. The final relations are:

$$X_1 = \left(\frac{LG_1}{G_1 - G_2}\right)\left(\frac{L_1}{L_2}\right) \tag{4-7}$$

and

$$X_2 = \left(\frac{LG_2}{G_1 - G_2}\right)\left(\frac{L_2}{L_1}\right) \tag{4-8}$$

where X_1 and X_2 are the distances (in feet) to the right and left from A and B; they cannot exceed L_1 and L_2, respectively.

4-8. Vertical Curve to Pass Through Fixed Point

In the office work of grade-line design it is often necessary to find the length of a vertical curve that will join given tangents and will pass through a fixed point. The fixed point P (Fig. 4-4) may be a road intersection or the minimum clearance over a culvert or rock outcrop. Given values will usually be the gradients G_1 and G_2, the elevation and stationing of the fixed point P, and the elevation and stationing either of the beginning of the vertical curve or of the vertex.

For any given set of conditions, the unknown length of curve may always be found by substituting known values in formula 4-3. Careful attention must be paid to algebraic signs.

Example

Given: $G_1 = -4.2\%$; $G_2 = +1.6\%$; Sta. $P = 17+00$; Elev. $P = 614.00$.

- *Case A—Beginning of Curve Known.* Sta. = 13+00; Elev. = 624.53. $y = aX^2 + G_1X$, where $y = 614.00 - 624.53$ and $a = [100(G_2 - G_1)]/2L$. Substituting,

$$-10.53 = \left(\frac{580}{2L}\right)(4)^2 - (4.2)(4)$$

from which

$$6.27L = 4640$$

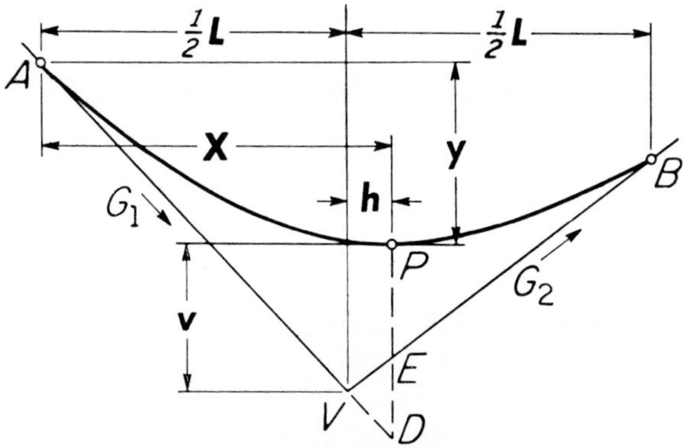

Figure 4-4 Parabola through point P.

and

$$L = 740 \text{ ft}$$

• *Case B—Vertex of Curve Known*. Sta. = 16+70; Elev. = 608.99. In this case

$$y = \frac{G_1 L}{200} + (614.00 - 608.99).$$

Or

$$\frac{-4.2L}{200} + 5.01 = \left(\frac{580}{2L}\right)\left(\frac{L}{200} + 0.3\right)^2 - (4.2)\left(\frac{L}{200} + 0.3\right)$$

which reduces to $L^2 - 744.83L + 3600 = 0$.

A rapid slide-rule solution for L is performed by first converting the quadratic to the form $L(L - 744.83) = -3600$. Then, by trial, $L = 740$. (This is the same curve as in Case A.)

The need for solving a quadratic equation may be eliminated by using the following alternate solution.*

From the rule of offsets and the construction shown in Fig. 4-4,

$$\frac{DP}{EP} = \frac{(\frac{1}{2}L + h)^2}{(\frac{1}{2}L - h)^2}$$

which reduces to

$$\sqrt{\frac{DP}{EP}} = \frac{L + 2h}{L - 2h}$$

But

$$DP = V - \frac{G_1 h}{100} \quad \text{and} \quad EP = V - \frac{G_2 h}{100}$$

Each term in these expressions is known. To simplify, therefore, replace

$$\sqrt{\frac{V - \dfrac{G_1 h}{100}}{V - \dfrac{G_2 h}{100}}}$$

by the constant C. Substituting and solving for L gives

$$L = \frac{2h(C + 1)}{C - 1} \tag{4-9}$$

* From suggestion by Max Kurtz, P. E., Brooklyn, N.Y.

In this example,

$$C = \sqrt{\frac{5.01 - \dfrac{(-4.2)(30)}{100}}{5.01 - \dfrac{(1.6)(30)}{100}}} = 1.1765$$

Therefore

$$L = \frac{2 \times 30 \times 2.1765}{0.1765} = 740 \text{ ft}$$

as before.

In formula 4-9, h is positive when the station of P exceeds the station of V and negative when the reverse is true. *For crest vertical curves*, C is found from

$$C = \sqrt{\frac{V + \dfrac{G_1 h}{100}}{V + \dfrac{G_2 h}{100}}}$$

● *Case C—Fixed Point is Highest or Lowest Point.* In this practical case the elevation of the turning point (such as the clearance over the invert of a culvert) and the vertex of the curve are known. Although the stationing of the turning point can be scaled approximately from the profile, it is assumed to be unknown and is not used in finding the required curve.

If the position of the vertex is the same as in Case B and the turning point is at elevation 613.28, then the value of X_t in stations is $LG_1/100(G_1 - G_2)$ (from 4-6). By substitution,

$$\frac{-4.2L}{200} + 4.29 = \left(\frac{580}{2L}\right)\left(\frac{-4.2L}{-580}\right)^2 - (4.2)\left(\frac{-4.2L}{-580}\right)$$

which reduces to

$$0.0058L = 4.29$$

and

$$L = 740 \text{ ft}$$

4-9. Reversed Vertical Curve

In highway location where the terrain is rolling, and also on interchange ramps, it is often necessary to insert reversed vertical curves. Figure 4-5 shows a simple example of this problem.

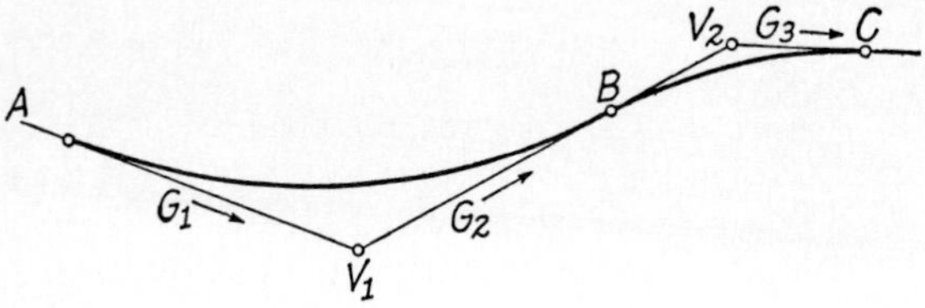

Figure 4-5 Reversed vertical curve.

A large number of cases are possible, the conditions depending on the particular combination of given values. Some cases have little practical value. However, all of them may be solved by proper use of formula 4-1. In one practical situation the given values will be the gradient G_1 and the elevation and stationing of points A, B, and C. A numerical example of this case is given in problem 4-6.

4-10. Laying Out Parabola by Taping

Though it is inconvenient to stake a horizontal parabola by the deflection-angle method, laying out such a curve by tape alone is very simple.

Figure 4-6 shows two taping methods applied to the general case of an unequal-tangent parabola. The "middle-ordinate" method (shown by solid lines) is the simpler in field work and arithmetic; it requires only division by 2. The "tangent-offset" method (shown dotted) is based upon the rule of offsets.

One variation of the middle-ordinate method follows: Set stakes at the desired controlling points A, V, and B. Measure AV and VB; divide by 2 and set the midpoints V_1 and V_2. Measure V_1V_2; divide by 2 and set the point M on the axis of the parabola (VM may be measured, but it is not essential). Then measure AM and MB; divide by 2 and set the midpoints C_1 and C_2. Set two more points, M_1 and M_2, on the parabola at the midpoints of C_1V_1 and C_2V_2 (C_1V_1 and C_2V_2 both equal $\frac{1}{2}MV$). If more points are required to define the curve, measure the chords AM_1, M_1M, MM_2, and M_2B; divide by 2 and set four more points by middle ordinates as before. When conditions

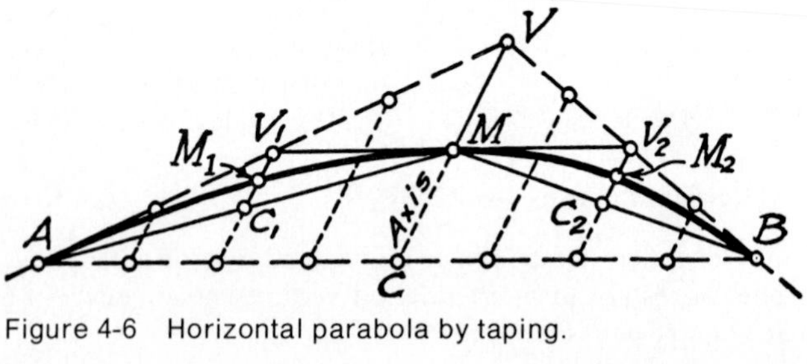

Figure 4-6 Horizontal parabola by taping.

make the long chord AB inaccessible, the middle-ordinate method is particularly suitable.

In the tangent-offset method the long chord AB is divided into an even number of equal parts. The tangents AV and VB are also divided into equal parts, their number being the same as on the long chord. Then CV is measured and bisected, and the point M on the axis of the parabola is thus located. Offsets to the parabola from the points on the tangents are measured in the direction of the corresponding points on the long chord; the distances are simple fractions of VM, computed according to the rule of offsets (see section 4-2). In approximate work, points on the long chord may be dispensed with, except for C; the offsets from the tangents are computed as before, but are aligned by eye.

PROBLEMS

4-1. Compute grade elevations to hundredths at all full stations on the following equal-tangent vertical curves. Work each case by two methods: (1) chord gradients, and (2) the algebraic method (section 4-4).

CASE	STA. V	ELEV. V	G_1 (%)	G_2 (%)	L (ft)
(a)	35 + 20	128.63	+4.8	−3.6	700
(b)	63 + 50	275.84	+5.0	+1.4	600
(c)	74 + 65	318.26	−1.26	−5.0	850
(d)	21 + 10	440.63	−3.6	+4.8	700
(e)	45 + 70	582.68	+1.4	+5.0	600
(f)	62 + 80	504.17	−5.0	−1.26	850

4-2. Calculate and check independently the curve elevation at the specified plus point on the following curves in problem 4-1:

CASE	PLUS POINT	ANSWER	CASE	PLUS POINT	ANSWER
(a)	35 + 50	121.41	(d)	19 + 05	449.27
(b)	62 + 25	268.67	(e)	44 + 18	—
(c)	75 + 31	312.13 (or 0.12)	(f)	64 + 75	—

4-3. Compute the station and elevation of the highest point on the curve in problem 4-1(a) and the lowest point on the curve in problem 4-1(d). *Answers:* Problem 4-1(a), 35+70 and 121.43; problem 4-1(d), 20+60 and 447.83.

4-4. Unequal-tangent vertical curve. *Given:* Sta. $V = 63+50$; elev. $V = 572.46$; $G_1 = -3.0\%$; $G_2 = +5.0\%$; $L_1 = 300$ ft; $L_2 = 500$ ft. *Find:* the external VM, and the station and elevation of the lowest point on the curve. *Answers:* $VM = 7.5$ ft; sta. of lowest point $= 62+30$, elev. $=$ 578.76.

4-5. Given the following data for six vertical curves:

	G_1	G_2	STA. A	ELEV. A	STA. V	ELEV. V
(a)	-2.0%	$+3.1\%$	28 + 00	450.06	—	—
(b)	-3.8%	-0.54%	56 + 50	719.86	—	—
(c)	-2.7%	$+4.3\%$	—	—	24 + 25	587.32
(d)	$+1.5\%$	-2.5%	—	—	18 + 60	426.18
(e)	-2.2%	$+4.6\%$	—	—	75 + 00	327.51
(f)	$+2.5\%$	-1.2%	—	—	45 + 25	212.47

Calculate L (to the nearest foot) for curves which will also pass through these points:

(a) Sta. P = 33+00; elev. P = 447.82. *Answer:* 822 ft.
(b) Sta. P = 60+00; elev. P = 708.50. *Answer:* 1029 ft.
(c) Sta. P = 25+00; elev. P = 594.76. *Answer:* 752 ft.
(d) Sta. P = 17+50; elev. P = 423.64. *Answer:* 526 ft.
(e) Turning point P_t, elev. 332.68. *Answer:* 695 ft.
(f) Turning point P_t, elev. 207.66. *Answer:* ___.

4-6. Given the following data for two reversed vertical curves:

	STA. AND ELEV. OF A	STA. AND ELEV. OF B	STA. AND ELEV. OF C	G_1
(a)	53 + 00 265.40	61 + 00 269.40	65 + 00 275.40	-1.0%
(b)	0 + 00 621.41	— —	17 + 00 612.06	$+1.0\%$

In curve (a) find: G_2, G_3, and the separate values of $2a$. *Answers:* $G_2 = +2.0\%$; $G_3 = +1.0\%$; $2a = +0.375$ and -0.25% per sta.

In curve (b) the reversed curves must have the same length, and $2a$ for the crest curve must be twice its values for the sag curve. Find: G_2, G_3, the values of $2a$, and the station and elevation of B. *Answers:* $G_2 = -1.48\%$; $G_3 = -0.24\%$; $2a$ (crest) $= -0.292\%$ per sta.; (sag) $= +0.146\%$ per sta.; sta. B = 25+50; elev. B = 613.08.

Chapter 5
Spirals

5-1. Foreword

In high-speed operation over alignment on which the curves are circular arcs, an abrupt change from a straight path to a circular path is required at the T.C. of the curve. It is obviously impossible to make this change instantaneously. Smooth, safe operation around railroad and highway curves requires a *gradual transition* between the uniform operating conditions on tangents and the different (but also uniform) operating conditions on circular curves. Any curve inserted to provide such a transition is called an "easement" curve.

Figure 5-1 shows a simple curve $A'E'B'$. The only way in which easement curves can be inserted at the tangents, while still preserving the radius R, is to shift the curve inward to a position represented by the parallel curve KEK'. It is impossible, however, to use a circular arc having the same length as the original circular arc. The portions KC and $C'K'$ must be deducted in order to provide room for the easement curves AC and $C'B$.

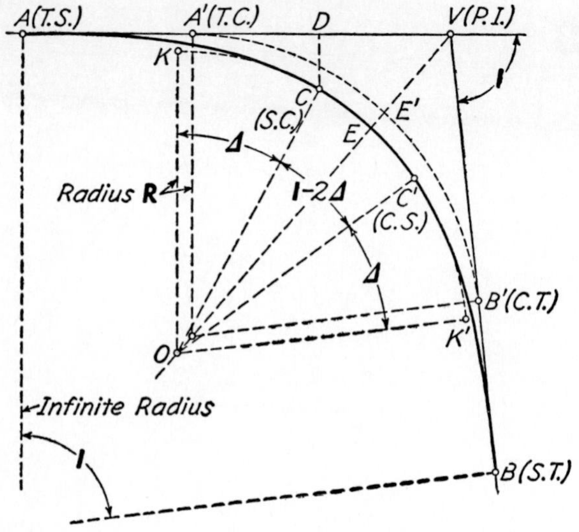

Figure 5-1 Simple curve with equal spirals.

The easement curve AC is tangent to the initial tangent at A, at which point its radius of curvature is infinite. (The tangent may also be thought of as a curve of infinite radius.) At successive points along AC, the radius of curvature decreases until it becomes equal to R at point C, where the easement AC and the circular arc CC' have a common center at O. Thus, instead of abrupt changes in direction at A' and B' on the original simple curve, there are now gradual transitions between the tangents and the simple curve CC' by means of the easement curves AC and $C'B$.

In Fig. 5-1 the curved layout starts at the T.S. (tangent to spiral) and ends at the S.T. (spiral to tangent). The *approach spiral* AC joins the circular arc at the S.C. (spiral to curve), and the circular arc joins the *leaving spiral* $C'B$ at the C.S. (curve to spiral). It should be observed that the total central angle I is unchanged. However, the central angle of the circular arc is less than that of the original simple curve $A'B'$ by the amount used up by the two spirals. Thus, the central angle I_c of the arc CC' equals $I - 2\Delta$. From theory to be developed, the spiraled curve AB will be shown to be exactly one spiral length longer than curve $A'B'$.

The effect of an easement curve is to introduce centrifugal force gradually, thus reducing shock to track and equipment on railroads and making high-speed "streamliner" operation attractive to passengers. Moreover, the easement curve tends to "build safety into the highways" by following the natural path of the vehicle between tangent and circular arc, in that way reducing the tendency to veer from the traffic lane.

An easement curve serves several incidental purposes, the most important of which is to provide a logical place for accomplishing the gradual change from zero to full superelevation. It also simplifies, on highway curves, the addition of the extra pavement width found to be needed on curves for mechanical and psychological reasons.

Finding a suitable easement curve is not difficult. On the contrary, the problem confronting engineers has been to decide which of several available forms should be selected. Many forms have been used. Some, including the cubic parabola, the lemniscate, and the clothoid, have definite mathematical equations; others, such as the Searles spiral (a multicompound curve) and the A.R.E.A. 10-chord spiral (see section 5-12), are empirical. Within the limits used in practice, all these easements give substantially the same curve on the ground. However, consideration of their relative merits from three important viewpoints—mathematical simplicity, adaptability to a variety of conditions, and ease of staking out in the field—has led most American engineers to favor the clothoid, or the spiral first investigated by the Swiss mathematician, Leonard Euler.

The clothoid (called the *Euler spiral*, the *American spiral*, or the *transition spiral*) is adopted in this book. For simplicity it will be referred to hereafter as "the spiral." All spiral tables (Tables XI through XVI-D) are based upon this spiral; some are set up with small corrections which enable rapid conversion from the arc definition to the chord definition of D on the circular arc, thus changing the spiral practically to the A.R.E.A. 10-chord form used in the past on many railroads. Therefore, the tables are adaptable to wide usage.

Although the spiral is a mathematically rich curve with applications in other fields of pure and applied science, its use in route surveying involves spiral parts that have clear graphical significance, simple formulas, and easily remembered analogies to the parts of a simple curve. Only a few relations require use of calculus in their derivations. Tabulations of spiral functions are given so completely in Part III that the actual calculation and staking of a spiral-curve layout requires only slightly more trigonometry and very little more field work than are needed for a simple curve. On the other hand, it is possible to compute a satisfactory transition spiral without referring to special tables of any kind (see section 5-11).

The adaptability of the spiral to a great variety of practical uses, especially in modern highway alignment, makes it advisable to treat the subject under three headings: simple spirals, combining spirals, and compound spirals.

A *simple spiral* connects and is tangent to a straight line and a circular arc. At the beginning of the spiral its radius of curvature is infinite; at the end, its radius equals that of the connecting arc. The most common application of the simple spiral is shown in Fig. 5-1. A special case is outlined in section 5-10.

A *combining spiral* connects and is tangent to two circular arcs having different radii of curvature. At each point of tangency the radius of curvature of the spiral equals that of the circular arc to which it connects. Such a spiral is often used to provide a smooth transition between the arcs of a compound curve, as in Fig. 5-7.

A *compound spiral* connects and is tangent to a straight line and a circular arc. In contrast to the simple spiral, however, the compound spiral *does not* have the same radius of curvature as the circular arc at the point of tangency. In addition, the compound spiral may have a finite radius of curvature where tangent to the straight line.

SIMPLE SPIRALS

5-2. Geometry of the Simple Spiral

The simple spiral obeys an exact law:

> The radius of the spiral at any point is inversely proportional to its length.

All the relations needed for computing and staking a spiraled curve stem from Fig. 5-1 and the law of the spiral. In contrast to a circular arc, the spiral is a curve of variable radius, or variable degree of curve. At any point on the spiral, however, the inverse relationship between R and D is still correctly represented by equation 2-10. Since the radius of the spiral is infinite at the T.S., its degree of curve at that point must be zero. But the law of the spiral shows an inverse relationship between the radius at any point and the distance to that point from the T.S. The following statement is therefore true:

> The degree of curve of the spiral increases at a uniform rate from zero at the T.S. to the degree D of the circular arc at the S.C.

The constant rate of increase in degree of curve per station along a spiral is represented by k. The basic formula for k is derived from its definition by dividing the total change in degree of curve, D, by the length of the spiral in stations. Since this length is $L_s/100$,

$$k = \frac{100D}{L_s} \tag{5-1}$$

The constant k is useful in a variety of problems. As an illustration, suppose a combining spiral is required in making the transition between arcs of a compound curve on which $D_L = 3°$ and $D_S = 9°$. In such a case a limitation would be placed on the sharpness of the transition by specifying, for example, that the value of k should not exceed 2° per station. This condition would be met by a 300-ft spiral ($k = 2°$) or by a 400-ft spiral ($k = 1\frac{1}{2}°$). These lengths are the latter portions of simple spirals that are 450 ft and 600 ft long.

An extremely important element of a spiral is its central angle, or *spiral angle*, Δ. For a curve of *constant* radius (a circular arc), equation 2-13 shows that the central angle equals the length of curve in stations times the degree of curve, or $I = LD/100$. Since the spiral is a curve of *uniformly changing* degree of curve, it follows that its central angle equals the length of spiral in stations times the *average* degree of curve, or

$$\Delta = \frac{L_s D}{200} \tag{5-2}$$

The foregoing central angle is exactly half that of a simple curve of the same length and degree (see formula 2-13).

The coordinates of the S.C. are very useful in computations involving simple spirals. Their values, which are $X = AD$ and $Y = DC$ (see Fig. 5-2), have been computed from exact relations derived in Appendix A and tabulated in the spiral tables of Part III; it is never necessary for the surveyor to compute them from their theoretical formulas.

In a special problem the coordinates X' and Y' may be needed. These are the coordinates of the T.S. with respect to the S.C. With the aid of Fig. 5-2 it may be deduced that

$$X' = X \cos \Delta + Y \sin \Delta \tag{5-3}$$

$$Y' = X \sin \Delta - Y \cos \Delta \tag{5-4}$$

The point K (the theoretical point where a tangent to the circular curve produced backward becomes parallel to the tangent AD) is known as the

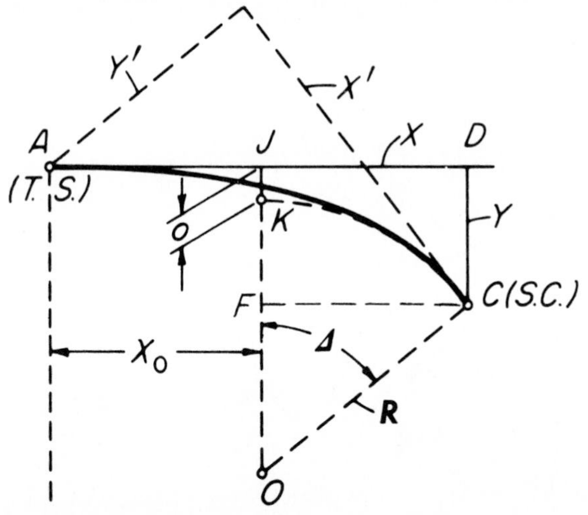

Figure 5-2 Spiral coordinates.

offset T.C. It is sometimes needed in the field; but is used more often in computations. If a perpendicular CF is dropped from C to OK, the coordinates of the offset T.C., which are $X_0 = AJ$ and $o = JK$, are found from $X_o = AD - CF$ and $o = CD - FK$, or

$$X_o = X - R \sin \Delta \tag{5-3a}$$

$$o = Y - R \text{ vers } \Delta \tag{5-4a}$$

The distance o is often called the "throw" (in Great Britain, the "shift"). It is the distance through which the circular curve must be moved inward in order to provide clearance for inserting the spiral. The shift $E'E$ at the middle of the curve (Fig. 5-1) is also called the throw. Obviously, $E'E = o \sec \frac{1}{2}I$. In any problem the particular distance referred to as the throw will be clear from the context.

Figure 5-3 represents a spiral on which P is any point located by the spiral angle δ, the radius r, and the length $AP = l$. Differentials being used,

$$d\delta = \frac{dl}{r}$$

However, from the law of the spiral,

$$r{:}R = L_s{:}l$$

Consequently,

$$d\delta = \frac{l\,dl}{R\,L_s}$$

By integration,

$$\delta = \frac{l^2}{2\,R\,L_s} \tag{5-5}$$

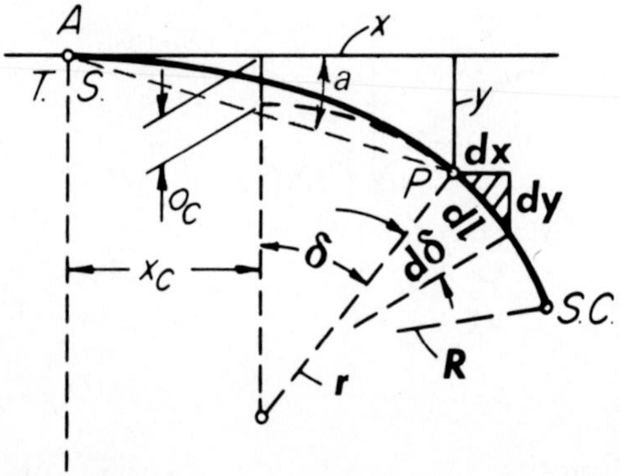

Figure 5-3

and

$$\Delta = \frac{L_s}{2R} \tag{5-6}$$

in which the angles are in radians.

When 5-5 is divided by 5-6 and the resulting equation is solved for δ, it is found that

$$\delta = \left(\frac{l}{L_s}\right)^2 \Delta \tag{5-7}$$

which expresses the following important property:

Spiral angles are directly proportional to the squares of the lengths from the T.S.

(On a simple curve, central angles are directly proportional to the first powers of the lengths from the T.C.)

Also, from Fig. 5-3,

$$\sin \delta = \frac{dy}{dl} = \delta \text{ (approximately)}$$

Therefore,

$$dy = \delta\, dl = \left(\frac{l^2}{2RL_s}\right) dl$$

By integration,

$$y = \frac{l^3}{6RL_s} \qquad \text{(5-8) Approx.}$$

This relation shows that:

Tangent offsets are closely proportional to the cubes of the lengths from the T.S.

Let a be the deflection angle in radians to any point P on a spiral. For the flat spirals used on modern alignment, a is almost equal to $\sin a$ and

$$\sin a = \frac{y}{l} \text{ (approximately)}$$

When the value in 5-8 is substituted for y,

$$a = \frac{l^2}{6RL_s} \qquad \text{(5-9) Approx.}$$

Hence, A, the deflection angle to the S.C., is

$$A = \frac{L_s}{6R} \qquad \text{(5-10) Approx.}$$

If relation 5-9 is divided by 5-10 and the resulting equation solved for a,

$$a = \left(\frac{l}{L_s}\right)^2 A \qquad \text{(5-11) Approx.}$$

This equation expresses another important property:

> Deflection angles are closely proportional to the squares of the lengths from the T.S.

(On a simple curve, deflection angles are exactly proportional to the first powers of the lengths from the T.C.)

It follows from the foregoing that

$$a = \tfrac{1}{3}\delta \qquad \text{(5-12) Approx.}$$

and

$$A = \tfrac{1}{3}\Delta \qquad \text{(5-13) Approx.}$$

Relations 5-12 and 5-13 are correct for most practical purposes. Theoretically, the relations produce values which are slightly too large. Should exact deflection angles be needed (as on a very long, sharp spiral), they are given in Table XV for any 10-chord spiral, and may be obtained quickly from Tables XVI, XVI-A, and XVI-B for a spiral staked with any number of chords up to 20. The derivation of the small correction which, if subtracted, would make equations 5-12 and 5-13 exact, is given in Appendix A.

Figure 5-4 shows the first half of Fig. 5-1. The original simple curve has been omitted, however, and certain construction lines have been added to aid in deriving necessary formulas. The triangle ABC in Fig. 5-4 is analogous to that formed by the vertex and tangent points of a simple curve. The simple curve has equal tangents and equal angles at the long chord; the spiral, on the contrary, cannot have equal local tangents or equal angles. From formula 5-13, the angle BCA must be almost exactly equal to $\tfrac{2}{3}\Delta$.

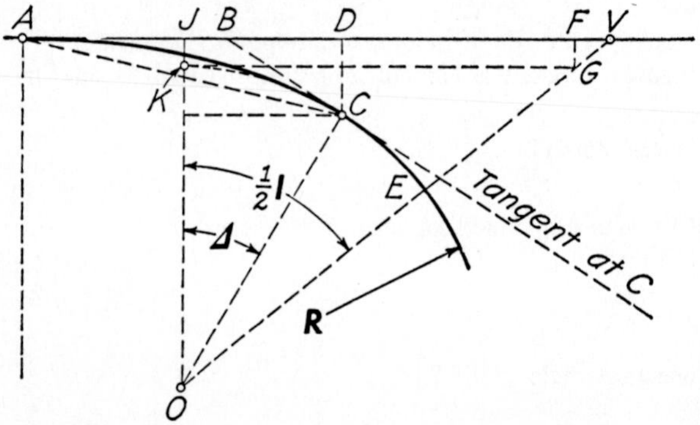

Figure 5-4

The three lengths AC, AB, and BC are occasionally useful in field work; they are called the *long chord* (*L.C.*), *long tangent* (*L.T.*), and *short tangent* (*S.T.*) of the spiral. When needed, their values are taken from tables, though they may readily be computed when X, Y, and Δ are known. For the flat spirals used on modern alignment, the L.T. and the S.T. are approximately in a 2:1 ratio.

5-3. Simple Curve with Spirals

Theoretical relations for laying out any spiral, once the T.S. has been located, were given in section 5-2. The T.S. is usually staked, as in a simple curve, by measuring the calculated tangent distance T_s from the P.I. The tangent distances will be equal in the usual case of equal spirals at the tangents. There is rarely any justification for using unequal spirals, except in realigning existing railroad track.

If equal spirals are assumed in Fig. 5-4 and a line drawn parallel to AV from K to G, T_s is made up of the three segments AJ, KG, and FV, which are X_0, $R \tan \frac{1}{2}I$, and $o \tan \frac{1}{2}I$. Therefore,

$$T_s = (R + o) \tan \tfrac{1}{2}I + X_0 \qquad (5\text{-}14)$$

or

$$T_s = T + X_0 + o \tan \tfrac{1}{2}I \qquad (5\text{-}15)$$

By the same construction, the external distance E_s (VE) may be divided into two segments EG and GV, which are R exsec $\frac{1}{2}I$ and $o \sec \frac{1}{2}I$. Therefore,

$$E_s = E + o \sec \tfrac{1}{2}I \qquad (5\text{-}16)$$

By means of a trigonometric conversion, this equation may be written in the form

$$E_s = (R + o) \operatorname{exsec} \tfrac{1}{2}I + o \qquad (5\text{-}17)$$

For the rare case of unequal spirals,

$$T_{s1} = T + X_{o1} + \frac{o_2 - o_1 \cos I}{\sin I} \qquad (5\text{-}15a)$$

and

$$T_{s2} = T + X_{o2} + \frac{o_1 - o_2 \cos I}{\sin I} \qquad (5\text{-}15b)$$

In these formulas subscripts 1 and 2 refer to the initial and the final spirals and the resulting unequal values of T_s. The last term in each formula may be positive or negative, depending on the magnitudes of o_1, o_2, and I. (See the end of section 5-20 for general versions of these formulas as applied to a completely-spiraled compound curve.)

Calculations and field work for a spiral follow the general pattern described in Chapter 2 for a simple curve; the variations are in details only. In the usual case, sta. P.I., I, and D are known. Briefly, sequence of the remaining work is as follows:

(a) Select L_s to fit the imposed conditions (section 5-15 contains reference to the choice of L_s).

(b) Calculate Δ from formula 5-2.

(c) Take X and Y from tables, and calculate X_0 and o from formulas 5-3a and 5-4a. (This is theoretical; in practice, X_0 and o also may be taken from tables.)

(d) Calculate T_s from formula 5-14 or formula 5-15.

(e) Calculate the stationing of the T.S., which is sta. P.I. $-$ T_s; stationing of the S.C., which is sta. T.S. $+$ L_s; stationing of the C.S., i.e., sta. S.C. $+$ $[100(I - 2\Delta)]/D$; and the stationing of the S.T., which is sta. C.S. $+$ L_s.

(f) Calculate deflection angles at selected points on the approach spiral, using formulas 5-13 and 5-11. (This is theoretical; for regularly spaced points on the spiral, tables or abbreviated relations may be used.)

(g) Set hubs at the T.S. and S.T. by measuring T_s from the P.I.

(h) Occupy the T.S. and stake an approach spiral to the S.C. by deflection angles. (L_s ordinarily equals the sum of the chords used to lay it out; corrections to chords are necessary only for fairly long chords near the end of long, sharp spirals.)

(i) Occupy the S.C. and backsight to the T.S. with $(\Delta - A)$ set off on the proper side of 0°; $\Delta - A$ is almost exactly equal to $2A$. Then, plunge the telescope and stake the simple curve a chosen check point by the usual methods. (The check point may be the C.S. or any point on the simple curve.)

(j) Occupy the S.T. and run in the leaving spiral to the C.S. (For regularly spaced points, the deflections are the same as those on the approach spiral.)

(k) Make a final check at the selected check point, which should preferably be near the middle of the simple curve. A setup at the C.S. is required if this is done, but on final location the resulting smooth junction at the C.S. justifies this method.

Conditions frequently warrant varying the foregoing procedure. For example, instead of following steps (h) and (j), the S.C. and the C.S. may be located (or checked) by any one of three other methods: (1) by measuring the X and Y coordinates; (2) by measuring the long tangent (L.T.), the angle Δ, and the short tangent (S.T.) of the spiral; (3) by measuring $\frac{1}{3}\Delta$ and the long chord (L.C.).

Tables may be used to expedite some of the operations just described.

Table XI: See explanation on page 354.

Table XII: See explanation on page 354.

Table XIII: This table is explained in section 5-9.

Table XIV: Discussed in section 5-10.

Table XV: This table gives exact deflection angles for any spiral up to $\Delta = 45^\circ$ staked with 10 equal chords.

If a table such as Table XV is unavailable, it is not necessary to use formulas 5-13 and 5-11 when points are spaced regularly. The following method is quicker:

Let a_1 = the deflection angle (in minutes) to the *first of n regular points.* Then

$$a_1 \text{ (in minutes)} = \frac{20\,\Delta \text{ (in degrees)}}{n^2} \tag{5-18}$$

where n = the number of equal chords on the spiral. Proof of this is left as an exercise by students; *see* problem 5-13 (a).

Calculate a_1 for given values of Δ and n. Then, to find the remaining deflection angles, simply multiply a_1 by the squares of the chord-point numbers. Thus, if Δ were 6° and n were 10, $a_1 = 1.2'$. The remaining deflections are $4\,a_1$, $9\,a_1$, $16\,a_1$, $25\,a_1$, and so forth. (Students should check these against Table XV.)

Table XVI: This table is an extension of the principle just described. The theory underlying deflections for a setup on spiral is outlined in section 5-7.

5-4. Locating Any Intermediate Point on Spiral

Though spiral is usually laid out with equal chords, the number commonly being 10, such a process does not serve all purposes. For example, on location prior to grading, earthwork estimates are made more rapidly if cross sections are taken at regular full stations and possibly half-stations. Furthermore, important "breaks" requiring cross sectioning may fall between regularly spaced points. During construction it may be necessary to set points on a spiral at trestle bents or on bridge piers. For these reasons it is convenient to have a simple formula for determining the deflection angle to any point at a distance l ft beyond the beginning of a spiral. The following relation serves this purpose:

$$a\text{(in minutes)} = \frac{kl^2}{1000} \tag{5-19}$$

The constant k may be computed from formula 5-1. It is recommended that an integral value of k be chosen in order to simplify the computation.

5-5. Field Notes

Notes for staking a spiraled curve are set up in continuous stationing according to a modification of the form in section 2-9. Deflection angles for the spirals should start with 0° at the T.S. and S.T. The comments in section 3-13 are pertinent.

FORM OF NOTES FOR SPIRALED CURVE

STA.	POINT	TOTAL DEFL.	FIELD PROCEDURE	CURVE DATA
47 + 15.8	⊙S.T.	0°00′	At S.T., orient by	$I = 14°10'$
46 + 55.8		0°04′	sighting to P.I.	$D = 3°$
45 + 95.8		0°14′	with 0°00′ on	P.I. = 43 + 31.1
45 + 35.8		0°32′	vernier. Make	$T_s = 387.5$
44 + 75.8		0°58′	final check at	$L_s = 300$
44 + 15.8	C.S.	1°30′	C.S.	
44 + 15.8	C.S.	2°35′	At S.C., orient by	
44 + 00		2°21′	sighting to T.S.	
43 + 50		1°36′	with 3°00′ on	
43 + 00		0°51′	proper side of	
42 + 50		0°06′	vernier.	
42 + 43.6	⊙S.C.	0°00′		
42 + 43.6	S.C.	1°30′	At T.S., orient by	
41 + 83.6		0°58′	sighting to P.I.	
41 + 23.6		0°32′	with 0°00′ on	
40 + 63.6		0°14′	vernier.	
40 + 03.6		0°04′		
39 + 43.6	⊙T.S.	0°00′		

In the accompanying form, the columns for calculated and magnetic bearings have been omitted in order to insert explanatory notes. Otherwise the form is a typical example of the left-hand page of a field book. Distances are here computed to tenths of a foot, and deflection angles taken to the nearest minute. In final location a higher degree of precision would be required.

In the foregoing example both spirals were run as five-chord spirals. The deflection angles could be computed by finding a_1 from formula 5-18 and multiplying by the squares of the chord-point numbers; or they could be taken directly from Table XV for every other chord point of the 10-chord spiral.

To illustrate how easily spiral deflections can be computed to fit any desired conditions, the following alternate notes are given for staking the approach spiral.

ALTERNATE NOTES FOR APPROACH SPIRAL

CASE A: 4-CHORD SPIRAL. USE FORMULA 5–18 AND TABLE XVI.			CASE B: STAKES AT REGULAR FULL STATIONS. USE FORMULA 5–19.		
STA.	POINT	TOTAL DEFL.	STA.	POINT	TOTAL DEFL.
42 + 43.6	S.C.	1°30′	42 + 43.6	S.C.	1°30′
41 + 68.6		0°51′	42 + 00		1°06′
40 + 93.6		0°22′	41 + 00		0°24′
40 + 18.6		0°06′	40 + 00		0°03′
39 + 43.6	T.S.	0°00′	39 + 43.6	T.S.	0°00′

5-6. Theory of the Osculating Circle

In Fig. 5-5, P is any point on a spiral just as in Fig. 5-3. The circular arc drawn tangent to the spiral at point P has a radius r equal to the radius of curvature of the spiral at the point of tangency. Any such tangential arc is called an *osculating circle.* From a law of the spiral, the osculating circle at any point must lie inside the spiral toward the T.S. and outside the spiral toward the S.C.

At the T.S. a spiral has an infinitely long radius of curvature and a degree of curve equal to zero. Therefore, the osculating circle at the T.S. is a straight line coinciding with the main tangent. But any circle is a curve of constant degree, whereas a spiral is a curve of uniformly changing degree. It follows, as an important principle, that:

> A spiral departs in both directions from any osculating circle at the same rate as from the tangent at the T.S.

This statement, known as the "theory of the osculating circle," is valid only for rate of departure in degree of curvature. As commonly used, however, the theory is also assumed to apply to rates of departure in angular direction and offset. These are approximations which, fortunately, are close enough for the relatively flat spirals used on primary highway and railroad alignment. But where sharp spirals are used, as sometimes occurs on highway-interchange roadways, the approximations must be replaced by mathematical exactness. It is more appropriate to develop these mathematical relations in connection with the theory of the combining spiral. This is done in section 5-17 and Appendix B.

In order to conform to the theory of the osculating circle, any complete spiral is bisected by its throw o, and vice versa. That is, line JK and spiral

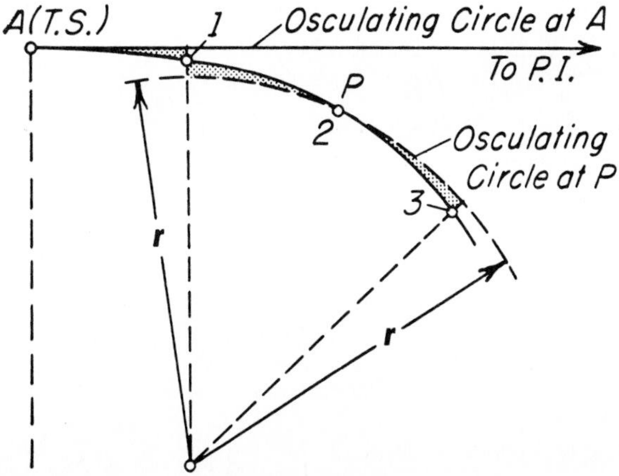

Figure 5-5 Osculating-circle theory.

AC (Fig. 5-2) bisect each other. Since tangent offsets are closely proportional to the cubes of the spiral lengths, this means that

$$o = \frac{Y}{4} \qquad \text{(5-20) Approx.}$$

Another useful relation is found by substituting $L_s^2/6\,R$ for Y (from formula 5-8), giving

$$o = \frac{L_s^2}{24\,R} \qquad \text{(5-21) Approx.}$$

The mutual bisection of spiral and throw is also true for any portion of a spiral. In Fig. 5-5, for example, assume that P is the second of several regularly-spaced chord points on a spiral. The offsets to chord point 1 from the osculating circles at A and P must be equal, and each is equal to the offset at chord point 3.

An interesting consequence of the foregoing principles is that the three shaded portions of Fig. 5-5 are equal in area. This property has practical value in computing areas of highway pavement on spiraled curves. (See section 8-21.)

5-7. Transit Setups on Spiral

Field conditions sometimes require a transit setup on a spiral. The osculating-circle principle supplies the basis for locating any other point on a spiral from such a setup.

In Fig. 5-6 assume that the spiral has been staked as far as point t from a setup at the T.S., and that obstructions to the line of sight prevent further staking from that setup. The problem is how to orient the transit at point t

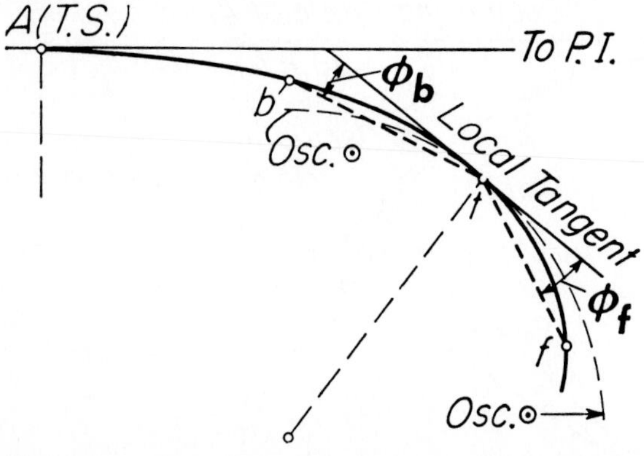

Figure 5-6 Transit setup on a spiral.

and stake the rest of the spiral from the new setup. Point b is any point on the spiral (including the T.S.) to which a backsight is taken in order to orient the transit. Point f is any point on the spiral beyond point t (including the S.C.) which is to be located by means of a foresight from t. The angle ϕ_b is the orientation angle turned between the backsight to b and the local tangent, and ϕ_f is the deflection angle to the foresight point f. Spiral lengths from the T.S. to these points are denoted by l_b, l_t, and l_f.

Angle ϕ_b equals the deflection angle for a circular arc having a length of $(l_t - l_b)$ and a degree of curve equal to that of the osculating circle *minus* the angular departure between the osculating circle and the spiral in the distance tb. Since D is the degree of curve of the spiral at the S.C., the degree of the osculating circle at t equals $(l_t/L_s)\,D$. In general,

$$\phi_b = \frac{1}{2}\left(\frac{l_t - l_b}{100}\right)\left(\frac{l_t D}{L_s}\right) - \frac{1}{3}\left(\frac{l_t - l_b}{L_s}\right)^2 \Delta$$

(Refer to formulas 2-11, 5-7, and 5-12.) When D is replaced by its equivalent, $200\Delta/L_s$, this expression may be simplified to

$$\phi_b = (2l_t + l_b)(l_t - l_b)\frac{\Delta}{3L_s^{\,2}} \qquad \text{(5-22) Approx.}$$

Angle ϕ_f equals the deflection angle to the same osculating circle for a length of $(l_f - l_t)$ *plus* the angular departure between the osculating circle and the spiral in the distance tf. That is

$$\phi_f = \frac{1}{2}\left(\frac{l_f - l_t}{100}\right)\left(\frac{l_t D}{L_s}\right) + \frac{1}{3}\left(\frac{l_f - l_t}{L_s}\right)^2 \Delta$$

which may be simplified to

$$\phi_f = (l_f + 2l_t)(l_f - l_t)\frac{\Delta}{3L_s^{\,2}} \qquad \text{(5-23) Approx.}$$

The total deflection angle ϕ from backsight to foresight equals $\phi_b + \phi_f$. When formulas 5-22 and 5-23 are added, the simplified result is

$$\phi = (l_b + l_t + l_f)(l_f - l_b)\frac{\Delta}{3L_s^{\,2}} \qquad \text{(5-24) Approx.}$$

Where stakes on a spiral are spaced regularly and a setup is required on a regular chord point, the foregoing formulas may be expressed in terms of chord point numbers. Since L_s equals n equal chords and the length l of any portion of the spiral sighted over equals L_s/n times the chord point number, it follows that

$$\phi_b(\text{in minutes}) = (2n_t + n_b)(n_t - n_b)\frac{20\Delta(\text{in degrees})}{n^2} \qquad \text{(5-25) Approx.}$$

Also

$$\phi_f(\text{in minutes}) = (n_f + 2n_t)(n_f - n_t)\frac{20\Delta(\text{in degrees})}{n^2} \qquad \text{(5-26) Approx.}$$

and

$$\phi(\text{in minutes}) = (n_b + n_t + n_f)(n_f - n_b)\frac{20\Delta(\text{in degrees})}{n^2} \qquad \text{(5-27) Approx.}$$

The second terms in the six foregoing formulas represent the spiral lengths sighted over in performing the indicated operations. In the last three formulas the final term is the deflection angle in minutes to the first regular chord point (see formula 5-18); the products of the first two terms yield the coefficients given in Table XVI. These formulas give accurate results except for long sights over long sharp spirals, where the effect of the slight approximation in formula 5-12 may become noticeable. Exact results may be obtained by applying the corrections found in Tables XVI-C and XVI-D, as explained in section 5-17 and Appendix C.

Example

Given: $D_a = 10°00'$; $L_s = 900$ ft. Spiral to be staked using fifteen 60-ft chords. Assume that obstructions require setups at chord points 5 and 12 in addition to the T.S.

PRELIMINARY COMPUTATIONS

$$\Delta = \frac{L_s D}{200} = \frac{900 \times 10}{200} = 45° \qquad a_1 = \frac{20\Delta}{n_2} = \frac{20 \times 45}{15^2} = 4'$$

If this spiral were staked entirely from one setup at the T.S., the deflections to the distant chord points would have to be *decreased* by slight corrections. The theory behind these corrections is outlined in Appendix A; corrections are listed in Table XVI-A. Table XVI-A shows that these corrections exceed $\frac{1}{2}'$ when the ratio of spiral length sighted to length L_s is 0.7 or greater. In this example the correction for chord point 12 would be $-74''$; for the 900-ft-long sight to the S.C., it would be 286″, or almost 5′. Obviously, one beneficial effect of setting up at points 5 and 12 is the elimination of such corrections; thus, the maximum correction is 17″ for the sight from point 5 to point 12 (see Appendix C for method of computing this correction).

Regardless of the location of transit setups used in staking a spiral, the actual chords taped should (theoretically) be slightly shorter than their nominal arc lengths. For most practical purposes, L_s equals the sum of the nominal chords used to stake the spiral. Chord corrections can ordinarily be

COMPUTATION OF DEFLECTIONS

SIGHT AT POINT	COEFFICIENT (TABLE XVI)	× a_1	= DEFLECTION TO SPIRAL	COMMENTS
Transit at T.S. (Orient by sighting along tangent with 0°00′ on vernier.)				
1	1	4′	0°04′	These same deflections obtainable from Table XV at every other chord point by entering Table with
2	4	4′	0°16′	
3	9	4′	0°36′	
4	16	4′	(64′) 1°04′	
5	25	4′	(100′) 1°40′	
			$\Delta = \left(\frac{5}{15}\right)^2 \times 45° = 5°$	
Transit at Point 5 (Orient by backsight to T.S. with 3°20′ on vernier.)				
0 (T.S.)	50	4′	(200′) 3°20′	Orientation angle equals twice the deflection from the T.S. Plunge telescope after backsight and set deflections for points 6 to 12 on the other side of 0°00′.
6	16	4′	(64′) 1°04′	
7	34	4′	(136′) 2°16′	
8	54	4′	(216′) 3°36′	
9	76	4′	(304′) 5°04′	
10	100	4′	(400′) 6°40′	
11	126	4′	(504′) 8°24′	
12	154	4′	(616′) 10°16′	
Transit at Point 12 (Orient by backsight to point 5 with 13°32′ on vernier.)				
5	203	4′	(812′) 13°32′	Plunge telescope after backsight and set deflections for points 13 to 15 on the other side of 0°00′.
13	37	4′	(148′) 2°28′	
14	76	4′	(304′) 5°04′	
15 (S.C.)	117	4′	(468′) 7°48′	

neglected; they are necessary only for fairly long chords near the end of a long sharp spiral. In the preceding example the position of the S.C. would overrun its theoretical position by only 0.04 ft if the spiral were staked entirely from the T.S. with fifteen 60-ft chords. If necessary to allow for this small discrepancy, each of the last four chords could be taped as 59.99 ft long. Chord corrections may be found with the aid of Tables II and IV.

5-8. Fitting Spiraled Curve to Specified E_s or T_s

As in the case of a simple curve, it may be necessary to fit a spiraled curve to definite field conditions such as a specified E_s or T_s. There are voluminous tables* giving convenient combinations of D and L_s which approximate a specified value for E_s or T_s; the exact values of E_s and T_s for a selected combination are obtained by interpolation.

The tables in this book may also be used to select a spiral-curve layout fitting given conditions.

Example

Given: $I = 27°00'$; E_s = approximately 34 ft. Determine suitable values of D and L_s.

First use Table VIII to obtain a fairly close value of D. Thus,

$$D = \frac{163}{34} = 4.8° \text{ (roughly)}$$

Considerations of the relation $\Delta = L_s D/200$ shows two unknowns. Consequently, there are any number of combinations of Δ and L_s which will fit the conditions. A glance at Table XI might suggest a suitable combination.

Suppose that L_s is assumed to be 250 ft. Then

$$\Delta = \frac{250}{200} \times 4.8° = 6.0° \text{ (tentatively)}$$

From Table XII, $o = 0.00872 \times 250 = 2.2$ (approx.).

A close value of $R + o$ is then found from formula 5-17.

Thus,

$$R + o = \frac{34 - 2.2}{0.02842} = 1119$$

Therefore, $R = 1117$ and $D = 5°08'$ (by Table I).

If D were chosen to be 5°, the 250-ft spiral is one listed in "selected spirals" of Table XI. For this combination, E_s would differ slightly from 34 ft, but a rigid requirement permitting no deviation from a fixed E_s is rarely met (it would require an odd value of either D or L_s).

If the approximate value of T_s were fixed, the first trial can also be made by using Table VIII. This gives a rough value of D, but allows selection of a suitable value for L_s. Since o is very small in comparison with R, formula 5-14 can be expressed approximately as $T_s = T + \frac{1}{2}L_s$, which can be solved for T.

Suppose that T_s is fixed at approximately 400 ft in this example. From Table VIII.

$$D = \frac{1376}{400} = 3.4° \text{ (roughly)}$$

* Barnett, Joseph, *Transition Curves for Highways*, Washington, D.C. U.S. Government Printing Office, 1938.

Trying L_s as 250 ft, $T = T_s - \frac{1}{2}L_s = 400 - 125 = 275$ ft (approx.); and (by Table VIII) $D = 1376/275 = 5°$ as before.

5-9. Spiraled-Curve Formulas with Radius as Parameter

In connection with the use of even-radius curves (section 2-12) it is convenient to express formulas for needed spiraled-curve parts in terms of the radius of the circular arc.

In Fig. 5-4, certain distances may be related to R by means of coefficients N, P, M, and S, which have the following significance:

$N\,R =$ the distance $AD = X$

$P\,R =$ the distance $DC = Y$

$M\,R =$ the distance $OJ = R + o$

$S\,R =$ the distance $AJ = X_0$

Values of these coefficients for $R = 1$ have been computed for various assumed values of Δ (see Table XIII). When these possibilities are substituted in the following relations, parts needed for computing and staking the layout result:

$$X = R\,N \tag{5-28}$$

$$Y = R\,P \tag{5-29}$$

$$T_s = R(M \tan \tfrac{1}{2}I + S) \tag{5-30}$$

$$E_s = R(M \sec \tfrac{1}{2}I - 1) = R[M \operatorname{exsec} \tfrac{1}{2}I + (M - 1)] \tag{5-31}$$

$$L_s = 2\,R\,\Delta \text{ (radians)} \tag{5-6}$$

In applying this procedure to practical problems in the highway field, a "proportion rule" has been suggested* for choosing L_s. The "proportion of transition" p is the ratio of the length of the two spirals to the total length of the spiraled curve. It may be proved that

$$\Delta = \left[\frac{p}{2(2 - p)}\right] I \tag{5-32}$$

It is not possible (except by coincidence) to have integral values of R, L_s, and Δ (in degrees); one, at least, must be an odd value. When it is desired to use Table XIII without interpolation, the odd variable is L_s.

As an illustration the example of section 5-8 is solved by this method. For a trial, assume that $p =$ about 0.6 (evidence shows that the natural value

* Leeming, J. J., "The General Principles of Highway Transition Curve Design," *Transactions, ASCE*, Vol. 113, 1948, pp. 877ff.

of p ranges between 0.4 and 0.7 for high-speed operation on highways). Then

$$\Delta = \left[\frac{0.6}{2(2 - 0.6)}\right] \times 27^\circ = 5.8^\circ$$

Round off to 6° for convenience in using Table XIII.

From formula 5-31, $R = 1130$ (approx.). Use $R = 1100$, which is listed in Table V.

Finally $2\,\Delta = 12^\circ = 0.20944$ rad. (Table VI) and $L_s = 0.20944 \times 1100 = 230.4$ ft.

If L_s were rounded to a convenient value (say 250 ft), then Δ would be the odd value, but the circular arc could still be staked readily by means of the deflections listed in Table V. When the chord definition of D is used, Δ is obtained from 5-32; R_c, from 5-31; D_c, from Table I; and L_s, from 5-2.

The coefficients in formulas 5-28 through 5-31 apply whether or not a central circular arc is present; that is, they also can be used for the double-spiral curve described in the next article.

5-10. Double-Spiral Curve

Some engineers advocate making curves transitional throughout by using a double spiral, which consists of two equal spirals placed end to end with the curvature changing from increasing to decreasing at the middle of the layout. This simplifies the selection of a curve that will have a specified value of E_s or T_s; since $\Delta = \frac{1}{2}I$, it follows that $T_s = X + Y \tan \frac{1}{2}I$ and $E_s = Y \sec \frac{1}{2}I$.

Table XIV gives values of T_s and E_s for $L_s = 1$. Multiplying the proper tabulated values by the actual length L_s gives T_s and E_s for a double-spiral curve having the given value of I.

Obviously, a double-spiral curve of given I is longer than a spiraled simple curve having the same D; hence there is a much greater length over which operating conditions—centrifugal force and superelevation, in particular—are variable. For this reason, the all-transitional curve should be used only in exceptional cases, at least on highway work.

5-11. Calculating Spiral Without Special Tables

It is possible to compute a spiral without reference to special tables (if an engineer will do this occasionally, he will understand the spiral better). The method suggested is based upon certain approximations the significance of which may be seen by examining Fig. 5-2 in connection with Table XI. All calculations may be made by slide rule.

Procedure Recommended for Flat Spirals (*up to* $\Delta = 5^\circ$)

(a) Compute Y from $Y = L_s \sin \frac{1}{3}\Delta$

(b) Calculate X from $X = L_s - Y^2/2\,L_s$

(c) Compute o from $o = \frac{1}{4}Y$

(d) Compute X_0 from $X_0 = \frac{1}{2}X$
(e) Calculate deflections by applying the principle they are proportional to the squares of the distances.

If the foregoing method is applied to a 200-ft spiral joining a 3° curve, the computed values are: $Y = 3.49$, $X = 199.97$, $o = 0.87$, and $X_0 = 99.98$. For all pratical purposes these are the same as the values listed in Table XI.

As Δ increases beyond 5°, more significant differences arise between the exact and approximate values of these four spiral parts. The greatest discrepancy will be found in the value of X. However, X is not often needed in the field. Values of o and X_0 have more practical use, for they are needed to obtain T_s by formula 5-15. The following table shows the relatively small errors that would result from using the approximate method even when Δ is as large as 15°.

GIVEN		VALUES BY APPROX. METHOD	EXACT VALUES FROM TABLE XI $D_a = 10°$	$D_c = 10°$
$L_s = 300$	$Y =$	26.1	26.05	26.05
$D = 10°00'$	$X =$	298.86	297.95	297.95
$(\Delta = 15°00')$	$o =$	6.52	6.53	6.51
	$X_0 =$	149.43	149.66	149.47

5-12. The A.R.E.A. 10-Chord Spiral

The 10-chord spiral of the American Railway Engineering Association has been used by many American railroads since about 1912. This spiral is an approximation of equation 5-8 in which L_s is measured by 10 equal chords instead of around the spiral itself. It is commonly used in connection with the chord definition of D.

The spiral in this book may be converted practically to the A.R.E.A. spiral by applying the corrections marked with an asterisk, as explained in Part III on page 354. (See Appendix A for the source of those corrections.) Consequently, there is no need to master the details of the A.R.E.A. spiral or to use tables prepared for it alone, as demonstrated by the following comparison in the case of an exceptionally sharp railroad curve:

	D_c	L_s	X	Y	X_0	o	A
Spiral in this book	10°	300	297.95	26.05	149.47	6.51	4°59.8′
A.R.E.A. spiral	10°	300	297.96	26.05	149.48	6.51	4°59.8′

5-13. Laying Out Spiral by Taping

A spiral may be staked by using tangent offsets, chord offsets, or middle ordinates. The operations in the field are similar to those described in section

2-15. However, the calculations differ slightly from those for a simple curve owing to the variable curvature on a spiral.

Tangent offsets for selected points on the spiral are computed from the cube law (section 5-2). It is sufficiently accurate for the flat spirals used on modern highway and railroad alignment to assume that the spiral and the throw o bisect each other, and that $o = Y/4$ (equation 5-20). The throw is computed from the relation $L_s^2/24\ R$ (equation 5-21), or is taken from tables, after which offsets from the tangent to the midpoint of the spiral (at equidistant points) are found from the principle that offsets are proportional to cubes of the distances. The same offsets are then used to locate the second half of the spiral by measuring them radially from the circular arc (osculating circle) produced backward from the S.C. For example, the five offsets required to locate a 10-chord spiral are found by multiplying the throw by 0.004, 0.032, 0.108, 0.256, and 0.500. The advantage of this method, in comparison with measuring all offsets from the tangent, is that the measurements usually come well within the limits of the graded roadbed.

Chord offsets and middle ordinates are approximately proportional to lengths from the T.S. Needed values can be obtained by substituting $(D_1+D_2)/2$ for D in the various simple-curve offset formulas, D_1 and D_2 being the degrees of curve of a spiral at the ends of a particular chord. The values of C.O. and M.O. in Table I facilitate the computation.

5-14. Parallel Spirals

For the same reasons given in section 2-16 it is often necessary to stake an offset curve parallel to a spiral along the center line. This matter is more complex than on circular curve alignment. It is also closely related to the subjects of edge lengths and widening of highway pavements. For these reasons parallel spirals are discussed in detail in section 8-20 as a special curve problem in highway design.

5-15. Length of Spirals

A spiral need not have a particular length, but it should be at least long enough for the transition to be made safely and comfortably. Design speed and the rate of attaining superelevation are controlling factors in this respect.

On railroads, practice has been fairly well standardized for a number of years, though the operation of high-speed streamliners added new problems. Chapter 9 contains examples and recommendations.

Practice on highways is not so definite as is that on railroads. Lack of standardization is due to the later adoption of spirals, more diverse operating conditions, and greater number of administrative units involved. Recent practice is outlined in Chapter 8.

COMBINING SPIRALS

5-16. Combining Spiral-Method for Flat Spirals

The difference in curvature at the C.C. of a compound curve may be abrupt enough to justify the insertion of a combining spiral. In Fig. 5-7 the combining spiral *AC* is tangent to the curves having radii R_S and R_L; and it has the same radius of curvature as the circular arc at each point of tangency. That is, *AC* is a portion of a simple spiral cut to fit as a transition between curves of degree D_S and D_L. With D_S and D_L known, an additional value—usually the length *AC* but sometimes the offset *JK*—must be assumed. The problem is then determinate; only one combining spiral will satisfy the given conditions.

If the combining spiral is fairly flat (true central angle less than 15°), the theory of the osculating circle is satisfactory. According to this theory the offset or "throw" *JK* is the value of *o* for a spiral of length *AC* and terminal degree $D_S - D_L$. It also follows that, for all practical purposes, *JK* and *AC* bisect each other. Accordingly, the problem of calculating deflections for running in the spiral *AC* is identical with that of calculating the deflections for staking the remainder of a spiral from a setup at any point, as explained in section 5-7.

If the length of the spiral *AC* be designated by l_s, its *true* central angle equals the sum of the central angles for the simple curves produced, that is, for the arcs *AJ* and *KC*. From formula 2-13 the true central angle Δ_t is

$$\Delta_t = \left(\frac{l_s}{2}\right)\left(\frac{D_S}{100}\right) + \left(\frac{l_s}{2}\right)\left(\frac{D_L}{100}\right)$$

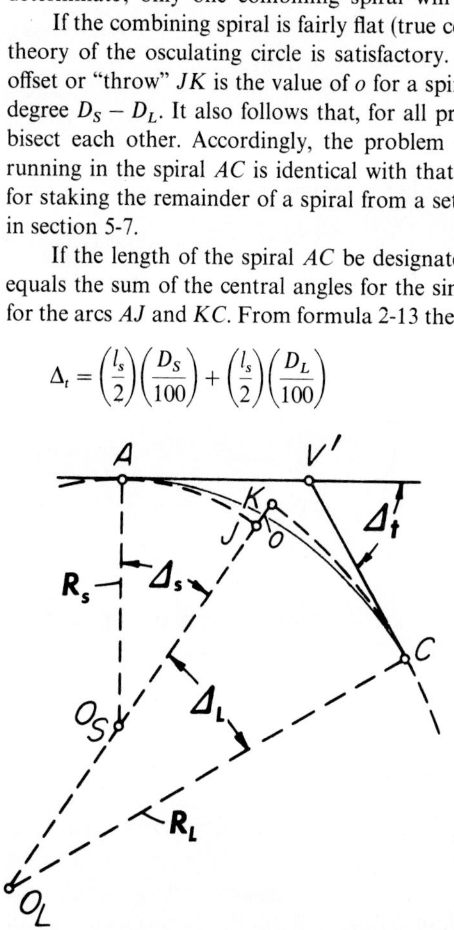

Figure 5-7 Combining spiral by osculating-circle theory.

which reduces to

$$\Delta_t = \left(\frac{l_s}{200}\right)(D_S + D_L) = \Delta_S + \Delta_L \tag{5-33}$$

In computing spiral deflection angles from a local tangent at A or C, the *nominal* value of the central angle Δ_N is used. This is defined as the difference between the true central angle of the combining spiral and that of the osculating circle at either end. Since the osculating circle at A has a central angle of $l_s D_S/100$ in the distance l_s,

$$\Delta_N = \frac{l_s D_S}{100} - \Delta_t$$

which reduces to

$$\Delta_N = \frac{l_s}{200}(D_S - D_L) = \Delta_S - \Delta_L \tag{5-34}$$

Table XI may be used for combining spirals by substituting l_s, Δ_N, and $(D_S - D_L)$ for the headings L_s, Δ, and D.

Example

Given: $D_S = 7°$; $D_L = 2°$; $AC = l_s = 300$ ft. Here $D_S - D_L = 5°$; consequently, $o = 3.27$ (Table XI).

From formula 5-1 the value of k is

$$\frac{100(7° - 2°)}{300} = 1\tfrac{2}{3}° \text{ per sta.}$$

which means that l_s is the last 300 ft of a simple spiral that is $(7°/k) \times 100 =$ 420 ft long.

From formula 5-33, $\Delta_t = 13°30'$; and from formula 5-34 (or Table XI), $\Delta_N = 7°30'$.

In computing spiral deflection angles, use Δ_N for Δ in formula 5-18. Thus, if the spiral AC were to be divided into four equal parts,

$$a_1 = \frac{20 \times 7.5°}{4^2} = 9.375'$$

and the deflections to the spiral from the osculating circle at C are a_1, $4\,a_1$, $9\,a_1$, and $16\,a_1$. These are added to the deflections from the tangent to the corresponding points on the osculating circle, namely, 0°45′, 1°30′, 2°15′, and 3°00′. The final deflections are: 0°54′, 2°08′, 3°39′, and 5°30′. For running in the spiral from the sharper arc (setup at A), the required deflections from the tangent would be 2°28′, 4°38′, 6°28′, and 8°00′. *Note:* This spiral is one of three used in problem 5-11(a).

5-17. Combining Spiral-Method for Sharp Spirals

In contrast to the assumptions shown in Fig. 5-7, the derivations in Appendix B prove that the correct throw is not JK. Instead, the throw occurs along a radial line which is rotated slightly toward the flatter end of the spiral. This rotation is shown (greatly exaggerated) in Fig. 5-8. The vertex V' is preserved, as is the total change in direction between the local tangents. In consequence, the combining spiral takes position $A'C'$ instead of position AC, as in the osculating-circle theory. The true throw is $J'K'$, or p_a.

In Appendix B it is shown that β is always larger than Δ_S by a quantity C_b, where

$$C_b = 0.03655\Delta_S\Delta_N(\Delta_S - \Delta_N)\{1 + 121[24\Delta_S(\Delta_S - \Delta_N) - 7\Delta_N{}^2]10^{-8}\} \qquad (5\text{-}35)$$

in which Δ_S and Δ_N are in degrees and C_b is in seconds.

In a practical problem, C_b may be obtained quickly from Table XVI-C

Example

Given: $D_S = 20°$; $D_L = 10°$; $A'C' = l_s = 400$ ft. Formulas 5-33 and 5-34 yield $\Delta_t = 60°$, $\Delta_N = 20°$, and $\Delta_S = 40°$. The value of C_b from Table XVI-C is 596″ or 9′56″. Accordingly, $\beta = 40°09'56''$ and $\alpha = 19°50'04''$.

Appendix B also shows that p_a is always less than the throw o by a quantity C_p, where

$$C_p = 0.2215 l_s\Delta_S\Delta_N(\Delta_S - \Delta_N)10^{-7} \qquad (5\text{-}36)$$

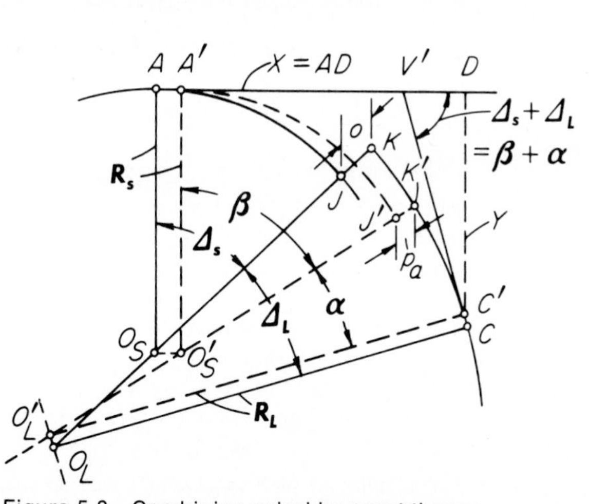

Figure 5-8 Combining spiral by exact theory.

in which C_p and l_s are in feet and Δ_S and Δ_N are in degrees. Table XVI-C also contains values of C_p based on $l_s = 1000$ ft. In this example the table yields $C_p = 0.4 \times 0.354 = 0.142$ ft. If Table XII is now entered with $\Delta = \Delta_N = 20°$, the value of o is found to be $400 \times 0.02896 = 11.584$ ft. Therefore $p_a = 11.584 - 0.142 = 11.442$ ft.

Derivation of the correction to be applied to a deflection angle obtained by the osculating-circle theory is likewise given in Appendix B. The correction, C_ϕ, for the end of a combining spiral is

$$C_\phi = 0.012185\Delta_S\Delta_N(\Delta_S - \Delta_N) + 0.3535\Delta_S{}^3\Delta_N(\Delta_S - 2\Delta_N)10^{-6} + 0.2946\Delta_S\Delta_N{}^3(181\Delta_S - 61\Delta_N)10^{-8} + 0.00309\Delta_N{}^3 + 0.00228\Delta_N{}^5 10^{-5} \cdots \tag{5-37}$$

in which C_ϕ is in seconds and Δ_S and Δ_N are in degrees.

If Δ_S is zero the combining spiral reverts to a simple spiral; the first three terms in formula 5-37 disappear and Δ_N becomes Δ. Formula 5-37 is then the same as formula A-6 (Appendix A); that is C_ϕ becomes C_s. In a numerical problem these first three terms comprise the greater portion of the total correction. It is convenient to group these terms and denote them as C_a. Thus,

$$C_a = 0.012185\Delta_S\Delta_N(\Delta_S - \Delta_N) + 0.3535\Delta_S{}^3\Delta_N(\Delta_S - 2\Delta_N)10^{-6} + 0.2946\Delta_S\Delta_N{}^3(181\Delta_S - 61\Delta_N)10^{-8} \cdots \tag{5-38}$$

Summarizing: C_s gives the correction for a simple spiral; $C_\phi = C_a + C_s$ is the correction for a combining spiral.

In general, *when sighting toward the sharper end of a combining spiral*, the corrected deflection, ϕ_c (see Fig. B-1 in Appendix B), is expressed by

$$\phi_c = \phi_f - C_\phi = \phi_f - (C_a + C_s) \tag{5-39}$$

where ϕ_f is found from formula 5-23 (see also Fig. 5-6).

Similarly, *when sighting toward the flatter end of a combining spiral*,

$$\phi_c = \phi_b + C_\phi = \phi_b + (C_a + C_s) \tag{5-40}$$

where ϕ_b is found from formula 5-22.

In working numerical problems, formula 5-37 is rather cumbersome to use. This objection is eliminated by recourse to Tables XVI-C and XVI-D, which yield the separate values of C_a and C_s quickly. Appendix C contains detailed examples showing convenient forms of computation for deflection angles to a series of points along a combining spiral from a transit setup at any point on the spiral.

5-18. Computations for Field Layout

The computations for staking a combining spiral are based either on the approximate method of section 5-16, or on the exact method of section 5-17.

The choice is affected by sharpness of the spiral and by the accuracy required in a particular situation.

Suppose a more searching examination is made of the example in section 5-16. In this 300-ft spiral, $\Delta_t = 13°30'$, $\Delta_S = 10°30'$, and $\Delta_N = 7°30'$. By interpolation in Tables XVI-C and XVI-D, $C_b = 9''$, $C_p = 0.002$ ft, and C_ϕ (to end of spiral) $= 4''$. These corrections are small enough to neglect, and the osculating-circle method of section 5-16 is satisfactory.

On the other hand, in the sharper spiral of the example in section 5-17, $\Delta_t = 60°$, and the corresponding corrections are $C_b = 9'56''$, $C_p = 0.142$ ft, and C_ϕ (to end of spiral) $= 3'45''$. To neglect corrections of such magnitude would result in significant errors in field layout; therefore, the exact method of section 5-17 should be used.

In field layout it is convenient to stake the vertex V' as a transit setup on the control traverse. The spiral ends are then located by measuring local tangents $V'A$ and $V'C$ (or $V'A'$ and $V'C'$). Although the spiral could be staked completely from a transit setup at either end, it is better to lay it out in part from both ends. This reduces the lengths of sights and may make the corrections to deflection angles negligible.

If the osculating-circle theory is used, the local tangents can be found from the traverse $O_SAV'CO_LO_S$ (Fig. 5-8), with north assumed in direction O_SA. This yields

$$V'C = \frac{R_S + (R_L - R_S - o)\cos\Delta_S - R_L\cos(\Delta_S + \Delta_L)}{\sin(\Delta_S + \Delta_L)} \tag{5-41}$$

and

$$V'A = R_L\sin(\Delta_S + \Delta_L) - (R_L - R_S - o)\sin\Delta_S - V'C\cos(\Delta_S + \Delta_L) \tag{5-42}$$

As a check on the computation, tangent $V'A$ may be verified from

$$V'A = \frac{R_L - (R_L - R_S - o)\cos\Delta_L - R_S\cos(\Delta_S + \Delta_L)}{\sin(\Delta_S + \Delta_L)} \tag{5-43}$$

This formula comes from setting Σ *departure* = zero with north assumed in direction CV'.

If the exact theory is used, the local tangents come from traverse $O'_SA'V'C'O'_LO'_S$. The resulting alternative formulas are

$$V'C' = \frac{R_S + (R_L - R_S - p_a)\cos\beta - R_L\cos(\Delta_S + \Delta_L)}{\sin(\Delta_S + \Delta_L)} \tag{5-41a}$$

and

$$V'A' = R_L\sin(\Delta_S + \Delta_L) - (R_L - R_S - p_a)\sin\beta - V'C'\cos(\Delta_S + \Delta_L) \tag{5-42a}$$

or

$$V'A' = \frac{R_L - (R_L - R_S - p_a)\cos\alpha - R_S\cos(\Delta_S + \Delta_L)}{\sin(\Delta_S + \Delta_L)} \tag{5-43a}$$

As a further check on computation of the tangents,

$$V'C' = Y\csc(\Delta_S + \Delta_L) \tag{5-44}$$

and

$$V'A' = X - Y\cot(\Delta_S + \Delta_L) \tag{5-45}$$

in which X and Y come from formulas B-7 and B-8, Appendix B.

5-19. Practical Example of Combining Spiral

The importance of computing a sharp spiral by exact theory is not yet commonly understood. This is a natural consequence of the historical development of curves in route alignment. Spiraled curves on railroads are invariably flat. In most cases the spirals used on primary highway alignment have also been flat enough for satisfactory computation by the osculating-circle theory. Spirals used on the Interstate Highway System have sometimes been so sharp as to warrant computation by exact theory. This anomaly stems from the high unit cost of the Interstate System and its interchanges when located in mountainous terrain or urban areas. The great flexibility of compound curves and combining spirals permits fitting alignment to terrain conditions to save many thousands of dollars in grading costs or land damages.

There are instances where sharp combining spirals on the Interstate System have been computed by the osculating-circle theory, only to find the work would not check in the field. Designers have used this theory because of its simplicity and long heritage of success, without realizing that the small approximations involved may grow into large errors in sharp spirals.

An example of such a case occurred at an interchange ramp to Interstate Highway 77, where a sharp combining spiral was used to reduce the cost of grading in rough terrain. In this instance a 300-ft combining spiral was chosen to connect curves having radii of approximately 150 and 1600 ft. The approximate values of Δ_S and Δ_N were 57° and 52°. In performing the original calculations the osculating-circle method of section 5-16 was assumed to apply. Exact values of the resulting tangent distances were $V'C = 210.33$ and $V'A = 123.29$ ft.

In the field it was found that very rough terrain prevented staking the spiral. However, the vertex V' was located and the ends of the spiral staked by measuring the tangents $V'A$ and $V'C$. Following rough grading it was discovered that a 300-ft spiral would not fit between the hubs at A and C. In particular, the deflection angle across the spiral did not check by about

13′ and the curved distance was too long by approximately 4 ft. Eventually, the reason for the misfit was traced to use of the osculating-circle theory rather than the exact method of computation outlined in section 5-17. By this time, however, construction up to and beyond the spiral had progressed so far it was advisable to preserve points *A* and *C*. Consequently, the solution utilized was to proportion out the error in angle and distance to produce a smooth curve suitable for paving. The effect of this mistake is that instead of a true combining spiral exactly 300 ft long between points *A*′ and *C*′ there exists a pseudo spiral that is about 304 ft long between points *A* and *C*. The exact data for this case are given in problem 5-10.

5-20. Completely Spiraled Compound Curve

Figure 5-9 shows the geometric relations between the nonspiraled compound curve *ACB* and the same curve provided with three transition spirals. The spirals themselves are omitted from the sketch, and the offsets needed to accomodate them are exaggerated for the purpose of clarity. For this case, in which the layout begins with the flatter arc, point *a* is the offset T.C. for the first spiral. Its throw *A'a* is denoted by o_L. The corresponding throw for the third spiral is *B'b*, denoted by o_S. The throw o_C for the *combining spiral* between the two arcs is *JK*. Section 5-16 is assumed to apply.

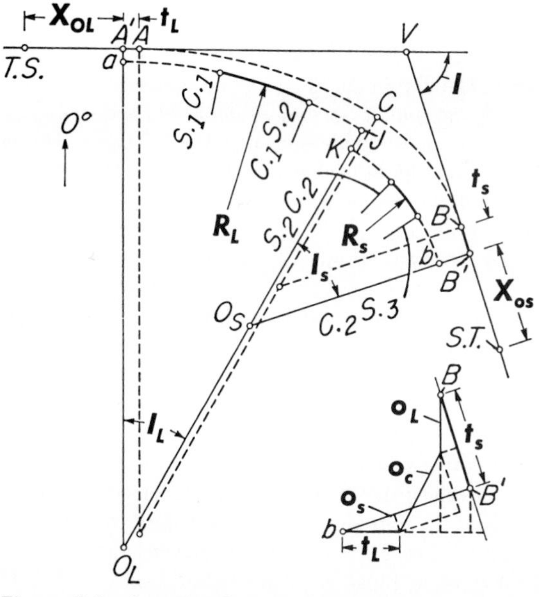

Figure 5-9 Completely-spiraled compound curve.

Assume the given values to be sta. V, I, I_L, R_L, R_S, and the lengths (or the k-values) of the spirals. The problem is to find the stationing of all curve points (T.S., $S._1C._1$, $C._1S._2$, and so forth) so that the layout may be computed and staked. The best solution directly from the given data is to find VB' and $A'V$ by the traverse method. Use the traverse $O_LA'VB'O_SO_L$, and follow the procedure developed in section 3-5. (Note the similarity between the following tabulation and that on page 50.)

LENGTH	AZIMUTH	LATITUDE	DEPARTURE
$R_L + o_L$	$0°$	$R_L + o_L$	0
$A'V$	$90°$	0	$A'V$
VB'	$90° + I$	$-VB' \sin I$	$VB' \cos I$
$R_S + o_S$	$180° + I$	$-(R_S + o_S) \cos I$	$-(R_S + o_S) \sin I$
$R_L - R_S - o_C$,	$180° + I_L$	$-(R_L - R_S - o_C) \cos I_L$	$-(R_L - R_S - o_C) \sin I_L$

From Σ *latitudes*,

$$VB' = \frac{(R_L + o_L) - (R_S + o_S) \cos I - (R_L - R_S - o_C) \cos I_L}{\sin I} \tag{5-46}$$

Then from Σ *departures*,

$$A'V = (R_S + o_S) \sin I + (R_L - R_S - o_C) \sin I_L - VB' \cos I \tag{5-47}$$

The tangent distances T_{sL} and T_{sS} are $X_{oL} + A'V$ and $VB' + X_{oS}$. Stations of the curve points are found in the usual way by subtracting T_{sL} from sta. V to obtain sta. T.S. and adding successive spiral and arc lengths to sta. T.S. It is necessary to use the principles brought out in section 5-16 in computing the true central angles of the circular arcs.

If the layout begins with the sharper arc, formulas 5-46 and 5-47 remain the same. The only difference is that sta. T.S. equals sta. V minus T_{sS}, and, of course, the stationing increases in the direction opposite to that in Fig. 5-9 (see also Fig. 5-7).

This problem may be solved by another method that is particularly convenient if the given values pertain to a nonspiraled compound curve already computed. In such a case the tangent distances ($T_L = AV$ and $T_S = VB$) are known from appropriate formulas derived in section 3-5. If it is now required to modify the curve to insert three spirals, X_{oL} and X_{oS} are found readily from Tables XI or XII. The additional corrections needed to convert T_L and T_S to T_{sL} and T_{sS} are the distances AA' and BB', denoted by t_L and t_S.

The arcs AC and CB of the nonspiraled curve are shifted to their locations aJ and Kb on the spiraled curves as the result of movements of translation involving the throws o_L, o_C, and o_S. The combined effects of these translations yield the distances t_L and t_S, which may be written by means of the enlarged sketch in Fig. 5-9 as

$$t_L = \frac{o_S - o_L \cos I - o_C \cos I_S}{\sin I} \tag{5-48}$$

and

$$t_S = \frac{o_L - o_S \cos I + o_C \cos I_L}{\sin I} \tag{5-49}$$

Both t_L and t_S will normally be positive quantities. If either is negative, points A and A' (or B and B') are interchanged in Fig. 5-9, and the *t-correction* must be subtracted. One advantage of this method is that the terms in formulas 5-48 and 5-49 are so small that the t-corrections may be found by slide-rule calculation.

If the combining spiral is omitted, o_C equals zero and the formulas yield the t-corrections either for a compound curve with spirals at the main tangents only or for a simple curve with unequal spirals. (Compare the resulting t-correction formulas with the final terms in formulas 5-15a and 5-15b.) Furthermore, for a simple curve with equal spirals, $o_L = o_S = o$, and $t_L = t_S$. In this case the t-correction becomes $(o)[(1 - \cos I)/\sin I]$, which equals $o \tan \frac{1}{2}I$ (see formula 5-15).

COMPOUND SPIRALS

5-21. Definition and Fields of Use

In a special situation, a better fit with terrain can be obtained by permitting a simple spiral to have a different radius of curvature than that of the circular arc at the point of tangency. At the S.C. the condition is analogous to that at the C.C. of a compound curve. The spiral in this case is called a *compound spiral*, and the difference in degree of curvature between spiral and circular arc at the point of tangency is called *degree of compoundancy*. Some degree of compoundancy may also be allowed at the beginning of the spiral where it is tangent to the straight line.

Compound spirals provide smoother alignment along interchange ramps where restricted physical conditions would otherwise require multicompound curves. They may also be used on open highways in locations where a small degree of compoundancy is not objectionable.

5-22. Computation of Compound Spirals

The simplest case of a compound spiral is a simple spiral with compoundancy at the S.C. (denoted as $S.C_c$, in such a case). For example, suppose that preliminary alignment consisted of a 300-ft simple spiral to an 8° curve. If closer examination showed that tight field conditions could be met better by using a 10° curve while preserving the same alignment along the spiral, the original spiral could still be used with 2° of compoundancy at the S.C.$_c$. No special operations would be required in computing the spiral; the procedure would follow that for any simple spiral.

PROBLEMS

5-1. Given the following simple spirals. Compute values of X, Y, X_0, o, L.T., S.T., and L.C. with the aid of Table XII. Where applicable use Table XI to verify answers.
(a) $D_a = 2°00'$; $L_s = 150$ (d) $D_c = 8°00'$; $L_s = 200$
(b) $D_a = 6°15'$; $L_s = 400$ (e) $D_c = 9°15'$; $L_s = 400$
(c) $D_a = 24°00'$; $L_s = 500$ (f) $D_c = 25°00'$; $L_s = 420$

5-2. Assume simple curves with equal spirals corresponding to those in problem 5-1. For each layout, compute values of T_s and E_s using formulas 5-14 and 5-17. Check the results by Tables VIII, IX, and X where applicable. Given values of I are:
(a) $I = 14°43'$ (c) $I = 150°00'$ (e) $I = 57°24'$
(b) $I = 35°18'$ (d) $I = 41°27'$ (f) $I = 130°45'$
Partial answers: (b) $T_s = 493.67$, $E_s = 52.90$;
(d) $T_s = 371.93$, $E_s = 52.07$; (e) $T_s = 544.43$, $E_s = 99.03$

5-3. Prepare sets of field notes (see section 5-5) for staking certain of the curves in problem 5-2 with setups at the T.S., S.C., and S.T. On approach spirals calculate five equal chords; on leaving spirals and circular arcs compute full and half-stations. Carry deflections to seconds. Use answers to problems 5-1 and 5-2 where needed. (b) Sta. P.I. = 86+10.54; (d) Sta. P.I. = 106+44.18; (e) Sta. P.I. = 185+83.24

5-4. Prepare deflections for staking the spiral in problem 5-1(c) as a 20-chord spiral. Use the osculating circle theory of section 5-6 and set up the table as in section 5-7.
(a) Whole spiral staked forward from the T.S.
(b) Entire spiral staked backward from the S.C.
(c) Spiral staked forward from setups at the T.S. and chord points 8 and 15.
(d) Spiral staked backward from setups at the S.C. and chord points 15 and 8.

5-5. For the following simple curves with equal spirals, find all spirals listed in Table XI that conform to the stated conditions. Use the method of section 5-8.
(a) $I = 28°42'$; $E_S = 50 \pm 6$; $L_s = 200$ to 400. *Answers:* $D_a = 3°30'$ with $L_s = 200$, 250, 300, or 350.
(b) $I = 28°42'$; $T_s = 450 \pm 10$; $L_s = 200$ to 400.
(c) $I = 40°00'$; $E_s = 150 \pm 10$; $L_s = 200$ to 400.

(d) $I = 40°00'$; $T_s = 700 \pm 15$; $L_s = 200$ to 400. *Answers:* $D_a = 3°30'$ with $L_s = 200$; $D_a = 4°00'$ with $L_s = 350$.

5-6. Use the method of section 5-9 to find a simple curve with equal spirals in which L_s is an exact multiple of 50 ft, R is a value listed in Table V, and p is approximately 0.5.

(a) $I = 18°42'$; E_s = approximately 27 ft. *Answers:* $L_s = 200$ ft with $R = 1900$ ft.

(b) $I = 38°00'$; T_s = approximately 580 ft. *Answers:* $L_s = 300$ ft with $R = 1300$ ft.

(c) Compute deflections for staking the approach spiral in (a) as a five-chord spiral. *Answers:* 0°02′25″, 0°09′39″, 0°21′43″, 0°38′36″, and 1°00′19″.

(d) Compute deflections for staking the approach spiral in (b) by setting full and half-stations, assuming that sta. T.S. = 60+20. *Answers:* 0°01′19″, 0°09′24″, 0°24′50″, 0°47′36″, 1°17′43″, and 2°12′13″.

5-7. A simple curve has $I = 56°36'$ and $D_a = 5°$.

(a) Replace it by a double spiral having the same radius at its midpoint. Find: L_s, T_s, and E_s. *Answers:* $L_s = 1132$; $T_s = 1203.30$; $E_s = 208.02$.

(b) How much longer is the double spiral of problem (a) than the original simple curve? Is the result a coincidence or is it required by theory? Prove your conclusion.

(c) Replace the simple curve by a double spiral passing through the same midpoint. Find: L_s, T_s, and E_s. *Answers:* $L_s = 846.54$; $T_s = 899.86$; $E_s = 155.56$.

5-8. Using the approximate method of section 5-11, compute Y, X, o, and X_o for the spirals in problems 5-1(a), (b), (d), and (e). Compare the results with those in problem 5-1.

5-9. Prepare deflections for staking the spiral in problem 5-1(c) as a 20-chord spiral. Use the exact theory of section 5-17 and Appendix *B*, and set up tables as in Appendix *C*.

(a) Same as 5-4(a). See Appendix *C*, Case I, and example.

(b) Same as 5-4(b). See Appendix *C*, Case II, and example.

(c) Same as 5-4(c). See Appendix *C*, Case III, and example.

5-10. Refer to Fig. 5-8 and the example in section 5-19. The given values were $R_s = 150$ ft; $R_L = 1581.02$ ft; $l_s = 300$ ft.

(a) Find the values of $V'A$ and $V'C$ assuming the osculating circle theory to apply. *Answers:* $V'A = 123.29$; $V'C = 210.33$.

(b) Find the deflection angle across the spiral (transit at C, sight at A) by solving the triangle $V'AC$. *Answer:* 22°19′49″.

(c) To find the error caused by the osculating circle theory, compute the exact deflection angle ϕ_c, across the 300-ft combining spiral $A'C'$ by formula 5-39 aided by Tables XVI-C and XVI-D. *Answer:* 22°32′38″. Error in osculating circle theory = 0°12′49″.

(d) Find the tangents $V'A'$ and $V'C'$. *Answers:* $V'A' = 122.75$; $V'C' = 206.62$.

(e) Find the approximate length of the pseudo spiral AC. *Answer:* 304.2 ft.

5-11. Completely-spiraled compound curves. Find the stationing of all curve points.

	FIRST ARC		SECOND ARC		EACH
STA. V	D	I	D	I	SPIRAL
(a) 42 + 24.08	$2°(D_a)$	23°48′	$7°00'(D_a)$	48°02′	$k = 1\frac{2}{3}°$
(b) 56 + 34.87	$1\frac{2}{3}°(D_c)$	$33°57\frac{1}{2}'$	$6°00'(D_c)$	$41°23\frac{1}{2}'$	150 ft.

Answers: (a) T.S. = 27+95.49; $S._1C._1$ = 29+15.49; $C._1S._2$ = 38+95.49; $S._2C._2$ = 41+95.49; $C._2S._3$ = 45+21.68; S.T. = 49+41.68. (b) T:S. = 36 + 20.75; $S._1C._1$ = 37 + 70.75; $C._1S._2$ = 56 + 58.25; $S._2C._2$ = 58 + 08.25; $C._2S._3$ = 63 + 48.11; S.T. = 64 + 98.11.

5-12. Refer to the multicompound curve in problem 3-3. Replace this curve by an equivalent simple spiral having its T.S. on the same initial tangent and its S.C. at the same tangent point B. Compute L_s, T_S, T_L, and sta. T.S., assuming sta. A = 51+83.25. *Answers:* $L_s = 770.26$; $T_S = 286.98$; $T_L = 546.49$; sta. T.S. = 52+36.44.

5-13. Derive:

(a) Formula 5-18 from the basic relations in section 5-2.

(b) Formula 5-19 from the basic relations in section 5-2.

(c) Formulas 5-30 and 5-31.

(d) Formula 5-32.

(e) The approximate formulas in section 5-11, using a sketch to show the nature of approximations made.

Chapter 6
Earthwork

6-1. Foreword

Payment for grading on highway and railroad construction is usually based on a bid price per cubic yard for *excavation measured in place as computed from survey notes.* The unit price ordinarily includes: hauling excavated material (*cut*) from within the limits of the roadway or moving in other material (*borrow*) from outside areas; building the embankments (*fill*) to specified form; disposing of surplus material (*waste*); and performing such operations as forming earth shoulders, trimming slopes, and preparing the subgrade for railroad ballast or for highway base or surface courses.

Separate unit prices for different types of material excavated may be used. There are advantages, however, in reducing the number of classifications to two—"rock excavation" and "common excavation"—or even to a single type called "unclassified excavation."

Fill quantities are important in grade-line design, though they are not paid for directly in the usual contract. However, on projects consisting wholly of embankment—such as levees—the payment is based on a unit price for fill as computed from survey notes.

6-2. Earthwork Operations

The operations included under this heading depend on the nature and magnitude of the project.

A. On major highway projects where photogrammetric methods are used, the usual operations in sequence are:

1. Office work—taking cross-section measurements by photogrammetric techniques along a selected route center line prior to construction.
2. Office work—obtaining quantities by electronic computer methods from data obtained in 1.
3. Field work—setting construction stakes from measurements obtained in 1 and from alignment data.
4. Miscellaneous—making added measurements and computations needed for determining final pay quantities.

B. On secondary roads and most single-track railroad projects where photogrammetric methods are not used, the usual operations in sequence are:

1. Office work—making preliminary estimates of grading quantities along a selected route center line drawn on a contour map.
2. Field work—setting cross-section stakes along the route after the alignment is staked.
3. Office work—computing grading quantities from data obtained in 2.
4. Miscellaneous—making added measurements and computations needed for determining final pay quantities.

6-3. Types of Cross Sections

The determination of earthwork quantities is based on office or field cross sections taken in a specified manner. On highway and railroad work, cross sections are vertical and at right angles to the survey center line. Every section is an area bounded by the original and the graded surface, the latter being defined by its side slopes, shoulders, subgrade, median strip, and drainage ditches.

Figure 6-1 shows a portion of a graded roadbed passing from fill to cut (side ditches are omitted for simplicity). The sketch illustrates several cross sections, the types depending on their shape and the number of measurements used to determine them.

At any cross section a rod reading is always taken at the center-line stake. Two additional readings are usually taken at the intersections of the side slopes and the ground surface; if stakes are driven at these points they

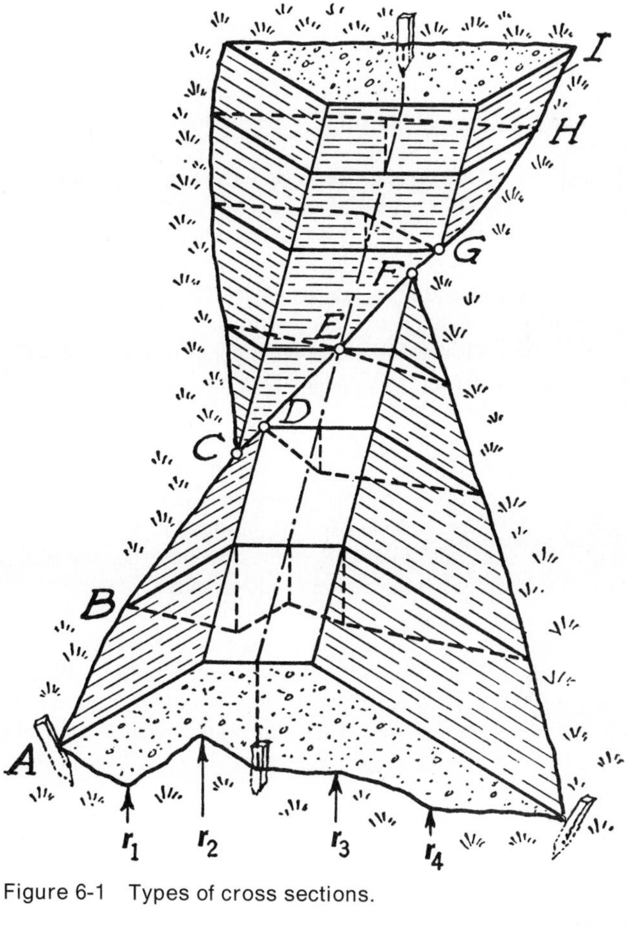

Figure 6-1 Types of cross sections.

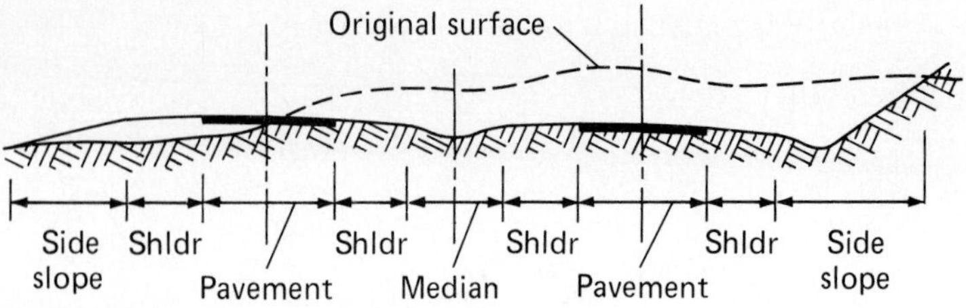

Figure 6-2

are called *slope stakes.* Additional rod readings are taken where conditions require them.

The section at *H* is a regular *three-level section* in cut, so named because three rod readings are used to fix it—one at the center stake and two others at the slope-stake locations. In railroad and two-lane highway construction, this type (in cut or fill) occurs more often than any other.

The sections at *D* and *G* are special cases of a three-level section, each having a *grade point* (point *D* or *G*) at one corner.

The section at *B* is a *five-level section* in fill. This is a modification of a three-level section in which two additional readings are taken directly below (or above) the edges of the base.

The section at *A* is an *irregular section* in fill. A large number of rod readings are required to fix it—in the case shown, there are four readings at points *r* in addition to the three at the slope stakes and center stake.

The section at *E* is a *side-hill section,* having cut on one side and fill on the other side of a grade point at *E.* In the case illustrated the grade point is on the center line, but in general it may fall anywhere between the edges of the base.

The section at *I* is a *level section,* so designated because the ground is level transversely and only one rod reading at the center line is sufficient.

A major rural highway, such as on the Interstate System, will usually have four or more lanes, a swaled median, wide shoulders, and drainage ditches at the toe of cut side slopes. Figure 6-2 shows a cross section fitting this description.

6-4. Location of Cross Sections

For convenience in calculations and field work, cross sections are usually taken at each full-station (or half-station) stake on the survey center line. They are also taken at curve points and additional plus-points where important "breaks" in the topography occur. Where grading is very heavy or where unit costs are high, as in rock excavation, cross sections are taken at closer intervals.

If the transition between cut and fill occurs on a side hill, as many as five cross sections may be needed. In Fig. 6-1 these sections are located at *C*, *D*, *E*, *F*, and *G*. Theoretically, complete cross sections are not necessary at *C* and *F*, but their stationing is needed to locate the apexes of the pyramids having triangular end bases at *E*. Thus, the cross sections at the transition are reduced to three: (1) at the fill-base grade point *D*, (2) at the center-line grade point *E*, and (3) at the cut-base grade point *G*. The points *C* and *F* are often so close to *D* and *G* they are omitted from the notes and the apexes of the transition pyramids assumed to fall at *D* and *G*. Where the three sections at the transition are very close together, the grade contour *DEG* is assumed to be at right angles to the center line; there are then wedged-shaped solids on either side of the grade contour. See page 127.

6-5. Special Formulas for End Areas

All end areas that are bounded by straight lines can be found by the coordinate method (section 6-6). However, the simple geometric areas defined in section 6-3 occur so frequently that it is convenient to have special formulas for them.

Figure 6-3 shows the areas at certain cross sections in Fig. 6-1. The common notation is spread among the several sketches. The distance c is always the vertical distance (cut or fill) between ground and grade at the center line, and h_l (or h_r) is the vertical distance between ground and grade at the slope stake. Distances between ground and grade at other points are denoted by c_l and c_r in cut, and by f_l and f_r in fill (as at section *B*). The inclination s of the side slopes is expressed by the ratio of horizontal distance to vertical distance (as unity).

The horizontal distance from the survey center line to any slope stake is

$$d_l \text{ (or } d_r) = \tfrac{1}{2}b + s\,h_l \text{ (or } s\,h_r) \tag{6-1}$$

The area of a *level section* (as at *I*) is

$$A_L = c\left(\frac{b + D}{2}\right) \tag{6-2}$$

or

$$A_L = c(b + c\,s) \tag{6-3}$$

The area of a *regular three-level section* (as at *H*) is found by adding the areas of the two cross-hatched triangles to the areas of the two triangles having the common base c. Thus,

$$A_3 = \tfrac{1}{2}c(d_l + d_r) + \tfrac{1}{4}b(h_l + h_r)$$

Substitution of H for $h_l + h_r$ and D for $d_l + d_r$ reduces the relation to

$$A_3 = \frac{c\,D}{2} + \frac{b\,H}{4} \tag{6-4}$$

Another convenient formula for the area of a *regular three-level section* is found by extending the side slopes to an intersection at the center line to form a triangle, called the *grade triangle*, below (or above) the base. The dimensions of the grade triangle are constant until the base or slope changes; its fixed area is $b^2/4\,s$. Consequently, the "grade-triangle formula" is

$$A_3 = \frac{D}{2}\left(c + \frac{b}{2\,s}\right) - \frac{b^2}{4\,s} \tag{6-5}$$

Formula 6-5 is slightly more convenient than 6-4 for computing a long series of regular three-level sections in cut or fill, owing to the constant terms $b/2\,s$ and $b^2/4\,s$.

The *special three-level section* having a grade point at the ground surface (as at D or G, Fig. 6-1) is also determined by formula 6-4. One of the four triangles disappears; therefore, $D = d_l$ (or d_r) $+ \tfrac{1}{2}b$, and $H = h_l$ or h_r. The grade-triangle formula also applies if properly modified, but its use is not recommended at the transition between cut and fill.

At a *side-hill section* (as at E) the end areas for cut and fill are kept separate. Obviously, both are triangles. In the general case, with the grade point not at the center line, each area is

$$A_T = \tfrac{1}{2}\,w\,h \tag{6-6}$$

where w is the actual base width of the triangle. At section E in Fig. 6-3, $w = b/2$.

The area of a *five-level section* (as at B) is found by combining the indicated triangles having common bases. The final relation is

$$A_5 = \tfrac{1}{2}(c\,b + f_l\,d_l + f_r\,d_r) \tag{6-7}$$

If the section is in cut, c_l and c_r are substituted for f_l and f_r.

6-6. End Areas by Coordinates

The area of an irregular section, such as A in Fig. 6-1, is best found by a coordinate method. The procedure takes the origin of coordinates at the

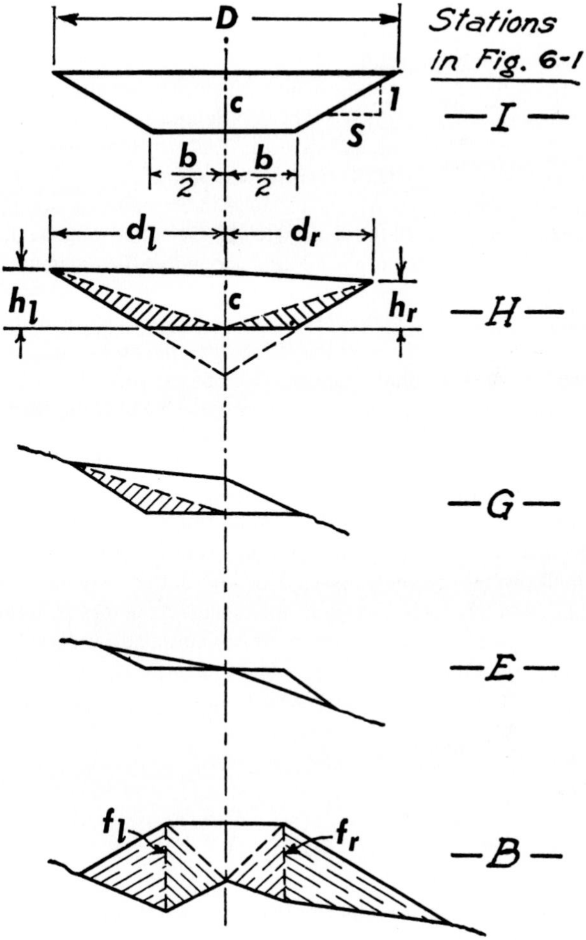

Figure 6-3

center of the base and the coordinates of the corners of the area written in clockwise order starting with point A and repeating the coordinates of A to close the figure. This is shown as the first form on page 120.

From theory developed in plane surveying, the area of a closed figure whose coordinates are set up in the first form is equal to one-half the difference between the algebraic sums of the products indicated by full diagonal lines and those indicated by dotted diagonal lines. In this form careful attention must be paid to algebraic signs.

The possibility of errors in signs can be avoided by writing the coordinates as in the second form on page 120. The coordinates start at the center of the base as origin, and then proceed *clockwise around the left portion* of the section and *counter-clockwise around the right portion* in the form of a figure 8 on its side. *Algebraic signs are omitted.* All products indicated by the solid diagonal lines have the same sign; all those indicated by the dotted diagonal lines have the opposite sign. As before, the end area is one-half the difference between the sums of the two sets of products.

On large rural highway projects the cross sections approximate Fig. 6-2 and most end areas will be irregular. Moreover, the cross-section data will have been obtained by photogrammetric methods. In such cases the data are used as input to electronic computer programs which use the theory of coordinates to produce the end areas.

Machine calculation of areas is done best with coordinates set up in the second form. In using the machine, multiply the figures connected by the solid diagonal lines, accumulating the products in the proper register. Then reverse-multiply the figures connected by the dotted diagonal lines, thus subtracting those products. The result remaining on the register is the *double area.* For cut areas, the sum of the products indicated by dotted diagonal lines is larger than the sum of the products indicated by full diagonal lines; for fill areas, the reverse is true. When using the calculating machine the products making up the larger sum are set up first.

Ordinate

$$\frac{-h_l}{-d_l} \times \frac{0}{-\frac{1}{2}b} \times \frac{0}{+\frac{1}{2}b} \times \frac{-h_r}{+d_r} \times \frac{-f_4}{+d_4} \times \frac{-f_3}{+d_3} \times \frac{-c}{0} \times \frac{-f_2}{-d_2} \times \frac{-f_1}{-d_1} \times \frac{-h_l}{-d_l}$$

Abscissa

Double Area by Coordinates—First Form

℄

$$\frac{0}{0} \times \frac{c}{0} \times \frac{f_2}{d_2} \times \frac{f_1}{d_1} \times \frac{h_l}{d_l} \times \frac{0}{\frac{b}{2}} \times \frac{0}{0} \times \frac{c}{0} \times \frac{f_3}{d_3} \times \frac{f_4}{d_4} \times \frac{h_r}{d_r} \times \frac{0}{\frac{b}{2}} \times \frac{0}{0}$$

Double Area by Coordinates—Second Form

Some computers prefer to use machine calculation for all areas, even the standard types represented by formulas 6-2 through 6-7. This is done by a generalization of the method just described. In all cases omit algebraic signs. For areas lying entirely to the left of the survey center line, use the coordinates in clockwise order; for areas to the right, use counter-clockwise order of the coordinates. It makes no difference which coordinate is used first, so long as it is repeated at the end, that is, so long as the traverse is closed. For side-hill sections (Fig. 6-5) use the figure 8 form of coordinates described for section *A*, starting at the grade point separating the cut and fill portions.

6-7. Compound and Other Irregular Sections

In locations where rock lies between ground and grade, a *compound section* occurs, as in Fig. 6-4. If there are different unit prices for rock and earth ("common") excavation, it is necessary to determine the quantities of each material. This is done by first cross sectioning the original surface, and then doing the same to the rock surface after stripping or by probing through earth to rock.

In modern highway construction the trend is toward denoting all excavation as "unclassified." There is only one unit price, regardless of whether the material excavated is ledge rock, loose rock, earth of any type, or a combination of such materials with boulders and (as on much reconstruction) old pavement, building foundations, railroad ties, or miscellaneous scrap deposited in dumps. Even in a clear-cut situation such as in Fig. 6-4, the contractor may prefer to expedite the work by drilling through the earth cover and blasting the underlying rock without having to clean the rock surface for separate cross sectioning. Obviously, the unclassified specification speeds the work and eliminates arguments between contractor and engineer as to the pay classification of material excavated.

Railroad roadbeds on curves are not usually crowned or banked; the superelevation is adjusted in the rock ballast. Earthwork quantities in the drainage ditches are usually computed as a separate item.

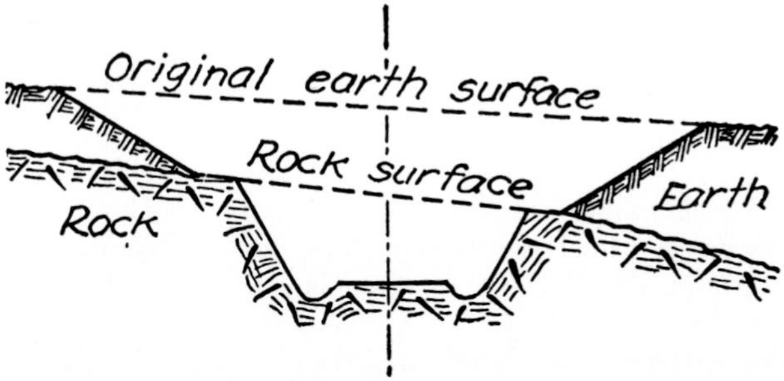

Figure 6-4 Compound section.

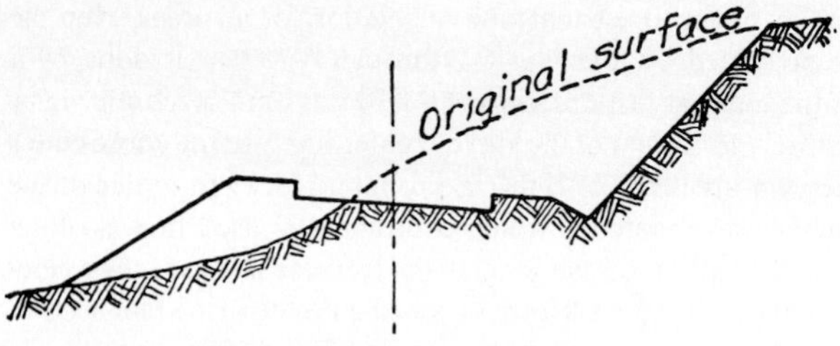

Figure 6-5 Side-hill section on curve.

On highway roadbeds the subgrade may be crowned on tangents and is usually banked parallel to the surface on curves. Moreover, the drainage ditches and earth shoulders are usually considered part of the cross-sectional area. (See Fig. 6-5.) The resulting irregular areas may be found by the coordinate method (possibly employing a calculating machine or an electronic computer) or by graphical methods.

6-8. End Areas by Graphical Methods

End areas, no matter how irregular, are easily found by plotting them to scale and running a planimeter around the boundaries. This method is widely used in highway work, especially if ditches and shoulders are part of the cross section proper. It is particularly adapted to projects on which a permanent graphical record of the cross sections is desired, as in tunnel construction through rock. Another distinct advantage of plotted sections is their value in studying the effects of minor changes in alignment or grade elevation.

In order to obtain precision consistent with the field work, areas are plotted to a fairly large scale, usually 1 in. = 10 ft or larger; consequently, the file of cross section sheets is voluminous.

Another graphical method, which is especially useful for shallow areas plotted accurately to scale, is to mark off on the edge of a paper strip the continuous summation of verticals at 1-ft intervals across the area. Each vertical is the area of a trapezoidal strip 1 ft wide; the total length, applied against the scale of the cross section, is the desired end area. A special scale (*Avol rule*) can replace the paper strip.

6-9. Field Cross Sectioning

The term *cross sectioning* is loosely used to include any vertical and horizontal measurements made on a transverse section. On routes such as railroads, highways, and canals, two methods of cross sectioning are in use: (1) cross-section leveling, and (2) slope staking.

Cross-section leveling is used when end areas are to be determined graphically, as by planimeter. It is also the method which must be used to obtain whatever cross profiles are needed in the office work of grade-line design. The field process is essentially that of profile leveling, the difference being that intermediate foresights are taken at breaks in the transverse profile in addition to breaks along the survey center line. The left-hand page of the notes is similar to one in differential leveling, but the notes run up the page. On the right-hand page the cross section notes are entered in the form of fractions, the numerator of each fraction giving the rod reading (intermediate foresight) and the denominator showing the transverse distance to the rod from the survey center line.

Form of Notes for Cross-Section Leveling

Portion of left-hand page		*Right-hand page*							
Station	**H.I.**		**L**		**℄**		**R**		
		473.4	469.6	468.4	467.9	467.3	467.2	465.1	464.5
(T.C.) 71+76.2	....	−0.8	3.0	4.2	4.7	5.3	5.4	7.5	8.1
		40	26	12	0	12	18	32	40
		470.5	468.3	466.9	466.4	465.8	465.6	464.1	462.8
71+00	472.58	2.1	4.3	5.7	6.2	6.8	7.0	8.5	9.8
		40	20	12	0	12	16	24	40
					H.I. = 472.6				

To obtain occasional rod readings on ground higher then the H.I., the hand level is a useful supplement to the engineer's level; rod readings in such cases are recorded as negative; they must be *added* to the H.I. elevation (see the extreme left reading at sta. 71 + 76.2 in the notes).

The form of notes represents a portion of the left-hand page and complete notes for two cross sections on the right-hand page.

Slope staking is a special form of leveling used only after the grade line and form of cross section have been decided upon. The cross sections are not usually plotted; either areas or volumes are computed directly from the field notes.

This method may be employed solely for the purpose of obtaining data for calculating volumes without actually setting slope stakes. It is also used preceding construction where grading is heavy and slope stakes are needed to control the work.

The process of slope staking at any cross section consists of finding and recording the positions where the designed side slopes will intersect the ground surface, and recording breaks in the transverse ground profile between slope stakes. The work may be done in connection with profile leveling over the located line or as a separate process afterward. The latter is a simpler and speedier method.

Slope staking in mountainous terrain is best accomplished by using EDM equipment with a computer program that delivers the elevation and offset of slope stakes on each digitized cross section.

When profile levels have already been run, the slope-staking party is provided with the ground and subgrade elevations at each station stake. Leveling is done with an engineer's level, a hand level, a level board, or a combination of these instruments. The form of notes resembles that used in cross section leveling. But the numerator of the fraction, instead of being the rod reading, is the difference between the ground-surface elevation and the center-line grade elevation at the particular station. Cuts are designated *C* or +; fills, *F* or −. No record is kept of the actual rod readings used in the process. Also, the extreme left-hand and right-hand entries in the notes are for the slope-stake locations.

The form below is a record of slope-staking notes for sections along the two-lane road shown in Fig. 6-1.

In the example illustrated, elevations will have been supplied for the full stations only, since the exact locations of the grade points are not known in

SLOPE-STAKING NOTES—FIRST EXAMPLE

STA.	SURFACE ELEV.	GRADE ELEV.	L						R
(*I*) 63 + 00	570.4	562.84	C7.6/31.4			C7.6			C7.6/31.4
(*H*) 62 + 00	568.4	561.64	C7.0/30.5			C6.8			C6.4/29.6
(*G*) 61 + 48	565.1	561.02	C4.8/27.2			C4.1			0/20
(*E*) 61 + 36	560.9	560.87	C2.6/23.9			0			F2.0/18.0
(*D*) 61 + 25	557.1	560.74	0/15			F3.6			F5.8/23.7
(*B*) 61 + 00	554.6	560.44	F5.4/23.1		F6.9/15.0	F5.8	F6.7/15.0		F8.2/27.3
(*A*) 60 + 00	551.2	559.24	F7.2/25.8	F8.8/20.0	F7.0/14.7	F8.0	F8.1/18.0	F9.3/24.0	F9.4/29.1

Bases: 40 ft in cut; 30 ft in fill. Side slopes: $1\frac{1}{2}$; 1; Gradient: +1.2%.

advance. Grade points are found by trial in the field, their grade elevations computed, and their surface elevations obtained by leveling from any convenient known point. Note that stations *C* and *F* were omitted. These points are assumed to fall at *D* and *G*, as explained in section 6-4.

The process of finding the slope-stake locations in the field is a matter of trial and error. It is one of those procedures for which a detailed numerical illustration of the method appears more complicated than it actually is. Briefly, the process follows:

1. Record the cut or fill at the center found by taking the difference between the given elevations.
2. Observe whether the ground slopes up or down transversely, *estimate* the cut or fill h at the probable slope-stake location, and calculate the corresponding distance d from the center line by the relation $d = \frac{1}{2}b + sh$.
3. Take a rod reading at the computed distance and find the *actual* h.
4. If the actual and estimated values of h differ by more than 0.1 ft, make a new estimate, being guided by the first result; then repeat the process until d and h satisfy formula 6-1. Distances and depths are determined to the nearest tenth.

Slope staking is done rapidly by the foregoing method. With a little experience many of the points are located close enough for the purpose on the first trial. More than two trials are seldom needed.

Slope stakes, if set, are driven aslant. The cut or fill is marked on the side facing the center line, and the stationing is marked on the back.

If profile levels have not been run or if it is desired to verify them in slope staking, the engineer's level is used to carry continuous elevations along the line by usual leveling methods. In this case it is convenient to use a device known as the *grade rod* (see Fig. 6-6), which is the imaginary reading on a

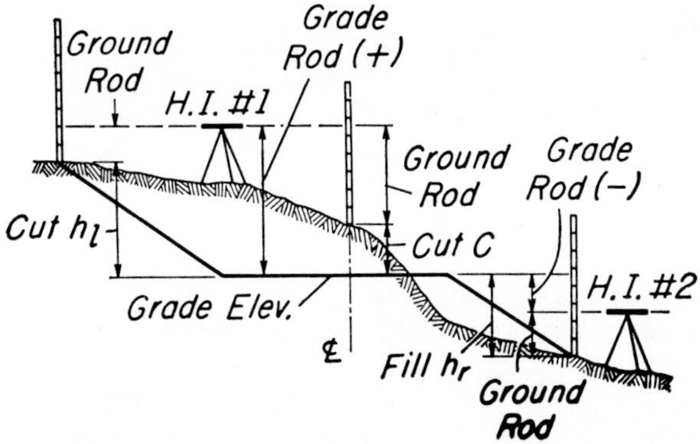

Figure 6-6

rod held on the finished subgrade. If the line of sight is below subgrade elevation, the rod is assumed to be read downward and the grade rod given a negative sign. Various combinations are possible, the sign of the result depending on the type of section (cut or fill) and relative elevations of instrument and subgrade. The following rules always give correct results if algebraic signs are strictly observed:

1. H.I. minus grade elevation equals grade rod.
2. Grade rod minus ground rod equals cut or fill at the ground rod (the sign + signifies cut; the sign − indicates fill).

If desired, grade-rod readings may be entered in a separate column in the leveling notes on the left-hand page of the field book, with slope-staking notes on the right-hand page. Once the center cut or fill has been determined, the slope-staking process is the same as that previously described.

In brush or on steep slopes the field work may be expedited by using a hand level or a level board to supplement the engineer's level.

The second example of slope-staking notes on page 127 shows irregular ground with more-complex transitions between excavation and embankment. Included also are wedge-shaped solids either side of a grade contour at right angles to the center line.

6-10. Volume by Average End Areas

Except where the solid between cross sections is a pyramid (as between *E* and *F* in Fig. 6-1), it is usually considered a prism whose right cross-sectional area is the average of the end areas. For sections having areas of A_1 and A_2 ft^2 and L ft apart, the average-end-area formula for volume in cubic yards is

$$V_e = \frac{L}{27}\left(\frac{A_1 + A_2}{2}\right) \tag{6-8}$$

This formula is exact only when the end areas are equal. For other cases it usually gives volumes slightly larger than their true values. If it were to be applied to a pyramid, for example, the error would be the maximum and equal to 50% of the correct volume. In practice, however, the total error in a long line is rarely more than 2%. Also, calculation of the errors or corrections (see section 6-13) is much more complicated than determining the average-end-area volumes themselves. In consequence, the average-end-area method is almost always used, and invariably ruled to apply in the absence of specifications to the contrary.

In applying the average-end-area formula the simplest method is to add the end areas (determined by calculation or by planimeter) and multiply their sum by $L/54$. Table XIX facilitates the process.

SLOPE-STAKING NOTES—SECOND EXAMPLE

STA.	SURFACE ELEV.	GRADE ELEV.	L						R
92+00	891.6	874.63	C18.1/61.2	C20.0/40.3	C20.1/25.0	C17.0	C17.9/17.8	C15.1/36.8	C10.2/45.4
91+52	875.6	875.59	0/25	0/20		0		0/20	0/25
91+00	864.5	876.63	F6.9/33.8			F12.1			F11.4/42.8
90+50	867.3	877.63	F8.2/36.4			F10.3			F10.1/40.2
90+05	875.3	878.53	0/20			F3.2			F12.8/45.6
89+73	879.2	879.17	C16.3/57.6			0			F12.2/44.4
89+59	885.6	879.45	C17.4/59.8			C6.1		0/14	F8.3/36.6
89+43	879.8	879.77	C10.2/45.4			0			F7.3/34.6
89+08	873.7	880.47	0/20			F6.8			F16.1/52.2
88+68	881.3	881.27	C16.8/58.6			0			F12.9/45.8
88+30	890.9	882.03	C24.0/73.0			C8.9			0/25

Bases: 50 ft in cut; 40 ft in fill. Side slopes: 2:1; Gradient −2%.

6.11. Example of Earthwork Calculation

The tabulation on page 128 gives the results of earthwork calculations for the notes on page 124. Areas were computed from formulas given in section 6-4; volumes from the average-end-area formula (except for the two pyramids).

The number of significant figures used in computing the areas and volumes is inconsistent with the principles set forth in sections 2-17 and 2-18. Both the rod readings and measured distances are round numbers; consequently, the computed volumes are reliable to only two significant figures.

Yet it is conventional practice to consider the recorded measurements as exact numbers and to compute volumes to the nearest cubic yard (or to the nearest $\frac{1}{10}$ yd³ as in this example).

6.12. Earthwork Tables

Table XIX is especially adapted to highway work or other projects on which cross-sectional areas are obtained graphically. Two other tables in Part III are useful in earthwork computations:

Table XVII gives cubic yards per 100 ft for level sections having various base widths and side-slope ratios. It is very useful in making preliminary estimates of grading quantities by scaling center cuts and fills from a projected paper location.

Table XVIII gives cubic yards per 50 ft for triangular prisms having various widths and heights. It is the most convenient earthwork table for general purposes, since practically all solids met in route surveying may be broken up into constituent triangular prisms.

If w is the width (base) of any triangle and h the height (altitude), the volume in cubic yards for a triangular prism 50 ft long is

$$V_{50\text{ ft}} = \frac{50}{54} w\, h \tag{6-9}$$

Comparison with the average-end-area formula shows that the volume between two end sections with areas A_1 and A_2 spaced 100 ft apart is equal to the sum of the volumes of two 50-ft prisms having the given right cross-sectional areas.

	END AREAS		VOLUMES (yd³)	
STA.	FORMULA	ft²	CUT	FILL
63 + 00	6–3	390.64		
			1350.0	
62 + 00	6–4	338.34		
			465.2	
61 + 48	6–4	144.76		
			37.9	2.2
61 + 36	6–6	C 26.00 F 15.00		
			3.5	26.1
61 + 25	6–4	113.16		
				171.9
61 + 00	6–7	258.15		
				1104.3
60 + 00	coordinates	338.16		
		Totals =	1856.6	1304.5

Even more generally, if the area of any type of section is converted into the form $A = \frac{1}{2} \times$ (the product of two quantities), the volume in cubic yards between sections 100 ft apart may be found simply by adding two values taken from Table XVIII, each one found by entering the table with the proper given quantities. If the distance between the stations is less than 100 ft, the sum of the tabulated quantities is multiplied by the ratio of the actual spacing to 100. By means of this principle, volumes may be computed directly from slope-staking notes without separate computation of areas.

The following computations show the use of Table XVIII in determining three of the volumes previously found (section 6-11) by the average-end-area method.

STA.	FORMULA CONVERTED TO FORM $\frac{1}{2}wh$	TABLE XVIII: ENTRIES	TABLE XVIII: yd^3 per 50 ft	yd^3 CUT BETWEEN STATIONS
63 + 00	$\frac{1}{2}(2c)(b + cs)$	$w = 2c = 15.2$ $h = b + cs = 51.4$	723.4	
				1350.0
62 + 00	$\frac{1}{2}cD + \frac{1}{2}\frac{b}{2}(H)$	$w = c = 6.8$ $h = D = 60.1$ $w = \frac{b}{2} = 20$ $h = H = 13.4$	626.6	
				465.2
61 + 48	$\frac{1}{2}cD + \frac{1}{2}\frac{b}{2}(h_l)$	$w = c = 4.1$ $h = D = 47.2$ $w = \frac{b}{2} = 20$ $h = h_l = 4.8$	268.1	
				37.9
61 + 36	$\frac{1}{2}wh_l$	$w = \frac{b}{2} = 20$ $h = h_l = 2.6$	48.1	

In entering Table XVIII either of the given quantities may be taken as the height. By properly shifting the decimal point all the separate values making up a given product can be taken from the same line in the table. For example, at sta. 63 the given quantities are 15.2 and 51.4.

Enter left-hand column with 15.2

Take out under 5 (×10)	703.7
Take out under 1	14.1
Take out under 4 (×0.1)	5.6
Sum = yd^3 per 50 ft =	723.4

At sta. 62, instead of entering the table with 20.0 and taking out the three values under 1($\times$ 10), 3, and 4($\times$ 0.1), it is quicker to enter with 13.4 and get the result in one operation under 2($\times$ 10). Any slight error caused by multiplying a tabular value by 10 can be eliminated, if desired, by adding proper values. For example, a total found under 2($\times$ 10) is obtained more accurately by adding the tabular values under 6, 7, and 7.

6-13. Prismoidal Volumes and Corrections

As noted in section 6-10, the average-end-area formula usually gives volumes slightly larger than their true values. When a precise value is necessary—and the field measurements are refined enough to warrant it—the solid between cross sections is considered to be a *prismoid* rather than an average-end-area prism.

The prismoidal formula for volume in cubic yards is

$$V_p = \frac{L}{27}\left(\frac{A_1 + 4\,A_m + A_2}{6}\right) \tag{6-10}$$

where A_m is the area of a section midway between A_1 and A_2 and the other terms have the same meanings as in formula 6-8.

In route surveying the prismoidal formula applies to any solid generated by a straight line passing around the sides of plane parallel end-bases. Accordingly, it fits warped-surface solids as well as plane-surface solids, provided that the warp is continuous between the ends. The formula also applies to a wide variety of solids seldom found in earthwork calculations, such as the frustums of prisms, cylinders, and cones.

Owing to the need for computing the area of the midsection A_m, direct determination of volumes from the basic prismoidal formula is inconvenient. It is easier to apply a *prismoidal correction* C_p to the average-end-area volume. By definition,

$$V_e - C_p = V_p \tag{6-11}$$

When the values given by formulas 6-8 and 6-10 are substituted in formula 6-11 and the resulting formula is reduced, the general prismoidal-correction formula is

$$C_p = \frac{L}{3 \times 27}(A_1 - 2\,A_m + A_2) \tag{6-12}$$

More convenient working formulas for solids commonly met in practice are found by calculating A_m in terms of the given dimensions of A_1 and A_2 and substituting in formula 6-12. (*Note:* A_m is not the mean of A_1 and A_2, but its dimensions are the means of corresponding dimensions at the end sections.) For a solid having triangular end areas the result is

$$C_T = \frac{L}{12 \times 27}(w_1 - w_2)(h_1 - h_2) \tag{6-13}$$

Formula 6-13 can be made to fit any type of end area by dividing it into triangles. However, the prevalence of three-level sections makes the following formula valuable:

$$C_3 = \frac{L}{12 \times 27}(D_1 - D_2)(c_1 - c_2) \tag{6-14}$$

Although formula 6-14 is derived from the dimensions of three-level sections, it also fits a solid having level-section end areas and a solid having a triangular section at one end.

The prismoidal correction is applied with the sign indicated in formula 6-11; that is, it is normally *subtracted* from the average-end-area volume. In case the sign of C_T or C_3 should happen to be negative (rare, but possible), the prismoidal correction is added.

The corrections to the three volumes computed in section 6-12, in cubic yards, are:

$$\text{Sta. } 62+00 \text{ to } 63+00\text{: } C_3 = \frac{100}{12 \times 27}(62.8 - 60.1)(7.6 - 6.8) = 0.7$$

$$\text{Sta. } 61+48 \text{ to } 62+00\text{: } C_3 = \frac{52}{12 \times 27}(60.1 - 47.2)(6.8 - 4.1) = 5.6$$

$$\text{Sta. } 61+36 \text{ to } 61+48\text{: } C_3 = \frac{12}{12 \times 27}(47.2 - 23.9)(4.1 - 0.0) = 3.5$$

Prismoidal corrections can also be determined by means of Table XVIII. The tabulated values come from formula 6-9, which may be written in the following general form:

$$V_{50\text{ ft}} = \frac{50}{54} \times (\text{product of two quantities})$$

When $L = 100$ the prismoidal-correction formulas 6-13 and 6-14 may also be written as follows:

$$C_p = (\tfrac{1}{3})(\tfrac{50}{54}) \times (\text{product of two quantities})$$

Consequently, the prismoidal correction for sections 100 ft apart is one-third the value found by entering Table XVIII with $(w_1 - w_2)$ and $(h_1 - h_2)$, or $(D_1 - D_2)$ and $(c_1 - c_2)$, as the given height and width. The three corrections previously computed are verified by Table XVIII to be:

$$\tfrac{1}{3} \times 2.00 = 0.7$$

$$\tfrac{1}{3} \times 0.52 \times 32.25 = 5.6$$

$$\tfrac{1}{3} \times 0.12 \times 88.5 = 3.5$$

The foregoing corrections are 0.05%, 1.2%, and 9.2% of the respective average-end-area volumes. It is evident that prismoidal corrections are insignificant, except at transitions between cut and fill. Since, normally, these locations account for only a small percentage of the total yardage, it is obvious that volumes determined by the average-end-area method are adequate for all but rare situations.

6-14. Correction for Curvature

Where conditions warrant calculation of prismoidal corrections, they may also justify correcting the prismoidal volumes on curves for the slight error involved in assuming the center line to be straight. On curves, cross sections are taken radially. The true volume between two such sections is a curved solid with plane, nonparallel ends, as portrayed by Fig. 6-7(a). But when curvature is ignored, the computed volume is that represented by Fig. 6-7(b). *The curvature correction is the difference between these volumes, that is,*

$$C_c = V_{\text{Computed}} - V_{\text{True}} \tag{6-15}$$

Figure 6-7(c) represents a typical cross section at a station on a curve. The center of gravity of the total end area is at point G, located at a distance e from the survey center line. The volume generated by revolving the end area is the product of the area and the length of path described by its center of gravity (theorem of Pappus). Obviously, if the end area were shaped so that G fell on the survey center line, the curvature correction is zero.

Let A denote the total end area of the section in Fig. 6-7(c). The curvature correction per station is

$$A \times 1 \text{ sta.} - A(R - e)\frac{\pi}{180}D^\circ$$

Since $(\pi/180)RD^\circ$ equals 1 station, the curvature correction is $(\pi/180)AeD^\circ$ per station. When A is in square feet and e is in feet, this reduces to

$$C_c = \frac{AeD^\circ}{1550} \text{ yd}^3 \text{ per sta.} \tag{6-16}$$

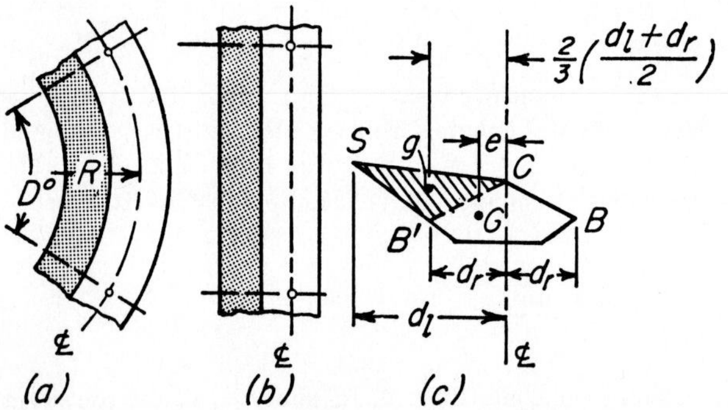

Figure 6-7

Three-level sections occur so often that it is convenient to have a special version of formula 6-16 in which A and e are replaced by the notation used in three-level sections. Figure 6-7(c) is a three-level section in cut with the slope stake S on the inside of the curve. If B' is drawn on the inner side slope at the same elevation as slope stake B, the nonshaded portion of the end area is symmetrical about the survey center line and there is no curvature correction for that portion. The remaining shaded area $SB'C$ has its center of gravity at g, which is two-thirds the distance from C to the midpoint of SB'.

The curvature correction per station is

$$\text{Area } SB'C \times 1 \text{ sta.} - \text{Area } SB'C \left[R - \tfrac{1}{3}(d_l + d_r)\right] \frac{\pi}{180} D^\circ$$

Since $(\pi/180)RD^\circ$ equals 1 station, the curvature correction is

$$\text{Area } SB'C \; \tfrac{1}{3}(d_l + d_r) \frac{\pi}{180} D^\circ \text{ per sta.}$$

But Area $SB'C$ equals $\frac{1}{2}(\frac{1}{2}b + sc)(h_l - h_r)$ ft^2, where $(h_l - h_r)$ is the difference between the slope-stake heights, and always used as a plus quantity. Therefore,

$$C_c = (\tfrac{1}{2}b + sc)(h_l - h_r)(d_l + d_r)\left(\frac{D^\circ}{9300}\right) \text{yd}^3 \text{ per sta.} \qquad (6\text{-}17)$$

For irregular sections, C_c may be found by plotting the sections to scale, drawing for each an equivalent three-level section by estimation, and scaling the values needed in formula 6-17. If this method is not considered accurate enough (as it may not be for highly-eccentric sections in rock), the following procedure may be applied: An irregular section may be divided into triangles; each triangular area may be multiplied by the distance from its center of gravity to a vertical axis at the survey center line; and the algebraic sum of the products may be divided by the total area of the irregular section. The result is the *eccentricity* e of the total cross section, or the distance from the survey center line to the center of gravity G (see Fig. 6-7). This is the familiar method of moments.

The curvature correction is applied as indicated in formula 6-15. The sign of C_c, however, may be positive or negative. If the end area is unsymmetrical about the survey center line and has excess area on the *inside* of the curve, C_c has a positive sign and is *subtracted* from the volume as computed by ignoring the curvature. Where the excess area is on the *outside* of the curve, the curvature correction is *added*. These rules for the sign of the curvature correction apply to both excavation and embankment and hold true regardless of the direction of curvature.

In applying either of the formulas to find the curvature correction for a solid between two different cross sections L ft apart, the results are averaged and multiplied by the ratio of L to 100. For example, if a 10° curve to the left is assumed, calculations for the curvature correction between stations 61+48 and 62+00 in the notes in section 6-9 give:

Sta. 62+00 . . . $C_c = 1.17$ yd³ per sta.

Sta. 61+48 . . . $C_c = 6.37$

$$\text{Sum} = 7.54 \text{ yd}^3 \text{ per sta.}$$

$$\text{Av} = 3.77 \text{ yd}^3 \text{ per sta.}$$

$$C_c = 0.52 \times 3.77 = 2.0 \text{ yd}^3 \text{ to be subtracted}$$

6-15. Earthwork Volumes by Grading Contours

The use of freeway design, in which double roadways are often at different levels separated by a median of varying width, complicates not only the construction staking but also calculation of grading quantities. Good results can be obtained by substituting a contour grading plan for the usual voluminous set of cross-section sheets. This is a device adapted from landscape grading.

In essence, "contour grading" consists of superimposing contour lines of the proposed construction on an existing contour map, thereby forming a series of horizontal areas bounded by closed contours. The areas are planimetered, and the volumes of the horizontal slices of earthwork are determined by the average-end-area method.

A detailed description of contour grading is omitted here since the subject is well covered in many plane surveying books (for example, *Surveying* by Moffitt and Bouchard, section 16-14).

6-16. Borrow Pits

When the quantity of material within the theoretical limits of excavation is not enough to make the fills, it is necessary to provide additional material, termed *borrow*. It is most convenient to obtain borrow by widening the cuts adjacent to the fills where the material is needed. When this can be done within the right-of-way limits (and without interfering with existing or planned structures), it has the added advantages of permitting wider shoulders (on highways), "daylighting" curves, reducing slope erosion, and minimizing snow drifting on the traveled way.

Material taken from borrow pits adjacent to the main construction may be measured by extending the regular cross-sections and adding intermediate ones where necessary. The work is conveniently done by the cross-section-leveling method.

Borrow pits located away from the route are cross-sectioned independently of the survey stationing. A convenient method is to stake out over the area a system of rectangles referenced to points outside the limits of the work. By leveling at the stakes before and after excavation, data are obtained from which to compute the volume of borrow taken from the pit.

Figure 6-8 shows a borrow-pit area over which 28 squares were originally staked out. The cross-hatched line represents the limits of the excavation. Squares are of such size that no important breaks, either in the original ground surface or in the pit floor, are assumed to lie between the corners of squares or between the edge of the excavation and the nearest interior corner. Those squares falling completely within the excavation are outlined by a heavy line. Within that line each square excavated to the pit floor is the volume of a truncated square prism. Square 7, for example, has the surface corner points *b*, *c*, *d*, and *e*; after excavation, corresponding points on the pit floor are *b*′, *c*′, *d*′, and *e*′ (see Fig. 6-9). The volume of the resulting prism is the product of the right cross-sectional area *A* and the average of the four corner heights *bb*′, *cc*′, *dd*′ and *ee*′. In cubic yards,

$$V_7 = \frac{A}{4 \times 27}(bb' + cc' + dd' + ee') \qquad \text{exactly}$$

Each similar complete prism might be computed by the preceding method. However, when a number of such prisms adjoin one another, it is quicker to use the following relation which gives the total volume of *any number* of complete prisms:

$$V = \frac{A}{4 \times 27}(\Sigma h_1 + 2\,\Sigma h_2 + 3\,\Sigma h_3 + 4\,\Sigma h_4) \qquad (6\text{-}18)$$

In formula 6-18, *A* is the right cross-sectional area of the unit rectangular prism, *not* the total area of all the complete prisms; h_1 is a corner height found in only one prism; h_2 is one common to two prisms; h_3 is one common to three prisms; and h_4 is one common to four prisms. For example, *ee*′ is an h_1, *dd*′ is an h_2, and *cc*′ is an h_4.

The total borrow-pit quantity also includes the wedge-shaped volumes lying between the complete prisms and the limits of the excavation.

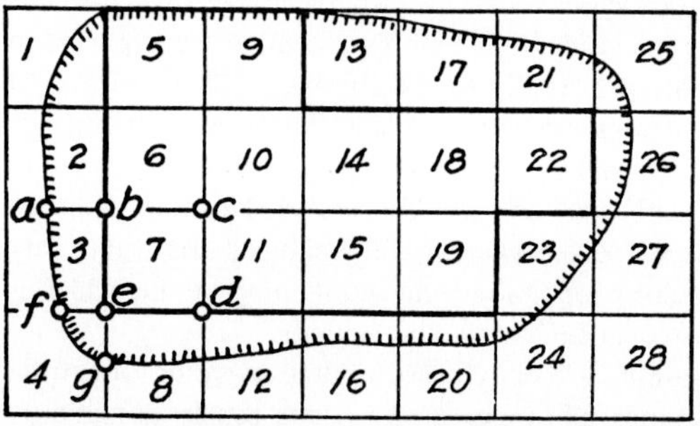

Figure 6-8 Borrow pit.

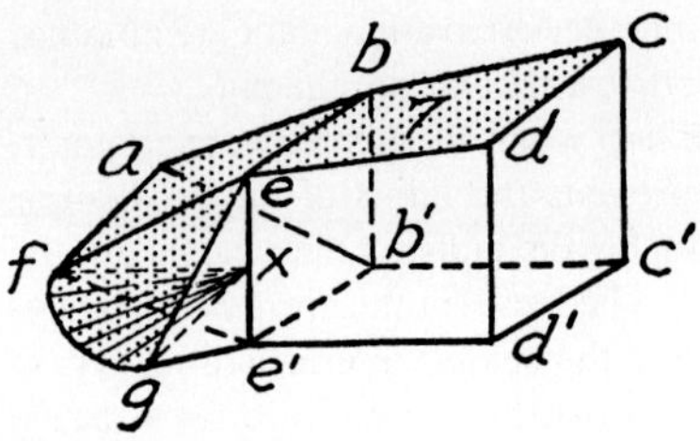

Figure 6-9

The portion of square 3, Fig. 6-8, excavated to the near face of prism 7 is shown in Fig. 6-9 to be a wedge-shaped mass with the cutting edge *fa* and the trapezoidal base *ebb'e'*. For all practical purposes its volume is one-half the product of the area of the base and the average of the horizontal distances *ab* and *fe*.

At a corner the portion of the square excavated is composed approximately of two quarter-cones, base to base. For example, as shown in Fig. 6-9, the mass in square 4 has one quarter-cone with base *fxg* and altitude *xe*, and another quarter-cone with the same base but with altitude *xe'*. The radius *r* of the circular base may be taken as the average of *fx* and *gx*. Consequently, the volume at the corner equals $\frac{1}{3}(\pi r^2/4) \times ee'$ ft^3. If the height *ee'* is designated by *h*, the volume reduces to approximately $r^2 h/103$ yd^3.

6-17. Shrinkage, Swell, and Settlement

On many routes, one object of the paper-location study is to design the grade line so that total cut within the limits of the work will equal total fill. If it is assumed economical to haul all excavated material to embankments, borrow and waste are eliminated. Attainment of this ideal is prevented by many factors, one of which is the uncertainty regarding shrinkage or swell of the material.

Shrinkage denotes the fact—commonly noticed—that 1 yd^3 of earth as measured by cross sectioning before excavation will occupy less than a cubic yard of space when excavated, hauled to an embankment, and compacted in place. This difference is due principally to the combined effects of loss of material during hauling and compaction to a greater than original density by the heavy equipment used in making the embankment. Shrinkage is small in the case of granular materials such as sand and gravel; larger in ordinary earth containing appreciable percentages of silt, loam, or clay; and very high (possibly as much as 30%) for shallow cuts containing humus which is discarded as being unsuitable for building embankments.

Since payment for grading is usually based upon excavation quantities (see section 6-1), the shrinkage allowance in grade-line design is made by adding a percentage to the calculated fill quantities.

Swell is the term used in referring to a condition which is the reverse of shrinkage. It occurs rarely, and then usually in the case of broken rock blasted from solid beds and mixed with little, if any, earth in making embankments.

Swell is apt to be fairly uniform for a given material. Shrinkage, however, varies not only with changes in the soil constituents but also with fluctuations in moisture content when compacted and the type of construction equipment used. Consequently, a percentage allowance assumed in design may eventually prove to be 5% or more in error. A common shrinkage allowance is 10 to 15% for ordinary earth having little material unsuitable for fills.

The term *settlement* refers to subsidence of the completed embankment. It is due to slow additional compaction under traffic and to gradual plastic flow of the foundation material beneath the embankment. On railroad fills, small settlement can be corrected by tamping more ballast beneath the ties as routine maintenance work. In highway construction, new fills are sometimes built higher than the designed subgrade elevation and the placing of permanent pavement is deferred until most of the settlement has taken place. With modern construction methods, however, involving good foundations and compaction at optimum mositure content, settlement of fills is rarely serious.

DISTRIBUTION ANALYSIS

On projects in which embankments are built from material excavated and hauled from cuts within the limits of the right-of-way, mere calculation of separate cut and fill quantities does not provide enough information. The distribution of the earthwork, which involves the quantity, direction, and distance hauled, is also important both in planning the work and in computing extra payment in case the contract contains an overhaul clause.

6-18. Haul, Free Haul, and Overhaul

The word *haul* has several definitions. In earthwork analyses, however, it means either the distance over which material is moved or (in a more technical sense) the product of volume and distance moved, the units being station-yards.

The contract sometimes contains a clause providing extra payment for hauling material a distance greater than a specified amount, known as the *limit of free haul*. In this case there is one unit price, *per cubic yard*, for earth excavation and another unit price, *per station-yard*, for *overhaul*. The former price includes hauling within the free-haul limit and forming the embankments either inside or outside that limit (see section 6-1). Short hauls are never averaged with those longer than the free-haul limit; therefore, there is no need to calculate the station-yards of *free haul*.

6-19. Limit of Economic Haul

With an overhaul clause in effect, there is obviously a certain distance beyond which the cost of overhaul exceeds the cost of excavation without overhaul.

This *limit of economic haul* equals the limit of free haul plus the quotient found by dividing the unit price for borrow (or for excavation, if there is no separate price for borrow) by the unit price for overhaul. Thus, if the free-haul limit were 1000 ft and unit prices for excavation and overhaul were \$2.40 per yd^3 and \$0.20 per sta.-yd, respectively, the limit of economic haul would be $10 + (2.40/0.20) = 22$ stations.

6-20. Balance Points

The principal problem in making a distribution analysis is locating the stationing of *balance points* between which excavation equals fill plus shrinkage allowance. On a small job the primary balance points may be found by making separate subtotals of the cuts and corrected fills, balance points being located where the two subtotals are equal. On important work this method is inadequate. It does not fix intermediate balance points; neither does it give data for computing overhaul, nor show how to schedule the work. More detailed analyses may be made by the *station-to-station method* or by the *mass-diagram method.*

Regardless of how complete an analysis is made, it is fairly common practice to show balance points on the plans, together with estimated quantities of cut, fill, borrow, and waste. It is advisable to label such balance points as "approximate," in order to avoid controversy with the contractor in case the balance points should prove to be in error because of variable shrinkage.

6-21. Station-to-Station Method

Making a distribution analysis by the station-to-station method is a numerical process. The steps are illustrated by Fig. 6-10 which shows a portion of the profile along a route center line. A grade point G is first located in the notes. Then balance points A and A', a distance apart equal to the limit of economic haul, are found by adding computed cuts and fills (plus shrinkage allowance) in opposite directions from G. Balance points L and L', spaced at the limit of free haul, are located similarly.

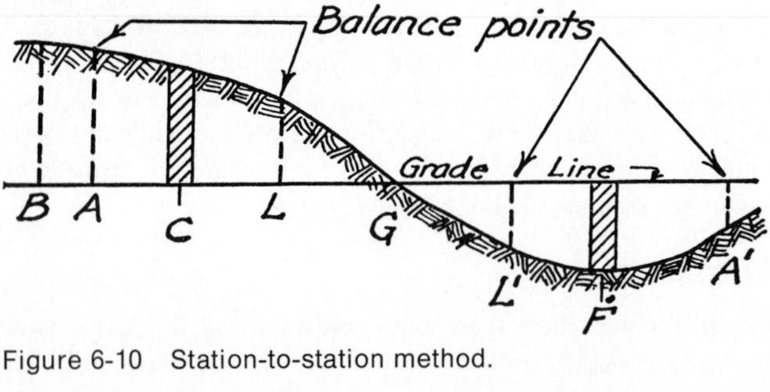

Figure 6-10 Station-to-station method.

Excavation between A and L, which just equals that portion of the fill between L' and A', is subject to payment for overhaul. The average distance over which that excavation is hauled is assumed to be the distance between the center of gravity of the cut mass and the center of gravity of the fill mass. Deducting the free-haul limit LL' from that distance and multiplying by the yardage hauled gives the overhaul in station-yards.

In this method the center of gravity of each cut solid and each fill solid between adjacent cross sections (usually one station apart) is assumed to lie midway between the sections. Overhaul on each solid is the product of its volume and the distance between its center of gravity and the beginning of the free-haul limit. Thus, in Fig. 6-10, C is at the center of gravity of the individual cross-hatched cut volume and F at the center of gravity of the indicated fill volume. The overhaul on the cut is its volume times CL; the overhaul on the fill is its volume times $L'F$. The total overhaul is the sum of the products found by multiplying each volume of cut between A and L by the distance between its center of gravity and station L, plus those products found by multiplying each volume of fill (plus shrinkage) between L' and A' by the distance between its center of gravity and station L'.

If B is an economic balance point following an earlier grade point, the quantity of excavation between B and A is not used in making embankment; it represents *waste*.

6-22. Mass-Diagram Method

Though the numerical method just described is quite simple and rapid, it is not adapted to making a broad study of grading operations by analyzing the effects upon the overall economy produced by various shifts in balance points. This is best done by a semigraphic method in which the *mass diagram* is used.

The earthwork mass-diagram is a continuous graph of net cumulative yardage. It is plotted with stations as abscissas and algebraic sums of cut and fill as ordinates. Customarily, a cut volume is given as a plus sign: a fill volume (plus shrinkage allowance) is given a minus sign.

Haul in station-yards is measured by areas on the mass diagram. In Fig. 6-11, suppose 1 yd^3 of excavation at A on the profile is moved X stations to A' in the embankment. The haul is obviously X station-yards, which is shown graphically on the mass diagram by the cross-hatched trapezoidal area. If the remaining excavation between A and G is moved to the embankment between G and A', the haul for each cubic yard is shown on the mass diagram as a stack of trapezoidal areas above the one indicated. The total haul in station-yards between A and A' is the area aga'.

The profile illustrated in Fig. 6-10 is repeated (to reduced scale) in the upper sketch of Fig. 6-12. Directly below is a representation of the corresponding mass diagram.

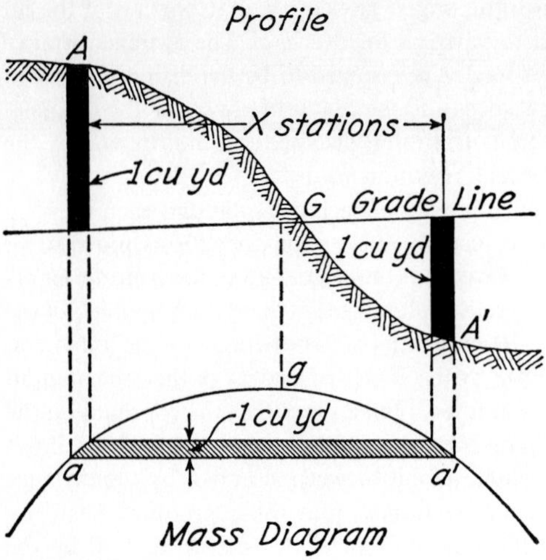

Figure 6-11

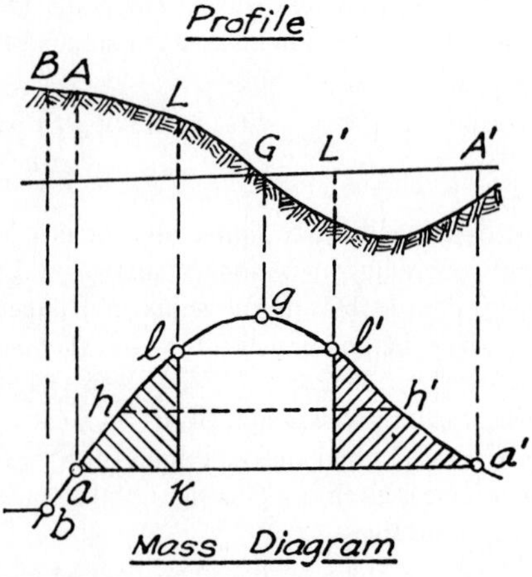

Figure 6-12

By reference to Fig. 6-12 the following characteristics of a mass diagram are apparent:

(a) Any horizontal line (as *aa'*) intersecting the mass diagram at two points is a *balance line;* total cut and total fill are equal between the stations at the intersections (as *A* and *A'*).

(b) Any ordinate between two balance lines (as *kl*) is a measure of the yardage between the stations at the extremities of the balance lines that is, between *A* and *L* or *L'* and *A'*. Stated more generally, the vertical distance between two points on the diagram (as *a* and *b*) is a measure of the yardage between the corresponding stations.

(c) The highest point of a loop (as at *g*) indicates a change from cut to fill (in the direction of the stationing); conversely, the lowest point represents a change from fill to cut. Such points may not fall exactly at the stationing of center-line grade points if there is a side-hill transition (represented by sta. *E* in Fig. 6-1).

(d) The area between the diagram and any balance line is a measure of the haul in station-yards between stations at the extremities of the balance line. If this area is divided by the maximum ordinate between the balance line and mass diagram, the result equals the average haul in stations. In Fig. 6-12 the area bounded by aga' measures the total haul between A and A'; that bounded by lgl' measures the total haul between L and L'. Since the latter is free haul, as is also the station-yards represented by the rectangle kl', the difference between total haul and free haul is the *overhaul* between A and A'. This overhaul is represented by the two cross-hatched areas.

When the portions of a mass diagram on the sides of two related overhaul areas are fairly smooth (as al and $l'a'$), even though not straight, the sum of the two areas may be found by drawing a horizontal line midway between the two balance lines, deducting the free-haul distance from its length, and multiplying the difference by the ordinate between the balance lines. For example, hh' bisects kl. The points h and h' are approximately at the centers of gravity of the volumes between A and L and L' and A'. Consequently, the overhaul from A to A' is $(hh' - ll') \times kl$.

If the mass diagram is very irregular between balance lines, overhaul may be determined either by (a) planimeter or (b) the method of moments. In (a) the overhaul is found directly by planimetering the areas representing overhaul and applying necessary scale factors to convert areas to station-yards. If needed, the distance to the center of mass of the yardage overhauled could be found by dividing the overhaul by the volume. Thus, in Fig. 6-12, the station of the center of mass of the yardage between A and L is sta. $A +$ area alk/kl.

In (b), the method of moments, each separate volume is multiplied by its distance from a selected station, and the sum of the products is divided by the sum of the volumes. The result is the distance from the selected station to the center of mass. As in any other method, *overhaul* = *yardage* × (*distance between centers of mass* − *free-haul distance*).

Other useful principles in making a distribution analysis by mass diagram are illustrated by Fig. 6-13, which represents the profile and mass diagram of a continuous section of line.

Balance lines equal in length to the limit of economic haul (as aa' and cc') are first drawn in the larger loops.

Between a' and c the most economical position of the balance line is at bb', which is drawn so that $bb_1 = b_1b'$ with neither segment longer than the limit

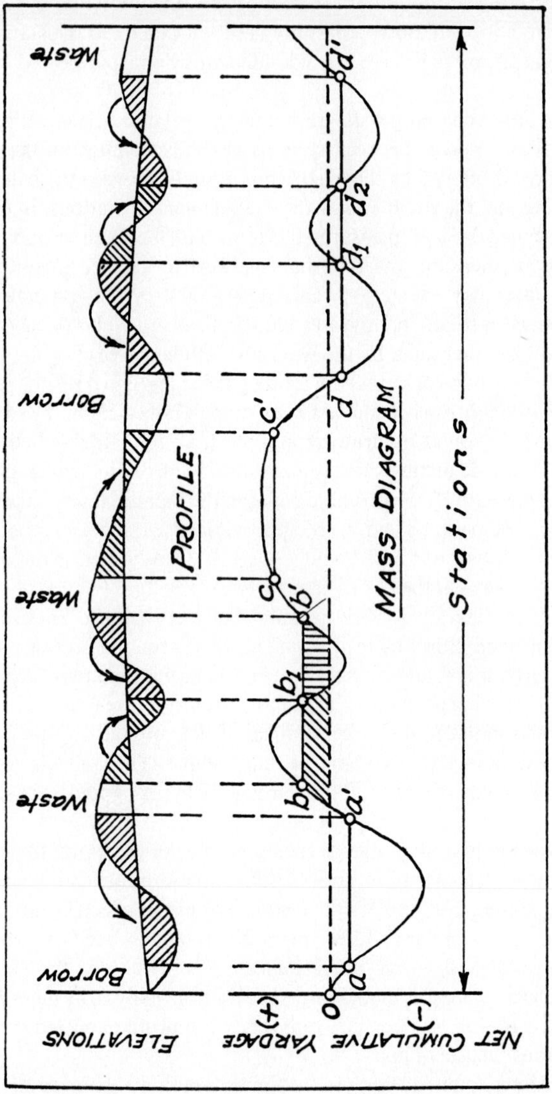
ELEVATIONS
NET CUMULATIVE YARDAGE
(+)
(-)
PROFILE
MASS DIAGRAM
Stations
Borrow
Waste
Waste
Borrow
Waste
o
a
a'
b
b1
b'
c
c'
d
d1
d2
d'

Figure 6-13

of economic haul. That this is the best position may be shown by imagining bb' lowered to coincide with the horizontal plotting axis. There would be no change in the total waste; the waste at b would be decreased by the increase at b'. However, the total haul would be increased by the area shown cross-hatched diagonally and would be decreased by the area cross-hatched vertically. Since these areas have equal bases bb_1 and b_1b', there is a net increase in area, or haul. Shifting the balance line higher than bb' would obviously have the same effect.

The balance line dd' is adjusted so that $(dd_1 + d_2d' - d_1d_2)$ is equal to the limit of economic haul and no segment is greater than that limit. An analysis similar to that made for bb' would prove that raising or lowering dd' from the position shown also increases the cost.

In general, the most economical position for a balance line cutting *any even number of loops* is that for which the sum of the segments cutting convex loops equals the sum of the segments cutting concave loops, no segment being longer than the limit of economic haul. The most economical position for a balance line cutting *any odd number of loops* is that for which the sum of the segments cutting convex (or concave) loops less the sum of the segments in loops of opposite form equals the limit of economic haul, no segment being longer than that limit.

Theoretically, the foregoing principles are unaffected by the length of free haul. For example, if the alternate positions of balance line bb' produced segments longer than the free-haul limit, overhaul would be increased with consequent increase in the payment to the contractor. If the balance lines were shorter than the free-haul limit, there would be no actual payment for overhaul in either case. Nevertheless, the total haul in station-yards would be increased, thus adding to the contractor's cost of doing the work and, possibly, influencing him to submit slightly higher bid prices.

In drawing balance lines, one note of caution should be mentioned: *adjacent balance lines must not overlap*. The effect is to use part of the mass diagram twice—an obvious impossibility except by borrowing an extra mass of earthwork measured by the distance between the overlapping balance lines.

Figures 6-14 and 6-15 show two solutions for a case not found in Fig. 6-13. This is the case in which there is an intermediate loop that is not cut by a balance line equal in length to the limit of economic haul.

In both solutions, AB is the limit of economic haul and CD is the free-haul distance. The total overhaul in Fig. 6-14 is the sum of the two numbered cross-hatched triangles on the mass diagram (found in the usual way) *plus* the overhaul (shaded) on yardage K, which equals $(XY - CD) \times K$ station-yards.

In the solution shown by Fig. 6-15, the total overhaul is the sum of the four numbered cross-hatched triangles *plus* the overhaul (shaded) on yardage K', which equals $(X'Y' - CD) \times K'$ station-yards.

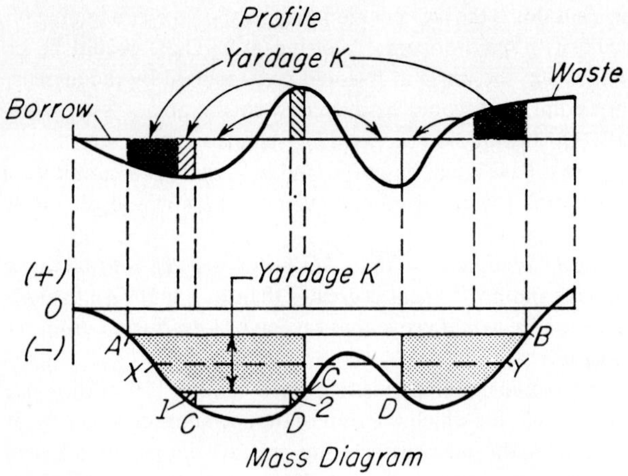

Figure 6-14 Two-way hauling from intermediate cut.

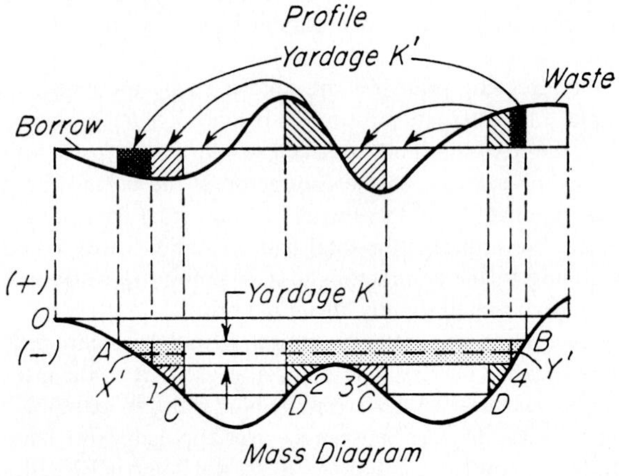

Figure 6-15 Unidirectional hauling.

Theoretically, the solution in Fig. 6-15 is the more economical one because it has less overhaul, more free haul, and the same amounts of borrow and waste. Yet, in practice, the solution in Fig. 6-14 might be preferred because of the two-way hauling and the shorter haul distances.

Making a distribution analysis by mass diagram is not the purely mechanical process implied in the preceding discussion. Factors other than obtaining theoretical maximum economy enter into the planning of grading operations. For example, on steep grades the contractor prefers loaded hauls to be down hill. Moreover, he may prefer to haul more of a particular cut in a certain direction than is indicated on the plans. Again, there may be one fill that acts as a bottle-neck. Building it ahead of schedule, possibly by using extra borrow or longer hauls than those theoretically needed, may save time and money. These preferences may be realized by exercising good judgment in altering the theoretical balance lines. The result may be submission of lower bid prices. Even if the bid is not lower, the contractor is better satisfied—a condition that should produce a better job and friendlier relations with the owners.

Even if a grading contract contains no overhaul clause (this practice is becoming more common), the mass diagram is still very useful in the work of grade-line design. Approximate balance points are shown on the final plans to indicate the grading schedule to the contractor. It is then his responsibility to calculate or estimate the hauls and adjust his bid prices for excavation to cover their cost.

PROBLEMS

6-1. Compute yardages of cut and fill between stations specified by the instructor. Use formula 6-8 (except for the pyramids) and check the results using Table XIX.

6-2. Use Table XVIII to obtain yardages of cut and fill between stations specified by your instructor.

6-3. What condition must exist in formulas 6-13 or 6-14 in order for the prismoidal volume to (a) exceed the average-end-area volume, and (b) equal the average-end-area volume? Which solids in the text figures conform to either of these conditions?

6-4. Compute the prismoidal corrections for the solids between the following stations, and state whether corrections should be added to, or subtracted from, the average-end-area volumes.

(a) 88+30 to 88+68 (cut). *Answer:* Subtract 41.1 yd^3.
(b) 88+68 to 89+08 (fill). *Answer:* Subtract 22.2 yd^3.
(c) 89+43 to 89+59 (fill). *Answer:* Add 0.2 yd^3.
(d) 90+05 to 90+50 (fill). *Answer:* Subtract 10.8 yd^3.
(e) 91+00 to 91+52 (fill). *Answer:* Subtract 32.4 yd^3.
(f) 91+52 to 92+00 (cut). *Answer:* Subtract 42.3 yd^3.

6-5. Compute the prismoidal corrections for any (or all) of the solids in problem 6-4, using Table XVIII as explained in section 6-13.

6-6. Use formula 6-10 to compute the values of V_p for any (or all) of the solids in problem 6-4. Compare the results with those obtained by combining problems 6-1 (or 6-2) and 6-4 (or 6-5).

6-7. Assume that all sections lie on a curve to the right. By inspection of the notes, determine the sign of the curvature correction for each solid.

* *Note.* Problems 6-1 through 6-8 refer to the notes on page 127.

Table A. FREE-HAUL LIMIT 2400 ft.

STA.	NET CUMULATIVE YARDAGE	STA.	NET CUMULATIVE YARDAGE	STA.	NET CUMULATIVE YARDAGE
0	0	15	8400	31	−550
1	320	16	7480	32	−400
2	660	17	6230	33	−540
3	1050	18	4250	34	−1160
4	1670	19	2100	35	−3600
5	2500	20	0	36	−5180
6	3100	21	−1370	37	−5650
7	4050	22	−4100	38	−5900
8	5160	23	−5350	39	−5580
9	6420	24	−6030	40	−4950
10	7750	25	−6200	41	−4160
11	9100	26	−6050	42	−3100
12	9380	27	−5600	43	−1900
12 + 70	9500	28	−4900	44	0
13	9400	29	−3680	45	2100
14	9120	30	−2150	46	3800

Cost of excavation \$2.50 per yd^3; borrow \$2.00 per yd^3; overhaul \$0.25 per sta-yd.

Table B. FREE-HAUL LIMIT 2400 ft.

STA.	NET CUMULATIVE YARDAGE	STA.	NET CUMULATIVE YARDAGE	STA.	NET CUMULATIVE YARDAGE
12	0	28	4500	44	−4000
13	1260	29	6510	45	−4980
14	4260	30	7350	46	−5390
15	6490	31	8160	47	−5420
16	7740	32	8380	48	−5110
17	8460	33	8150	49	−4520
18	9000	34	8010	50	−3700
19	9240	35	7640	51	−2100
20	9360	36	7000	52	+20
21	9000	37	6000	53	+2000
22	7750	38	4980	54	+2800
23	6100	39	3160	55	+3320
24	3500	40	1500	56	+4010
25	2320	41	−200	57	+4550
26	2200	42	−1760	58	+4910
27	2520	43	−3310	59	+4840

Cost of excavation \$3.75 per yd^3; borrow \$3.00 per yd^3; overhaul \$0.40 per sta-yd.

6-8. Assume that all sections lie on a 10°00′ curve to the right. Determine the curvature corrections for any (or all) of the solids in problem 6-4. Indicate sign. *Partial answers:* (a) add 25.6 yd^3; (b) subtract 10.9 yd^3.

6-9. Plot a mass diagram from the data in *Table A* or *Table B*, using a horizontal scale of 1 in. = 5 sta. and a vertical scale of 1 in. = 5000 yd^3. (An $8\frac{1}{2} \times 11$ in. sheet of cross-section paper will suffice.) Show a hypothetical profile above the diagram, as in Fig. 6-13. Establish balance points and compute the cost of grading. On the profile show the separate amounts of waste, borrow, and yardage excavated; indicate the disposition of excavation by arrows.

Part II
PRACTICAL APPLICATIONS

Chapter 7
Special Curve Problems

7-1. Foreword

In a subject as utilitarian as route surveying, there is hardly a strict division between basic principles and practical applications. Though these headings are used in this book, many practical features have already been referred to in Part I; moreover, some problems involving additional theory will be found in Part II. Nevertheless, Part I is complete in itself; it does not require this additional material in order to understand and apply the theory to any practical problem met in route surveying.

Part II contains specific applications of basic theory to some common problems and survey procedures found in practice. It is not a detailed compilation of instructions covering field and office work; education is well taken care of in the manuals published by most state highway departments and various other organizations to guide their party chiefs. Part II is in the nature of a supplement to such manuals. Technical knowledge of route surveying being assumed, the explanations are briefer than those given in Part I and the simpler proofs are omitted.

In practice, special curve problems occur in such great variety that it is not possible to include a large number of them in the space allotted to this chapter. Doing so, if possible, would have questionable merit, since the "textbook type" of problem is less apt to occur than some perplexing variation of it. To serve the purpose, a few more common problems will be described and general methods of approach outlined. These, combined with a thorough grounding in the basic principles of Part I, should enable an engineer to develop the skill needed for solving any special curve problem.

7-2. Methods of Solution

In solving special curve problems there are four general methods of approach: (1) an exact geometric or trigonometric solution, (2) a cut-and-try calculation method, (3) a graphical method, and (4) a cut-and-try field method.

Generally the first method is preferred. However, if the solution cannot be found or is very complicated, cut-and-try calculation often provides a fairly quick solution. In case a problem is not readily solvable by either of these methods, the unknowns may sometimes be scaled from a careful drawing; the scale must be fairly large to give adequate precision. Some problems are adaptable to cut-and-try solution in the field (see section 3-14 for an example previously described). Additional examples of these methods of solution follow.

OBSTACLE PROBLEMS

7-3. P.I. Inaccessible

• *Simple Curve.* This problem (Fig. 7-1) occurs frequently. The degree of curve is known, and the curve points *A* and *B* must be located.

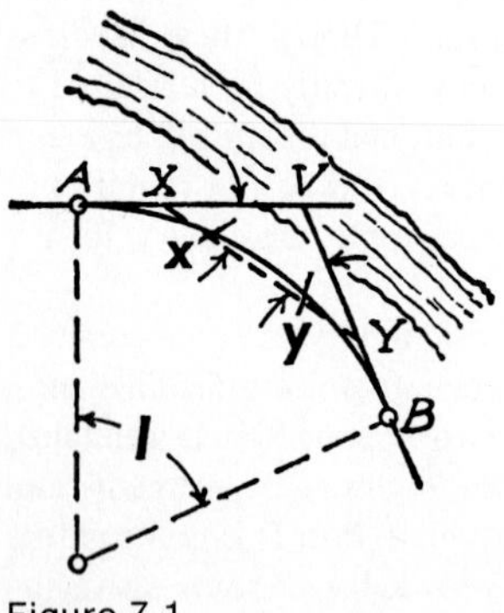

Figure 7-1

Find any convenient line XY cutting the established tangents.

Measure distance XY and deflection angles x and y. Angle $I = x + y$.

Calculate XV and VY by using the sine law. Subtract their values from T (or T_s), giving the required distances to the beginning and end of the curve.

● *Compound Curves, Fig. 7-2.* In this case the cutoff line is chosen so as to establish the T.C. and C.T. at X and Y. One more variable must be known. Assume R_S, since the maximum D is often fixed by specifications. Thus, the four known values are XY, R_S, and the angles x and y.

Use the closed traverse shown by heavy lines and solve by the traverse method described in section 3-5.

Angle $I = x + y$

From Σ *latitudes:*

$$R_L - XY \sin x - R_S \cos I - (R_L - R_S) \cos I_L = 0$$

If the hint preceding equation 3-2 in section 3-5 is used, this relation reduces to

$$\text{vers } I_L = \frac{XY \sin x - R_S \text{ vers } I}{R_L - R_S} \tag{7-1}$$

Similarly, the relation based on Σ *departures* reduces to

$$\sin I_L = \frac{XY \cos x - R_S \sin I}{R_L - R_S} \tag{7-2}$$

When equation 7-1 is divided by 7-2, the result is

$$\tan \tfrac{1}{2} I_L = \frac{XY \sin x - R_S \text{ vers } I}{XY \cos x - R_S \sin I} \tag{7-3}$$

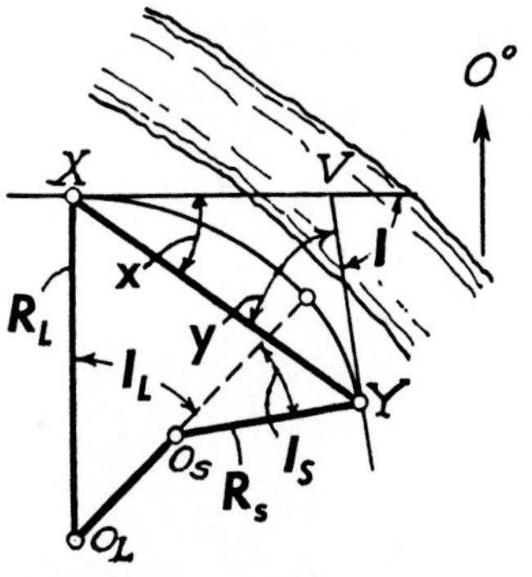

Figure 7-2

Solve equation 7-3 for I_L. Then obtain $R_L - R_S$ from equation 7-1 (or 7-2). Finally, $I_S = I - I_L$.

Use the same method if R_L were assumed; the final relations differ only in signs and subscripts.

7-4. T.C. or C.T. Inaccessible

In Fig. 7-3, assume *B* to be the C.T. The problem is to stake the computed curve and to check the work for alignment and stationing.

Set a check point *F* on the forward tangent by measuring *VF*, using right-angle offsets to get around the obstacle. Sta. F = Sta. C.T. + $VF - T$.

Stake the curve from *A* to a station (as *P*) from which a sight parallel to the forward tangent would clear the obstacle.

Occupy *P* and deflect angle $I - (a/2)$ (that is, angle *I* minus the tabulated deflection for the station occupied), thereby placing the line of sight parallel to the tangent. Set point *E* on this line by measuring PE = Sta. F − Sta. B + $R \sin (I - \alpha)$.

Occupy *E*, turn 90° from *EP*, and check the distance and direction to *F*. The offset *EF* should equal R vers $(I - a)$. (If *EF* is small, it can be computed by slide rule from formula 2-30, in which arc *PB* is used for *s*.)

While the transit is at *A* (or *P*), set stakes on the curve for cross sectioning between *P* and the obstacle, leaving one stake at the plus point marking the beginning of the obstacle.

If *B* were the T.C., run in the part *AP* of the curve backward from the C.T. The rest of the process is similar in principle to that just described.

If the curve is spiraled and the S.T. or T.S. is made inaccessible, by an obstacle at *position 1* in Fig. 7-4, the field procedure is the same but the computations differ somewhat.

In the general case, assume the line of sight from the C.S., pointed parallel to the forward tangent, is cut off by the obstacle at the S.T. As before, run in the layout to a station *P* on the circular arc, *a* being the central angle (twice the deflection difference) between *P* and the C.S. at *C*.

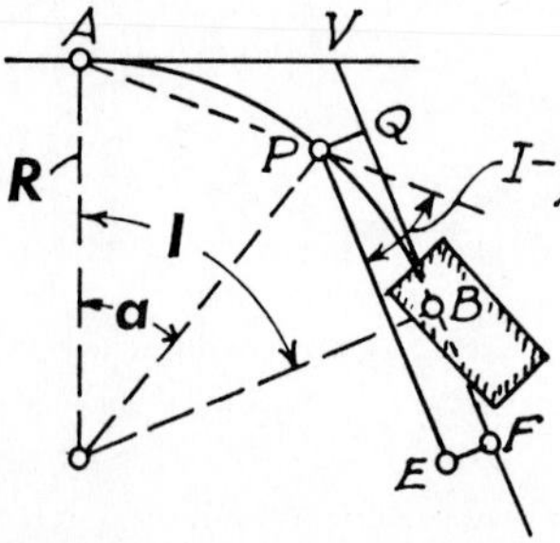

Figure 7-3

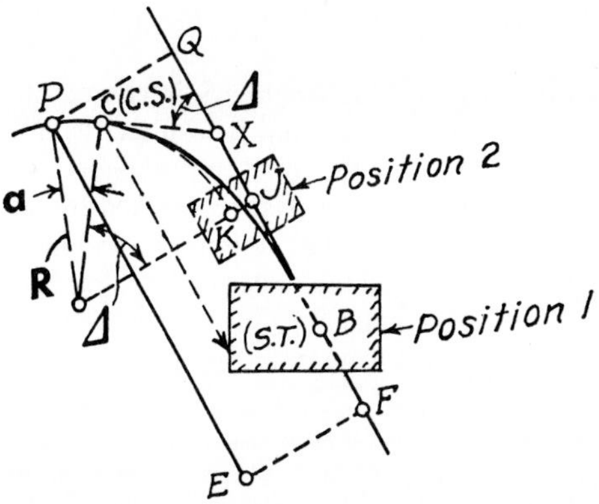

Figure 7-4

Occupy P, turn the line of sight parallel to the forward tangent, and measure the distance PE to any convenient point beyond the obstacle.

Point K is the *offset* C.T. ($KJ = o$; $BJ = X_0$).

Offset $PQ = EF = o + \frac{7}{8}(PK)^2 D$, the value for PK being taken equal to the difference in stationing between P and C plus $\frac{1}{2} L_s$. If the offset is large, compute it from the relation $PQ = o + R \text{ vers } (a + \Delta)$.

For checking out on the tangent, the relation is:

$$\text{sta. } F = \text{sta. } P + PC + CB + BF$$

or

$$\text{sta. } F = \text{sta. } P + \frac{100\,a}{D} + L_s + [PE - X_0 - R \sin (a + \Delta)]$$

If both the P.I. and T.C. (or C.T.) are inaccessible, a combination of the foregoing procedures will provide the solution.

7-5. Obstacle on Curve

For methods of bypassing obstacles preventing sights to curve points or obstacles on the curve itself, see sections 2-11 and 2-15 along with Figs. 2-10 and 2-14.

If the obstacle cuts the spiral and tangent but does not obstruct the S.T. (or T.S.), as at *position 2* in Fig. 7-4, first run in the layout to the C.S. (or S.C.). Then set up at the C.S. (or S.C.) and locate stakes on the spiral which are needed for cross sectioning between the setup and the obstacle. Finally, check the position of the S.T. (or T.S.) and direction of the main tangent, by measuring the spiral coordinates *X* and *Y* and turning the necessary right angles.

An alternate procedure, which may be used if the obstacle interferes with the sight from *C* pointed parallel to the main tangent, is to place the line of sight on the local tangent at *C* and measure the spiral short tangent to a point *X* on the main tangent. The field work can then be checked by occupying point *X*, deflecting angle Δ from a backsight to point *C*, and measuring (possibly using right-angle offsets around the obstacle) the spiral long tangent to check the S.T. (or T.S.) of the layout.

7-6. Obstacle on Tangent

For the general method of bypassing an obstacle on a tangent at a point unaffected by a curve, see section 12-8.

Examples of obstacles spanning both a tangent and a curve are given in sections 7-4 and 7-5.

Special problems affecting tangent distances of curves have infinite variety. Figure 7-5 represents a case in which it was required to run a spur track to a proposed warehouse. Conditions (not shown) fixed the warehouse

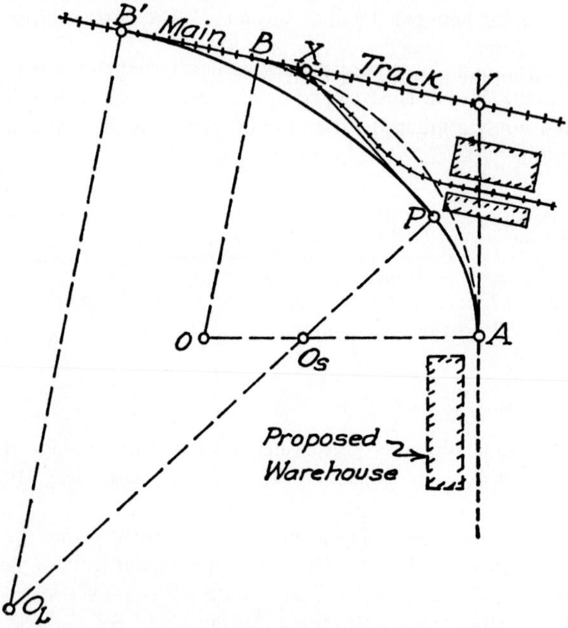

Figure 7-5

location as indicated. It was impossible to run in a simple curve *AB* without interfering with existing buildings and a turnout track. The solution was a compound curve *APB'*. See problem 7-9 for the numerical values involved.

CHANGE-OF-LOCATION PROBLEMS

After part of the paper-location alignment has been staked, desirable minor adjustments often become apparent (see section 1-12). Frequently the data obtained and the stakes already set for the original curve can be used to advantage in making the revision. A few common cases coming under this heading are outlined in the following articles. More extensive adjustments are best made by locating the new P.I. and staking the revised curve irrespective of the original layout.

If a line has been staked beyond an alignment change, it is not necessary to renumber all the stakes following the junction. Instead a "station equation" is written in the notes and on the stake marking the junction. For example, the station equation "75+92.74 back = 75+60.21 forward" not only shows where the alignment change terminates but also that the revision saves 32.53 ft of distance. A mistake in taping which is discovered after stakes have been driven and marked is treated in the same way.

7-7. Practical Suggestions

Skill in solving change-of-location problems does not come from memorizing certain "textbook solutions" but is developed by identifying the key steps in those solutions and applying them to the unusual problems that arise in practice. The following practical hints, numbered for ease of reference, are helpful.

• *Hint 1*. Draw a careful sketch, which is not necessarily to scale. Exaggerate small distances to make their effects clear. Preserve right angles. Do not make other angles close to 90°; otherwise, a special case might result. If only one graphical solution is possible when the known data are used, the problem is definite and determinate.

• *Hint 2*. If the problem involves a revision of some kind, such as the shift of a tangent, the solution must contain that desired revision. Try to connect established revisions to fixed points on the original layout by simple geometric construction, especially by triangles. Considering a triangle containing curve centers or vertices often leads to the solution.

• *Hint 3*. A planned linear revision may often be expressed as the difference between a fixed part of the original layout and a similar unknown part of the revised scheme. Also, an unknown linear revision may equal the difference between similar fixed parts of the two layouts.

• *Hint 4.* If a point is common to both layouts, perpendiculars dropped from that point to tangents or radii frequently disclose the key to the solution.

• *Hint 5.* Although there is only one correct set of numerical answers to a definite problem, there may be several correct geometric solutions. If a certain solution cannot produce adequate precision, determine the reason for the lack of precision and search for a better solution.

• *Hint 6.* If a solution by means of simple construction cannot be found, recall that a solution by traverse (section 3-5) or by the missing data method (section 3-6), may be applicable.

7-8. Simple Curve; New Parallel Tangent; Same *D*

Assume that the forward tangent is to be shifted outward parallel to itself a small distance p in order to reduce grading or improve the approach to the next curve.

In Fig. 7-6, the skew shift obviously equals $AA' = OO' = VV' = BB'$. From a triangle at any one of these positions (see Hint 2, section 7-7),

$$\text{skew shift} = \frac{p}{\sin I} \tag{7-4}$$

Tape the skew shift from A and B to locate hubs at the new curve points A' and B'.

For a tangent shifted inward, use $A'B'$ as the original curve.

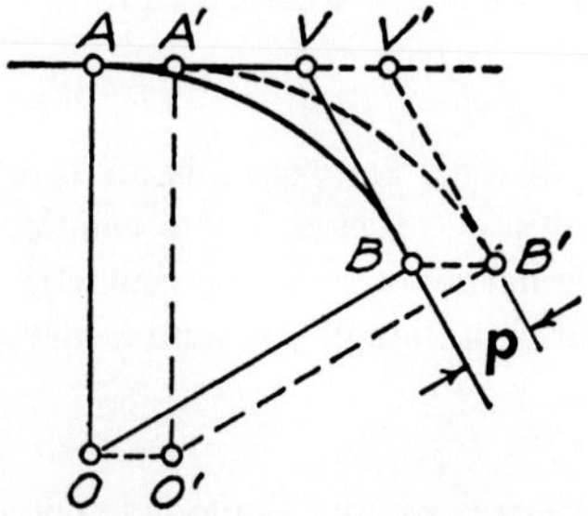

Figure 7-6

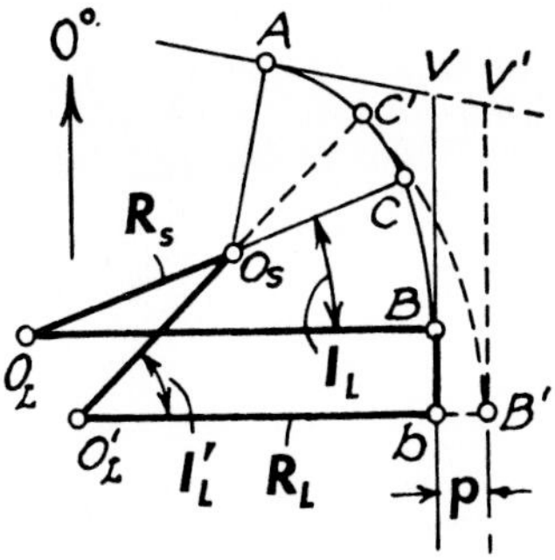

Figure 7-7

7-9. Compound Curve; New Parallel Tangent; Same *D*'s

This is the same problem as the one in section 7-8, except that the curves are computed at C and C' (Fig. 7-7).

Although a solution by construction is possible (Hints 3 and 4 being used), it may not be readily apparent. Therefore, use the traverse method (see Hint 6).

In the closed traverse $bBO_LO_sO_L'b$,

$$\Sigma \textit{ departures} = -R_L + (R_L - R_S)\cos I_L - (R_L - R_S)\cos I_L' + (R_L - p) = 0$$

from which

$$\cos I_L' = \cos I_L - \frac{p}{R_L - R_S} \qquad (7\text{-}5)$$

The distance bB needed for locating the new C.T. at B' is found by setting Σ *latitudes* equal to zero and reducing. The result is

$$bB = (R_L - R_S)(\sin I_L' - \sin I_L) \qquad (7\text{-}6)$$

(Observe that equation 7-6 also comes directly from Hints 3 and 4 by dropping perpendiculars from O_S to the radii.)

There are four variations of this problem, the solution depending on whether the layout starts with the sharper or flatter arc and on the direction (outward or inward) of the tangent shift. The final relations contain different signs and subscripts.

7-10. Simple Curve; Parallel Tangent; Same T.C.

In contrast to the situation in section 7-8, assume that the original T.C. must be preserved, thereby requiring a curve of new *D*, as indicated in Fig. 7-8.

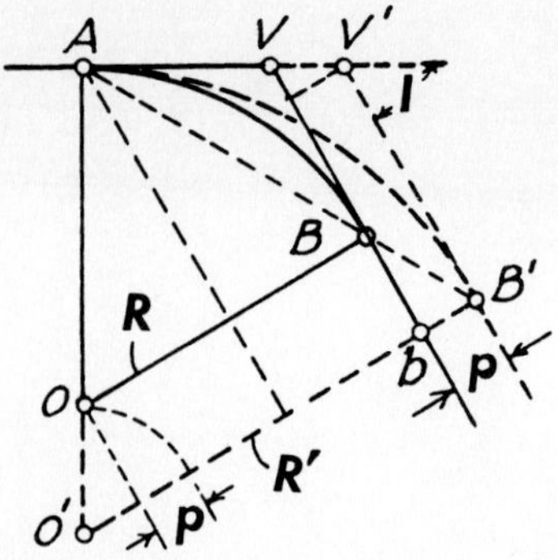

Figure 7-8

From the triangle at the vertex (Hint 2), VV' = the skew shift = $p/\sin I'$ and the new tangent distance T' equals $T + VV'$.

Then D' is found from Table VIII, or, if preferred (Hints 4 and 3), from

$$R' = R + \frac{p}{\text{vers } I} \tag{7-7}$$

For setting the new C.T. at B', notice that B' must lie on AB produced. Since angle bBB' equals $\frac{1}{2}I$,

$$BB' = \frac{p}{\sin \frac{1}{2}I} \tag{7-8}$$

When the tangent is shifted inward, use AB' as the original curve.

7-11. Simple Curve; New C.T. Opposite Original C.T.

If conditions do not permit moving the C.T. forward (as in Fig. 7-8), it may be kept on the same radial line opposite the original position, as in Fig. 7-9.

For the triangle at the vertex (Hint 2), $VK = p \cot I$, and the new tangent distance T' equals $T - VK$.

Then D' is found from Table VIII, or (if preferred) $BB' = BX - B'X$ (Hint 3). That is, $p = (R - R')$ exsec I, from which

$$R' = R - \frac{p}{\text{exsec } I} \tag{7-9}$$

For setting the new T.C. at A', notice that AA' *does not* equal VV'. Find AA' from the triangle at the centers (Hint 2). This triangle gives

$$AA' = (R - R') \tan I \tag{7-10}$$

When the tangent is shifted inward, use $A'B'$ as the original curve.

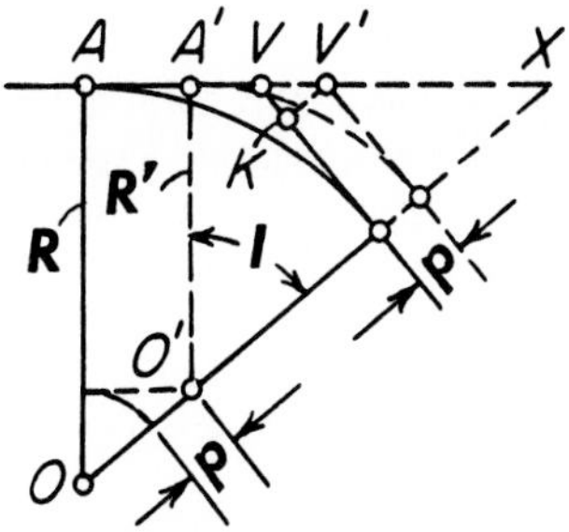

Figure 7-9

7-12. Simple Curve; New Direction from C.T.

Changing the direction of the forward tangent after a curve has been staked may place the alignment on more favorable ground than by shifting the tangent parallel to its original position. Figure 7-10 represents a case in which the tangent is swung inward through a measured angle a, the C.T. at B being preserved. The new central angle I' equals $I + a$.

Using Hint 4, drop perpendiculars (not shown) from B to the tangent AV produced and to the radii OA and $O'A'$. Then, by inspection, R' vers $I' = R$ vers I, and

$$R' = \frac{R \text{ vers } I}{\text{vers } I'} \tag{7-11}$$

Also, from Hint 3,

$$AA' = R \sin I - R' \sin I' \tag{7-12}$$

If preferred, solve triangle VBV' (Hint 2) for T'. Thus,

$$T' = \frac{T \sin I}{\sin I'} \tag{7-13}$$

Then, obtain D' from Table VIII.

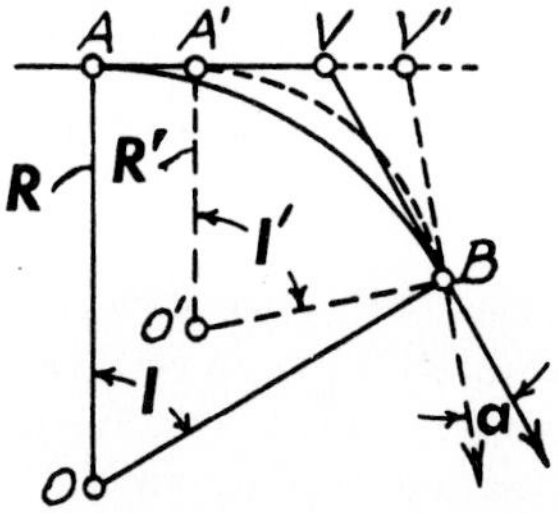

Figure 7-10

7-13. Modification for Spiraled Curve

In the preceding examples illustrating change-of-location problems, the curves were not spiraled. Spiraling complicates field adjustments to a certain extent. The best general method is to locate on the ground the positions of the offset T.C.'s and the offset P.I., thereby converting the problem to one involving unspiraled curves.

In a simple problem it may not be necessary actually to stake the offset T.C.; but the calculations must then take the spiral into account. (See Fig. 7-4 for an example.)

RELOCATION PROBLEMS

Major relocations of existing highways are continually being made in order to bring them into conformity with modern standards. To a lesser extent, some large sections of railroad line are being relocated, especially in mountainous terrain where the amount of traffic originally expected did not warrant low grades and expensive alignment. In such work little if any use is made of the existing alignment records.

Minor relocations, both on highways and railroads, are even more common; they will probably continue to outrank major relocation projects in total mileage and construction cost. The shorter the relocation the more convenient it becomes to tie the survey work closely to the original alignment. Survey and design problems are closely related to those requiring new right-of-way and abandoning old right-of-way.

Two typical minor relocation problems are outlined in the following articles. References to some major projects are given in Chapter 9.

7-14. Replacement of Broken-Back Curve

A "broken-back" curve consists of two curves in the same direction separated by a tangent shorter than the sum of the distances needed to run out the superelevation. Formerly, such alignment was often used for reasons of economy, but present standards rarely justify the practice. When the entire layout is visible, it is very unsightly; even though obscured, it is apt to be dangerous (on highways).

Elimination of the tangent between curves is a common relocation problem. Occasionally it can be done by inserting a single curve between the outer tangents of existing curves.

A more general method is illustrated in Fig. 7-11, which shows the original tangent BC separating curves with centers at O_S and O_L. A new curve with its center at O is sprung between points A and D on the existing curves, thus forming a three-centered compound curve (section 3-9). The problem is to locate points A and D.

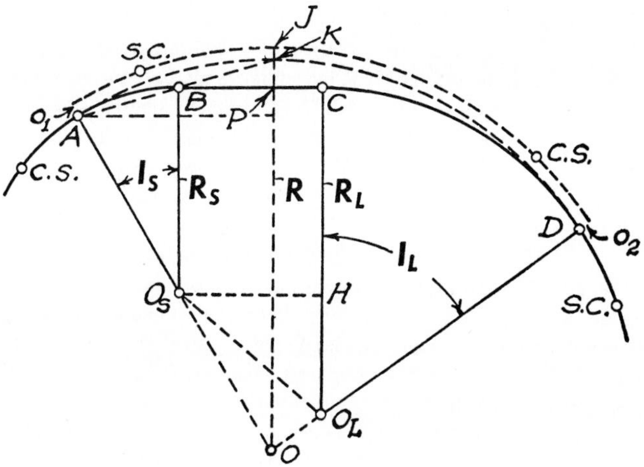

Figure 7-11

If a value is assumed for the radius R of the new curve, the positions of A and D can be found by solving the right triangle O_SHO_L and the triangle O_SOO_L, and then calculating the angles I_S and I_L, which locate A and D.

The maximum offset PK between the original and revised layouts should be computed to see if it falls within the limit permitted by topography.

$$PK = (R - R_S) \text{ vers } I_S \tag{7-14}$$

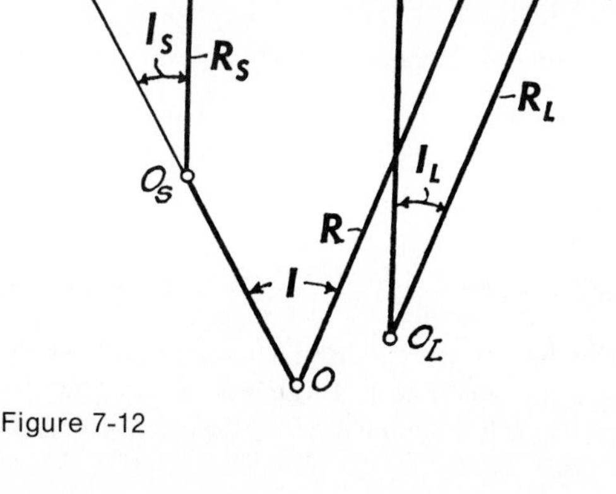

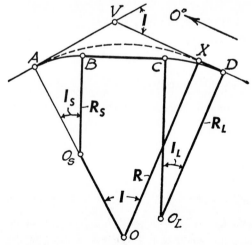

Figure 7-12

If spirals are required between the curves, the problem then becomes a special case of inserting spirals between arcs of a compound curve. The center O is preserved and the radius of the new curve is increased to $R + KJ$, where $KJ = o_1 = o_2$. The two spirals usually have different lengths, since the differences between the degrees of the new and original curves will rarely be the same. The theory of the process is outlined in sections 5-16 and 5-20.

Figure 7-12 illustrates how a change in specifications sometimes complicates a problem. The existing layout is a broken-back curve similar to that in Fig. 7-11, but field conditions require that the new curve start at A (the T.C. of the original shorter-radius curve) and end on the existing forward tangent at an unknown point X. The problem is to find a new radius R and distance DX. No solution is possible by merely solving two triangles, as was the case in Fig. 7-11. To avoid more-intricate construction, the traverse method is used. (Hint 6, section 7-7).

The closed traverse $DXOO_SBCO_LD$ is imagined to be oriented so that direction DX is 0° azimuth (Rule 2, section 3-5). The data for the traverse are then as follows:

SIDE	LENGTH	AZIMUTH
DX	DX	0°
XO	R	270°
OO_S	$R - R_S$	$90° - I$
O_SB	R_S	$90° - I_L$
BC	BC	$180° - I_L$
CO_L	R_L	$270° - I_L$
O_LD	R_L	90°

Setting Σ *departures* equal to 0 and reducing gives

$$R = \frac{R_S(\cos I_L - \cos I) + BC \sin I_L + R_L \operatorname{vers} I_L}{\operatorname{vers} I} \qquad (7\text{-}15)$$

Similarly, setting Σ *latitudes* equal to 0 gives

$$DX = (R_L - R_S)\sin I_L - (R - R_S)\sin I + BC\cos I_L \qquad (7\text{-}16)$$

As in Fig. 7-11, the maximum offset is expressed by formula 7-14.

The foregoing formulas are solved quickly by a calculator and Table XX, in which all needed natural functions appear in the same table. (See problem 7-5.)

7-15. Replacement of Reverse Curve

Figure 7-13 shows an existing reverse curve ACB which is to be replaced by a new simple curve starting at the same T.C. at A. The problem is to find the a radius R and distance XB. The following solution is by the traverse method

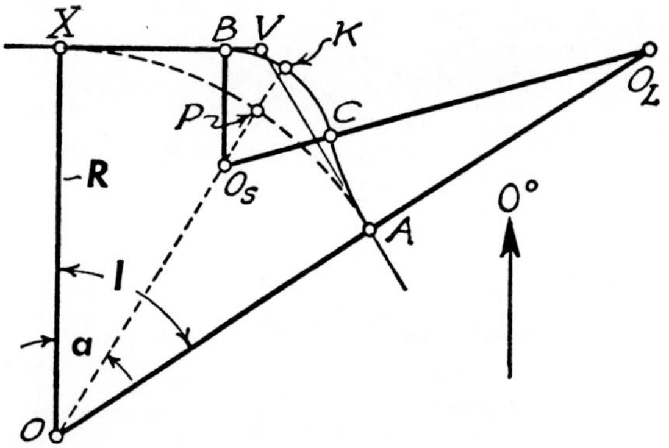

Figure 7-13

employed in Fig. 7-12, 0° azimuth being taken parallel to OX in the closed traverse $OXBO_SO_LO$. Angle $I = I_S - I_L$.

Setting Σ *latitudes* equal to 0 and reducing gives

$$R = \frac{R_L(\cos I - \cos I_S) + R_S \text{ vers } I_S}{\text{vers } I} \tag{7-17}$$

Setting Σ *departures* equal to 0 and reducing gives

$$XB = (R_L + R)\sin I - (R_L + R_S)\sin I_S \tag{7-18}$$

To find the maximum offset PK, first calculate angle a from the relation

$$\tan a = \frac{XB}{R - R_S} \tag{7-19}$$

Then,

$$\text{Offset } PK = (R - R_S) \text{ exsec } a \tag{7-20}$$

(See problem 7-6.)

7-16. General Method for Major Relocation

Even in the case of a long relocation supplanting several curves and their intervening tangents, it is useful to tie the survey to the existing alignment and compute the resulting closed traverse by means of the old alignment data on file. This not only gives a check on the field work without extra surveying but also provides the coordinates needed for drawing a map of the two layouts.

Relocations often result in surprisingly simple alignment. Figure 7-14 shows a case in which a single 1° curve and two tangents replaced seventeen curves having total central angles of more than 900°.

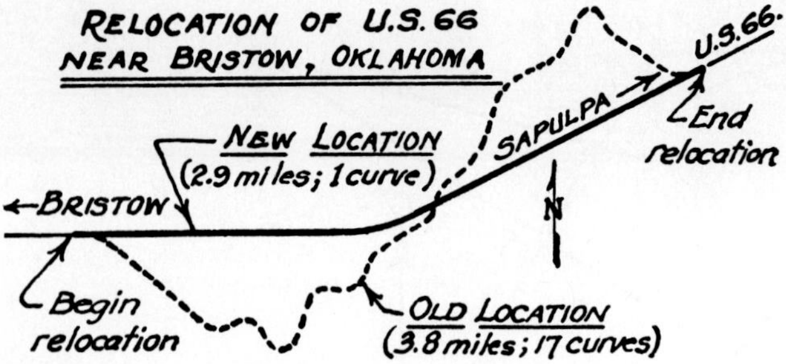

Figure 7-14

MISCELLANEOUS PROBLEMS

7-17. Curve Through Fixed Point

A curve can be forced to pass through (or close to) a fixed point either by trial methods or by an exact geometric solution. Where the P.I. and directions of the tangents have been established, an exact geometric solution is better.

Figure 7-15 shows this situation, P being the point fixed by ties to the P.I. The tie measurements may be the distances VP' and $P'P$ or the distance VP and the angle a. Either pair of values can be obtained from the other by computation. The problem is to find the unknown radius R, after which all remaining data are easily computed.

The key to a trigonometric solution is the triangle OPV (the interior angles of which are denoted by o, p, and v). Although only one side (PV) and one angle (v) are known, the relation between the two unknown sides is fixed, therefore, the triangle can be solved.

The ratio of the unknown sides OP and OV is $R:R \sec \frac{1}{2}I$, which equals $\cos \frac{1}{2}I$. Therefore, the law of sines can be used in the key triangle, giving

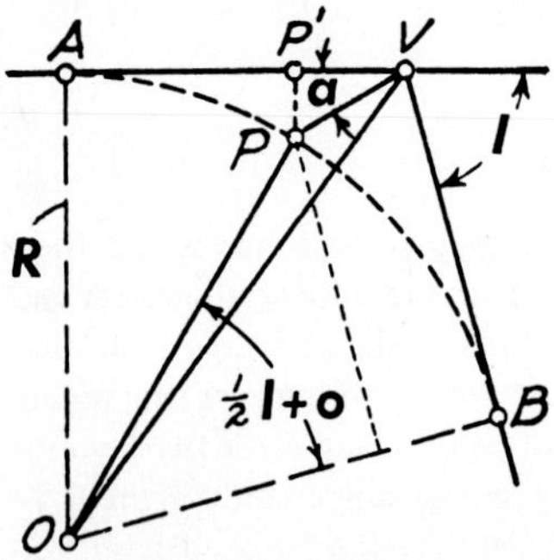

Figure 7-15

$\sin p : \sin v = OV : OP$, from which

$$\sin p = \frac{\sin v}{\cos \frac{1}{2}I} \tag{7-21}$$

Since all angles of the triangle are now known, R may be found from the sine law. However, if angles v and o are very small, R may not be calculable to sufficient precision owing to the rapid variation in the sines of the angles (see Hint 5, section 7-7). In this case, solve for R from the relation

$$R = \frac{PV \sin (I + a)}{\text{vers } (\frac{1}{2}I + o)} \tag{7-22}$$

As a check, $P'P$ must equal $R \text{ vers } (\frac{1}{2}I - o)$.

7-18. Intersection of Straight Line and Curve

Figure 7-16 represents a more general case of Fig. 7-15 in which a straight line XP' cuts the tangent AV of a given curve. The problem is to locate the intersection of the line and the curve at P. Among other places this problem occurs in right-of-way work, as in defining the corner of the piece of property shown shaded.

All curve data are known and the survey notes also provide the angle a and the distance AP' (or $P'V$).

This information being given, the problem could be converted to the preceding one by drawing another tangent (not shown) from P' to the curve, thereby making P' correspond to point V in Fig. 7-15. However, the traverse method is more direct.

If 0° azimuth is assumed in the direction PP', setting Σ *departures* equal to 0 in the traverse $PP'AOP$ gives

$$\cos (a + c) = \frac{R \cos a - P'A \sin a}{R} \tag{7-23}$$

from which the central angle c and the stationing of P can be computed.

From Σ *latitudes* $= 0$,

$$PP' = R \sin a + P'A \cos a - R \sin (a + c) \tag{7-24}$$

The solution can also be found by the method of section 3-6 as follows:

1. Assume 0°00′ azimuth in the direction shown on Fig. 7-16. Take point O as the origin of coordinates.
2. Using the traverse program find coordinates of point P' from the traverse OAP'.
3. With the bearing-distance intersect program compute the coordinates of P' from bearing of $P'P$ and distance OP ($=R$).
4. By the inverse program calculate the distance $P'P$.

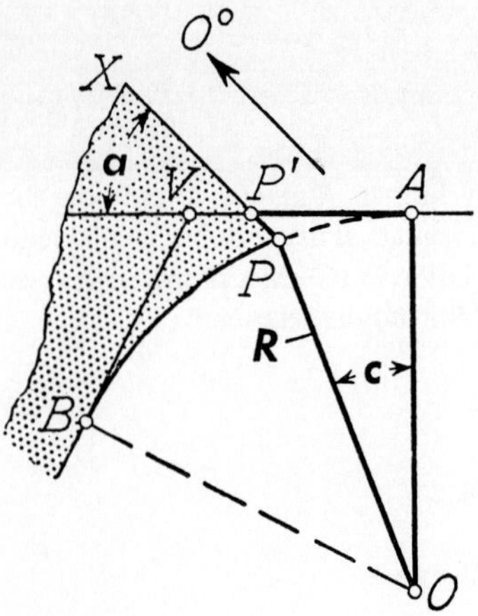

Figure 7-16

This is the type of problem readily solved by graphical methods. For example, if AP' is known, the exact tangent offset at P' is computed and laid off at right angles to a line representing the tangent AV. A large scale should be used (say 1 in. = 1 ft). Then, at a point 10 ft closer to A, another tangent offset is computed and laid off parallel to the previous one and 10 in. away. A spline fitted to three such points is used for drawing the curve, and the line XP' extended to the intersection at P. Finally, the required distance $P'P$ is scaled to the nearest 0.01 in., the ground distance thus being determined with an error hardly greater than 0.01 ft.

In a field problem, contrasted with one met in the office, the point P may be found by setting two points close together on the curve either side of where the straight line comes through. Point P is then located by "string intersection."

PROBLEMS

7-1. Simple curve with P.I. inaccessible (Fig. 7-1). *Find:* Sta. T.C. (A), sta. C.T. (B), and distance YB.

(a) Sta. $X = 67+00$; $XY = 748.63$; $x = 34°00'$; $y = 32°21'$; $D_c = 5°18'$. *Answers:* sta. T.C. = 64 + 30.31; sta. C.T. = 76 + 82.20; $YB = 249.99$.

(b) Sta. $X = 75+18.27$; $XY = 366.28$; $x = 32°30'$; $y = 23°22'$; $D_a = 6°00'$. *Answers:* sta. T.C. = 71 + 87.45; sta. C.T. = 81 + 18.56; $YB = 268.56$.

(c) Sta. $X = 117+50$; $XY = 2{,}100$; $x = 23°38'45''$; $y = 52°09'30''$; $R = 3000$. *Answers:* sta. T.C. = 111 + 25.01; sta. C.T. = 150 + 94.11; $YB = 1466.91$.

7-2. Compound curve with P.I. inaccessible (Fig. 7-2). *Find:* Sta. C.T. (Y), I_L, L_S, and the missing curve.

(a) Sta. T.C. (X) = 71 + 63.75; XY = 1,176.24; $x = 35°28'40''$; $y = 47°10'20''$; D_s (chord def.) = 9°00′. *Answers:* sta. C.T. = 84+40.14; $I_L = 42°34'03''$; $I_S = 40°04'57''$; D_L (chord def.) = 5°07′20″.

(b) Sta. T.C. (X) = 83 + 25; XY = 698.27; $x = 24°42'$; $y = 33°59'$; D_s (arc def.) = 12°00′. *Answers:* sta. T.C. = 90+51.84; $I_L = 30°51'$; $I_S = 27°50'$; D_L (arc def.) = 6°14′01″.

7-3. Change-of-location problem. *Find:* station of the new C.T. and distances needed to set hubs at the new T.C. and C.T. by taping from their original positions.

(a) Fig. 7-6: Sta. A = 36 + 18.52; $I = 41°34'$; $D_a = 8°00'$; $p = 5$. *Answers:* sta. new C.T. (B') = 41 + 45.64; tape skew shift = 7.54 from A and B to A' and B'.

(b) Fig. 7-8: Sta. A = 71 + 92.44; $I = 29°18'30''$; $D_a = 6°00'$; $p = 8$. *Answers:* sta. new C.T. (B') = 77 + 12.88; tape BB' = 31.62.

(c) Fig. 7-9: Sta. A = 63 + 50; $I = 52°25'$; $R = 2000$; $p = 12$. Answers: sta. new C.T. = 81 + 86.90; tape AA' = 24.38.

(d) Fig. 7-10: Sta. A = 106 + 04.27; $I = 28°17'40''$; $D_c = 1°30'$; $a = 2°41'10''$. *Answers:* sta. new C.T. = 124+97.87; tape AA' = 163.87.

7-4. Broken-back curve (Fig. 7-11). *Find:* I_S, I_L, PK, and increased length of the relocation between points A and D.

(a) $R_S = 500$; $R_L = 800$; $R = 1000$; $BC = 385.47$. *Answers:* $I_S = 28°47'54''$; $I_L = 46°18'22''$; $PK = 61.84$; increased length = 27.48.

(b) $D_S = 4°00'$; $D_L = 2°30'$; $D_C = 1°20'$; $BC = 257.80$. This is railroad track; chord def. of D applies. (From the answers, notice that the track could be jacked to its revised position without having to add rails.) *Answers:* $I_S = 2°48'31''$; $I_L = 3°21'26''$; $PK = 3.44$; increased length = 0.12.

7-5. Broken-back-curve (Fig. 7-12). *Find:* D of the new simple curve, the maximum offset, and increased length of the relocation between points A and D.

(a) $D_S = 5°30'$; $D_L = 3°24'$ (both chord def.); $I_S = 30°10'15''$; $I_L = 23°26'45''$; $BC = 901.17$. *Answers:* New $D_C = 2°47'23''$; max. offset = 137.09; increased length = 50.71.

(b) $D_S = 20°00'$; $D_L = 10°30'$ (both arc def.); $I_S = 34°24'$; $I_L = 32°10'$; $BC = 350.25$. *Answers:* New $D_A = 8°39'12''$; max. offset = 65.69; increased length = 30.48.

7-6. Replacement of reverse curve (Fig. 7-13). *Find:* R of the new simple curve, PK, and the decreased length of the relocation between points X and A.

(a) $R_L = 1400$; $R_S = 1000$; $I_L = 31°00'$; $I_S = 62°00'$. *Answers:* R = 7514.40; PK = 453.32; decreased length = 246.08.

(b) $D_L = 8°30'$; $D_S = 16°00'$ (both arc def.); $I_L = 26°44'$; $I_S = 65°13'$.

Answers: $R = 2085.9$ ($D = 2°44'49''$); $PK = 168.06$; decreased length = 101.48.

7-7. Curve through fixed point (Fig. 7-15). *Find:* D_A, sta. T.C. (A), and sta. P. Verify D_A by trial calculation.

(a) Sta. P.I. = 19+27.45; $I = 38°46'30''$; $VP = 96.17$; $a = 12°24'45''$. *Answers:* $D_a = 7°26'02''$; sta. A = 16 + 56.22; sta. P = 18 + 35.13.

(b) Sta. P.I. = 81+63.42; $I = 58°10'$; $VP' = 490.69$; $PP' = 26.40$. *Answers:* $D_A = 4°12'24''$; sta. A = 74+05.87; sta. P = 76+74.47.

7-8. Intersection of line and curve (Fig. 7-16). *Find:* $P'P$, sta. P, and the coordinates of P by the method of Art. 7-18.

Sta. T.C. (A) = 81+92.47; coordinates of V = N 6055.22, E 5409.63; bearings AV = S 52°42'20''W, VB = S 8°08'40''E, XP' = S 80°37'10''E; lengths AV (T) = 587.28, AP' = 480.27, $R = 1000$. *Answers:* $P'P = 115.41$; sta. P = 86+05.17; coordinates of P = N 6101.25, E 5608.63.

(NOTE: The remaining problems are "original" in the sense that no special formulas for solving them appear in this book. Suggestions in Sections 7-2 and 7-7 are pertinent. Problems are given in the order of increasing difficulty.)

7-9. Compound curve (Fig. 7-5). Arc AP consisted of four stations of 10°00' curve (chord def.); distance $PX = 580.34$; angle $VXP = 40°02'$. Find D_L (chord def.). *Answer:* $D_L = 3°35'50''$.

7-10. A reverse curve begins with the flatter arc. *Given:* $I_L = 93°32'$; $I_S = 31°54'$; $D_L = 4°30'$; $D_S = 7°30'$ (both arc def.). Assume the curve is replaced by a 3°00' (arc def.) simple curve joining the same tangents. Find: distances $A'A$ and $B'B$ between the T.C.'s and C.T.'s of the two layouts. *Answers:* $A'A = 729.41$; $B'B = 862.90$.

7-11. A 400-ft vertical curve joining gradients of −4.00% and +6.00% ends at elev. 127.86. The 6.00% gradient must be lowered parallel to itself by 2.5 ft, but drainage requirements preclude any horizontal shift in the low point of the curve. Find the length of the revised vertical curve and elevation of its low point. *Answers:* L = 650 ft; low point = 122.66.

7-12. Connect the center lines of two parallel roadways 300 ft apart with a crossover consisting of equal turnout curves of 1000-ft radius separated by a 300-ft tangent. Find the central angle of the turnout curves and overall distance required for the layout. *Answers:* 24°15'58''; 1095.45 ft.

7-13. A reverse curve between parallel tangents has angle between tangents at P.I. = 23°04'30''; T_L (P.I. to T.C.) = 3220.16; T_S (P.I. to C.T.) = 591.84; sta. T.C. = 265+87.23. Conditions require that both arcs have exactly the same radius. Find: R, sta. of point of reversal, and sta. of C.T. *Answers:* $R = 2163.30$; sta. point of reversal = 275+30.76; sta. C.T. = 293+45.53.

Chapter 8
Curve Problems in Highway Design

8-1. Foreword

Because of the public nature of highway traffic, highway curves have a greater effect upon safety of operation than do curves on railroad lines. Railways, operating over fixed track on private right-of-way, are able to control the volume and spacing of traffic and enforce slow orders on dangerous curves. Such restrictions are not practicable on the public highway. Consequently, it is necessary to "build safety into the highways" by proper location and design. It would be difficult enough to meet this requirement if conditions were static, but the continuous improvements in vehicle design and highway construction, both of which encourage ever-increasing speeds, make safety the highway designer's paramount engineering problem—as yet unsolved in several important respects.

The importance and variety of curve problems in highway design warrant devoting a separate chapter to aspects of these problems not fully covered in the preceding chapters. Though certain physical and geometric principles are reasonably well established, the numerical recommendations controlling design are subject to periodic revision. Illustrative examples are taken from recent research or current design "policies" or "standards." Revision of policies and standards can be expected as conditions change and research discloses facts not yet known.

For many years the various elements entering highway-alignment design had been fixed largely by rule-of-thumb methods, and there was little agreement among State highway departments. This difference in practice was partly due to the mushroom growth of traffic, which forced the highway engineer to concentrate upon meeting the resulting demands quickly by any methods that seemed adequate at the time. However, the principal reason was lack of basic research concerning the human and mechanical factors that contribute to safe operation at high speeds.

An important step toward correcting this condition was taken in 1937 by the American Association of State Highway Officials (AASHO) with the formation of a "Committee on Planning and Design Policies." Within a few years seven brochures on various aspects of geometric design were approved by the member States and played a major role in replacement of rule-of-thumb methods by scientific design based on research. In 1954 this material was revised, expanded, and brought up to date in a single volume entitled "A Policy on Geometric Design of Rural Highways." This volume, known as the "Blue Book" was a monumental work; it was further revised and updated in 1965.

More recently the term AASHO was changed to AASHTO (American Association of State Highway and Transportation Officials). In this chapter the numerical recommendations for design can be considered to come from the AASHO "Blue Book" of 1965, since no new revision of this policy is expected before the 1980s. (For amplification of some of this material and an extensive list of bibliographical references supporting the basic research, see the Fourth edition of this book.)

SIGHT DISTANCE

8-2. Speeds

Speed and sight distance are closely related. Several definitions of speed are used. *Overall travel speed* is the speed over a specified section of highway, being the distance divided by the overall travel time. The term *running speed* refers to the distance divided by the time the vehicle is in motion. In either case, the *average speed* for all traffic, or component thereof, is the summation of distances divided by the summation of running (or overall travel) times. The most useful concept of speed is *design speed*, which is the maximum safe speed that can be maintained over a specified section of highway when conditions are so favorable that the design features of the highway govern.

8-3. Definitions

Two definitions of sight distance are in use, known as "stopping" and "passing."

Stopping sight distance is the shortest distance required for a vehicle traveling at the assumed design speed to stop safely before reaching a stationary object in its path. At horizontal curves and crest vertical curves, height of the driver's eye is assumed to be 3.75 ft, and height of the object taken as 6 in. At no point on a highway should the sight distance be less than the stopping value.

Passing sight distance on a tangent is the shortest distance required for a vehicle safely to pull out of a traffic lane, pass a vehicle traveling in the same direction, and return to the correct lane without interfering either with the overtaken vehicle or opposing traffic. At horizontal curves and crest vertical curves, passing sight distance is the length of road that must visibly be free of obstructions in order to permit a vehicle moving at the design speed to pass a slower-moving vehicle. For these cases the height of eye is taken as 3.75 ft; height of object 4.5 ft. Highways on which passing must be accomplished in lanes that may be occupied by opposing traffic should be provided with frequent safe passing sections having a sight distance not less than the passing value for the assumed design speed.

Sight distances on overlapping horizontal and vertical curves are determined independently for each type of curvature. The critical sight distance at any point is then taken as the smaller of the two.

8-4. Stopping Sight Distance

Stopping sight distance is the sum of two distances: (1) that traversed during perception plus brake reaction time; (2) that required for stopping after brakes are applied.

Numerous scientifically controlled tests have been made to determine perception time and brake reaction time. As might be expected, the results vary according to vehicle speed, age and natural aptitude of the driver, and the conditions accompanying the test. Brake reaction time is assumed to be 1 s, this having been found to be sufficient for most drivers; perception time is selected as slightly greater than that required by most drivers, and assumed to be 1.5 s.

For a speed of V mph and perception plus brake reaction time of t s, total reaction distance in feet is

$$D_r = 1.47\, V\, t \qquad (8\text{-}1)$$

(The conversion factor 1.47 may be recalled more readily by means of the exact relation: 60 mph = 88 ft per s.)

Braking distance can be determined from fundamental principles of mechanics. The force causing a vehicle to stop after application of the brakes

Table 8-1. MINIMUM STOPPING SIGHT DISTANCE—WET PAVEMENTS

DESIGN SPEED (mph)	ASSUMED SPEED FOR CONDITION (mph)	PERCEPTION PLUS REACTION		COEFFICIENT OF FRICTION	BRAKING DISTANCE ON LEVEL (ft)	STOPPING SIGHT DISTANCE	
		TIME (s)	DIST. (ft)			COMPUTED (ft)	ROUNDED FOR DESIGN (ft)
	V	t	D_r	f	D_b	$D_r + D_b$	
30	28	2.5	103	0.36	73	176	200
40	36	2.5	132	0.33	131	263	275
50	44	2.5	161	0.31	208	369	350
60	52	2.5	191	0.30	300	491	475
65	55	2.5	202	0.30	336	538	550
70	58	2.5	213	0.29	387	600	600
75	61	2.5	224	0.28	443	667	675
80	64	2.5	235	0.27	506	741	750

equals *mass times deceleration*, or

$$F = M\,a = \frac{W}{g}\,a$$

If the coefficient of friction f is assumed to be uniform during deceleration, $F = W f$; hence, $a = f\,g$. Since the distance traversed in decelerating from a velocity v to rest is $v^2/2a$, the braking distance equals $v^2/2fg$. When g is 32.2 ft per s^2, and speed is converted to V in miles per hour, the braking distance in feet reduces to

$$D_b = \frac{V^2}{30f} \qquad (8\text{-}2)$$

Actually the coefficient of friction is not constant during deceleration, but assuming it to be so introduces no error if the proper equivalent uniform value is assumed to fit the speed in effect at the beginning of the operation. The coefficients of friction used apply to normal clean wet pavements that are free of mud, snow, or ice.

In Table 8-1, design speeds of 75 and 80 mph are applicable only to highways with full control of access. The speed for wet conditions is taken to be slightly less than the design speed so the greater proportion of traffic, traveling at yet lower speeds, will enjoy an additional safety factor. If the full design speed were used along with the coefficients of friction for dry pavements (almost double the tabulated values), the required stopping sight distances would be somewhat less than those for the assumed wet conditions. Therefore, the critical design values are those in Table 8-1.

Theoretically, stopping distances are affected slightly by grades. If G is the percent grade divided by 100, the formula for braking distance becomes

$$D_b = \frac{V^2}{30\,(f \pm G)} \qquad (8\text{-}3)$$

In practice the sight distance is usually longer on downgrades than on upgrades, a fact that automatically compensates for the greater braking distances on downgrades required by formula 8-3. Exceptions would be one-way lanes on divided highways having independent profiles for the two roadways.

8-5. Passing Sight Distance

In the "Blue Book" the minimum passing sight distance for *two-lane highways* is the sum of four distances which are shown in Fig. 8-1:

d_1 = distance traversed during the preliminary delay period (the distance traveled during perception and reaction time and during the initial acceleration to the point of encroachment on the left lane).

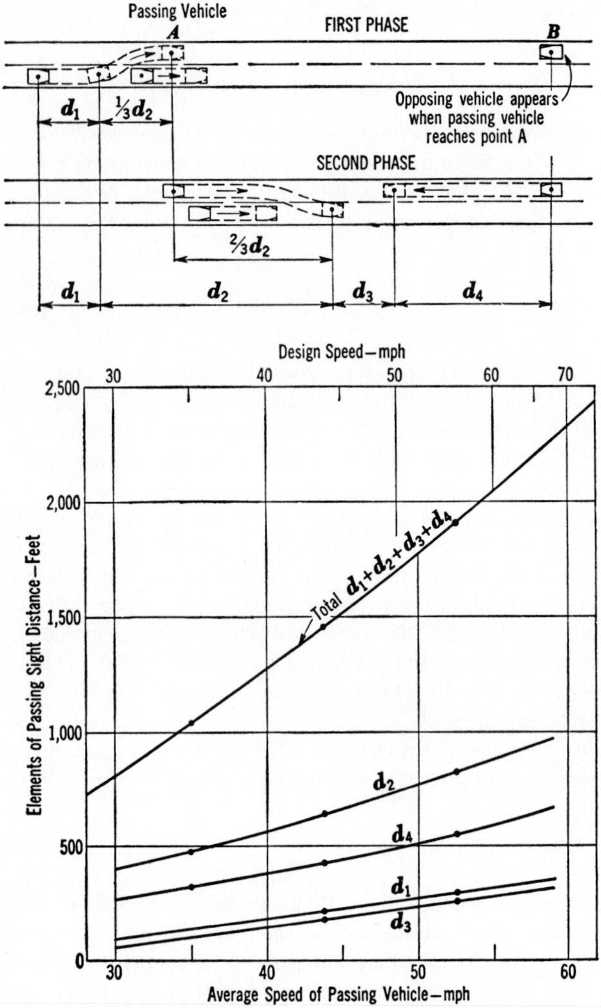

Figure 8-1 Passing sight distance (two-lane highways).

d_2 = distance traveled while the passing vehicle occupies any part of the left lane.

d_3 = distance between the passing vehicle at the end of its maneuver and the opposing vehicle.

d_4 = distance traversed by an opposing vehicle for two-thirds of the time the passing vehicle occupies the left lane, or $\frac{2}{3}d_2$.

Table 8-2. ELEMENTS OF SAFE PASSING SIGHT DISTANCE—TWO-LANE HIGHWAYS

SPEED GROUP (mph) AVERAGE PASSING SPEED (mph)	30–40 34.9	40–50 43.8	50–60 52.6	60–70 62.0
Initial maneuver:				
a = av acceleration, mph per s	1.40	1.43	1.47	1.50
t_1 = time, s	3.6	4.0	4.3	4.5
d_1 = distance traveled, ft	145	215	290	370
Occupation of left lane:				
t_2 = time, s	9.3	10.0	10.7	11.3
d_2 = distance traveled, ft	475	640	825	1030
Clearance length:				
d_3 = distance traveled, ft	100	180	250	300
Opposing vehicle:				
d_4 = distance traveled, ft	315	425	550	680
Total distance, $d_1 + d_2 + d_3 + d_4$, ft	1035	1460	1915	2380

The preliminary delay period d_1 is a complex one. For purposes of analysis it may be broken into two components: (1) time for perception and reaction, and (2) an interval during which the driver accelerates his vehicle from the trailing speed $V - m$ to the passing speed V at the point of encroachment on the left lane. The distance traversed is expressed as

$$d_1 = 1.47\, t_1 \left(V - m + \frac{at_1}{2} \right) \tag{8-4}$$

The distance traveled while the passing vehicle occupies the left lane is

$$d_2 = 1.47\, Vt_2 \tag{8-5}$$

Reference to Fig. 8-1 shows that during the first part of the passing maneuver the driver can still return to the right lane if he sees an opposing vehicle. From experience, this "uncommitted" distance is about $\frac{1}{3}d_2$. Since the opposing and passing vehicles are assumed to be traveling at the same speed, $d_4 = \frac{2}{3}d_2$.

Basic data used in establishing the design curves in Fig. 8-1 are summarized in Table 8-2. The values of a, t_1, t_2, d_3, and average passing speed come from a report on extensive field observations of driver behavior during passing maneuvers. The average value of m is taken as 10 mph.

Table 8-3 contains a summary of passing sight distances determined from the foregoing analysis.

Table 8-3. MINIMUM PASSING SIGHT DISTANCE

DESIGN SPEED (mph)	ASSUMED SPEEDS		MINIMUM PASSING SIGHT DISTANCE (ft)	
	PASSED VEHICLE (mph)	PASSING VEHICLE (mph)	FIG. 8-1	ROUNDED
30	26	36	1090	1100
40	34	44	1480	1500
50	41	51	1840	1800
60	47	57	2140	2100
65	50	60	2310	2300
70	54	64	2490	2500
75	56	66	2600	2600
80	59	69	2740	2700

HORIZONTAL ALIGNMENT

8-6. Superelevation Theory

Figure 8-2 shows the forces W (weight of vehicle) and F (centrifugal force) acting through the center of gravity c of a vehicle traveling at a speed v around a curve of radius R, when the pavement is superelevated at an angle θ with the horizontal ($\theta = \tan^{-1} e$).

In order to simplify the analysis, the two forces are resolved into their components normal and parallel to the pavement. The resultant of the forces

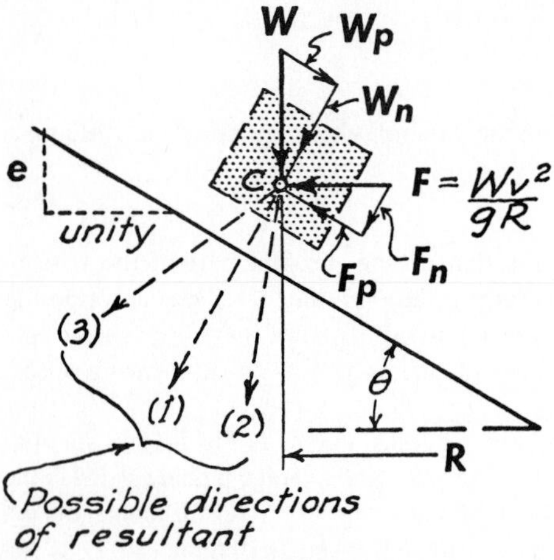

Figure 8-2 Superelevation theory.

must take one of three possible general directions:

1. When $W_p = F_p$, the resultant is perpendicular to the pavement and no centrifugal sensation is felt by the occupants of the vehicle. The speed which produces this effect is called "equilibrium speed."
2. When $W_p > F_p$, the resultant is inclined to the pavement down the slope. Consequently, there is a tendency for the vehicle to slide inward, and this effect is resisted by a lateral force acting up the slope at the contact between the wheels and the road surface. Obviously, there is also a clockwise overturning moment causing the vehicle to tilt inward.
3. When $W_p < F_p$, the resultant is inclined to the pavement up the slope. The effects are then opposite to those in (2); the resisting lateral force acts down the slope and the tilt is outward.

At equilibrium speed, $W_p = F_p$, or

$$W \sin \theta = \frac{W v^2}{g R} \cos \theta$$

Solving for θ gives the following basic equilibrium formula:

$$\tan \theta = e = \frac{v^2}{g R} \tag{8-6}$$

The equilibrium formula is expressed in more usable form by replacing v in feet per second by V in miles per hour and by substituting 32.16 ft per s^2 for g. The result is

$$e = \frac{0.067 V^2}{R} = \frac{V^2}{15 R} \tag{8-7}$$

or

$$e = \frac{V^2 D_a}{85{,}700} \tag{8-8}$$

When $W_p \neq F_p$, the magnitude of the tendency for the vehicle to move laterally may be denoted by the term f, which is the ratio of the lateral component of the resultant to its normal component. If all forces are assumed to have positive signs, the value of f is

$$f = \frac{F_p - W_p}{F_n + W_n} = \frac{F \cos \theta - W \sin \theta}{F \sin \theta + W \cos \theta}$$

Since the superelevation angles used in practice are small (θ is rarely permitted to exceed 7°), it is sufficiently accurate to assume that $F \sin \theta = 0$, this term being minute in comparison with $W \cos \theta$. Therefore, f is approximately equal to

$$\frac{F \cos \theta - W \sin \theta}{W \cos \theta} = \frac{F}{W} - \tan \theta$$

which reduces to

$$f = \frac{v^2}{g R} - e \tag{8-9}$$

The resulting working relations are:

when $W_p > F_p$ (or V is less than equilibrium speed),

$$e - f = \frac{V^2}{15 R} = \frac{V^2 D_a}{85{,}700} \tag{8-10}$$

when $W_p < F_p$ (or V exceeds equilibrium speed),

$$e + f = \frac{V^2}{15 R} = \frac{V^2 D_a}{85{,}700} \tag{8-11}$$

It should be observed that f is necessarily proportional to, and opposite in direction to, the resisting side friction which automatically comes into action at other than equilibrium speed.

In solving problems for various operating conditions, confusion in signs may be avoided by considering the terms in the foregoing equations to be positive quantities that are proportional to the actual forces acting on the vehicle. Since the forces themselves act in definite directions, these same directions may be given to the terms e, f, and $V^2/15\ R$. Figure 8-3 and the accompanying analysis illustrate this procedure.

The term f has been called the "lateral ratio" in research done by Leeming in Great Britain, and this appears to be the most logical expression. In the United States, f has been termed "unbalanced centrifugal ratio," "cornering ratio" (at General Motors Proving Ground), "unbalanced side friction," and "side friction factor." Because of the widespread use of the last expression, it

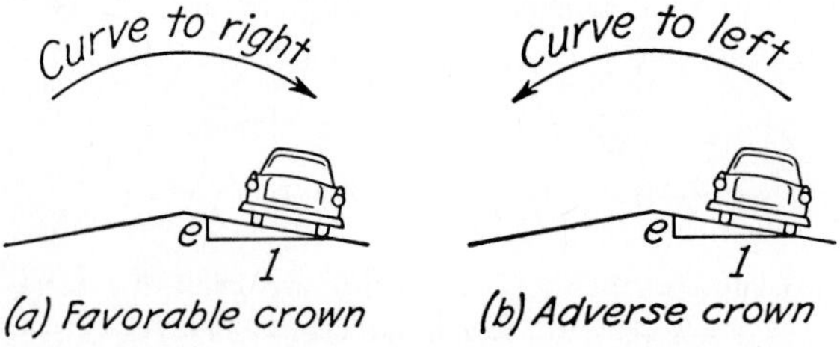

Figure 8-3

FIG. 8-3(a)—FAVORABLE CROWN (1) $V =$ Equilibrium speed Superelevation counterbalances the centrifugal force. There is no side friction.	$\frac{V^2}{15R} = e$ (Formula 8-7)
(2) $V <$ Equilibrium speed Side friction and centrifugal force assist each other in counterbalancing excessive superelevation.	$f + \frac{V^2}{15R} = e$ (Formula 8-10)
(3) $V >$ Equilibrium speed Side friction and superelevation assist each other in counterbalancing centrifugal force.	$\frac{V^2}{15R} = e + f$ (Formula 8-11)
FIG. 8-3(b)—ADVERSE CROWN Only one possible case regardless of speed. Side friction alone holds car on road.	$f = e + \frac{V^2}{15R}$

will be adopted in this book—with the warning that it should not be confused with the "coefficient of friction" as understood in dynamics.

8-7. Dynamics of Vehicle Operation on Curves

On a curve having accurately built constant superelevation, there is a particular speed, known as the *equilibrium speed* or "handsoff" speed, at which a car steers itself around the curve. At this speed $W_p = F_p$, and the value of f is zero.

At other than equilibrium speed, equations 8-10 and 8-11 seem to indicate that safe operation around a given curve is entirely within the control of the driver. He has only to adopt any desired speed and rely upon the automatic development of whatever value of f is needed to make up for the lack of balance between e and $V^2/15\,R$. However, the matter is not as simple as this. On the contrary, it is very complex, and further research is required before curve design for high-speed operation can be placed upon a sound scientific basis.

When $W_p \neq F_p$, the car tends to creep out of the traffic lane. To offset this tendency the driver must exert force at the steering wheel and must steer slightly toward or away from the center of curvature, the direction depending on whether V is greater or less than equilibrium speed. As a result, each pair of tires must "slip" across the surface at a definite angle between the path of travel and the longitudinal axis of the wheels. Figure 8-4 illustrates the normal condition when V exceeds the equilibrium speed.

Front and rear slip angles are rarely equal. When the front slip angle is greater than the rear slip angle, the car is said to be "understeering," as in Fig. 8-4; when the reverse is true, the car is "oversteering." Whether a car

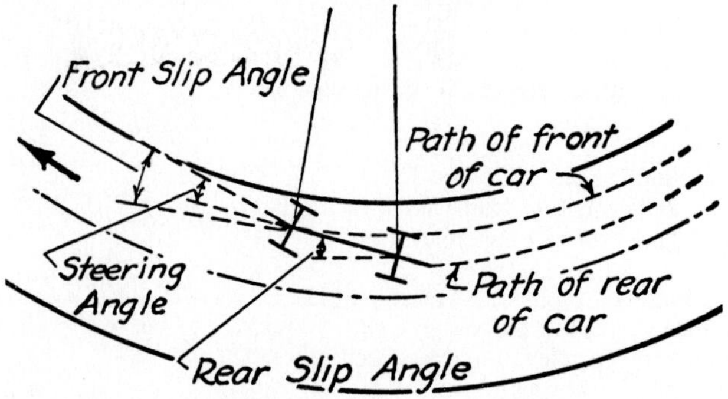

Figure 8-4 Understeering action.

understeers or oversteers depends principally on its design and partly on factors within the control of the operator. For example, research at the General Motors Proving Ground shows that it is possible to make a car either highly oversteering or highly understeering by merely varying front and rear tire pressures within certain limits.

The understeering car is somewhat more stable and susceptible to control than the oversteering type. Perhaps this is true because an increase in speed on a curve requires an *increase* in the steering angle of the understeering car (an operation that is instinctively natural), whereas a *decrease* in the steering angle is necessary in the case of the oversteering car. Above a certain critical speed the front wheels of the understeering car start to slide off the road, but by careful braking the driver can generally regain control and return to a fairly fixed course at a speed below the critical value. It is more difficult to hold an oversteering car on a fixed path even at moderate speeds. Above a certain critical speed the rear end of the car starts to slide off the road, and any slight application of brakes is apt to put the car into a spin.

During the time a car develops certain slip angles in rounding a curve, there is some tilt or "body roll." The roll angle is a linear function of f, at least up to the limits of f considered safe. Roll angles are not large. Tests show that, when $f = 0.20$, body-roll angles vary between 1.8° and 3.5°, the value depending on the make and model year of the car. Though body roll has less effect upon a car's general "road-ability" than do the slip angles, it is a factor that must be allowed for in accurate use of the ball bank indicator (see section 8-8).

8-8. Side Friction Factors

The value of f at which side skidding is imminent depends principally upon the speed of the vehicle, condition of the tires, and characteristics of the roadway surface.

An important problem in curve design—especially on curves to be marked with safe-speed signs—is to determine the percentage of the maximum side friction that can be utilized safely by the average driver. The resulting values of f used in design should give posted speeds that have an ample margin of safety even when the pavement is wet. Furthermore, when the posted speed is exceeded, the added unbalanced centrifugal force should be enough to produce an uncomfortable sensation and an instinctive reduction in speed.

The simplest device yet developed for determining maximum safe speeds and their relation to side friction factors is the ball bank indicator, apparently first used by the Missouri State Highway Department in 1937. It consists of a sealed curved glass tube containing liquid and a steel ball slightly smaller than the bore of the tube. When the indicator is mounted on the dash by means of rubber suction cups, the ball is free to roll transversely under the influence of the forces acting upon it. The liquid produces enough damping effect to hold the ball fairly steady, even when the car is driven around a curve at high speed on a slightly rough surface. Readings are taken on a scale graduated in degrees with the 0° mark at the center of the tube.

Where the ball bank indicator is to be used, it is first mounted on the dash with the ball at the 0° reading when the car is in a stationary level position. Obviously, all observers who are to be in the car during the test run must be in their assigned positions when the indicator is set in place.

The indicator is used for two purposes: (1) to determine the ball bank angle and side friction factor at the maximum comfortable speed on a particular curve, and (2) to determine the speed on a curve required to produce a specified ball bank angle.

In the first use, the body roll ρ of the test car is determined by stopping the car on the curve and reading the ball bank angle β. When the car is stationary, $\rho = \beta - \theta$, as shown in Fig. 8-5. (Obviously the superelevation θ

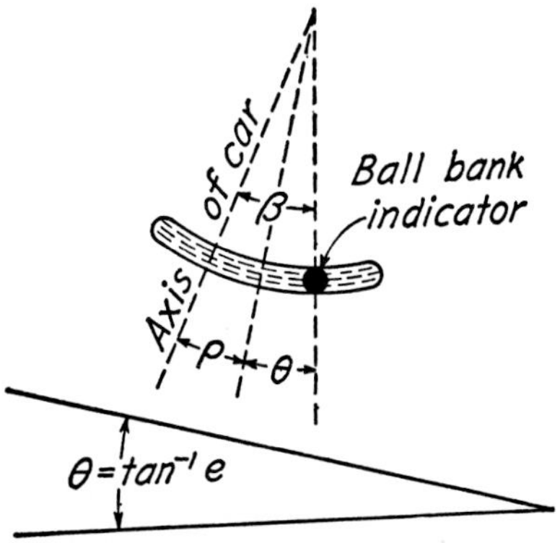

Figure 8-5 Ball bank indicator—car stopped.

must be measured.) Then trial runs are made around the curve at various constant speeds until that speed is reached which first produces an uncomfortable centrifugal sensation. Simultaneous readings of speed and ball bank angle are then made, and the averages taken. Thus, one of the desired values, β, is measured by direct observation.

To find the desired side friction factor f, the following analysis, based on Fig. 8-6, is made:

When a car rounds a superelevated curve at equilibrium speed, the resultant of the forces is perpendicular to the pavement surface; there is no body roll and the ball bank indicator reads 0°. At a speed greater than equilibrium speed, side friction comes into play, acting inward. Were it not for the body roll, the ball bank angle would be practically a direct function of this side friction. However, the outward body roll tilts the car toward the horizontal, thereby increasing the reading of the ball bank indicator. The condition is shown in Fig. 8-6, from which

$$\beta - \rho = \tan^{-1} \frac{v^2}{g R} - \theta \tag{8-12}$$

But equation 8-9 shows that

$$f = \frac{v^2}{g R} - e$$

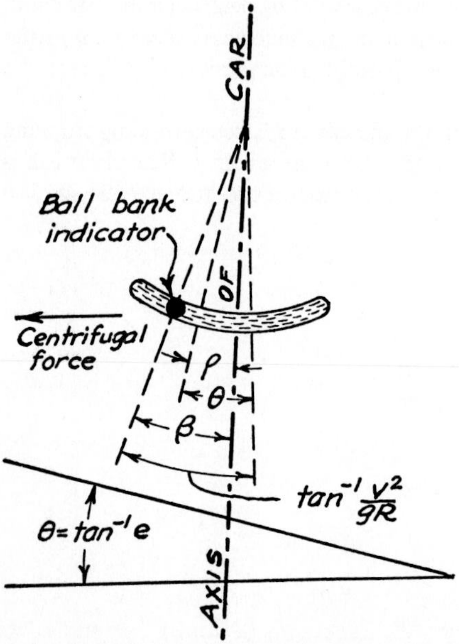

Figure 8-6 Ball bank indicator when $V >$ equilibrium speed.

or

$$\tan^{-1} f = \tan^{-1}\left(\frac{v^2}{g\,R} - e\right)$$

Because of the small angles involved, it is sufficiently accurate to assume that

$$\tan^{-1} f = \tan^{-1}\frac{v^2}{g\,R} - \tan^{-1} e = \tan^{-1}\frac{v^2}{g\,R} - \theta$$

Substituting from equation 8-12 gives

$$\tan^{-1} f = \beta - \rho \qquad \text{(8-13) Approx.}$$

The work may be checked by assuming that ρ is zero and solving equation 8-12 for β, thus obtaining the theoretical ball bank angle on the assumption of zero body roll. The difference between the calculated and observed ball bank angles is the body roll, which should agree with the value measured while the car is stationary on the superelevated curve. For any one test car there should be a linear relation between the values ρ and f.

This first use of the ball bank indicator represents its application in research. After the ball bank angle corresponding to the maximum safe speed has been determined, the second use of the indicator is in connection with establishing posted speeds on curves. In this application it is not necessary to measure the radius or the superelevation. The speed of the test car is gradually increased until the specified ball bank angle is reached, and this maximum speed is recorded.

The practical value of the ball bank indicator rests in its simplicity and in the fact that there is a surprisingly close agreement among the various States using it in regard to the numerical value, 10°, most widely adopted to indicate maximum safe speed.

The values of f summarized in Fig. 8-7 embody the principle of a greater safety factor at high speeds. A straight-line relation assumed for curve design represents a reasonable compromise based on the safe values recommended by various investigators; it was purposely kept lower at the low design speeds in order to compensate for the tendency of drivers to overdrive on highways with low design speeds.

Table 8-4 gives recommended side friction factors corresponding to the straight-line relation in Fig. 8-7. The ball bank angles were found by calculation, average body roll angles reported by General Motors Proving Ground being used. The final ball bank angles are accurate enough for determining posted speeds on curves, even though test-car roll angles differ somewhat from those listed. In the range of speeds—45 to 55 mph—within which the safe-speed signs are most frequently used, the recommended side friction factors give ball bank angles very close to the 10° figure used by most states.

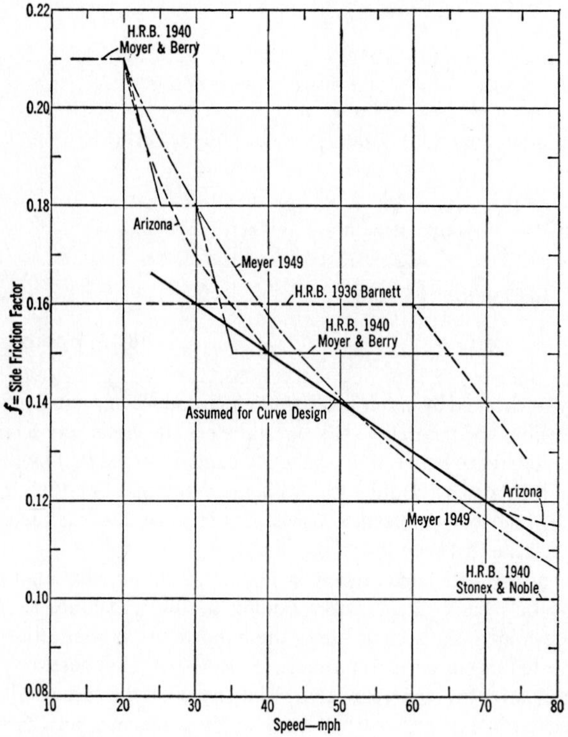

Figure 8-7

Table 8-4. RECOMMENDED VALUES FOR USE IN DESIGN

DESIGN SPEED V (mph)	SIDE FRICTION FACTOR f	$\tan^{-1} f$ OR $(\beta - \rho)^\circ$	BODY ROLL ANGLE ρ°	BALL BANK ANGLE β°
20	0.17	9.6	3.0	12.6
25	0.165	9.4	2.7	12.1
30	0.16	9.1	2.5	11.6
35	0.155	8.8	2.4	11.2
40	0.15	8.5	2.2	10.7
45	0.145	8.3	2.1	10.4
50	0.14	8.0	2.0	10.0
55	0.135	7.7	1.9	9.6
60	0.13	7.4	1.8	9.2
65	0.125	7.1	1.7	8.8
70	0.12	6.8	1.6	8.4
80	0.11	6.3	1.5	7.8
90	0.10	5.7	1.4	7.1
100	0.09	5.2	1.3	6.5

8-9. Maximum Superelevation Rates

Because of the presence of both slow-moving and fast traffic, and the variations in weather conditions over the seasons, it is impossible to design a highway having superelevation ideal for all traffic at all times. Safety is the paramount consideration. This requires a fairly low superelevation rate, with the result that road speeds are usually greater than the equilibrium speed.

When the speed is less than the equilibrium value, the resultant (Fig. 8-2) acts inward and the driver must steer slightly away from the center of curvature in order to maintain a true course in the traffic lane. The effect is somewhat like the action of an oversteering car. Since steering outward on a curve is not a natural operation, there is a tendency for slow-moving vehicles to "edge in" toward the shoulder or toward the inner traffic lane. Moreover, if the side friction factor is greater than about 0.05 when the road is icy, vehicles may slide inward despite the driver's efforts to steer a true course. On the other hand, if the superelevation rate is too low, design speeds on curves are limited

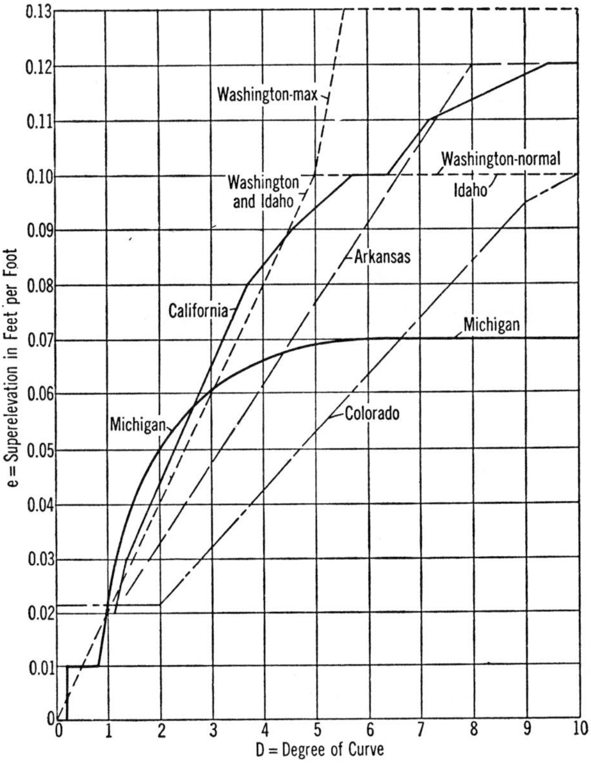

Figure 8-8

by the safe side friction factors to values less than considered practicable in modern design. Therefore, a compromise between these conflicting requirements is necessary.

There is fairly general agreement that a superelevation rate of 0.10 (approximately $1\frac{1}{4}$ in. per ft) is about the maximum that should be used in regions where snow and ice are encountered. Where exceptionally adverse winter conditions are likely to prevail for several months, a maximum superelevation rate of 0.07 or 0.08 is recommended. A maximum as high as 0.13 is used in localities free from snow or ice.

Figure 8-8 shows superelevation practice in several States when it was customary to ignore design speed. This was the traditional method. Since publication of the "Blue Book" an increasing number of States vary superelevation with design speed. In general the same maximum superelevation is used at all speeds, but at the higher design speeds the maximum is reached on flatter curves.

A number of different maximum superelevation rates are used, varying from 0.06 for urban design to 0.12 for high-speed rural highways.

8-10. Maximum Degree of Curve

The use of any recommended maximum rate of superelevation in combination with a particular design speed results in a maximum degree (or minimum radius) of curve. If sharper curvature were used with the stated design speed, either the superelevation rate or the side friction factor would have to be increased beyond recommended safe limits. Thus, this maximum degree of curve is a significant value in alignment design.

Table 8-5 lists limiting values of curvature for four maximum rates of superelevation. The final values are consistent with formula 8-11 in which are substituted the safe side friction factors from Table 8-4.

The relationships given by formula 8-11 are shown graphically in Fig. 8-9 by the straight lines for each design speed. Superimposed on the graph are the values of maximum curvature, as taken from Table 8-5.

8-11. Superelevation Rate over Range of Curvature

It is not necessary to superelevate very flat curves; the normal crown is carried around the curve unchanged. (This matter is treated in section 8-12.) However, it is not logical to change abruptly from zero superelevation on a very flat curve to maximum superelevation at some arbitrary value of *D* or *R*. There should be a transition range of curvature within which maximum superelevation increases in a rational manner from zero to the full value permitted by climatic conditions. Figure 8-8 indicates how some States have handled this matter.

Table 8-5. MAXIMUM CURVATURE

DESIGN SPEED V (mph)	MAXIMUM e	MAXIMUM f	TOTAL $e+f$	MINIMUM R (ft)	MAXIMUM D (deg)
30	0.06	0.16	0.22	273	21.0
40	0.06	0.15	0.21	508	11.3
50	0.06	0.14	0.20	833	6.9
60	0.06	0.13	0.19	1263	4.5
65	0.06	0.13	0.19	1483	3.9
70	0.06	0.12	0.18	1815	3.2
75	0.06	0.11	0.17	2206	2.6
80	0.06	0.11	0.17	2510	2.3
30	0.08	0.16	0.24	250	22.9
40	0.08	0.15	0.23	464	12.4
50	0.08	0.14	0.22	758	7.6
60	0.08	0.13	0.21	1143	5.0
65	0.08	0.13	0.21	1341	4.3
70	0.08	0.12	0.20	1633	3.5
75	0.08	0.11	0.19	1974	2.9
80	0.08	0.11	0.19	2246	2.5
30	0.10	0.16	0.26	231	24.8
40	0.10	0.15	0.25	427	13.4
50	0.10	0.14	0.24	694	8.3
60	0.10	0.13	0.23	1043	5.5
65	0.10	0.13	0.23	1225	4.7
70	0.10	0.12	0.22	1485	3.9
75	0.10	0.11	0.21	1786	3.2
80	0.10	0.11	0.21	2032	2.8
30	0.12	0.16	0.28	214	26.7
40	0.12	0.15	0.27	395	14.5
50	0.12	0.14	0.26	641	8.9
60	0.12	0.13	0.25	960	6.0
65	0.12	0.13	0.25	1127	5.1
70	0.12	0.12	0.24	1361	4.2
75	0.12	0.11	0.23	1630	3.5
80	0.12	0.11	0.23	1855	3.1

As a result of careful analysis of the dynamic factors involved in four different methods of approach to this problem, "it was concluded that a parabolic form, with the horizontal distance governing, represents a practical distribution [of superelevation] over the range of curvature." This is the type of relationship proposed in the first edition of this textbook as "a rational suggestion for general design."

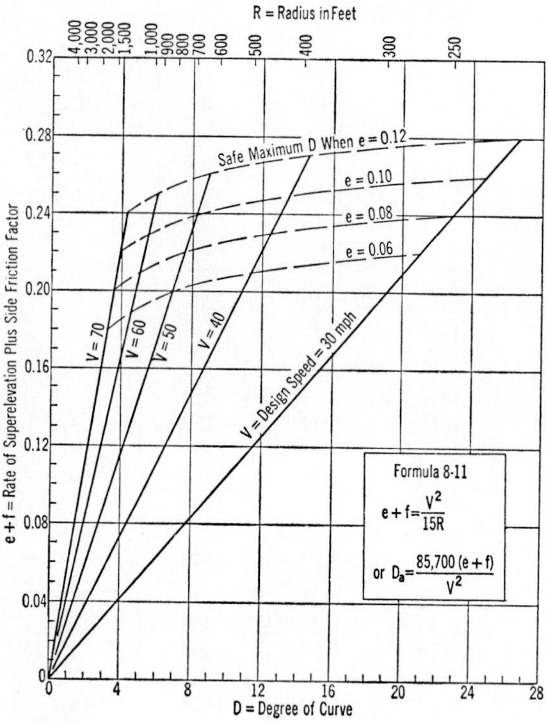

Figure 8-9

8-12. Superelevation Rates for Design

Figure 8-10 shows recommended design superelevation rates for the case where the maximum rate of superelevation equals 0.10. The maximum values of curvature at $e = 0.10$ correspond to those in Table 8-5. On curves flatter than the maximum, the superelevation rates lie on parabolic curves of the form recommended in section 8-11. In practice similar design curves could be constructed for other maximum superelevation rates, such as those given in Table 8-5.

On two-lane highways, if the pavement cross section normally used on tangents is carried around a horizontal curve unchanged, traffic entering a curve to the right has the benefit of some favorable superelevation from the crowned pavement. On the other hand, traffic entering a curve to the left meets an adverse crown, as in Fig. 8-3(b). This leads to a consideration of the maximum curvature for which the normal crowned cross section is suitable.

As a general rule, it is recommended that the minimum rate of superelevation on any curve (except at a reverse transition) should be about 0.012

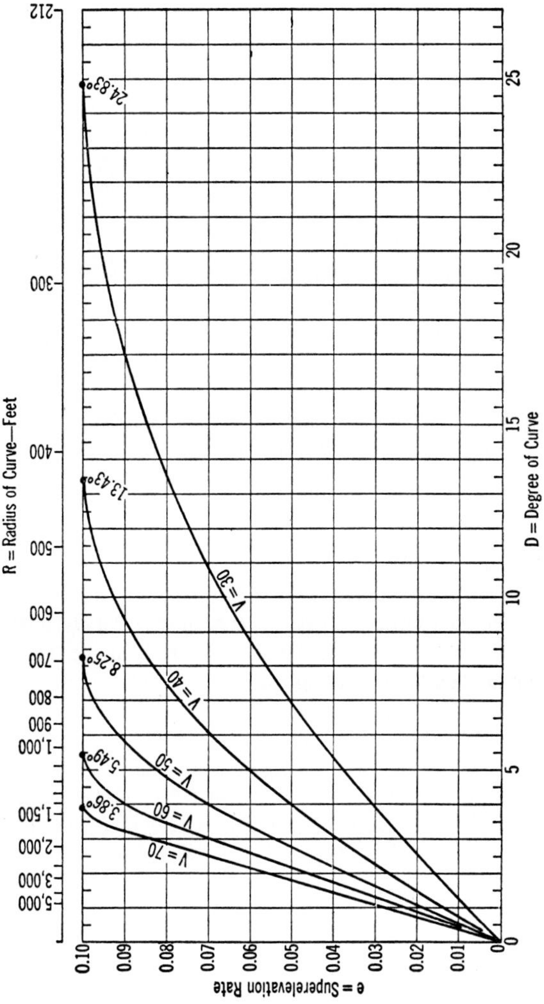

Figure 8-10 Design superelevation rates, $e_{max} = 0.10$.

Table 8-6. MAXIMUM CURVATURE FOR NORMAL CROWN SECTION

DESIGN SPEED V (mph)	AVERAGE RUNNING SPEED (mph)	MAXIMUM D	MINIMUM R (ft)	RESULTING SIDE FRICTION FACTOR f WHEN ADVERSE CROWN $e = 0.012$	
				AT DESIGN SPEED	AT RUNNING SPEED
30	28	1°21′	4,250	0.026	0.024
40	36	0°48′	7,160	0.027	0.024
50	44	0°32′	10,810	0.027	0.024
60	52	0°23′	14,690	0.028	0.024
70	58	0°18′	19,100	0.029	0.024

and that the particular value should correspond to the average rate of cross slope used on tangents.

With an average adverse cross slope of 0.012, the corresponding degree of curve (rounded) for each design speed in Fig. 8-10 is shown in the third column of Table 8-6. Observe that the resulting side friction required to counteract both adverse superelevation and centrifugal force is very small. Obviously, if the curves were made sharper, the required side friction factors would increase, and a point would be reached where a favorable slope across the entire pavement would be desirable.

It is recommended that a plane slope across the pavement should be used wherever a curve is sharp enough to require a superelevation rate in excess of about 0.02. This practical limit corresponds to degrees of curve ranging from 2°30′ at 30 mph to 0°30′ at 80 mph. For curves between these values and those in Table 8-6, a compromise could be made, in the interest of construction economy, by rotating the normal crown slightly toward the inside of the curve. However, a change to a plane slope across both lanes would be preferable, at least at the higher design speeds.

Table 8-7 shows the resulting design superelevation rates for the case in which $e = 0.10$. It is the tabular form of Fig. 8-10. Similar tables for three other values of maximum e are found in the *AASHO Policy*. The basis for selecting the tabulated runoff or spiral lengths is discussed in sections 8-13 and 8-15.

8-13. Length of Spiral

The purposes served by an easement curve and the reasons for choosing the spiral as the easement were stated succinctly in section 5-1.

Safe operation at high speeds requires that curves be designed to fit natural driver-vehicle behavior. It is obviously impossible, when traveling at any appreciable speed, to change instantaneously from a straight to a circular path at the T.C. of an unspiraled curve. On such alignment the driver makes his own transition as a matter of necessity, usually by starting to steer toward the curve in advance of the T.C. In so doing, there is bound to be some deviation from the traffic lane. If the curve is sharp or if the speed is high, the deviation may result in dangerous encroachment on the shoulder or on the adjacent traffic lane (see Fig. 8-11).

Though the operational and aesthetic advantages of spirals are generally recognized, their adoption in the United States (since about 1925) has been very gradual; as of 1969 more than one-half of the State highway departments used them. Among the reasons for this are the cumbersome, highly-mathematical treatments often presented, which result in the belief that tedious computations and awkward field work are inherent in the use of the spiral; the mistaken belief that a spiral of particular length is required for each different value of curve radius or of design speed; and inertia—a reluctance to change existing rule-of-thumb practices.

It is hoped that availability of modern computers and a simple presentation of the basic geometry of the spiral in Chapter 5, in which easily remembered analogies to the corresponding parts of a circular curve are emphasized, will assist in dispelling the bugaboo of mathematical complexity. Moreover, the various spiral tables in Part III are so complete and so readily adapted to basing the circular arc either upon the radius, or upon the chord or the arc definition of degree of curve, that the required trigonometry and field work are no more complicated for calculating and staking of a spiral-curve layout than for an unspiraled curve.

The belief that proper length of a spiral is closely calculable adds to the impression of mathematical complexity. This belief is probably based upon early research done by Shortt in the field of railroad practice. It is significant that Shortt's research was on *unsuperelevated* transition curves on *railway track*. Briefly, the result of his work seemed to indicate that the length of transition for comfortable operation was a definitely calculable variable. The reasoning was as follows: At constant speed, the length of time required

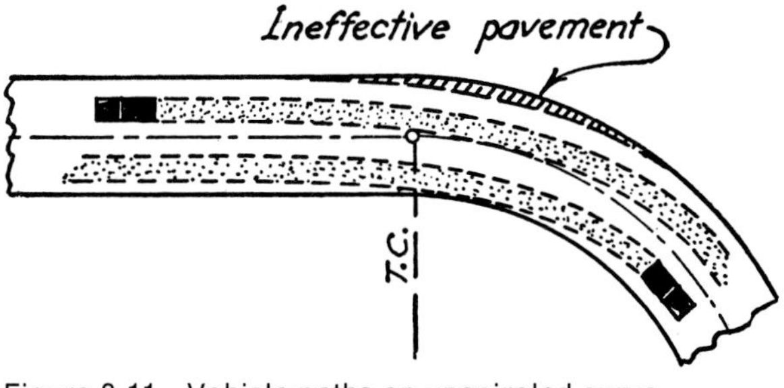

Figure 8-11 Vehicle paths on unspiraled curve.

Table 8-7. DESIGN VALUES FOR RATE OF ELEVATION (e) AND MINIMUM LENGTH OF RUNOFF OR SPIRAL CURVE

		$V = 30$			$V = 40$			$V = 50$			$V = 60$			$V = 70$		
			L			L			L			L			L	
D	R	e	TWO-LANE	FOUR-LANE	e	TWO-LANE	FOUR-LANE	e	TWO-LANE	FOUR-LANE	e	TWO-LANE	FOUR-LANE	e	TWO-LANE	FOUR-LANE
0°15′	22918′	NC	0	0	NC	0	0	NC	0	0	NC	0	0	RC	200	200
0°30′	11459′	NC	0	0	NC	0	0	RC	150	150	RC	175	175	RC	200	200
0°45′	7639′	NC	0	0	RC	125	125	RC	150	150	0.018	175	175	0.020	200	200
1°00′	5730′	NC	0	0	RC	125	125	0.018	150	150	0.022	175	175	0.028	200	200
1°30′	3820′	RC	100	100	0.020	125	125	0.027	150	150	0.034	175	175	0.042	200	200
2°00′	2865′	RC	100	100	0.027	125	125	0.036	150	150	0.046	175	190	0.055	200	250
2°30′	2292′	0.020	100	100	0.033	125	125	0.045	150	160	0.059	175	240	0.069	210	310
3°00′	1910′	0.024	100	100	0.038	125	125	0.054	150	190	0.070	190	280	0.083	250	370

3°30′	1637′	0.027	100	100	0.045	125	140	0.063	150	230	0.081	220	330	0.096	290	430
4°00′	1432′	0.030	100	100	0.050	125	160	0.070	170	250	0.090	240	360	0.100	300	450
5°00′	1146′	0.038	100	100	0.060	130	190	0.083	200	300	0.099	270	400	D_{max} = 3.9°		
6°00′	955′	0.044	100	120	0.068	140	210	0.093	220	330	0.100	270	400			
7°00′	819′	0.050	100	140	0.076	160	240	0.097	230	350	D_{max} = 5.5°					
8°00′	716′	0.055	100	150	0.084	180	260	0.100	240	360						
9°00′	637′	0.061	110	160	0.089	190	280	0.100	240	360						
10°00′	573′	0.065	120	180	0.093	200	290	D_{max} = 8.3°								
11°00′	521′	0.070	130	190	0.096	200	300									
12°00′	477′	0.074	130	200	0.098	210	310									
13°00′	441′	0.078	140	210	0.099	210	310									
14°00′	409′	0.082	150	220	0.100	210	320									
16°00′	358′	0.087	160	240	D_{max} = 13.4°											
18°00′	318′	0.093	170	250												
20°00′	286′	0.096	170	260												
22°00′	260′	0.099	180	270												
24°00′	239′	0.100	180	270												
24.8°	231′	0.100	180	270												
		D_{max} = 24.8°														

NOTES. NC = normal crown section. RC = remove adverse crown, superelevate at normal crown slope. Spirals desirable but not as essential above heavy line. Lengths rounded in multiples of 25 or 50 ft permit simpler calculations.

to traverse an easement curve of length L_s is L_s/v s. On the curve, the acceleration toward the center is v^2/R ft per s^2. Consequently, the average (assumed constant) rate of change in centripetal acceleration is $v^3/R\,L_s$ ft per s^3. Shortt denoted this constant rate by C. Converting v to V in miles per hour and solving for L_s gives the basic Shortt formula, which is

$$L_s = \frac{3.15\ V^3}{R\ C} \tag{8-14}$$

Shortt concluded from these experiments on unsuperelevated railroad track that a value for C of 1 ft per s^3 was the maximum that would go unnoticed.

In transplanting the Shortt formula to highway practice, it has been customary to overlook the possible effect of superelevation and also to use a larger value of C. Tentative suggestions that 2 might be a suitable value of C for highways led to the following version of the Shortt formula:

$$L_s = \frac{1.58\ V^3}{R} \tag{8-15}$$

Unfortunately, equation 8-15 has sometimes been used quite literally, as though to deviate from it would represent a departure from correct spiral theory as applied to highways. Arbitrary use of this equation involves two fallacies. One is neglect of the superelevation. Actually, the presence of superelevation cancels out part or all of the centrifugal force (as far as its effect upon comfort is concerned), thereby invalidating any mathematical relation that is based upon *unsuperelevated* curves. Review of the geometry of a spiral shows that it is a curve of *uniformly increasing degree.* If superelevation on a curve is *increased uniformly*, it follows from basic superelevation theory (section 8-6) that the side friction factor f will be zero when traversing the spiral at the so-called equilibrium speed for the circular arc. In other words, it would be theoretically possible for a car to steer itself on a true course around the complete spiral-curve layout without the driver touching the wheel.

It may be shown that the correct form of the Shortt formula for *superelevated* curves is

$$L_s = \frac{1}{C}\left(\frac{v^3}{R}\cos\theta - v\,g\sin\theta\right) \tag{8-16}$$

Since the values of θ used in practice are small, it is close enough to assume that $\cos\theta = 1$ and $\sin\theta = \tan\theta = e$. When these substitutions are made and v is converted to V in miles per hour, equation 8-16 becomes

$$L_s = \frac{3.15\ V}{C}\left(\frac{V^2}{R} - 15\,e\right) \tag{8-17}$$

When $e = 0$, formula 8-17 reduces to the Shortt formula for unsuperelevated curves; for this reason it has been called the "modified Shortt formula."

In practice speed on the circular arc is usually greater than the equilibrium value, resulting in a uniform value of f that must fit equation 8-11. Both comfort and safety at high speed make it desirable to approach f at a *uniform* rate along some form of easement curve. The spiral, when banked at a uniformly increasing rate, produces this condition.

A simpler form of equation 8-17 for the case in which V is greater than the equilibrium value is found by substituting $15(e + f)$ for V^2/R. The result is

$$L_s = \frac{47\ V f}{C} \qquad (8\text{-}18)$$

In applying any formula for length of spiral it is important to consider the rotational effect. Though an extremely short spiral to a superelevated curve will result in no centrifugal sensation (if traversed at equilibrium speed), yet the rotational change about the axis of the car may be too rapid for comfort or safety.

Long years of operating experience on spiraled superelevated railroad track have led the American Railway Engineering Association to recommend that minimum spiral lengths be based upon attaining superelevation across standard-gauge track at a desirable maximum rate of $1\frac{1}{4}$ in. per s of time. Highway and railroad operation are by no means analogous; but, in the absence of research on the motor-vehicle rotational rate at which discomfort begins, there is logic in tentative adoption of the same rate used successfully on railroads.

A rise of 1.25 in. per s across the track gauge of 4 ft $8\frac{1}{2}$ in. is equivalent to about 0.022 ft per ft per s. Consequently,

$$L_s = 1.47\ V\left(\frac{e}{0.022}\right)$$

or

$$L_s = 67\ e\ V \qquad (8\text{-}19)$$

The two formulas for minimum desirable spiral length are separately inconsistent for certain conditions. Equation 8-18 gives values which are too short when f is small ($L_s = 0$ at equilibrium speed), whereas equation 8-19 gives values which are too short when e is small ($L_s = 0$ for unsuperelevated curves). However, by using each formula within its proper range, the resulting minimum spiral lengths will produce neither unsafe values of f nor an uncomfortable rate of angular rotation.

The spiral length at which the change from equation 8-18 to 8-19 occurs is found by equating the two values of L_s. Thus,

$$\frac{47\,V f}{C} \text{ governs choice of minimum } L_s \text{ when } e < 0.7\frac{f}{C}$$

$$67\,e\,V \text{ governs choice of minimum } L_s \text{ when } e > 0.7\frac{f}{C}$$

The value of e is found from the relation

$$\frac{V^2 D_a}{85{,}700} - f \left(\text{or } \frac{V^2}{15\,R} - f\right).$$

Rotational change, rather than side friction and centripetal acceleration, is likely to govern minimum spiral lengths on the open highway. This is because e will exceed $0.7(f/C)$ wherever the actual superelevation rate is greater than about 60% of the recommended maximum for the given degree of curve and design speed.

Table 8-8 shows values of spiral length on the sharpest curves recommended when $e = 0.10$, as determined from the several formulas just described. The variations serve to emphasize that there is no basis for insisting on great precision in calculating lengths of spirals for design.

The practical value of the rotational-change method of choosing minimum spiral lengths is well-illustrated in the case of the Pennsylvania Turnpike. The spirals on this highway (Table 8-10) were designed on the basis of a rotational change of about 0.02 ft per ft per s—practically the value used in deriving equation 8-19. The generally favorable operating characteristics of these spirals, as observed on high-speed tests, are undoubtedly due in no small measure to the constant (and comfortable) rate of rotational change experienced on successive curves.

Objection to this method of selecting spiral lengths has been made on the grounds that it does not necessarily produce a constant value of C. The impression that C must be constant on a highway spiral—as it must of

Table 8-8. CALCULATED LENGTHS OF SPIRAL FOR MAXIMUM CURVATURE

BASIS $e = 0.10$	L_s = MINIMUM LENGTH OF SPIRAL IN ft FOR DESIGN SPEED, mph, OF:				
	30	40	50	60	70
Shortt formula, $C = 1$	370	470	570	650	730
Shortt formula, $C = 2$	185	235	285	325	365
Modified Shortt formula, $C = 1$	230	285	330	370	400
Modified Shortt formula, $C = 1.35$	170	210	250	275	295
Meyer, $L_s = eV$	200	270	335	400	470

necessity be in the case of a spiral on railroad track—is another fallacy which has been disclosed by research. This assumption is equivalent to stating that the driver of a motor vehicle who is free to choose his own path in making the transition from a tangent to a circular curve always turns his wheel (1) at a constant rate on any given transition and (2) at the same rate on all transitions.

Careful field tests and statistical analyses have verified the first part of this assumption. However, C has been found to vary between such wide limits that there is no "natural rate of turning" the wheel, such as is represented by $C = 2$ or by any other constant value. It appears that safety and comfort depend mainly on the size of f; and that a driver slows down to reduce f, not to reduce C. Relatively large values of C (far in excess of 2 ft per s^3) were frequently recorded as producing an "imperceptible" degree of discomfort, as long as the accompanying values of f were moderate (in the neighborhood of the design values suggested in Fig. 8-7).

The length of spiral may also be made equal to the distance required for superelevation runoff. Runoff length is determined by the rate at which the pavement cross slope is changed, or rotated, subject to some modification on the basis of appearance as viewed by the driver. Thus, the resulting formula for runoff (and spiral) length has the same general form as 8-19. This subject is treated in greater detail in section 8-15.

8-14. Minimum Curvature for Use of Spirals

Neither superelevation nor spirals are required on extremely flat curves. It appears logical to use spirals approaching all curves on which a plane cross slope is used over the entire pavement width. In section 8-12 the superelevation requiring such design was fixed at the rate $e = 0.02$ or greater. However, at this limiting rate a selected spiral has such a small o-distance (offset, or throw, at the offset T.C.) that the spiral is not needed. It could be omitted without hampering the ability of a driver to keep well within the traffic lane; if used, it would serve principally as a graceful method of changing from the normal crown to the superelevated cross section.

In view of the fact that minimum curvature for use of spirals must be set arbitrarily, the control values selected were taken as rounded values obtained from Fig. 8-10 at the points where $e = 0.03$, approximately. The resulting recommendations are given in Table 8-9. Throws in the last column are obtainable from Table XI. Minimum lengths of spirals approximate the distances traveled in 2 s at the design speeds.

8-15. Length of Superelevation Runoff

Superelevation runoff is the general term denoting the transition from the normal crown section on a tangent to the fully superelevated section on a

Table 8-9. MINIMUM DEGREE OF CURVE FOR USE OF SPIRALS

DESIGN SPEED (mph)	MINIMUM CURVATURE	ASSUMED MINIMUM LENGTH OF SPIRAL (ft)	CALCULATED *o*-DISTANCE (ft)
30	3°30′	100	0.25
40	2°15′	125	0.26
50	1°45′	150	0.29
60	1°15′	175	0.26
70	1°15′	200	0.36

curve. This transition should be spread over a long enough distance and be so positioned with respect to the horizontal curve that driving safety is "built in" to the design.

There is no completely rational method of determining length of runoff. It has been made equal to the length of spiral, as calculated from one of the formulas in section 8-13. If this method is used, the most logical formula would be that one having the same general form as 8-19, since this formula is based on restricting the rate of angular change in superelevation.

To possess a pleasing appearance as viewed by the driver, the edge profiles should not appear to be distorted. Control of runoff length from this standpoint has also been used; the numerical controls must, of course, be empirical. In the "Blue Book" suggested values are as follows:

Design speed, mph	30	40	50	60	70
Max. relative slope in % gradient between edges of two-lane pavement	1.33	1.14	1.00	0.89	0.80

On spiraled curves there is no sound basis for using different lengths of spiral and runoff. Simplicity in construction is gained by using identical lengths. Since length of runoff is applicable to all superelevated curves, whether spiraled or not, it is concluded that runoff lengths, as determined from the foregoing appearance controls, should also be used for minimum lengths of spirals. On four-lane highways the runoff lengths are taken to be 1.5 times the lengths for two-lane highways, purely on an empirical basis. The resulting spiral and runoff lengths are listed in Table 8-7. It should be emphasized that these are *minimum* values; high-type alignment or the attainment of proper pavement drainage may justify the use of greater lengths.

8-16. Methods of Attaining Superelevation

The transition from the normal crowned section on a tangent to the fully superelevated section on a curve should be pleasing in appearance and inherently safe and comfortable for the operation of vehicles at the highway design speed. In addition, it should be relatively simple to calculate and stake out the transition.

Spiraled curves may be superelevated by the method shown in Fig. 8-12. In this method the normal profile grade of the center line is unchanged. The outer lane between sections *a-a* and *b-b* is gradually warped from the normal crowned section to a straight *level* section at the T.S.; beyond the T.S. the section is rotated at a uniform rate about the survey center line until it reaches full superelevation at the S.C. (section *d-d*). The normal profile grade of the inner edge of the pavement is continued as far as section *c-c*. Between sections *b-b* and *c-c* the normal convex crown (if any) on the inner lane is gradually converted to a straight *inclined* section at *c-c*, where the rate of superelevation equals that on the outside lane. (The pavement areas over which the crown is taken out are shown cross hatched in the plan view.) Between sections *c-c* and *d-d* there is a uniformly increasing one-way bank across both lanes. The same method is used on a curve to the left and on the leaving spiral.

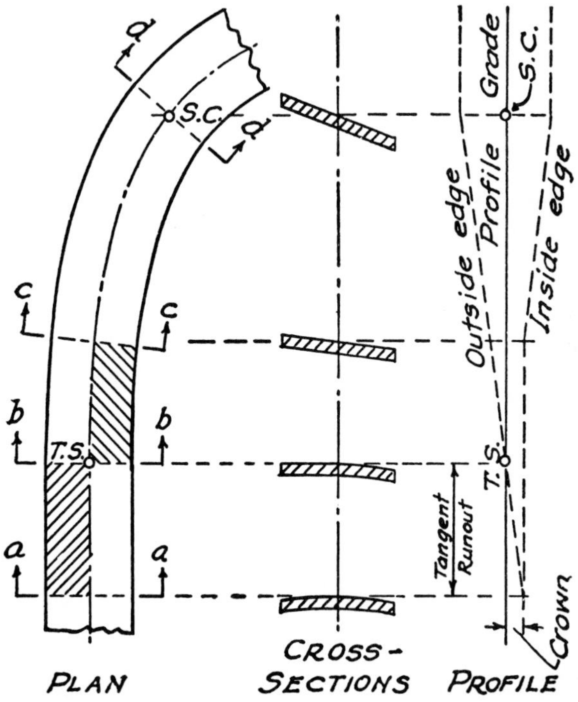

Figure 8-12 Attaining superelevation at spiraled curve.

In Fig. 8-12 the edge profiles are shown as straight lines merely to illustrate the basic design; the edge breaks would actually be rounded in construction. Some agencies obtain the effect of short vertical curves by eye adjustments of the stakes or forms; others insert true vertical curves at the breaks; and a few States use reversed vertical curves. Graphical determination of edge profiles by means of splines is an excellent and economical method in office design.

The preceding method of rotating a section about the center line should not be adhered to rigidly. Practical considerations, as well as aesthetics, are poorly served by such stereotyped design. Drainage conditions, for example, may not permit depressing the inside edge of the pavement by the amount required by this method. In such a case the pavement can be rotated about the normal profile grade of the inside edge or about a line a short distance from that edge and parallel to it. Where summit vertical curves and horizontal curves overlap, rotation about the normal profile grade of the outside edge may be the obvious method of reducing the unsightly hump produced along that edge by either of the preceding methods. No one method is best for all situations; each case should be studied individually.

Multilane highways having a median strip present an especially difficult problem in runoff design. Here, consideration of all important factors—aesthetics, drainage conditions, economics of grading, riding comfort, and safety—is necessary in arriving at a harmonious solution.

The design finally adopted on the original section of the Pennsylvania Turnpike is especially instructive in these respects. On this highway the rather narrow median strip (10 ft wide) and frequent curves—aggregating $50\frac{1}{2}$ miles in the 160-mile distance—were principally responsible for the decision to keep the edges of the paved roadway nearest the median strip in the same horizontal plane at all times. Each 24-ft roadway slopes away from these edges at a rate of $\frac{1}{8}$ in. per ft. Consequently, on tangents there is surface drainage over the roadways from the median strip. This disadvantage, however, is offset by the simpler method of runoff design made possible. The method is essentially that of Fig. 8-12, adapted to rotation about the edges closest to the median strip. Figure 8-13 shows the runoff details for a curve to the left; specific curve data are given in Table 8-10.

It should be observed that the level-inclined section, corresponding to *b-b* in Fig. 8-12, occurs in advance of the T.S. in Fig. 8-13, and that the inclined section corresponding to *c-c* in Fig. 8-12 occurs at the T.S. in Fig. 8-13. These modifications are required by the one-way bank over the roadways on tangents.

Observations of high-speed driving over spiraled curves built with and without a tangent runout on the outer lane indicate that the runout is a desirable feature of superelevation runoff design. In tests on the Pennsylvania Turnpike the runout was found to be "a pronounced aid in entering horizontal curvature. This is quite noticeable at night during adverse visibility conditions. The car appears to steer itself into the curve before the operator is aware of

Table 8-10. PENNSYLVANIA TURNPIKE SPIRALS

DEGREES OF CURVE	RATE OF SUPERELEVATION PER ft OF WIDTH (in.)	LENGTH OF SPIRAL (ft)	TANGENT RUNOUT (ft)
1°45′	$\frac{5}{16}$	150	260
2°00′	$\frac{3}{8}$	150	210
2°15′	$\frac{7}{16}$	200	210
2°30′	$\frac{9}{16}$	200	164
2°45′	$\frac{11}{16}$	250	162
3°00′	$\frac{3}{4}$	250	160
3°15′	$\frac{13}{16}$	280	158
3°30′	$\frac{7}{8}$	300	157
4°00′	1	350	154
4°30′	$1\frac{1}{16}$	370	148
5°00′	$1\frac{3}{16}$	410	146
5°15′	$\frac{3}{16}$	400	145
5°30′	$1\frac{3}{16}$	400	141

its presence. This quality of self-steering eliminates the element of surprise when entering curves traveling at speeds in excess of visibility requirements during adverse weather."

Unspiraled circular curves are superelevated by various rule-of-thumb methods. Obviously, no method can be completely rational, since it is impossible to have full superelevation between the T.C. and the C.T. (where it belongs) without placing the runoff entirely on the tangent (where it does not belong). On the other hand, runoff cannot be accomplished completely on

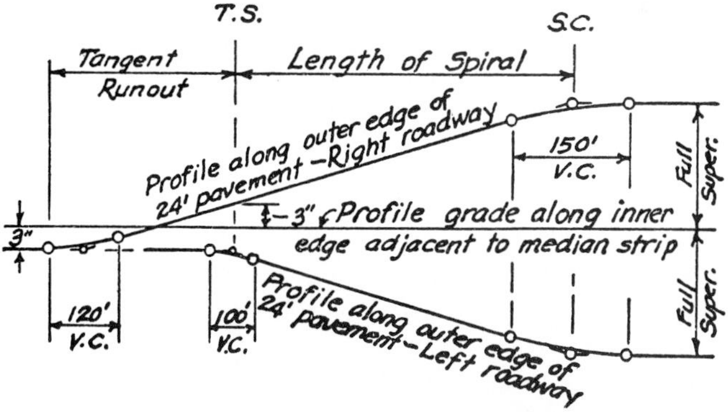

Figure 8-13 Runoff design on Pennsylvania Turnpike.

the curve without having inadequate and variable superelevation over a substantial portion of the distance. No agencies are known to follow the latter procedure, although several States place all the runoff on the tangent. The method adopted is usually a compromise, in which the runoff starts on the tangent and ends at a point some distance beyond the T.C.

In the foregoing method there is invariably a section of tangent at each end of the curve over which the cross section varies from the normal crown on tangents to a one-way bank at the T.C. and C.T., where the rate is between seven-tenths and nine-tenths of the full superelevation value. There is no standard practice as to details. For example, some State highway departments provide full superelevation a fixed distance beyond the T.C., regardless of the length of runoff. The distance may or may not coincide with the point at which full widening (section 8-18) is attained. Other States provide full superelevation beyond the T.C. at a variable distance equal to a certain fraction of the length of runoff. Rotation may be about the center line or about any other line parallel to the center line; the modifying circumstances that need to be considered are the same as those in the case of spiraled curves.

The objection inherent in all methods of designing superelevation runoff for unspiraled curves is the inevitable violation of sound dynamic principles. In traversing that portion of the runoff on the entering tangent, the driver—if he is to maintain a straight course—must steer against the gradually increasing superelevation. However, as soon as he reaches the T.C. of a curve involving the usual combination of an understeering car and a speed greater than the equilibrium value, he must steer toward the center of curvature (see section 8-7). This reversal in steering direction is neither natural nor obtainable instantaneously. Consequently, during the approach to the T.C. the vehicle usually creeps toward the shoulder or toward the inner lane. Near the T.C. it traverses a reverse curve in getting back into the traffic lane. Upon leaving the curve similar effects are produced.

The effects just described are not particularly important where vehicles are operated at low speeds. But at high design speeds they may have consequences serious enough to justify universal adoption of center-line spiraling. In fact, one of the conclusions resulting from carefully-instrumented high-speed tests on the Pennsylvania Turnpike was: "the use of spirals in modern highway design is imperative if inherent safety is to be provided."

8-17. Pavement Widening on Curves

The practice of widening pavements on sharp curves is well established, although there is little uniformity among State highway departments as to the amount of widening required.

When a vehicle travels at equilibrium speed around a curve, the rear wheels track inside the front wheels by an amount equal to $R - \sqrt{R^2 - L^2}$ (see Fig. 8-14). At other than equilibrium speed, the rear wheels track further

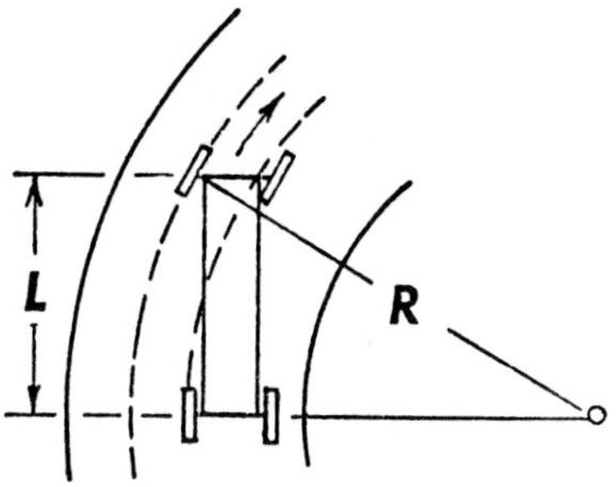

Figure 8-14

in or out, the positions depending on whether the speed is less or greater than the equilibrium value. When the speed is high, though still within safe limits, the rear wheels may even track outside the front wheels (see Fig. 8-4). There is no way of determining the exact amount of extra lane widening required to compensate for the nontracking effect, except at equilibrium speed, since it depends on the particular slip angles developed by each vehicle. However, measurements of actual wheel paths on two-lane highways show that drivers sense the need for greater clearance between opposing traffic, and that they instinctively increase the clearance when the speed is higher or the curvature sharper.

One of the earliest formulas giving the recommended total extra width for a two-lane highway was

$$w = 2[R - \sqrt{R^2 - L^2}] + \frac{35}{\sqrt{R}} \tag{8-20}$$

Formula 8-20, with L taken as 20 ft, has been used by several agencies. The expression is not entirely rational, however, since the first term applies only at equilibrium speed and the second is purely empirical. Other simpler empirical formulas give results fully as satisfactory. Among these are:

$$w = \sqrt[3]{D}, \quad w = \frac{D}{10} + 1, \quad \text{and} \quad w = \frac{D}{5}$$

It is customary to limit the widening for a two-lane pavement to between 1.5 and 6.0 ft, approximately, there being no widening on curves flatter than 5° or 6°. There are many exceptions to these values, however. Some States determine the widening only to the nearest foot; others work as close as the nearest 0.1 ft.

In an attempt to rationalize the subject, the "Blue Book" contains an analysis in which four factors are included in the formula for widening. These are (1) track width of the design vehicle (a single-unit truck or bus); (2) lateral clearance per vehicle; (3) width of front overhang of the vehicle on the inner lane; and (4) an extra width allowance that depends on sharpness of the curve and design speed. Accompanying the analysis is the recommendation that design values for widening should be multiples of $\frac{1}{2}$ ft and the minimum value should be 2 ft. On this basis no widening is required on two-lane pavements that are normally 24 ft wide or greater. In much modern alignment design no extra widening is used, since high design speeds limit curvature to 5° or 6° and traffic lanes are commonly 11 or 12 ft wide.

8-18. Transition to Widened Section

Theoretically, the full extra widening should continue for the entire length of a circular arc. It is relatively easy to do this with spiraled curves. If curves are not spiraled, however, it is impossible to carry the full widening from the T.C. to the C.T. without introducing undesirable kinks in the edge of the pavement.

Spiraled curves may be widened by the methods shown in Figs. 8-15 and 8-16. In Fig. 8-15 the total widening is placed on the inside of the curve. Widening begins at the T.S., reaches the full amount at the S.C., and tapers off from the C.S. to the S.T. in like manner. At intermediate points on the transition the *widenings are proportional to the distances from the T.S.* Forms for the curve at the inside edge of the roadway are located on radial offsets from the survey center line at distances equal to $\frac{1}{2}b$ plus the proportional amounts of the widening.

In this method the curve of the inside edge is not a true spiral. However, the transition is smooth except at the S.C., where the slight break may be remedied by eye adjustment of the stakes or forms. The alternative would be to calculate a separate spiral for the inside edge having a throw equal to w plus the throw of the center-line spiral, and having a radius R' equal to $R - \frac{1}{2}b - w$. The length of the edge spiral is computed as in the example on page 208. This procedure, preferred by some engineers, results in a spiral

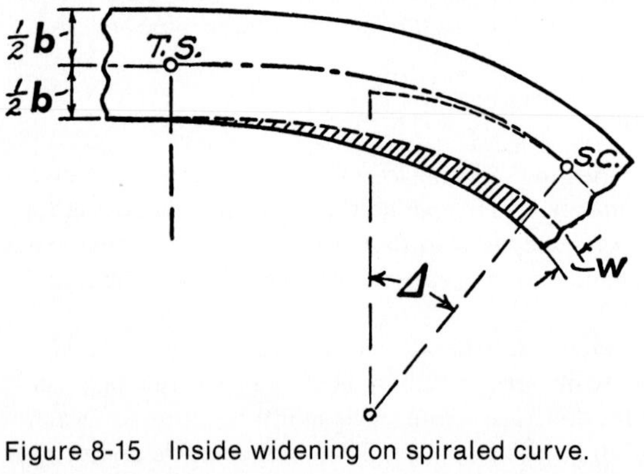

Figure 8-15 Inside widening on spiraled curve.

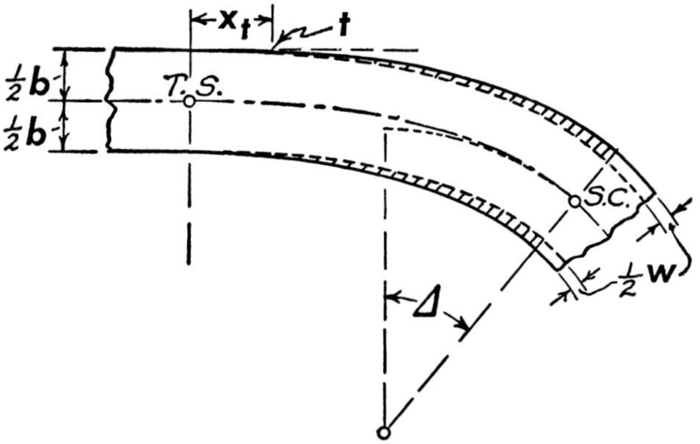

Figure 8-16 Equal lane widening on spiraled curve.

much longer than that on the center line; it also complicates the field work somewhat. Figure 8-17 illustrates the layout; see also problem 8-9.

In Fig. 8-16 the total widening is divided equally between the inside and outside edges, and is distributed along the transition by the proportion method used in Fig. 8-15. If the spirals are long enough to attain the superelevation properly, the breaks in the edges of the pavement at the S.C. are hardly noticeable; in case they are apparent at all, they can be rectified by eye adjustment of the forms.

The slight reverse curve in the outer edge of the pavement near the T.S. may be avoided by starting the outside widening at the point where the tangent produced intersects the widened pavement, as point t in Fig. 8-16. The distance from the T.S. to t is approximately equal to

$$x_t = L_s \sqrt{\frac{w}{2\,Y}} \tag{8-21}$$

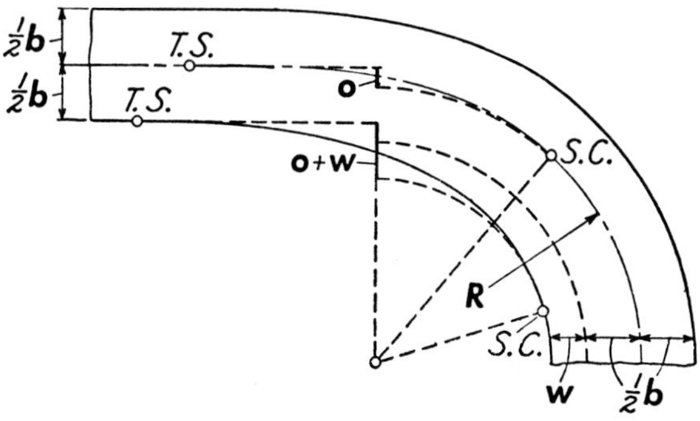

Figure 8-17 Spiraled center line and inside edge.

Theoretically, there is a slight break at point t, but it is imperceptible. Moreover, the small loss in widening between t and the T.S. is negligible. For example, on a 10° curve for which $L_s = 400$ ft and $w = 2$ ft, $x_t = 59$ ft and the widening at point t is only 0.15 ft.

When curves are spiraled, the method of Fig. 8-16 has considerable justification, particularly in the case of a two-lane concrete pavement having a longitudinal joint on the survey center line. For one thing, staking is simplified. In addition, traffic on the outside lane is provided with the same extra widening that is given to the inner lane, and the tendency to edge across the longitudinal joint is thereby reduced. On sharp curves requiring center striping, there is no unsightly and confusing deviation between the striping and the center joint.

Unspired circular curves are widened at the inside edge of the pavement and approached by some sort of easement curve, such as a spiral (Fig. 8-18). Theoretically, a perfectly smooth transition results by using a spiral that has a throw equal to w and that terminates in a circular curve having a radius R' equal to $R - \frac{1}{2}b - w$. There are several methods to determine the value of L_s. The simplest assumes that $Y = 4$ times the throw (equation 5-20). Then,

$$L_s = \sqrt{24\, R'o} \quad \text{(from equation 5-21)}$$

after which

$$\Delta \text{ (in radians)} = \frac{L_s}{2\, R'} \quad \text{(from equation 5-6)}$$

Example

Given: $D_a = 10°$; $\frac{1}{2}b = 12$ ft; $w = 2$ ft.

$$L_s = \sqrt{24 \times 558.958 \times 2} = 163.8 \text{ ft}, \quad \text{and} \quad \Delta = 8.395°$$

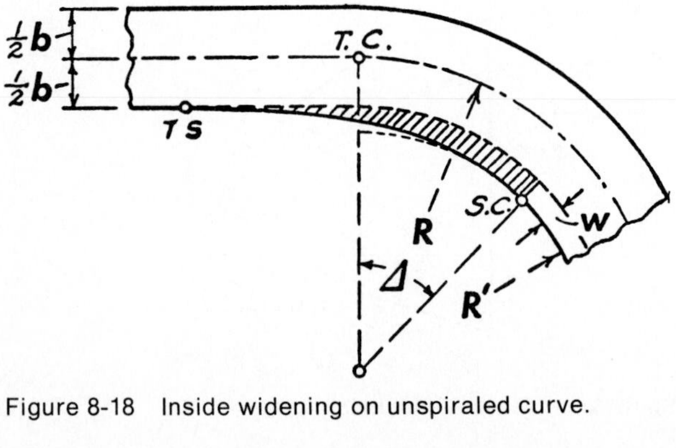

Figure 8-18 Inside widening on unspiraled curve.

On a very sharp curve having a large value of Δ, the approximation $Y = 4\,o$ may not give sufficient accuracy. In this case, calculate the coefficient M (section 5-9) and determine Δ from Table XIII. Thus, in the preceding example,

$$M = \frac{R' + w}{R'} = \frac{560.96}{558.96} = 1.003578$$

From Table XIII, by interpolation, $\Delta = 8.397^\circ$. Then, from formula 5-6, $L_s = 163.8$ ft as before. (In this example the two methods agree.) Or, from Table XII,

$$L_s = \frac{2.0}{0.01221} = 163.8 \text{ ft}$$

The method of Fig. 8-18 leaves a strip of ineffective pavement at the outside edge near the T.C. (see Fig. 8-11). In order to compensate for this, the outside edge may be independently spiraled to produce a layout in which the survey center line is a simple curve but the edges of the pavement are true spirals. (The State of Indiana used this practice for many years on curves having $D = 5^\circ$ and greater.) One method of selecting the spirals uses the following procedure, illustrated with the aid of Fig. 8-19:

1. Choose a value for the throw o of the outside spiral. In practice, o is usually about $\frac{1}{2}w$.
2. Calculate L_s (and Δ) for the outside spiral, as previously described, using $R' = R + \frac{1}{2}b - o$.
3. Compute L_s (and Δ) for the inside spiral, using a throw equal to $o + w$ and a radius $R' = R - \frac{1}{2}b - o - w$.
4. Examine the two spiral lengths and, if necessary, adjust the value of o (and possibly of w) to give more suitable spirals.

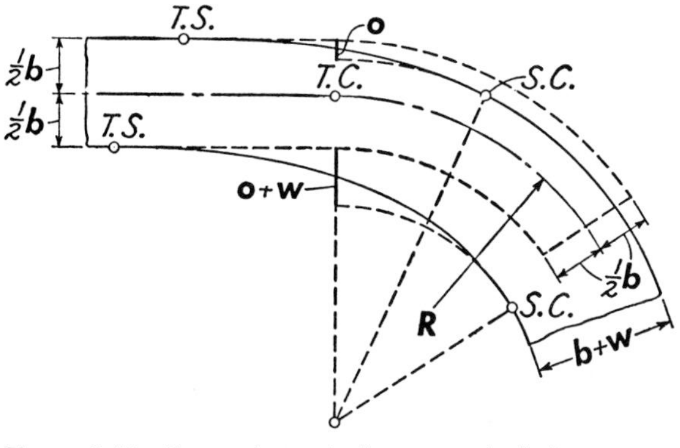

Figure 8-19 True edge spirals on unspiraled curve.

When the inner and outer edges are separately spiraled, the S.C.'s of the two spirals will not be opposite each other. Moreover, the inside spiral must be longer than that on the outside edge, because of the greater throw (see problem 8-7).

An objection to using a spiral for the widening transition to unspiraled curves—aside from the fact that each combination of R, b, and w requires a different spiral—is that only one-half the desired widening is attained at the T.C. of the circular arc. Furthermore, if the circular curve is short, there may not be distance enough beyond the T.C. in which to reach full widening unless an exceedingly short spiral is used. Accordingly, most State highway departments that have not yet adopted center-line spiraling use transitions which reach full widening on the inside edge at some relatively small fixed distance beyond the T.C., regardless of the values of R, b, and w. The length of the transition is also a fixed distance, although some States vary the length with the degree of curve according to a rule-of-thumb procedure. The edge transition itself cannot be a spiral; usually it is a curve approximating a parabola. On pavement work, the eye adjustment of the forms required to produce a smooth curve near the point of maximum widening is likely to be quite extensive. This matter is treated in greater detail in section 8-19.

Voluminous tables worked up by some State highway departments for staking widening transitions and superelevation runoffs on unspiraled curves—in contrast to the simple procedures based upon Figs. 8-12, 8-15, and 8-16—are themselves strong arguments for adoption of center-line spiraling.

8-19. Edge Lengths

It is often necessary to determine the curved length of the inner or outer edge of a pavement or of a line a certain distance from either edge. This problem occurs in connection with estimating the length of curb or guard rail and in staking an offset curve parallel to the center line.

Unwidened circular arcs present no special difficulty. This form of the problem is treated in detail in section 2-16.

In the case of a circular arc on which inside widening is accomplished by means of a true spiral (Fig. 8-18), the length of each spiraled edge is found from the basic spiral-length formula, $L_s = 2\,\Delta\,R'$ (equation 5-6), Δ being in radians and R' being the radius of the fully widened edge, that is, $R' = R - \frac{1}{2}b - w$. Total length of the curved inside edge between the T.S. and S.T. of the spirals is $(I + 2\,\Delta)\ (R - \frac{1}{2}b - w)$, whereas length of the curved outside edge is $I(R + \frac{1}{2}b)$.

Modification of the preceding case, in which the true spiral is replaced by a rule-of-thumb transition reaching full widening a short distance beyond the T.C., does not permit any mathematically simple method of calculating edge lengths accurately. Since high precision is rarely required, approximate

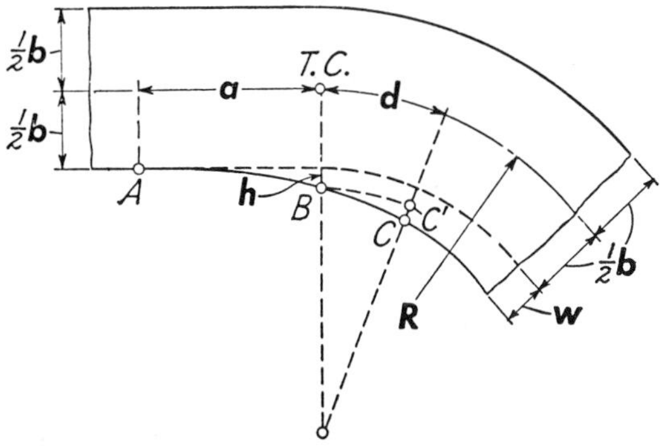

Figure 8-20

methods usually suffice. Thus, in Fig. 8-20, full widening is attained at C, the distances a, d, h, and w being given. The length of curved edge AB is approximately equal to the chord distance, or

$$AB = a + \frac{h^2}{2\,a} \text{ (approx.)}$$

Also, the curve BC is approximately equal to the arc BC', or

$$BC = BC' = \left(\frac{R - \frac{1}{2}b - h}{R}\right) d \text{ (approx.)}$$

Curved edge lengths along an unwidened curve which is spiraled along the center line depend on the method of aligning the edges. If true spirals are used for the edge curves, their lengths equal $2\,\Delta(R \pm \frac{1}{2}b)$ and the total curved edge lengths equal $(I + 2\,\Delta)(R \pm \frac{1}{2}b)$. However, the edge spirals do not quite begin, or end, on radial lines through the T.S. and S.C. Moreover, the pavement width is not exactly uniform between those points. Simplicity is gained and the effects of true spirals are closely approximated by keeping a uniform width of pavement. That is, the edges of the pavement are on radial offsets from the center line at distances of $\frac{1}{2}b$ (as in Fig. 8-15, but without the widening).

The lengths of the resulting edge curves between points radially opposite the T.S. and S.C. are found as follows: In Fig. 5-3, $dl = r d\delta$, and the corresponding relations for the inside and outside edges are

$$dl_i = (r - \tfrac{1}{2}b)d\delta \quad \text{and} \quad dl_o = (r + \tfrac{1}{2}b)d\delta$$

Consequently, $dl_i = dl - \frac{1}{2}bd\delta$ and $dl_o = dl + \frac{1}{2}bd\delta$. But from formula 5-7,

$$\delta = \left(\frac{l}{L_s}\right)^2 \Delta, \quad \text{from which} \quad d\delta = \frac{2\Delta l dl}{L_s^{\,2}}$$

After substitution, therefore,

$$dl_i = dl - \frac{b\Delta l dl}{L_s^2} \quad \text{and} \quad dl_o = dl + \frac{b\Delta l dl}{L_s^2}$$

By integration

$$l_i = l - \left(\frac{b\Delta}{2}\right)\left(\frac{l}{L_s}\right)^2 \tag{8-22}$$

and

$$l_o = l + \left(\frac{b\Delta}{2}\right)\left(\frac{l}{L_s}\right)^2 \tag{8-23}$$

For the complete edge curve, $l = L_s$, $l_i = L_i$, and $l_o = L_o$. Consequently (Fig. 8-21),

$$L_i = L_s - \tfrac{1}{2}b\Delta \tag{8-24}$$

and

$$L_o = L_s + \tfrac{1}{2}b\Delta \tag{8-25}$$

in which Δ is in radians.

If the pavement is widened on the inside by the method shown in Fig. 8-15, an analysis similar to the foregoing yields

$$L_i = L_s - \Delta(\tfrac{1}{2}b + \tfrac{2}{3}w) \tag{8-26}$$

Where the widening is divided equally between the inside and outside lanes, as in Fig. 8-16, the edge lengths are given by

$$L_i = L_s - \Delta(\tfrac{1}{2}b + \tfrac{1}{3}w) \tag{8-27}$$

and

$$L_o = L_s + \Delta(\tfrac{1}{2}b + \tfrac{1}{3}w) \tag{8-28}$$

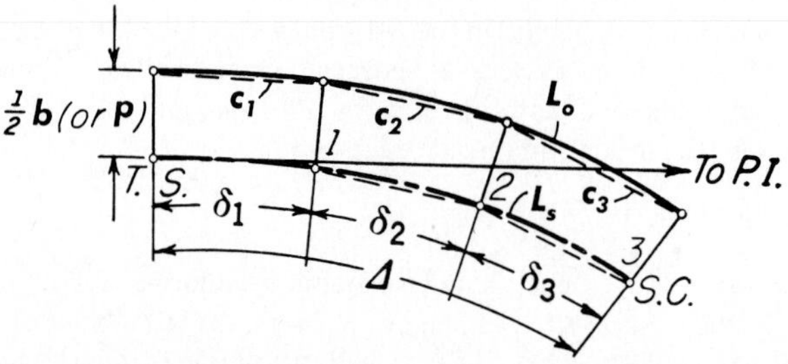

Figure 8-21 Offset curve parallel to spiral.

8-20. Staking Offset Curve Parallel to a Spiral

An offset curve parallel to a spiral may be staked either by deflection angles or by offsets (see Chapter 5). In either method it is necessary to find the chords to be taped. To illustrate the theory involved, Fig. 8-21 shows a center-line spiral L_s divided into three equal parts. It is required to stake the parallel offset curve L_o by setting points radially opposite the corresponding points on the center line. In contrast to the analogous problem at a circular curve (section 2-15), the chords along the offset curve are not the same length because of the increasing rate of curvature along the spiral and consequent increase in the values of δ.

If the offset curve is divided into n equal parts, formula 8-23 shows that

$$c_1 = \frac{L_s}{n} + \left(\frac{b\Delta}{2}\right)\left(\frac{1}{n}\right)^2$$

Also

$$c_2 = \frac{2L_s}{n} + \left(\frac{b\Delta}{2}\right)\left(\frac{2}{n}\right)^2 - c_1,$$

which may be reduced to

$$c_2 = c_1 + 2\left(\frac{b\Delta}{2}\right)\left(\frac{1}{n}\right)^2$$

Similarly

$$c_3 = c_2 + 2\left(\frac{b\Delta}{2}\right)\left(\frac{1}{n}\right)^2, \ldots, \text{etc.}$$

To generalize, substitute the offset p for $\frac{1}{2}b$. The formulas applicable to parallel offset curves outside or inside a center-line spiral then become

$$\left.\begin{aligned} c_1 &= \frac{L_s}{n} \pm i \\ c_2 &= c_1 \pm 2i \\ c_3 &= c_2 \pm 2i \\ &\cdots \\ c_n &= c_{n-1} \pm 2i \end{aligned}\right\} \qquad (8\text{-}29)$$

where the increment i equals $(p/n^2)(\Delta°/57.296)$.

The actual chords to be taped will differ slightly from nominal chords (given by formulas 8-29) only near the end of a long sharp spiral. Where this is the case, the chord corrections can be found with the aid of Tables II and III. Though the parallel offset curves have the same central angle, Δ, as the center-line spiral, the relation $A = \frac{1}{3}\Delta$ is not valid for these curves. This is because the offset curves are not true spirals; consequently, deflection angles cannot be taken from Table XV.

Correct deflection angles may be computed by first finding the coordinates of the end of an offset curve relative to its beginning. For the outer curve these coordinates are $x_o = X + p \sin \Delta$ and $y_o = Y + p \text{ vers } \Delta$, where X and Y refer to the center-line spiral. The corresponding coordinates for the inner curve are $x_i = X - p \sin \Delta$ and $y_i = Y - p \text{ vers } \Delta$. The total deflection angle to the end of an offset curve from a setup at its beginning is

$$A_o = \tan^{-1} \frac{y_o}{x_o} \left(\text{or } A_i = \tan^{-1} \frac{y_i}{x_i} \right)$$

Finally, the deflection angles to points along an offset curve can be computed on the assumption that they are closely proportional to the squares of the lengths from its beginning (formula 5-11).

If the center-line spiral has been staked and it becomes necessary to set parallel edge curves, the required points may be located by offsets (section 5-13). Each point is at a constant offset p from the center-line stake and also at the computed chord distance ahead of the offset stake previously set.

8-21. Pavement Areas on Curves

The area of pavement along any unwidened curve equals pavement width times length of curve. For a circular curve, $A = bL$, where L is the arc length between T.C. and C.T.

At an unwidened spiraled curve (see lightly-shaded portion of Fig. 8-22),

$$A = b\,[2L_s + (I - 2\Delta)R] \tag{8-30}$$

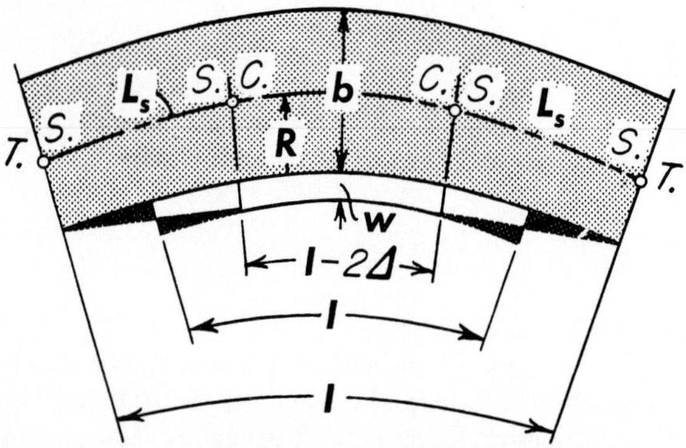

Figure 8-22 Pavement areas at spiraled curve.

where I and Δ are in radians; or

$$A = b\left[2L_s + \frac{100(I - 2\Delta)}{D}\right] \tag{8-30a}$$

where I and Δ are in degrees. In other words, the presence of spirals does not affect the basic relation for pavement area so long as the pavement width is uniform.

If the pavement is widened as in Fig. 8-15, the total extra area along the inside of a curve with equal spirals consists of a curved strip of uniform width w and two equal curved strips of variable width. From the osculating-circle principle (Fig. 5-5), the four heavily-shaded areas in Fig. 8-22 are approximately equal. Therefore, the total extra area due to widening is

$$A_w = wI(R - \tfrac{1}{2}b - \tfrac{1}{2}w) \qquad \text{(8-31) Approx.}$$

where I is in radians. If slightly higher precision is needed, a theoretically correct relation for the area of the strip of variable width (A_{vw}) may be found by expressing dA as the difference between two sectors. Then by integration,

$$A_{vw} = \tfrac{1}{2}wL_s - \Delta(\tfrac{1}{3}bw + \tfrac{1}{4}w^2) \tag{8-32}$$

where Δ is in radians. The second term in this formula is usually small enough to be neglected at the flat curves used on high-speed alignment. Omission of this term indicates that the area approximates a triangle.

8-22. Sight Distances on Horizontal Curves

Where a building, wall, cut slope, or other obstruction is located at the inside of a curve, the designer must consider the possible effect of the obstruction on the sight distance. Figure 8-23 shows this situation for the case in which the sight line AB is shorter than the curve. The driver's eye at A is assumed to be at the center of the inside lane. Although chord AB is the actual line of sight,

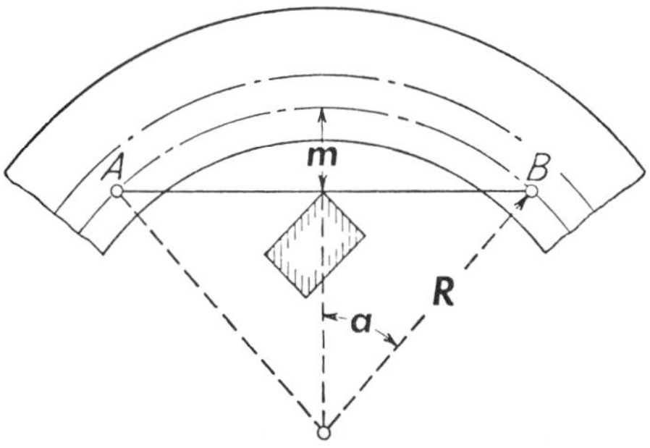

Figure 8-23 Sight distance on horizontal curve.

stopping sight distance S_H is taken as the arc AB because this is the travel distance available for stopping in order to avoid hitting an object at B.

In the following analysis, R, D, and L refer to the center of the inside lane, not to the survey center line. If the notation on Fig. 8-23 is used, $m = R$ vers a. But $a{:}D = \frac{1}{2}S_H{:}100$, and R is approximately equivalent to $5730/D$. Hence

$$m = \frac{5730}{D} \operatorname{vers} \frac{S_H D}{200} \tag{8-33}$$

Also,

$$S_H = \frac{R}{28.65} \cos^{-1} \frac{R - m}{R} \tag{8-34}$$

Figure 8-24 is a design chart showing the required middle ordinates, at various degrees of curve, needed to satisfy the stopping sight distances given in Table 8-1. Where the obstruction is a cut slope, criteria for height of eye and object (section 8-3) can be approximated by using a height of 2.0 ft at the point where m is measured.

If the sight distance is longer than the curve, the following approximate formulas can be used:

$$m = \frac{L(2S_H - L)}{8\,R} \qquad \text{(8-35) Approx.}$$

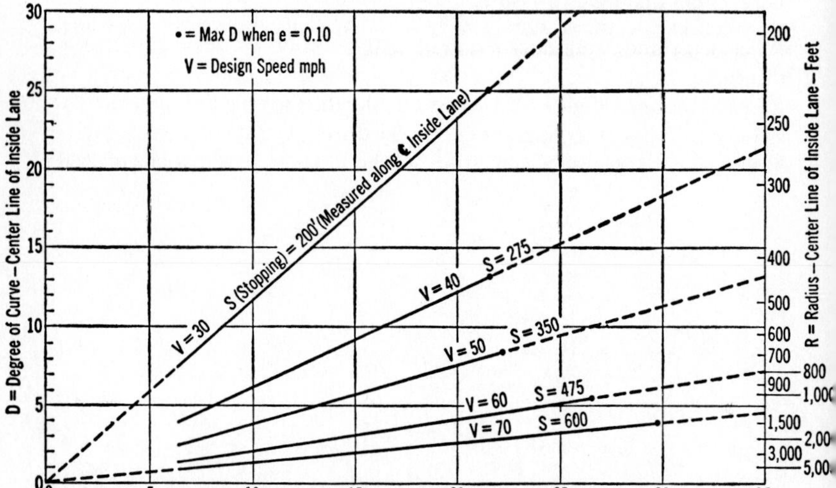

Figure 8-24

and

$$S_H = \frac{4\,R\,m}{L} + \frac{L}{2} \qquad \text{(8-36) Approx.}$$

These formulas can also be used with the passing sight distances given in Table 8-3. However, this application is of little value except on long flat curves.

Instead of using the foregoing methods, a designer may prefer to scale sight distances from the plans. Since high precision is unnecessary, the procedure is to place a straightedge on the survey center line (at the station for which the sight distance is to be determined) and tangent to the obstruction. The sight distance is then taken as the difference in stationing between the points where the straightedge intersects the center line. When scaled, it is suggested that S_H be recorded to the nearest 50 ft when less than 1000 ft and to the nearest 100 ft when greater than 1000 ft.

VERTICAL ALIGNMENT

8-23. Sight Distances at Crest Vertical Curves

At a crest vertical curve the sight distance is considered to be the horizontal projection of the line-of-sight for the assumed conditions. Figure 8-25 shows the situation where the sight distance S is less than the length L of the vertical curve, or $S < L$. Distances h_1 and h_2, representing height of the driver's eye and height of the object, respectively, are vertical offsets to the tangent sight line. In the general case, $h_1 \neq h_2$, and the sight line is not parallel to the chord joining the ends of the parabola.

From the *rule of offsets* in section 4-2 and formula 4-4,

$$h_1 : \frac{AL}{800} = S_1{}^2 : \left(\frac{L}{2}\right)^2$$

(In sight distance formulas, A is used as a positive number equal to the change

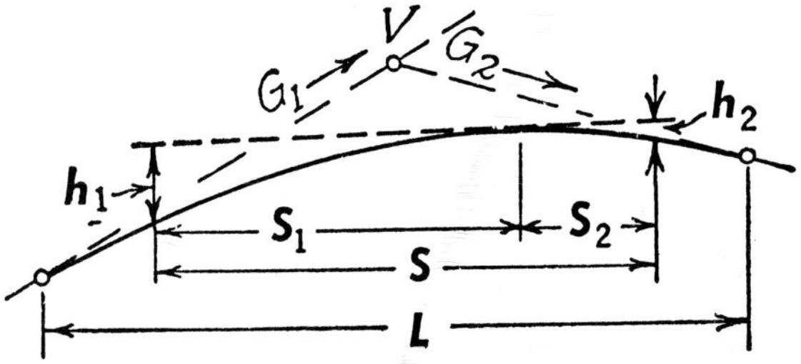

Figure 8-25 Sight distance on vertical curve, where $S < L$.

in gradient from G_1 to G_2.) Therefore,

$$S_1 = \sqrt{\frac{200\, L\, h_1}{A}}$$

Similarly,

$$S_2 = \sqrt{\frac{200\, L\, h_2}{A}}$$

The result obtained by substituting S for $S_1 + S_2$ and solving for L is the following general relation:

When $S < L$,

$$L = \frac{A\, S^2}{200(\sqrt{h_1} + \sqrt{h_2})^2} \tag{8-37}$$

In the "Blue Book" the values in the case of *stopping sight distance* are $h_1 = 3.75$ ft and $h_2 = 6$ in. In the case of *passing sight distance*, $h_1 = 3.75$ ft and $h_2 = 4.5$ ft. (See section 8-3.) Substituting these values in formula 8-37 gives the following practical relations:

When $S < L$,

(*nonpassing conditions*)

$$L = \frac{A\, S^2}{1398} \tag{8-38}$$

(*passing conditions*)

$$L = \frac{A\, S^2}{3295} \tag{8-39}$$

Figure 8-26 shows the situation where $S > L$. In the general case $h_1 \neq h_2$ and the sight line *ad* is not parallel to the chord joining the ends of the parabola.

The problem is to find the slope of the sight line that will make the distance *ad* a minimum. Let g represent the difference between the gradient of

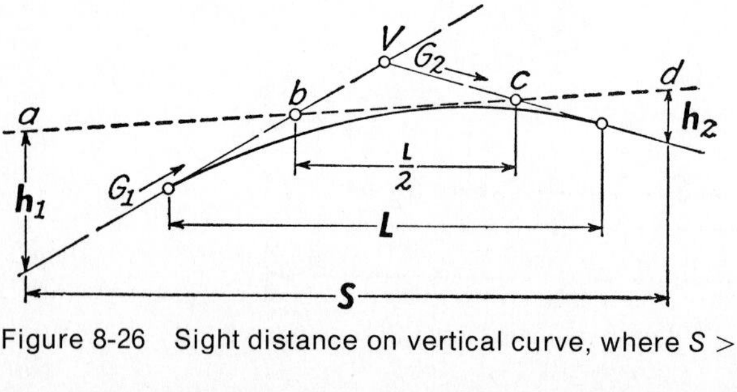

Figure 8-26 Sight distance on vertical curve, where $S > L$.

the sight line and the gradient G_1. Then $A - g$ will be the difference between the gradient of the sight line and gradient G_2. Use is also made of the following property of the parabola: If a tangent to the parabola is drawn between the main tangents, the horizontal projection of the intercept cut off on this new tangent by the main tangents is equal to one-half the horizontal projection of the long chord of the parabola; that is, the horizontal projection of bc is equal to $\frac{1}{2}L$.

By definition, S equals the sum of the horizontal projections of distances ab, bc, and cd. Consequently, the general expression for sight distance where $S > L$ is

$$S = \frac{100\,h_1}{g} + \frac{L}{2} + \frac{100\,h_2}{A - g} \qquad \text{(A)}$$

For the sight distance to be a minimum, $dS/dg = 0$, or

$$\frac{dS}{dg} = \frac{-100\,h_1}{g^2} + \frac{100\,h_2}{(A - g)^2} = 0$$

Solving for g gives

$$g = \frac{A\sqrt{h_1 h_2} - h_1\,A}{h_2 - h_1} \qquad \text{(B)}$$

The result of substituting the value of g from equation B in equation A and solving for L is the following general relation:

When $S > L$,

$$L = 2\,S - \frac{200(\sqrt{h_1} + \sqrt{h_2})^2}{A} \qquad \text{(8-40)}$$

For the adopted values of h_1 and h_2 the practical relations are:

When $S > L$,

(nonpassing conditions)

$$L = 2\,S - \frac{1398}{A} \qquad \text{(8-41)}$$

(passing conditions)

$$L = 2\,S - \frac{3295}{A} \qquad \text{(8-42)}$$

Figure 8-27 shows minimum lengths of crest vertical curves needed to provide stopping sight distances at various design speeds. The value of K, or the length of curve required to effect a 1% change in A (section 4-2), is a simple expression of the design control. For convenience, theoretical values of K required to fit the stopping distances in Table 8-1 have been rounded to integral values; lengths of vertical curves computed from the rounded values of K are plotted as heavy solid lines in Fig. 8-27.

Where $S > L$ the theoretical minimum lengths become zero for small values of A because the sight line passes over the crest of the vertical curve. However, good practice in design calls for inserting a vertical curve at all changes in vertical alignment. As an approximation of current practice, the minimum length of vertical curve is taken as about three times the design speed. This approximation is represented by the heavy vertical lines at the lower left of Fig. 8-27.

The heights of eye and object used in developing Fig. 8-27 are different from those appearing in the 1956 Blue Book. These changes are the result of later research as to the effect of lower vehicles on crest sight distances.

Drainage requirements may affect the maximum lengths of vertical curves. If pavements are curbed, experience indicates the desirability of attaining a minimum longitudinal grade of 0.35% at a point about 50 ft from the crest. This corresponds to a K-value of 143 ft for a change of 1% in A; the resulting line is plotted in Fig. 8-27 as the *drainage maximum.* Special attention to drainage should be given for combinations below and to the right of this line.

Some attempts have been made to introduce headlight sight distance as a design control for stopping sight distance at crest vertical curves, the height

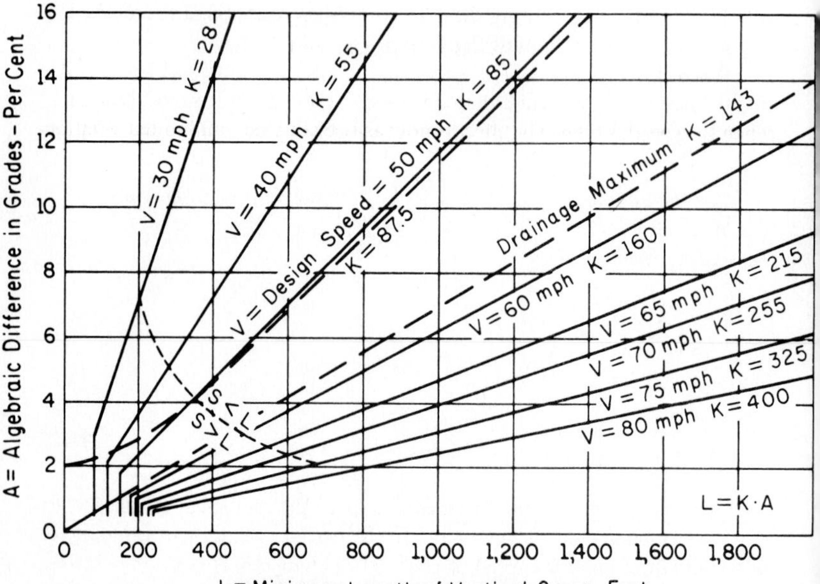

Figure 8-27 Design controls for crest vertical curves.

of headlight h_1 being taken as 2.0 ft and height of object h_2 as 4 in. or more. Obviously these conditions demand much longer vertical curves, but this requirement is considered to be unnecessary in view of the lower running speeds used in night driving.

8-24. Sight Distances at Sag Vertical Curves

There is no generally accepted basis for establishing the lengths of sag vertical curves. Four criteria have been used: (1) headlight-beam distance, (2) rider comfort, (3) drainage control, and (4) general appearance.

Headlight-beam distance, as used by the Pennsylvania Turnpike Commission, is represented by Figs. 8-28 and 8-29. With the aid of these figures it is not difficult to derive the practical relations, which are:

When $S < L$,

$$L = \frac{A\,S^2}{400 + 3.5\,S} \tag{8-43}$$

When $S > L$,

$$L = 2\,S - \frac{400 + 3.5\,S}{A} \tag{8-44}$$

Figure 8-30 shows lengths of sag vertical curves conforming to the preceding formulas, in which values of S are the stopping distances in Table 8-1. The K-values are rounded as in Fig. 8-27.

When a vehicle is traversing a sag vertical curve, centrifugal and gravitational force act in the same direction. As a result there is some discomfort when the speed is high. The effect is not easily evaluated, but limited attempts at its measurement have led to the general conclusion that riding on sag

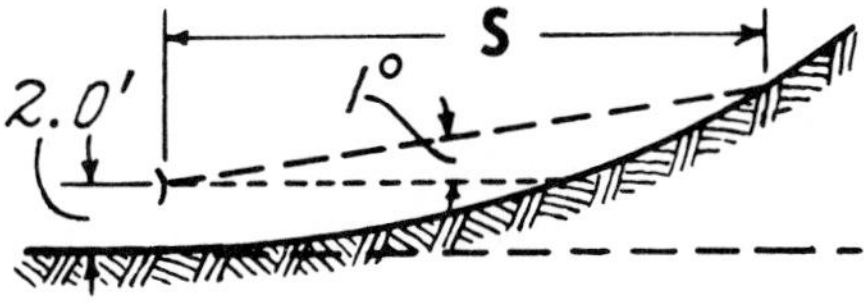

Figure 8-28 Headlight beam distance, where $S < L$.

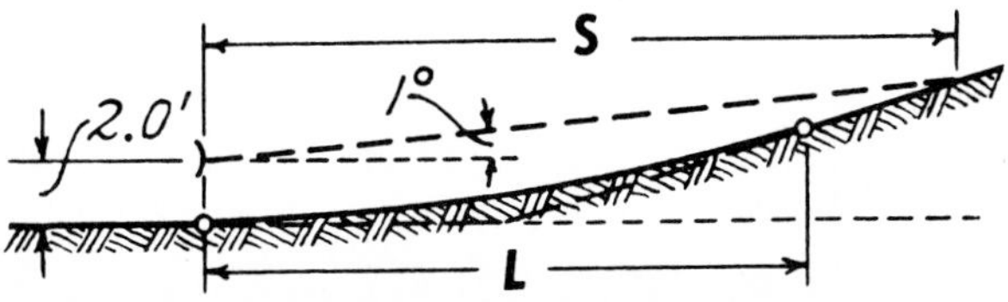

Figure 8-29 Headlight beam distance, where $S > L$.

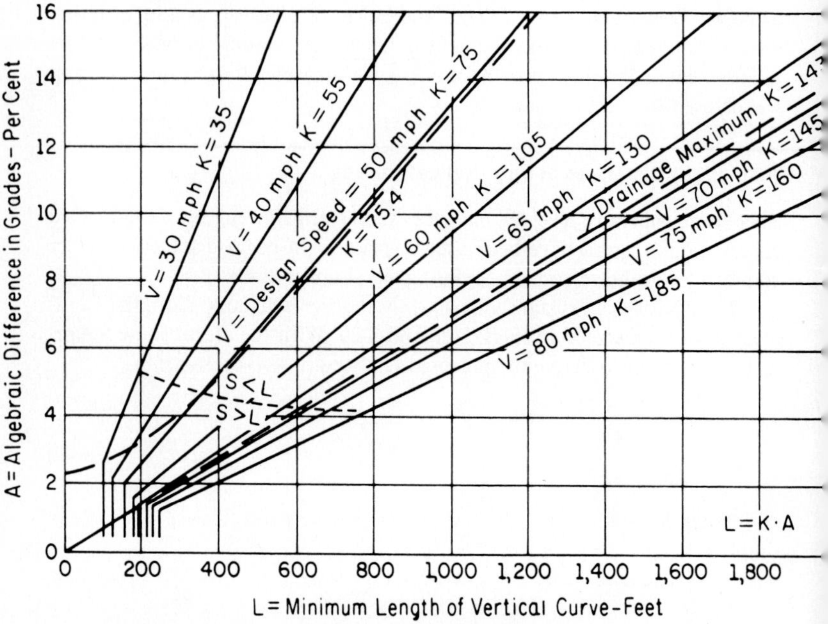

Figure 8-30 Design controls for sag vertical curves.

vertical curves is comfortable when the centripetal acceleration does not exceed 1 ft per s^2. A mathematical approximation of this criterion can be derived with the aid of Fig. 8-31.

Because of the small gradients used in practice, the curve *AB* may be taken as a circular arc and the distance *VB* as approximately $\frac{1}{2}L$. In the

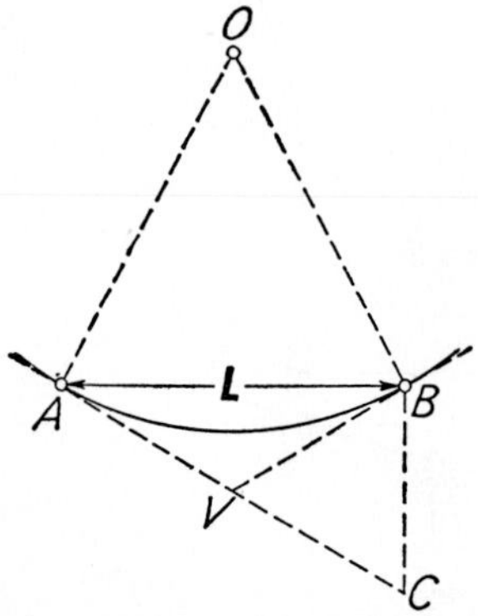

Figure 8-31

similar triangles AOB and CVB,

$$AB:OA = BC:VB \quad \text{or} \quad L:R = \frac{A\,L}{200}:\frac{L}{2} \text{ (approximately)}$$

Thus, $L = A\,R/100$. But $v^2/R = 1$ ft per s^2. Substituting for R and converting v to V in miles per hour gives

$$L = \frac{A\,V^2}{46.5} \tag{8-45}$$

Lengths of sag vertical curves satisfying this comfort factor are about 75% of those required by the conditions for headlight beam distance.

The criterion for drainage is the same as on a crest vertical curve; that is, K should not exceed 143 ft for a change of 1% in A. This criterion is plotted in Fig. 8-30 as the *drainage maximum.* The maximum length for drainage on a sag curve exceeds the minimum length required by headlight beam distance up to design speeds of about 70 mph—a fortunate circumstance.

Several rule-of-thumb relationships have been used to ensure that a sag curve will have satisfactory appearance. One of the simplest requires that L be at least 100 A. In Fig. 8-30 this corresponds to a design speed of almost 60 mph, based on headlight-beam distance.

Of the four criteria for the minimum length of a sag vertical curve, that based on headlight-beam distance appears to be the most logical. This is the conclusion in the "Blue Book", in which the K-values have been rounded off for convenience in design, as represented by the heavy solid lines in Fig. 8-30.

8-25. Sight Distances at Interchanges

Sight distances at interchanges should be at least as long as the stopping distances listed in Table 8-1. When a sag vertical curve occurs at an underpass, the overhead structure may shorten the sight distance otherwise obtainable. However, if the length of the sag curve conforms with that in Fig. 8-30, the structure does not shorten the sight distance below the minimum required for stopping. This is true even though the sight distance is measured from a height of eye of 6 ft (truck driver) to a height of object of 1.5 ft (tail light) and the vertical clearance is the recommended minimum of 16 ft (adopted 1969). The practical relations corresponding to these conditions are:

When $S < L$,

$$L = \frac{A\,S^2}{9800} \tag{8-46}$$

When $S > L$,

$$L = 2\,S - \frac{9800}{A} \tag{8-47}$$

In the case of dual highways it may be desirable to check the distance available for *passing* at an undercrossing without ramps. This can be done either by using the preceding formulas or by scaling from the profile.

Limited sight distance is more likely to occur when an interchange is located at a horizontal curve. Ordinarily, the lateral clearance to bridge rails at an overpass or to abutments at an underpass is not enough to permit use of maximum curvature for the design speed. Use the formulas for S_H in section 8-22 and see problem 8-14.

DESIGN PRINCIPLES

Exact adherence to the specific controls outlined in foregoing articles will not guarantee attainment of the best location. Experience, judgment, and the observance of recognized principles of good design are also necessary. When geometric controls are applied, certain general principles should also be observed. The significance and importance of many of the following principles are brought out vividly by photographs.

8-26. General Controls for Horizontal Alignment

Important principles relating to horizontal alignment are as follows:

1. Alignment between location control points (see section 10-2) should be as directional as possible. Long, flowing curves fitted to the topography (Fig. 8-32) are better than long tangents that slash through the terrain in an artificial manner.
2. Closely spaced short curves are poor design. Such unsightly kinks as broken-back and reverse curves may be converted into more pleasing alignment in several ways. (See sections 7-14 and 7-15.)
3. Small changes in direction should not be accomplished by means of the sharpest curve permitted by the design speed. The maximum degree of curve is permissible only with large central angles and at critical locations in general.
4. Curves, unless very flat, are to be avoided on long, high embankments.
5. Consistent alignment design is the ideal sought (see section 1-5). In case the topography requires a reduction in design speed, the change—reflected in reduced sight distances increased curvature, and shorter distances between curves—are best made gradually over a distance of several miles. Moreover, conspicuous warning signs are advisable to show that such a change is in progress.
6. Horizontal and vertical alignment are best studied together. Models are particularly useful in designing intricate interchanges.

8-27. General Controls for Vertical Alignment

When considering vertical alignment, some important principles are the following:

Figure 8-32 Merritt Parkway in Connecticut. (Photo by Josef Scaylea, A. Devaney, Inc., N.Y.)

1. A smooth-flowing profile with long vertical curves is preferable to a profile with numerous breaks and short grades.
2. Care should be taken to avoid sag vertical curves on comparatively straight horizontal alignment. These produce "hidden dips" and are serious hazards, especially during passing maneuvers.

3. In general, long steep grades may be broken to the advantage of traffic by placing the steepest grade at the bottom of the ascent.
4. Steep grades should be reduced through important intersections at grade, in order to minimize hazards to turning traffic.
5. Unnatural and unsightly design should be avoided. Among these defects are crest vertical curves on embankments and sag vertical curves in cuts; broken-back vertical curves; and the presence of numerous minor undulations in grade line so located that they are visible to the driver.
6. On important highways carrying a large percentage of commercial traffic, economic studies of motor-truck performance relative to grades should accompany the office work of grade-line design.

8-28. Combination of Horizontal and Vertical Alignment

Where there is a combination of horizontal and vertical curvature the following principles should be kept in mind.

1. It is not essential to separate horizontal and vertical curves; in general, the alignment is more natural and more pleasing in appearance if they are combined—subject to the limitations which follow in 2 and 3.
2. A change in horizontal alignment should preferably be made at a sag vertical curve where the change in direction is readily apparent to the driver. However, the horizontal curve should be flat, in order to avoid the distorted appearance caused by foreshortening.
3. If horizontal curvature at a crest vertical curve cannot be avoided, the change in direction should precede the change in profile.
4. Relatively small savings in cost of right-of-way or of grading should not be an excuse for the insertion of short sections of substandard design. On most locations in rural areas the cost of these relatively permanent elements of the highway is less than that of the pavement and other shorter-lived appurtenances. Therefore, it is unwise to reduce the built-in safety of some sections of a highway, and invite almost certain early obsolescence, for the sole purpose of reducing the cost by a small percentage.
5. As noted in section 1-5, "the aim of good location should be the attainment of consistent conditions with a proper balance between curvature and grade." Balanced design, everywhere consistent with the chosen design speed, is the ideal to be constantly sought. Straight alignment obtained at the expense of long, steep grades, or excessive curvature inserted to follow the grade contour closely, are both poor designs. The best design is a compromise in which safety, economics, and aesthetics are sensibly blended.

PROBLEMS

8-1. Compute the side friction factor f developed on the following curves, each of which has favorable crown. Is f within the safe maximum recommended by AASHTO?

(a) $R = 5000$ ft; $e = 0.10$; $V = 55$ mph. *Answer:* 0.06; yes.
(b) $D_a = 5°00'$; $e = 0.06$; $V = 60$ mph. *Answer:* 0.15; no.
(c) $R = 1200$ ft; $e = 0.08$; $V = 60$ mph. *Answer:* 0.12; yes.
(d) $D_a = 1°40'$; $e = 0.08$; $V = 60$ mph. *Answer:* 0.01; yes.

8-2. If the actual coefficient of sliding friction on icy pavement is 0.05, theoretically within what range of speeds could a vehicle be operated without sliding inward or outward on the curves in problem 8-1? *Answers:* (a) 61 to 106 mph; (b) 13 to $43\frac{1}{2}$ mph; (c) 23 to $48\frac{1}{2}$ mph; (d) 39 to $81\frac{1}{2}$ mph.

8-3. Compute design speeds at the following curves on the Pennsylvania Turnpike consistent with values of f recommended by AASHTO (Fig. 8-7 and Table 8-4). See Table 8-10 for the superelevation rates. (a) $D_a = 2°00'$; (b) $D_a = 3°00'$; (c) $D_a = 4°00'$; (d) $D_a = 5°00'$. *Partial answers:* (a) 78 mph; (b) 72 mph.

8-4. A typical section of the Massachusetts Turnpike is a six-lane dual highway. On tangents, each inner lane has a 2% cross slope toward the median; outer lanes, a 2% slope toward the shoulders. If this section were continued around a 3500-ft radius curve without rotating the pavement, for what maximum speeds would the lanes be safe? Take f from Table 8-4. *Answers:* 82 mph with favorable crown; 72 mph with adverse.

8-5. Compute the length of true spiral on the inside edge of the following spiraled curves assuming that the inside lane is widened as in Fig. 8-17: (a) $D_a = 5°00'$; $\frac{1}{2}b = 10$ ft; $w = 2$ ft; $L_s = 300$ ft. (b) $R = 500$ ft; $\frac{1}{2}b =$ 10 ft; $w = 2$ ft; $L_s = 200$ ft. *Answers:* (a) 379 ft; (b) 250 ft.

8-6. Compute the length of true spiral on the inside edge of the following unspiraled curves assuming that the inside lane is widened as in Fig. 8-18. Compare answers with those in problem 8-5.

(a) Data same as problem 8-5(a). *Answer:* 233.3 ft.
(b) Data same as problem 8-5(b). *Answer:* 153.1 ft.

8-7. Compute the length of true spirals on the inside and outside edges of the curves in problem 8-6 assuming that the inside lane is widened as in Fig. 8-19. Use $\frac{1}{2}w$ as the throw of the outside spiral. *Answers:* (a) inside 286 ft, outside 166 ft; (b) inside 187 ft, outside 111 ft.

8-8. Compute the edge lengths of the curves in problem 8-5 between points radially opposite the T.S. and S.C. assuming that widening is omitted. *Answers:* (a) $L_i = 298.69$ ft; $L_o = 301.31$ ft. (b) $L_i = 198$ ft; $L_o = 202$ ft.

8-9. Compute the inside-edge lengths of the curves in problem 8-5 between points radially opposite the T.S. and S.C. assuming that the widening is applied as in Fig. 8-15. *Answers:* (a) 298.5 ft; (b) 197.7 ft.

8-10. Compute the edge lengths of the curves in problem 8-5 between points radially opposite the T.S. and S.C. assuming that the widening is divided equally between the lanes as in Fig. 8-16. *Answers:* (a) $L_i =$ 298.60 ft; $L_o = 301.40$ ft. (b) $L_i = 197.9$ ft; $L_o = 202.18$ ft.

8-11. Compute the chords needed to locate offset curves parallel to the spirals in problem 8-5. (As a check, the sum of the chords should equal the value found by formula 8-24 or 8-25 in which p is substituted for $\frac{1}{2}b$.) Spiral: (a) $n = 6$, $p = 60$ ft outside. (b) $n = 4$, $p = 60$ ft inside.

8-12. Compute the total pavement area in square yards between T.S. and S.T. of the following unwidened spiraled curves:

(a) $I = 45°32'$; $D_a = 5°00'$; $\frac{1}{2}b = 10$ ft; $L_s = 300$ ft. *Answer:* 2690.4 yd^2.

(b) $I = 51°17'$; $R = 500$ ft; $\frac{1}{2}b = 10$ ft; $L_s = 200$ ft. *Answer:* 1439.0 yd^2.

8-13. Compute the total extra pavement area in square yards caused by widening the curves in problem 8-12 by the method of Fig. 8-15. Use formula 8-31; check by formula 8-32. Curve (a): $w = 2$ ft. *Answers:* 200.4 yd^2 approx.; 200.7 yd^2 exact. Curve (b) $w = 2$ ft.

8-14. The design for a two-lane highway provides for 11-ft lanes and 8-ft shoulders. Overhead bridge abutments are located 2 ft beyond the shoulders. At an abutment along a 5°00′ curve,

(a) What is S_H if $I = 30°00'$? *Answer:* 376 ft.

(b) What is the recommended maximum design speed in (a)? *Answer:* 52 mph.

(c) Would the foregoing speed be limited by the bridge abutment or side friction, considering e to be 0.08? *Answer:* the abutment.

8-15. Determine the recommended minimum lengths of crest and sag vertical curves where

(a) $A = 5.00\%$; $V = 50$ mph. *Answers:* Crest 438 ft; sag 377 ft.

(b) $A = 4.75\%$; $V = 70$ mph. *Answers:* Crest 1223 ft; sag 684 ft.

8-16. At which of the vertical curves in problem 8-15 would difficulty with surface drainage probably occur, assuming the pavements to have edge curbs? *Answer:* (b).

8-17. Compute the stopping sight distances at the following vertical curves:

(a) Crest curve: $L = 400$ ft; $A = 3.00\%$. *Answer:* 433 ft.

(b) Crest curve: $L = 600$ ft; $A = 3.00\%$. *Answer:* 529 ft.

(c) Sag curve: $L = 600$ ft; $A = 5.00\%$. *Answer:* 513 ft.

(d) Sag curve: $L = 400$ ft; $A = 4.00\%$. *Answer:* 444 ft.

8-18. Test the statement in the third sentence of section 8-25 by computing sight distances for the sag curves in problem 8-17(c) and (d) by the proper formula in section 8-25.

Chapter 9
Some Special Railroad Curve Problems

9-1. Foreword

The purpose of this chapter is to present a few examples of curve theory and surveying procedures applied to the field of railway surveying. More detailed descriptions of some subjects, particularly track layout and maintenance (turnouts, connecting tracks, string lining, and so forth), will be found in railway surveying handbooks, notably *Railroad Curves and Earthwork*, by Allen, or *Field Engineering*, by Searles, Ives, and Kissam.

Figure 9-1 shows an example of railroad location in the Rocky Mountains near Blacktail, Montana. The pictured alignment includes a simple 8°00′ curve with a total central angle of 152°40′ and a total length, including two 180-ft spirals, of 2088 ft. On the adjoining tangents the grade is 1.80%. The grade is decreased 0.04% per degree to compensate for friction between wheels and curved track, making the actual grade 1.48%. Shown is a 106-car freight train powered by a 5400-hp diesel locomotive at the head and a 4050-hp helper at the rear.

Figure 9-1 (Courtesy Great Northern Railway.)

9-2. Superelevation

Fundamentally, superelevation theory is the same on railways as on highways. Figure 8-2 and the equations developed in section 8-6 are valid for both types of operation.

The equilibrium formula for superelevation of railroad track* is

$$E = 0.0007\ V^2 D \tag{9-1}$$

in which E is the superelevation, in inches, of the outer rail. This relation corresponds to formula 8-8, and is found by substituting $E \div 59.5$ for e. ("Standard gauge" of track is 4 ft $8\frac{1}{2}$ in., but E is measured with respect to center to center of rails, that is, 4 ft $11\frac{1}{2}$ in.)

According to the A.R.E.A. Manual:

> If it were possible to operate all classes of traffic at the same speed on a curve, the ideal condition for smooth riding and minimum rail wear would be obtained by elevating for equilibrium. However, curved track must handle several classes of traffic operating at various speeds, which results in slow trains causing excessive wear on the inside rail and high-speed trains causing accelerated wear on the outside rail.
>
> Safety and comfort limit the speed with which a passenger train may negotiate a curve. Any speed which gives comfortable riding on a curve is well within the limits of safety. Experience has shown that the conventional baggage cars, passenger coaches, diners, and Pullman cars will ride comfortably around a curve at a speed which will require an elevation about 3 inches higher for equilibrium. Equipment designed with large center bearings, roll stabilizers, and outboard swing hangers can negotiate curves comfortably at greater than 3 inches unbalanced elevation because there is less car body roll. It is suggested that where complete passenger trains are equipped with cars utilizing the foregoing refinements that a lean test be made on the equipment to determine the amount of body roll. If the roll angle is less than 1°30′, experiments indicate that cars can negotiate curves comfortably at $4\frac{1}{2}$ inches unbalanced elevation.
>
> The inner rail should preferably maintained at grade.

Using these recommendations, the A.R.E.A. formula for superelevation based on maximum speed becomes

$$E = 0.0007(\text{max. } V)^2\ D - 3 \tag{9-2}$$

Formula 9-2 is analogous to equation 8-11.

Table 9-1 gives the equilibrium elevation E for various values of D_c. The 5-mph increments represent general practice for use on speed-limit signs.

9-3. Spirals

In contrast to the situation with regard to the use of spirals on highways (section 8-13), spirals have been used on railroad track since about 1880. (A

* 1956 *Manual*, American Railway Engineering Association.

Table 9-1. *E* = EQUILIBRIUM ELEVATION FOR VARIOUS SPEEDS ON CURVES.

DEGREE OF CURVE	V = SPEED (mph)															
D_c	10	20	30	35	40	45	50	55	60	65	70	75	80	85	90	95	100
0°30′	0.04	0.14	0.32	0.43	0.56	0.71	0.88	1.06	1.26	1.48	1.72	1.97	2.24	2.53	2.84	3.16	3.50
1°00′	0.07	0.28	0.63	0.87	1.12	1.42	1.75	2.12	2.52	2.96	3.43	3.94	4.48	5.06	5.67	6.32	7.00
1°30′	0.11	0.42	0.95	1.29	1.68	2.13	2.63	3.18	3.78	4.44	5.15	5.91	6.72	7.59	8.51	9.48	10.50
2°00′	0.14	0.56	1.26	1.72	2.24	2.84	3.50	4.24	5.04	5.92	6.86	7.88	8.96	10.12	11.34	12.46	
2°30′	0.18	0.70	1.58	2.14	2.80	3.54	4.38	5.29	6.30	7.39	8.58	9.84	11.20				
3°00′	0.21	0.84	1.89	2.57	3.36	4.25	5.25	6.35	7.56	8.87	10.29	11.81					
3°30′	0.25	0.98	2.21	3.00	3.92	4.96	6.13	7.41	8.82	10.35							
4°00′	0.28	1.12	2.52	3.43	4.48	5.67	7.00	8.47	10.08								
5°00′	0.35	1.40	3.15	4.29	5.60	7.09	8.75	10.59									
6°00′	0.42	1.68	3.78	5.15	6.72	8.51	10.50										
7°00′	0.49	1.96	4.41	6.00	7.84	9.92											
8°00′	0.56	2.24	5.04	6.86	8.96	11.34											
9°00′	0.63	2.52	5.67	7.72	10.08												
10°00′	0.70	2.80	6.30	8.58	11.20												
11°00′	0.77	3.08	6.93	9.43													
12°00′	0.84	3.36	7.56	10.29													

E, in inches $= 0.0007V^2D$

concise history of the use of spirals is given in *Proceedings, A.R.E.A.*, Vol. 40, 1939, pp. 172–174.)

As a result of long years of experience in operating over spiraled superelevated curves, American railroads almost invariably base minimum spiral length upon the rate of rotational change. For many years the A.R.E.A. had recommended that spiral length be based on attaining superelevation across standard-gauge track at a desirable maximum rate of 1.25 in. per s. Such a rate yields the expression $L_s = 1.17\ EV$. This equation is the same as 8-19 except for different notation.

As the result of a committee report* on "Length of Railway Transition Spiral—Analysis and Running Tests," the recommendation for desirable spiral length has been changed to read

$$L_s = 1.63\ E_u V \tag{9-3}$$

in which E_u is the unbalanced elevation in inches.

Where speed is designed for passenger equipment which has car-body roll controlled by special designs, the recommended desirable length is found from

$$L_s = 62\ E \tag{9-3a}$$

As for highways, the superelevation is run out uniformly over the spiral. The slight vertical curves in the outer rail at the beginning and end of the spiral are taken care of automatically by flexibility of the rail.

Track spirals are staked either by deflection angles or offsets. The basic theory is fully covered in Chapter 5. Section 5-13 contains practical suggestions for applying the offset methods.

9-4. String Lining

In spite of ballast and rail braces, tracks on curves tend to creep slowly out of line. This creeping is due principally to the unbalanced lateral forces caused by operation at other than equilibrium speed. Other contributing factors are rapid deceleration during emergency stops and, possibly, temperature expansion and contraction. Track once irregularly out of line becomes progressively worse, owing to the variable impact produced by moving trains.

The trend toward higher train speeds in both freight and passenger operation makes it more important than ever to maintain curved track continuously in good alignment. This can be done either by the deflection-angle method or by *string lining*. The latter method has so many obvious advantages that it has rapidly superseded the former method.

Briefly, string lining consists in shifting the track in or out along the circular curve until equal middle ordinates are obtained at equal chords.

* *Proceedings, A.R.E.A.*, Vol. 65, 1964, pp. 91–129.

Theoretically, the chord used may be of any length; but to obtain good control on main-line track it should be between 50 and 80 ft. Many engineers use a 62-ft chord. This is the value recommended by the A.R.E.A.; it is a convenient length, and also produces the useful relation that the degree of curve is numerically equal to the middle ordinate in inches (see equation 2-31).

Equipment consists simply of a tape, a strong fish line or a fine wire, and a scale for measuring the middle ordinates. Some engineers use wooden or metal templates which are held against the rail head; the string passes through holes or slots a fixed distance from the rail. With such devices, it is necessary to deduct the fixed distance from the measured middle ordinates, or to use a special scale with an offset zero point.

The procedure involves (1) preliminary field work, (2) calculation, and (3) track shifting in accordance with the approved calculations.

Preliminary field work

1. Locate the T.C. by sighting along the gage side of the outer rail. Make a keel mark on the inside of the rail head at this point and mark it sta. 0 on the web.
2. Mark sta. -1 similarly, 31 ft back from sta. 0 on the tangent. Then mark stations, 1, 2, 3, etc. at 31-ft intervals along the outer rail until the last station is beyond the end of the curve.
3. Stretch the line taut between the keel marks at stations -1 and 1. Measure and record the middle ordinate at sta. 0. In similar manner, stretch the cord between stations 0 and 2 and measure the middle ordinate at sta. 1. Continue this process until the middle ordinates become zero.

Calculation is based upon four simple rules. The first comes from the fact that the middle ordinate m is proportional to D. Since $D = 100\ I/L$, it follows that $\Sigma m \propto I$. This relation may be expressed by the following rule:

Rule 1. For any chord length the sum of the middle ordinates on a curve between given tangents is constant.

The other rules come from Fig. 9-2, in which the solid line represents the outer rail of curved track badly out of line, and the dotted line shows the correct position. (The scale is greatly exaggerated in order to make the relations clear.)

The offset distance between the original and final positions of the track at any station is called the *throw*. Track moved outward in revising its position is given a positive throw; for track moved inward, the throw is negative. Thus, the throw at sta. 1 is negative and is numerically equal to the distance *ab*.

The *error* at any station is found from the following relation: *error = original middle ordinate minus revised middle ordinate* (algebraically). In the sketch the error at sta. 1 is $ac - bd$.

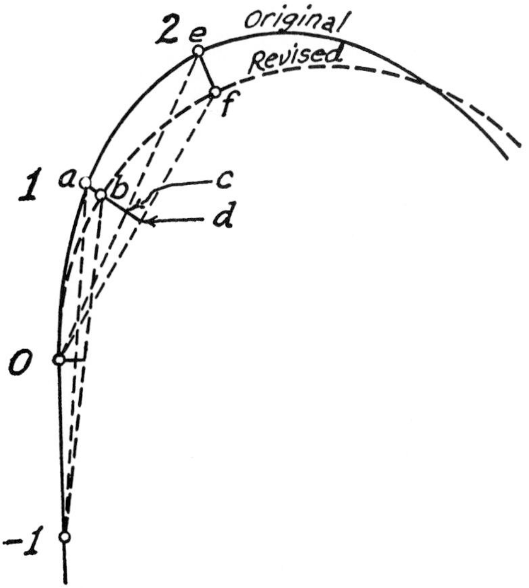

Figure 9-2

It is obvious that the first throw will occur at sta. 1. Since the middle ordinates at successive stations are practically parallel, the throw at sta. 1 is twice the error at sta. 0; both have negative signs. This relation may be expressed by the following rule:

Rule 2. At the first station at which a throw occurs, the half-throw ($\frac{1}{2}t$) equals the error at the preceding station.

The throw at sta. 2 is *ef* (negative sign); and for all practical purposes *ef* = twice *cd*. The length *cd* may be written in the following form:

$$\underset{(1)}{-cd} = \underset{(2)}{(-\tfrac{1}{2}ab)} + \underset{(3)}{(-\tfrac{1}{2}ab)} + \underbrace{[(ab + bc) - bd]}_{(4)}$$

Term (1) is the half-throw at sta. 2.

Term (2) is the half-throw at sta. 1.

Term (3) is the error at sta. 0.

Term (4) is the error at sta. 1.

From the foregoing relation, the following rule may be stated:

Rule 3. The half-throw at any station equals the half-throw at the preceding station plus the algebraic sum of the errors up to and including the preceding station.

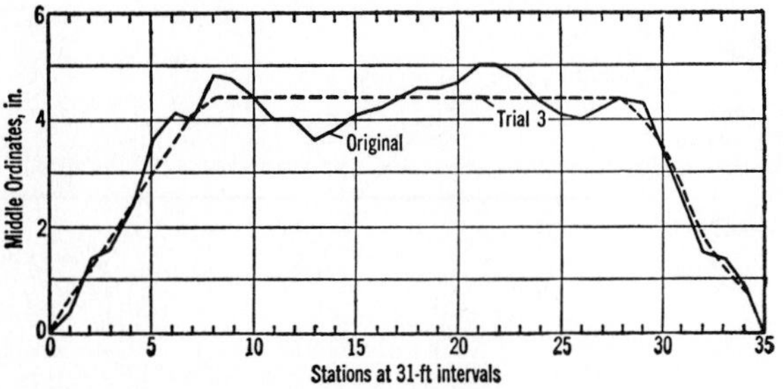

Figure 9-3

In solving a string-lining problem, it is helpful to plot a graph of the measured middle ordinates. Figure 9-3 shows such a graph for a spiraled curve very badly out of line. The curve is too flat near sta. 13 and too sharp near sta. 22. The original middle ordinates were measured to tenths of inches at stations 31 feet apart; their values (with the decimal point omitted for simplicity) are given in Table 9-2, column (2).

For perfect alignment the middle ordinates on the circular curve must be constant, and those on the spirals must change uniformly. Moreover, the half-throw at the final station must be zero; otherwise, the forward tangent will be shifted parallel to itself by an amount equal to the full throw.

Trial 1 is shown in Table 9-2; it approximates trial 3, shown in Fig. 9-3. The middle ordinate for a 4.4° curve between stations 7 and 28 was assumed to be the average of the existing middle ordinates on the circular arc.

Column (3) contains the revised middle ordinates. The middle ordinates on the spirals were adjusted so that the sum of the revised middle ordinates is equal to the original sum. Rule 1 requires this relation.

The errors and their algebraic sums were next calculated, and the arithmetic automatically checked by the zero value at the foot of column (5).

Then the half-throws were calculated from Rules 2 and 3 and entered in column (6). Arrows indicate the additions.

Up to about sta. 17, it appeared that a fairly close solution might be found on the first trial. After that, however, the half-throws became excessive. Instead of starting over when this happens, it is a good plan to continue col. (6) to completion; otherwise, a large number of trials might have to be made before finding a solution giving zero half-throw at the final station.

In the illustrative example the result of the first trial was modified by a method which guarantees a check on the second trial. The method is based upon the following rule:

> *Rule 4.* The effect upon the half-throw at any station caused by a change in middle ordinate at any preceding station equals the product of the change in middle ordinate and the difference in stationing; the sign of the product is opposite to the sign of the change in the middle ordinate.

Table 9-2. STRING-LINING CALCULATIONS

		TRIAL 1				TRIAL 2			
STATION (1)	ORIGINAL m (2)	REVISED m (3)	ERROR (4)	SUM OF ERRORS (5)	HALF-THROW (6)	REVISED m (7)	ERROR (8)	SUM OF ERRORS (9)	HALF-THROW (10)
−1	0	0	0	0	0	0	0	0	0
0	0	0	0	0	0	0	0	0	0
1	4	6	−2	−2	0	5	−1	−1	0
2	14	12	+2	0	−2	11	+3	+2	−1
3	16	19	−3	−3	−2	18	−2	0	+1
4	23	25	−2	−5	−5	24	−1	−1	+1
5	36	31	+5	0	−10	30	+6	+5	0
6	41	38	+3	+3	−10	37	+4	+9	+5
7	40	44	−4	−1	−7	44	−4	+5	+14
8	48	44	+4	+3	−8	44	+4	+9	+19
9	47	44	+3	+6	−5	44	+3	+12	+28
10	44	44	0	+6	+1	44	0	+12	+40
11	40	44	−4	+2	+7	44	−4	+8	+52
12	40	44	−4	−2	+9	44	−4	+4	+60
13	36	44	−8	−10	+7	44	−8	−4	+64
14	38	44	−6	−16	−3	44	−6	−10	+60
15	41	44	−3	−19	−19	44	−3	−13	+50
16	42	44	−2	−21	−38	44	−2	−15	+37
17	44	44	0	−21	−59	44	0	−15	+22
18	46	44	+2	−19	−80	44	+2	−13	+7
19	46	44	+2	−17	−99	44	+2	−11	−6
20	47	44	+3	−14	−116	44	+3	−8	−17
21	50	44	+6	−8	−130	44	+6	−2	−25
22	50	44	+6	−2	−138	44	+6	+4	−27
23	48	44	+4	+2	−140	44	+4	+8	−23
24	44	44	0	+2	−138	44	0	+8	−15
25	41	44	−3	−1	−136	44	−3	+5	−7
26	40	44	−4	−5	−137	44	−4	+1	−2
27	42	44	−2	−7	−142	44	−2	−1	−1
28	44	44	0	−7	−149	44	0	−1	−2
29	43	38	+5	−2	−156	39	+4	+3	−3
30	34	32	+2	0	−158	33	+1	+4	0
31	24	25	−1	−1	−158	26	−2	+2	+4
32	15	19	−4	−5	−159	21	−6	−4	+6
33	14	12	+2	−3	−164	12	+2	−2	+2
34	9	6	+3	0	−167	7	+2	0	0
35	0	0	0	0	−167	0	0	0	0
Sum	1231	1231				1231			

Table 9-2. *(Continued)*

	TRIAL 3			
STATION (11)	REVISED m (12)	ERROR (13)	SUM OF ERRORS (14)	HALF-THROW (15)
−1	0	0	0	0
0	0	0	0	0
1	7	−3	−3	0
2	13	+1	−2	−3
3	18	−2	−4	−5
4	24	−1	−5	−9
5	30	+6	+1	−14
6	35	+6	+7	−13
7	41	−1	+6	−6
8	44	+4	+10	0
9	44	+3	+13	+10
10	44	0	+13	+23
11	44	−4	+9	+36
12	44	−4	+5	+45
13	44	−8	−3	+50
14	44	−6	−9	+47
15	44	−3	−12	+38
16	44	−2	−14	+26
17	44	0	−14	+12
18	44	+2	−12	−2
19	44	+2	−10	−14
20	44	+3	−7	−24
21	44	+6	−1	−31
22	44	+6	+5	−32
23	44	+4	+9	−27
24	44	0	+9	−18
25	44	−3	+6	−9
26	44	−4	+2	−3
27	44	−2	0	−1
28	44	0	0	−1
29	40	+3	+3	−1
30	35	−1	+2	+2
31	26	−2	0	+4
32	18	−3	−3	+4
33	12	+2	−1	+1
34	8	+1	0	0
35	0	0	0	0
Sum	1231			

Revision—Trial 1 to 2

STA.	CHANGE IN m	CHANGE IN $\frac{1}{2}t$ AT STA. 35
1	−1	+34
2	−1	+33
3	−1	+32
4	−1	+31
5	−1	+30
6	−1	+29
Sum	−6	+189
29	+1	−6
30	+1	−5
31	+1	−4
32	+2	−6
33	0	0
34	+1	−1
Sum	+6	−22
Net	0	+167

Revision—Trial 2 to 3

CHANGE IN $\frac{1}{2}t$ STA.	CHANGE IN m	CHANGE IN $\frac{1}{2}t$ AT STA. 35	13	22
1	+2	−68	−24	−42
2	+2	−66	−22	−40
6	−2	+58	+14	+32
7	−3	+84	+18	+45
Sum	−1	+8	−14	−5
29	+1	−6		
30	+2	−10		
32	−3	+9		
34	+1	−1		
Sum	+1	−8		
Net	0	0		

The modification of trial 1 is shown at the top of page right of trial 3. It should be observed that the small changes were made entirely on the spirals. The middle ordinates were adjusted in a way to make their net change zero (Rule 1) and at the same time produce the required change of +167 units in the half-throw at sta. 35. The resulting half-throws are given in Column (10).

Trial 2 might be considered an acceptable solution, provided there are no objects which might interfere with the fairly large throws between stations 10 and 15.

Any number of solutions will give zero throw at the end of the curve. The best solution is the one having the smallest intermediate throws, consistent with specified clearances and smooth curvature.

Trial 3 shows a second solution to the illustrative example. In revising trial 2, an attempt was made to decrease the throws near sta. 13 without increasing those near sta. 22 too much. The tabulation at the bottom of the page to the right of trial 3 shows how this was done and the zero throw at the end checked before detailed calculations for trial 3 were performed. The resulting half-throws are given in column (15). By using trial 3 instead of trial 2, the full throw is reduced from a maximum of 12.8 in. to 10.0 in.

It is possible to obtain further improvement in this example by continuing the foregoing process, especially if the middle ordinates are expressed to the nearest 0.05 in. (This is suggested as a profitable exercise for students.)

String-lining problems are more complicated if it is necessary to hold the track fixed at certain points, such as at frogs, bridges, or station platforms. In such cases, zero throws are entered at the proper stations and the middle ordinates adjusted to produce the required result. Numerous restrictions on throws make it difficult to obtain perfectly smooth track.

Track shifting in conformity with the throws finally approved is controlled by setting suitable line (and, possibly, grade) stakes.

Stout tacked line stakes are driven between the ties opposite each station, or as close thereto as permitted by the position of ties. On double-track roadbeds, stakes are eliminated by setting tacks on the ties of the parallel track.

Some engineers prefer to set the line stakes on the revised center line. Track in the shifted position is then checked by means of the usual track gage. Instead of being centered, the line stakes can be set level with the base of the rail at a distance such that, when the track is shifted, the rail base will be a constant distance, for example, 1 ft, from the tack. On a curve requiring large throws, it may also be necessary to set grade stakes for adjusting the rails to proper superelevation.

9-5. Spiraling Existing Curves

In earlier years of railway surveying a variety of track realignment problems arose in connection with spiraling existing track originally laid out as simple

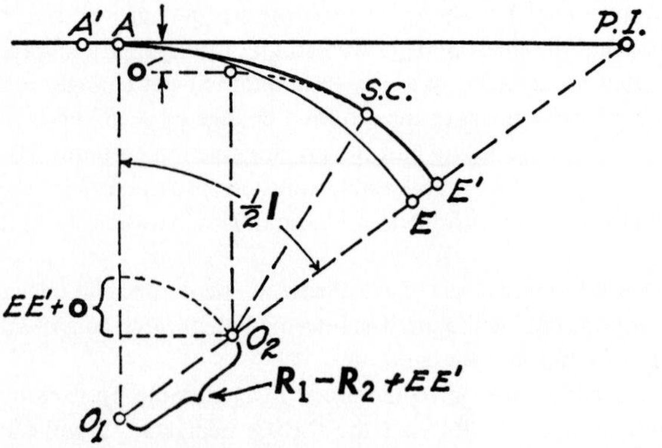

Figure 9-4 Spiraling an existing curve.

curves. Since such problems are much less common now, only one typical case will be illustrated.

Figure 9-4 shows half of an existing circular curve, *AE*. It is necessary to introduce spirals in such a way as to minimize shifting of the track and at the same time to keep the new and old track lengths practically the same. This facilitates shifting the track to its new position. The spiraled half-curve is shown at *A'E'*. The throw *EE'* at the center of the curve is usually restricted to a maximum value of 10 to 12 in.

Obviously, the new simple curve, with radius R_2, must be somewhat sharper than the original one. The selected value of L_s and the final value of R_2 must fit the following relation:

$$(R_1 - R_2 + EE') \text{ vers } \tfrac{1}{2}I = EE' + o \tag{9-4}$$

Any number of combinations of R_2 and L_s may be found. A suggested procedure follows:

1. Select a trial value of D_2 slightly greater than the original degree of curve.
2. Select a practical spiral length not less than that required by equation 9-3; and calculate the resulting value of o from the relation $L_s^2/24R_2$ or from Tables XI or XII.
3. By a trial calculation, determine the value of EE' needed to balance equation 9-4. If EE' is greater than the permitted maximum, go through the same process with new values of D_2 and L_s properly chosen to bring EE' within the required limit.
4. After satisfactory values of D_2 and L_s have been obtained, calculate the difference in length between the original and revised alignments. This difference should be figured between points common to both layouts, namely, the T.S. and the S.T. of the new alignment.

In other problems, it may be necessary to hold a certain portion of the circular curve in its original position, such as on a bridge, trestle, or high embankment. In this case, it is necessary to compound the curve with slightly sharper arcs in order to obtain the needed clearance for inserting the spirals. As in the previous case, any number of combinations of R_2 and L_s will fit the conditions.

9-6. Track Layouts

Railway track layouts involving surveying operations in their location are exceedingly complex. Included under this heading are:

1. *Turnouts*
   (a) Simple *split-switch* turnouts from straight track
   (b) Turnouts from curved track
   (c) Double turnouts, involving *three-throw* and *tandem* split switches
2. *Crossovers*
   (a) Between parallel straight tracks (see Fig. 9-5)
   (b) Between parallel curved tracks
3. *Crossings*
   (a) Straight or curved track
   (b) Combination crossings, or *slip switches*
4. *Connecting tracks* from turnout to:
   (a) Diverging track
   (b) Another turnout, such as at *wye tracks*
   (c) Parallel siding
5. *Yard layouts*
   (a) Complex combinations of the foregoing
   (b) Various arrangements of *ladder tracks*

Each of these layouts involves a multitude of features, including switches, frogs, guard rails, operating devices, rail braces, and fasteners of various kinds.

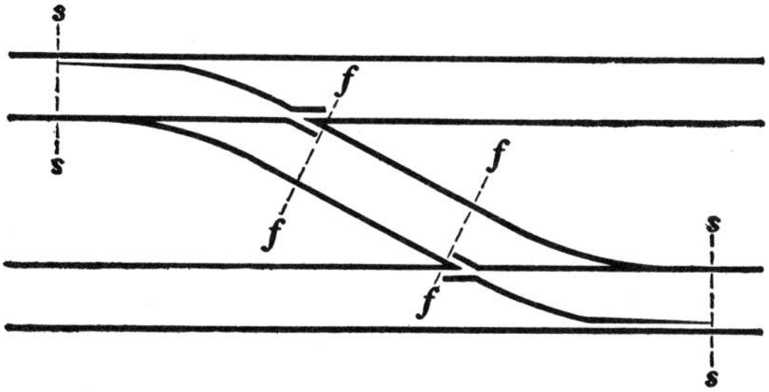

Figure 9-5

One of the layouts frequently used is a crossover between parallel straight tracks, a simplified diagram of which is shown in Fig. 9-5.

The layout involves two turnouts, each of which includes the two switch rails, the frog, and the sections of curved and straight track shown located between lines *s-s* and *f-f*. The crossover tracks are straight between lines *f-f*, which are located at the heels of the frogs.

The best source of information concerning track details is the portfolio of *Trackwork Plans of the A.R.E.A.*

9-7. Relocations

Though the principal trunk railroad lines in the United States that can be justified in the foreseeable future have already been located, relocations are continually being made in order to bring substandard sections of line into conformity with modern requirements.

A typical relocation problem takes the form of flattening a single sharp curve formerly requiring operation under a slow order. If economic considerations do not eventually force the relocation, a serious wreck of a fast passenger train usually does.

Major relocations are invariably made for economic reasons. The changes include (either separately or in combination) reduction in distance, grade, or rate of curvature. The objectives generally sought are: (1) increased speed, in order to reduce running time; (2) greater riding comfort for passengers; (3) operating savings; (4) increased train tonnage, either as a result of grade reduction or from conservation of momentum formerly wasted through braking at the approach to slow-order curves.

A complete treatment of the economics of railway location and operation is beyond the scope of a *Route Surveying* textbook. Excellent analyses of these subjects will be found in *Proceedings, A.R.E.A.*, Vol. 39 (1938), pp. 439–560, and in *Proceedings, A.R.E.A.*, Vol. 45 (1944), pp. 25–44. See also Chapter 16 in the latest A.R.E.A. *Manual.*

Major relocations involve surveying operations fully as complex and interesting as those met in the original location of trunk lines. In view of the existence of thousands of miles of track having curvature and grades which limit economical operation at high speeds, there is every reason to believe that such major relocations will continue to be made for many years. The railroads—stimulated by competition from other forms of transportation, as well as by competition from each other—are apparently aware of the necessity for making many improvements in the alignment of main-line track.

References to a few examples of relocation and new location follow (it is suggested that they be made required reading assignments for students):

1. "The Eight-Mile Cascade Tunnel, Great Northern Railway," *Transactions, ASCE*, Vol. 96 (1932), pp. 915–1004.
2. "Two Important Tunnels Built in 1945."
   (1) The 3,015-Foot Bozeman Pass Tunnel in Montanta (Northern Pacific), *Railway Age*, Vol. 120, No. 19, May 11, 1946, pp. 952–955.
   (2) The 2,550-Foot Tennessee Pass Tunnel in Colorado (Denver & Rio Grande Western), *Railway Age*, Vol. 120, No. 21, May 25, 1946, pp. 1056–1058.
3. "Frisco Line Changes Improve Operation" (12-mile relocation in Missouri, involving grade reduction from 2.3 to 1.27% and curvature reduction amounting to more than 1046°), *Railway Age*, Vol. 121, No. 7, Aug. 17, 1946, pp. 281–284.
4. "Rock Island Completes Relocation in Iowa" (about 88 miles of relocation, involving elimination of 1020 ft of rise and fall, and curvature reduction of 2900°), *Engineering News-Record*, Vol. 139, No. 22, Nov. 27, 1947, pp. 726–730.
5. "Grade and Line Revisions Lower Operating Costs on Missouri Pacific" (two relocations involving grade reduction from maximum of 2.45% to 1.25% compensated, distance saving of 3 miles, and curvature reduction of almost 900°), *Civil Engineering*, March, 1949.
6. "Short-cut Through Missouri" (71-mile relocation on C.B. & Q. reduces Chicago-Kansas City run by 22 miles and passenger schedules by 5 hours. Has high design standards, such as maximum grade of 0.8% and maximum curvature of 1°. Aerial photographs used in planning the location). *Engineering New-Record*, Vol. 145, No. 19, Nov. 9, 1950, pp. 32–35.
7. "Crews Race to Complete Port, Railroad" (193-mile ore-handling railroad built through rugged terrain in Quebec Province. About half of line is curve track with few curves sharper than $5\frac{1}{2}°$. Construction to high standards of curvature and grade required 14 million yd^3 of excavation, half of it rock. Extensive use made of aerial photogrammetry). *Engineering News-Record*, Vol. 165, No. 24, Dec. 15, 1960, pp. 34–40.
8. "New Rails Cross Oregon Trail to Reach Mine" (76-mile railroad spur built in Wyoming to main line criteria. Location surveyed and designed by combining traditional methods with use of aerial photographs and electronic computer). *Engineering News-Record*, Vol. 166, No. 24, June 15, 1961, pp. 44–46.
9. "Southern Pacific Finishes 78-Mile Cut-Off Line at Cost of $22 Million" (huge job required new construction methods and equipment). *Railway Age*, Vol. 163, July 17, 1967, pp. 22–23 and 26–29.
10. "Railroad Track Relocation Saves Time, Money." Relocation by the Santa Fe Railway at the famous Cajon Pass (California) replaced two

10°00′ curves (one with central angle of 134°00′) and eight other sharp reverse curves with six flatter curves, the maximum being 4°00′. *Civil Engineering*, Vol. 44, No. 5, May, 1974, pp. 86–87.

PROBLEMS

9-1. Given a nonspiraled railroad curve. The track is to be realigned with equal spirals to permit a maximum speed of 50 mph. Use the method of section 9-5 to obtain a layout in which EE' is less than 10 in. and the difference in length is less than 3 in.

(a) Existing $D_c = 4°00'$, $I = 21°24'$, and sta. T.C. = 87 + 19.34. *Answer* (one of several possible combinations): New $D_c = 4°30'$; $L_s = 250$ ft; $EE' = 0.74$ ft; sta. T.S. = 86 + 24.09; revised length = 0.06 ft longer than original.

(b) Existing $D_c = 5°00'$, $I = 36°28'$, and sta. T.C. = 124 + 18.53. *Answer* (one of several possible combinations): New $D_c = 5°30'$; $L_s = 350$ ft; $EE' = 0.36$ ft; sta. T.S. = 122 + 76.46; revised length = 0.45 ft shorter than original.

9-2. Given a nonspiraled railroad curve with $D_c = 3°00'$, and $I = 27°16'$; maximum speed allowed with 3 in. unbalanced elevation is 50 mph. The central 200 ft of track is on a trestle. Revise the curve by compounding and spiraling in a way to avoid shifting track on the trestle (see last paragraph in section 9-5). *Solution:* D_c of first and third arcs = 3°10′; $L_s = 270$ ft; unchanged length of 3°00′ curve on trestle = 210.21 ft; sta. T.S. revised length = 0.13 ft shorter than original.

Chapter 10
Reconnaissance for Route Location

10-1. Foreward

In this and the following two chapters, the basic system of route surveying and design as outlined in section 1-8 will be expanded and discussed. The shaded blocks on p. 246 indicate the portion of the basic system covered in Chapter 10.

Because of the scope and purpose of this book, the chapter will concentrate on various methods of gathering information needed for a route reconnaissance study. Basic as well as more sophisticated techniques are discussed, offering the designer a spectrum of methods from which to choose. The first part of the chapter concentrates on field methods of reconnaissance surveying, while the later presents photogrammetric methods. Basic photogrammetric terms needed for understanding later chapters are included.

10-2. Reconnaissance

Reconnaissance is collecting information on broad areas of land between termini for the purpose of route planning and corridor selection. Approximate horizontal positions of natural and human-made features together with

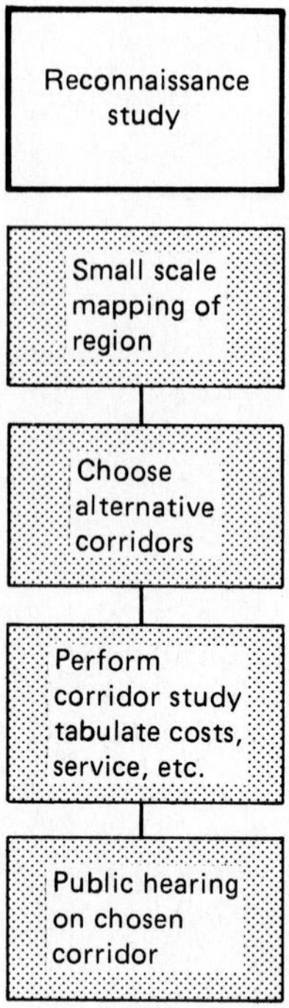

the approximate shape of the ground surface must be made available to the locating engineer so that alternate locations can be identified and studied. The term *reconnaissance surveys* refer to the various methods of collecting and displaying the horizontal and vertical positions of features in the broad area of land under study. In addition, a planning study must consider a multitude of other items such as existing soils, geologic structures, drainage patterns, ecology, sociological impact of the route, land usage, land ownership, and other identified location controls.

The proposed route must pass through or near certain controlling points. Some primary controls will have been fixed in the project conception; others, including most secondary ones, will be revealed by office studies of the maps and whatever field reconnaissance is needed to verify questionable points. Typical location controls include:

Primary Controls
The termini.
Important intermediate traffic centers.
Unique mountain passes, tunnel sites, or major stream crossings.

Secondary Controls

Minor intermediate markets or production centers.
Water courses.
Crossings of existing railroads or important highways.
Swampy areas.
Areas subject to snow or rock slides.
Areas involving costly land damages.
Topography in general, as it affects the economical attainment of desirable grades and curvature.

10-3. Methods of Reconnaissance

Reconnaissance surveys can be achieved by a broad spectrum of methods with each having its value in a particular situation. Low-cost methods of reconnaissance are used on many projects in which corridor selection is not critical, whereas the more sophisticated methods apply to routes for which mistakes in corridor selection might cause excessive construction, operating, or maintenance costs.

The horizontal position of features are determined by field reconnaissance, a stadia survey of the area, use of existing maps, single oblique photos, aerial mosaics, and single vertical photos. The land elevation and shape is determined by visual inspection, use of existing maps, rough transit-stadia mapping, or stereoscopic study of vertical aerial photos with simple parallax measurements. All these methods are explained in more detail in sections to follow on reconnaissance surveys.

Existing soil conditions are determined from available soil maps, local knowledge of engineers and residents, field inspection, or by remote sensing. Gathering information on soils is termed a *soil survey* (see section 10-7). The purpose of a soil survey in the planning stage is to identify relative effects of the soils on suitability of various corridors being considered.

Subsoil and geologic conditions can be investigated to determine their relative effects on the corridors. Knowledge of existing subsoil and geology is obtained from selected soil borings, existing geologic maps, gravimetric surveys, local knowledge of engineers and residents, and by remote sensing.

Drainage patterns along with surface and groundwater conditions are vitally important to engineers in the planning stage. Watershed areas and drainage patterns can be determined from existing topographic maps, field inspections, stereoscopic examination of aerial photos, or by infrared photography. Groundwater histories and conditions are secured from records kept by the U.S. Geological Survey, the U.S. Soil Conservation Service, and data kept by local water management districts.

Ecological and environmental effects are now essential in evaluation of alternate corridors. Most transportation authorities are required to file *impact statements* that analyze each proposed route's effect on the region's environment. To perform these tasks, specially trained biologists, foresters, and environmentalists are consulted regularly. In urban areas noise effects on surrounding communities must be evaluated as part of the planning study.

The locating engineer must have definite knowledge of existing land use in the subject area so the proposed route will be planned to enhance development of suburban areas and aid in inner city renewal. Land use is defined from existing land-use maps, field inspections, interpretation of aerial photography, and most recently by automatic scanning of remote sensing imagery. In addition, the engineer must know state and local land plans for the area so the proposed route will be consistent with community goals.

Sociologists recognize that highways and railroads have a profound effect on the sociological structure of a community. For urban routes, the engineer must consider the impact of alternate corridors on people in the community.

Land ownership often becomes an important factor in locating and selecting alternate corridors. During the planning phase, the engineer should develop an *ownership map* showing the various land parcels existing in the area along with parcel owners. The map must be developed from courthouse research, data furnished by a local abstractor, or sometimes the records of a local land surveyor.

Most investigations mentioned are beyond the scope of this book. However, continued study of the methods in texts on highway engineering and the articles referenced at the end of this chapter is recommended.

10-4. Field Reconnaissance of the Land

Typically, conception of a project is followed first by a careful study of the best available maps and then by field reconnaissance of the terrain between proposed termini. During early years of railroad expansion in the United States, it was often necessary to reconnoiter relatively unexplored regions without aid of maps. Directions were determined by pocket compass; elevations, by barometer; slopes, by clinometer; and distances, by pedometer or timing the gait of saddle horses. Though such devices are still useful at times, they have been made nearly obsolete by the growing file of good maps and improvements in the science of photogrammetry. In fact, it is becoming increasingly apparent that, except for minor projects, aerial-surveying methods may almost entirely supplant the older methods of field reconnaissance.

10-5. Stadia Traverse

In remote regions that are inadequately mapped and on projects for which aerial-surveying methods are not justified, reconnaissance may not definitely disclose the best general route. In such cases a stadia traverse should be run along each of the possible locations.

Needed measurements are made as rapidly as possible, consistent with the required accuracy. Transit stations are far apart. Single deflection angles need be read no closer than the nearest 10′, though it is advisable to check the resulting bearings by compass in order to guard against blunders. A few intermediate shots along the traverse line may be needed to give data for plotting a profile of the traverse. Only enough side shots (supplemented by sketches) need be taken to give approximate positions and elevations of secondary controls.

Especially good judgment is needed in conducting this type of stadia survey; if the survey is entrusted to inexperienced personnel, it usually reverts to an unnecessarily detailed and time-consuming topographical survey.

The resulting maps, plotted by protractor and scale, really serve as high-grade reconnaissance; if the work is well done, they permit the definite discard of certain routes and go far toward fixing closely the best general route to be followed in more precise preliminary surveys.

10-6. Use of Existing Maps

Today, most of the country has been mapped to a small scale by public agencies such as the U.S. Geological Survey. The U.S.G.S. "quadrangle series" of maps now are available for most areas at a scale of 1:24000, or 1 in. equals 2000 ft. Such small scale maps are largely used today in planning routes but have limited further use in design because of their small scale. The maps, called *quad sheets*, are constantly being updated and revised by aerial methods of U.S.G.S.

10-7. Soil Surveys

Comprehensive soil surveys, rarely needed in railroad location, have become standard practice among progressive state highway departments. Such surveys are superseding the former types of soil surveys which were often restricted to borings at bridge sites and cursory examination of the route for surface indications of snow slides or unstable side slopes.

Modern alignment standards often require that routes traverse topographic features formerly avoided; as a result, long heavy fills and deep cuts—sometimes through bedrock—are frequently necessary. Since the cost of a modern highway may exceed one million dollars per mile, possible savings in construction and maintenance costs justify using fairly expensive methods of determining pertinent information about surface and subsurface soil conditions.

Soils investigations for highways continue to grow more comprehensive in scope. This phase of highway engineering requires close cooperation with the soil physicist, the geologist, and even the seismologist. Modern tendency is to go beyond merely making auger borings along the proposed route and classifying the soil samples in a laboratory. Instead, past geologic history of the area is investigated. From these studies, area soil maps are prepared which show the soil "pattern"—landforms, types of soil deposits, swamp areas, drainage conditions, and related information. A proposed route traversing the region covered by an area soil map is then the subject of a preliminary report which shows the relationship of the soils to the engineering considerations of alignment, grade, drainage, and grading and compaction processes. Availability of materials for borrow, subbase, or concrete aggregates is also indicated. The report may suggest alignment changes, and also may contain specific recommendations regarding the extent of subsurface exploration needed to answer detailed questions for design and construction.

In glaciated regions, erratic depths to bedrock may justify use of seismic methods of subsurface exploration. Massachusetts, for example, has developed a procedure in which a geologic "strip" map is prepared and the locations where seismic studies are recommended, such as at deep cuts and bridge sites, are shown on this map. Extensive use was made of seismic profiles in obtaining quantity estimates on the Massachusetts Turnpike. Seismic methods have become less costly with the development of a lightweight seismograph, the sound waves for which are generated by the impact of a sledge hammer on a metal plate instead of by an explosive charge.

Complete treatment of the subject of modern highway soil surveys is beyond the scope of a text on Route Surveying. For detailed information, see any modern text on Highway Engineering.

AERIAL RECONNAISSANCE METHODS

10-8. Photogrammetry for Reconnaissance

A photo shows an infinite amount of detail. For this reason, aerial photos have superseded field methods of reconnaissance on all but the very simplest of route locations. Photogrammetry is used to such a great extent in route location that the same route may require as many as seven different photographic flights at appropriate times during the reconnaissance, location, design, construction, and completion of a project. Typical flights may be used for

- area reconnaissance to identify corridors to study
- a reconnaissance of route alternatives for comparison and selection
- aerotriangulation (see section 11-15) to supply control points for corridor mapping
- mapping of the corridor for alignment design
- measurement of cross sections for earthwork quantities before construction

mapping of the construction limits to large scale for final design
measurement of cross sections of the completed construction for final quantities

The first two flights in the list indicate the reconnaissance may occur in two steps. In the first step, small-scale photos will enable the designer to select important controls and all feasible route corridors which may be from one-quarter mile to a mile in width. The second reconnaissance step is comparison of route corridors and selection of the most promising one. In some instances, the choice will become apparent during the first step, but in more difficult cases it may be necessary to prepare small-scale topographic maps (1 in. equals 400 ft to 1 in. equals 1000 ft) so approximate quantities can be computed for comparison of corridors.

10-9. Photogrammetry Definitions

Definitions of photogrammetric terms are given in this article. For more complete definitions, consult the *Manual of Photogrammetry* published by the American Society of Photogrammetry and complete texts on the subject.

Photogrammetry is the science or art of obtaining reliable measurements by means of photography. The subject is subdivided into *terrestrial photogrammetry* and *aerial photogrammetry*. In terrestrial photogrammetry photos are taken from one or more ground stations; in aerial photogrammetry, from an airplane in flight. Aerial photogrammetry utilizes *vertical photographs*, those taken with the camera pointed down along the vertical; and *oblique photographs*, those not taken vertical but on a definite tilt angle to the ground. The angle between the true vertical and camera axis is called *tilt* of the photo. In vertical photos, a small amount of accidental tilt is usually present due to motion of the airplane and camera. In oblique photographs, tilt angles may be as large as 50°. The process of removing accidental tilt from near vertical photos is called *rectification* (see Fig. 10-1).

An important property of a photo is its *scale*, the ratio of a photo distance to a corresponding ground distance. Photo scale is determined by the camera *focal length* and its *flying height* above the terrain. Scale of the developed negative is termed the photograph's *negative scale.*

A *flight strip* is a succession of overlapping aerial photographs taken along a single course called the *flight line. Endlap* is the percentage that each photo overlaps the adjacent photo in a flight strip. Sixty to eighty percent endlap is usual in practice. Overlap of adjacent flight lines, called *sidelap* is usually about 30%.

A *mosaic* is an assemblage of aerial photos trimmed and matched to form a continuous photographic representation of the earth's surface. If the photos are matched without reference to ground control points, the resulting

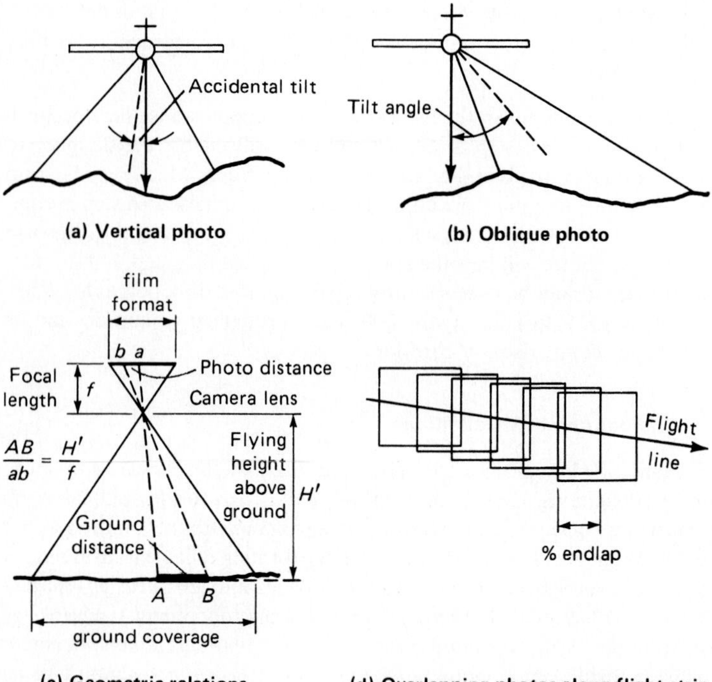

(a) Vertical photo (b) Oblique photo

(c) Geometric relations (d) Overlapping photos along flight strip

Figure 10-1

map is an *uncontrolled mosaic;* if they have first been brought to a uniform scale and fitted to ground control stations, the map is a *controlled mosaic.*

The photographs used may be *contact prints,* (made with the negatives in contact with sensitized photographic paper); *ratio prints,* (scales have been changed from the negatives by enlargement or reduction); or *stereoscopic* pairs (stereo pairs), in which two photos of the same area are taken to permit stereoscopic examination. Ratio prints used as maps are called *photomaps.*

As stereo pairs are viewed through a stereoscope, the viewer sees a three-dimensional model of the ground surface called a *stereo model. Parallax* is the total distance a ground point moves with respect to the focal plane from one photo to the next. Parallax measurements of points in the stereo model allow heights to be computed.

A *planimetric map* shows the horizontal position of selected natural and cultural features, whereas a *topographic map* also shows relief in measurable form, usually by contours. The term *base map* refers to a map containing only essential features of the land and is used as a background for later drafting of special features of interest. A base map may be used many times throughout a route location, design, and construction project.

10-10. The Important Geometric Facts of Aerial Photos

To the novice in photogrammetry, a "vertical" photograph appears to be a perfect map. However, four significant differences exist between the perspective photo projection and the orthographic map projection that cause problems for those wishing to use a photo as a map: (1) odd negative scales, (2) varying scales caused by relief, (3) relief displacement, and (4) tilt.

First, scale of an aerial negative as it comes from the camera is an odd value not easily used. To produce a photo to a scale of 1 in. equals 2000 ft. with a 6-in. focal length camera, the flying height must be 12,000 ft above the ground surface (not above sea level). It becomes apparent that to produce a 9 × 9 in. contact print to a scale, the camera must be flown at exact height above the ground, which is impossible. Therefore, a vertical 9 × 9 in. photo cannot be made to a precise scale unless the developed negative is enlarged or reduced photographically. The term *negative scale* refers to the scale of the 9 × 9 in. negative as it comes from the camera—it is usually an odd scale value.

Second, the ground has relief, so the flying height above a certain point will differ from that at a second point, making the scale between images near the first point different from the scale near the second. Therefore, only vertical photos of perfectly flat ground will come out of the camera at a constant scale throughout giving reliable measurements on the resulting print.

Relief displacement is the outward shifting of a photo image due to the height of the ground point producing the image. The top and base of a flag pole have a single map position. However, due to the perspective camera view, the pole will appear to "lean" on the photo with the top being displaced outward from the base in a direction away from the photo's center. This relative displacement of two points having the same map position but different elevations is termed relief displacement causing buildings to lean on photographs—especially the buildings at the edge of a photo. Near the photo center, relief displacement is nil because the camera sees the pole's top and base superimposed upon each other as a single point. If an aerial photo is to be used as a photomap, relief displacement is a bad feature because distances cannot be scaled between images of points of different elevations. Referring to Figure 10-2, flat ground surfaces at left are photographed in correct shape (square) but at a larger scale than flat surfaces on the right. The datum grid is reproduced at a smaller scale yet, called the datum scale of the photo. Straight lines on the ground, level or sloping, become straight lines on the photo. Sloping surfaces are distorted in shape. Each ground point is displaced on the photo from its datum position by relief displacement (d) depending on the radial distance from the photo center (r), height above the datum of the ground point (h), and flying height of the photo (H). The center of the photo is not displaced.

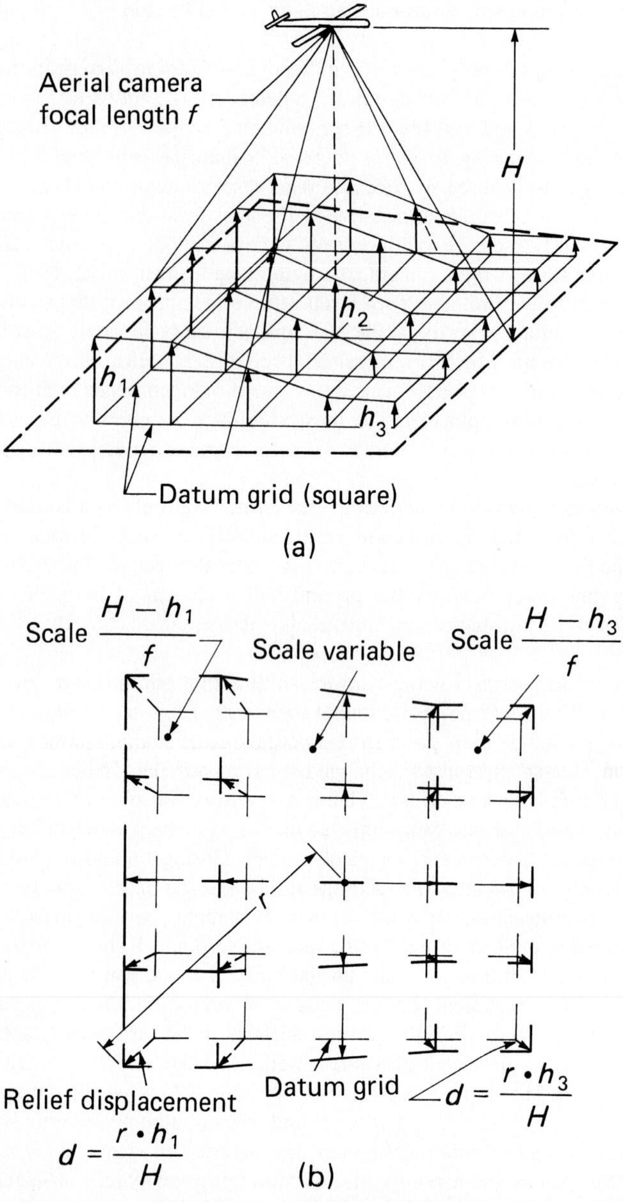

Figure 10-2 Scale Distortions caused by relief. (a) Ground surface with relief above datum plane. (b) Scale distortions caused by relief displacement of ground points from the datum position.

Relief displacement can be reduced, but not completely eliminated by photographing only very flat land, flying at a large height over rolling land, or by using a 12-in. focal length camera. The later method is popular in producing photomaps because the 12-in. camera photographs only the middle one-fourth of the ground area that a 6-in. camera would—the ground area containing little relief displacement.

Tilt of the camera at the instant of exposure also will hinder use of a photograph as a map because it causes scale distortions. On an oblique photo that is purposely tilted, objects near the horizon photograph at a smaller scale than near the camera. The same effect takes place to a much lesser degree on vertical photos containing minor accidental tilts. To remove the effect of tilt, the photos must be *rectified* by a projection process in the darkroom that varies the relation between the plane of the negative in the projector and the photographic print until the identical relation that existed at the instant of exposure between the film plane in the camera and the ground datum plane is duplicated.

For example, if a household camera is used to take a picture of the vertical face of a square building from an oblique angle, the plane of the building face will not be parallel with the plane of the film at the instant of exposure. The resulting picture distorts the square building face into an oblong shape. And if the relative position between the planes of a screen and the slide in the projector is adjusted until the building appears square on the screen, the angular relation between projector and screen is now identical to the relation that existed between camera and building face. The photo has now been rectified. A print of the resulting squared projected building face would be a *rectified print*. In addition, the scale of that rectified building face on the screen could be changed to any desired ratio by moving the screen closer to the projector for a smaller scale, or away for a larger one. Therefore, as a normal part of the rectification process, the resulting print is made to any desired enlarged scale.

Modern rectifier-enlarging projectors can expand aerial negatives up to nine times the negative scale. Referring to Figure 10-3, an aerial sheet is prepared on a rectifier-enlarger. The easel is tilted using the handwheels until the projected images of control points match the same points plotted to the desired scale of the print. A mylar sheet with a photographic coating is then placed on the easel and exposed for the required time. The exact angular relation that existed between camera and ground is recreated.

10.11. Aerial Mosaics for Reconnaissance

In making an *uncontrolled* mosaic, contact prints covering the area to be studied are trimmed and assembled by matching like images, then fastened to a rigid or flexible backing. If the mosaic is for temporary field use the prints are mounted on linen or other material that will permit it to be rolled

Figure 10-3 A rectifier-enlarger. In the darkroom, an aerial sheet is prepared on a rectifier-enlarger. The easel is tilted using the handwheels until the projected images of control points match points plotted to the desired scale of the print. A mylar sheet with a photographic coating is then placed on the easel and exposed for the required time. The exact angular relation that existed between camera and ground is recreated. (Courtesy WILD-Heerbrugg.)

up. For this purpose semimatte prints are preferred because they take pencil lines readily and are not scratched as easily as glossy prints. For more permanent use, properly matched glossy prints can be stapled or pasted to a rigid backing. If desired, the assembly may be photographed to preserve one or more copies of the complete map, after which the mosaic is dismantled so that the contact prints may be used for stereoscopic study.

Such a mosaic is relatively inexpensive and, though subject to errors because of scale variations and displacements, it is extremely valuable for reconnaissance and miscellaneous uses.

A *controlled* mosaic is assembled by bringing the photographs to a uniform scale, correcting them for tilt, and fitting them in their correct relative positions. This procedure requires locating control points by ground surveying methods.

Control stations, properly distributed over the area, are first selected with the aid of a stereoscope. They should be definite points easily recognized on the photographs and accessible on the ground. Buildings, fence corners, or road intersections usually serve this purpose. Preparatory to planning the ground control surveys, the selected control points are marked on each photograph by a circled prick point. Ground control surveys consist of suitable triangulation, traverse, and level circuits from which the coordinates and elevations of the control points can be computed. Ground control is costly and should therefore be no more precise or extensive than is required for the purpose.

In making the finished mosaic, the contact prints are ratioed (brought to the same predetermined scale), rectified (corrected for tilt), and fitted on a base board to the plotted ground control points. There are several methods of doing this, all of which are highly technical and require special equipment. Only the central part of each photograph is used in compiling the mosaic, and the trimmed edges are feathered on the underside. In addition, prints having the same tone, or degree of exposure, are selected. The finished mosaic then has the appearance of a single large photograph.

Following each flight of an area, an *index mosaic* is constructed to allow the user to select the photos covering a certain ground area. Each aerial photo is identified by a number printed on its corner. In making an index mosaic, photos are laid up with features matching and identification number of each photo visible. The resulting composite is rephotographed and enlarged prints of the index mosaic made for future use in locating particular photos.

10-12. Oblique Aerial Photography for Reconnaissance

Aerial oblique photos can be utilized to illustrate topography, land use, drainage, soil condition, traffic, and other data for highway location and design. Today, when public hearings are essential in the process of determining route alternatives, oblique aerial photos are invaluable as a medium for showing in perspective each proposed solution of the highway engineering problems. Features to be included in the oblique photo should be anything identifiable which will give persons viewing the oblique photos orientation while they examine the artist's delineation of the proposed route.

An angle of obliquity should be specified for each photo. The horizon will appear on 9 × 9 in. photos if the angle to the camera axis is made 56° from the vertical for a 6-in. camera. Angles of obliquity between 25 and 40° usually provide the best views.

An advantage of oblique photos is that large areas of land can be displayed on one print for illustration purposes. The obvious disadvantage is lack of consistent scale for measuring purposes.

Land areas can be mapped to a scale from an oblique photo by a perspective grid drafted on the photo. Features are interpolated between grid points of the perspective grid and plotted on a map sheet containing rectangular grid lines to desired scale.

10-13. Stereoscopic Study of Aerial Photos

Most aerial photography is done by flying parallel flight lines across an area. During the flights, photographs are taken at intervals such that adjacent photographs overlap approximately 30% at the sides and more than 60% in the flight direction. This ensures that the center (principal point) in each photograph will appear in the adjacent picture taken in the line of flight, thus providing what is called "stereoscopic overlap." By properly orienting the overlapping photographs (stereopairs) and viewing them through a stereoscope, the process known as "stereoscopic fusion" takes place. In this process there is a vivid mental impression of the terrain in three dimensions. A simple demonstration of stereoscopic fusion is shown in Fig. 10-4. In effect, two positions of the camera lens several thousand feet apart are substituted for the observer's eyes. In the resulting image (known as the "stereo model"), relative heights of hills and structures, depths of canyons, and slopes of terrain are determinable. Used in this way, the old principle of stereoscopic vision has become probably the most important basic tool for studying the manifold problems of route location.

Not only can stereovision give the engineer a three-dimensional view of the ground, (see Fig. 10-5) but measurements of image parallax on a stereo

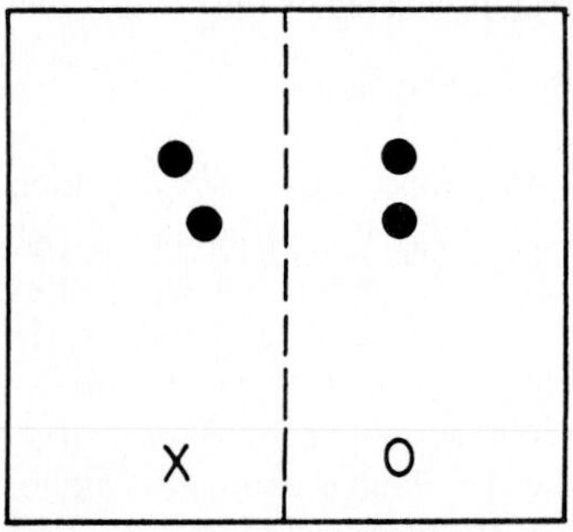

Figure 10-4 Stereoscopic fusion. Hold a card at right angles to the page and along the dotted line. Closing one eye at a time, adjust the head so that the letter X (but not O) can be seen with the left eye, and conversely with the right eye. Then open both eyes and focus them beyond the page. The four dots above the letters will fuse and appear as two. Moveover, the lower dot will seem to be floating in space relative to the upper dot. To prove that all four dots are in the image, notice the letters X and O are superimposed.

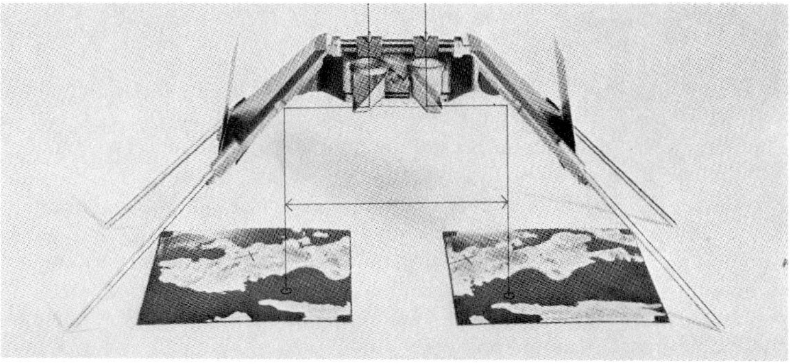

Figure 10-5 A mirror stereoscope. Images on the right photo are viewed with the right eye, and the left photo is viewed with the left. The effect is to place each eye at the space location on the left and right exposure stations. The mind perceives an optical model of the ground. (Courtesy WILD-Heerbrugg.)

pair permits the calculation of heights, elevations, and approximate grades. Any text on photogrammetry gives details of this valuable tool for reconnaissance.

10-14. Summary of Advantages and Limitations of Aerial Photography

To summarize the information in this chapter, it is apparent that the outstanding advantages of aerial photography for reconnaissance are:

1. The larger area and wider route corridors covered by the photographs give greater flexibility in route location and practically ensure that a better location has not been overlooked.
2. Practically all the studies preceding construction can be made without encroaching on private property or arousing premature fears in regard to the extent of property damage. Land speculation is thereby reduced. Moreover, eventual acquisition of right-of-way is expedited because property owners can see clearly on the photographs the effects of the takings.
3. Overall survey and mapping costs may be considerably less than by ground methods.
4. Aerial photographs have many auxiliary uses and contain more information about a variety of significant features than the engineer can obtain by ground methods except at greatly increased cost.

Photogrammetry for purposes of route location is not without its limitations. Clarity in photographs requires good atmospheric conditions—freedom from clouds, mist, smoke, and severe haze. In some parts of the world such ideal conditions may occur only one or two days in a year.

SUPPLEMENTAL READINGS

Davies, J. H. 1976. Remote Sensing of Atmospheric Pollutants to Assess Environmental Impact of Highway Projects. *Transportation Research Record* Bulletin 594: 1–5.

Deloach, W. C. 1973. Remote Sensing Applications to Environmental Analysis. *Highway Research Record* Bulletin 452: 1–9.

Hawkes III, T. Wallace and Brown, Drew A. 1973. Application of Aerial Mapping to Development of Highways. *Highway Research Record* Bulletin 452: 10–18.

Kiefer, R. W. 1972. Sequentical Aerial Photography and Imagery for Soil Studies. *Highway Research Record* Bulletin 421: 85–92.

Manual of Photogrammetry, 3rd ed. 1966. American Society of Photogrammetry, Washington, D.C., 1966.

Melhorn, W. N. and Keller, E. A. 1973. Landscape Aesthetics Numerically Determined: Applications to Highway Corridor Selection. *Highway Research Record* Bulletin 452: 1–9.

Moffitt, F. H. and Mikhail, E. M. *Photogrammetry*, 3rd ed., New York: Harper & Row.

Smedes, H. W., Turner, A. K., and Reed Jr, J. C. 1976. Oblique Airphotos for Mapping. Educating Users and Enhancing Public Participation in Environmental Planning. *Transportation Research Record* Bulletin 594: 1–5.

Specifications for Aerial Surveys and Mapping by Photogrammetric Methods for Highways, Reference Guide, the Photogrammetry for Highways Committee of the American Society of Photogrammetry, published by the U.S. Department of Transportation, Federal Highway Administration, 1968.

Tanguay, M. G. and Miles, R. D. 1970. Multispectral Data Interpretation for Engineering Soils Mapping. *Highway Research Record* Bulletin 319: 58–70.

Chapter 11
Base Line Surveys and Corridor Mapping

11-1. Foreword

As indicated by the shaded blocks on p. 262, this chapter discusses the various methods of gathering the information needed in an alignment study and design. The chapter's first part deals with methods for the preliminary traverse, introducing electronic measuring techniques as well as the older transit and tape methods. The later part of the chapter covers methods of mapping the corridor starting with field and finishing with photogrammetric procedures. Refer to Chapter 10 on reconnaissance for a logical introduction and the basic photogrammetric terms.

11-2. Alignment Design and Preliminary Surveys

Following reconnaissance, a locating engineer must design the exact horizontal alignment within the chosen corridor. In addition, an approximate vertical location must be selected along with the correct horizontal alignment

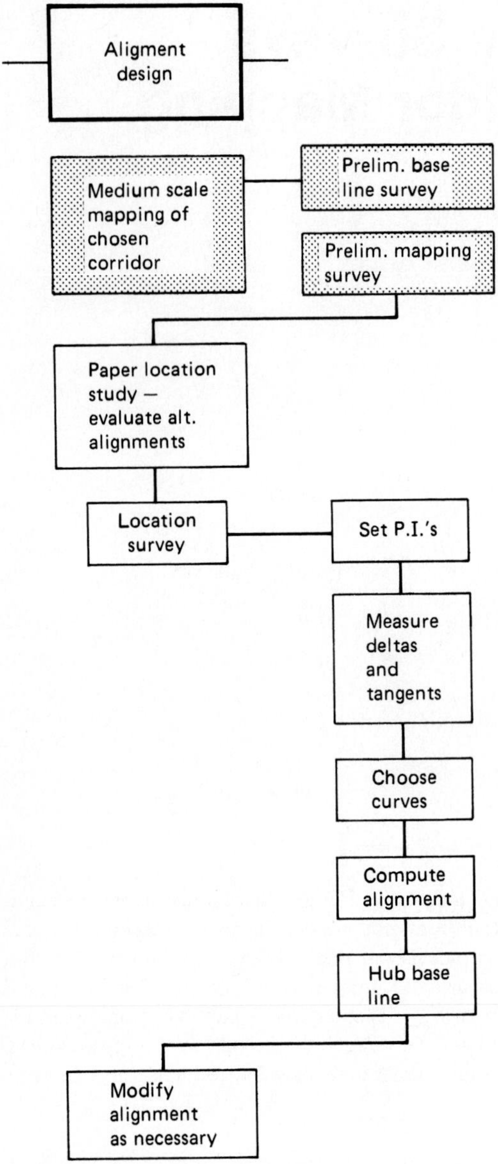
Aligment design
Medium scale mapping of chosen corridor
Prelim. base line survey
Prelim. mapping survey
Paper location study — evaluate alt. alignments
Location survey
Set P.I.'s
Measure deltas and tangents
Choose curves
Compute alignment
Hub base line
Modify alignment as necessary

because of their interdependence in design. Preliminary limits of required right-of-way are determined so that property acquisitions can begin prior to construction. This phase of route location is termed *alignment design.*

For making the many decisions in alignment design, the primary tool is a precise topographic map of the corridor drawn to a medium scale such as 1 in. equals 100 ft or 1 in. equals 200 ft. Corridor mapping can be done either by field or photogrammetric methods. Today, photogrammetric methods are used almost exclusively on new alignments to examine large corridor widths. On minor realignments within old corridors, field mapping methods may be useful. To provide horizontal control points for mapping, a base line is run in a random direction, and bench marks established throughout the corridor. Establishment of a field base line and subsequent corridor mapping are generally termed *preliminary surveys.* The sole purpose of a preliminary survey is to provide the engineer with necessary information for choosing horizontal and vertical route alignments.

11-3. Organizing Field Work for the Preliminary Survey

Prior to starting extensive field work, a locating engineer preferably should make a field examination of the general route recommended in the reconnaissance report. Before going into the field, he must become acquainted with the general objectives of the project. Moreover, he should be informed as to how much information can be divulged to curious property owners along the route.

A great variety of useful purposes are accomplished in this preliminary field examination. Among these are:

Permission to trespass on private property for survey purpose, with whatever qualifications are imposed by the owner.

Identification of convenient bench marks described in Federal or State publications.

Determination of best access to the work by automobile or other means of transportation.

Making of special notes regarding secondary controls.

Taken together, the information acquired should enable the locating engineer to select needed field and office personnel and determine efficient methods of performing the various surveying operations.

11-4. Transit-and-Tape Traverse

Whenever the reconnaissance narrows the location down to a fairly definite route, a careful transit-and-tape traverse, called the "base line" or "*P*-line," is run joining the various location controls. This traverse then serves not only as an accurate framework for plotting topographic details but also as a convenient means of transferring the paper location to the ground and checking its accuracy.

The *P*-line is a continuous traverse located as close as convenient to the expected position of the final location. Transit stations are marked by substantial tacked hubs which are driven flush, carefully referenced, and identified by guard stakes. Angles are at least doubled, the telescope being reversed on the second angle in order to eliminate instrumental errors. To guard against blunders in reading angles the magnetic bearing of the forward line should be compared with the calculated bearing before leaving any transit station. Accuracy of the traverse should be checked whenever convenient by tying to monuments of the State Plane Coordinate System.

Horizontal distances measured with a 100-ft (or longer) steel tape, usually are recorded to the nearest hundredth of a foot. Station stakes are set at regular 100-ft stations, at intermediate plus points where profile breaks occur, and possibly, at fence lines and highways crossing the center line.

11-5. Levels

If the survey is over terrain permitting the transit party to make rapid progress, it is often worthwhile to start two 2-man level parties at the time at which the transit party begins. One, the bench level party, carries elevations from the nearest bench mark and sets additional bench marks every 1000 ft or so along the general route. It is not necessary for this party to follow the exact path taken by the transit party; swampy areas and steep slopes, for example, should be bypassed wherever possible. In fact, the best plan is for the bench level party to work ahead of the transit party along a general route specified by the locating engineer. This party can then close its levels back to the starting point each half-day without falling behind the transit party.

The second level group, the profile level party, takes its initial backsight on a bench mark established near station *P* 0+00 by the bench group. It follows immediately behind the transit party, taking rod readings on the ground at every station stake and intermediate plus point. In proceeding along the line, a check is made on bench marks previously set by bench leveling. Thus, there is no need for the profile level group to close its levels back; it is always close to the transit party, where the locating engineer frequently needs the information on elevations to plan the positions of forward transit stations.

Bench-mark elevations determined by the profile level party should be compared with their adjusted values each day. The adjusted elevations should be noted in the field book and used on all subsequent work.

11-6. The EDMI-Theodolite Geodetic Base Line

As an alternative to the transit-tape traverse discussed in section 11-4, the EDMI-theodolite base line is used on projects requiring more accuracy. The geodetic base line has the following characteristics:

1. Angles are measured with a theodolite.
2. Distances are determined with an *electronic distance measuring instrument* (*EDMI*).
3. Stations are widely spaced.
4. Geodetic horizontal control stations serve as terminal points.
5. The base line is located outside the anticipated limits of construction so that monuments will be preserved.
6. Each station consists of a brass marker set in a concrete monument.
7. All computations are performed on the State Plane Coordinate System.
8. Traverse adjustment is executed by the least squares method.
9. Base line layout and field methods are designed according to national specifications for a chosen order of accuracy. The latest set of specifications published by the National Geodetic Survey in 1974 is shown in Table 11-1.

11-7. The Electronic Distance Measuring Instrument

In the late 1950s basic research led to the development of the first electronic distance measuring instruments which have since revolutionized all phases of surveying. Two basic instruments, the geodimeter and the tellurometer, were independently developed in different parts of the world within several years of each other.

The geodimeter, employing visible light, was evolved in Sweden by a physicist, Dr. Bergstrand, as a by-product from his basic research into the speed of light. The first geodimeter used in this country, the Model 2, weighed over 200 pounds. In the middle 1960s, the visible light source was replaced by a laser beam and later models of the geodimeter were refined into smaller and lighter electronic packages. In the late 1960s further basic research led to development of the gallium arsenide diode that generates large amounts of infrared radiation. Several companies quickly incorporated the infrared light source into light-weight instruments of shorter ranges, that is, 1 to 2 miles. Since introduction of the short-range infrared EDMI in the early 1970s, several other companies have built competing models of ever smaller dimensions and lighter weight. The various existing models using visible light, laser light, and infrared radiation, that still employ the basic measuring system developed by Dr. Bergstrand, are collectively termed *lightwave instruments.*

The second basic measuring instrument, the tellurometer, was produced in the late 1950s by a research group in South Africa under a government funded contract. The tellurometer differs from a geodimeter by using radio waves, called *microwaves*, instead of light. Due to the characteristics of radio

Table 11-1. TRAVERSE SPECIFICATIONS

CLASSIFICATION	FIRST-ORDER	SECOND-ORDER		THIRD-ORDER	
		CLASS I	CLASS II	CLASS I	CLASS II
RECOMMENDED SPACING OF PRINCIPAL STATIONS	Network stations 10–15 km Other surveys seldom less than 3 km.	Principal stations seldom less than 4 km except in metropolitan area surveys where the limitation is 0.3 km.	Principal stations seldom less than 2 km except in metropolitan area surveys where the limitation is 0.2 km.	Seldom less than 0.1 km in tertiary surveys in metropolitan area surveys. As required for other surveys.	
HORIZONTAL DIRECTIONS OR ANGLES					
Instrument	0″.2	0″.2 or 1″.0	0″.2 or 1″.0	1″.0	1″.0
Number of observations	16	8 or 12[a]	6 or 8[a]	4	2
Rejection limit from mean	4″	4″ or 5″	4″ or 5″	5″	5″
LENGTH MEASUREMENTS					
Standard error	1 part in 600,000	1 part in 300,000	1 part in 120,000	1 part in 60,000	1 part in 30,000
RECIPROCAL VERTICAL ANGLE OBSERVATIONS					
Number of and spread between observations	3 D/R—10″	3 D/R—10″	2 D/R—10″	2 D/R—10″	2 D/R—20″
Number of stations between known elevations	4–6	6–8	8–10	10–15	15–20

ASTRO AZIMUTHS					
Number of courses between azimuth checks	5–6	10–12	15–20	20–25	30–40
No. of obs. per night	16	16	12	8	4
No. of nights	2	2	1	1	1
Standard error	0″.45	0″.45	1″.5	3″.0	8″.0
Azimuth closure at azimuth check point not to exceed	1″.0 per station or 0.2″ $\sqrt{N}$	1″.5 per station or 3″ $\sqrt{N}$. Metropolitan area surveys seldom to exceed 2″.0 per station or 3″ $\sqrt{N}$	2″.0 per station or 6″ $\sqrt{N}$. Metropolitan area surveys seldom to exceed 4″.0 per station or 8″ $\sqrt{N}$	3″.0 per station or 10″ $\sqrt{N}$. Metropolitan area surveys seldom to exceed 6″.0 per station or 15″ $\sqrt{N}$	8″ per station or 30″ $\sqrt{N}$
POSITION CLOSURE					
after azimuth adjustment	0.04 m $\sqrt{K}$ or 1:100,000	0.08 m $\sqrt{K}$ or 1:50,000	0.2 m $\sqrt{K}$ or 1:20,000	0.4 m $\sqrt{K}$ or 1:10,000	0.8 m $\sqrt{K}$ or 1:5000

[a] May be reduced to 8 and 4, respectively, in metropolitan areas.

Source: *Classification, Standards of Accuracy, and General Specifications of Geodetic Control Surveys*, NGS, Rockville, Md., 1974.

waves, radiation cannot be "bounced" from a mirror-type reflector, but must be received and retransmitted by a remote unit on the other end of the line being measured. The master and remote units are identical in construction and require operators at each end of the line. The master unit receives the retransmitted radiation and performs a phase comparison on multiple frequencies to determine the distance. Microwave instruments have the advantage of voice communication between the operators on a special broadcast channel of the radio waves being used.

The latest developments in EDMI's have integrated the distance measuring function with a theodolite for simultaneous measurement of angle and distance. Such equipment is termed "total station" because of the ability to measure distance, horizontal angles, and vertical angles from one pointing of the instrument. Several such instruments have an automatic pendulum that senses the angle of vertical inclination of the line-of-sight and breaks the

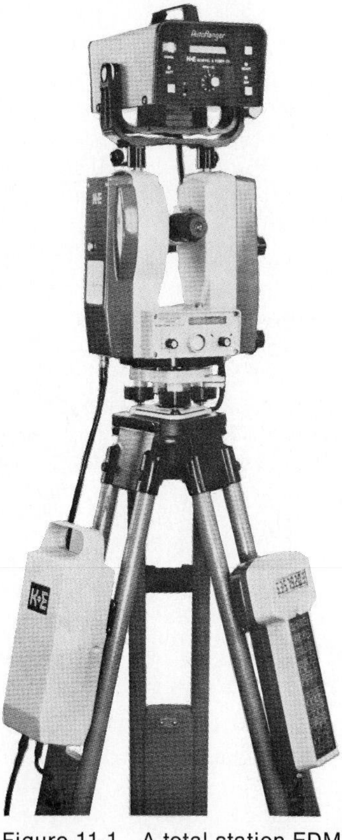

Figure 11-1 A total station EDMI. (Courtesy K & E Company.)

measured slope distance into horizontal and vertical components of distance automatically. The components are computed and displayed in digital form. With the "total station" instrument, elevations can be computed along the traverse together with horizontal angle and distance (see Fig. 11-1).

11-8. Doppler and Inertial Surveying Systems

Currently being developed are two methods of performing base line surveys that should have many applications in route surveying.

The *doppler positioning* system uses signals sent from navigational satellites orbiting earth. A special radio receiver with an antenna is erected over a survey point. As an orbiting satellite approaches the receiver's position and transmitted radio waves are "compressed" because of the satellite's velocity (like sound waves of an approaching train whistle) causing a slight increase in the frequency of received signals. Likewise, as the satellite moves away from the receiver, radio signals are "stretched" and the received frequency is reduced. The receiver measures increase and decrease in signal frequency as the satellite passes which permits the computation of latitude, longitude, and elevation of the survey point using the known orbit geometry of the satellite. Recent tests have shown that multiple satellite passes determine positions of survey points with accuracies approaching 1 m. Improvements in this survey system will surely provide electronic position measurement comparable with traditional methods.

Another measuring tool that may find wide application in route surveying is the *inertial system*. Three accelerometers are mounted inside a truck and aligned along north-south, east-west, and vertical coordinate axes by gyroscopes. The truck is positioned over a point of known horizontal and vertical position. As the truck is driven in the general direction of the survey, the accelerometers measure the acceleration-time function. An on-board computer performs two successive numerical integrations of the acceleration-time function to obtain the distance-time relation. Therefore, the computer can constantly compute and record north-south and east-west coordinates as well as elevation of the truck *as it moves*. Preliminary tests show that "traverses" can be run to close out on widely-spaced control points with an accuracy approaching 1 in 20,000.

11-9. Methods of Mapping the Corridor

Mapping of topographic features within the route corridor follows completion of the preliminary base line. The final map may take three basic forms: a drafted line map from field measurements; a photomap consisting of an actual photo; or a drafted line map made from measurement of a stereo model.

Field surveys locate the horizontal positions of features by *transit-tape*, *stadia*, or *hand level*. Elevation and shape of the corridor ground surface is generally determined by the cross-section method as discussed in Chapter 6. The preliminary base line and bench marks are used as control points for eventual plotting of details. Measurements are generally recorded by a topographic party. In the office, a line map is drawn to the desired scale from the field notes. Plotting is generally done by hand, but computers are now performing map drafting.

Photogrammetric methods of mapping produce either a scaled photomap or a line map produced from aerial photos. Photomaps are made by the general process of enlargment (see section 10-10), rectification (see section 10-10), and orthophotography (see section 11-17). Line maps are obtained from aerial photos by measuring an optical model of the land created in a stereoplotter. Line maps are hand drafted, scribed, or prepared on a computer-controlled plotter. These processes will be discussed in sections to follow.

11-10. Topography by Transit, Tape, and Level

Transit and tape topographic surveys are commonly used on secondary rural routes and on minor realignment jobs within old corridors. Station stakes set on the preliminary traverse are used as the mapping base line. Horizontal points to be mapped are tied in by the *plus-offset method*. Simple devices such as *right-angled prisms* are used to find the base line point that lies at the foot of a perpendicular from the mapped point. Once found, the station plus and offset are measured. Notes, consisting of a sketch of points in proper relation to the base line, are constructed by the party chief. The map is usually hand drafted to the desired scale.

Elevations and contours are determined by the cross-section method. At each base-line station stake, a cross section is taken covering the corridor limits. Resulting elevations and offsets are plotted on the planimetric map, and contours found by interpolation between ground "break points." Usual field equipment includes a dumpy or automatic level, and cloth tape.

11-11. Topography by Stadia

The method to be used for taking topography depends on the character of the terrain, and selected contour interval. A scale of 1 in. = 400 ft with 10-ft contours, is about the smallest useful combination. A better one is 1 in. = 200 ft, with 5-ft contours. A scale of 1 in. = 100 ft, with 5-ft contours, is also popular.

Under some circumstances the stadia method of taking topography is the most efficient one. This may be the case in open regions permitting unobscured sights, especially if a wide strip of topography is required. In this method, no station stakes are set between traverse stations, and profile leveling is unnecessary. All field work (except bench leveling) is done by one large party. It is a good practice to have a separate computer and field draftsman in the party, so the notes can be reduced and plotted as field work progresses. This method requires skilled personnel and careful supervision by the locating engineer.

11-12. Topography by Hand Level

Wherever a narrow strip of accurate topography is required through a region covered with brush or timber, the method of taking topography with hand level, rod, and tape is almost essential. In this method the topographic party is supplied with the ground elevation at each station stake, as determined by the profile leveling. Locations of contours on lines at right angles to the survey center line are then determined by the following method:

A perpendicular to the traverse is first established at each station, either by estimation, or with the aid of a pocket compass, an optical square, or cross staff. In timber or brush the transverse lines are kept reasonably straight by ranging through with three flags or range poles.

Location of the first regular contour on a transverse line is determined by hand leveling from the known ground elevation at the station stake. A forked stick cut for a 5-ft height of eye is a convenient hand level support. To illustrate the process, assume that the center-line ground elevation is 673.2 and locations of 5-ft contours are required along rising ground to one side of the center line. The levelman, resting the hand level in the fork of the 5-ft stick, directs the rodman out until the hand level reads 3.2 on the rod. This reading locates the 675-ft contour, the distance to which, say 36 ft, is measured with a metalic tape. If notes are kept (somewhat similar to cross-section leveling, section 6-9, the entry is recorded as 675/36. The levelman then continues out past the rodman until he reads 10.0 on the rod, and the distance beyond the previous point, say 42 ft, is measured. The corresponding entry would be 680/78. If distances between contours are too great for hand-level readings, intermediate readings are taken at shorter distances by the same step-by-step process until the desired contour is reached.

The hand-level method is surprisingly accurate. With a little experience, levels may be carried 400 ft from the center line and checked back with an error of less than 0.5 ft. Since each new cross line starts with a correct center-line elevation, there is no cumulative error.

In wooded terrain taking topography by hand leveling is faster than by any other method giving comparable accuracy. Only a thin gap need be cut through brush. Trees never need be cut, for the tape and sight line can be offset around them by eye without introducing serious error, considering scale of the map.

In addition to locating contours, the topographic party locates all buildings, property lines, highways, streams, rock outcrops, and other physical features likely to affect location.

Though some engineers prefer to record notes of contour locations (supplemented by sketches in the field book), a method which is usually more accurate and much faster is to plot the topography in the field at once on special field sheets. These are usually strips of cross-section drawing paper mounted on a topographer's sketch board about 12 × 18 in. in size. Sheets are prepared in advance by drawing the survey center line straight and continuous through traverse stations, provided the angles are small (possibly less than 5°). Where the deflection angles are larger, distortion of plotted topography is reduced by repeating the traverse station after leaving a 1-in. gap in the center line. A still better method—one that eliminates all distortion at large horizontal angles is to cut adjoining sheets on the proper bevel through the plotted position of the transit station and mount them on the sketch board so the center line is an exact reproduction of the traverse line. Lines on the cross-section paper are thus parallel and perpendicular to the center line, thereby being more convenient for plotting in the field. The final step in preparing the fields sheets is to record stations and elevations along one edge.

Particular advantage of the preceding method, aside from eliminating voluminous notes, is that contours and all other topographic details are plotted immediately in the field, where faithfulness of the reproduction is readily apparent. Moreover, field plotting to the subsequent map scale gives the topographer a clear idea of the degree of precision needed in the field work. He is better able to decide whether cross lines at some stations might logically be omitted, and if contours along additional lines (such as on lines parallel to the center line or on ridge lines and valley lines making an angle with the center line) are necessary to complete the topography of difficult sections.

11-13. Planimetric Maps by Photogrammetry

A planimetric map shows accurate positions of natural and cultural features such as watercourses, forests, highways, and buildings, and is constructed from aerial photographs. The first step is to plot control points on the map manuscript from computed coordinates. Additional points whose coordinates were determined by aerotriangulation, called *PUG* points (see section 11-15), are also plotted by their coordinates. A minimum of two horizontal and three vertical control points are needed to orient a stereo model. Of course, additional points are desired to ensure against blunders and provide a check on the orientation.

Positions of details are traced from the stereo model (see Fig. 11-2) to the map sheet using complex optical and mechanical instruments called "stereoplotters." Two adjacent photos are printed as transparencies on glass plates called "diapositives," and each is mounted in a projector. Some plotters can accommodate three or more diapositives by having additional projectors, but the "two-projector" plotter is most common. Through *direct*

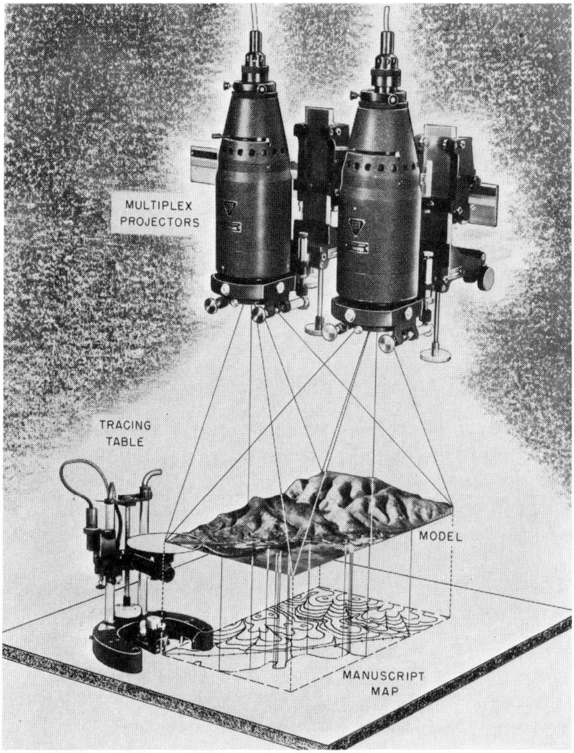

Figure 11-2 The stereo model. Shown is an artist's conception of the stereo model as seen by the plotter operator. (Courtesy U.S. Geological Survey.)

optical projection onto a tracing table, or by projecting the photos through an *optical train* to a binocular eyepiece, a viewer can see an exactly scaled "stereo model" of the overlapping area (see Fig. 11-3).

These instruments employ the principle of a floating mark. As the model is viewed, each eye will see a small dot. The dots, when adjusted properly, fuse into a single one, called the floating mark, which will appear in the viewer's stereovision either above or below the ground surface. Using a thumb or foot wheel, the operator can move the floating mark up or down to coincide with the ground surface or any other desired level. In addition, adjusting the elevation of the floating mark is mechanically converted into a reading on some type of counter wheel that displays the elevation of the floating mark.

Planimetric features are traced from the model by adjusting the floating mark height so it is in apparent contact with the road edge, building, stream edge, or tree line being mapped. Then the operator manually moves the floating

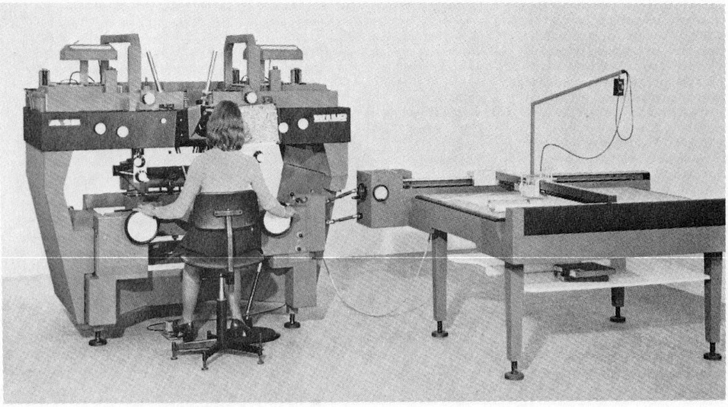

Figure 11-3 Optical train plotter. The operator views the model through an optical train of lenses and prisms and moves the measuring mark with handwheels. The coordinatograph at right is mechanically connected to the motion of the measuring mark and plots the manuscript. (Courtesy WILD-Heerburgg.)

mark along the desired line in the model. Motion of the mark is mechanically converted to movement of a pencil point on the map sheet. The pencil and map sheet are located directly under the tracing table on direct projection plotters. Optical train plotters have the pencil mechanically linked to the motion of the floating mark so plotting is done on a *coordinatograph* located directly beside the plotter within operator's view.

The manuscript map is edited and completed by the operator during plotting. In general, outlines of regular features such as buildings are initially plotted irregularly because of the operator limitations in following a precise line with the floating mark. Using straight edges and templates, the operator must "square up" buildings and smooth out other lines that are meant to be regular lines. Some current plotters have the ability to "square up" features automatically using a small computer that controls the plotting pencil after the operator traces the corners and buildings outlines. In addition, these semiautomatic plotters will draft standard mapping symbols such as center lines, fence lines, and trees.

11-14. Contours by Photogrammetric Methods

A stereoscopic plotting instrument may also be manipulated to measure differences in elevation between points. In principle, the first step is to adjust the floating mark until it rests on a vertical control point. The operator moves the floating mark horizontally to a point whose elevation is desired and makes the mark rise or sink until it is at the same elevation as the ground point. Finally the elevation is read from the counter.

To plot a contour, the desired elevation is preset on the elevation counter. The floating mark is moved until it appears to rest on the ground surface of the stereo model, and the plotting pencil dropped to start plotting. The ground surface is read by the operator, moving the mark, keeping contact with the ground. Obviously, the pencil continuously plots the contour. (See Fig. 11-4.)

The feasibility of plotting contours by photogrammetry is limited by dense foliage and very flat land. In sparse ground foliage, the operator of the stereoplotter will see random "spots" of ground and can measure their elevation. Contours are then drafted on the map by interpolation between spots. In very flat land, the operator cannot accurately judge which way to move the floating mark when continuously plotting a contour. This problem is solved by taking spot elevations on a grid pattern covering the flat land. If

(*a*)

Figure 11-4 Reproduction of an aerial photograph, (a) and a photogrammetric map, (b). (Courtesy Ammann International, Inc.)

CITY OF KEOKUK, IOWA

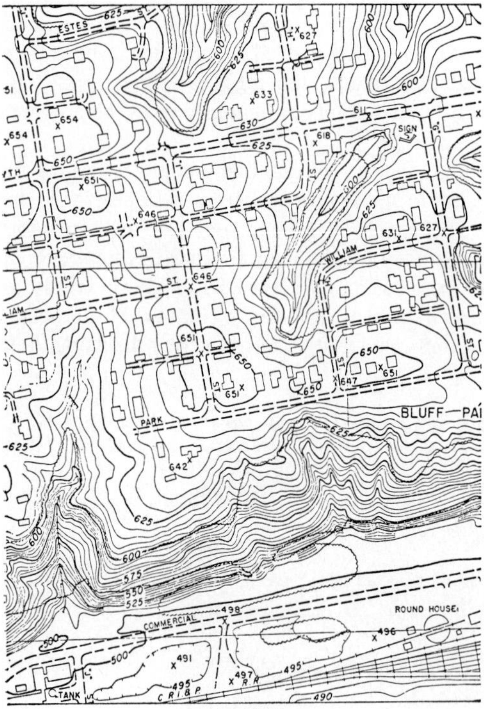

(*b*)

Figure 11-4 (*continued*)

desired, contours can be drafted by interpolation between the spot elevations taken.

11-15. Aerotriangulation as an Aid to Mapping

Aerotriangulation is the process of determining the *x*, *y*, and *z* coordinates of targeted points solely from photographic measurements using widely spaced ground control points of known position and elevation. Referring to Fig. 11-5, a strip of photos is shown "bridging" between ground control points. Basically, aerotriangulation achieves a densification of control points without costly field measurements.

Aerotriangulation can be accomplished three ways: by mechanical bridging, using a line of individual projectors on a long plotting surface; by semianalytic methods, consisting of computations following the measurement

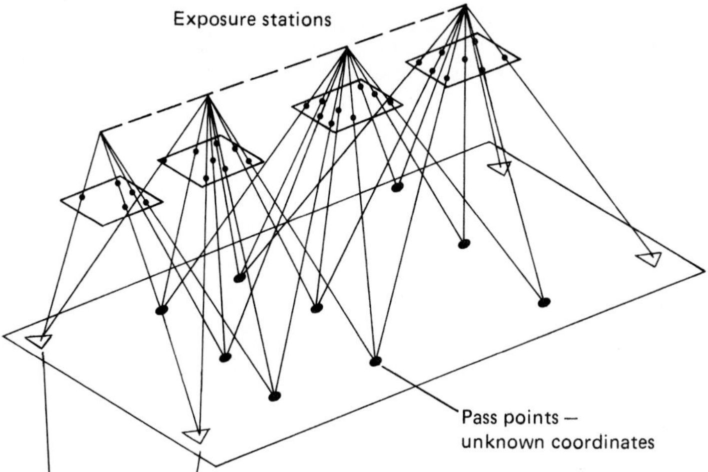

Figure 11-5 Aerotriangulation.

of model coordinates by a precise plotter; or by full-analytic computations from measurements of photo-coordinates using a *comparator*. A comparator is a device that measures precise coordinates on a photograph. In the analytic method photographic coordinates of control and pass points are measured. Since the ground point, photo point, and exposure station must be colinear, two equations can be written for each point measured. A large system of equations is then solved for unknown coordinates (x, y, and z) of pass points. Aerotriangulation aids mapping by providing many control coordinates and elevations needed to orient stereo models. For a more complete study of these methods, consult the textbooks on photogrammetry listed at the end of this chapter.

The technology, science, and equipment existing today allow an engineer to measure ground distances and elevations within 0.2 ft strictly using photographs and the process of aerotriangulation. Such possibilities have opened up new methods of route location and design as discussed in Chapter 13.

11-16. Enlarged Rectified Photomaps

A photomap is an actual portion of a photo enlarged and printed to desired scale. Enlarging and rectification take place in one operation as discussed in section 10-10. If ground relief is minimal, distortions caused by relief displacement will be small, and the photo can be used as a planimetric map.

The main drawback of a photomap in alignment design is the lack of elevations and contours. To solve the problem, contours are drafted on the photomap from cross sections taken in the field, or added from stereoplotter measurements. The principal advantage of the photomap is that all planimetric photo details are available for choosing the best alignment through built-up areas.

11-17. Orthophotomaps

An improvement over conventional stereoscopic photogrammetry took place about 1956 with the development of equipment to produce the equivalent of an orthographic photograph, that is a uniform-scale photograph free from distortions due to tilt and relief.

At about the same time a different system of producing topographic maps based on orthophotography was developed. The product, called the *photo-contour map* (Fig. 11-6), is not a mosaic with contours superimposed. Because of the method by which the photographic perspective is rectified, it is an orthographic projection at the same scale as the contour plotting. The finished map is a photo copy on which contours are shown as black or white lines, or a combination of both. Tone of the photograph usually determines which is preferable. The photo-contour map is practically self-checking. Inaccurate work in compiling is revealed by obvious mismatches in contours and planimetric features. Inspection of Fig. 11-6 shows that all contours properly track the visible drains and roads.

In general, photo-contour maps cost more than topographic maps made by conventional photogrammetric methods. However, where a proposed route location passes through areas having dense planimetric detail, they may be less costly because of saving in drafting time.

11-18. Scribing

The majority of base maps made for route location studies are inked on a transparent material such as vellum paper or drafting film. The process of *scribing* is an alternate method of map drafting with distinct advantages over ink or pencil drafting (see Fig. 11-7).

Scribing is the scraping of a special opaque paint-like coating away from one side of a clear plastic sheet along a desired line instead of putting an ink line on plastic as in the case of traditional drafting. The line forms a clear "window" that contrasts with the opaque coating of the sheet. First, a scribe sheet is placed over the map manuscript on a light table causing lines to be traced to show through. Then a "graver" is selected that will produce a line of desired width and the features are traced for that line width. The finished sheet is called a *scribed negative* because map features show as clear lines on an opaque background. This scribed negative is then transformed into the

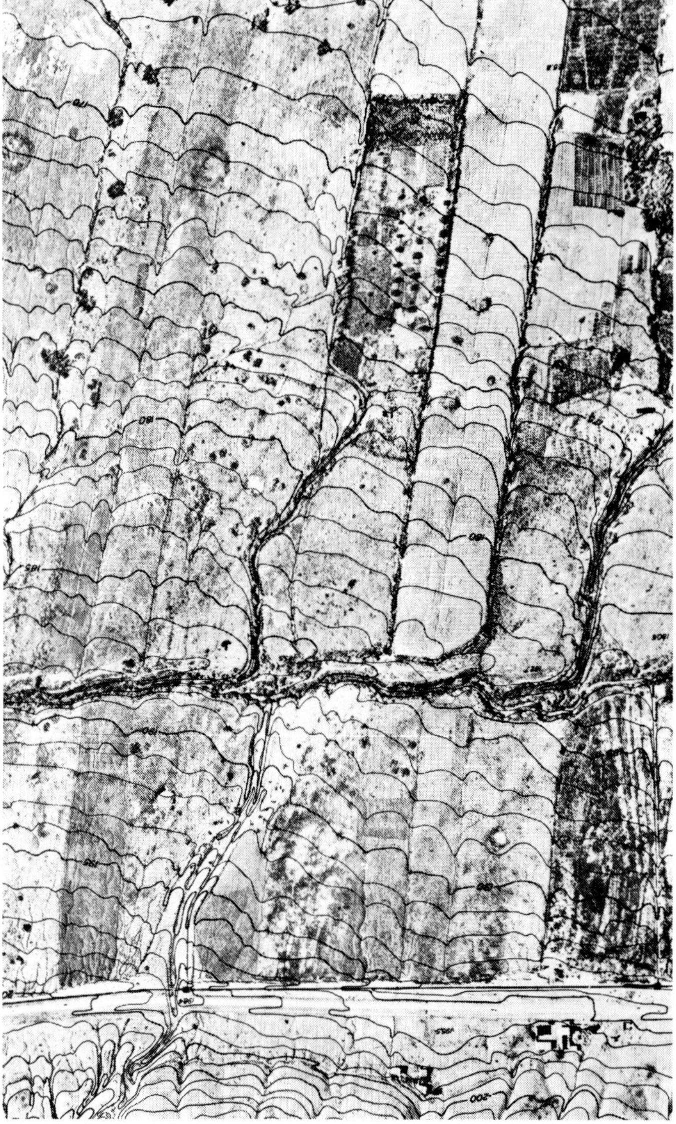

Figure 11-6 Photo-contour map. (Courtesy Aero Service Corp.)

(a)

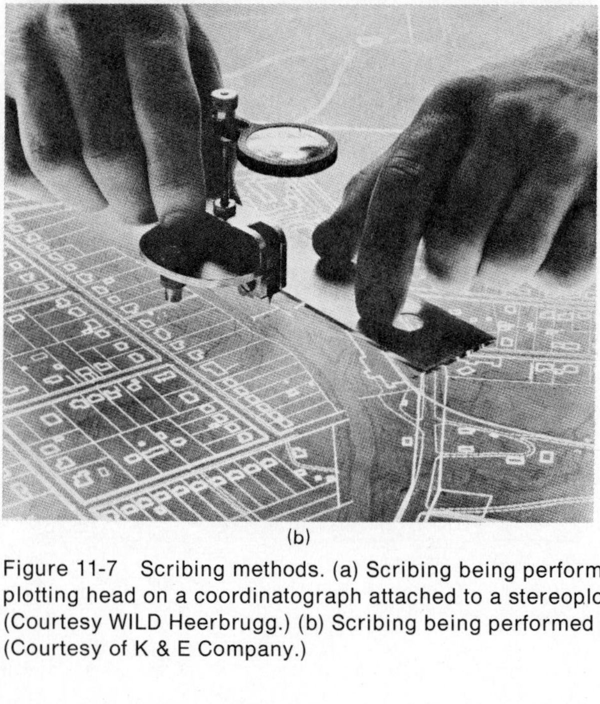

(b)

Figure 11-7 Scribing methods. (a) Scribing being performed by the plotting head on a coordinatograph attached to a stereoplotter. (Courtesy WILD Heerbrugg.) (b) Scribing being performed by hand. (Courtesy of K & E Company.)

finished positive map by contact printing on photographic drafting film in the darkroom. The exposed positive is developed for the final map.

Scribing's advantages over pen and ink drafting are that the finished negative is more durable with hard usage; the work can be done by low-paid employees with little training; revisions or corrections can be made by simply painting over the area to be revised and then recut; and the scribed lines are more consistent in width and quality. Disadvantages are (1) the materials cost more and (2) an extra photographic process is necessary to transform the negative into a positive map.

11-19. Automatic Mapping Systems

All mapping systems previously discussed produce either a photomap or a hand-drafted line map. Automatic mapping systems offer an engineer the opportunity to speed the mapping process and reduce errors inherent in plotting by hand-drafting methods. Automatic mapping systems have two main components: the digitizing unit and the computer-controlled plotter.

A map digitizing unit is an electromechanical device that transforms map points into x, y, and z coordinates and stores them on some useful medium such as computer cards, punch paper tape, or magnetic tape. An existing map can be digitized point by point using a *digitized drafting table*, containing a rectangular system of wires imbedded beneath the drafting surface and electronic cursor held in the operator's hand. As the cursor is centered on a map point, wires in the table sense the cursor's position and coordinates. When the operator presses a button on the cursor, the unit permanently stores coordinates on punch cards, punch tape, or magnetic tape.

Points in a stereo model can be digitized in x, y, and z by attaching three electronic encoders to two horizontal and one vertical motion of the floating mark. Outputs of the digitizers are fed into a recording unit located within reach of the operator. The record unit has a keyboard with which the operator enters numerical codes for various types of objects being plotted such as buildings, road edges, and contours. Features can be traced point by point in the stereo model digitized and coded. For continuous plotting of contours, the digitizing unit can be set to automatically record the x, y, and z coordinates of points at preset time intervals, for example one point per second. As a result, the whole stereo model consisting of buildings, roads, trees, utilities, and contours, is digitized into permanent storage on cards or tape.

The computer-controlled plotter consists of a pen-holder that is driven by a computer. Two types of plotter configurations are in use, (1) the drum type and (2) the flat bed type. A drum type mounts drafting material such as paper or drafting film on a cylinder of approximately 1 ft diameter. On flat bed plotters, the material is secured to a flat surface of dimensions up to 4 by 4 ft. The digitized and stored points are "read" into the plotter's computer,

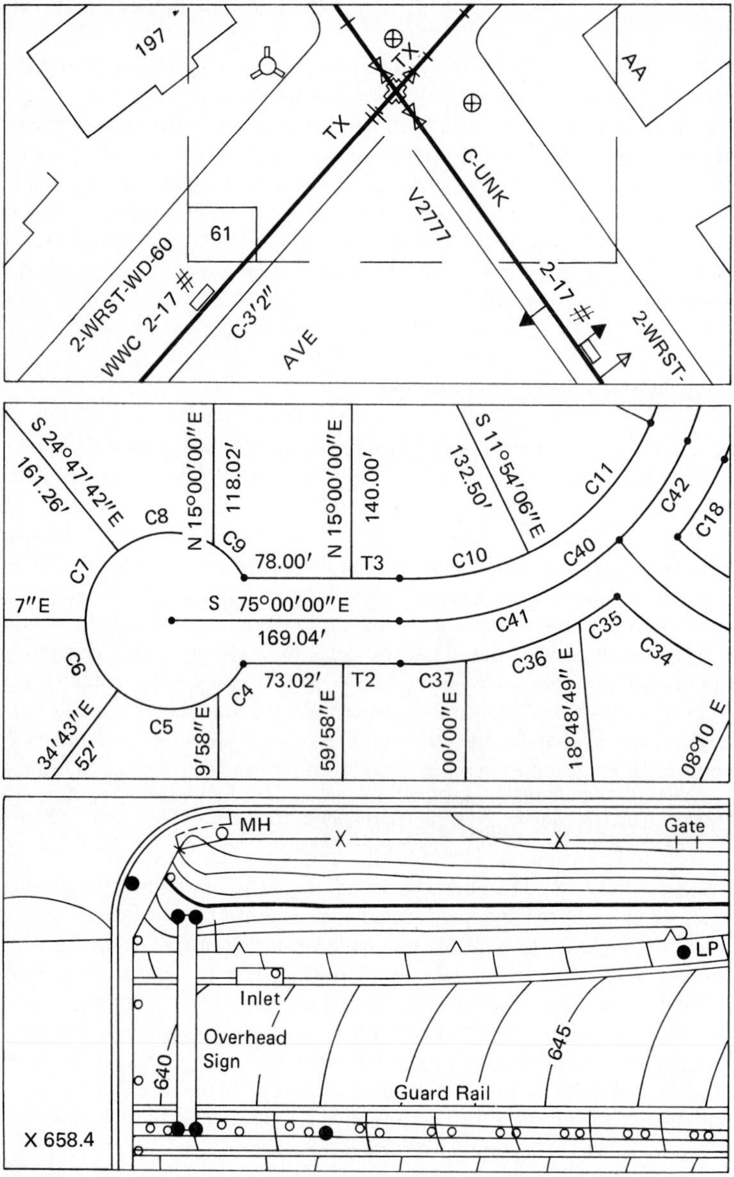

Figure 11-8 Maps drawn by automatic methods. The three example maps have been drawn by different plotter systems, the top by Calcomp Inc., the middle one by Wang Systems Inc., and the lower by an optical system. (Source: Kneeland A. Godfrey, Jr., "Making Maps by Computer," *Civil Engineering*, Feb. 1977, p. 42. Reprinted with permission of the author.

and following instructions of the operator, the mechanical pen-holder reproduces the map features under computer control. In some plotters, four different size drafting pens are mounted in the holder permitting the operator to choose the desired pen size for a particular feature.

Perhaps best map quality is produced by a photographic plotting process in which drafting pens are replaced by a controlled light source and the map sheet by photosensitized paper or film. A computer-driven pen-holder moves the beam of light, thereby exposing the map features on the photographic emulsion. The finished map is then developed by a standard photographic darkroom process. (See Figure 11-8.)

Advantages of automatic mapping are obvious. Once digitized, the finished map can be produced at any scale merely by changing instructions to the computer prior to plotting. Total time required for the drafting is reduced to a fraction of that required by hand methods. To edit or revise a map, the coordinates are entered into the computer replacing the old ones. A revised map is then completely redrawn.

SUPPLEMENTAL READINGS

Aquilar, A. M. 1975. Evaluation of Automatic Cartographic Systems for Engineering Mapping Functions. *Journal of the Surveying and Mapping Division, A.S.C.E.* 101, pp. 17–26.

Brown, D. C. 1976. Doppler Positioning by the Short Arc Method, a paper presented to the International Geodetic Symposium on Satellite Doppler Positioning, published by Geodetic Services Incorporated, Melbourne, Fla.

Cartographic Scribing Materials, Instruments and Techniques, American Congress on Surveying and Mapping, Technical Monograph No. CA-3, June 1970.

Derenyi, E. E. and Maarek, A. 1973. Photogrammetric Control Extension for Route Design. *Journal of the Surveying and Mapping Division A.S.C.E.* 100: 49–62.

Hou, M. C. Y. 1973. Aerotriangulation Precision Attainable for Highway Photogrammetry. *Journal of the Surveying and Mapping Division A.S.C.E.* 100: 7–14.

Karara, H. M. and Marks, G. W. 1970. Analytical Aerial Triangulation for Highway Location and Design. *Highway Research Record* Bulletin 319: 1–2.

MacLeod, M. H. and Turner, J. B. 1973. Semiautomated Large Scale Mapping. *Highway Research Record* Bulletin 452: 40–44.

Making Maps by Computer. 1977. *Civil Engineering—A.S.C.E.*, Vol. 47 No. 2.

Moffitt, F. H. and Mikhail, E. M. 1979. *Photogrammetry*, 3rd ed. New York, Harper and Row.

Saxena, N. K. 1975. Electro-Optical Short-Range Surveying Instruments. *Journal of the Surveying and Mapping Division, A.S.C.E.* 101: 137–147.

Winikka, C. C. and Morse, Jr, S. A. 1974. Orthophotoquads for Arizona Land-Use Mapping and Planning. *Journal of the Surveying and Mapping Division, A.S.C.E.* 100: 1–6.

Wolf, P. R. 1974. *Elements of Photogrammetry*. New York: McGraw-Hill.

Chapter 12 Basic Methods in Alignment and Roadway Design

12-1. Foreword

The outline on p. 285 indicates coverage of this chapter which emphasizes surveying techniques. Alternate systems, taking more advantage of precise photogrammetric mapping, are presented in Chapter 13. Refer to Chapters 2 through 9 for technical information and methods of performing the steps noted.

12-2. Alignment and Roadway Design

After the corridor for a proposed route is chosen in reconnaissance study and mapped to a medium scale in the preliminary survey, it can be designed by the engineer in two steps, *alignment design* and *roadway design.* Alignment design fixes the exact horizontal geometry of the route's *base line*, also termed the *location line* (*L-line*). Roadway design, often termed *final design*, applies a chosen *typical section*, or *template*, to the selected alignment to determine the exact extent of all items of construction within construction limits.

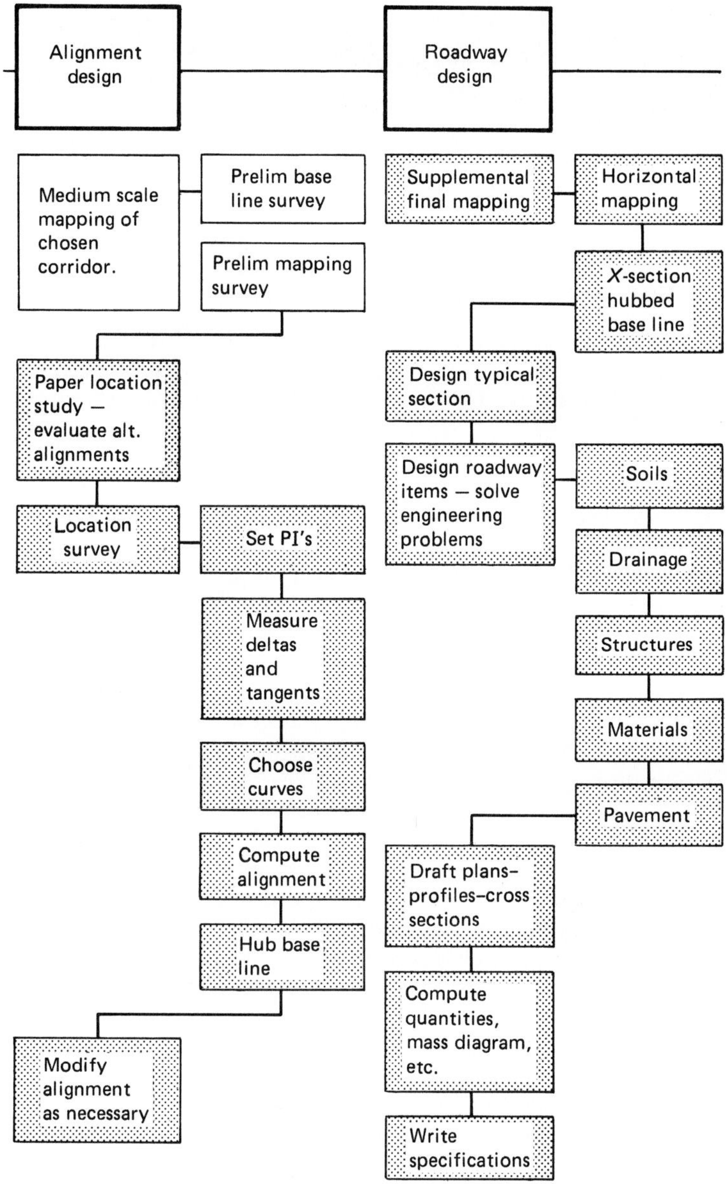
Alignment design
Roadway design
Medium scale mapping of chosen corridor.
Prelim base line survey
Prelim mapping survey
Paper location study — evaluate alt. alignments
Location survey
Set PI's
Measure deltas and tangents
Choose curves
Compute alignment
Hub base line
Modify alignment as necessary
Supplemental final mapping
Horizontal mapping
X-section hubbed base line
Design typical section
Design roadway items — solve engineering problems
Soils
Drainage
Structures
Materials
Pavement
Draft plans–profiles–cross sections
Compute quantities, mass diagram, etc.
Write specifications

The base line is some chosen line of the roadway such as the center line of pavement for two lane roads, center line of the median for divided highways, edge of the pavement for interchange ramps, or the center line of rails in railroads. Once designed, the base line is made up of tangents, curves, and spirals of mathematically fixed size and relationship. Stationing of the route is expressed by distance along the base line. Positions of roadway features are described by offsets from stationed base line points.

12-3. Tools for Alignment Design—Visual Inspection, Topographic Map, the Digital Terrain Model

The most basic method of alignment design is visual inspection in which the route's alignment is chosen completely in the field. For low speed roads, such as those in a park, one valid method of alignment design is to simply drive a vehicle at a comfortable pace along a curving path taking advantage of natural scenery. The track of the vehicle is then staked at the points of curvature thereby dividing the track into a series of curves and tangents. Additional stakes are placed on curves and tangents at regular intervals. The road is constructed by following the set stakes. Vertical alignment follows the natural ground surface. It is obvious that the particular horizontal and vertical alignments chosen are not critical to the road's cost, maintenance, or functioning. However, this is not the case for higher types of routes in which the engineer is forced to meet many restrictions of curvature, sight distance, and grade through densely populated terrain containing horizontal and vertical controls. To meet the restrictions, massive earth moving and purchase of high-priced land may be necessary, causing the route's cost to skyrocket. Thus the engineer is forced by economics to spend additional time planning and studying alternative alignments in an attempt to find the "best" solution.

The tool used by engineers for years in route alignment studies is the *medium-scale topographic map* (scale 1 in. equals 100 ft) to supplement visual inspection. As trial lines are graphically constructed on the map, their intersection with contour lines are scaled and an approximate profile of the existing ground along the trial alignment is plotted to guide the choice of vertical alignments. Likewise, cross lines are constructed on the map, and their intersections with contour lines scaled to produce existing ground cross sections at regular intervals. As vertical alignment is tentatively chosen, the typical section is used to plot proposed cross sections, which are employed for earthwork calculations. The alignment study results in the chosen *paper location* (*the L-line*), a graphical representation of the route. The paper location procedure is discussed further in section 12-4.

To reduce time taken for plotting profiles and cross sections of trial alignments, a *Digital Terrain Model* can be effectively used. A digital terrain model is a numerical representation of the ground surface within a route corridor consisting of stored *xyz coordinates* of a large number of points in a computer forming a "model" of the ground surface through which trial alignments may be passed and studied. Alignment design using a digital terrain model is discussed further in section 12-5.

12-4. Paper-Location Procedure

The preliminary surveys and subsequent office work result in a topographic map of a strip of territory, varying possibly from 400 to 1000 ft in width, in which the ultimate location is expected to lie.

Finding a satisfactory location having suitable curves and grades is not usually a difficult task; it is largely a technical process involving patience and quite a bit of routine scaling and calculation. But finding the *best* location requires something more than drafting-room technique. As stated in section 1-5, "To produce a harmonious balance between curvature and grade, and to do it economically, requires the engineer to possess broad experience, mature judgment, and a thorough knowledge of the objectives of the project."

It is obviously impossible to write a set of rules which, if followed, will inevitably produce the best location. The set which follows is merely a suggested office procedure that will be principally of value to the novice; the intangible ingredients, skill and judgment, grow with the locator's experience.

(*Note:* In the technique to be described, splines may be substituted for thread and curve templates.)

1. Using the preliminary map, set small pins at the termini and also near the intermediate controls. Stretch a fine thread around the pins.
2. Examine the terrain along the thread line, and set additional pins (angle points) where they apparently are needed between location controls.
3. Scale the elevations at the pins and the distances between them, thus giving approximate grades.
4. Place transparent circular-curve templates tangent to the thread. Shift pins and change templates until a trial alignment is obtained which fits the alignment specifications and appears to be reasonable in so far as gradients and probable earthwork quantities are concerned.
5. Pencil the trial alignment lightly but precisely, and station it continuously along the tangents and curves by stepping around it with dividers.
6. Plot the ground profile of the trial alighment to an exaggerated vertical scale.
7. Using pins and thread, establish a tentative grade line on the profile that fits the specifications for grades and appears to produce a reasonable balance of earthwork. Circular-curve templates can be used for vertical curves. In estimating earthwork balance from center cuts

and fills, make approximate allowance for the fact that the graded roadbed is wider in cuts than on fills and that a certain percentage will be added to the fill quantities for shrinkage.

8. Examine the trial alignment and grade line together. Go over the alignment station by station, visualizing the finished roadbed if built as indicated. The need for drainage structures and the probable maintenance difficulties dealing with drainage, slides, and snow drifting should be examined. Then make whatever changes will obviously improve operating and maintenance characteristics without changing earthwork quantities appreciably.
9. Make an earthwork estimate, using the following column headings:

	CENTER	CUBIC YARDS	
STATION	HEIGHT	CUT	FILL = *%

(*Insert suitable shrinkage factor—possibly from 10 to 15% for purposes of estimate). In the station column, enter in proper order each 100-ft station, the station and plus of each high or low point, and the station and plus of each grade point. Scale center heights from the profile at positions midway between the points entered in the station column; designate the values *C* or *F*. Take earthwork quantities from the proper table of *level sections* (Table XVII, Part III). Do not forget to reduce tabulated quantities for fractional distances or to add the shrinkage percentage to the fill quantities.

10. Enter the subtotals for each increment of cut and fill. If conditions appear to warrant striving for an approximate balance of quantities along certain grading sections, see how close the balance comes. Observe also the relative sizes of adjoining cuts and fills and the approximate distances between their centers of gravity. For convenience, note these numerical values on the profile.
11. Reexamine the tentative alignment and grade line in the light of the numerical values obtained in the first earthwork estimate. Make any minor changes in the alignment or the grade line, or both, which will reduce the pay quantities and improve the balance and distribution of earthwork. Do not erase the first trial; merely use a different style of line for the revised portions. Scale new center heights along the revised portions and determine the new quantities.
12. Repeat step 11 until the location appears to be the best one possible.

The paper location finally established by the foregoing process should not be accepted rigidly as final. Though the earthwork estimate is based upon level sections, the errors introduced by transverse slopes tend to cancel. In a long line, level-section quantities will usually be within 5% of the true values found later by cross sectioning.

12-5. Alignment Design with the Digital Terrain Model

The digital terrain model (DTM) system, initially developed at Massachusetts Institute of Technology, is a system that results in maximum integration of photogrammetry and computer in highway work. Figure 12-1 gives a pictorial concept of a terrain model stored in a computer.

A digital terrain model is "built" within the computer in several ways: by field cross sections, digitizing elevations from an existing map, or by directly digitizing coordinates and elevations from the stereo model of a stereoplotter. Using field cross sections, the horizontal position (x and y) of each rod shot and ground elevation (z) is entered and stored thereby forming the terrain model. An existing contour map is digitized by plotting a coordinate grid on the map and measuring coordinates of desired features. Elevation of each grid point is found by interpolation between adjacent contours. The x, y, and z coordinates are then entered in the computer to form the digital terrain model.

The best integration of computer and photogrammetry is achieved by digitizing the stereo model directly without plotting the medium-scale map. Figure 12-2 shows a commercial system for digitizing stereoplotter output. It consists of a Kelsh plotter modified so the horizontal and vertical movements of the tracing table are registered by encoder assemblies as the table is constrained to move along the horizontal guide bar (cross-section bar). Encoders convert tracing table positions to a digital coded output. When the floating point is on the ground at a desired place, the plotter operator steps on a foot switch that stores horizontal coordinates and elevation on a computer-input medium such as punch tape, punch cards, or magnetic tape.

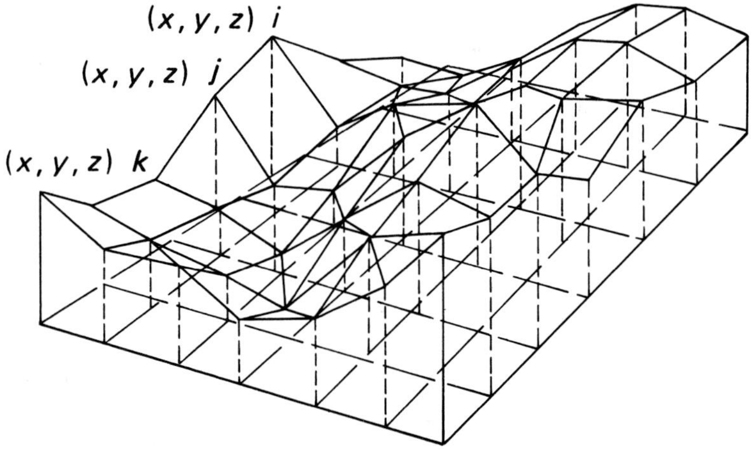

Figure 12-1 Sketch of a digital terrain model.

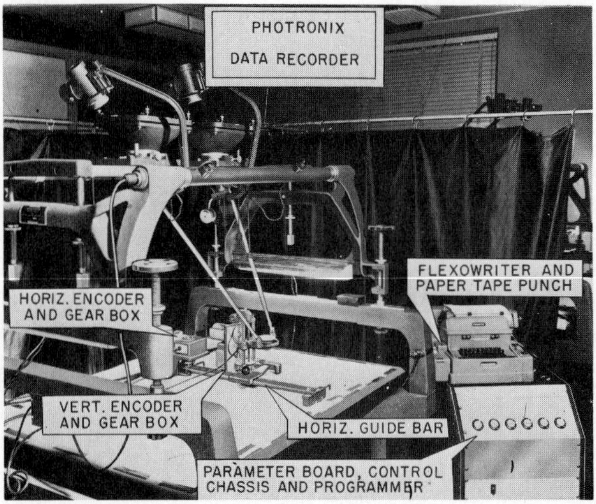

Figure 12-2 Commercial system for digitizing stereoplotter output. (Courtesy Photronix, Inc.)

Coordinates are subsequently read into the computer to form the terrain model.

After the terrain model is finished, an engineer will study many alternate trial alignments by introducing desirable ones and a chosen typical section into the computer. The computer mathematically passes the desired location through the terrain model and computes existing cross sections, proposed cross sections, and earthwork quantities. Cross sections may be output in digital form consisting of stations, offsets, and elevations listed on paper, or the computer will draw the cross sections directly on a computer-controlled plotter.

Many engineers believe there is no good substitute for a topographic map on which the designer can lay trial lines and visualize many other matters pertaining to the design. However, automatic readout of terrain data directly from a stereo model theoretically eliminates the need for a map.

In studying the aesthetics and visual appeal of proposed trial alignments, *interactive graphics* have begun to play a large role. Under federally funded research programs, several States have developed interactive graphic systems that allow an engineer to view on a CRT (cathode ray tube) display a perspective of the proposed route from the viewpoint of the driver. Studies have shown that highway safety is increased when the driver sees a smooth, continuous strip of pavement ahead. Sudden discontinuities in the appearance of approaching pavement are caused by an inappropriate combination of horizontal and vertical pavement curvature. Interactive graphics, tied into a digital terrain model, allows the engineer to actually "drive" the proposed route and view the sequence of scenes on the CRT display. Faults and discontinuities in the approaching pavement can be corrected by modifying the alignment design parameters.

12-6. Alignment Computation—Computer Applications—COGO

Once the desired alignment is fixed on paper, the next job is to describe it mathematically by calculating the base line parameters.

Two general approaches exist for fixing the chosen alignment geometry. The first and more traditional approach is to perform a *location survey* during which the proposed alignment is staked in the field and the precise base line geometry field measured. Location survey procedures are discussed in subsequent sections 12-7 and 12-8. A second approach, formulated and published by Jack E. Leisch, eliminates the need for field staking at the design stage by relying on precision corridor mapping. This second approach is presented in Chapter 13.

The location survey approach admits that the mapping may not be perfect, and therefore a field location is desirable before proceeding with final design. Additional information may be found upon field staking that escaped initial mapping requiring small changes in alignment. The second approach is based on the completeness and accuracy of aerial photos and medium scale mapping performed photogrammetrically. Finalization of alignment is done without fear of missing a key land feature affecting design.

In either of the two approaches, the electronic computer has become a tool that frees the engineer of a multitude of routine calculations needed to finalize the alignment mathematically. When first applied to route location, the digital computer merely replaced the desk calculator in performing familiar routine computations. As the spectacular time savings became known, computer programs were developed for all phases of horizontal and vertical alignment—straight and curved, for structural design, traffic problems, soils analyses, and administrative procedures. And the use of computers mushroomed. By 1961, 49 of the 53 member departments of AASHTO were using computers in highway work; their use is now universal.

The financial burden of program development has been eased by the formation of "user groups" composed of highway departments and consulting firms using the same make and model of computer. In order to encourage exchange of programs, the Federal Highway Administration has a *Computer Program Library* that serves as an agency for the receipt and distribution of programs used in all fields of highway operation.

A valuable approach called COGO (from COordinate GeOmetry) was introduced in 1961. COGO freed engineers from having to force a solution to an established form. It gave him control over the form of input/output and the sequence of mathematical operations, without requiring costly program investment. Since then, COGO has become one of several subsystems in the

Integrated Civil Engineering System (ICES). Information on available ICES systems is circulated at intervals.

12-7. The Location Survey

The principal purpose of the location survey is to transfer the paper location, called the "*L*-line," to the ground. This may be done quite accurately if the preliminary map is based upon a transit-tape traverse plotted by coordinates.

The first step is to scale coordinates of the *L*-line P.I.'s and from them compute bearings of the tangents and distances between P.I.'s. If the map is carefully drawn to a scale of 1 in. = 100 ft, these computed bearings and lengths are usually reliable to the nearest minute and foot. The bearings and lengths are useful not only in making ties to the *P*-line (preliminary traverse line) but also in checking the location field work.

The *L*-line should never be run by turning the calculated angles and measuring the calculated distances continuously from beginning to end. Instead, each *L*-line tangent should be tied independently to the *P*-line and run to a string intersection with the adjacent *L*-line tangent, at which point the exact central angle of the curve is then measured.

An excellent tie exists wherever an *L*-line tangent crosses the *P*-line. The *P*-line stationing of the crossing can be scaled closely enough for the purpose if the angle of intersection is large; otherwise, it may be computed by coordinates. A hub at the intersection is then located by sighting and measuring from the most convenient *P*-line hub, after which the *L*-line tangent is projected in both directions by setting up at the intersection hub and turning off the calculated angle.

L-line tangents may sometimes be tied to the *P*-line by right-angle offsets. Another method is to produce a *P*-line course for a scaled or calculated distance to an intersection with the *L*-line. Considerable ingenuity and field experience are needed in establishing ties rapidly and accurately.

Great care should be taken to ensure the straightness of the *L*-line tangents. The lines should always be produced forward through P.O.T.'s by double centering. Obstacles on the tangents may be bypassed by the method described in section 12-8.

After the P.I.'s are located and the central angles measured, the curves and spirals are run in as described in Part I. Stationing is continuous along the tangents and curves. Some minor adjustments in alignment are usually made during or following the curve staking. (See section 1-12). These adjustments may involve some of the special curve problems described in Chapter 7.

It is a good plan to run profile levels over the staked *L*-line to compare the actual ground profile with the paper-location ground profile. The degree of "fit" is a measure of topographic accuracy.

If it is suspected that the earthwork estimate may be in error, owing possibly to the prevalence of side-hill sections, the line may be cross sectioned.

The more accurate earthwork quantities may then justify some slight grade line revisions without necessarily changing the alignment. Further analysis of earthwork distribution can be made by means of a mass diagram (see section 6-22).

The location survey also includes ties to property lines and existing improvements, as well as a variety of measurements needed for the design of miscellaneous structures.

12-8. Bypassing Obstacles on Tangents

Where location tangents (or other straight survey lines) are produced through woods, large trees often obstruct the line. To avoid felling them without authorization, and also to save time, the best recourse is to bypass them by a small-angle deflection-angle traverse.

Figure 12-3 and the accompanying notes illustrate the process. Point A is a transit station (P.O.T.) on the location tangent. It is assumed that large trees obstruct the line beyond A. Consequently, a small deflection angle is turned to the right and a hub B set at any convenient distance. At B it does not prove possible to get back on the tangent; therefore, the auxiliary traverse is continued through convenient openings between the trees until point D is reached, after which the tangent is resumed at E.

It is convenient to adopt a systematic form of notes that will indicate the distance from the tangent as well as the measurements needed to get back on line. In the form suggested, for example, the algebraic sum of the products of distances and angles in minutes must be zero up to a point on the tangent. Moreover, since the reciprocal of the sine of 1′ is 3440, the offset at any point is found by dividing the proper algebraic sum by that constant. The sign of any product is determined by the angle that the course makes with the tangent. An angle to the right gives a plus product; one to the left, minus.

In the example given, it is assumed that a deflection angle of 20′ R at D is a convenient direction toward the tangent. The required distance to a P.O.T. at E is, therefore, $4060 \div 17 = 238.82$ ft. At E the deflection required to place the line of sight on the correct tangent produced is obviously 17′ R.

If the angle between any auxiliary course and the tangent is kept below 1° or 2°, a negligible error results from assuming that the length along the auxiliary traverse is the same as the distance along the tangent. In the foregoing example the true tangent distance between A and E is only 0.06 ft shorter than the traverse distance of 1698.82 ft. This small correction is found quickly by slide rule, if required, by summing the products of the distances and the versines of the angles with the tangent.

It is not usual to move station stakes back on the correct tangent. However, if required, it may be done by eye with the aid of the offsets in the last column of the notes.

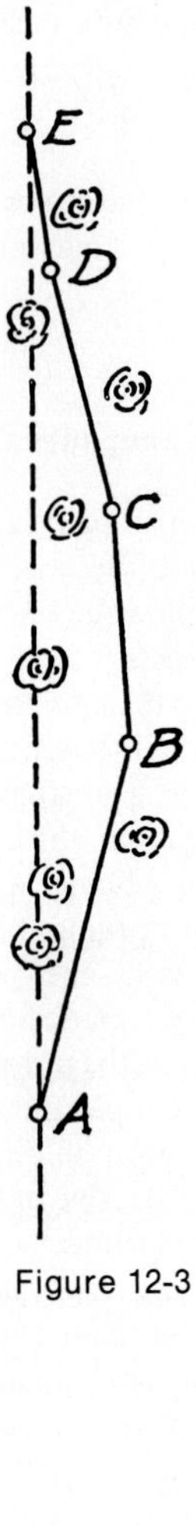

Figure 12-3

NOTES FOR FIG. 12-3

				PRODUCTS		
STA.	DEFL.	ANGLE WITH TANGENT	DIST.	ANGLE X DIST.	ALG. SUM	OFFSET FROM TANGENT
A	35′ *R*					zero
		35′ *R*	640	+22,400		
B	42′ *L*				+22,400	6.5 *R*
		7′ *L*	400	−2,800		
C	30′ *L*				+19,600	5.7 *R*
		37′ *L*	420	−15,540		
D	20′ *R*				+4,060	1.2 *R*
		17′ *L*	238.82	−4,060		
E	17′ *R*				zero	zero
		zero				
Tang.						

12-9. Roadway Design—Preparation of Plans

Office procedure in design and preparation of plans for a major highway project involves a multitude of operations. Some are quite routine and may be done by subprofessional members of the design team; others require specialized training and experience.

State highway departments usually follow certain "standards" with regard to methods of design, sizes of drawing sheets, arrangement of work, and forms for estimating quantities. No one scheme is best for all projects; a great deal depends on the size and type of project and on the personnel available. However, on Federal-Aid projects certain specifications relative to size of drawing sheet and form of layout must be followed.

The final objective of the office work is to prepare a cost estimate and complete set of plans showing clearly all information needed (1) by the engineers in laying out the lines and grades to be used by the contractor in building the project, (2) by the contractor in estimating the nature and extent of all work to be performed, in order that he may prepare his bid, and (3) by the legal agents to assist in preparing right-of-way descriptions and other data connected with land takings and easements.

A detailed description of office design methods not only would be too voluminous for inclusion in this book but also would encroach upon subjects more properly treated in a study of highway engineering. Consequently, only an outline of conventional office routine is given to show the relation between survey work and design.

Supplied with the reconnaissance report and all alignment data from the location survey, designers carry out these steps:

1. *Cross-section base line.* Field measure the precise profile and cross sections of the alignment as staked during the location survey.

2. *Final mapping.* Perform final mapping of the precise location of features within construction limits. Mapping may be selective depending on the quality of preliminary mapping and may include profiles of all intersecting streets and drainage systems. In general, anything not included in the medium scale mapping or not located accurately must be included in the final mapping.

3. *Right-of-way surveys.* Make right-of-way surveys in the field. As soon as approximate construction limits are known, choose a right-of-way width, establish property lines of parcels to be taken, and construct land descriptions.

4. *Design typical sections.* These are dimensioned drawings showing roadway cross sections of standard portions of the project. Included are width, thickness, and crown of pavements; shoulder widths and slopes; positions of ditches, side slopes, curbs, median strips, guard rails, and other construction details.

5. *Prepare location base maps.* Usually these are done on a series of 22 × 36 in. Federal Aid Sheets, which show the profile as well as plan. Common scales are 1 in. equals 100 ft for rural routes and 1 in. equals 40 ft or

50 ft for urban areas. The plan shows survey base lines, topographic details, and all alignment data. The profile depicts the ground line.

6. *Plot cross sections.* Ground cross sections, used for earthwork calculation and grade-line design, are plotted directly from cross section leveling notes. A common scale is 1 in. equals 10 ft.

7. *Establish final profile grade.* Finalize grade-line design considering relative importance of economy of construction, balance of earthwork quantities, property damage, sight distances, safety of operation, drainage and soil conditions, aesthetics, environment, adaptability to future property development and future grade separations.

8. *Draw cross-section plans.* Proposed roadway cross sections are drawn on ground cross-section sheets in conformity with designed profile grades. These sections show the pay lines for excavation. Widening and superelevation are allowed for.

9. *Finish plan-profile sheet concurrent with final design.* Determine the exact location of all roadway items such as curbs, sidewalks, lighting, signalization, turnouts, striping, signing, sodding, seeding, and so forth. Draft all items on the plan-profile, cross-section sheets.

10. *Make special detail drawings.* Include detail drawings of all types of drainage structures, retaining walls, and complicated interchanges and intersections.

11. *Prepare right-of-way plans.* Property maps of all parcels to be acquired are drafted showing locations, owners' names, and ties to existing and proposed right-of-ways.

12. *Estimate quantities.* Make detailed estimates of quantities of paving, grading, and other construction work. A *Quantity Summary* is the basis for the engineer's cost estimate and an aid to contractors in preparing bids.

13. *Prepare specifications.* Detailed general and special provisions are written relating to contract conditions, submission of bids, prosecution of work, construction details, and methods of measurement and payment.

12-10. Construction Surveys

Briefly, the principal survey work related to new construction includes the following operations.

1. Reestablishing the Final Location

Checking and referencing key points, such as P.I.'s, T.C.'s, C.T.'s, and occasional intermediate points on long tangents, so that they are quickly available during construction.

Resetting enough station stakes on curves to control clearing of the right-of-way.

Checking bench-mark elevations and setting convenient new bench marks in locations where they are not likely to be disturbed during construction.

2. Setting Construction Stakes

Cross-sectioning the line after clearing (and just ahead of grading operations), together with setting slope stakes wherever needed, setting line and grade stakes (and in some cases, batter boards) for appurtenant structures, such as buildings, bridge piers, culverts, and bridges; finishing stakes for completing cuts and fills to exact grade; stakes for borrow pits; and stakes for right-of-way fences.

3. Periodic Quantity Measurements

Measurements, calculations, and estimates of work done to serve as basis of monthly payments to contractor, as well as for progress reports to headquarters.

4. Final Measurements

Final cross sections for calculation of total grading pay quantities.

"As-built" measurements of all work to serve as basis for final payment, as well as for preparation of "record" plans.

Monumenting curve points.

5. Property Surveys

Making all measurements needed for preparing legal descriptions of easements and land acquired by purchase or through condemnation proceedings.

Setting right-of-way monuments.

In the conventional layout method, tacked line stakes, marked with station and offset, are set no more than 50 ft apart on lines offset from the construction base lines. Their elevations are determined and recorded for future use in setting grade stakes. After the right-of-way has been cleared, a double line of slope stakes or "rough grading" stakes is set at 50-ft intervals. Finishing stakes are necessary for the final operations of side-slope trimming, subgrade preparation, and setting forms for paving. After the grading has been completed, "blue-topped" line and grade stakes are set on the subgrade near enough to the work to permit placing forms using a short grade board.

In mountainous terrain, where grading is very heavy and there are the complications of variable slopes and benches, the customary method of setting construction slope stakes is very clumsy. Instead, a "traverse method" may be used to great advantage. Because of its specialized application, the traverse method will not be described.

Freeway design, in which double roadways are often at different levels separated by a median of varying width, complicates not only construction staking but also calculation of grading quantities. Good results can be obtained by substituting a contour grading plan for the usual voluminous set of cross-section sheets. In essence, "contour grading" consists of superimposing contour lines of proposed construction on the existing contour map, thereby forming a series of areas bounded by closed contours. The areas are planimetered, and the volumes of horizontal slices of earthwork are determined by the average-end-area method. This procedure is subject to further refinement and greater accuracy if partial contour intervals are taken into account. As a result of some time studies, it is estimated that earthwork calculations, together with the drafting and survey operations, can save about 40% in man-hours. Even this saving is small, however, compared with that resulting from use of electronic computers.

Warped surfaces at intersections require specially worked out staking arrangements in order to produce smooth riding surfaces.

Record plans of all work "as built" are prepared as construction proceeds. Since pavement is usually paid for on the basis of surface area, final measurement of the project length is somewhat greater than horizontal survey measurement.

SUPPLEMENTAL READINGS

Beilfuss, C. W. 1973. Interactive Graphics in Highway Engineering. *Highway Research Record* Bulletin 455: 1–10.

Berrill, J. B. and Feeser, L. J. 1973. Photo-Computer Plot Montages for Highway Design. *Highway Research Record* Bulletin 437: 1–8.

Computer User Groups. 1974. *Civil Engineering*, Vol. 44, No. 6:51–53.

Howard, B. E., Brammick, Z., and Shaw, J. F. B. June, 1968. Optimum Curvature Principle in Highway Routing. *Journal of the Highway Division, A.S.C.E.* 94, No. HW1.

Maxwell, D. A. 1970. Mathematical Surface Approximation of the Terrain. *Highway Research Record* Bulletin 319: 16–29.

McNoldy, C. E. 1970. Highway Location and Design Utilizing Photogrammetric Terrain Data. *Highway Research Record* Bulletin 319: 30–39.

Schultz, V. H. 1971. Evaluation of Photogrammetric Cross Sections for Earthwork Payment. *Highway Research Record* Bulletin 375: 1–7.

Smith, B. L. and Holmes, R. C. 1973. Highway Design Models: Scales and Uses. *Highway Research Record* Bulletin 437: 23–31.

Smith, B. L., Yotter, E. B., and Murphy, J. S. 1971. Alignment Coordination in Highway Design. *Highway Research Record* Bulletin 371.

Suhrbier, J. H. and Roberts, P. O. 1965. Engineering of Location—Selection and Evaluation of Trial Grade Lines by Electronic Digital Computer. *Highway Research Record* Bulletin 83: 88–113.

Chapter 13
Alternate Route Survey and Design Systems

13-1. Foreward

Basic procedures in surveying, alignment design, and layout are described in Chapters 11 and 12. These procedures comprise a time tested method successful in route design. However, along with the advent of photogrammetry and electronic distance measuring have come opportunities to create new route survey and design systems. Today, many consulting engineers and transportation organizations are choosing various components to form systems that match their individual traditions, requirements, and equipment. The newer procedures rely heavily on use of the computer, photogrammetry, electronic measurement, and a coordinated base line. One modern system, originated and published by Jack E. Leisch in articles listed at the end of this chapter, streamlines alignment design by eliminating the location survey layout until required for construction. Additional time saving techniques are incorporated to speed the alignment computation phase. The method will be discussed in this chapter.

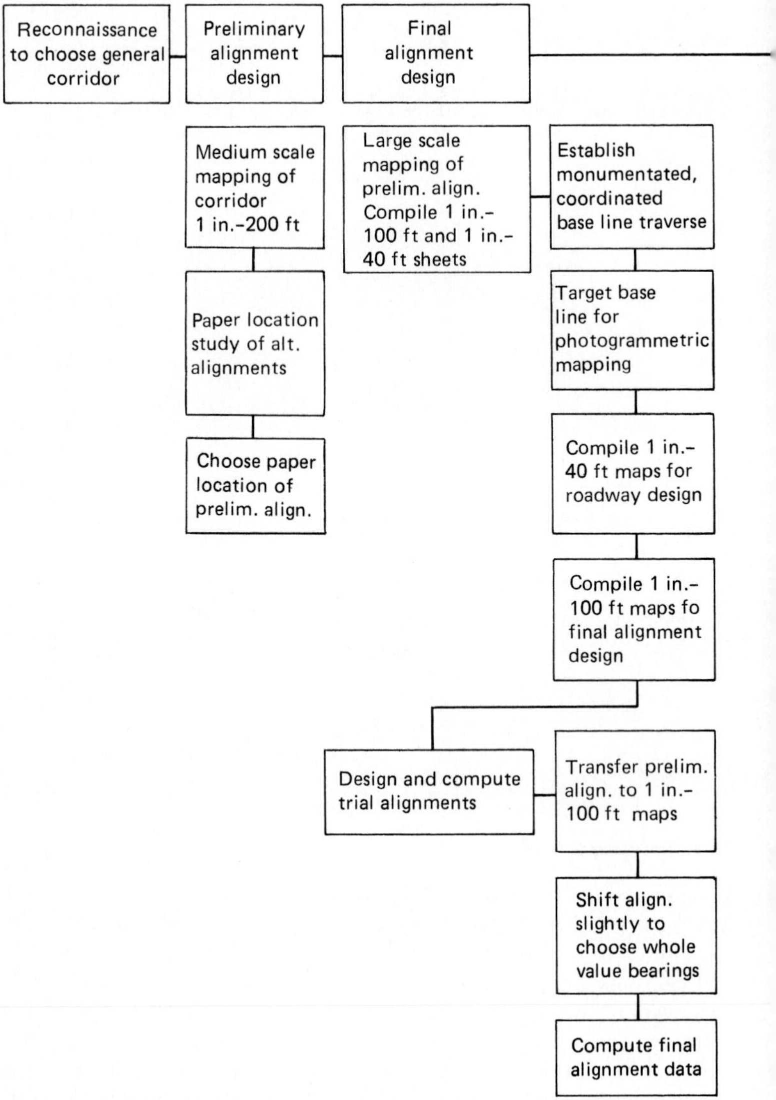

Figure 13-1 A coordinate based route survey and design system.

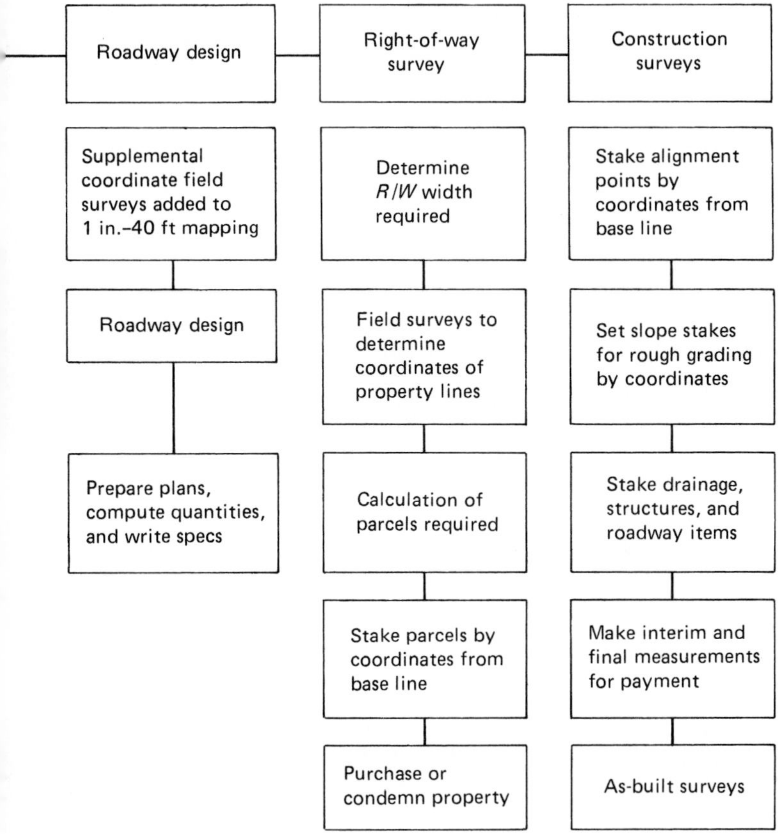

Figure 13-1 (*continued*)

Many other combinations of the components are also valid depending on the type of route being designed. For other route survey systems used in Florida and Illinois, see Supplemental Readings at the end of this chapter.

13-2. Sequence of Operations

New techniques involve more than technical changes in executing the several steps in design and stakeout. The sequence of the steps is also changed significantly, as may be seen in Fig. 13-1 when compared to the basic system outlined in Fig. 1-1.

A key change is deferring alignment stakeout until after final design and plan preparation, to reduce difficulty in running a location survey through built-up urban areas. Eliminating the location survey as a part of alignment design requires complete reliance upon the large-scale, precision aerial photos and maps.

13-3. Initial Mapping

Initial photogrammetric mapping involves preparation of small-scale base maps to a scale normally 1 in. = 200 ft with 2- or 5-ft contours. These maps cover areas large enough to permit disclosing all feasible alternative alignments.

13-4. Paper Location of the Center Line

Alternative alignments are drawn on the base maps, following conventional location procedures of Chapter 12. Splines bent to fit the numerous controlling points, are useful in developing flowing curves.

Since the purpose of this step is to select the best of several possible alignments for final design, each is subjected to detailed analysis. This includes drawing longitudinal profiles to which suitable grade lines are fitted; plotting cross sections for estimates of grading; examination of aesthetic values and operational characteristics, including safety considerations; and compilation of adequate road-user and overall construction costs.

13-5. Final Mapping

The larger-scale base maps needed for final design are also prepared by photogrammetric methods. Control for this mapping depends on first locating an accurate, coordinated base line, which is chosen tentatively with the aid of the small-scale maps used in paper location of the center line. The base line is modified, as conditions require, when staked in the field. Enough intervisible control points on the base line are monumented to permit staking critical points along the center and right-of-way lines before, during, and after construction. Electronic measurement of distance is essential on the base-line survey if traffic movements are present.

The topographic base map for final alignment design is drawn to a scale of 1 in. = 100 ft, with coordinate grid, control base line, and spot elevations superimposed. Larger-scale base maps, up to 1 in. = 40 ft, are prepared for development of other plan details.

13-6. Calculation of Alignment

Final alignment may now be computed without reliance on field stakeouts. For use in a large office, time is saved by preparing numerous special design aids. Among these are tables and templates. Such devices are used in addition to the usual route surveying tables and circular curve templates.

Table 13-1 is an example for circular curves. A large number of similar tables would be needed covering other ranges of central angle and different degrees of curve, and three-centered and multicentered curves. Table XI (Part III in this book) is already a "special" table for simple spirals.

Table 13-1. DESIGN TABLE FOR CIRCULAR CURVES

I°	T	L	L.C.	E	M
20	202.057	400.000	397.973	17.678	17.409
21	212.384	420.000	417.656	19.515	19.188
22	222.744	440.000	437.304	21.448	21.054
23	233.140	460.000	456.920	23.476	23.004
24	243.573	480.000	476.501	25.601	25.041
25	254.045	500.000	496.046	27.823	27.163
26	264.556	520.000	515.552	30.142	29.370
27	275.111	540.000	535.019	32.561	31.662
28	285.710	560.000	554.447	35.081	34.038
29	296.356	580.000	573.831	37.701	36.500
30	307.048	600.000	593.172	40.423	39.046

$D = 5^\circ$ ($R = 1145.92$) $I = 20^\circ$ to 30°

Design tables giving properties of combining spirals are particularly useful in fixing ramp alignments at interchanges. These tables are best built-up around numerous convenient lengths of combining spirals joining various combinations of D_S and D_L.

Design tables serve no purpose without special templates or *transparencies*. These are cut to a scale of 1 in. = 100 ft. Each transparency portrays several of the values listed in the appropriate design table, as shown by Fig. 13-2.

With these design aids available, final alignment design and calculation proceeds according to the following steps:

1. Transfer the selected center line to the 100-scale base map. Before drawing the curves adjust each spline slightly to a suitable degree of curve by template, as shown in Fig. 13-3. Then pencil the complete "initial alignment."
2. Revise the alignment slightly by substituting a closely fitted mathematical line. Start with the coordinates of a known point near the beginning of the alignment, draw the tangents, and extend them to tentative P.I.'s. Compute the bearing of the initial tangent from coordinates. Then round the bearing to the nearest whole degree and replot to conform.

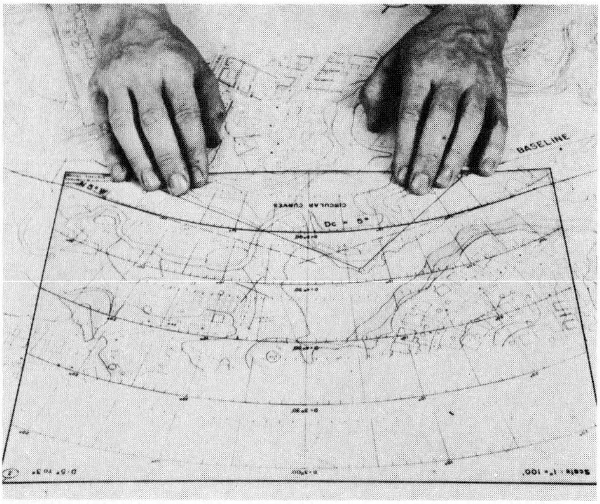

Figure 13-2

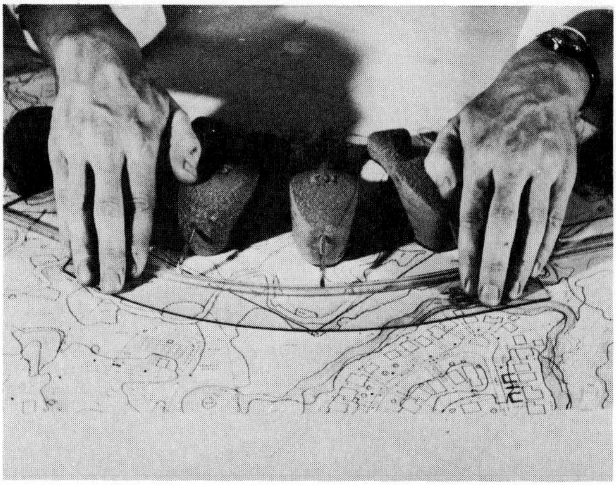

Figure 13-3

3. Repeat step 2 for each successive tangent, thus yielding whole-degree *I*-values for curves found in tables similar to Table 13-1.
4. Check the selection of degree of curve and the acceptability of its slightly altered position by applying the appropriate transparency (Fig. 13-3).
5. Where spirals are specified apply the same circular-curve transparencies with *o*-offsets taken from Table XI.

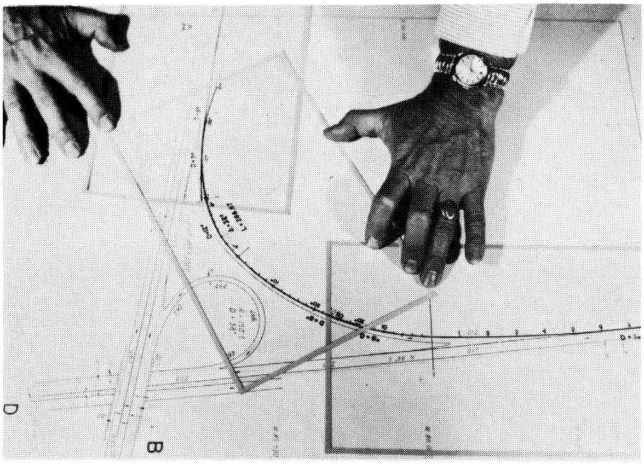

Figure 13-4

6. Compute the coordinates of the adjusted P.I.'s and the stationing of all curve points on the final alignment.

Similar design aids can be used to advantage in alignment design at interchange ramps. The transparencies portraying spirals, combining spirals, and three-centered curves are used singly or in combination to best fit the requirements. Figure 13-4 shows three transparencies being adjusted to yield a suitable ramp alignment. By simply overlaying the transparencies on the base map, without actually drawing the curves, pertinent data can be picked off and taken from design tables for entry as input to an electronic computer program for calculation of ramp alignment.

13-7. Final Design and Plans

Under circumstances requiring these new techniques, final plans for cross sections, drainage, right-of-way limits, interchanges, and other structures are normally developed with the aid of photogrammetric base maps plotted to a scale of 1 in. = 40 ft. The final alignment is plotted on these larger-scale maps by use of the coordinate system. Some supplementary ground surveys may be needed to supply details not adequately disclosed by the photogrammetric maps.

13-8. Stakeout of Alignment

The key to efficient stakeout of the designed center line is the accurately coordinated and monumented base line. In contrast to traditional procedure, the P.I.'s are rarely used in setting points on curves. Instead, such points are set by angle and distance measurements from control stations on the base line.

Proper control in stakeout of the center line requires the location of points at intervals not much over 500 ft. For this purpose the T.C.'s and C.T.'s are supplemented by points on tangents (P.O.T.'s) and points on curves (P.O.C.'s). The coordinated base line rarely contains enough well-located stations for setting all control points on the center line. Thus, it is necessary to monument numerous auxiliary stations along the base line.

In preparation for field work, the electronic computer is used to deliver lists of coordinates of base-line stations and center-line points, in addition to the distances and azimuths to center-line points from two or more base-line stations.

A control point on the center line is set by intersecting directions from two stations on the base line and checked by measuring the distance from one of the stations. Obviously, other combinations of measurements may be used.

For construction purposes the short sections of center line between control points can be filled in by conventional survey methods. The surveys need only be to third-order accuracy, since they start and close on accurately coordinated points previously set from the base line. Where traffic interferes with taping along the center line, points in these short sections can be set by triangulation from base-line stations. In Fig. 13-5, stations on the base line as initially staked are shown by open triangles; auxiliary stations, by solid triangles. The dotted lines illustrate how some points on the center line are set and checked by ties from the base line.

The difficult problem of slope staking in mountainous terrain is accomplished in California by using a computer program that delivers the elevation and offset of slope stakes on each digitized cross section. The printout also includes the grid azimuths and vertical angles from base-line stations to the slope-stake locations. In the field, slope stakes are set by "triangulation" at the intersection of horizontal and vertical lines of sight from two or more theodolites. No time-consuming linear measurements are required.

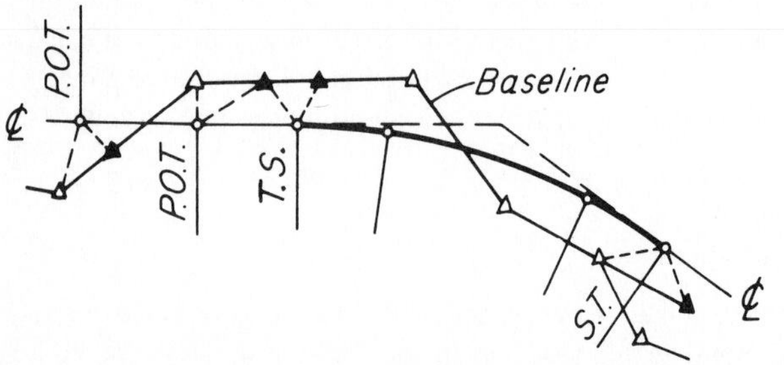

Figure 13-5

SUPPLEMENTAL READINGS

Beazley, J. S. April, 1970. Modern Day Surveys for Highway Location and Design. *Journal of the Surveying and Mapping Division, A.S.C.E.* 96.

Leisch, J. E. 1959. Developing Tools and Techniques for Geometric Design of Highways. *Proceedings, Highway Research Board* 38: 321–352.

Leisch, J. E. 1966. New Techniques in Alinement Design and Stakeout. *Journal of the Highway Division, A.S.C.E.* 92: No. HW1.

Leisch, J. E. 1977. Communicative Aspects in Highway Design. *Transportation Research Record* Bulletin 631: 64–67.

Stipp, D. W. 1968. Trigonometric Construction Staking. *Surveying and Mapping, Journal of the American Congress on Surveying and Mapping* 28: 437–446.

Chapter 14
Surveys for Other Routes

14-1. Foreword

Reference to the broad definition of *transportation* stated in section 1-2 suggests that the following additional types of transportation may involve surveying operations similar to those already described for railroads and highways:

1. Transportation (transmission) of power and messages by means of overhead tower or pole lines, or by lines in underground conduits.
2. Transportation of liquids and gases through closed conduits under pressure, such as pipe lines for water, gasoline, oil, and natural gas; through closed conduits by means of gravity, such as sewers and aqueducts; and through open channels, such as canals and flumes.
3. Transportation of materials (sand, gravel, stone, or selected borrow) to the site of large construction projects, by means of cableways and belt conveyors.

Whenever any of the foregoing are projects of considerable magnitude and involve termini a fairly long distance apart, the required surveying operations can properly be included in the term *route surveying.*

Special types of surveys are necessary in the case of tunnel location and construction. These are noted briefly in section 14-6.

14-2. Similarity to Railroad and Highway Surveys

Surveys for all routes of transportation and communication are similar in general respects to those described in Chapters 10 through 13 for railroads and highways. Because all routes have certain location controls, in fitting the line to those controls the natural sequence of field and office work approximates that outlined in section 1-8. Particular differences that do occur are caused by requirements peculiar to a specific type of route. An engineer acquainted with railroad or highway surveying should have no difficulty in adapting his knowledge to surveys for other routes, once he knows the uses to which the surveys will be put. Surveys for some other routes are described briefly in the succeeding sections.

14-3. Transmission-Line Surveys

Location of a power transmission line is controlled less by topography than other types of routes. Power loss due to voltage drop is proportional to the length of the conductor; consequently, high-tension transmission lines run as directly as possible from generating station to substation. Changes in direction, where required by intermediate controls, are made at angle towers instead of along curves. A trunk telephone or telegraph line is usually located within the right-of-way of a highway or of a railroad, in which case it must be followed.

Unless aerial photographs are used, the field and office work for transmission-line location involves, after a study of available maps, the following operations:

1. Reconnaissance to locate intermediate features to be avoided, such as buildings, cemeteries, extensive swamps, stands of heavy timber, and particularly valuable improved land; and for location of intermediate controls fixing points on the line, such as the most advantageous crossings of important highways, railroads, and streams.

2. A transit-and-tape (or stadia) traverse. The traverse can be either a preliminary line or the final center line, the selection depending on difficulty of the problem. In staking long straight sections between intermediate controls, the deflection-angle method of bypassing obstacles on tangents (described in section 12-8) is particularly useful. Contours are not located; however, all topographic features and right-of-way data are measured with respect to the traverse.

3. Levels sufficient in extent to aid in locating the towers. These may be merely "spot" elevations or a complete profile along the traverse line. On final location in difficult terrain, it is advisable to take levels along two lines, one on each side of the center line, to obtain proper conductor clearance when spotting positions of towers.

4. Office studies, including the features common to all route location: drawing the map, describing right-of-way easements, estimating quantities and cost, and preparing specifications. A special problem in transmission-line design is the location of towers. This location work may be done with the aid of special transparent templates, as described after step 5.

5. Construction surveys. These are relatively simple on transmission-line construction, since there is practically no grading. Stakes are needed only for clearing the right-of-way and for building tower footings. However, the surveyor's assistance is also valuable in planning other details related to construction, such as spotting cable reels and locating suitable dead-end and pulling points.

A convenient method of spotting tower locations on the profile is to use a transparent template, the lower edge of which is cut to the curve (approximately a parabola) that will be taken by the conductor cables. Obviously the curve must be modified to fit the profile scale. Two other curves are inscribed on the template parallel to the curve of the lower edge. The axial distance from the lower edge to the middle curve equals the maximum cable sag for a particular span; that from the middle curve to the upper curve equals the specified minimum ground clearance.

The template is used as shown in Fig. 14-1. First a point X is located at a suitable position for a tower; then the lower edge of the template is placed on this point and moved until the middle curve touches the ground line. The other point Y at which the lower edge of the template intersects the ground line is the possible location of the next tower. One template fits a considerable range of spans with sufficient accuracy.

After the towers at certain controlling points have been located, establishing intermediate towers is a matter of cut and try; the object is to cover the greatest length of line with the least number of towers.

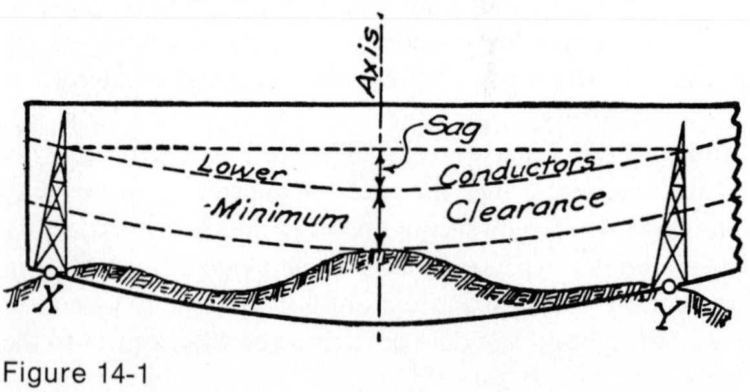

Figure 14-1

On important new transmission-line work, aerial photographs are now commonly used, at least for reconnaissance, and sometimes serve for all phases of the survey work except final staking. The growing file of available aerial photographs often makes this method feasible where it was formerly prevented by economic considerations.

14-4. Surveys for Pressure Pipe Lines and Underground Conduits

Surveys to locate long pressure pipe lines are almost as simple as those for transmission lines. In fact, the descriptions contained in steps 1 and 2 in section 14-3 apply also to pipe-line surveys. However, since pressure pipe lines are usually located underground, greater attention is paid to foundation conditions and especially to avoiding costly rock excavation and frequent stream crossings. Accessibility to power for operating booster pumping stations is also an important intermediate control.

Grades and undulations in the profile are relatively unimportant, especially on small-diameter steel pipe lines; consequently, detailed profile levels may be omitted and replaced by spot elevations at proposed pumping stations and the high and low points along the line.

On construction, line stakes are more important than grade stakes. In fact, grade stakes for steel pipe lines may be needed only at pumping stations and at crossings of highways, railroads, and streams. Along intervening sections, at least in easy terrain, several sections of pipe are welded together on the ground before being laid in the relatively uniform ditch dug by the trenching machine.

Large reinforced-concrete pipe lines require much more careful attention to undulations in the profile, since there is a practical limit to the change in direction possible at each joint, and beyond that limit special pipe sections are necessary.

As in the case of transmission lines, right-of-way for a pressure pipe line usually takes the form of easements for its construction and operation.

Aerial surveys are particularly useful in the location of long pipe lines. Only mosaics are used, since contours are not essential. As a rule, a stereoscopic study of the photographs will give enough information for the preliminary location. After this the line may be "walked over" prior to deciding upon the final location. It is possible, however, to rely upon aerial photographs to an even greater extent (Refs. 1 and 2). For example, in building one of the longest pipe lines from the southwestern part of the United States to the industrial middle west, no surveyor went on the job until the sections of pipe were ready to be laid; yet, the right-of-way agents completed much of their work before that time.

Surveys for underground conduits containing power lines on private right-of-way are much the same as those for pressure pipe lines.

Underground communication circuits are commonly placed in conduits located beneath the highway pavement. Access for maintenance is by means of manholes. Coaxial cables used for telephone, broadcast, and television circuits may be drawn through existing conduits. However, they are also placed directly in a shallow ploughed trench beside the highway. In neither case is any extensive survey work required.

14-5. Surveys for Construction at the Hydraulic Gradient

Surveys for hydraulic construction in which flow is by gravity require very careful attention to elevations, owing to the flat grades used. If the flow is in an open channel, as in a canal, the alignment may have to be circuitous in order to obtain proper velocity and avoid costly grading. A more direct alignment is possible if the construction is below the ground surface, such as in the case of a grade-line tunnel, aqueduct, or sewer.

Surveys for surface construction may be identical with those for railroad location, the principal modification being that a narrower strip of topography will suffice. This condition is necessary to keep the gradients between relatively narrow limits.

Stations on the preliminary traverse are kept close to the final location by setting them near a "grade contour." A grade contour is the line on the ground (starting at a controlling point) along which the grade changes at the rate best suited to the construction. In locating the stations, it is obviously necessary that the leveling keep up with the transit work. In simple irrigation-ditch construction in easy terrain, a line may be located on the grade contour in the field by tape and level, without need for a transit.

On a contour map the grade contour is found by starting at a controlling point and stepping from contour to contour with dividers set at a distance equal to the contour interval divided by the desired rate of grade. The closer the final alignment follows the grade contour, the smaller will be the grading quantities.

Since economy of grading is an important factor in canal construction, careful cross sections are taken at short intervals. Construction surveys for canals are very similar to those for highways, but all stakes must be set on offset lines.

There are cases in which aerial surveys have been used in studies for canal location—for example, on the proposed Florida Barge Canal. They have also been used in studies for levees and dikes to control meandering rivers. Difficulty of access for ground-survey parties is an important consideration favoring use of aerial-survey methods.

For gravity-flow structures below the ground surface, it is entirely suitable to use railroad surveying methods, supplemented by adequate subsurface exploration. A most important aspect of the office studies for such construction is to decide whether cut-and-cover or tunnel construction is the better. Often a combination of the two provides the most economical solution.

14-6. Tunnel Surveys

In mountainous terrain, it is sometimes necessary to use tunnels on route alignment. Surveying operations for locating tunnels vary greatly in complexity. Preliminary studies are best made by using aerial photography, especially in regions which have experienced earth movements. Even detailed field studies may not disclose old earthquake faults, but good photographs quickly reveal them. As an example, some topographically-favorable tunnel sites considered for Interstate Highway 70 under the Continental Divide, west of Denver, were found to follow major fault zones (Ref. 3), a fact which resulted in the choice of a different location (Ref. 4).

The final alignment of a short tunnel can be fixed by locating a transit line on the ground directly over the tunnel. As a rule, however, an indirect precise traverse is necessary. In the case of subaqueous tunnels or long tunnels to be driven through rugged mountain ranges, triangulation control must be used. This is a subject outside the scope of route surveying.

Traverse or triangulation control provides only the data for calculating the tunnel alignment; elevations must be determined by careful spirit leveling between the proposed portals.

Locating the portals, adits, and shafts by means of accurate control surveys is only one of the surveyor's important tasks. His work in controlling accuracy of the tunnel driving is fully as important; it must be done with the highest precision, for it cannot be verified conclusively until the headings are holed through—and then it is too late to make adjustments.

Surveying for tunnels driven through rock involves specialized operations not found in other types of route surveying. Among these are:

1. Carrying the alignment down shafts by means of heavy bobs damped in oil and suspended from piano wires. The equipment also includes lateral adjusting devices for the sheaves and scales for measuring wires' swing.
2. Transferring the alignment from the wires to plumb bobs suspended from riders, or "skyhooks," mounted on scales attached to the roof of the tunnel.
3. Extending the alignment into the tunnel on "spads" driven in plugged holes in the roof. The transitman usually works on a suspended platform, out of the way of the muck cars.
4. Carrying the alignment to the working face, or "painting the heading," for locating drill holes.

5. Transferring grade down shafts by means of weighted tapes or by taping down elevator guides.
6. Carrying temporary grade into the tunnel by means of inverted rod readings on the wood plugs in which the spads are driven.
7. Cross sectioning twice; first for locating "tights" (points needing trimming), and finally for obtaining permanent graphical records of the sections and computing pay yardage and overbreak.

Several ingenious devices for cross sectioning have been used, such as pantographs and "sunflowers." The latter are designed to locate breaks in the tunnel cross section by polar coordinates.

Since 1965 several different commercial laser instruments (Ref. 5) have become available for accurate line-and-grade control in the construction of tunnels and underwater pipelines.

References

1. MacDonald, G. E. 1956. Surveys and Maps for Pipelines. *Transactions, A.S.C.E.* 121: 121–134.
2. Guss, P. June, 1961. Aerial Photography Aids Pipeline Location. *Civil Engineering* 31, No. 6: 48–51.
3. Mitcham, T. W. Aug. 1961. Tunnel-Site Selection by Use of Aerial Photography. *Civil Engineering* 31, No. 8: 64.
4. Road Tunnel Will Pierce Divide. 1961. *Engineering NewsRecord* 167, No. 7: 25.
5. Bengston, D. 1968. Construction Control with Lasers. *Civil Engineering* 38, No. 4: 72–74.

SUPPLEMENTAL READINGS

Cooney, A. 1970. Laser Alignment Techniques in Tunneling. *Journal of the Surveying and Mapping Division A.S.C.E.* 96.

Jenik, E. C. 1977. Computerized Mapping and Record Systems for Utilities. *Transportation Research Record* Bulletin 631: 64–67.

Peterson, E. W. and Frobenius, P. 1973. Tunnel Survey and Tunneling Machine Control. *Journal of the Surveying and Mapping Division, A.S.C.E.* 99, No. SU1; 21–38.

Shields, J. M. May, 1971. Channel Rectification Engineering Survey Program. *Journal of the Surveying and Mapping Division, A.S.C.E.* 97.

List of Tables

Table I. RADII, DEFLECTIONS, OFFSETS, ETC.

CHORD DEFINITION OF DEGREE OF CURVE* (D_c)	ARC DEFINITION OF DEGREE OF CURVE* (D_a)
MATHEMATICAL RELATIONS	MATHEMATICAL RELATIONS
1. $\sin \frac{1}{2}D_c = \frac{50}{R}$, . . . , exactly	1. $D_a = \frac{5729.58}{R}$, . . . , exactly
2. $\text{C.O.} = \frac{100^2}{R} = 200 \sin \frac{1}{2}D_c$ exactly	2. $\text{C.O.} = 4\,R \sin^2 \frac{1}{2}D_a$ exactly
3. T.O. = $\frac{1}{2}$C.O., . . . , exactly	3. T.O. = $\frac{1}{2}$C.O., . . . , exactly
4. M.O. = $\frac{1}{8}$C.O., . . . , approx.	4. M.O. = $\frac{1}{8}$C.O., . . . , approx.

Notes

(1) For values of D_a less than 12°, the C.O. (not tabulated) may be taken equal to the tabulated C.O. for the corresponding value of D_c. Exact values of C.O. are tabulated for both definitions of D, where D equals or exceeds 12°.

(2) For values of D_c and D_a less than 12°, relation 4 will give M.O. without perceptible error. Beyond $D_c = 12°$, the exact values of M.O. are tabulated.

(3) For any value of D, the ratio M.O./C.O. is numerically the same for both definitions of D. Thus, if D_a were 52°, the M.O. equals (11.54/87.67)(84.69), which is 11.17.

(4) To obtain *c.o.*, *t*, and *m.o.* for a chord or arc shorter than 100 ft, multiply C.O., T.O., and M.O. by (chord or arc/100)^2.

* See sections 2-4 and 2-15.

TABLE I.—RADII, DEFLECTIONS, OFFSETS, ETC.

DEGREE OF CURVE D	DEFL. PER FT OF STA. (MIN)	CHORD DEFINITION			ARC DEFINITION	
		RADIUS R	LOG R	C.O. 1 STA.	RADIUS R	LOG R
0° 0'		Infinite	Infin.		Infinite	Infin.
1'	0.005	343775.	5.536274	0.03	343775.	5.536274
2'	0.01	171887.	5.235244	0.06	171887.	5.235244
3'	0.015	114592.	5.059153	0.09	114592.	5.059153
4'	0.02	85943.7	4.934214	0.12	85943.7	4.934214
5'	0.025	68754.9	4.837304	0.15	68754.9	4.837304
6'	0.03	57295.8	4.758123	0.17	57295.8	4.758123
7'	0.035	49110.7	4.691176	0.20	49110.7	4.691176
8'	0.04	42971.8	4.633184	0.23	42971.8	4.633184
9'	0.045	38197.2	4.582031	0.26	38197.2	4.582031
10'	0.05	34377.5	4.536274	0.29	34377.5	4.536274
11'	0.055	31252.3	4.494881	0.32	31252.2	4.494881
12'	0.06	28647.8	4.457093	0.35	28647.8	4.457093
13'	0.065	26444.2	4.422331	0.38	26444.2	4.422331
14'	0.07	24555.4	4.390146	0.41	24555.3	4.390146
15'	0.075	22918.3	4.360183	0.44	22918.3	4.360183
16'	0.08	21485.9	4.332154	0.47	21485.9	4.332154
17'	0.085	20222.1	4.305825	0.49	20222.0	4.305825
18'	0.09	19098.6	4.281002	0.52	19098.6	4.281001
19'	0.095	18093.4	4.257521	0.55	18093.4	4.257520
20'	0.1	17188.8	4.235244	0.58	17188.7	4.235244
21'	0.105	16370.2	4.214055	0.61	16370.2	4.214055
22'	0.11	15626.1	4.193852	0.64	15626.1	4.193851
23'	0.115	14946.8	4.174547	0.67	14946.7	4.174546
24'	0.12	14324.0	4.156064	0.70	14323.9	4.156063
25'	0.125	13751.0	4.138335	0.73	13751.0	4.138334
26'	0.13	13222.1	4.121302	0.76	13222.1	4.121300
27'	0.135	12732.4	4.104911	0.79	12732.4	4.104910
28'	0.14	12277.7	4.089117	0.81	12277.7	4.089116
29'	0.145	11854.3	4.073877	0.84	11854.3	4.073876
30'	0.15	11459.2	4.059154	0.87	11459.2	4.059153
31'	0.155	11089.6	4.044914	0.90	11089.5	4.044912
32'	0.16	10743.0	4.031125	0.93	10743.0	4.031124
33'	0.165	10417.5	4.017762	0.96	10417.4	4.017760
34'	0.17	10111.1	4.004797	0.99	10111.0	4.004795
35'	0.175	9822.18	3.992208	1.02	9822.13	3.992206
36'	0.18	9549.34	3.979973	1.05	9549.29	3.979971
37'	0.185	9291.25	3.968074	1.07	9291.21	3.968072
38'	0.19	9046.75	3.956493	1.11	9046.70	3.956490
39'	0.195	8814.78	3.945212	1.13	8814.73	3.945209
40'	0.2	8594.42	3.934216	1.16	8594.37	3.934214
41'	0.205	8384.80	3.923493	1.19	8384.75	3.923490
42'	0.21	8185.16	3.913027	1.21	8185.11	3.913025
43'	0.215	7994.81	3.902808	1.25	7994.76	3.902805
44'	0.22	7813.11	3.892824	1.28	7813.06	3.892821
45'	0.225	7639.49	3.883065	1.31	7639.44	3.883061
46'	0.23	7473.42	3.873519	1.34	7473.36	3.873516
47'	0.235	7314.41	3.864179	1.37	7314.35	3.864176
48'	0.24	7162.03	3.855036	1.39	7161.97	3.855033
49'	0.245	7015.87	3.846082	1.43	7015.81	3.846078
50'	0.25	6875.55	3.837308	1.45	6875.49	3.837304
51'	0.255	6740.74	3.828708	1.48	6740.68	3.828704
52'	0.26	6611.12	3.820275	1.51	6611.05	3.820270
53'	0.265	6486.38	3.812002	1.54	6486.31	3.811998
54'	0.27	6366.26	3.803885	1.57	6366.20	3.803880
55'	0.275	6250.51	3.795916	1.60	6250.45	3.795911
56'	0.28	6138.90	3.788091	1.63	6138.83	3.788086
57'	0.285	6031.20	3.780404	1.66	6031.14	3.780399
58'	0.29	5927.22	3.772851	1.69	5927.15	3.772846
59'	0.295	5826.76	3.765427	1.72	5826.69	3.765422

TABLE I.—RADII, DEFLECTIONS, OFFSETS, ETC.

DEGREE OF CURVE D	DEFL. PER FT OF STA. (MIN)	CHORD DEFINITION			ARC DEFINITION	
		RADIUS R	LOG R	C.O. 1 STA.	RADIUS R	LOG R
1° 0'	0.3	5729.65	3.758128	1.75	5729.58	3.758123
1'	0.305	5635.72	3.750950	1.77	5635.65	3.750944
2'	0.31	5544.83	3.743888	1.80	5544.75	3.743882
3'	0.315	5456.82	3.736939	1.83	5456.74	3.736933
4'	0.32	5371.56	3.730100	1.86	5371.48	3.730094
5'	0.325	5288.92	3.723367	1.89	5288.84	3.723360
6'	0.33	5208.79	3.716737	1.92	5208.71	3.716730
7'	0.335	5131.05	3.710206	1.95	5130.97	3.710199
8'	0.34	5055.59	3.703772	1.98	5055.51	3.703765
9'	0.345	4982.33	3.697432	2.01	4982.24	3.697425
10'	0.35	4911.15	3.691183	2.03	4911.07	3.691176
11'	0.355	4841.98	3.685023	2.07	4841.90	3.685015
12'	0.36	4774.74	3.678949	2.09	4774.65	3.678941
13'	0.365	4709.33	3.672959	2.12	4709.24	3.672951
14'	0.37	4645.69	3.667051	2.15	4645.60	3.667042
15'	0.375	4583.75	3.661221	2.18	4583.66	3.661213
16'	0.38	4523.44	3.655469	2.21	4523.35	3.655460
17'	0.385	4464.70	3.649792	2.24	4464.61	3.649783
18'	0.39	4407.46	3.644189	2.27	4407.37	3.644179
19'	0.395	4351.67	3.638656	2.30	4351.58	3.638647
20'	0.4	4297.28	3.633194	2.33	4297.18	3.633184
21'	0.405	4244.23	3.627799	2.35	4244.13	3.627789
22'	0.41	4192.47	3.622470	2.39	4192.37	3.622460
23'	0.415	4141.96	3.617206	2.41	4141.86	3.617196
24'	0.42	4092.66	3.612005	2.44	4092.56	3.611995
25'	0.425	4044.51	3.606866	2.47	4044.41	3.606855
26'	0.43	3997.49	3.601787	2.50	3997.38	3.601775
27'	0.435	3951.54	3.596766	2.53	3951.43	3.596755
28'	0.44	3906.64	3.591803	2.56	3906.53	3.591791
29'	0.445	3862.74	3.586896	2.59	3862.64	3.586884
30'	0.45	3819.83	3.582044	2.62	3819.71	3.582031
31'	0.455	3777.85	3.577245	2.65	3777.74	3.577232
32'	0.46	3736.79	3.572499	2.68	3736.68	3.572486
33'	0.465	3696.61	3.567804	2.71	3696.50	3.567791
34'	0.47	3657.29	3.563160	2.73	3657.18	3.563146
35'	0.475	3618.80	3.558564	2.76	3618.68	3.558550
36'	0.48	3581.10	3.554017	2.79	3580.99	3.554003
37'	0.485	3544.19	3.549517	2.82	3544.07	3.549502
38'	0.49	3508.02	3.545063	2.85	3507.91	3.545048
39'	0.495	3472.59	3.540654	2.88	3472.47	3.540638
40'	0.5	3437.87	3.536289	2.91	3437.75	3.536274
41'	0.505	3403.83	3.531968	2.94	3403.71	3.531952
42'	0.51	3370.46	3.527690	2.97	3370.34	3.527673
43'	0.515	3337.74	3.523453	3.00	3337.62	3.523437
44'	0.52	3305.65	3.519257	3.03	3305.53	3.519241
45'	0.525	3274.17	3.515101	3.05	3274.04	3.515085
46'	0.53	3243.29	3.510985	3.08	3243.16	3.510968
47'	0.535	3212.98	3.506908	3.11	3212.85	3.506890
48'	0.54	3183.23	3.502868	3.14	3183.10	3.502850
49'	0.545	3154.03	3.498866	3.17	3153.90	3.498847
50'	0.55	3125.36	3.494900	3.20	3125.22	3.494881
51'	0.555	3097.20	3.490970	3.23	3097.07	3.490951
52'	0.56	3069.55	3.487075	3.26	3069.42	3.487056
53'	0.565	3042.39	3.483215	3.29	3042.25	3.483195
54'	0.57	3015.71	3.479389	3.32	3015.57	3.479369
55'	0.575	2989.48	3.475596	3.35	2989.34	3.475576
56'	0.58	2963.72	3.471836	3.37	2963.58	3.471816
57'	0.585	2938.39	3.468109	3.40	2938.25	3.468087
58'	0.59	2913.49	3.464413	3.43	2913.34	3.464392
59'	0.595	2889.01	3.460749	3.46	2888.86	3.460727

TABLE I.—RADII, DEFLECTIONS, OFFSETS, ETC.

DEGREE OF CURVE D	DEFL. PER FT OF STA. (MIN)	CHORD DEFINITION			ARC DEFINITION	
		RADIUS R	LOG R	C.O. 1 STA.	RADIUS R	LOG R
2° 0'	0.6	2864.93	3.457115	3.49	2864.79	3.457093
1'	0.605	2841.26	3.453511	3.52	2841.11	3.453488
2'	0.61	2817.97	3.449937	3.55	2817.83	3.449914
3'	0.615	2795.06	3.446392	3.58	2794.92	3.446369
4'	0.62	2772.53	3.442876	3.61	2772.38	3.442852
5'	0.625	2750.35	3.439388	3.64	2750.20	3.439364
6'	0.63	2728.52	3.435928	3.66	2728.37	3.435903
7'	0.635	2707.04	3.432495	3.69	2706.89	3.432470
8'	0.64	2685.89	3.429089	3.72	2685.74	3.429064
9'	0.645	2665.08	3.425710	3.75	2664.92	3.425684
10'	0.65	2644.58	3.422356	3.78	2644.42	3.422330
11'	0.655	2624.39	3.419029	3.81	2624.23	3.419002
12'	0.66	2604.51	3.415727	3.84	2604.35	3.415700
13'	0.665	2584.93	3.412449	3.87	2584.77	3.412422
14'	0.67	2565.65	3.409197	3.90	2565.48	3.409169
15'	0.675	2546.64	3.405968	3.93	2546.48	3.405940
16'	0.68	2527.92	3.402763	3.96	2527.75	3.402735
17'	0.685	2509.47	3.399582	3.98	2509.30	3.399553
18'	0.69	2491.29	3.396424	4.01	2491.12	3.396395
19'	0.695	2473.37	3.393289	4.04	2473.20	3.393259
20'	0.7	2455.70	3.390176	4.07	2455.53	3.390145
21'	0.705	2438.29	3.387085	4.10	2438.12	3.387055
22'	0.71	2421.12	3.384016	4.13	2420.95	3.383985
23'	0.715	2404.19	3.380969	4.16	2404.02	3.380938
24'	0.72	2387.50	3.377943	4.19	2387.32	3.377911
25'	0.725	2371.04	3.374938	4.22	2370.86	3.374905
26'	0.73	2354.80	3.371954	4.25	2354.62	3.371921
27'	0.735	2338.78	3.368990	4.28	2338.60	3.368956
28'	0.74	2322.98	3.366046	4.30	2322.80	3.366012
29'	0.745	2307.39	3.363122	4.33	2307.21	3.363087
30'	0.75	2292.01	3.360217	4.36	2291.83	3.360183
31'	0.755	2276.84	3.357332	4.39	2276.65	3.357297
32'	0.76	2261.86	3.354466	4.42	2261.68	3.354430
33'	0.765	2247.08	3.351618	4.45	2246.89	3.351582
34'	0.77	2232.49	3.348789	4.48	2232.30	3.348753
35'	0.775	2218.09	3.345797	4.51	2217.90	3.345942
36'	0.78	2203.87	3.343187	4.54	2203.68	3.343149
37'	0.785	2189.84	3.340412	4.57	2189.65	3.340374
38'	0.79	2175.98	3.337655	4.60	2175.79	3.337617
39'	0.795	2162.30	3.334916	4.62	2162.10	3.334877
40'	0.8	2148.79	3.332193	4.65	2148.59	3.332154
41'	0.805	2135.44	3.329488	4.68	2135.25	3.329448
42'	0.81	2122.26	3.326799	4.71	2122.07	3.326759
43'	0.815	2109.24	3.324127	4.74	2109.05	3.324086
44'	0.82	2096.39	3.321471	4.77	2096.19	3.321430
45'	0.825	2083.68	3.318832	4.80	2083.48	3.318790
46'	0.83	2071.13	3.316208	4.83	2070.93	3.316166
47'	0.835	2058.73	3.313600	4.86	2058.53	3.313557
48'	0.84	2046.48	3.311008	4.89	2046.28	3.310964
49'	0.845	2034.37	3.308431	4.92	2034.17	3.308387
50'	0.85	2022.41	3.305869	4.94	2022.20	3.305825
51'	0.855	2010.59	3.303323	4.97	2010.38	3.303278
52'	0.86	1998.90	3.300791	5.00	1998.69	3.300745
53'	0.865	1987.35	3.298274	5.03	1987.14	3.298228
54'	0.87	1975.93	3.295771	5.06	1975.72	3.295725
55'	0.875	1964.64	3.293283	5.09	1964.43	3.293236
56'	0.88	1953.48	3.290809	5.12	1953.27	3.290761
57'	0.885	1942.44	3.288349	5.15	1942 23	3.288301
58'	0.89	1931.53	3.285902	5.18	1931.32	3.285854
59'	0.895	1920.75	3.283470	5.21	1920.53	3.283421

TABLE I.—RADII, DEFLECTIONS, OFFSETS, ETC.

DEGREE OF CURVE D	DEFL. PER FT OF STA. (MIN)	CHORD DEFINITION			ARC DEFINITION	
		RADIUS R	LOG R	C.O. 1 STA.	RADIUS R	LOG R
3° 0'	0.9	1910.08	3.281051	5.24	1909.86	3.281001
1'	0.905	1899.53	3.278646	5.26	1899.31	3.278595
2'	0.91	1889.09	3.276253	5.29	1888.87	3.276203
3'	0.915	1878.77	3.273874	5.32	1878.55	3.273823
4'	0.92	1868.56	3.271508	5.35	1868.34	3.271456
5'	0.925	1858.47	3.269155	5.38	1858.24	3.269102
6'	0.93	1848.48	3.266814	5.41	1848.25	3.266761
7'	0.935	1838.59	3.264486	5.44	1838.37	3.264432
8'	0.94	1828.82	3.262170	5.47	1828.59	3.262116
9'	0.945	1819.14	3.259867	5.50	1818.91	3.259812
10'	0.95	1809.57	3.257576	5.53	1809.34	3.257520
11'	0.955	1800.10	3.255296	5.56	1799.87	3.255240
12'	0.96	1790.73	3.253029	5.58	1790.49	3.252973
13'	0.965	1781.45	3.250774	5.61	1781.22	3.250716
14'	0.97	1772.27	3.248530	5.64	1772.03	3.248472
15'	0.975	1763.18	3.246297	5.67	1762.95	3.246239
16'	0.98	1754.19	3.244077	5.70	1753.95	3.244018
17'	0.985	1745.29	3.241867	5.73	1745.05	3.241808
18'	0.99	1736.48	3.239669	5.76	1736.24	3.239609
19'	0.995	1727.75	3.237481	5.79	1727.51	3.237421
20'	1.0	1719.12	3.235305	5.82	1718.87	3.235244
21'	1.005	1710.57	3.233140	5.85	1710.32	3.233078
22'	1.01	1702.10	3.230985	5.88	1701.85	3.230922
23'	1.015	1693.72	3.228841	5.90	1693.47	3.228778
24'	1.02	1685.42	3.226707	5.93	1685.17	3.226644
25'	1.025	1677.20	3.224584	5.96	1676.95	3.224520
26'	1.03	1669.06	3.222472	5.99	1668.81	3.222407
27'	1.035	1661.00	3.220369	6.02	1660.75	3.220303
28'	1.04	1653.01	3.218277	6.05	1652.76	3.218210
29'	1.045	1645.11	3.216195	6.08	1644.85	3.216128
30'	1.05	1637.28	3.214122	6.11	1637.02	3.214055
31'	1.055	1629.52	3.212060	6.14	1629.26	3.211991
32'	1.06	1621.84	3.210007	6.17	1621.58	3.209938
33'	1.065	1614.22	3.207964	6.19	1613.96	3.207894
34'	1.07	1606.68	3.205930	6.22	1606.42	3.205860
35'	1.075	1599.21	3.203906	6.25	1598.95	3.203835
36'	1.08	1591.81	3.201892	6.28	1591.55	3.201820
37'	1.085	1584.48	3.199886	6.31	1584.21	3.199814
38'	1.09	1577.21	3.197890	6.34	1576.95	3.197817
39'	1.095	1570.01	3.195903	6.37	1569.75	3.195830
40'	1.1	1562.88	3.193925	6.40	1562.61	3.193851
41'	1.105	1555.81	3.191956	6.43	1555.54	3.191881
42'	1.11	1548.80	3.189996	6.46	1548.53	3.189921
43'	1.115	1541.86	3.188045	6.49	1541.59	3.187969
44'	1.12	1534.98	3.186103	6.51	1534.71	3.186026
45'	1.125	1528.16	3.184169	6.54	1527.89	3.184091
46'	1.13	1521.40	3.182244	6.57	1521.13	3.182165
47'	1.135	1514.17	3.180327	6.60	1514.43	3.180248
48'	1.14	1508.06	3.178419	6.63	1507.78	3.178339
49'	1.145	1501.48	3.176519	6.66	1501.20	3.176438
50'	1.15	1494.95	3.174627	6.69	1494.67	3.174546
51'	1.155	1488.48	3.172744	6.72	1488.20	3.172661
52'	1.16	1482.07	3.170868	6.75	1481.79	3.170786
53'	1.165	1475.71	3.169001	6.78	1475.43	3.168918
54'	1.17	1469.41	3.167142	6.81	1469.12	3.167058
55'	1.175	1463.16	3.165291	6.83	1462.87	3.165206
56'	1.18	1456.96	3.163447	6.86	1456.67	3.163362
57'	1.185	1450.81	3.161612	6.89	1450.53	3.161526
58'	1.19	1444.72	3.159784	6.92	1444.43	3.159697
59'	1.195	1438.68	3.157963	6.95	1438.39	3.157876

TABLE I.—RADII, DEFLECTIONS, OFFSETS, ETC.

DEGREE OF CURVE D	DEFL. PER FT OF STA. (MIN)	CHORD DEFINITION			ARC DEFINITION	
		RADIUS R	LOG R	C.O. 1 STA.	RADIUS R	LOG R
4° 0'	1.2	1432.69	3.156151	6.98	1432.39	3.156063
1'	1.205	1426.74	3.154346	7.01	1426.45	3.154257
2'	1.21	1420.85	3.152548	7.04	1420.56	3.152458
3'	1.215	1415.01	3.150758	7.07	1414.71	3.150667
4'	1.22	1409.21	3.148975	7.10	1408.91	3.148884
5'	1.225	1403.46	3.147200	7.13	1403.16	3.147108
6'	1.23	1397.76	3.145431	7.15	1397.46	3.145339
7'	1.235	1392.10	3.143670	7.18	1391.80	3.143577
8'	1.24	1386.49	3.141916	7.21	1386.19	3.141822
9'	1.245	1380.92	3.140170	7.24	1380.62	3.140074
10'	1.25	1375.40	3.138430	7.27	1375.10	3.138334
11'	1.255	1369.92	3.136697	7.30	1369.62	3.136600
12'	1.26	1364.49	3.134971	7.33	1364.19	3.134873
13'	1.265	1359.10	3.133251	7.36	1358.79	3.133153
14'	1.27	1353.75	3.131539	7.39	1353.44	3.131440
15'	1.275	1348.45	3.129833	7.42	1348.14	3.129734
16'	1.28	1343.18	3.128134	7.45	1342.87	3.128034
17'	1.285	1337.96	3.126442	7.47	1337.64	3.126341
18'	1.29	1332.77	3.124756	7.50	1332.46	3.124654
19'	1.295	1327.63	3.123077	7.53	1327.32	3.122974
20'	1.3	1322.53	3.121404	7.56	1322.21	3.121300
21'	1.305	1317.46	3.119738	7.59	1317.14	3.119633
22'	1.31	1312.43	3.118078	7.62	1312.12	3.117972
23'	1.315	1307.45	3.116424	7.65	1307.13	3.116318
24'	1.32	1302.50	3.114777	7.68	1302.18	3.114669
25'	1.325	1297.58	3.113136	7.71	1297.26	3.113028
26'	1.33	1292.71	3.111501	7.74	1292.39	3.111392
27'	1.335	1287.87	3.109872	7.76	1287.55	3.109762
28'	1.34	1283.07	3.108249	7.79	1282.74	3.108139
29'	1.345	1278.30	3.106632	7.82	1277.97	3.106521
30'	1.35	1273.57	3.105022	7.85	1273.24	3.104910
31'	1.355	1268.87	3.103417	7.88	1268.54	3.103304
32'	1.36	1264.21	3.101818	7.91	1263.88	3.101705
33'	1.365	1259.58	3.100225	7.94	1259.25	3.100111
34'	1.37	1254.98	3.098638	7.97	1254.65	3.098523
35'	1.375	1250.42	3.097057	8.00	1250.09	3.096941
36'	1.38	1245.89	3.095481	8.03	1245.56	3.095365
37'	1.385	1241.40	3.093912	8.06	1241.06	3.093974
38'	1.39	1236.94	3.092347	8.08	1236.60	3.092229
39'	1.395	1232.51	3.090789	8.11	1232.17	3.090670
40'	1.4	1228.11	3.089236	8.14	1227.77	3.089116
41'	1.405	1223.74	3.087689	8.17	1223.40	3.087566
42'	1.41	1219.40	3.086147	8.20	1219.06	3.086025
43'	1.415	1215.09	3.084610	8.23	1214.75	3.084487
44'	1.42	1210.82	3.083079	8.26	1210.47	3.082955
45'	1.425	1206.57	3.081553	8.29	1206.23	3.081429
46'	1.43	1202.36	3.080033	8.32	1202.01	3.079908
47'	1.435	1198.17	3.078518	8.35	1197.82	3.078392
48'	1.44	1194.01	3.077008	8.38	1193.66	3.076881
49'	1.445	1189.88	3.075504	8.40	1189.53	3.075376
50'	1.45	1185.78	3.074005	8.43	1185.43	3.073876
51'	1.455	1181.71	3.072511	8.46	1181.36	3.072381
52'	1.46	1177.66	3.071022	8.49	1177.31	3.070891
53'	1.465	1173.65	3.069538	8.52	1173.29	3.069406
54'	1.47	1169.66	3.068059	8.55	1169.30	3.067927
55'	1.475	1165.70	3.066585	8.58	1165.34	3.066452
56'	1.48	1161.76	3.065116	8.61	1161.40	3.064982
57'	1.485	1157.85	3.063653	8.64	1157.49	3.063517
58'	1.49	1153.97	3.062194	8.67	1153.61	3.062057
59'	1.495	1150.11	3.060740	8.69	1149.75	3.060603

TABLE I.—RADII, DEFLECTIONS, OFFSETS, ETC.

DEGREE OF CURVE D	DEFL. PER FT OF STA. (MIN)	CHORD DEFINITION			ARC DEFINITION	
		RADIUS R	LOG R	C.O. 1 STA.	RADIUS R	LOG R
5° 0'	1.5	1146.28	3.059290	8.72	1145.92	3.059153
1'	1.505	1142.47	3.057846	8.75	1142.11	3.057707
2'	1.51	1138.69	3.056407	8.78	1138.33	3.056267
3'	1.515	1134.94	3.054972	8.81	1134.57	3.054831
4'	1.52	1131.21	3.053542	8.84	1130.84	3.053400
5'	1.525	1127.50	3.052116	8.87	1127.13	3.051974
6'	1.53	1123.82	3.050696	8.90	1123.45	3.050553
7'	1.535	1120.16	3.049280	8.93	1119.79	3.049135
8'	1.54	1116.52	3.047868	8.96	1116.15	3.047723
9'	1.545	1112.91	3.046462	8.99	1112.54	3.046315
10'	1.55	1109.33	3.045059	9.01	1108.95	3.044912
11'	1.555	1105.76	3.043662	9.04	1105.38	3.043514
12'	1.56	1102.22	3.042268	9.07	1101.84	3.042119
13'	1.565	1098.70	3.040880	9.10	1098.32	3.040729
14'	1.57	1095.20	3.039495	9.13	1094.82	3.039343
15'	1.575	1091.73	3.038115	9.16	1091.35	3.037963
16'	1.58	1088.28	3.036740	9.19	1087.89	3.036587
17'	1.585	1084.85	3.035368	9.22	1084.46	3.035215
18'	1.59	1081.44	3.034002	9.25	1081.05	3.033847
19'	1.595	1078.05	3.032639	9.28	1077.66	3.032483
20'	1.6	1074.68	3.031281	9.31	1074.30	3.031124
21'	1.605	1071.34	3.029927	9.33	1070.95	3.029769
22'	1.61	1068.01	3.028577	9.36	1067.62	3.028418
23'	1.615	1064.71	3.027231	9.39	1064.32	3.027071
24'	1.62	1061.43	3.025890	9.42	1061.03	3.025729
25'	1.625	1058.16	3.024552	9.45	1057.77	3.024390
26'	1.63	1054.92	3.023219	9.48	1054.52	3.023056
27'	1.635	1051.70	3.021890	9.51	1051.30	3.021726
28'	1.64	1048.49	3.020565	9.54	1048.09	3.020400
29'	1.645	1045.31	3.019244	9.57	1044.91	3.019078
30'	1.65	1042.14	3.017927	9.60	1041.74	3.017760
31'	1.655	1039.00	3.016614	9.62	1038.59	3.016446
32'	1.66	1035.87	3.015305	9.65	1035.47	3.015136
33'	1.665	1032.76	3.013999	9.68	1032.36	3.013829
34'	1.67	1029.67	3.012698	9.71	1029.27	3.012527
35'	1.675	1026.60	3.011401	9.74	1026.19	3.011229
36'	1.68	1023.55	3.010107	9.77	1023.14	3.009935
37'	1.685	1020.51	3.008818	9.80	1020.10	3.008644
38'	1.69	1017.49	3.007532	9.83	1017.08	3.007357
39'	1.695	1014.50	3.006250	9.86	1014.08	3.006074
40'	1.7	1011.51	3.004972	9.89	1011.10	3.004795
41'	1.705	1008.55	3.003698	9.92	1008.14	3.003520
42'	1.71	1005.60	3.002427	9.94	1005.19	3.002248
43'	1.715	1002.67	3.001160	9.97	1002.26	3.000980
44'	1.72	999.762	2.999897	10.00	999.345	2.999715
45'	1.725	996.867	2.998637	10.03	996.448	2.998455
46'	1.73	993.988	2.997381	10.06	993.568	2.997198
47'	1.735	991.126	2.996129	10.09	990.705	2.995944
48'	1.74	988.280	2.994880	10.12	987.858	2.994695
49'	1.745	985.451	2.993635	10.15	985.028	2.993448
50'	1.75	982.638	2.992393	10.18	982.213	2.992206
51'	1.755	979.840	2.991155	10.21	979.415	2.990967
52'	1.76	977.060	2.989921	10.23	976.632	2.989731
53'	1.765	974.294	2.988690	10.26	973.866	2.988499
54'	1.77	971.544	2.987463	10.29	971.115	2.987271
55'	1.775	968.810	2.986239	10.32	968.379	2.986045
56'	1.78	966.091	2.985018	10.35	965.659	2.984824
57'	1.785	963.387	2.983801	10.38	962.954	2.983606
58'	1.79	960.698	2.982587	10.41	960.264	2.982391
59'	1.795	958.025	2.981377	10.44	957.590	2.981179

TABLE I.—RADII, DEFLECTIONS, OFFSETS, ETC.

DEGREE OF CURVE D	DEFL. PER FT OF STA. (MIN)	CHORD DEFINITION			ARC DEFINITION	
		RADIUS R	LOG R	C.O. 1 STA.	RADIUS R	LOG R
6° 0'	1.8	955.366	2.980170	10.47	954.930	2.979971
2'	1.81	950.093	2.977766	10.53	949.654	2.977565
4'	1.82	944.877	2.975375	10.58	944.436	2.975173
6'	1.83	939.719	2.972998	10.64	939.275	2.972793
8'	1.84	934.616	2.970633	10.70	934.170	2.970426
10'	1.85	929.569	2.968282	10.76	929.121	2.968072
12'	1.86	924.576	2.965943	10.82	924.126	2.965731
14'	1.87	919.637	2.963616	10.87	919.184	2.963402
16'	1.88	914.750	2.961303	10.93	914.294	2.961086
18'	1.89	909.915	2.959001	10.99	909.457	2.958783
20'	1.9	905.131	2.956711	11.05	904.670	2.956490
22'	1.91	900.397	2.954434	11.11	899.934	2.954211
24'	1.92	895.712	2.952168	11.16	895.247	2.951943
26'	1.93	891.076	2.949915	11.22	890.608	2.949687
28'	1.94	886.488	2.947673	11.28	886.017	2.947442
30'	1.95	881.946	2.945442	11.34	881.474	2.945209
32'	1.96	877.451	2.943223	11.40	876.976	2.942988
34'	1.97	873.002	2.941015	11.45	872.525	2.940778
36'	1.98	868.598	2.938819	11.51	868.118	2.938579
38'	1.99	864.238	2.936633	11.57	863.756	2.936391
40'	2.0	859.922	2.934459	11.63	859.437	2.934214
42'	2.01	855.648	2.932295	11.69	855.161	2.932048
44'	2.02	851.417	2.930142	11.75	850.927	2.929892
46'	2.03	847.228	2.928000	11.80	846.736	2.927748
48'	2.04	843.080	2.925869	11.86	842.585	2.925616
50'	2.05	838.972	2.923747	11.92	838.475	2.923490
52'	2.06	834.904	2.921637	11.98	834.405	2.921377
54'	2.07	830.876	2.919536	12.04	830.374	2.919274
56'	2.08	826.886	2.917446	12.09	826.381	2.917181
58'	2.09	822.934	2.915365	12.15	822.427	2.915098
7° 0'	2.1	819.020	2.913295	12.21	818.511	2.913025
2'	2.11	815.144	2.911234	12.27	814.632	2.910961
4'	2.12	811.303	2.909183	12.33	810.789	2.908908
6'	2.13	807.499	2.907142	12.38	806.983	2.906864
8'	2.14	803.731	2.905111	12.44	803.212	2.904830
10'	2.15	799.997	2.903089	12.50	799.476	2.902805
12'	2.16	796.299	2.901076	12.56	795.775	2.900790
14'	2.17	792.634	2.899073	12.62	792.108	2.898784
16'	2.18	789.003	2.897079	12.67	788.474	2.896787
18'	2.19	785.405	2.895094	12.73	784.874	2.894800
20'	2.2	781.840	2.893118	12.79	781.306	2.892821
22'	2.21	778.307	2.891151	12.85	777.771	2.890852
24'	2.22	774.806	2.889193	12.91	774.267	2.888891
26'	2.23	771.336	2.887244	12.96	770.795	2.886939
28'	2.24	767.897	2.885303	13.02	767.354	2.884996
30'	2.25	764.489	2.883371	13.08	763.944	2.883061
32'	2.26	761.112	2.881448	13.14	760.563	2.881135
34'	2.27	757.764	2.879534	13.20	757.213	2.879218
36'	2.28	754.445	2.877627	13.25	753.892	2.877309
38'	2.29	751.155	2.875730	13.31	750.600	2.875408
40'	2.3	747.894	2.873840	13.37	747.336	2.873516
42'	2.31	744.661	2.871959	13.43	744.101	2.871631
44'	2.32	741.456	2.870086	13.49	740.894	2.869756
46'	2.33	738.279	2.868221	13.55	737.714	2.867888
48'	2.34	735.129	2.866363	13.60	734.561	2.866028
50'	2.35	732.005	2.864514	13.66	731.436	2.864176
52'	2.36	728.909	2.862673	13.72	728.336	2.862332
54'	2.37	725.838	2.860840	13.78	725.263	2.860496
56'	2.38	722.793	2.859014	13.84	722.216	2.858667
58'	2.39	719.774	2.857196	13.89	719.194	2.856846

TABLE I.—RADII, DEFLECTIONS, OFFSETS, ETC.

DEGREE OF CURVE D	DEFL. PER FT OF STA. (MIN)	CHORD DEFINITION			ARC DEFINITION	
		RADIUS R	LOG R	C.O. 1 STA.	RADIUS R	LOG R
8° 0'	2.4	716.779	2.855385	13.95	716.197	2.855033
2'	2.41	713.810	2.853583	14.01	713.226	2.853227
4'	2.42	710.865	2.851787	14.07	710.278	2.851428
6'	2.43	707.945	2.849999	14.13	707.355	2.849638
8'	2.44	705.048	2.848219	14.18	704.456	2.847854
10'	2.45	702.175	2.846446	14.24	701.581	2.846078
12'	2.46	699.326	2.844679	14.30	698.729	2.844309
14'	2.47	696.499	2.842921	14.36	695.900	2.842547
16'	2.48	693.696	2.841169	14.42	693.094	2.840792
18'	2.49	690.914	2.839424	14.47	690.311	2.839045
20'	2.5	688.156	2.837687	14.53	687.549	2.837304
22'	2.51	685.419	2.835956	14.59	684.810	2.835570
24'	2.52	682.704	2.834232	14.65	682.093	2.833843
26'	2.53	680.010	2.832515	14.71	679.397	2.832123
28'	2.54	677.338	2.830805	14.76	676.722	2.830410
30'	2.55	674.686	2.829102	14.82	674.068	2.828704
32'	2.56	672.056	2.827405	14.88	671.435	2.827004
34'	2.57	669.446	2.825715	14.94	668.822	2.825311
36'	2.58	666.856	2.824032	15.00	666.230	2.823624
38'	2.59	664.286	2.822355	15.05	663.658	2.821944
40'	2.6	661.736	2.820685	15.11	661.105	2.820270
42'	2.61	659.205	2.819021	15.17	658.572	2.818603
44'	2.62	656.694	2.817363	15.23	656.059	2.816941
46'	2.63	654.202	2.815712	15.29	653.564	2.815288
48'	2.64	651.729	2.814067	15.34	651.088	2.814640
50'	2.65	649.274	2.812428	15.40	648.632	2.811998
52'	2.66	646.838	2.810796	15.46	646.193	2.810362
54'	2.67	644.420	2.809169	15.52	643.773	2.808732
56'	2.68	642.021	2.807549	15.58	641.371	2.807109
58'	2.69	639.639	2.805935	15.63	638.986	2.805492
9° 0'	2.7	637.275	2.804327	15.69	636.620	2.803880
2'	2.71	634.928	2.802724	15.75	634.271	2.802275
4'	2.72	632.599	2.801128	15.81	631.939	2.800675
6'	2.73	630.286	2.799538	15.87	629.624	2.799081
8'	2.74	627.991	2.797953	15.92	627.326	2.797494
10'	2.75	625.712	2.796374	15.98	625.045	2.795911
12'	2.76	623.450	2.794801	16.04	622.780	2.794335
14'	2.77	621.203	2.793234	16.10	620.532	2.792764
16'	2.78	618.974	2.791673	16.16	618.300	2.791199
18'	2.79	616.760	2.790117	16.21	616.084	2.789640
20'	2.8	614.563	2.788566	16.27	613.883	2.788086
22'	2.81	612.380	2.787021	16.33	611.699	2.786538
24'	2.82	610.214	2.785482	16.39	609.530	2.784995
26'	2.83	608.062	2.783948	16.45	607.376	2.783457
28'	2.84	605.926	2.782420	16.50	605.237	2.781926
30'	2.85	603.805	2.780897	16.56	603.114	2.780399
32'	2.86	601.698	2.779379	16.62	601.005	2.778878
34'	2.87	599.607	2.777867	16.68	598.911	2.777362
36'	2.88	597.530	2.776360	16.74	596.831	2.775851
38'	2.89	595.467	2.774858	16.79	594.766	2.774346
40'	2.9	593.419	2.773361	16.85	592.715	2.772846
42'	2.91	591.384	2.771870	16.91	590.678	2.771351
44'	2.92	589.364	2.770383	16.97	588.655	2.769861
46'	2.93	587.357	2.768902	17.03	586.646	2.768376
48'	2.94	585.364	2.767426	17.08	584.651	2.766897
50'	2.95	583.385	2.765955	17.14	582.669	2.765422
52'	2.96	581.419	2.764489	17.20	580.700	2.763952
54'	2.97	579.466	2.763028	17.26	578.745	2.762487
56'	2.98	577.526	2.761572	17.32	576.803	2.761028
58'	2.99	575.599	2.760120	17.37	574.874	2.759573

TABLE I.—RADII, DEFLECTIONS, OFFSETS, ETC.

DEGREE OF CURVE D	DEFL. PER FT OF STA. (MIN)	CHORD DEFINITION			ARC DEFINITION	
		RADIUS R	LOG R	C.O. 1 STA.	RADIUS R	LOG R
10° 0'	3.0	573.686	2.758674	17.43	572.958	2.758123
2'	3.01	571.784	2.757232	17.49	571.054	2.756667
4'	3.02	569.896	2.755796	17.55	569.163	2.755237
6'	3.03	568.020	2.754364	17.61	567.285	2.753801
8'	3.04	566.156	2.752937	17.66	565.419	2.752370
10'	3.05	564.305	2.751514	17.72	563.565	2.750944
12'	3.06	562.466	2.750096	17.78	561.723	2.749522
14'	3.07	560.638	2.748683	17.84	559.894	2.748106
16'	3.08	558.823	2.747274	17.89	558.076	2.746693
18'	3.09	557.019	2.745870	17.95	556.270	2.745285
20'	3.1	555.227	2.744471	18.01	554.475	2.743882
22'	3.11	553.447	2.743076	18.07	552.692	2.742483
24'	3.12	551.678	2.741686	18.13	550.921	2.741089
26'	3.13	549.920	2.740300	18.18	549.161	2.739700
28'	3.14	548.174	2.738918	18.24	547.412	2.738314
30'	3.15	546.438	2.737541	18.30	545.674	2.736933
32'	3.16	544.714	2.736169	18.36	543.947	2.735557
34'	3.17	543.001	2.734800	18.42	542.231	2.734185
36'	3.18	541.298	2.733436	18.47	540.526	2.732817
38'	3.19	539.606	2.732077	18.53	538.832	2.731453
40'	3.2	537.924	2.730721	18.59	537.148	2.730094
42'	3.21	536.253	2.729370	18.65	535.475	2.728739
44'	3.22	534.593	2.728023	18.71	533.812	2.727388
46'	3.23	532.943	2.726681	18.76	532.159	2.726041
48'	3.24	531.303	2.725342	18.82	530.516	2.724699
50'	3.25	529.673	2.724008	18.88	528.884	2.723360
52'	3.26	528.053	2.722677	18.94	527.262	2.722026
54'	3.27	526.443	2.721351	19.00	525.649	2.720696
56'	3.28	524.843	2.720029	19.05	524.047	2.719370
58'	3.29	523.252	2.718711	19.11	522.454	2.718048
11° 0'	3.3	521.671	2.717397	19.17	520.871	2.716730
2'	3.31	520.100	2.716087	19.23	519.297	2.715417
4'	3.32	518.539	2.714781	19.28	517.733	2.714106
6'	3.33	516.986	2.713479	19.34	516.178	2.712799
8'	3.34	515.443	2.712181	19.40	514.633	2.711497
10'	3.35	513.909	2.710887	19.46	513.097	2.710199
12'	3.36	512.385	2.709596	19.52	511.569	2.708905
14'	3.37	510.869	2.708310	19.57	510.051	2.707614
16'	3.38	509.363	2.707027	19.63	508.542	2.706327
18'	3.39	507.865	2.705748	19.69	507.042	2.705044
20'	3.4	506.376	2.704473	19.75	505.551	2.703765
22'	3.41	504.896	2.703202	19.81	504.068	2.702490
24'	3.42	503.425	2.701934	19.86	502.595	2.701218
26'	3.43	501.962	2.700671	19.92	501.129	2.699950
28'	3.44	500.507	2.699410	19.98	499.672	2.698685
30'	3.45	499.061	2.698154	20.04	498.224	2.697425
32'	3.46	497.624	2.696901	20.10	496.784	2.696168
34'	3.47	496.195	2.695652	20.15	495.353	2.694914
36'	3.48	494.774	2.694407	20.21	493.929	2.693665
38'	3.49	493.361	2.693165	20.27	492.514	2.692418
40'	3.5	491.956	2.691926	20.33	491.107	2.691176
42'	3.51	490.559	2.690692	20.38	489.708	2.689937
44'	3.52	489.171	2.689460	20.44	488.316	2.688701
46'	3.53	487.790	2.688233	20.50	486.933	2.687469
48'	3.54	486.417	2.687008	20.56	485.557	2.686241
50'	3.55	485.051	2.685788	20.62	484.190	2.685015
52'	3.56	483.694	2.684570	20.67	482.830	2.683974
54'	3.57	482.344	2.683357	20.73	481.477	2.682576
56'	3.58	481.001	2.682146	20.79	480.132	2.681361
58'	3.59	479.666	2.680939	20.85	478.795	2.680149

TABLE I.—RADII, DEFLECTIONS, OFFSETS, ETC.

DEGREE OF CURVE D	DEFL. PER FT OF STA. (MIN)	CHORD DEFINITION				ARC DEFINITION		
		RADIUS R	LOG R	C.O. 1 STA.	M.O. 1 STA.	RADIUS R	LOG R	C.O. 1 STA.
12° 0'	3.6	478.339	2.679735	20.91	2.62	477.465	2.678941	20.87
10'	3.65	471.810	2.673767	21.19	2.66	470.924	2.672951	21.15
20'	3.7	465.459	2.667881	21.48	2.69	464.560	2.667042	21.44
30'	3.75	459.276	2.662074	21.77	2.73	458.366	2.661213	21.73
40'	3.8	453.259	2.656345	22.06	2.77	452.335	2.655460	22.02
50'	3.85	447.395	2.650691	22.35	2.80	446.461	2.649783	22.30
13° 0'	3.9	441.684	2.645111	22.64	2.84	440.737	2.644179	22.59
10'	3.95	436.117	2.639603	22.93	2.88	435.158	2.638647	22.88
20'	4.0	430.690	2.634164	23.22	2.91	429.718	2.633184	23.16
30'	4.05	425.396	2.628794	23.51	2.95	424.413	2.627789	23.45
40'	4.1	420.233	2.623490	23.80	2.99	419.237	2.622460	23.74
50'	4.15	415.194	2.618251	24.09	3.02	414.186	2.617196	24.03
14° 0'	4.2	410.275	2.613075	24.37	3.06	409.256	2.611995	24.31
10'	4.25	405.473	2.607962	24.66	3.10	404.441	2.606855	24.60
20'	4.3	400.782	2.602908	24.95	3.13	399.738	2.601775	24.89
30'	4.35	396.200	2.597914	25.24	3.17	395.143	2.596755	25.17
40'	4.4	391.722	2.592978	25.53	3.20	390.653	2.591791	25.46
50'	4.45	387.345	2.588097	25.82	3.24	386.264	2.586884	25.74
15° 0'	4.5	383.065	2.583272	26.11	3.28	381.972	2.582031	26.03
10'	4.55	378.880	2.578501	26.39	3.31	377.774	2.577232	26.32
20'	4.6	374.786	2.573783	26.68	3.35	373.668	2.572486	26.60
30'	4.65	370.780	2.569116	26.97	3.39	369.650	2.567791	26.89
40'	4.7	366.859	2.564500	27.26	3.42	365.718	2.563146	27.17
50'	4.75	363.022	2.559933	27.55	3.46	361.868	2.558550	27.46
16° 0'	4.8	359.265	2.555415	27.83	3.50	358.099	2.554003	27.74
10'	4.85	355.585	2.550944	28.12	3.53	354.407	2.549502	28.03
20'	4.9	351.981	2.546519	28.41	3.57	350.790	2.545048	28.31
30'	4.95	348.450	2.542140	28.70	3.61	347.247	2.540638	28.60
40'	5.0	344.990	2.537806	28.99	3.64	343.775	2.536274	28.89
50'	5.05	341.598	2.533516	29.27	3.68	340.371	2.531952	29.16
17° 0'	5.1	338.273	2.529268	29.56	3.72	337.034	2.527673	29.45
10'	5.15	335.013	2.525062	29.85	3.75	333.762	2.523437	29.73
20'	5.2	331.816	2.520898	30.14	3.79	330.553	2.519241	30.02
30'	5.25	328.689	2.516774	30.42	3.82	327.404	2.515085	30.31
40'	5.3	325.604	2.512690	30.71	3.86	324.316	2.510968	30.59
50'	5.35	322.585	2.508645	31.00	3.90	321.285	2.506890	30.87
18° 0'	5.4	319.623	2.504638	31.29	3.94	318.310	2.502850	31.16
10'	5.45	316.715	2.500668	31.57	3.97	315.390	2.498847	31.44
20'	5.5	313.860	2.496736	31.86	4.01	312.522	2.494881	31.72
30'	5.55	311.056	2.492839	32.15	4.04	309.707	2.490951	32.00
40'	5.6	308.303	2.488978	32.44	4.08	306.942	2.487056	32.29
50'	5.65	305.599	2.485152	32.72	4.12	304.225	2.483195	32.58
19° 0'	5.7	302.943	2.481361	33.01	4.16	301.557	2.479369	32.86
10'	5.75	300.333	2.477603	33.30	4.19	298.934	2.475576	33.14
20'	5.8	297.768	2.473878	33.58	4.23	296.357	2.471816	33.42
30'	5.85	295.247	2.470186	33.87	4.26	293.825	2.468087	33.71
40'	5.9	292.770	2.466526	34.16	4.30	291.334	2.464392	33.99
50'	5.95	290.334	2.462897	34.44	4.34	289.886	2.460727	34.27
20° 0'	6.0	287.939	2.459300	34.73	4.37	286.479	2.457093	34.55
10'	6.05	285.583	2.455733	35.02	4.41	284.111	2.453488	34.83
20'	6.1	283.267	2.452195	35.30	4.45	281.783	2.449914	35.12
30'	6.15	280.988	2.448688	35.59	4.48	279.492	2.446369	35.40
40'	6.2	278.746	2.445209	35.87	4.52	277.238	2.442852	35.68
50'	6.25	276.541	2.441759	36.16	4.56	275.020	2.439364	35.96
21° 0'	6.3	274.370	2.438337	36.45	4.59	272.837	2.435903	36.24
10'	6.35	272.234	2.434943	36.73	4.63	270.689	2.432470	36.52
20'	6.4	270.132	2.431576	37.02	4.67	268.574	2.429064	36.80
30'	6.45	268.062	2.428235	37.30	4.70	266.492	2.425684	37.08
40'	6.5	266.024	2.424921	37.59	4.74	264.442	2.422330	37.37
50'	6.55	264.018	2.421633	37.88	4.78	262.423	2.419002	37.65

TABLE I.—RADII, DEFLECTIONS, OFFSETS, ETC.

DEGREE OF CURVE D	DEFL. PER FT OF STA. (MIN)	CHORD DEFINITION				ARC DEFINITION		
		RADIUS R	LOG R	C.O. 1 STA.	M.O. 1 STA.	RADIUS R	LOG R	C.O. 1 STA.
22° 0'	6.6	262.042	2.418371	38.16	4.81	260.435	2.415700	37.93
10'	6.65	260.098	2.415134	38.45	4.85	258.477	2.412422	38.21
20'	6.7	258.180	2.411922	38.73	4.89	256.548	2.409169	38.49
30'	6.75	256.292	2.408734	39.02	4.92	254.648	2.405940	38.77
40'	6.8	254.431	2.405571	39.30	4.96	252.775	2.402735	39.05
50'	6.85	252.599	2.402431	39.59	5.00	250.930	2.399553	39.33
23° 0'	6.9	250.793	2.399315	39.87	5.04	249.112	2.396395	39.61
10'	6.95	249.013	2.396222	40.16	5.07	247.320	2.393259	39.89
20'	7.0	247.258	2.393151	40.44	5.11	245.553	2.390145	40.16
30'	7.05	245.529	2.390103	40.73	5.14	243.812	2.387055	40.44
40'	7.1	243.825	2.387077	41.01	5.18	242.095	2.383985	40.72
50'	7.15	242.144	2.384074	41.30	5.22	240.402	2.380938	41.00
24° 0'	7.2	240.487	2.381091	41.58	5.26	238.732	2.377911	41.28
10'	7.25	238.853	2.378130	41.87	5.29	237.086	2.374905	41.56
20'	7.3	237.241	2.375190	42.15	5.33	235.462	2.371921	41.84
30'	7.35	235.652	2.372270	42.44	5.37	233.860	2.368956	42.11
40'	7.4	234.084	2.369371	42.72	5.40	232.280	2.366012	42.39
50'	7.45	232.537	2.366492	43.00	5.44	230.721	2.363087	42.67
25° 0'	7.5	231.011	2.363633	43.29	5.48	229.183	2.360183	42.95
10'	7.55	229.506	2.360794	43.57	5.51	227.665	2.357279	43.23
20'	7.6	228.020	2.357974	43.86	5.55	226.168	2.354430	43.50
30'	7.65	226.555	2.355173	44.14	5.59	224.689	2.351582	43.78
40'	7.7	225.108	2.352391	44.42	5.62	223.230	2.348753	44.06
50'	7.75	223.680	2.349627	44.71	5.66	221.790	2.345942	44.33
26° 0'	7.8	222.271	2.346882	44.99	5.70	220.368	2.343149	44.60
30'	7.95	218.150	2.338755	45.84	5.81	216.210	2.334877	45.43
27° 0'	8.1	214.183	2.330785	46.69	5.92	212.207	2.326759	46.26
30'	8.25	210.362	2.322967	47.54	6.03	208.348	2.318790	47.08
28° 0'	8.4	206.678	2.315295	48.38	6.14	204.628	2.310964	47.91
30'	8.55	203.125	2.307764	49.23	6.25	201.038	2.303278	48.72
29° 0'	8.7	199.696	2.300370	50.08	6.36	197.572	2.295725	49.55
30'	8.85	196.385	2.293108	50.92	6.47	194.223	2.288301	50.36
30° 0'	9.0	193.185	2.285974	51.76	6.58	190.986	2.281001	51.18
32° 0'	9.6	181.398	2.258632	55.13	7.03	179.049	2.252973	54.42
34° 0'	10.2	171.015	2.233035	58.47	7.47	168.517	2.226644	57.62
36° 0'	10.8	161.803	2.208988	61.80	7.92	159.155	2.201820	60.79
38° 0'	11.4	153.578	2.186328	65.11	8.37	150.778	2.178339	63.93
41° 0'	12.3	142.773	2.154645	70.04	9.04	139.746	2.145339	68.55
44° 0'	13.2	133.473	2.125395	74.92	9.72	130.218	2.114669	73.09
48° 0'	14.4	122.930	2.089657	81.35	10.63	119.366	2.076881	78.98
52° 0'	15.6	114.058	2.057128	87.67	11.54	110.184	2.042119	84.69
57° 0'	17.1	104.787	2.020307	95.43	12.70	100.519	2.002248	91.55
64° 0'	19.2	94.354	1.974760	106.0	14.34	89.525	1.951943	100.6
72° 0'	21.6	85.065	1.929751	117.6	16.25	79.577	1.900790	110.0
82° 0'	24.6	76.213	1.882027	131.2	18.69	69.873	1.844309	120.3
95° 0'	28.5	67.817	1.831339	147.5	22.00	60.311	1.780399	131.1
115° 0'	34.5	59.284	1.772941	168.7	27.43	49.822	1.697425	141.8

NOTE: The odd values of D between 30° and 115° are those whose arc-definition radii vary approximately from 190 feet to 50 feet by 10-ft intervals. For curves having exactly these radii, see Table V.

TABLE II.—LENGTHS OF ARCS AND TRUE CHORDS

D	CHORD DEFINITION OF D				ARC DEFINITION OF D		
	ARC FOR	TRUE CHORDS			TRUE CHORDS		
	1 STA.	1/10 STA.	1/4 STA.	1/2 STA.	1/4 STA.	1/2 STA.	1 STA.
1°	100.001	10	25	50	25	50	100
2°	100.005	10	25	50	25	50	100
3°	100.011	10	25	50	25	50	99.99
4°	100.020	10	25	50.01	25	50	99.98
5°	100.032	10	25.01	50.01	25	50	99.97
6°	100.046	10.01	25.01	50.02	25	50	99.95
7°	100.062	10.01	25.02	50.02	25	50	99.94
8°	100.081	10.01	25.02	50.03	25	49.99	99.92
9°	100.103	10.01	25.02	50.04	25	49.99	99.90
10°	100.127	10.01	25.03	50.05	25	49.98	99.88
11°	100.154	10.02	25.04	50.06	25	49.98	99.85
12°	100.183	10.02	25.04	50.07	25	49.98	99.82
13°	100.215	10.02	25.05	50.08	25	49.97	99.79
14°	100.249	10.02	25.06	50.09	25	49.97	99.75
15°	100.286	10.03	25.07	50.11	25	49.96	99.72
16°	100.326	10.03	25.08	50.12	25	49.96	99.68
17°	100.368	10.04	25.09	50.14	25	49.95	99.63
18°	100.412	10.04	25.10	50.16	24.99	49.95	99.59
19°	100.460	10.04	25.11	50.17	24.99	49.94	99.54
20°	100.510	10.05	25.12	50.19	24.99	49.94	99.49
21°	100.562	10.06	25.13	50.21	24.99	49.93	99.44
22°	100.617	10.06	25.14	50.23	24.99	49.92	99.39
23°	100.675	10.07	25.16	50.25	24.99	49.92	99.33
24°	100.735	10.07	25.17	50.27	24.99	49.91	99.27
25°	100.798	10.08	25.19	50.30	24.99	49.90	99.21
26°	100.863	10.08	25.20	50.32	24.99	49.89	99.14
27°	100.931	10.09	25.22	50.35	24.99	49.88	99.08
28°	101.002	10.10	25.23	50.38	24.98	49.88	99.01
29°	101.075	10.11	25.25	50.40	24.98	49.87	98.94
30°	101.152	10.11	25.27	50.43	24.98	49.86	98.86
32°	101.312	10.13	25.31	50.49	24.98	49.84	98.71
34°	101.482	10.15	25.35	50.56	24.98	49.82	98.54
36°	101.664	10.16	25.39	50.62	24.97	49.79	98.36
38°	101.857	10.18	25.43	50.69	24.97	49.77	98.18
41°	102.166	10.21	25.51	50.81	24.97	49.73	97.88
44°	102.500	10.25	25.59	50.94	24.96	49.69	97.56
48°	102.986	10.30	25.70	51.12	24.95	49.63	97.10
52°	103.516	10.35	25.82	51.31	24.95	49.57	96.60
57°	104.246	10.42	25.98	51.59	24.94	49.49	95.93
64°	105.394	10.53	26.26	52.01	24.92	49.35	94.88
72°	106.896	10.68	26.61	52.57	24.90	49.18	93.55
82°	109.073	10.90	27.12	53.38	24.87	48.94	91.68
95°	112.445	11.23	27.91	54.63	24.82	48.58	88.93
115°	118.992	11.88	29.44	57.03	24.74	47.93	84.04

Chord Definition of D

For degrees of curve not listed obtain excess arc per station approximately by interpolation, or exactly to 3 decimal places (up to D = 15°) from:

$$\text{Excess} = 0.00127\ D^2$$

Arc Definition of D

For degrees of curve not listed obtain chord deficiency per station approximately by interpolation, or exactly to 2 decimal places (up to D = 30°) from:

$$\text{Deficiency} = 0.00127\ D^2$$

Tables III & IV. CORRECTION COEFFICIENTS FOR SUBCHORDS

CHORD DEFINITION OF D

The following table may be used to obtain true lengths of subchords not listed in Table II.

For any degree of curve the small correction to be added to the nominal length in order to obtain the true length is almost a constant percentage of the excess of arc for 1 station on a curve of that degree. The maximum correction is required for a nominal subchord about 57.5 ft long.

NOMINAL SUBCHORD	RATIO OF CHORD CORRECTION TO EXCESS ARC PER STATION
5	0.050
10	0.099
15	0.147
20	0.192
25	0.234
30	0.273
35	0.307
40	0.336
45	0.359
50	0.375
55	0.383
60	0.384
65	0.375
70	0.357
75	0.328
80	0.288
85	0.236
90	0.171
95	0.093
100	0

Example

Given: a 20° curve. Required: true subchord for 75-ft nominal length.

SOLUTION

Excess arc per sta. = 0.510 (from Table II). Corr = 0.510 × 0.328 = 0.17. *Add corr. to nominal length* giving true subchord = 75.17.

For subchords not listed, interpolate for correction coefficients.

ARC DEFINITION OF D

The following table may be used to obtain true lengths of subchords not listed in Table II.

For any degree of curve the small correction to be subtracted from an arc length in order to obtain the true chord length is almost a constant percentage of the chord deficiency for a 100-ft arc on a curve of that degree. These percentages vary approximately as the cubes of the arc lengths.

LENGTH OF ARC	RATIO OF CHORD CORRECTION TO CHORD DEFICIENCY FOR 100-ft ARC
5	0.0001
10	0.0010
15	0.0034
20	0.0080
25	0.016
30	0.027
35	0.043
40	0.064
45	0.091
50	0.125
55	0.166
60	0.216
65	0.275
70	0.343
75	0.422
80	0.512
85	0.614
90	0.729
95	0.857
100	1

Example

Given: a 25° curve. Required: true subchord for 75-ft arc.

SOLUTION

Chord def. per sta. = 0.79 (from Table II). Corr. = 0.79 × 0.422 = 0.33. *Subtract corr. from arc* giving true subchord = 74.67.

For arcs not listed, note that correction coefficients are proportional to the cubes of the arcs.

TABLE V.—EVEN-RADIUS CURVES. DEFLECTIONS AND CHORDS

RADIUS FT	DEFL. MIN PER FT OF ARC	DEFLECTIONS FOR ARCS OF			CHORDS FOR ARCS OF		
		10 FT	25 FT	100 FT	10 FT	25 FT	100 FT
50	34.377	5°43.78'	14°19.44'	57°17.75'	9.98	24.74	84.15
60	28.648	4°46.48'	11°56.20'	47°44.79'	9.99	24.82	88.82
70	24.555	4°05.55'	10°13.88'	40°55.53'	9.99	24.87	91.73
80	21.486	3°34.86'	8°57.15'	35°48.59'	10	24.90	93.62
90	19.099	3°10.99'	7°57.46'	31°49.86'	10	24.92	94.94
100	17.189	2°51.89'	7°09.72'	28°38.87'	10	24.93	95.89
110	15.626	2°36.26'	6°30.65'	26°02.61'	10	24.95	96.59
120	14.324	2°23.24'	5°58.10'	23°52.39'	10	24.96	97.13
130	13.222	2°12.22'	5°30.55'	22°02.21'	10	24.96	97.55
140	12.278	2°02.78'	5°06.94'	20°27.77'	10	24.96	97.88
150	11.459	1°54.59'	4°46.48'	19°05.92'	10	24.97	98.16
160	10.743	1°47.43'	4°28.58'	17°54.30'	10	24.97	98.38
170	10.111	1°41.11'	4°12.77'	16°51.10'	10	24.97	98.57
180	9.549	1°35.49'	3°58.73'	15°54.93'	10	24.98	98.72
190	9.047	1°30.47'	3°46.17'	15°04.67'	10	24.98	98.85
		25 FT	50 FT		25 FT	50 FT	
200	8.594	3°34.86'	7°09.72'	14°19.44'	24.98	49.87	98.96
225	7.639	3°10.99'	6°21.97'	12°43.94'	24.99	49.90	99.18
250	6.875	2°51.89'	5°43.78'	11°27.55'	24.99	49.92	99.34
275	6.250	2°36.26'	5°12.52'	10°25.05'	24.99	49.93	99.45
300	5.730	2°23.24'	4°46.48'	9°32.96'	24.99	49.94	99.54
325	5.289	2°12.22'	4°24.44'	8°48.88'	24.99	49.95	99.61
350	4.911	2°02.78'	4°05.55'	8°11.11'	25	49.96	99.66
375	4.584	1°54.59'	3°49.18'	7°38.37'	25	49.96	99.70
400	4.297	1°47.43'	3°34.86'	7°09.72'	25	49.97	99.74
425	4.044	1°41.11'	3°22.22'	6°44.44'	25	49.97	99.77
450	3.820	1°35.49'	3°10.99'	6°21.97'	25	49.97	99.79
475	3.619	1°30.47'	3°00.93'	6°01.87'	25	49.98	99.82
500	3.438	1°25.94'	2°51.89'	5°43.77'	25	49.98	99.83
550	3.235	1°18.13'	2°36.26'	5°12.52'	25	49.98	99.86
600	2.865	1°11.62'	2°23.24'	4°46.48'	25	49.99	99.89
650	2.644	1°06.11'	2°12.22'	4°24.44'	25	49.99	99.90
700	2.455	1°01.39'	2°02.78'	4°05.55'	25	49.99	99.92
750	2.292	0°57.30'	1°54.59'	3°49.18'	25	50	99.93
800	2.149	0°53.71'	1°47.43'	3°34.86'	25	50	99.93
850	2.022	0°50.56'	1°41.11'	3°22.22'	25	50	99.94
900	1.910	0°47.75'	1°35.49'	3°10.99'	25	50	99.95
950	1.809	0°45.23'	1°30.47'	3°00.93'	25	50	99.95
1000	1.719	0°42.97'	1°25.94'	2°51.89'	25	50	99.96
1050	1.637	0°40.93'	1°21.85'	2°43.70'	25	50	99.96
1100	1.563	0°39.07'	1°18.13'	2°36.26'	25	50	99.96
1200	1.432	0°35.81'	1°11.62'	2°23.24'	25	50	99.97
1300	1.322	0°33.06'	1°06.11'	2°12.22'	25	50	99.97
1400	1.228	0°30.69'	1°01.39'	2°02.78'	25	50	99.98
1500	1.146	0°28.65'	0°57.30'	1°54.59'	25	50	99.98
1600	1.074	0°26.86'	0°53.72'	1°47.43'	25	50	99.98
1700	1.011	0°25.28'	0°50.56'	1°41.11'	25	50	99.99
1800	0.955	0°23.87'	0°47.75'	1°35.49'	25	50	99.99
1900	0.905	0°22.62'	0°45.23'	1°30.47'	25	50	100
2000	0.859	0°21.49'	0°42.97'	1°25.94'	25	50	100
2100	0.819	0°20.46'	0°40.93'	1°21.85'	25	50	100
2200	0.781	0°19.53'	0°39.07'	1°18.13'	25	50	100
2300	0.747	0°18.68'	0°37.37'	1°14.73'	25	50	100
2400	0.716	0°17.91'	0°35.81'	1°11.62'	25	50	100
2500	0.688	0°17.19'	0°34.38'	1°08.75'	25	50	100
2600	0.661	0°16.53'	0°33.06'	1°06.11'	25	50	100

TABLE V.—EVEN-RADIUS CURVES. DEFLECTIONS AND CHORDS

RADIUS FT	DEFL. MIN PER FT OF ARC	DEFLECTIONS FOR ARCS OF			CHORDS FOR ARCS OF		
		25 FT	50 FT	100 FT	25 FT	50 FT	100 FT
2700	0.637	0°15.91'	0°31.83'	1°03.66'	25	50	100
2800	0.614	0°15.35'	0°30.69'	1°01.39'	25	50	100
2900	0.593	0°14.82'	0°29.64'	0°59.27'	25	50	100
3000	0.573	0°14.32'	0°28.65'	0°57.30'	25	50	100
3100	0.555	0°13.86'	0°27.72'	0°55.45'	25	50	100
3200	0.537	0°13.43'	0°26.86'	0°53.71'	25	50	100
3300	0.521	0°13.02'	0°26.04'	0°52.09'	25	50	100
3400	0.506	0°12.64'	0°25.28'	0°50.55'	25	50	100
3500	0.491	0°12.28'	0°24.56'	0°49.11'	25	50	100
3600	0.477	0°11.94'	0°23.87'	0°47.75'	25	50	100
3700	0.465	0°11.61'	0°23.23'	0°46.46'	25	50	100
3800	0.452	0°11.31'	0°22.62'	0°45.23'	25	50	100
3900	0.441	0°11.02'	0°22.04'	0°44.07'	25	50	100
4000	0.430	0°10.74'	0°21.49'	0°42.97'	25	50	100
4100	0.419	0°10.48'	0°20.96'	0°41.92'	25	50	100
4500	0.382	0°09.55'	0°19.10'	0°38.20'	25	50	100
5000	0.344	0°08.59'	0°17.19'	0°34.38'	25	50	100
5500	0.313	0°07.81'	0°15.63'	0°31.25'	25	50	100
6000	0.286	0°07.16'	0°14.32'	0°28.65'	25	50	100
6500	0.264	0°06.61'	0°13.22'	0°26.44'	25	50	100
7000	0.246	0°06.14'	0°12.28'	0°24.56'	25	50	100
7500	0.229	0°05.73'	0°11.46'	0°22.92'	25	50	100
8000	0.215	0°05.37'	0°10.74'	0°21.49'	25	50	100
9000	0.191	0°04.77'	0°09.55'	0°19.10'	25	50	100
10000	0.172	0°04.30'	0°08.59'	0°17.19'	25	50	100

Notes

Degree of Curve (arc definition) = twice the deflection for a 100-ft arc. Deflections for even-radius curves or arcs not listed may be computed from:

$$\text{Defl. (in minutes)} = \frac{1718.873}{R} \times \text{Arc length}$$

Chords not listed may be obtained by interpolation or computed from:

$$\text{Chord} = 2\,R \sin \text{defl.}$$

Total length of curve may be determined by use of Table VI. See section 5-9 for selection of spirals for even-radius curves.

TABLE VI.—LENGTHS OF CIRCULAR ARCS; RADIUS=1
(Degrees, Minutes, and Seconds to Radians)

DEG	LENGTH	DEG	LENGTH	MIN	LENGTH	SEC	LENGTH
1	0.017 45 329	61	1.064 65 084	1	.000 29 089	1	.000 00 485
2	.034 90 659	62	.082 10 414	2	0 58 178	2	00 970
3	.052 35 988	63	.099 55 743	3	0 87 266	3	01 454
4	.069 81 317	64	.117 01 072	4	1 16 355	4	01 939
5	0.087 26 646	65	1.134 46 401	5	.001 45 444	5	.000 02 424
6	.104 71 976	66	.151 91 731	6	1 74 533	6	02 909
7	.122 17 305	67	.169 37 060	7	2 03 622	7	03 394
8	.139 62 634	68	.186 82 389	8	2 32 711	8	03 879
9	.157 07 963	69	.204 27 718	9	2 61 799	9	04 363
10	0.174 53 293	70	1.221 73 048	10	.002 90 888	10	.000 04 848
11	.191 98 622	71	.239 18 377	11	3 19 977	11	05 333
12	.209 43 951	72	.256 63 706	12	3 49 066	12	05 818
13	.226 89 280	73	.274 09 035	13	3 78 155	13	06 303
14	.244 34 610	74	.291 54 365	14	4 07 243	14	06 787
15	0.261 79 939	75	1.308 99 694	15	.004 36 332	15	.000 07 272
16	.279 25 268	76	.326 45 023	16	4 65 421	16	07 757
17	.296 70 597	77	.343 90 352	17	4 94 510	17	08 242
18	.314 15 927	78	.361 35 682	18	5 23 599	18	08 727
19	.331 61 256	79	.378 81 011	19	5 52 688	19	09 211
20	0.349 06 585	80	1.396 26 340	20	.005 81 776	20	.000 09 696
21	.366 51 914	81	.413 71 669	21	6 10 865	21	10 181
22	.383 97 244	82	.431 16 999	22	6 39 954	22	10 666
23	.401 42 573	83	.448 62 328	23	6 69 043	23	11 151
24	.418 87 902	84	.466 07 657	24	6 98 132	24	11 636
25	0.436 33 231	85	1.483 52 986	25	.007 27 221	25	.000 12 120
26	.453 78 561	86	.500 98 316	26	7 56 309	26	12 605
27	.471 23 890	87	.518 43 645	27	7 85 398	27	13 090
28	.488 69 219	88	.535 88 974	28	8 14 487	28	13 575
29	.506 14 548	89	.553 34 303	29	8 43 576	29	14 060
30	0.523 59 878	90	1.570 79 633	30	.008 72 665	30	.000 14 544
31	.541 05 207	91	.588 24 962	31	9 01 753	31	15 029
32	.558 50 536	92	.605 70 291	32	9 30 842	32	15 514
33	.575 95 865	93	.623 15 620	33	9 59 931	33	15 999
34	.593 41 195	94	.640 60 950	34	9 89 020	34	16 484
35	0.610 86 524	95	1.658 06 279	35	.010 18 109	35	.000 16 969
36	.628 31 853	96	.675 51 608	36	10 47 198	36	17 453
37	.645 77 182	97	.692 96 937	37	10 76 286	37	17 938
38	.663 22 512	98	.710 42 267	38	11 05 375	38	18 423
39	.680 67 841	99	.727 87 596	39	11 34 464	39	18 908
40	0.698 13 170	100	1.745 32 925	40	.011 63 553	40	.000 19 393
41	.715 58 499	101	.762 78 254	41	11 92 642	41	19 877
42	.733 03 829	102	.780 23 584	42	12 21 730	42	20 362
43	.750 49 158	103	.797 68 913	43	12 50 819	43	20 847
44	.767 94 487	104	.815 14 242	44	12 79 908	44	21 332
45	0.785 39 816	105	1.832 59 571	45	.013 08 997	45	.000 21 817
46	.802 85 146	106	.850 04 901	46	13 38 086	46	22 301
47	.820 30 475	107	.867 50 230	47	13 67 175	47	22 786
48	.837 75 804	108	.884 95 559	48	13 96 263	48	23 271
49	.855 21 133	109	.902 40 888	49	14 25 352	49	23 756
50	0.872 66 463	110	1.919 86 218	50	.014 54 441	50	.000 24 241
51	.890 11 792	111	.937 31 547	51	14 83 530	51	24 726
52	.907 57 121	112	.954 76 876	52	15 12 619	52	25 210
53	.925 02 450	113	.972 22 205	53	15 41 708	53	25 695
54	.942 47 780	114	.989 67 535	54	15 70 796	54	26 180
55	0.959 93 109	115	2.007 12 864	55	.015 99 885	55	.000 26 665
56	0.977 38 438	116	.024 58 193	56	16 28 974	56	27 150
57	0.994 83 767	117	.042 03 522	57	16 58 063	57	27 634
58	1.012 29 097	118	.059 48 852	58	16 87 152	58	28 119
59	1.029 74 426	119	.076 94 181	59	17 16 240	59	28 604
60	1.047 19 755	120	.094 39 510	60	17 45 329	60	29 089

TABLE VII.—MINUTES AND SECONDS IN DECIMALS OF A DEGREE

MIN	SECONDS							
	0	10	15	20	30	40	45	50
0	.0000	.0028	.0042	.0056	.0083	.0111	.0125	.0139
1	.0167	.0194	.0208	.0222	.0250	.0278	.0292	.0306
2	.0333	.0361	.0375	.0389	.0417	.0444	.0458	.0472
3	.0500	.0528	.0542	.0556	.0583	.0611	.0625	.0639
4	.0667	.0694	.0708	.0722	.0750	.0778	.0792	.0806
5	.0833	.0861	.0875	.0889	.0917	.0944	.0958	.0972
6	.1000	.1028	.1042	.1056	.1083	.1111	.1125	.1139
7	.1167	.1194	.1208	.1222	.1250	.1278	.1292	.1306
8	.1333	.1361	.1375	.1389	.1417	.1444	.1458	.1472
9	.1500	.1528	.1542	.1556	.1583	.1611	.1625	.1639
10	.1667	.1694	.1708	.1722	.1750	.1778	.1792	.1806
11	.1833	.1861	.1875	.1889	.1917	.1944	.1958	.1972
12	.2000	.2028	.2042	.2056	.2083	.2111	.2125	.2139
13	.2167	.2194	.2208	.2222	.2250	.2278	.2292	.2306
14	.2333	.2361	.2375	.2389	.2417	.2444	.2458	.2472
15	.2500	.2528	.2542	.2556	.2583	.2611	.2625	.2639
16	.2667	.2694	.2708	.2722	.2750	.2778	.2792	.2806
17	.2833	.2861	.2875	.2889	.2917	.2944	.2958	.2972
18	.3000	.3028	.3042	.3056	.3083	.3111	.3125	.3139
19	.3167	.3194	.3208	.3222	.3250	.3278	.3292	.3306
20	.3333	.3361	.3375	.3389	.3417	.3444	.3458	.3472
21	.3500	.3528	.3542	.3556	.3583	.3611	.3625	.3639
22	.3667	.3694	.3708	.3722	.3750	.3778	.3792	.3806
23	.3833	.3861	.3875	.3889	.3917	.3944	.3958	.3972
24	.4000	.4028	.4042	.4056	.4083	.4111	.4125	.4139
25	.4167	.4194	.4208	.4222	.4250	.4278	.4292	.4306
26	.4333	.4361	.4375	.4389	.4417	.4444	.4458	.4472
27	.4500	.4528	.4542	.4556	.4583	.4611	.4625	.4639
28	.4667	.4694	.4708	.4722	.4750	.4778	.4792	.4806
29	.4833	.4861	.4875	.4889	.4917	.4944	.4958	.4972
30	.5000	.5028	.5042	.5056	.5083	.5111	.5125	.5139
31	.5167	.5194	.5208	.5222	.5250	.5278	.5292	.5306
32	.5333	.5361	.5375	.5389	.5417	.5444	.5458	.5472
33	.5500	.5528	.5542	.5556	.5583	.5611	.5625	.5639
34	.5667	.5694	.5708	.5722	.5750	.5778	.5792	.5806
35	.5833	.5861	.5875	.5889	.5917	.5944	.5958	.5972
36	.6000	.6028	.6042	.6056	.6083	.6111	.6125	.6139
37	.6167	.6194	.6208	.6222	.6250	.6278	.6292	.6306
38	.6333	.6361	.6375	.6389	.6417	.6444	.6458	.6472
39	.6500	.6528	.6542	.6556	.6583	.6611	.6625	.6639
40	.6667	.6694	.6708	.6722	.6750	.6778	.6792	.6806
41	.6833	.6861	.6875	.6889	.6917	.6944	.6958	.6972
42	.7000	.7028	.7042	.7056	.7083	.7111	.7125	.7139
43	.7167	.7194	.7208	.7222	.7250	.7278	.7292	.7306
44	.7333	.7361	.7375	.7389	.7417	.7444	.7458	.7472
45	.7500	.7528	.7542	.7556	.7583	.7611	.7625	.7639
46	.7667	.7694	.7708	.7722	.7750	.7778	.7792	.7806
47	.7833	.7861	.7875	.7889	.7917	.7944	.7958	.7972
48	.8000	.8028	.8042	.8056	.8083	.8111	.8125	.8139
49	.8167	.8194	.8208	.8222	.8250	.8278	.8292	.8306
50	.8333	.8361	.8375	.8389	.8417	.8444	.8458	.8472
51	.8500	.8528	.8542	.8556	.8583	.8611	.8625	.8639
52	.8667	.8694	.8708	.8722	.8750	.8778	.8792	.8806
53	.8833	.8861	.8875	.8889	.8917	.8944	.8958	.8972
54	.9000	.9028	.9042	.9056	.9083	.9111	.9125	.9139
55	.9167	.9194	.9208	.9222	.9250	.9278	.9292	.9306
56	.9333	.9361	.9375	.9389	.9417	.9444	.9458	.9472
57	.9500	.9528	.9542	.9556	.9583	.9611	.9625	.9639
58	.9667	.9694	.9708	.9722	.9750	.9778	.9792	.9806
59	.9833	.9861	.9875	.9889	.9917	.9944	.9958	.9972

Table VIII. TANGENTS AND EXTERNALS FOR A 1° CURVE (Chord or Arc Definition).

EXPLANATION

CHORD DEFINITION OF DEGREE OF CURVE (D_c)	ARC DEFINITION OF DEGREE OF CURVE (D_a)
MATHEMATICAL RELATIONS	MATHEMATICAL RELATIONS
1. $\sin \frac{1}{2}D_c = \frac{50}{R}$, . . . , exactly	1. $D_a = \frac{5729.58}{R}$, . . . , exactly
1a. $D_c = \frac{5730}{R}$, . . . , approx.	1a. $D_a = \frac{5730}{R}$, . . . , approx.
2. $T = R \tan \frac{1}{2}I$, . . . , exactly	2. $T = R \tan \frac{1}{2}I$, . . . , exactly
2a. $T = \frac{T_{1^\circ}}{D_c}$, . . . , approx.	2a. $T = \frac{T_{1^\circ}}{D_a}$, . . . , exactly
3. $E = R \operatorname{exsec} \frac{1}{2}I$, . . . , exactly	3. $E = R \operatorname{exsec} \frac{1}{2}I$, . . . , exactly
3a. $E = \frac{E_{1^\circ}}{D_c}$, . . . , approx.	3a. $E = \frac{E_{1^\circ}}{D_a}$, . . . , exactly

Table VIII gives values of T and E for a 1° curve having various central angles. The values may be considered correct for both the chord definition and the arc definition of D, since the difference in R for $D = 1°$ is only seven units in the sixth significant figure. To find T and E for any given values of I and D, use relations 2a and 3a, as in the following example:

Example

Given: $I = 68°45'$, $D = 8°40'$. (*Note:* To avoid lack of precision in calculation, use D as $8\frac{2}{3} = \frac{26}{3}$, instead of 8.666 . . .)
For arc definition of D,

$$T = 3919.5 \times \tfrac{3}{26} = 452.25$$

$$E = 1212.3 \times \tfrac{3}{26} = 139.88$$

For chord definition of D,

T = Arc-definition value plus correction of 0.43 found from Table IX by interpolation; or $T = 452.68$

E = Arc-definition value plus correction of 0.14 found from Table X by interpolation; or $E = 140.02$

TABLE VIII.—TANGENTS AND EXTERNALS FOR A 1° CURVE
(Chord or Arc Definition)

'	I = 0°		I = 1°		I = 2°		I = 3°		I = 4°		'
	T	E	T	E	T	E	T	E	T	E	
0	0.0	0.0	50.0	0.2	100.0	0.9	150.0	2.0	200.1	3.5	0
1	0.8	0.0	50.8	0.2	100.8	0.9	150.9	2.0	200.9	3.5	1
2	1.7	0.0	51.7	0.2	101.7	0.9	151.7	2.0	201.8	3.6	2
3	2.5	0.0	52.5	0.2	102.5	0.9	152.5	2.0	202.6	3.6	3
4	3.3	0.0	53.3	0.2	103.3	0.9	153.4	2.1	203.4	3.6	4
5	4.2	0.0	54.2	0.3	104.2	0.9	154.2	2.1	204.3	3.7	5
6	5.0	0.0	55.0	0.3	105.0	1.0	155.0	2.1	205.1	3.7	6
7	5.8	0.0	55.8	0.3	105.8	1.0	155.9	2.1	205.9	3.7	7
8	6.7	0.0	56.7	0.3	106.7	1.0	156.7	2.1	206.8	3.7	8
9	7.5	0.0	57.5	0.3	107.5	1.0	157.5	2.2	207.6	3.8	9
10	8.3	0.0	58.3	0.3	108.3	1.0	158.4	2.2	208.4	3.8	10
11	9.2	0.0	59.2	0.3	109.2	1.0	159.2	2.2	209.3	3.8	11
12	10.0	0.0	60.0	0.3	110.0	1.1	160.0	2.2	210.1	3.9	12
13	10.8	0.0	60.8	0.3	110.8	1.1	160.9	2.3	210.9	3.9	13
14	11.7	0.0	61.7	0.3	111.7	1.1	161.7	2.3	211.8	3.9	14
15	12.5	0.0	62.5	0.3	112.5	1.1	162.5	2.3	212.6	3.9	15
16	13.3	0.0	63.3	0.4	113.3	1.1	163.4	2.3	213.4	4.0	16
17	14.2	0.0	64.2	0.4	114.2	1.1	164.2	2.4	214.3	4.0	17
18	15.0	0.0	65.0	0.4	115.0	1.2	165.0	2.4	215.1	4.0	18
19	15.8	0.0	65.8	0.4	115.9	1.2	165.9	2.4	215.9	4.1	19
20	16.7	0.0	66.7	0.4	116.7	1.2	166.7	2.4	216.8	4.1	20
21	17.5	0.0	67.5	0.4	117.5	1.2	167.6	2.4	217.6	4.1	21
22	18.3	0.0	68.3	0.4	118.4	1.2	168.4	2.5	218.4	4.2	22
23	19.2	0.0	69.2	0.4	119.2	1.2	169.2	2.5	219.3	4.2	23
24	20.0	0.0	70.0	0.4	120.0	1.3	170.1	2.5	220.1	4.2	24
25	20.8	0.0	70.8	0.4	120.9	1.3	170.9	2.5	220.9	4.3	25
26	21.7	0.0	71.7	0.4	121.7	1.3	171.7	2.6	221.8	4.3	26
27	22.5	0.0	72.5	0.5	122.5	1.3	172.6	2.6	222.6	4.3	27
28	23.3	0.0	73.3	0.5	123.4	1.3	173.4	2.6	223.4	4.4	28
29	24.2	0.1	74.2	0.5	124.2	1.3	174.2	2.6	224.3	4.4	29
30	25.0	0.1	75.0	0.5	125.0	1.4	175.1	2.7	225.1	4.4	30
31	25.8	0.1	75.8	0.5	125.9	1.4	175.9	2.7	226.0	4.5	31
32	26.7	0.1	76.7	0.5	126.7	1.4	176.7	2.7	226.8	4.5	32
33	27.5	0.1	77.5	0.5	127.5	1.4	177.6	2.8	227.6	4.5	33
34	28.3	0.1	78.3	0.5	128.4	1.4	178.4	2.8	228.5	4.6	34
35	29.2	0.1	79.2	0.5	129.2	1.5	179.2	2.8	229.3	4.6	35
36	30.0	0.1	80.0	0.6	130.0	1.5	180.1	2.8	230.1	4.6	36
37	30.8	0.1	80.8	0.6	130.9	1.5	180.9	2.9	231.0	4.7	37
38	31.7	0.1	81.7	0.6	131.7	1.5	181.7	2.9	231.8	4.7	38
39	32.5	0.1	82.5	0.6	132.5	1.5	182.6	2.9	232.6	4.7	39
40	33.3	0.1	83.3	0.6	133.4	1.6	183.4	2.9	233.5	4.8	40
41	34.2	0.1	84.2	0.6	134.2	1.6	184.2	3.0	234.3	4.8	41
42	35.0	0.1	85.0	0.6	135.0	1.6	185.1	3.0	235.1	4.8	42
43	35.8	0.1	85.8	0.6	135.9	1.6	185.9	3.0	236.0	4.9	43
44	36.7	0.1	86.7	0.7	136.7	1.6	186.7	3.0	236.8	4.9	44
45	37.5	0.1	87.5	0.7	137.5	1.7	187.6	3.1	237.6	4.9	45
46	38.3	0.1	88.3	0.7	138.4	1.7	188.4	3.1	238.5	5.0	46
47	39.2	0.1	89.2	0.7	139.2	1.7	189.2	3.1	239.3	5.0	47
48	40.0	0.1	90.0	0.7	140.0	1.7	190.1	3.2	240.1	5.0	48
49	40.8	0.1	90.8	0.7	140.9	1.7	190.9	3.2	241.0	5.1	49
50	41.7	0.2	91.7	0.7	141.7	1.8	191.7	3.2	241.8	5.1	50
51	42.5	0.2	92.5	0.7	142.5	1.8	192.6	3.2	242.6	5.1	51
52	43.3	0.2	93.3	0.8	143.4	1.8	193.4	3.3	243.5	5.2	52
53	44.2	0.2	94.2	0.8	144.2	1.8	194.2	3.3	244.3	5.2	53
54	45.0	0.2	95.0	0.8	145.0	1.8	195.1	3.3	245.2	5.2	54
55	45.8	0.2	95.8	0.8	145.9	1.9	195.9	3.3	246.0	5.3	55
56	46.7	0.2	96.7	0.8	146.7	1.9	196.7	3.4	246.8	5.3	56
57	47.5	0.2	97.5	0.8	147.5	1.9	197.6	3.4	247.7	5.3	57
58	48.3	0.2	98.3	0.8	148.4	1.9	198.4	3.4	248.5	5.4	58
59	49.2	0.2	99.2	0.9	149.2	1.9	199.2	3.5	249.3	5.4	59

TABLE VIII.—TANGENTS AND EXTERNALS FOR A 1° CURVE
(Chord or Arc Definition)

′	I = 5°		I = 6°		I = 7°		I = 8°		I = 9°		′
	T	E	T	E	T	E	T	E	T	E	
0	250.2	5.5	300.3	7.9	350.4	10.7	400.7	14.0	450.9	17.7	0
1	251.0	5.5	301.1	7.9	351.3	10.8	401.5	14.1	451.8	17.8	1
2	251.8	5.5	301.9	8.0	352.1	10.8	402.3	14.1	452.6	17.8	2
3	252.7	5.6	302.8	8.0	352.9	10.9	403.2	14.2	453.4	17.9	3
4	253.5	5.6	303.6	8.0	353.8	11.0	404.0	14.2	454.3	18.0	4
5	254.3	5.6	304.5	8.1	354.6	11.0	404.8	14.3	455.1	18.0	5
6	255.2	5.7	305.3	8.1	355.5	11.0	405.7	14.3	456.0	18.1	6
7	256.0	5.7	306.1	8.2	356.3	11.1	406.5	14.4	456.8	18.2	7
8	256.8	5.8	307.0	8.2	357.1	11.1	407.4	14.5	457.6	18.2	8
9	257.7	5.8	307.8	8.3	358.0	11.2	408.2	14.5	458.5	18.3	9
10	258.5	5.8	308.6	8.3	358.8	11.2	409.0	14.6	459.3	18.4	10
11	259.3	5.9	309.5	8.4	359.6	11.3	409.9	14.6	460.2	18.4	11
12	260.2	5.9	310.3	8.4	360.5	11.3	410.7	14.7	461.0	18.5	12
13	261.0	5.9	311.1	8.4	361.3	11.4	411.5	14.8	461.8	18.6	13
14	261.9	6.0	312.0	8.5	362.2	11.4	412.4	14.8	462.7	18.7	14
15	262.7	6.0	312.8	8.5	363.0	11.5	413.2	14.9	463.5	18.7	15
16	263.5	6.1	313.7	8.6	363.8	11.5	414.1	14.9	464.4	18.8	16
17	264.4	6.1	314.5	8.6	364.7	11.6	414.9	15.0	465.2	18.9	17
18	265.2	6.1	315.3	8.7	365.5	11.6	415.7	15.1	466.0	18.9	18
19	266.0	6.2	316.2	8.7	366.3	11.7	416.6	15.1	466.9	19.0	19
20	266.9	6.2	317.0	8.8	367.2	11.8	417.4	15.2	467.7	19.1	20
21	267.7	6.3	317.8	8.8	368.0	11.8	418.2	15.2	468.5	19.1	21
22	268.5	6.3	318.7	8.9	368.8	11.9	419.1	15.3	469.4	19.2	22
23	269.4	6.3	319.5	8.9	369.7	11.9	419.9	15.4	470.2	19.3	23
24	270.2	6.4	320.3	8.9	370.5	12.0	420.8	15.4	471.1	19.3	24
25	271.0	6.4	321.2	9.0	371.4	12.0	421.6	15.5	471.9	19.4	25
26	271.9	6.4	322.0	9.0	372.2	12.1	422.4	15.6	472.7	19.5	26
27	272.7	6.5	322.8	9.1	373.0	12.1	423.3	15.6	473.6	19.5	27
28	273.5	6.5	323.7	9.1	373.9	12.2	424.1	15.7	474.4	19.6	28
29	274.4	6.6	324.5	9.2	374.7	12.2	424.9	15.7	475.3	19.7	29
30	275.2	6.6	325.4	9.2	375.5	12.3	425.8	15.8	476.1	19.7	30
31	276.1	6.6	326.2	9.3	376.4	12.3	426.6	15.9	476.9	19.8	31
32	276.9	6.7	327.0	9.3	377.2	12.4	427.5	15.9	477.8	19.9	32
33	277.7	6.7	327.9	9.4	378.1	12.5	428.3	16.0	478.6	20.0	33
34	278.6	6.8	328.7	9.4	378.9	12.5	429.1	16.0	479.5	20.0	34
35	279.4	6.8	329.5	9.5	379.7	12.6	430.0	16.1	480.3	20.1	35
36	280.2	6.9	330.4	9.5	380.6	12.6	430.8	16.2	481.1	20.2	36
37	281.1	6.9	331.2	9.6	381.4	12.7	431.7	16.2	482.0	20.2	37
38	281.9	6.9	332.0	9.6	382.2	12.7	432.5	16.3	482.8	20.3	38
39	282.7	7.0	332.9	9.7	383.1	12.8	433.3	16.4	483.6	20.4	39
40	283.6	7.0	333.7	9.7	383.9	12.8	434.2	16.4	484.5	20.4	40
41	284.4	7.1	334.6	9.8	384.7	12.9	435.0	16.5	485.3	20.5	41
42	285.2	7.1	335.4	9.8	385.6	13.0	435.8	16.6	486.2	20.6	42
43	286.1	7.1	336.2	9.9	386.4	13.0	436.7	16.6	487.0	20.7	43
44	286.9	7.2	337.1	9.9	387.3	13.1	437.5	16.7	487.8	20.7	44
45	287.7	7.2	337.9	10.0	388.1	13.1	438.4	16.7	488.7	20.8	45
46	288.6	7.3	338.7	10.0	388.9	13.2	439.2	16.8	489.5	20.9	46
47	289.4	7.3	339.6	10.1	389.8	13.2	440.0	16.9	490.4	20.9	47
48	290.3	7.3	340.4	10.1	390.6	13.3	440.9	17.0	491.2	21.0	48
49	291.1	7.4	341.2	10.2	391.4	13.4	441.7	17.0	492.0	21.1	49
50	291.9	7.4	342.1	10.2	392.3	13.4	442.5	17.1	492.9	21.2	50
51	292.8	7.5	342.9	10.3	393.1	13.5	443.4	17.1	493.7	21.2	51
52	293.6	7.5	343.7	10.3	394.0	13.5	444.2	17.2	494.6	21.3	52
53	294.4	7.6	344.6	10.4	394.8	13.6	445.1	17.3	495.4	21.4	53
54	295.3	7.6	345.4	10.4	395.6	13.6	445.9	17.3	496.2	21.5	54
55	296.1	7.6	346.3	10.5	396.5	13.7	446.7	17.4	497.1	21.5	55
56	296.9	7.7	347.1	10.5	397.3	13.8	447.6	17.5	497.9	21.6	56
57	297.8	7.7	347.9	10.6	398.1	13.8	448.4	17.5	498.8	21.7	57
58	298.6	7.8	348.8	10.6	399.0	13.9	449.3	17.6	499.6	21.7	58
59	299.4	7.8	349.6	10.7	399.8	13.9	450.1	17.7	500.4	21.8	59

TABLE VIII.—TANGENTS AND EXTERNALS FOR A 1° CURVE
(Chord or Arc Definition)

'	I = 10°		I = 11°		I = 12°		I = 13°		I = 14°		'
	T	E	T	E	T	E	T	E	T	E	
0	501.3	21.9	551.7	26.5	602.2	31.6	652.8	37.1	703.5	43.0	0
1	502.1	22.0	552.5	26.6	603.1	31.6	653.7	37.2	704.4	43.1	1
2	503.0	22.0	553.4	26.7	603.9	31.7	654.5	37.3	705.2	43.2	2
3	503.8	22.1	554.2	26.7	604.7	31.8	655.3	37.4	706.1	43.3	3
4	504.6	22.2	555.1	26.8	605.6	31.9	656.2	37.5	706.9	43.4	4
5	505.5	22.3	555.9	26.9	606.4	32.0	657.0	37.6	707.7	43.5	5
6	506.3	22.3	556.7	27.0	607.3	32.1	657.9	37.6	708.6	43.7	6
7	507.2	22.4	557.6	27.1	608.1	32.2	658.7	37.7	709.4	43.8	7
8	508.0	22.5	558.4	27.2	609.0	32.3	659.6	37.8	710.3	43.9	8
9	508.8	22.6	559.3	27.2	609.8	32.4	660.4	37.9	711.1	44.0	9
10	509.7	22.6	560.1	27.3	610.6	32.4	661.3	38.0	712.0	44.1	10
11	510.5	22.7	561.0	27.4	611.5	32.5	662.1	38.1	712.8	44.2	11
12	511.4	22.8	561.8	27.5	612.3	32.6	662.9	38.2	713.7	44.3	12
13	512.2	22.8	562.6	27.6	613.2	32.7	663.8	38.3	714.5	44.4	13
14	513.0	22.9	563.5	27.6	614.0	32.8	664.6	38.4	715.4	44.5	14
15	513.9	23.0	564.3	27.7	614.9	32.9	665.5	38.5	716.2	44.6	15
16	514.7	23.1	565.2	27.8	615.7	33.0	666.3	38.6	717.1	44.7	16
17	515.6	23.1	566.0	27.9	616.5	33.1	667.2	38.7	717.9	44.8	17
18	516.4	23.2	566.8	28.0	617.4	33.2	668.0	38.8	718.7	44.9	18
19	517.2	23.3	567.7	28.1	618.2	33.3	668.9	38.9	719.6	45.0	19
20	518.1	23.4	568.5	28.1	619.1	33.3	669.7	39.0	720.4	45.1	20
21	518.9	23.5	569.4	28.2	619.9	33.4	670.5	39.1	721.3	45.2	21
22	519.8	23.5	570.2	28.3	620.8	33.5	671.4	39.2	722.1	45.3	22
23	520.6	23.6	571.1	28.4	621.6	33.6	672.2	39.3	723.0	45.4	23
24	521.4	23.7	571.9	28.5	622.4	33.7	673.1	39.4	723.8	45.5	24
25	522.3	23.8	572.7	28.6	623.3	33.8	673.9	39.5	724.7	45.6	25
26	523.1	23.8	573.6	28.6	624.1	33.9	674.8	39.6	725.5	45.8	26
27	524.0	23.9	574.4	28.7	625.0	34.0	675.6	39.7	726.4	45.9	27
28	524.8	24.0	575.3	28.8	625.8	34.1	676.5	39.8	727.2	46.0	28
29	525.6	24.1	576.1	28.9	626.7	34.2	677.3	39.9	728.1	46.1	29
30	526.5	24.1	576.9	29.0	627.5	34.3	678.1	40.0	728.9	46.2	30
31	527.3	24.2	577.8	29.1	628.3	34.4	679.0	40.1	729.8	46.3	31
32	528.2	24.3	578.6	29.1	629.2	34.4	679.8	40.2	730.6	46.4	32
33	529.0	24.4	579.5	29.2	630.0	34.5	680.7	40.3	731.4	46.5	33
34	529.8	24.4	580.3	29.3	630.9	34.6	681.5	40.4	732.3	46.6	34
35	530.7	24.5	581.2	29.4	631.7	34.7	682.4	40.5	733.1	46.7	35
36	531.5	24.6	582.0	29.5	632.6	34.8	683.2	40.6	734.0	46.8	36
37	532.4	24.7	582.8	29.6	633.4	34.9	684.1	40.7	734.8	46.9	37
38	533.2	24.8	583.7	29.7	634.2	35.0	684.9	40.8	735.7	47.0	38
39	534.0	24.8	584.5	29.7	635.1	35.1	685.8	40.9	736.5	47.1	39
40	534.9	24.9	585.4	29.8	635.9	35.2	686.6	41.0	737.4	47.3	40
41	535.7	25.0	586.2	29.9	636.8	35.3	687.4	41.1	738.2	47.4	41
42	536.6	25.1	587.0	30.0	637.6	35.4	688.3	41.2	739.1	47.5	42
43	537.4	25.1	587.9	30.1	638.5	35.5	689.1	41.3	739.9	47.6	43
44	538.2	25.2	588.7	30.2	639.3	35.6	690.0	41.4	740.8	47.7	44
45	539.1	25.3	589.6	30.3	640.2	35.7	690.8	41.5	741.6	47.8	45
46	539.9	25.4	590.4	30.3	641.0	35.7	691.7	41.6	742.5	47.9	46
47	540.8	25.5	591.3	30.4	641.8	35.8	692.5	41.7	743.3	48.0	47
48	541.6	25.5	592.1	30.5	642.7	35.9	693.4	41.8	744.2	48.1	48
49	542.5	25.6	592.9	30.6	643.5	36.0	694.2	41.9	745.0	48.2	49
50	543.3	25.7	593.8	30.7	644.4	36.1	695.1	42.0	745.8	48.3	50
51	544.1	25.8	594.6	30.8	645.2	36.2	695.9	42.1	746.7	48.5	51
52	545.0	25.9	595.5	30.9	646.1	36.3	696.7	42.2	747.5	48.6	52
53	545.8	25.9	596.3	30.9	646.9	36.4	697.6	42.3	748.4	48.7	53
54	546.7	26.0	597.2	31.0	647.7	36.5	698.4	42.4	749.2	48.8	54
55	547.5	26.1	598.0	31.1	648.6	36.6	699.3	42.5	750.1	48.9	55
56	548.3	26.2	598.8	31.2	649.4	36.7	700.1	42.6	750.9	49.0	56
57	549.2	26.3	599.7	31.3	650.3	36.8	701.0	42.7	751.8	49.1	57
58	550.0	26.3	600.5	31.4	651.1	36.9	701.8	42.8	752.6	49.2	58
59	550.9	26.4	601.4	31.5	652.0	37.0	702.7	42.9	753.5	49.3	59

TABLE VIII.—TANGENTS AND EXTERNALS FOR A 1° CURVE
(Chord or Arc Definition)

′	I = 15°		I = 16°		I = 17°		I = 18°		I = 19°		′
	T	E	T	E	T	E	T	E	T	E	
0	754.3	49.4	805.2	56.3	856.3	63.6	907.5	71.4	958.8	79.7	0
1	755.2	49.6	806.1	56.4	857.2	63.8	908.3	71.6	959.7	79.8	1
2	756.0	49.7	806.9	56.5	858.0	63.9	909.2	71.7	960.5	80.0	2
3	756.9	49.8	807.8	56.7	858.9	64.0	910.1	71.8	961.4	80.1	3
4	757.7	49.9	808.6	56.8	859.7	64.1	910.9	72.0	962.2	80.2	4
5	758.6	50.0	809.5	56.9	860.6	64.3	911.8	72.1	963.1	80.4	5
6	759.4	50.1	810.3	57.0	861.4	64.4	912.6	72.2	964.0	80.5	6
7	760.3	50.2	811.2	57.1	862.3	64.5	913.5	72.4	964.8	80.7	7
8	761.1	50.3	812.0	57.3	863.1	64.6	914.3	72.5	965.7	80.8	8
9	762.0	50.4	812.9	57.4	864.0	64.8	915.2	72.6	966.5	80.9	9
10	762.8	50.6	813.7	57.5	864.8	64.9	916.0	72.8	967.4	81.1	10
11	763.7	50.7	814.6	57.6	865.7	65.0	916.9	72.9	968.2	81.2	11
12	764.5	50.8	815.5	57.7	866.5	65.2	917.7	73.0	969.1	81.4	12
13	765.3	50.9	816.3	57.9	867.4	65.3	918.6	73.2	970.0	81.5	13
14	766.2	51.0	817.2	58.0	868.2	65.4	919.5	73.3	970.8	81.7	14
15	767.0	51.1	818.0	58.1	869.1	65.5	920.3	73.4	971.7	81.8	15
16	767.9	51.2	818.9	58.2	869.9	65.7	921.2	73.6	972.5	82.0	16
17	768.7	51.3	819.7	58.3	870.8	65.8	922.0	73.7	973.4	82.1	17
18	769.6	51.5	820.6	58.5	871.6	65.9	922.9	73.8	974.2	82.2	18
19	770.4	51.6	821.4	58.6	872.5	66.1	923.7	74.0	975.1	82.4	19
20	771.3	51.7	822.3	58.7	873.3	66.2	924.6	74.1	976.0	82.5	20
21	772.1	51.8	823.1	58.8	874.2	66.3	925.4	74.3	976.8	82.7	21
22	773.0	51.9	824.0	58.9	875.1	66.4	926.3	74.4	977.7	82.8	22
23	773.8	52.0	824.8	59.1	875.9	66.6	927.1	74.5	978.5	83.0	23
24	774.7	52.1	825.7	59.2	876.8	66.7	928.0	74.7	979.4	83.1	24
25	775.5	52.2	826.5	59.3	877.6	66.8	928.9	74.8	980.2	83.2	25
26	776.4	52.4	827.4	59.4	878.5	67.0	929.7	74.9	981.1	83.4	26
27	777.2	52.5	828.2	59.5	879.3	67.1	930.6	75.1	982.0	83.5	27
28	778.1	52.6	829.1	59.7	880.2	67.2	931.4	75.2	982.8	83.7	28
29	778.9	52.7	829.9	59.8	881.0	67.3	932.3	75.4	983.7	83.8	29
30	779.8	52.8	830.8	59.9	881.9	67.5	933.1	75.5	984.5	84.0	30
31	780.6	52.9	831.6	60.0	882.7	67.6	934.0	75.6	985.4	84.1	31
32	781.5	53.0	832.5	60.2	883.6	67.7	934.8	75.8	986.3	84.3	32
33	782.3	53.2	833.3	60.3	884.4	67.9	935.7	75.9	987.1	84.4	33
34	783.2	53.3	834.2	60.4	885.3	68.0	936.6	76.0	988.0	84.6	34
35	784.0	53.4	835.0	60.5	886.1	68.1	937.4	76.2	988.8	84.7	35
36	784.9	53.5	835.9	60.7	887.0	68.3	938.3	76.3	989.7	84.8	36
37	785.7	53.6	836.7	60.8	887.9	68.4	939.1	76.5	990.5	85.0	37
38	786.6	53.7	837.6	60.9	888.7	68.5	940.0	76.6	991.4	85.1	38
39	787.4	53.9	838.4	61.0	889.6	68.6	940.8	76.7	992.3	85.3	39
40	788.3	54.0	839.3	61.1	890.4	68.8	941.7	76.9	993.1	85.4	40
41	789.1	54.1	840.1	61.3	891.3	68.9	942.5	77.0	994.0	85.6	41
42	790.0	54.2	841.0	61.4	892.1	69.0	943.4	77.1	994.8	85.7	42
43	790.8	54.3	841.8	61.5	893.0	69.2	944.3	77.3	995.7	85.9	43
44	791.7	54.4	842.7	61.6	893.8	69.3	945.1	77.4	996.5	86.0	44
45	792.5	54.6	843.5	61.8	894.7	69.4	946.0	77.6	997.4	86.2	45
46	793.4	54.7	844.4	61.9	895.5	69.6	946.8	77.7	998.3	86.3	46
47	794.2	54.8	845.2	62.0	896.4	69.7	947.7	77.8	999.1	86.5	47
48	795.1	54.9	846.1	62.1	897.2	69.8	948.5	78.0	1000.0	86.6	48
49	795.9	55.0	846.9	62.3	898.1	70.0	949.4	78.1	1000.8	86.8	49
50	796.8	55.1	847.8	62.4	898.9	70.1	950.2	78.3	1001.7	86.9	50
51	797.6	55.2	848.6	62.5	899.8	70.2	951.1	78.4	1002.6	87.1	51
52	798.5	55.4	849.5	62.6	900.7	70.4	952.0	78.5	1003.4	87.2	52
53	799.3	55.5	850.3	62.8	901.5	70.5	952.8	78.7	1004.3	87.3	53
54	800.2	55.6	851.2	62.9	902.4	70.6	953.7	78.8	1005.1	87.5	54
55	801.0	55.7	852.0	63.0	903.2	70.8	954.5	79.0	1006.0	87.6	55
56	801.9	55.8	852.9	63.1	904.1	70.9	955.4	79.1	1006.9	87.8	56
57	802.7	56.0	853.7	63.3	904.9	71.0	956.2	79.2	1007.7	87.9	57
58	803.6	56.1	854.6	63.4	905.8	71.2	957.1	79.4	1008.6	88.1	58
59	804.4	56.2	855.5	63.5	906.6	71.3	958.0	79.5	1009.4	88.2	59

TABLE VIII.—TANGENTS AND EXTERNALS FOR A 1° CURVE
(Chord or Arc Definition)

'	I = 20°		I = 21°		I = 22°		I = 23°		'
	T	E	T	E	T	E	T	E	
0	1010.3	88.4	1061.9	97.6	1113.7	107.2	1165.7	117.4	0
1	1011.2	88.5	1062.8	97.7	1114.6	107.4	1166.6	117.6	1
2	1012.0	88.7	1063.7	97.9	1115.5	107.6	1167.4	117.7	2
3	1012.9	88.8	1064.5	98.1	1116.3	107.7	1168.3	117.9	3
4	1013.7	89.0	1065.4	98.2	1117.2	107.9	1169.2	118.1	4
5	1014.6	89.1	1066.2	98.4	1118.1	108.1	1170.1	118.3	5
6	1015.4	89.3	1067.1	98.5	1118.9	108.2	1170.9	118.4	6
7	1016.3	89.4	1068.0	98.7	1119.8	108.4	1171.8	118.6	7
8	1017.2	89.6	1068.8	98.8	1120.7	108.6	1172.7	118.8	8
9	1018.0	89.7	1069.7	99.0	1121.5	108.7	1173.5	118.9	9
10	1018.9	89.9	1070.6	99.2	1122.4	108.9	1174.4	119.1	10
11	1019.8	90.0	1071.4	99.3	1123.2	109.1	1175.3	119.3	11
12	1020.6	90.2	1072.3	99.5	1124.1	109.2	1176.1	119.5	12
13	1021.5	90.3	1073.1	99.6	1125.0	109.4	1177.0	119.6	13
14	1022.3	90.5	1074.0	99.8	1125.8	109.6	1177.9	119.8	14
15	1023.2	90.6	1074.9	100.0	1126.7	109.7	1178.7	120.0	15
16	1024.0	90.8	1075.7	100.1	1127.6	109.9	1179.6	120.2	16
17	1024.9	90.9	1076.6	100.3	1128.4	110.1	1180.5	120.3	17
18	1025.8	91.1	1077.5	100.4	1129.3	110.2	1181.3	120.5	18
19	1026.6	91.2	1078.3	100.6	1130.2	110.4	1182.2	120.7	19
20	1027.5	91.4	1079.2	100.7	1131.0	110.6	1183.1	120.9	20
21	1028.3	91.6	1080.0	100.9	1131.9	110.7	1183.9	121.0	21
22	1029.2	91.7	1080.9	101.1	1132.8	110.9	1184.8	121.2	22
23	1030.1	91.9	1081.8	101.2	1133.6	111.1	1185.7	121.4	23
24	1030.9	92.0	1082.6	101.4	1134.5	111.2	1186.6	121.6	24
25	1031.8	92.2	1083.5	101.6	1135.4	111.4	1187.4	121.8	25
26	1032.6	92.3	1084.4	101.7	1136.2	111.6	1188.3	121.9	26
27	1033.5	92.5	1085.2	101.9	1137.1	111.8	1189.2	122.1	27
28	1034.4	92.6	1086.1	102.0	1138.0	111.9	1190.0	122.3	28
29	1035.2	92.8	1086.9	102.2	1138.8	112.1	1190.9	122.5	29
30	1036.1	92.9	1087.8	102.3	1139.7	112.3	1191.8	122.6	30
31	1037.0	93.1	1088.7	102.5	1140.6	112.4	1192.6	122.8	31
32	1037.8	93.2	1089.5	102.7	1141.4	112.6	1193.5	123.0	32
33	1038.7	93.4	1090.4	102.8	1142.3	112.8	1194.4	123.2	33
34	1039.5	93.5	1091.3	103.0	1143.2	112.9	1195.2	123.3	34
35	1040.4	93.7	1092.1	103.2	1144.0	113.1	1196.1	123.5	35
36	1041.3	93.8	1093.0	103.3	1144.9	113.3	1197.0	123.7	36
37	1042.1	94.0	1093.9	103.5	1145.8	113.4	1198.9	123.9	37
38	1043.0	94.2	1094.7	103.6	1146.6	113.6	1198.7	124.1	38
39	1043.8	94.3	1095.6	103.8	1147.5	113.8	1199.6	124.2	39
40	1044.7	94.5	1096.4	104.0	1148.4	113.9	1200.5	124.4	40
41	1045.6	94.6	1097.3	104.1	1149.2	114.1	1201.3	124.6	41
42	1046.4	94.8	1098.2	104.3	1150.1	114.3	1202.2	124.8	42
43	1047.3	94.9	1099.0	104.5	1151.0	114.5	1203.1	124.9	43
44	1048.1	95.1	1099.9	104.6	1151.8	114.6	1203.9	125.1	44
45	1049.0	95.2	1100.8	104.8	1152.7	114.8	1204.8	125.3	45
46	1049.9	95.4	1101.6	104.9	1153.6	115.0	1205.7	125.5	46
47	1050:7	95.5	1102.5	105.1	1154.4	115.1	1206.6	125.7	47
48	1051.6	95.7	1103.4	105.3	1155.3	115.3	1207.4	125.8	48
49	1052.4	95.9	1104.2	105.4	1156.2	115.5	1208.3	126.0	49
50	1053.3	96.0	1105.1	105.6	1157.0	115.7	1209.2	126.2	50
51	1054.2	96.2	1105.9	105.8	1157.9	115.8	1210.0	126.4	51
52	1055.0	96.3	1106.8	105.9	1158.8	116.0	1210.9	126.6	52
53	1055.9	96.5	1107.7	106.1	1159.6	116.2	1211.8	126.7	53
54	1056.8	96.6	1108.5	106.3	1160.5	116.3	1212.6	126.9	54
55	1057.6	96.8	1109.4	106.4	1161.4	116.5	1213.5	127.1	55
56	1058.5	97.0	1110.3	106.6	1162.2	116.7	1214.4	127.3	56
57	1059.3	97.1	1111.1	106.7	1163.1	116.9	1215.3	127.5	57
58	1060.2	97.3	1112.0	106.9	1164.0	117.0	1216.1	127.6	58
59	1061.1	97.4	1112.9	107.1	1164.8	117.2	1217.0	127.8	59

TABLE VIII.—TANGENTS AND EXTERNALS FOR A 1° CURVE
(Chord or Arc Definition)

′	I = 24° T	I = 24° E	I = 25° T	I = 25° E	I = 26° T	I = 26° E	I = 27° T	I = 27° E	′
0	1217.9	128.0	1270.2	139.1	1322.8	150.7	1375.6	162.8	0
1	1218.7	128.2	1271.1	139.3	1323.7	150.9	1376.4	163.0	1
2	1219.6	128.4	1272.0	139.5	1324.5	151.1	1377.3	163.2	2
3	1220.5	128.5	1272.9	139.7	1325.4	151.3	1378.2	163.4	3
4	1221.4	128.7	1273.7	139.9	1326.3	151.5	1379.1	163.6	4
5	1222.2	128.9	1274.6	140.1	1327.2	151.7	1380.0	163.8	5
6	1223.1	129.1	1275.5	140.3	1328.1	151.9	1380.9	164.0	6
7	1224.0	129.3	1276.4	140.4	1328.9	152.1	1381.7	164.2	7
8	1224.8	129.5	1277.2	140.6	1329.8	152.3	1382.6	164.5	8
9	1225.7	129.6	1278.1	140.8	1330.7	152.5	1383.5	164.7	9
10	1226.6	129.8	1279.0	141.0	1331.6	152.7	1384.4	164.9	10
11	1227.5	130.0	1279.9	141.2	1332.5	152.9	1385.3	165.1	11
12	1228.3	130.2	1280.7	141.4	1333.3	153.1	1386.1	165.3	12
13	1229.2	130.4	1281.6	141.6	1334.2	153.3	1387.0	165.5	13
14	1230.1	130.6	1282.5	141.8	1335.1	153.5	1387.9	165.7	14
15	1230.9	130.7	1283.4	142.0	1336.0	153.7	1388.8	165.9	15
16	1231.8	130.9	1284.2	142.2	1336.8	153.9	1389.7	166.1	16
17	1232.7	131.1	1285.1	142.4	1337.7	154.1	1390.6	166.3	17
18	1233.6	131.3	1286.0	142.5	1338.6	154.3	1391.4	166.5	18
19	1234.4	131.5	1286.9	142.7	1339.5	154.5	1392.3	166.7	19
20	1235.3	131.7	1287.7	142.9	1340.4	154.7	1393.2	167.0	20
21	1236.2	131.8	1288.6	143.1	1341.2	154.9	1394.1	167.2	21
22	1237.0	132.0	1289.5	143.3	1342.1	155.1	1395.0	167.4	22
23	1237.9	132.2	1290.4	143.5	1343.0	155.3	1395.9	167.6	23
24	1238.8	132.4	1291.2	143.7	1343.9	155.5	1396.7	167.8	24
25	1239.7	132.6	1292.1	143.9	1344.8	155.7	1397.6	168.0	25
26	1240.5	132.8	1293.0	144.1	1345.6	155.9	1398.5	168.2	26
27	1241.4	132.9	1293.9	144.3	1346.5	156.1	1399.4	168.4	27
28	1242.3	133.1	1294.7	144.5	1347.4	156.3	1400.3	168.6	28
29	1243.2	133.3	1295.6	144.7	1348.3	156.5	1401.2	168.8	29
30	1244.0	133.5	1296.5	144.9	1349.2	156.7	1402.0	169.0	30
31	1244.9	133.7	1297.4	145.0	1350.0	156.9	1402.9	169.3	31
32	1245.8	133.9	1298.2	145.2	1350.9	157.1	1403.8	169.5	32
33	1246.6	134.1	1299.1	145.4	1351.8	157.3	1404.7	169.7	33
34	1247.5	134.2	1300.0	145.6	1352.7	157.5	1405.6	169.9	34
35	1248.4	134.4	1300.9	145.8	1353.6	157.7	1406.5	170.1	35
36	1249.3	134.6	1301.7	146.0	1354.4	157.9	1407.3	170.3	36
37	1250.1	134.8	1302.6	146.2	1355.3	158.1	1408.2	170.5	37
38	1251.0	135.0	1303.5	146.4	1356.2	158.3	1409.1	170.7	38
39	1251.9	135.2	1304.4	146.6	1357.1	158.5	1410.0	170.9	39
40	1252.8	135.4	1305.3	146.8	1358.0	158.7	1410.9	171.2	40
41	1253.6	135.5	1306.1	147.0	1358.8	158.9	1411.8	171.4	41
42	1254.5	135.7	1307.0	147.2	1359.7	159.1	1412.6	171.6	42
43	1255.4	135.9	1307.9	147.4	1360.6	159.3	1413.5	171.8	43
44	1256.3	136.1	1308.8	147.6	1361.5	159.5	1414.4	172.0	44
45	1257.1	136.3	1309.6	147.8	1362.4	159.7	1415.3	172.2	45
46	1258.0	136.5	1310.5	148.0	1363.2	159.9	1416.2	172.4	46
47	1258.9	136.7	1311.4	148.2	1364.1	160.1	1417.1	172.6	47
48	1259.7	136.9	1312.3	148.4	1365.0	160.4	1417.9	172.8	48
49	1260.6	137.0	1313.1	148.6	1365.9	160.6	1418.8	173.1	49
50	1261.5	137.2	1314.0	148.7	1366.8	160.8	1419.7	173.3	50
51	1262.4	137.4	1314.9	148.9	1367.6	161.0	1420.6	173.5	51
52	1263.2	137.6	1315.8	149.1	1368.5	161.2	1421.5	173.7	52
53	1264.1	137.8	1316.7	149.3	1369.4	161.4	1422.4	173.9	53
54	1265.0	138.0	1317.5	149.5	1370.3	161.6	1423.3	174.1	54
55	1265.9	138.2	1318.4	149.7	1371.2	161.8	1424.1	174.3	55
56	1266.7	138.4	1319.3	149.9	1372.0	162.0	1425.0	174.6	56
57	1267.6	138.5	1320.2	150.1	1372.9	162.2	1425.9	174.8	57
58	1268.5	138.7	1321.0	150.3	1373.8	162.4	1426.8	175.0	58
59	1269.4	138.9	1321.9	150.5	1374.7	162.6	1427.7	175.2	59

TABLE VIII.—TANGENTS AND EXTERNALS FOR A 1° CURVE (Chord or Arc Definition)

'	I = 28°		I = 29°		I = 30°		I = 31°		'
	T	E	T	E	T	E	T	E	
0	1428.6	175.4	1481.8	188.5	1535.3	202.1	1589.0	216.2	0
1	1429.4	175.6	1482.7	188.7	1536.1	202.4	1589.9	216.5	1
2	1430.3	175.8	1483.6	189.0	1537.0	202.6	1590.8	216.7	2
3	1431.2	176.1	1484.5	189.2	1537.9	202.8	1591.7	217.0	3
4	1432.1	176.3	1485.3	189.4	1538.8	203.0	1592.6	217.2	4
5	1433.0	176.5	1486.2	189.6	1539.7	203.3	1593.5	217.5	5
6	1433.9	176.7	1487.1	189.8	1540.6	203.5	1594.4	217.7	6
7	1434.8	176.9	1488.0	190.1	1541.5	203.7	1595.3	217.9	7
8	1435.6	177.1	1488.9	190.3	1542.4	204.0	1596.2	218.2	8
9	1436.5	177.3	1489.8	190.5	1543.3	204.2	1597.1	218.4	9
10	1437.4	177.6	1490.7	190.7	1544.2	204.4	1598.0	218.7	10
11	1438.3	177.8	1491.6	191.0	1545.1	204.7	1598.8	218.9	11
12	1439.2	178.0	1492.5	191.2	1546.0	204.9	1599.7	219.1	12
13	1440.1	178.2	1493.4	191.4	1546.9	205.1	1600.6	219.4	13
14	1441.0	178.4	1494.2	191.6	1547.8	205.4	1601.5	219.6	14
15	1441.8	178.6	1495.1	191.9	1548.7	205.6	1602.4	219.9	15
16	1442.7	178.9	1496.0	192.1	1549.6	205.8	1603.3	220.1	16
17	1443.6	179.1	1496.9	192.3	1550.4	206.1	1604.2	220.4	17
18	1444.5	179.3	1497.8	192.5	1551.3	206.3	1605.1	220.6	18
19	1445.4	179.5	1498.7	192.8	1552.2	206.5	1606.0	220.8	19
20	1446.3	179.7	1499.6	193.0	1553.1	206.8	1606.9	221.1	20
21	1447.2	180.0	1500.5	193.2	1554.0	207.0	1607.8	221.3	21
22	1448.1	180.2	1501.4	193.4	1554.9	207.2	1608.7	221.6	22
23	1448.9	180.4	1502.3	193.7	1555.8	207.5	1609.6	221.8	23
24	1449.8	180.6	1503.1	193.9	1556.7	207.7	1610.5	222.1	24
25	1450.7	180.8	1504.0	194.1	1557.6	207.9	1611.4	222.3	25
26	1451.6	181.0	1504.9	194.3	1558.5	208.2	1612.3	222.5	26
27	1452.5	181.2	1505.8	194.6	1559.4	208.4	1613.2	222.8	27
28	1453.4	181.5	1506.7	194.8	1560.3	208.7	1614.1	223.0	28
29	1454.3	181.7	1507.6	195.0	1561.2	208.9	1615.0	223.3	29
30	1455.1	181.9	1508.5	195.2	1562.1	209.1	1615.9	223.5	30
31	1456.0	182.1	1509.4	195.5	1563.0	209.4	1616.8	223.8	31
32	1456.9	182.3	1510.3	195.7	1563.9	209.6	1617.7	224.0	32
33	1457.8	182.5	1511.2	195.9	1564.8	209.8	1618.6	224.2	33
34	1458.7	182.8	1512.1	196.2	1565.7	210.1	1619.5	224.5	34
35	1459.6	183.0	1512.9	196.4	1566.6	210.3	1620.4	224.7	35
36	1460.5	183.2	1513.8	196.6	1567.5	210.5	1621.3	225.0	36
37	1461.4	183.4	1514.7	196.8	1568.4	210.8	1622.2	225.2	37
38	1462.2	183.6	1515.6	197.1	1569.2	211.0	1623.1	225.5	38
39	1463.1	183.9	1516.5	197.3	1570.1	211.2	1624.0	225.7	39
40	1464.0	184.1	1517.4	197.5	1571.0	211.5	1624.9	226.0	40
41	1464.9	184.3	1518.3	197.8	1571.9	211.7	1625.8	226.2	41
42	1465.8	184.5	1519.2	198.0	1572.8	212.0	1626.7	226.5	42
43	1466.7	184.7	1520.1	198.2	1573.7	212.2	1627.6	226.7	43
44	1467.6	185.0	1521.0	198.4	1574.6	212.4	1628.5	226.9	44
45	1468.5	185.2	1521.9	198.7	1575.5	212.7	1629.4	227.2	45
46	1469.3	185.4	1522.8	198.9	1576.4	212.9	1630.3	227.4	46
47	1470.2	185.6	1523.6	199.1	1577.3	213.1	1631.2	227.7	47
48	1471.1	185.8	1524.5	199.4	1578.2	213.4	1632.1	227.9	48
49	1472.0	186.1	1525.4	199.6	1579.1	213.6	1633.0	228.2	49
50	1472.9	186.3	1526.3	199.8	1580.0	213.9	1633.9	228.4	50
51	1473.8	186.5	1527.2	200.1	1580.9	214.1	1634.8	228.7	51
52	1474.7	186.7	1528.1	200.3	1581.8	214.3	1635.7	228.9	52
53	1475.6	187.0	1529.0	200.5	1582.7	214.6	1636.6	229.2	53
54	1476.5	187.2	1529.9	200.7	1583.6	214.8	1637.5	229.4	54
55	1477.3	187.4	1530.8	201.0	1584.5	215.1	1638.4	229.7	55
56	1478.2	187.6	1531.7	201.2	1585.4	215.3	1639.3	229.9	56
57	1479.1	187.8	1532.6	201.4	1586.3	215.5	1640.2	230.2	57
58	1480.0	188.1	1533.5	201.7	1587.2	215.8	1641.1	230.4	58
59	1480.9	188.3	1534.4	201.9	1588.1	216.0	1642.0	230.7	59

TABLE VIII.—TANGENTS AND EXTERNALS FOR A 1° CURVE
(Chord or Arc Definition)

′	I = 32°		I = 33°		I = 34°		I = 35°		′
	T	E	T	E	T	E	T	E	
0	1643.0	230.9	1697.2	246.1	1751.7	261.8	1806.6	278.1	0
1	1643.9	231.1	1698.1	246.3	1752.6	262.1	1807.5	278.3	1
2	1644.8	231.4	1699.0	246.6	1753.6	262.3	1808.4	278.6	2
3	1645.7	231.6	1699.9	246.9	1754.5	262.6	1809.3	278.9	3
4	1646.6	231.9	1700.8	247.1	1755.4	262.9	1810.2	279.2	4
5	1647.5	232.1	1701.7	247.4	1756.3	263.1	1811.1	279.4	5
6	1648.4	232.4	1702.6	247.6	1757.2	263.4	1812.1	279.7	6
7	1649.3	232.6	1703.5	247.9	1758.1	263.7	1813.0	280.0	7
8	1650.2	232.9	1704.5	248.1	1759.0	263.9	1813.9	280.3	8
9	1651.1	233.1	1705.4	248.4	1759.9	264.4	1814.8	280.5	9
10	1652.0	233.4	1706.3	248.7	1760.8	264.5	1815.7	280.8	10
11	1652.9	233.6	1707.2	248.9	1761.8	264.7	1816.6	281.1	11
12	1653.8	233.9	1708.1	249.2	1762.7	265.0	1817.6	281.4	12
13	1654.7	234.1	1709.0	249.4	1763.6	265.3	1818.5	281.7	13
14	1655.6	234.4	1709.9	249.7	1764.5	265.5	1819.4	281.9	14
15	1656.5	234.6	1710.8	250.0	1765.4	265.8	1820.3	282.2	15
16	1657.4	234.9	1711.7	250.2	1766.3	266.1	1821.2	282.5	16
17	1658.3	235.1	1712.6	250.5	1767.2	266.4	1822.1	282.8	17
18	1659.2	235.4	1713.5	250.7	1768.1	266.6	1823.1	283.0	18
19	1660.1	235.6	1714.4	251.0	1769.1	266.9	1824.0	283.3	19
20	1661.0	235.9	1715.3	251.3	1770.0	267.2	1824.9	283.6	20
21	1661.9	236.2	1716.3	251.5	1770.9	267.4	1825.8	283.9	21
22	1662.8	236.4	1717.2	251.8	1771.8	267.7	1826.7	284.2	22
23	1663.7	236.7	1718.1	252.0	1772.7	268.0	1827.6	284.4	23
24	1664.6	236.9	1719.0	252.3	1773.6	268.2	1828.6	284.7	24
25	1665.5	237.2	1719.9	252.6	1774.5	268.5	1829.5	285.0	25
26	1666.4	237.4	1720.8	252.8	1775.4	268.8	1830.4	285.3	26
27	1667.3	237.7	1721.7	253.1	1776.4	269.1	1831.3	285.6	27
28	1668.2	237.9	1722.6	253.3	1777.3	269.3	1832.2	285.8	28
29	1669.1	238.2	1723.5	253.6	1778.2	269.6	1833.2	286.1	29
30	1670.0	238.4	1724.4	253.9	1779.1	269.9	1834.1	286.4	30
31	1670.9	238.7	1725.3	254.1	1780.0	270.1	1835.0	286.7	31
32	1671.8	238.9	1726.2	254.4	1780.9	270.4	1835.9	287.0	32
33	1672.8	239.2	1727.2	254.7	1781.8	270.7	1836.8	287.2	33
34	1673.7	239.4	1728.1	254.9	1782.8	270.9	1837.8	287.5	34
35	1674.6	239.7	1729.0	255.2	1783.7	271.2	1838.7	287.8	35
36	1675.5	239.9	1729.9	255.4	1784.6	271.5	1839.6	288.1	36
37	1676.4	240.2	1730.8	255.7	1785.5	271.8	1840.5	288.4	37
38	1677.3	240.5	1731.7	256.0	1786.4	272.0	1841.4	288.6	38
39	1678.2	240.7	1732.6	256.2	1787.3	272.3	1842.4	288.9	39
40	1679.1	241.0	1733.5	256.5	1788.2	272.6	1843.3	289.2	40
41	1680.0	241.2	1734.4	256.8	1789.2	272.9	1844.2	289.5	41
42	1680.9	241.5	1735.3	257.0	1790.1	273.1	1845.1	289.8	42
43	1681.8	241.7	1736.3	257.3	1791.0	273.4	1846.0	290.0	43
44	1682.7	242.0	1737.2	257.6	1791.9	273.7	1847.0	290.3	44
45	1683.6	242.2	1738.1	257.8	1792.8	273.9	1847.9	290.6	45
46	1684.5	242.5	1739.0	258.1	1793.7	274.2	1848.8	290.9	46
47	1685.4	242.8	1739.9	258.3	1794.6	274.5	1849.7	291.2	47
48	1686.3	243.0	1740.8	258.6	1795.6	274.8	1850.6	291.5	48
49	1687.2	243.3	1741.7	258.9	1796.5	275.0	1851.5	291.7	49
50	1688.1	243.5	1742.6	259.1	1797.4	275.3	1852.5	292.0	50
51	1689.0	243.8	1743.5	259.4	1798.3	275.6	1853.4	292.3	51
52	1690.0	244.0	1744.4	259.7	1799.2	275.9	1854.3	292.6	52
53	1690.9	244.3	1745.4	259.9	1800.1	276.1	1855.2	292.9	53
54	1691.8	244.5	1746.3	260 2	1801.1	276.4	1856.2	293.2	54
55	1692.7	244.8	1747.2	260.5	1802.0	276.7	1857.1	293.4	55
56	1693.6	245.1	1748.1	260.7	1802.9	277.0	1858.0	293.7	56
57	1694.5	245.3	1749.0	261.0	1803.8	277.2	1858.9	294.0	57
58	1695.4	245.6	1749.9	261.3	1804.7	277.5	1859.8	294.3	58
59	1696.3	245.8	1750.8	261.5	1805.6	277.8	1860.8	294.6	59

TABLE VIII.—TANGENTS AND EXTERNALS FOR A 1° CURVE
(Chord or Arc Definition)

′	I = 36°		I = 37°		I = 38°		I = 39°		′
	T	E	T	E	T	E	T	E	
0	1861.7	294.9	1917.1	312.2	1972.9	330.1	2029.0	348.6	0
1	1862.6	295.1	1918.0	312.5	1973.8	330.5	2029.9	349.0	1
2	1863.5	295.4	1919.0	312.8	1974.7	330.8	2030.9	349.3	2
3	1864.4	295.7	1919.9	313.1	1975.7	331.1	2031.8	349.6	3
4	1865.4	296.0	1920.8	313.4	1976.6	331.4	2032.7	349.9	4
5	1866.3	296.3	1921.7	313.7	1977.5	331.7	2033.7	350.2	5
6	1867.2	296.6	1922.7	314.0	1978.5	332.0	2034.6	350.5	6
7	1868.1	296.9	1923.6	314.3	1979.4	332.3	2035.5	350.8	7
8	1869.0	297.1	1924.5	314.6	1980.3	332.6	2036.5	351.2	8
9	1870.0	297.4	1925.5	314.9	1981.3	332.9	2037.4	351.5	9
10	1870.9	297.7	1926.4	315.2	1982.2	333.2	2038.4	351.8	10
11	1871.8	298.0	1927.3	315.5	1983.1	333.5	2039.3	352.1	11
12	1872.7	298.3	1928.2	315.8	1984.1	333.8	2040.2	352.4	12
13	1873.7	298.6	1929.2	316.1	1985.0	334.1	2041.2	352.7	13
14	1874.6	298.9	1930.1	316.4	1985.9	334.4	2042.1	353.0	14
15	1875.5	299.2	1931.0	316.7	1986.9	334.7	2043.1	353.4	15
16	1876.4	299.4	1932.0	316.9	1987.8	335.0	2044.0	353.7	16
17	1877.4	299.7	1932.9	317.2	1988.7	335.3	2044.9	354.0	17
18	1878.3	300.0	1933.8	317.5	1989.7	335.6	2045.9	354.3	18
19	1879.2	300.3	1934.7	317.8	1990.6	335.9	2046.8	354.6	19
20	1880.1	300.6	1935.7	318.1	1991.5	336.2	2047.8	354.9	20
21	1881.0	300.9	1936.6	318.4	1992.5	336.6	2048.7	355.3	21
22	1882.0	301.2	1937.5	318.7	1993.4	336.9	2049.6	355.6	22
23	1882.9	301.5	1938.5	319.0	1994.3	337.2	2050.6	355.9	23
24	1883.8	301.7	1939.4	319.3	1995.3	337.5	2051.5	356.2	24
25	1884.7	302.0	1940.3	319.6	1996.2	337.8	2052.5	356.5	25
26	1885.7	302.3	1941.2	319.9	1997.1	338.1	2053.4	356.8	26
27	1886.6	302.6	1942.2	320.2	1998.1	338.4	2054.3	357.2	27
28	1887.5	302.9	1943.1	320.5	1999.0	338.7	2055.3	357.5	28
29	1888.4	303.2	1944.0	320.8	1999.9	339.0	2056.2	357.8	29
30	1889.4	303.5	1945.0	321.1	2000.9	339.3	2057.2	358.1	30
31	1890.3	303.8	1945.9	321.4	2001.8	339.6	2058.1	358.4	31
32	1891.2	304.1	1946.8	321.7	2002.8	339.9	2059.0	358.7	32
33	1892.1	304.3	1947.7	322.0	2003.7	340.2	2060.0	359.1	33
34	1893.1	304.6	1948.7	322.3	2004.6	340.6	2060.9	359.4	34
35	1894.0	304.9	1949.6	322.6	2005.6	340.9	2061.9	359.7	35
36	1894.9	305.2	1950.5	322.9	2006.5	341.2	2062.8	360.0	36
37	1895.8	305.5	1951.5	323.2	2007.4	341.5	2063.7	360.3	37
38	1896.7	305.8	1952.4	323.5	2008.4	341.8	2064.7	360.7	38
39	1897.7	306.1	1953.3	323.8	2009.3	342.1	2065.6	361.0	39
40	1898.6	306.4	1954.3	324.1	2010.2	342.4	2066.6	361.3	40
41	1899.5	306.7	1955.2	324.4	2011.2	342.7	2067.5	361.6	41
42	1900.4	307.0	1956.1	324.7	2012.1	343.0	2068.5	361.9	42
43	1901.4	307.2	1957.0	325.0	2013.0	343.3	2069.4	362.3	43
44	1902.3	307.5	1958.0	325.3	2014.0	343.7	2070.3	362.6	44
45	1903.2	307.8	1958.9	325.6	2014.9	344.0	2071.3	362.9	45
46	1904.1	308.1	1959.8	325.9	2015.9	344.3	2072.2	363.2	46
47	1905.1	308.4	1960.8	326.2	2016.8	344.6	2073.2	363.5	47
48	1906.0	308.7	1961.7	326.5	2017.7	344.9	2074.1	363.9	48
49	1906.9	309.0	1962.6	326.8	2018.7	345.2	2075.0	364.2	49
50	1907.9	309.3	1963.6	327.1	2019.6	345.5	2076.0	364.5	50
51	1908.8	309.6	1964.5	327.4	2020.5	345.8	2076.9	364.8	51
52	1909.7	309.9	1965.4	327.7	2021.5	346.1	2077.9	365.1	52
53	1910.6	310.2	1966.4	328.0	2022.4	346.5	2078.8	365.5	53
54	1911.6	310.5	1967.3	328.3	2023.4	346.8	2079.8	365.8	54
55	1912 5	310.8	1968.2	328.6	2024.3	347.1	2080.7	366.1	55
56	1913.4	311.0	1969.1	328.9	2025.2	347.4	2081.6	366.4	56
57	1914.3	311.3	1970.1	329.2	2026.2	347.7	2082.6	366.8	57
58	1915.3	311.6	1971.0	329.5	2027.1	348.0	2083.5	367.1	58
59	1916.2	311.9	1971.9	329.8	2028.0	348.3	2084.5	367.4	59

TABLE VIII.—TANGENTS AND EXTERNALS FOR A 1° CURVE
(Chord or Arc Definition)

'	I = 40°		I = 41°		I = 42°		I = 43°		'
	T	E	T	E	T	E	T	E	
0	2085.4	367.7	2142.2	387.4	2199.4	407.6	2257.0	428.5	0
1	2086.4	368.0	2143.2	387.7	2200.4	408.0	2257.9	428.9	1
2	2087.3	368.4	2144.1	388.0	2201.3	408.3	2258.9	429.2	2
3	2088.3	368.7	2145.1	388.4	2202.3	408.7	2259.9	429.6	3
4	2089.2	369.0	2146.0	388.7	2203.2	409.0	2260.8	429.9	4
5	2090.1	369.3	2147.0	389.0	2204.2	409.4	2261.8	430.3	5
6	2091.1	369.7	2147.9	389.4	2205.1	409.7	2262.7	430.6	6
7	2092.0	370.0	2148.9	389.7	2206.1	410.0	2263.7	431.1	7
8	2093.0	370.3	2149.8	390.0	2207.1	410.4	2264.7	431.3	8
9	2093.9	370.6	2150.8	390.4	2208.0	410.7	2265.6	431.7	9
10	2094.9	371.0	2151.7	390.7	2209.0	411.1	2266.6	432.0	10
11	2095.8	371.3	2152.7	391.0	2209.9	411.4	2267.6	432.4	11
12	2096.8	371.6	2153.6	391.4	2210.9	411.8	2268.5	432.8	12
13	2097.7	371.9	2154.6	391.7	2211.8	412.1	2269.5	433.1	13
14	2098.6	372.3	2155.5	392.1	2212.8	412.5	2270.5	433.5	14
15	2099.6	372.6	2156.5	392.4	2213.8	412.8	2271.4	433.8	15
16	2100.5	372.9	2157.4	392.7	2214.7	413.1	2272.4	434.2	16
17	2101.5	373.2	2158.4	393.1	2215.7	413.5	2273.4	434.5	17
18	2102.4	373.6	2159.3	393.4	2216.6	413.8	2274.3	434.9	18
19	2103.4	373.9	2160.3	393.7	2217.6	414.2	2275.3	435.2	19
20	2104.3	374.2	2161.2	394.1	2218.6	414.5	2276.2	435.6	20
21	2105.3	374.5	2162.2	394.4	2219.5	414.9	2277.2	435.9	21
22	2106.2	374.9	2163.2	394.7	2220.5	415.2	2278.2	436.3	22
23	2107.2	375.2	2164.1	395.1	2221.4	415.6	2279.1	436.7	23
24	2108.1	375.5	2165.1	395.4	2222.4	415.9	2280.1	437.0	24
25	2109.0	375.8	2166.0	395.7	2223.3	416.3	2281.1	437.4	25
26	2110.0	376.2	2167.0	396.1	2224.3	416.6	2282.0	437.7	26
27	2110.9	376.5	2167.9	396.4	2225.3	416.9	2283.0	438.1	27
28	2111.9	376.8	2168.9	396.8	2226.2	417.3	2284.0	438.4	28
29	2112.8	377.1	2169.8	397.1	2227.2	417.6	2284.9	438.8	29
30	2113.8	377.5	2170.8	397.4	2228.1	418.0	2285.9	439.2	30
31	2114.7	377.8	2171.7	397.8	2229.1	418.3	2286.9	439.5	31
32	2115.7	378.1	2172.7	398.1	2230.1	418.7	2287.8	439.9	32
33	2116.6	378.5	2173.6	398.4	2231.0	419.0	2288.8	440.2	33
34	2117.6	378.8	2174.6	398.8	2232.0	419.4	2289.8	440.6	34
35	2118.5	379.1	2175.5	399.1	2232.9	419.7	2290.7	441.0	35
36	2119.5	379.4	2176.5	399.5	2233.9	420.1	2291.7	441.3	36
37	2120.4	379.8	2177.4	399.8	2234.9	420.4	2292.7	441.7	37
38	2121.4	380.1	2178.4	400.1	2235.8	420.8	2293.6	442.0	38
39	2122.3	380.4	2179.4	400.5	2236.8	421.1	2294.6	442.4	39
40	2123.3	380.8	2180.3	400.8	2237.7	421.5	2295.6	442.7	40
41	2124.2	381.1	2181.3	401.2	2238.7	421.8	2296.5	443.1	41
42	2125.1	381.4	2182.2	401.5	2239.7	422.2	2297.5	443.5	42
43	2126.1	381.8	2183.2	401.8	2240.6	422.5	2298.5	443.8	43
44	2127.0	382.1	2184.1	402.2	2241.6	422.9	2299.4	444.2	44
45	2128.0	382.4	2185.1	402.5	2242.5	423.2	2300.4	444.6	45
46	2128.9	382.7	2186.0	402.9	2243.5	423.6	2301.4	444.9	46
47	2129.9	383.1	2187.0	403.2	2244.5	423.9	2302.3	445.3	47
48	2130.8	383.4	2187.9	403.5	2245.4	424.3	2303.3	445.6	48
49	2131.8	383.7	2188.9	403.9	2246.4	424.6	2304.3	446.0	49
50	2132.7	384.1	2189.9	404.2	2247.3	425.0	2305.2	446.4	50
51	2133.7	384.4	2190.8	404.6	2248.3	425.3	2306.2	446.7	51
52	2134.6	384.8	2191.8	404.9	2249.3	425.7	2307.2	447.1	52
53	2135.6	385.1	2192.7	405.2	2250.2	426.0	2308.1	447.4	53
54	2136.5	385.4	2193.7	405.6	2251.2	426.4	2309.1	447.8	54
55	2137.5	385.7	2194.6	405.9	2252.2	426.7	2310.1	448.2	55
56	2138.4	386.1	2195.6	406.3	2253.1	427.1	2311.1	448.5	56
57	2139.4	386.4	2196.5	406.6	2254.1	427.4	2312.0	448.9	57
58	2140.3	386.7	2197.5	407.0	2255.0	427.8	2313.0	449.3	58
59	2141.3	387.0	2198.5	407.3	2256.0	428.1	2314.0	449.6	59

TABLE VIII.—TANGENTS AND EXTERNALS FOR A 1° CURVE (Chord or Arc Definition)

′	I = 44°		I = 45°		I = 46°		I = 47°		′
	T	E	T	E	T	E	T	E	
0	2314.9	450.0	2373.3	472.1	2432.1	494.8	2491.3	518.2	0
1	2315.9	450.3	2374.3	472.5	2433.1	495.2	2492.3	518.6	1
2	2316.9	450.7	2375.3	472.8	2434.1	495.6	2493.3	519.0	2
3	2317.8	451.1	2376.2	473.2	2435.0	496.0	2494.3	519.4	3
4	2318.8	451.4	2377.2	473.6	2436.0	496.4	2495.3	519.8	4
5	2319.8	451.8	2378.2	474.0	2437.0	496.7	2496.3	520.2	5
6	2320.7	452.2	2379.2	474.3	2438.0	497.1	2497.3	520.6	6
7	2321.7	452.5	2380.1	474.7	2439.0	497.5	2498.3	521.0	7
8	2322.7	452.9	2381.1	475.1	2440.0	497.9	2499.3	521.4	8
9	2323.7	453.3	2382.1	475.4	2440.9	498.3	2500.2	521.8	9
10	2324.6	453.6	2383.1	475.8	2441.9	498.7	2501.2	522.2	10
11	2325.6	454.0	2384.0	476.2	2442.9	499.1	2502.2	522.6	11
12	2326.6	454.4	2385.0	476.6	2443.9	499.4	2503.2	523.0	12
13	2327.5	454.7	2386.0	477.0	2444.9	499.8	2504.2	523.4	13
14	2328.5	455.1	2387.0	477.3	2445.9	500.2	2505.2	523.7	14
15	2329.5	455.4	2388.0	477.7	2446.9	500.6	2506.2	524.1	15
16	2330.5	455.8	2388.9	478.1	2447.8	501.0	2507.2	524.5	16
17	2331.4	456.2	2389.9	478.5	2448.8	501.4	2508.2	524.9	17
18	2332.4	456.5	2390.9	478.8	2449.8	501.8	2509.2	525.3	18
19	2333.4	456.9	2391.9	479.2	2450.8	502.2	2510.2	525.7	19
20	2334.3	457.3	2392.8	479.6	2451.8	502.5	2511.2	526.1	20
21	2335.3	457.6	2393.8	480.0	2452.8	502.9	2512.2	526.5	21
22	2336.3	458.0	2394.8	480.3	2453.8	503.3	2513.2	526.9	22
23	2337.3	458.4	2395.8	480.7	2454.7	503.7	2514.1	527.3	23
24	2338.2	458.7	2396.8	481.1	2455.7	504.1	2515.1	527.7	24
25	2339.2	459.1	2397.7	481.5	2456.7	504.5	2516.1	528.1	25
26	2340.2	459.5	2398.7	481.9	2457.7	504.9	2517.1	528.5	26
27	2341.1	459.8	2399.7	482.2	2458.7	505.3	2518.1	528.9	27
28	2342.1	460.2	2400.7	482.6	2459.7	505.6	2519.1	529.3	28
29	2343.1	460.6	2401.7	483.0	2460.7	506.0	2520.1	529.7	29
30	2344.1	460.9	2402.6	483.4	2461.7	506.4	2521.1	530.1	30
31	2345.0	461.3	2403.6	483.8	2462.6	506.8	2522.1	530.5	31
32	2346.0	461.7	2404.6	484.1	2463.6	507.2	2523.1	530.9	32
33	2347.0	462.1	2405.6	484.5	2464.6	507.6	2524.1	531.3	33
34	2348.0	462.4	2406.6	484.9	2465.6	508.0	2525.1	531.7	34
35	2348.9	462.8	2407.5	485.3	2466.6	508.4	2526.1	532.1	35
36	2349.9	463.2	2408.5	485.7	2467.6	508.8	2527.1	532.5	36
37	2350.9	463.5	2409.5	486.0	2468.6	509.2	2528.1	532.9	37
38	2351.8	463.9	2410.5	486.4	2469.6	509.5	2529.1	533.3	38
39	2352.8	464.3	2411.5	486.8	2470.5	509.9	2530.1	533.7	39
40	2353.8	464.6	2412.4	487.2	2471.5	510.3	2531.1	534.1	40
41	2354.8	465.0	2413.4	487.6	2472.5	510.7	2532.1	534.6	41
42	2355.7	465.4	2414.4	487.9	2473.5	511.1	2533.1	535.0	42
43	2356.7	465.8	2415.4	488.3	2474.5	511.5	2534.1	535.4	43
44	2357.7	466.1	2416.4	488.7	2475.5	511.9	2535.0	535.8	44
45	2358.7	466.5	2417.4	489.1	2476.5	512.3	2536.0	536.2	45
46	2359.6	466.9	2418.3	489.5	2477.5	512.7	2537.0	536.6	46
47	2360.6	467.2	2419.3	489.8	2478.5	513.1	2538.0	537.0	47
48	2361.6	467.6	2420.3	490.2	2479.4	513.5	2539.0	537.4	48
49	2362.6	468.0	2421.3	490.6	2480.4	513.9	2540.0	537.8	49
50	2363.5	468.4	2422.3	491.0	2481.4	514.3	2541.0	538.2	50
51	2364.5	468.7	2423.2	491.4	2482.4	514.6	2542.0	538.6	51
52	2365.5	469.1	2424.2	491.7	2483.4	515.0	2543.0	539.0	52
53	2366.5	469.5	2425.2	492.1	2484.4	515.4	2544.0	539.4	53
54	2367.4	469.8	2426.2	492.5	2485.4	515.8	2545.0	539.8	54
55	2368.4	470.2	2427.2	492.9	2486.4	516.2	2546.0	540.2	55
56	2369.4	470.6	2428.2	493.3	2487.4	516.6	2547.0	540.6	56
57	2370.4	471.0	2429.1	493.7	2488.4	517.0	2548.0	541.0	57
58	2371.3	471.3	2430.1	494.1	2489.3	517.4	2549.0	541.4	58
59	2372.3	471.7	2431.1	494.4	2490.3	517.8	2550.0	541.8	59

TABLE VIII.—TANGENTS AND EXTERNALS FOR A 1° CURVE
(Chord or Arc Definition)

'	I = 48°		I = 49°		I = 50°		I = 51°		'
	T	E	T	E	T	E	T	E	
0	2551.0	542.2	2611.2	566.9	2671.8	592.3	2732.9	618.4	0
1	2552.0	542.6	2612.2	567.4	2672.8	592.8	2733.9	618.8	1
2	2553.0	543.0	2613.2	567.8	2673.8	593.2	2734.9	619.3	2
3	2554.0	543.5	2614.2	568.2	2674.8	593.6	2736.0	619.7	3
4	2555.0	543.9	2615.2	568.6	2675.8	594.0	2737.0	620.2	4
5	2556.0	544.3	2616.2	569.0	2676.9	594.5	2738.0	620.6	5
6	2557.0	544.7	2617.2	569.4	2677.9	594.9	2739.0	621.0	6
7	2558.0	545.1	2618.2	569.9	2678.9	595.3	2740.1	621.5	7
8	2559.0	545.5	2619.2	570.3	2679.9	595.8	2741.1	621.9	8
9	2560.0	545.9	2620.2	570.7	2680.9	596.2	2742.1	622.4	9
10	2561.0	546.3	2621.2	571.1	2681.9	596.6	2743.1	622.8	10
11	2562.0	546.7	2622.2	571.5	2682.9	597.1	2744.2	623.3	11
12	2563.0	547.1	2623.2	572.0	2684.0	597.5	2745.2	623.7	12
13	2564.0	547.5	2624.2	572.4	2685.0	597.9	2746.2	624.1	13
14	2565.0	547.8	2625.3	572.8	2686.0	598.3	2747.2	624.6	14
15	2566.0	548.3	2626.3	573.2	2687.0	598.8	2748.3	625.0	15
16	2567.0	548.8	2627.3	573.6	2688.0	599.2	2749.3	625.4	16
17	2568.0	549.2	2628.3	574.1	2689.0	599.6	2750.3	625.9	17
18	2569.0	549.6	2629.3	574.5	2690.1	600.1	2751.3	626.4	18
19	2570.0	550.0	2630.3	574.9	2691.1	600.5	2752.4	626.8	19
20	2571.0	550.4	2631.3	575.3	2692.1	600.9	2753.4	627.2	20
21	2572.0	550.8	2632.3	575.7	2693.1	601.4	2754.4	627.7	21
22	2573.0	551.2	2633.3	576.2	2694.1	601.8	2755.4	628.1	22
23	2574.0	551.6	2634.3	576.6	2695.2	602.2	2756.5	628.6	23
24	2575.0	552.0	2635.3	577.0	2696.2	602.7	2757.5	629.0	24
25	2576.0	552.4	2636.4	577.4	2697.2	603.1	2758.5	629.5	25
26	2577.0	552.9	2637.4	577.9	2698.2	603.5	2759.5	629.9	26
27	2578.0	553.3	2638.4	578.3	2699.2	604.0	2760.6	630.4	27
28	2579.0	553.7	2639.4	578.7	2700.2	604.4	2761.6	630.8	28
29	2580.0	554.1	2640.4	579.1	2701.3	604.8	2762.6	631.2	29
30	2581.0	554.5	2641.4	579.5	2702.3	605.3	2763.7	631.7	30
31	2582.0	554.9	2642.4	580.0	2703.3	605.7	2764.7	632.1	31
32	2583.0	555.3	2643.4	580.4	2704.3	606.1	2765.7	632.6	32
33	2584.0	555.7	2644.4	580.8	2705.3	606.6	2766.7	633.0	33
34	2585.0	556.2	2645.4	581.2	2706.4	607.0	2767.8	633.5	34
35	2586.0	556.6	2646.5	581.7	2707.4	607.4	2768.8	633.9	35
36	2587.0	557.0	2647.5	582.1	2708.4	607.9	2769.8	634.4	36
37	2588.0	557.4	2648.5	582.5	2709.4	608.3	2770.8	634.8	37
38	2589.0	557.8	2649.5	582.9	2710.4	608.8	2771.9	635.3	38
39	2590.0	558.2	2650.5	583.4	2711.5	609.2	2772.9	635.7	39
40	2591.1	558.6	2651.5	583.8	2712.5	609.6	2773.9	636.2	40
41	2592.1	559.0	2652.5	584.2	2713.5	610.1	2775.0	636.6	41
42	2593.1	559.5	2653.5	584.6	2714.5	610.5	2776.0	637.1	42
43	2594.1	559.9	2654.6	585.1	2715.5	610.9	2777.0	637.5	43
44	2595.1	560.3	2655.6	585.5	2716.6	611.4	2778.1	638.0	44
45	2596.1	560.7	2656.6	585.9	2717.6	611.8	2779.1	638.4	45
46	2597.1	561.1	2657.6	586.3	2718.6	612.2	2780.1	638.9	46
47	2598.1	561.5	2658.6	586.8	2719.6	612.7	2781.1	639.3	47
48	2599.1	561.9	2659.6	587.2	2720.6	613.1	2782.2	639.8	48
49	2600.1	562.4	2660.6	587.6	2721.7	613.6	2783.2	640.2	49
50	2601.1	562.8	2661.6	588.0	2722.7	614.0	2784.2	640.7	50
51	2602.1	563.2	2662.7	588.5	2723.7	614.4	2785.3	641.1	51
52	2603.1	563.6	2663.7	588.9	2724.7	614.9	2786.3	641.6	52
53	2604.1	564.0	2664.7	589.3	2725.7	615.3	2787.3	642.0	53
54	2605.1	564.4	2665.7	589.8	2726.8	615.8	2788.4	642.5	54
55	2606.1	564.9	2666.7	590.2	2727.8	616.2	2789.4	642.9	55
56	2607.1	565.3	2667.7	590.6	2728.8	616.6	2790.4	643.4	56
57	2608.1	565.7	2668.7	591.0	2729.8	617.1	2791.4	643.8	57
58	2609.1	566.1	2669.8	591.5	2730.9	617.5	2792.5	644.3	58
59	2610.1	566.5	2670.8	591.9	2731.9	618.0	2793.5	644.7	59

TABLE VIII.—TANGENTS AND EXTERNALS FOR A 1° CURVE
(Chord or Arc Definition)

′	I = 52°		I = 53°		I = 54°		I = 55°		′
	T	E	T	E	T	E	T	E	
0	2794.5	645.2	2856.7	672.7	2919.4	700.9	2982.7	729.9	0
1	2795.6	645.6	2857.7	673.1	2920.5	701.4	2983.7	730.3	1
2	2796.6	646.1	2858.8	673.6	2921.5	701.8	2984.8	730.8	2
3	2797.6	646.5	2859.8	674.1	2922.6	702.3	2985.8	731.3	3
4	2798.7	647.0	2860.9	674.5	2923.6	702.8	2986.9	731.8	4
5	2799.7	647.4	2861.9	675.0	2924.7	703.3	2988.0	732.3	5
6	2800.7	647.9	2862.9	675.5	2925.7	703.8	2989.0	732.8	6
7	2801.8	648.3	2864.0	675.9	2926.8	704.2	2990.1	733.3	7
8	2802.8	648.8	2865.0	676.4	2927.8	704.7	2991.1	733.8	8
9	2803.8	649.2	2866.1	676.9	2928.9	705.2	2992.2	734.3	9
10	2804.9	649.7	2867.1	677.3	2929.9	705.7	2993.3	734.8	10
11	2805.9	650.2	2868.2	677.8	2931.0	706.1	2994.3	735.3	11
12	2806.9	650.6	2869.2	678.3	2932.0	706.6	2995.4	735.7	12
13	2808.0	651.1	2870.2	678.7	2933.1	707.1	2996.5	736.2	13
14	2809.0	651.5	2871.3	679.2	2934.1	707.6	2997.5	736.7	14
15	2810.0	652.0	2872.3	679.7	2935.2	708.1	2998.6	737.2	15
16	2811.1	652.4	2873.4	680.1	2936.2	708.6	2999.6	737.7	16
17	2812.1	652.9	2874.4	680.6	2937.3	709.0	3000.7	738.2	17
18	2813.1	653.3	2875.5	681.1	2938.3	709.5	3001.8	738.7	18
19	2814.2	653.8	2876.5	681.5	2939.4	710.0	3002.8	739.2	19
20	2815.2	654.3	2877.5	682.0	2940.4	710.5	3003.9	739.7	20
21	2816.2	654.7	2878.6	682.5	2941.5	710.9	3004.9	740.2	21
22	2817.3	655.2	2879.6	682.9	2942.5	711.4	3006.0	740.7	22
23	2818.3	655.6	2880.7	683.4	2943.6	711.9	3007.1	741.2	23
24	2819.3	656.1	2881.7	683.9	2944.6	712.4	3008.1	741.7	24
25	2820.4	656.5	2882.8	684.3	2945.7	712.9	3009.2	742.1	25
26	2821.4	657.0	2883.8	684.8	2946.7	713.4	3010.3	742.6	26
27	2822.4	657.5	2884.8	685.3	2947.8	713.8	3011.3	743.1	27
28	2823.5	657.9	2885.9	685.7	2948.9	714.3	3012.4	743.6	28
29	2824.5	658.4	2886.9	686.2	2949.9	714.8	3013.5	744.1	29
30	2825.6	658.8	2888.0	686.7	2951.0	715.3	3014.5	744.6	30
31	2826.6	659.3	2889.0	687.2	2952.0	715.8	3015.6	745.1	31
32	2827.6	659.7	2890.1	687.6	2953.1	716.2	3016.6	745.6	32
33	2828.7	660.2	2891.1	688.1	2954.1	716.7	3017.7	746.1	33
34	2829.7	660.7	2892.2	688.6	2955.2	717.2	3018.8	746.6	34
35	2830.7	661.1	2893.2	689.0	2956.2	717.8	3019.8	747.1	35
36	2831.8	661.6	2894.3	689.5	2957.3	718.2	3020.9	747.6	36
37	2832.8	662.0	2895.3	690.0	2958.3	718.7	3022.0	748.1	37
38	2833.8	662.5	2896.3	690.5	2959.4	719.1	3023.0	748.6	38
39	2834.9	663.0	2897.4	690.9	2960.5	719.6	3024.1	749.1	39
40	2835.9	663.4	2898.4	691.4	2961.5	720.1	3025.2	749.6	40
41	2837.0	663.9	2899.5	691.9	2962.6	720.6	3026.2	750.1	41
42	2838.0	664.3	2900.5	692.3	2963.6	721.1	3027.3	750.6	42
43	2839.0	664.8	2901.6	692.8	2964.7	721.6	3028.4	751.1	43
44	2840.1	665.3	2902.6	693.3	2965.7	722.1	3029.4	751.6	44
45	2841.1	665.7	2903.7	693.8	2966.8	722.5	3030.5	752.1	45
46	2842.1	666.2	2904.7	694.2	2967.9	723.0	3031.6	752.6	46
47	2843.2	666.6	2905.8	694.7	2968.9	723.5	3032.6	753.1	47
48	2844.2	667.1	2906.8	695.2	2970.0	724.0	3033.7	753.6	48
49	2845.3	667.6	2907.9	695.7	2971.0	724.5	3034.8	754.1	49
50	2846.3	668.0	2908.9	696.1	2972.1	725.0	3035.8	754.6	50
51	2847.3	668.5	2910.0	696.6	2973.1	725.5	3036.9	755.1	51
52	2848.4	669.0	2911.0	697.1	2974.2	725.9	3038.0	755.6	52
53	2849.4	669.4	2912.1	697.6	2975.3	726.4	3039.0	756.1	53
54	2850.5	669.9	2913.1	698.0	2976.3	726.9	3040.1	756.6	54
55	2851.5	670.3	2914.2	698.5	2977.4	727.4	3041.2	757.1	55
56	2852.5	670.8	2915.2	699.0	2978.4	727.9	3042.2	757.6	56
57	2853.6	671.3	2916.3	699.5	2979.5	728.4	3043.3	758.1	57
58	2854.6	671.7	2917.3	699.9	2980.5	728.9	3044.4	758.6	58
59	2855.7	672.2	2918.4	700.4	2981.6	729.4	3045.4	759.1	59

TABLE VIII.—TANGENTS AND EXTERNALS FOR A 1° CURVE
(Chord or Arc Definition)

′	I = 56°		I = 57°		I = 58°		I = 59°		′
	T	E	T	E	T	E	T	E	
0	3046.5	759.6	3110.9	790.1	3176.0	821.4	3241.7	853.5	0
1	3047.6	760.1	3112.0	790.6	3177.1	821.9	3242.8	854.0	1
2	3048.6	760.6	3113.1	791.1	3178.2	822.4	3243.9	854.5	2
3	3049.7	761.1	3114.2	791.6	3179.3	823.0	3245.0	855.1	3
4	3050.8	761.6	3115.3	792.1	3180.4	823.5	3246.1	855.6	4
5	3051.9	762.1	3116.3	792.7	3181.4	824.0	3247.2	856.2	5
6	3052.9	762.6	3117.4	793.2	3182.5	824.5	3248.3	856.7	6
7	3054.0	763.1	3118.5	793.7	3183.6	825.1	3249.4	857.3	7
8	3055.1	763.6	3119.6	794.2	3184.7	825.6	3250.5	857.8	8
9	3056.1	764.1	3120.7	794.7	3185.8	826.1	3251.6	858.3	9
10	3057.2	764.6	3121.7	795.2	3186.9	826.7	3252.7	858.9	10
11	3058.3	765.1	3122.8	795.8	3188.0	827.2	3253.8	859.4	11
12	3059.3	765.6	3123.9	796.3	3189.1	827.7	3254.9	860.0	12
13	3060.4	766.1	3125.0	796.8	3190.2	828.3	3256.0	860.5	13
14	3061.5	766.6	3126.1	797.3	3191.3	828.8	3257.1	861.1	14
15	3062.6	767.1	3127.2	797.8	3192.4	829.3	3258.2	861.6	15
16	3063.6	767.6	3128.2	798.3	3193.5	829.9	3259.3	862.2	16
17	3064.7	768.1	3129.3	798.9	3194.5	830.4	3260.4	862.7	17
18	3065.8	768.6	3130.4	799.4	3195.6	830.9	3261.5	863.2	18
19	3066.8	769.2	3131.5	799.9	3196.7	831.4	3262.6	863.8	19
20	3067.9	769.7	3132.6	800.4	3197.8	832.0	3263.7	864.3	20
21	3069.0	770.2	3133.6	800.9	3198.9	832.5	3264.8	864.9	21
22	3070.1	770.7	3134.7	801.5	3200.0	833.0	3265.9	865.4	22
23	3071.1	771.2	3135.8	802.0	3201.1	833.6	3267.0	866.0	23
24	3072.2	771.7	3136.9	802.5	3202.2	834.1	3268.1	866.5	24
25	3073.3	772.2	3138.0	803.0	3203.3	834.6	3269.2	867.1	25
26	3074.4	772.7	3139.1	803.5	3204.4	835.2	3270.3	867.6	26
27	3075.4	773.2	3140.1	804.1	3205.5	835.7	3271.4	868.2	27
28	3076.5	773.7	3141.2	804.6	3206.6	836.2	3272.6	868.7	28
29	3077.6	774.2	3142.3	805.1	3207.7	836.8	3273.7	869.3	29
30	3078.7	774.7	3143.4	805.6	3208.8	837.3	3274.8	869.8	30
31	3079.7	775.2	3144.5	806.1	3209.9	837.8	3275.9	870.4	31
32	3080.8	775.7	3145.6	806.7	3210.9	838.4	3277.0	870.9	32
33	3081.9	776.3	3146.6	807.2	3212.0	838.9	3278.1	871.5	33
34	3082.9	776.8	3147.7	807.7	3213.1	839.5	3279.2	872.0	34
35	3084.0	777.3	3148.8	808.2	3214.2	840.0	3280.3	872.6	35
36	3085.1	777.8	3149.9	808.8	3215.3	840.5	3281.4	873.1	36
37	3086.2	778.3	3151.0	809.3	3216.4	841.1	3282.5	873.7	37
38	3087.2	778.8	3152.1	809.8	3217.5	841.6	3283.6	874.2	38
39	3088.3	779.3	3153.2	810.3	3218.6	842.1	3284.7	874.8	39
40	3089.4	779.8	3154.2	810.9	3219.7	842.7	3285.8	875.3	40
41	3090.5	780.3	3155.3	811.4	3220.8	843.2	3286.9	875.9	41
42	3091.6	780.9	3156.4	811.9	3221.9	843.7	3288.0	876.4	42
43	3092.6	781.4	3157.5	812.4	3223.0	844.3	3289.2	877.0	43
44	3093.7	781.9	3158.6	812.9	3224.1	844.8	3290.3	877.5	44
45	3094.8	782.4	3159.7	813.5	3225.2	845.4	3291.4	878.1	45
46	3095.9	782.9	3160.8	814.0	3226.3	845.9	3292.5	878.6	46
47	3096.9	783.4	3161.8	814.5	3227.4	846.4	3293.6	879.2	47
48	3098.0	783.9	3162.9	815.0	3228.5	847.0	3294.7	879.7	48
49	3099.1	784.4	3164.0	815.6	3229.6	847.5	3295.8	880.3	49
50	3100.2	784.9	3165.1	816.1	3230.7	848.1	3296.9	880.8	50
51	3101.2	785.5	3166.2	816.6	3231.8	848.6	3298.0	881.4	51
52	3102.3	786.0	3167.3	817.2	3232.9	849.1	3299.1	881.9	52
53	3103.4	786.5	3168.4	817.7	3234.0	849.7	3300.2	882.5	53
54	3104.5	787.0	3169.5	818.2	3235.1	850.2	3301.4	883.1	54
55	3105.6	787.5	3170.6	818.7	3236.2	850.8	3302.5	883.6	55
56	3106.6	788.0	3171.6	819.3	3237.3	851.3	3303.6	884.2	56
57	3107.7	788.5	3172.7	819.8	3238.4	851.8	3304.7	884.7	57
58	3108.8	789.1	3173.8	820.3	3239.5	852.4	3305.8	885.3	58
59	3109.9	789.6	3174.9	820.8	3240.6	852.9	3306.9	885.8	59

TABLE VIII.—TANGENTS AND EXTERNALS FOR A 1° CURVE
(Chord or Arc Definition)

'	I = 60°		I = 61°		I = 62°		I = 63°		'
	T	E	T	E	T	E	T	E	
0	3308.0	886.4	3375.0	920.1	3442.7	954.8	3511.1	990.2	0
1	3309.1	886.9	3376.1	920.7	3443.9	955.3	3512.3	990.8	1
2	3310.2	887.5	3377.3	921.3	3445.0	955.9	3513.4	991.4	2
3	3311.3	888.1	3378.4	921.9	3446.1	956.5	3514.6	992.0	3
4	3312.5	888.6	3379.5	922.4	3447.3	957.1	3515.7	992.6	4
5	3313.6	889.2	3380.6	923.0	3448.4	957.7	3516.9	993.2	5
6	3314.7	889.7	3381.8	923.6	3449.5	958.3	3518.0	993.8	6
7	3315.8	890.3	3382.9	924.1	3450.7	958.8	3519.2	994.4	7
8	3316.9	890.8	3384.9	924.7	3451.8	959.4	3520.3	995.0	8
9	3318.0	891.4	3385.1	925.3	3452.9	960.0	3521.5	995.6	9
10	3319.1	891.9	3386.3	925.8	3454.1	960.6	3522.6	996.2	10
11	3320.2	892.5	3387.4	926.4	3455.2	961.2	3523.8	996.8	11
12	3321.4	893.1	3388.5	927.0	3456.3	961.8	3524.9	997.5	12
13	3322.5	893.6	3389.6	927.6	3457.5	962.4	3526.1	998.1	13
14	3323.6	894.2	3390.8	928.1	3458.6	963.0	3527.2	998.7	14
15	3324.7	894.7	3391.9	928.7	3459.8	963.5	3528.4	999.3	15
16	3325.8	895.3	3393.0	929.3	3460.9	964.1	3529.5	999.9	16
17	3326.9	895.9	3394.1	929.9	3462.0	964.7	3530.7	1000.5	17
18	3328.0	896.4	3395.3	930.4	3463.2	965.3	3531.8	1001.1	18
19	3329.2	897.0	3396.4	931.0	3464.3	965.9	3533.0	1001.7	19
20	3330.3	897.5	3397.5	931.6	3465.4	966.5	3534.1	1002.3	20
21	3331.4	898.1	3398.6	932.2	3466.6	967.1	3535.3	1002.9	21
22	3332.5	898.7	3399.8	932.7	3467.7	967.7	3536.4	1003.5	22
23	3333.6	899.2	3400.9	933.3	3468.9	958.2	3537.6	1004.1	23
24	3334.7	899.8	3402.0	933.9	3470.0	968.8	3538.7	1004.7	24
25	3335.9	900.3	3403.1	934.5	3471.1	969.4	3539.9	1005.3	25
26	3337.0	900.9	3404.3	935.0	3472.3	970.0	3541.0	1005.9	26
27	3338.1	901.5	3405.4	935.6	3473.4	970.6	3542.2	1006.5	27
28	3339.2	902.0	3406.5	936.2	3474.6	971.2	3543.3	1007.1	28
29	3340.3	902.6	3407.7	936.8	3475.7	971.8	3544.5	1007.7	29
30	3341.4	903.2	3408.8	937.3	3476.8	972.4	3545.6	1008.3	30
31	3342.6	903.7	3409.9	937.9	3478.0	973.0	3546.8	1008.9	31
32	3343.7	904.3	3411.0	938.5	3479.1	973.6	3547.9	1009.5	32
33	3344.8	904.8	3412.2	939.1	3480.3	974.2	3549.1	1010.1	33
34	3345.9	905.4	3413.3	939.7	3481.4	974.8	3550.2	1010.7	34
35	3347.0	906.0	3414.4	940.2	3482.5	975.3	3551.4	1011.3	35
36	3348.1	906.5	3415.6	940.8	3483.7	975.9	3552.5	1012.0	36
37	3349.3	907.1	3416.7	941.4	3484.8	976.5	3553.7	1012.6	37
38	3350.4	907.7	3417.8	942.0	3486.0	977.1	3554.8	1013.2	38
39	3351.5	908.2	3418.9	942.5	3487.1	977.7	3556.0	1013.8	39
40	3352.6	908.8	3420.1	943.1	3488.2	978.3	3557.2	1014.4	40
41	3353.7	909.4	3421.2	943.7	3489.4	978.9	3558.3	1015.0	41
42	3354.8	909.9	3422.3	944.3	3490.5	979.5	3559.5	1015.6	42
43	3356.0	910.5	3423.5	944.9	3491.7	980.1	3560.6	1016.2	43
44	3357.1	911.1	3424.6	945.4	3492.8	980.7	3561.8	1016.8	44
45	3358.2	911.6	3425.7	946.0	3494.0	981.3	3562.9	1017.4	45
46	3359.3	912.2	3426.9	946.6	3495.1	981.9	3564.1	1018.1	46
47	3360.4	912.8	3428.0	947.2	3496.2	982.5	3565.2	1018.7	47
48	3361.6	913.3	3429.1	947.8	3497.4	983.1	3566.4	1019.3	48
49	3362.7	913.9	3430.3	948.3	3498.5	983.7	3567.5	1019.9	49
50	3363.8	914.5	3431.4	948.9	3499.7	984.3	3568.7	1020.5	50
51	3364.9	915.0	3432.5	949.5	3500.8	984.9	3569.9	1021.1	51
52	3366.0	915.6	3433.7	950.1	3502.0	985.5	3571.0	1021.7	52
53	3367.2	916.2	3434.8	950.7	3503.1	986.1	3572.2	1022.3	53
54	3368.3	916.7	3435.9	951.3	3504.3	986.7	3573.3	1022.9	54
55	3369.4	917.3	3437.1	951.8	3505.4	987.3	3574.5	1023.5	55
56	3370.5	917.9	3438.2	952.4	3506.6	987.9	3575.6	1024.2	56
57	3371.7	918.4	3439.3	953.0	3507.7	988.4	3576.8	1024.8	57
58	3372.8	919.0	3440.5	953.6	3508.8	989.0	3578.0	1025.4	58
59	3373.9	919.6	3441.6	954.2	3510.8	989.6	3579.1	1026.0	59

TABLE VIII.—TANGENTS AND EXTERNALS FOR A 1° CURVE
(Chord or Arc Definition)

'	I = 64°		I = 65°		I = 66°		I = 67°		'
	T	E	T	E	T	E	T	E	
0	3580.3	1026.6	3650.2	1063.9	3720.9	1102.2	3792.4	1141.4	0
1	3581.4	1027.2	3651.4	1064.5	3722.1	1102.8	3793.6	1142.1	1
2	3582.6	1027.8	3652.5	1065.2	3723.2	1103.5	3794.8	1142.7	2
3	3583.8	1028.5	3653.7	1065.8	3724.4	1104.1	3796.0	1143.4	3
4	3584.9	1029.1	3654.9	1066.4	3725.6	1104.8	3797.2	1144.0	4
5	3586.1	1029.7	3656.1	1067.0	3726.8	1105.4	3798.4	1144.7	5
6	3587.2	1030.3	3657.2	1067.7	3728.0	1106.0	3799.6	1145.4	6
7	3588.4	1030.9	3658.4	1068.3	3729.2	1106.7	3800.8	1146.0	7
8	3589.6	1031.6	3659.6	1068.9	3730.4	1107.3	3802.0	1146.7	8
9	3590.7	1032.2	3660.7	1069.6	3731.6	1108.0	3803.2	1147.3	9
10	3591.9	1032.8	3661.9	1070.2	3732.7	1108.6	3804.4	1148.0	10
11	3593.0	1033.4	3663.1	1070.8	3733.9	1109.2	3805.6	1148.7	11
12	3594.2	1034.0	3664.3	1071.5	3735.1	1109.9	3806.8	1149.3	12
13	3595.4	1034.7	3665.4	1072.1	3736.3	1110.5	3808.0	1150.0	13
14	3596.5	1035.3	3666.6	1072.8	3737.5	1111.2	3809.2	1150.7	14
15	3597.7	1035.9	3667.8	1073.4	3738.7	1111.8	3810.4	1151.3	15
16	3598.8	1036.5	3669.0	1074.0	3739.9	1112.5	3811.6	1152.0	16
17	3600.0	1037.1	3670.1	1074.7	3741.1	1113.1	3812.8	1152.7	17
18	3601.2	1037.8	3671.3	1075.3	3742.2	1113.8	3814.0	1153.4	18
19	3602.3	1038.4	3672.5	1076.0	3743.4	1114.4	3815.2	1154.0	19
20	3603.5	1039.0	3673.7	1076.6	3744.6	1115.1	3816.4	1154.7	20
21	3604.7	1139.6	3674.8	1077.2	3745.8	1115.8	3817.6	1155.4	21
22	3605.8	1040.2	3676.0	1077.9	3747.0	1116.4	3818.8	1156.0	22
23	3607.0	1040.9	3677.2	1078.5	3748.2	1117.1	3820.0	1156.7	23
24	3608.2	1041.5	3678.4	1079.1	3749.4	1117.7	3821.2	1157.3	24
25	3609.3	1042.1	3679.5	1079.7	3750.6	1118.4	3822.4	1158.0	25
26	3610.5	1042.7	3680.7	1080.4	3751.8	1119.1	3823.6	1158.7	26
27	3611.6	1043.3	3681.9	1081.0	3752.9	1119.7	3824.8	1159.3	27
28	3612.8	1044.0	3683.1	1081.6	3754.1	1120.4	3826.0	1160.0	28
29	3614.0	1044.6	3684.3	1082.3	3755.3	1121.0	3827.2	1160.6	29
30	3615.1	1045.2	3685.4	1082.9	3756.5	1121.7	3828.4	1161.3	30
31	3616.3	1045.8	3686.6	1083.5	3757.7	1122.3	3829.6	1162.0	31
32	3617.5	1046.4	3687.8	1084.2	3758.9	1123.0	3830.8	1162.7	32
33	3617.6	1047.1	3689.0	1084.8	3760.1	1123.6	3832.0	1163.3	33
34	3619.8	1047.7	3690.1	1085.5	3761.3	1124.3	3833.3	1164.0	34
35	3621.0	1048.3	3691.3	1086.1	3762.5	1124.9	3834.5	1164.7	35
36	3622.1	1048.9	3692.5	1086.7	3763.7	1125.6	3835.7	1165.4	36
37	3623.3	1049.5	3693.7	1087.4	3764.9	1126.2	3836.9	1166.1	37
38	3624.5	1050.2	3694.9	1088.0	3766.1	1126.9	3838.1	1166.7	38
39	3625.6	1050.8	3696.0	1088.7	3767.3	1127.5	3839.3	1167.4	39
40	3626.8	1051.4	3697.2	1089.3	3768.5	1128.2	3840.5	1168.1	40
41	3628.0	1052.0	3698.4	1089.9	3769.6	1128.9	3841.7	1168.8	41
42	3629.1	1052.7	3699.6	1090.6	3770.8	1129.5	3842.9	1169.4	42
43	3630.3	1053.3	3700.8	1091.2	3772.0	1130.2	3844.1	1170.1	43
44	3631.5	1053.9	3702.0	1091.9	3773.2	1130.8	3845.3	1170.8	44
45	3632.6	1054.5	3703.1	1092.5	3774.4	1131.5	3846.5	1171.4	45
46	3633.8	1055.2	3704.3	1093.1	3775.6	1132.2	3847.7	1172.1	46
47	3635.0	1055.8	3705.5	1093.8	3776.8	1132.8	3849.0	1172.8	47
48	3636.1	1056.4	3706.7	1094.4	3778.0	1133.5	3850.2	1173.5	48
49	3637.3	1057.1	3707.9	1095.1	3779.2	1134.1	3851.4	1174.1	49
50	3638.5	1057.7	3709.0	1095.7	3780.4	1134.8	3852.6	1174.8	50
51	3639.7	1058.3	3710.2	1096.3	3781.6	1135.5	3853.8	1175.5	51
52	3640.8	1058.9	3711.4	1097.0	3782.8	1136.1	3855.0	1176.2	52
53	3642.0	1059.6	3712.6	1097.6	3784.0	1136.8	3856.2	1176.8	53
54	3643.2	1060.2	3713.8	1098.3	3785.2	1137.4	3857.4	1177.5	54
55	3644.3	1060.8	3715.0	1098.9	3786.4	1138.1	3858.6	1178.2	55
56	3645.5	1061.4	3716.1	1099.6	3787.6	1138.8	3859.8	1178.9	56
57	3646.7	1062.0	3717.3	1100.2	3788.8	1139.4	3861.1	1179.6	57
58	3647.8	1062.7	3718.5	1100.9	3790.0	1140.1	3862.3	1180.2	58
59	3649.0	1063.3	3719.7	1101.5	3791.2	1140.7	3863.5	1180.9	59

TABLE VIII.—TANGENTS AND EXTERNALS FOR A 1° CURVE (Chord or Arc Definition)

'	I = 68°		I = 69°		I = 70°		I = 71°		'
	T	E	T	E	T	E	T	E	
0	3864.7	1181.6	3937.9	1222.7	4011.9	1265.0	4086.9	1308.2	0
1	3865.9	1182.3	3939.1	1223.4	4013.2	1265.7	4088.2	1308.9	1
2	3867.1	1183.0	3940.3	1224.1	4014.4	1266.4	4089.4	1309.7	2
3	3868.3	1183.6	3941.6	1224.8	4015.7	1267.1	4090.7	1310.4	3
4	3869.6	1184.3	3942.8	1225.5	4016.9	1267.8	4091.9	1311.2	4
5	3870.8	1185.0	3944.0	1226.2	4018.2	1268.5	4093.2	1311.9	5
6	3872.0	1185.7	3945.2	1226.9	4019.4	1269.3	4094.5	1312.6	6
7	3873.2	1186.4	3946.5	1227.6	4020.6	1270.0	4095.7	1313.4	7
8	3874.4	1187.0	3947.7	1228.3	4021.9	1270.7	4097.0	1314.1	8
9	3875.6	1187.7	3948.9	1229.0	4023.1	1271.4	4098.2	1314.9	9
10	3876.8	1188.4	3950.2	1229.7	4024.4	1272.1	4099.5	1315.5	10
11	3878.0	1189.1	3951.4	1230.4	4025.6	1272.8	4100.8	1316.3	11
12	3879.3	1189.8	3952.6	1231.1	4026.9	1273.5	4102.0	1317.1	12
13	3880.5	1190.4	3953.9	1231.8	4028.1	1274.3	4103.3	1317.8	13
14	3881.7	1191.1	3955.1	1232.5	4029.4	1275.0	4104.5	1318.5	14
15	3882.9	1191.8	3956.3	1233.2	4030.6	1275.7	4105.8	1319.2	15
16	3884.1	1192.5	3957.5	1233.9	4031.8	1276.4	4107.1	1320.0	16
17	3885.3	1193.2	3958.8	1234.6	4033.1	1277.1	4108.3	1320.7	17
18	3886.6	1193.8	3960.0	1235.3	4034.3	1277.9	4109.6	1321.4	18
19	3887.8	1194.5	3961.2	1236.0	4035.6	1278.6	4110.9	1322.2	19
20	3889.0	1195.2	3962.5	1236.7	4036.8	1279.3	4112.1	1322.9	20
21	3890.2	1195.9	3963.7	1237.4	4038.1	1280.0	4113.4	1323.6	21
22	3891.4	1196.6	3964.9	1238.1	4039.3	1280.7	4114.6	1324.4	22
23	3892.6	1197.2	3966.2	1238.8	4040.6	1281.5	4115.9	1325.1	23
24	3893.9	1197.9	3967.4	1239.5	4041.8	1282.2	4117.2	1325.9	24
25	3895.1	1198.6	3968.6	1240.2	4043.1	1282.9	4118.4	1326.6	25
26	3896.3	1199.3	3969.9	1240.9	4044.3	1283.6	4119.7	1327.3	26
27	3897.5	1200.0	3971.1	1241.6	4045.6	1284.3	4121.0	1328.1	27
28	3898.7	1200.6	3972.3	1242.3	4046.8	1285.1	4122.2	1328.8	28
29	3900.0	1201.3	3973.6	1243.0	4048.1	1285.8	4123.5	1329.6	29
30	3901.2	1202.0	3974.8	1243.7	4049.3	1286.5	4124.8	1330.3	30
31	3902.4	1202.7	3976.0	1244.4	4050.6	1287.2	4126.0	1331.0	31
32	3903.6	1203.4	3977.3	1245.1	4051.8	1287.9	4127.3	1331.8	32
33	3904.8	1204.1	3978.5	1245.8	4053.1	1288.6	4128.6	1332.5	33
34	3906.1	1204.8	3979.7	1246.5	4054.3	1289.3	4129.8	1333.3	34
35	3907.3	1205.4	3981.0	1247.2	4055.6	1290.0	4131.1	1334.0	35
36	3908.5	1206.1	3982.2	1248.0	4056.8	1290.8	4132.4	1334.7	36
37	3909.7	1206.8	3983.4	1248.7	4058.1	1291.5	4133.6	1335.5	37
38	3910.9	1207.5	3984.7	1249.4	4059.3	1292.2	4134.9	1336.2	38
39	3912.2	1208.2	3985.9	1250.1	4060.6	1292.9	4136.2	1337.0	39
40	3913.4	1208.9	3987.2	1250.8	4061.8	1293.7	4137.4	1337.7	40
41	3914.6	1209.6	3988.4	1251.5	4063.1	1294.3	4138.7	1338.4	41
42	3915.8	1210.3	3989.6	1252.2	4064.3	1295.1	4140.0	1339.2	42
43	3917.1	1211.0	3990.9	1252.9	4065.6	1295.8	4141.2	1339.9	43
44	3918.3	1211.7	3992.1	1253.6	4066.8	1296.5	4142.5	1340.7	44
45	3919.5	1212.3	3993.3	1254.3	4068.1	1297.2	4143.8	1341.4	45
46	3920.7	1213.0	3994.6	1255.1	4069.3	1298.0	4145.0	1342.1	46
47	3921.9	1213.7	3995.8	1255.8	4070.6	1298.7	4146.3	1342.9	47
48	3923.2	1214.4	3997.1	1256.5	4071.9	1299.4	4147.6	1343.6	48
49	3924.4	1215.1	3998.3	1257.2	4073.1	1300.2	4148.8	1344.4	49
50	3925.6	1215.8	3999.5	1257.9	4074.4	1300.9	4150.1	1345.1	50
51	3926.8	1216.5	4000.8	1258.6	4075.6	1301.6	4151.4	1345.8	51
52	3928.1	1217.2	4002.0	1259.3	4076.9	1302.4	4152.7	1346.6	52
53	3929.3	1217.9	4003.3	1260.0	4078.1	1303.1	4153.9	1347.3	53
54	3930.5	1218.6	4004.5	1260.7	4079.4	1303.8	4155.2	1348.1	54
55	3931.7	1219.2	4005.7	1261.4	4080.6	1304.5	4156.5	1348.8	55
56	3933.0	1219.9	4007.0	1262.2	4081.9	1305.3	4157.7	1349.6	56
57	3934.2	1220.6	4008.2	1262.9	4083.1	1306.0	4159.0	1350.3	57
58	3935.4	1221.3	4009.5	1263.6	4084.4	1306.7	4160.3	1351.1	58
59	3936.7	1222.0	4010.7	1264.3	4085.7	1307.5	4161.6	1351.8	59

TABLE VIII.—TANGENTS AND EXTERNALS FOR A 1° CURVE
(Chord or Arc Definition)

'	I = 72°		I = 73°		I = 74°		I = 75°		'
	T	E	T	E	T	E	T	E	
0	4162.8	1352.6	4239.7	1398.0	4317.6	1444.6	4396.5	1492.4	0
1	4164.1	1353.3	4241.0	1398.8	4318.9	1445.4	4397.8	1493.2	1
2	4165.4	1354.1	4242.3	1399.5	4320.2	1446.2	4399.2	1494.0	2
3	4166.7	1354.8	4243.6	1400.3	4321.5	1447.0	4400.5	1494.8	3
4	4167.9	1355.6	4244.9	1401.1	4322.8	1447.8	4401.8	1495.6	4
5	4169.2	1356.3	4246.2	1401.8	4324.1	1448.5	4403.1	1496.4	5
6	4170.5	1357.1	4247.5	1402.6	4325.4	1449.3	4404.5	1497.3	6
7	4171.8	1357.8	4248.8	1403.4	4326.8	1450.1	4405.8	1498.1	7
8	4173.0	1358.6	4250.0	1404.2	4328.1	1450.9	4407.1	1498.9	8
9	4174.3	1359.3	4251.3	1404.9	4329.4	1451.7	4408.4	1499.7	9
10	4175.6	1360.1	4252.6	1405.7	4330.7	1452.5	4409.8	1500.5	10
11	4176.9	1360.8	4253.9	1406.5	4332.0	1453.3	4411.1	1501.3	11
12	4178.1	1361.6	4255.2	1407.3	4333.3	1454.1	4412.4	1502.1	12
13	4179.4	1362.3	4256.5	1408.0	4334.6	1454.9	4413.8	1502.9	13
14	4180.7	1363.1	4257.8	1408.8	4335.9	1455.7	4415.1	1503.7	14
15	4182.0	1363.8	4259.1	1409.6	4337.2	1456.4	4416.4	1504.5	15
16	4183.2	1364.6	4260.4	1410.4	4338.5	1457.2	4417.7	1505.4	16
17	4184.5	1365.3	4261.7	1411.2	4339.9	1458.0	4419.1	1506.2	17
18	4185.8	1366.1	4263.0	1411.9	4341.2	1458.8	4420.4	1507.0	18
19	4187.1	1366.8	4264.3	1412.7	4342.5	1459.6	4421.7	1507.8	19
20	4188.4	1367.6	4265.6	1413.5	4343.8	1460.4	4423.1	1508.6	20
21	4189.6	1368.4	4266.9	1414.3	4345.1	1461.2	4424.4	1509.4	21
22	4190.9	1369.1	4268.2	1415.0	4346.4	1462.0	4425.7	1510.2	22
23	4192.2	1369.9	4269.5	1415.8	4347.7	1462.8	4427.0	1511.0	23
24	4193.5	1370.6	4270.7	1416.6	4349.0	1463.6	4428.4	1511.8	24
25	4194.8	1371.4	4272.0	1417.3	4350.4	1464.4	4429.7	1512.6	25
26	4196.0	1372.2	4273.3	1418.1	4351.7	1465.2	4431.0	1513.5	26
27	4197.3	1372.9	4274.6	1418.9	4353.0	1466.0	4432.4	1514.3	27
28	4198.6	1373.7	4275.9	1419.7	4354.3	1466.8	4433.7	1515.1	28
29	4199.9	1374.4	4277.2	1420.4	4355.6	1467.6	4435.0	1515.9	29
30	4201.2	1375.2	4278.5	1421.2	4356.9	1468.4	4436.4	1516.7	30
31	4202.4	1376.0	4279.8	1422.0	4358.2	1469.2	4437.7	1517.5	31
32	4203.7	1376.7	4281.1	1422.8	4359.6	1470.0	4439.0	1518.3	32
33	4205.0	1377.5	4282.4	1423.5	4360.9	1470.8	4440.4	1519.2	33
34	4206.3	1378.2	4283.7	1424.3	4362.2	1471.6	4441.7	1520.0	34
35	4207.6	1379.0	4285.0	1425.1	4363.5	1472.4	4443.0	1520.8	35
36	4208.8	1379.8	4286.3	1425.9	4364.8	1473.2	4444.4	1521.6	36
37	4210.1	1380.5	4287.6	1426.7	4366.1	1474.0	4445.7	1522.4	37
38	4211.4	1381.3	4288.9	1427.4	4367.5	1474.8	4447.0	1523.3	38
39	4212.7	1382.0	4290.2	1428.2	4368.8	1475.6	4448.4	1524.1	39
40	4214.0	1382.8	4291.5	1429.0	4370.1	1476.4	4449.7	1524.9	40
41	4215.3	1383.6	4292.8	1429.8	4371.4	1477.2	4451.1	1525.7	41
42	4216.5	1384.3	4294.1	1430.6	4372.7	1478.0	4452.4	1526.5	42
43	4217.8	1385.1	4295.4	1431.3	4374.1	1478.8	4453.7	1527.4	43
44	4219.1	1385.8	4296.7	1432.1	4375.4	1479.6	4455.1	1528.2	44
45	4220.4	1386.6	4298.0	1432.9	4376.7	1480.4	4456.4	1529.0	45
46	4221.7	1387.4	4299.3	1433.7	4378.0	1481.2	4457.7	1529.8	46
47	4223.0	1388.1	4300.6	1434.5	4379.3	1482.0	4459.1	1530.6	47
48	4224.3	1388.9	4301.9	1435.2	4380.6	1482.8	4460.4	1531.5	48
49	4225.5	1389.6	4303.2	1436.0	4382.0	1483.6	4461.7	1532.3	49
50	4226.8	1390.4	4304.6	1436.8	4383.3	1484.4	4463.1	1533.1	50
51	4228.1	1391.2	4305.9	1437.6	4384.6	1485.2	4464.4	1533.9	51
52	4229.4	1391.9	4307.2	1438.4	4395.9	1486.0	4465.8	1534.8	52
53	4230.7	1392.7	4308.5	1439.1	4387.3	1486.8	4467.1	1535.6	53
54	4232.0	1393.4	4309.8	1439.9	4388.6	1487.6	4468.4	1536.4	54
55	4233.3	1394.2	4311.1	1440.7	4389.9	1488.4	4469.8	1537.2	55
56	4234.6	1395.0	4312.4	1441.5	4391.2	1489.2	4471.1	1538.1	56
57	4235.9	1395.7	4313.7	1442.3	4392.5	1490.0	4472.5	1538.9	57
58	4237.1	1396.5	4315.0	1443.0	4393.9	1490.8	4473.8	1539.7	58
59	4238.4	1397.2	4316.3	1443.8	4395.2	1491.6	4475.2	1540.6	59

TABLE IX.—CORRECTIONS FOR TANGENTS*
(Chord Definition of *D*)

After dividing $T_1°$ (Table VIII) by D_c add the correction.

I	Degree of Curve (D_c)											I
	2°	3°	4°	5°	6°	8°	10°	12°	14°	16°	20°	
5°	.00	.00	.01	.01	.02	.02	.03	.03	.04	.05	.06	5°
10°	.01	.02	.02	.03	.04	.05	.06	.07	.08	.10	.13	10°
15°	.02	.03	.04	.05	.06	.08	.10	.11	.13	.16	.19	15°
20°	.02	.03	.05	.06	.08	.10	.13	.15	.18	.21	.25	20°
25°	.03	.04	.06	.08	.10	.13	.16	.20	.23	.26	.32	25°
30°	.03	.04	.07	.08	.11	.15	.19	.23	.27	.30	.39	30°
35°	.04	.06	.08	.10	.13	.18	.22	.27	.32	.37	.46	35°
40°	.05	.08	.10	.13	.15	.21	.26	.32	.37	.42	.53	40°
45°	.05	.08	.12	.15	.18	.24	.30	.35	.42	.48	.61	45°
50°	.05	.09	.13	.16	.19	.26	.33	.40	.47	.54	.68	50°
52°	.05	.09	.14	.18	.21	.29	.36	.42	.49	.57	.71	52°
54°	.05	.10	.14	.18	.22	.30	.37	.45	.52	.59	.74	54°
56°	.06	.10	.15	.19	.23	.31	.38	.46	.54	.61	.78	56°
58°	.06	.11	.15	.19	.24	.32	.40	.48	.56	.64	.81	58°
60°	.06	.11	.16	.20	.25	.33	.42	.50	.58	.67	.84	60°
62°	.07	.12	.16	.21	.26	.34	.44	.53	.61	.70	.87	62°
64°	.07	.12	.17	.21	.26	.35	.45	.54	.63	.72	.90	64°
66°	.07	.13	.18	.22	.27	.37	.47	.56	.66	.75	.95	66°
68°	.07	.13	.18	.24	.28	.38	.49	.58	.68	.79	.98	68°
70°	.08	.14	.20	.25	.30	.40	.51	.61	.71	.82	1.02	70°
72°	.08	.15	.21	.26	.31	.42	.53	.63	.74	.84	1.06	72°
74°	.08	.15	.21	.26	.32	.43	.54	.65	.76	.88	1.10	74°
76°	.08	.15	.22	.27	.33	.45	.56	.68	.79	.91	1.14	76°
78°	.09	.16	.22	.28	.34	.46	.58	.70	.82	.94	1.18	78°
80°	.10	.17	.24	.30	.37	.49	.61	.73	.85	.98	1.23	80°

TABLE X.—CORRECTIONS FOR EXTERNALS*
(Chord Definition of *D*)

After dividing $E_1°$ (Table VIII) by D_c add the correction.

I	Degree of Curve (D_c)											I
	2°	3°	4°	5°	6°	8°	10°	12°	14°	16°	20°	
10°	.00	.00	.00	.00	.00	.00	.00	.00	.00	.00	.00	10°
20°	.00	.00	.00	.00	.01	.01	.01	.01	.02	.02	.02	20°
30°	.01	.01	.02	.02	.02	.03	.03	.03	.04	.04	.06	30°
40°	.01	.01	.02	.03	.03	.04	.05	.06	.07	.08	.10	40°
50°	.01	.02	.03	.04	.05	.06	.08	.09	.10	.12	.14	50°
55°	.02	.03	.04	.05	.06	.08	.10	.11	.13	.15	.18	55°
60°	.02	.03	.04	.06	.07	.09	.11	.13	.16	.18	.22	60°
65°	.03	.04	.05	.07	.08	.11	.14	.16	.19	.22	.27	65°
70°	.03	.05	.07	.08	.10	.13	.16	.19	.22	.26	.32	70°
75°	.04	.06	.08	.09	.11	.15	.19	.23	.27	.30	.38	75°

* See page 334 for explanation and example.

Tables XI and XII. Spiral Tables—General Explanation

Partial Notation

- D — Degree of central circular curve (D_a = arc definition of D; D_c = chord definition of D).
- L_s — Length of spiral curve, in feet.
- Δ — Central angle of the spiral, or spiral angle.
- X, Y — Coordinates (abscissa and ordinate) of the S.C. referred to the T.S. as origin and to the initial tangent as X-axis.
- X_0, o — Coordinates (abscissa and ordinate) of the offset T.C., which is the point where a tangent to the circular curve produced backward becomes parallel to the tangent at the T.S.
- L.T. — "Long tangent" of the spiral.
- S.T. — "Short tangent" of the spiral.
- L.C. — "Long chord" of the spiral.

(See Chapter 5 for complete theory and notation)

Table XI. This table gives spiral parts for various selected values of D and L_s. *It may be used for both chord definition and arc definition of D.* The definition of D affects only the coordinates of the offset T.C.; correct values of these coordinates may be obtained by observing the following:

1. *When arc definition of D is used,* select values of o and X_0 from sub-columns headed D_a.
2. *When chord definition of D is used,* subtract the corrections in the columns headed * from the arc-definition values of o and X_0.

Table XII. This table may be used to obtain spiral parts for *any combination* of D and L_s up to $\Delta = 90°$. Proceed as follows:

1. *When arc definition of D is used,* enter table with given value of Δ (interpolating if necessary) and multiply the tabulated coefficients by the value of L_s.
2. *When chord definition of D is used,* calculate spiral parts as above. Then, since only the coordinates of the offset T. C. are affected by the definition of D, correct the calculated values of o and X_0 by subtracting the products of D and the coefficients in the columns headed by an asterisk.

Example

$$D = 12; \quad L_s = 320; \quad \Delta = 19.2.$$

For arc definition,

$$o = 320 \times 0.02781 = 8.90$$

$$X_0 = 320 \times 0.49813 = 159.40$$

TABLE XI.—SELECTED SPIRALS

L_s	Δ	o	X_0	X	Y	L.T.	S.T.	L.C.
				D = 0° 30′				
100	0°15.0′	0.03	50.00	100.00	0.14	66.67	33.33	100.00
125	0 18.8	0.06	62.50	125.00	0.23	83.33	41.67	125.00
150	0 22.5	0.08	75.00	150.00	0.33	100.00	50.00	150.00
200	0 30.0	0.15	100.00	200.00	0.58	133.33	66.67	200.00
250	0 37.5	0.23	125.00	250.00	0.91	166.67	83.33	250.00
300	0 45.0	0.31	150.00	299.99	1.31	200.00	100.00	300.00
350	0 52.5	0.44	175.00	349.99	1.78	233.34	116.67	350.00
400	1 00.0	0.58	200.00	399.99	2.33	266.67	133.34	399.99
450	1 07.5	0.74	225.00	449.98	2.95	300.01	150.01	449.99
500	1 15.0	0.91	250.00	499.98	3.64	333.34	166.67	499.99
550	1 22.5	1.10	274.99	549.97	4.40	366.68	183.34	549.99
600	1 30.0	1.31	299.99	599.96	5.24	400.01	200.01	599.98
700	1 45.0	1.78	349.99	699.93	7.13	466.69	233.35	699.97
800	2 00.0	2.33	399.98	799.90	9.31	533.37	266.70	799.96
900	2 15.0	2.94	449.98	899.86	11.78	600.05	300.05	899.94
1000	2 30.0	3.64	499.97	999.81	14.54	666.73	333.39	999.92
				D = 0° 40′				
100	0°20.0′	0.05	50.00	100.00	0.18	66.67	33.33	100.00
125	0 25.0	0.08	62.50	125.00	0.30	83.33	41.67	125.00
150	0 30.0	0.11	75.00	150.00	0.44	100.00	50.00	150.00
200	0 40.0	0.20	100.00	200.00	0.78	133.33	66.67	200.00
250	0 50.0	0.30	125.00	249.99	1.21	166.67	83.33	250.00
300	1 00.0	0.44	150.00	299.99	1.75	200.00	100.00	300.00
350	1 10.0	0.59	175.00	349.99	2.38	233.34	116.67	349.99
400	1 20.0	0.78	200.00	399.98	3.10	266.67	133.34	399.99
450	1 30.0	0.98	225.00	449.97	3.93	300.01	150.01	449.99
500	1 40.0	1.21	249.99	499.96	4.85	333.35	166.68	499.98
550	1 50.0	1.47	274.99	549.95	5.87	366.69	183.35	549.98
600	2 00.0	1.75	299.99	599.93	6.98	400.03	200.02	599.97
700	2 20.0	2.38	349.98	699.88	9.50	466.71	233.37	699.95
800	2 40.0	3.10	399.97	799.83	12.41	533.39	266.72	799.92
900	3 00.0	3.93	449.96	899.75	15.70	600.09	300.08	899.89
1000	3 20.0	4.85	499.94	999.66	19.39	666.79	333.44	999.85
				D = 0° 50′				
100	0°25.0′	0.06	50.00	100.00	0.24	66.67	33.33	100.00
125	0 31.3	0.09	62.50	125.00	0.38	83.33	41.67	125.00
150	0 37.5	0.14	75.00	150.00	0.55	100.00	50.00	150.00
200	0 50.0	0.24	100.00	200.00	0.97	133.33	66.67	200.00
250	1 02.5	0.38	125.00	249.99	1.52	166.67	83.34	250.00
300	1 15.0	0.55	150.00	299.99	2.18	200.00	100.00	299.99
350	1 27.5	0.74	175.00	349.98	2.97	233.34	116.67	349.99
400	1 40.0	0.97	199.99	399.97	3.88	266.68	133.34	399.99
450	1 52.5	1.23	224.99	449.95	4.91	300.02	150.02	449.98
500	2 05.0	1.51	249.99	499.93	6.06	333.36	166.69	499.97
550	2 17.5	1.83	274.99	549.91	7.33	366.70	183.36	549.96
600	2 30.0	2.18	299.98	599.89	8.73	400.04	200.04	599.95
700	2 55.0	2.97	349.97	699.82	11.88	466.73	233.39	699.92
800	3 20.0	3.88	399.95	799.73	15.51	533.43	266.75	799.88
900	3 45.0	4.91	449.93	899.63	19.63	600.14	300.12	899.83
1000	4 10.0	6.06	499.91	999.47	24.23	666.85	333.50	999.77

For chord definition,

$$o = 8.90 - 12 \times 0.0041 = 8.85$$

$$X_0 = 159.40 - 12 \times 0.0240 = 159.11$$

TABLE XI.—SELECTED SPIRALS

L_s	Δ	o	X_0: D_a	X_0: *	X	Y	L.T.	S.T.	L.C.
D = 1°									
100	0°30.0′	0.07	50.00	....	100.00	0.29	66.67	33.33	100.00
125	0 37.5	0.11	62.50	....	125.00	0.45	83.33	41.67	125.00
150	0 45.0	0.16	75.00	....	150.00	0.65	100.00	50.00	150.00
200	1 00.0	0.29	100.00	....	199.99	1.16	133.34	66.67	200.00
250	1 15.0	0.45	125.00	....	249.99	1.82	166.67	83.34	249.99
300	1 30.0	0.65	150.00	....	299.98	2.62	200.01	100.01	299.99
350	1 45.0	0.89	174.99	....	349.97	3.56	233.34	116.68	349.99
400	2 00.0	1.16	199.99	....	399.95	4.65	266.68	133.35	399.98
450	2 15.0	1.47	224.99	....	449.93	5.89	300.02	150.02	449.97
500	2 30.0	1.82	249.98	....	499.91	7.27	333.37	166.70	499.96
550	2 45.0	2.20	274.98	....	549.87	8.80	366.71	183.37	549.94
600	3 00.0	2.62	299.97	....	599.84	10.47	400.06	200.05	599.93
700	3 30.0	3.56	349.94	....	699.74	14.25	466.76	233.42	699.89
800	4 00.0	4.65	399.94	.01	799.61	18.61	533.47	266.79	799.82
900	4 30.0	5.89	449.91	.01	899.44	23.55	600.19	300.18	899.76
1000	5 00.0	7.27	499.87	.01	999.24	29.07	666.93	333.58	999.66
D = 1° 15′									
100	0°37.5′	0.09	50.00	....	100.00	0.36	66.67	33.33	100.00
125	0 46.9	0.14	62.50	....	125.00	0.57	83.33	41.67	125.00
150	0 56.2	0.20	75.00	....	150.00	0.82	100.00	50.00	150.00
200	1 15.0	0.36	100.00	....	199.99	1.45	133.34	66.67	200.00
250	1 33.8	0.57	124.99	....	249.98	2.27	166.67	83.34	244.99
300	1 52.5	0.82	149.99	....	299.97	3.27	200.01	100.01	299.99
350	2 11.3	1.11	174.99	....	349.95	4.45	233.35	116.68	349.98
400	2 30.0	1.45	199.99	....	399.92	5.82	266.69	133.36	399.97
450	2 48.8	1.84	224.98	....	449.89	7.36	300.04	150.03	449.95
500	3 07.5	2.27	249.97	....	499.85	9.09	333.39	166.71	499.93
550	3 26.3	2.75	274.97	.01	549.80	11.00	366.74	183.40	549.91
600	3 45.0	3.27	299.96	.01	599.74	13.09	400.09	200.08	599.89
700	4 22.5	4.45	349.93	.01	699.59	17.81	466.81	233.46	699.82
800	5 00.0	5.82	399.90	.01	799.39	23.26	533.54	266.86	799.73
900	5 37.5	7.36	449.86	.01	899.13	29.43	600.30	300.28	899.62
1000	6 15.0	9.08	499.80	.01	998.80	36.33	667.08	333.71	999.48
D = 1° 30′									
100	0°45.0′	0.11	50.00	....	100.00	0.44	66.67	33.33	100.00
125	0 56.2	0.17	62.50	....	125.00	0.68	83.33	41.67	125.00
150	1 07.5	0.25	75.00	....	149.99	0.98	100.00	50.00	150.00
200	1 30.0	0.44	100.00	....	199.99	1.75	133.34	66.67	199.99
250	1 52.5	0.68	124.99	....	249.97	2.73	166.68	83.34	249.99
300	2 15.0	0.98	149.99	....	299.95	3.93	200.02	100.01	299.98
350	2 37.5	1.33	174.99	....	349.93	5.34	233.36	116.69	349.97
400	3 00.0	1.74	199.98	....	399.89	6.98	266.71	133.37	399.95
450	3 22.5	2.20	224.97	.01	449.85	8.83	300.06	150.05	449.93
500	3 45.0	2.72	249.96	.01	499.78	10.90	333.41	166.74	499.90
550	4 07.5	3.30	274.95	.01	549.71	13.19	366.77	183.42	549.87
600	4 30.0	3.92	299.94	.01	599.63	15.70	400.13	200.12	599.84
700	5 15.0	5.35	349.90	.01	699.41	21.36	466.88	233.52	699.74
800	6 00.0	6.98	399.86	.01	799.12	27.90	533.64	266.94	799.61
900	6 45.0	8.83	449.79	.01	898.75	35.31	600.44	300.40	899.44

* Subtract from tabulated value for D_a when chord definition of D is used.

TABLE XI.—SELECTED SPIRALS

			X_0						
L_s	Δ	o	D_a	*	X	Y	L.T.	S.T.	L.C.
				D = 1° 45′					
100	0°52.5′	0.13	50.00	....	100.00	0.51	66.67	33.33	100.00
125	1 05.6	0.20	62.50	....	125.00	0.80	83.33	41.67	125.00
150	1 18.8	0.29	75.00	....	149.99	1.15	100.00	50.00	150.00
200	1 45.0	0.51	100.00	....	199.98	2.36	133.34	66.67	199.99
250	2 11.3	0.80	124.99	....	249.96	3.18	166.68	83.35	249.98
300	2 37.5	1.15	149.99	.01	299.94	4.58	200.02	100.02	299.97
350	3 03.8	1.56	174.98	.01	349.90	6.23	233.37	116.70	349.96
400	3 30.0	2.03	199.98	.01	399.85	8.14	266.72	133.38	399.93
450	3 56.3	2.57	224.96	.01	449.79	10.30	300.07	150.07	449.91
500	4 22.5	3.18	249.95	.01	499.71	12.72	333.43	166.76	499.87
550	4 48.8	3.85	274.93	.01	549.61	15.39	366.80	183.46	549.83
600	5 15.0	4.58	299.92	.01	599.50	18.32	400.18	200.16	599.78
700	6 07.5	6.24	349.87	.01	699.20	24.92	466.95	233.59	699.65
800	7 00.0	8.14	399.80	.02	798.81	32.54	533.75	267.05	799.47
900	7 52.5	10.30	449.71	.02	898.31	41.18	600.60	300.55	899.25
				D = 2°					
100	1°00.0′	0.15	50.00	....	100.00	0.58	66.67	33.33	100.00
125	1 15.0	0.23	62.50	....	124.99	0.91	83.34	41.67	125.00
150	1 30.0	0.33	75.00	....	149.99	1.31	100.00	50.00	150.00
200	2 00.0	0.58	100.00	.01	199.98	2.33	133.34	66.67	199.99
250	2 30.0	0.91	124.99	.01	249.95	3.64	166.68	83.35	249.98
300	3 00.0	1.30	149.99	.01	299.92	5.24	200.03	100.03	299.96
350	3 30.0	1.78	174.98	.01	349.87	7.13	233.38	116.71	349.94
400	4 00.0	2.32	199.97	.01	399.80	9.30	266.73	133.40	399.91
450	4 30.0	2.94	224.96	.01	449.72	11.78	300.10	150.09	449.88
500	5 00.0	3.64	249.94	.01	499.62	14.54	333.47	166.79	499.83
550	5 30.0	4.40	274.92	.01	549.49	17.59	366.84	183.50	549.77
600	6 00.0	5.23	299.89	.02	599.34	20.93	400.23	200.21	599.71
700	7 00.0	7.13	349.82	.02	698.96	28.48	467.03	233.67	699.54
800	8 00.0	9.30	399.74	.02	798.44	37.18	533.88	267.16	799.30
				D = 2° 15′					
100	1°07.5′	0.16	50.00	....	100.00	0.65	66.67	33.33	100.00
125	1 24.4	0.26	62.50	....	124.99	1.02	83.34	41.67	125.00
150	1 41.2	0.37	74.99	....	149.99	1.47	100.00	50.00	149.99
200	2 15.0	0.65	99.99	.01	199.97	2.62	133.34	66.68	199.99
250	2 48.8	1.02	124.99	.01	249.94	4.09	166.69	83.35	249.97
300	3 22.5	1.47	149.98	.01	299.90	5.89	200.04	100.03	299.95
350	3 56.3	2.00	174.97	.01	349.83	8.02	233.39	116.72	349.93
400	4 30.0	2.62	199.96	.01	399.75	10.47	266.75	133.41	399.89
450	5 03.8	3.31	224.94	.01	449.65	13.24	300.12	150.11	449.84
500	5 37.5	4.09	249.92	.02	499.52	16.35	333.50	166.82	499.79
550	6 11.3	4.94	274.90	.02	549.36	19.78	366.89	183.54	549.72
600	6 45.0	5.89	299.86	.02	599.17	23.54	400.29	200.26	599.63
700	7 52.5	8.02	349.78	.03	698.69	32.03	467.13	233.75	699.42
800	9 00.0	10.46	399.67	.03	798.03	41.82	534.02	267.30	799.12

* Subtract from tabulated value for D_a when chord definition of D is used.

TABLE XI.—SELECTED SPIRALS

L_s	Δ	o	X_0 D_a	*	X	Y	L.T.	S.T.	L.C.
				D=2° 30′					
100	1°15.0′	0.18	50.00	....	100.00	0.72	66.67	33.33	100.00
125	1 33.8	0.28	62.50	....	124.99	1.14	83.34	41.67	125.00
150	1 52.5	0.41	75.00	.01	149.98	1.64	100.01	50.01	149.99
200	2 30.0	0.73	99.99	.01	199.96	2.91	133.35	66.68	199.98
250	3 07.5	1.13	124.99	.01	249.93	4.54	166.69	83.36	249.97
300	3 45.0	1.63	149.98	.01	299.87	6.54	200.04	100.04	299.94
350	4 22.5	2.22	174.97	.01	349.80	8.91	233.40	116.73	349.91
400	5 00.0	2.91	199.95	.02	399.70	11.63	266.77	133.43	399.86
450	5 37.5	3.68	224.93	.02	449.56	14.72	300.15	150.14	449.81
500	6 15.0	4.54	249.90	.02	499.40	18.16	333.54	166.86	499.73
550	6 52.5	5.49	274.87	.02	549.21	21.98	366.94	183.59	549.65
600	7 30.0	6.55	299.83	.02	598.97	26.15	400.36	200.33	599.54
700	8 45.0	8.90	349.73	.03	698.37	35.57	467.24	233.86	699.27
800	10 00.0	11.62	399.59	.03	797.57	46.44	534.18	267.45	798.92
				D=2° 45′					
100	1°22.5′	0.20	50.00	....	99.99	0.80	66.67	33.34	100.00
125	1 43.2	0.31	62.50	.01	124.99	1.25	83.34	41.67	125.00
150	2 03.8	0.45	75.00	.01	149.98	1.80	100.01	50.01	149.99
200	2 45.0	0.80	99.99	.01	199.95	3.20	133.35	66.68	199.98
250	3 26.3	1.25	124.98	.01	249.91	5.00	166.70	83.36	249.96
300	4 07.5	1.80	149.97	.01	299.85	7.20	200.06	100.05	299.93
350	4 48.8	2.45	174.96	.02	349.75	9.79	233.42	116.75	349.89
400	5 30.0	3.20	199.94	.02	399.63	12.79	266.80	133.45	399.84
450	6 11.3	4.05	224.91	.02	449.47	16.19	300.19	150.17	449.77
500	6 52.5	5.00	249.88	.02	499.28	19.98	333.58	166.90	499.68
550	7 33.8	6.04	274.84	.03	549.03	24.17	367.00	183.64	549.58
600	8 15.0	7.20	299.79	.03	598.76	28.76	400.43	200.40	599.45
700	9 37.5	9.80	349.67	.03	698.02	39.12	467.36	233.96	699.13
				D=3°					
100	1°30.0′	0.22	50.00	.01	99.99	0.87	66.67	33.34	100.00
125	1 52.5	0.34	62.50	.01	124.99	1.36	83.34	41.67	124.99
150	2 15.0	0.49	75.00	.01	149.98	1.96	100.01	50.01	149.99
200	3 00.0	0.87	99.99	.01	199.95	3.49	133.35	66.68	199.98
250	3 45.0	1.36	124.98	.01	249.89	5.45	166.70	83.37	249.95
300	4 30.0	1.96	149.97	.02	299.81	7.85	200.06	100.06	299.92
350	5 15.0	2.67	174.95	.02	349.71	10.68	233.44	116.76	349.87
400	6 00.0	3.49	199.93	.02	399.56	13.95	266.82	133.47	399.80
450	6 45.0	4.42	224.89	.03	449.37	17.65	300.22	150.20	449.72
500	7 30.0	5.46	249.86	.03	499.14	21.79	333.63	166.94	499.62
550	8 15.0	6.60	274.81	.03	548.86	26.36	367.06	183.70	549.49
600	9 00.0	7.85	299.75	.03	598.52	31.36	400.52	200.47	599.34
700	10 30.0	10.68	349.61	.04	697.66	42.66	467.49	234.08	698.96

* Subtract from tabulated value for D_a when chord definition of D is used.

TABLE XI.—SELECTED SPIRALS

D = 3° 30′

L_s	Δ	o	X_0 D_a	X_0 *	X	Y	L.T.	S.T.	L.C.
100	1°45.0′	0.25	50.00	.01	99.99	1.02	66.67	33.34	100.00
125	2 11.3	0.40	62.50	.01	124.98	1.59	83.34	41.67	124.99
150	2 37.5	0.57	74.99	.01	149.97	2.29	100.01	50.01	149.99
200	3 30.0	1.02	99.99	.02	199.93	4.07	133.36	66.69	199.97
250	4 22.5	1.59	124.98	.02	249.85	6.36	166.72	83.38	249.94
300	5 15.0	2.29	149.96	.02	299.75	9.16	200.09	100.08	299.89
350	6 07.5	3.12	174.93	.03	349.60	12.46	233.47	116.97	349.82
400	7 00.0	4.07	199.90	.03	399.40	16.27	266.88	133.52	399.73
450	7 52.5	5.15	224.86	.03	449.15	20.59	300.30	150.27	449.63
500	8 45.0	6.36	249.80	.04	498.84	25.41	333.74	167.04	499.48
550	9 37.5	7.69	274.74	.04	548.45	30.73	367.21	183.83	549.31
600	10 30.0	9.16	299.66	.05	597.99	36.56	400.71	200.64	599.10
700	12 15.0	12.45	349.46	.05	696.80	49.72	467.79	234.35	698.58

D = 4°

L_s	Δ	o D_a	o *	X_0 D_a	X_0 *	X	Y	L.T.	S.T.	L.C.
100	2°00.0′	0.29	....	50.00	.01	99.99	1.16	66.67	33.34	99.99
125	2 30.0	0.45	....	62.50	.01	124.98	1.82	83.34	41.67	124.99
150	3 00.0	0.65	....	74.99	.02	149.96	2.62	100.01	50.01	149.98
200	4 00.0	1.16	....	99.98	.02	199.90	4.65	133.37	66.70	199.96
250	5 00.0	1.82	....	124.97	.03	249.81	7.27	166.73	83.39	249.92
300	6 00.0	2.62	....	149.95	.03	299.67	10.46	200.11	100.10	299.85
350	7 00.0	3.56	....	174.91	.04	349.48	14.24	233.52	116.83	349.77
400	8 00.0	4.65	....	199.87	.04	399.22	18.59	266.94	133.58	399.65
450	9 00.0	5.89	....	224.82	.05	448.89	23.52	300.39	150.35	449.50
500	10 00.0	7.26	....	249.74	.05	498.48	29.02	333.87	167.15	499.32
550	11 00.0	8.79	.01	274.66	.06	547.98	35.11	367.38	183.98	549.10
600	12 00.0	10.46	.01	299.56	.06	597.37	41.75	400.92	200.84	598.83
650	13 00.0	12.27	.01	324.44	.07	646.66	48.98	434.51	217.74	648.51

D = 4° 30′

L_s	Δ	o D_a	o *	X_0 D_a	X_0 *	X	Y	L.T.	S.T.	L.C.
100	2°15.0′	0.33	....	50.00	.01	99.98	1.31	66.67	33.34	99.99
125	2 48.8	0.51	....	62.50	.02	124.97	2.04	83.34	41.67	124.99
150	3 22.5	0.74	....	74.99	.02	149.95	2.94	100.02	50.02	149.98
200	4 30.0	1.31	....	99.98	.03	199.88	5.23	133.38	66.71	199.95
250	5 37.5	2.04	....	124.96	.03	249.76	8.18	166.75	83.41	249.89
300	6 45.0	2.94	....	149.93	.04	299.58	11.77	200.15	100.13	299.81
350	7 52.5	4.01	....	174.89	.04	349.34	16.02	233.56	116.88	349.71
400	9 00.0	5.23	....	199.84	.05	399.02	20.91	267.01	133.65	399.56
450	10 07.5	6.62	.01	224.77	.06	448.60	26.45	300.49	150.45	449.37
500	11 15.0	8.17	.01	249.68	.06	498.08	32.64	334.01	167.28	499.14
550	12 22.5	9.88	.01	274.57	.07	547.44	39.47	367.57	184.15	548.86
600	13 30.0	11.76	.01	299.45	.08	596.68	46.94	401.17	201.07	598.52
650	14 37.5	13.79	.01	324.30	.08	645.78	55.05	434.82	218.02	648.12

* Subtract from tabulated value for D_a when chord definition of D is used.

TABLE XI.—SELECTED SPIRALS

		o		X_0						
L_s	Δ	D_a	*	D_a	*	X	Y	L.T.	S.T.	L.C.
D = 5°										
100	2°30.0′	0.36	....	50.00	.02	99.98	1.45	66.67	33.34	99.99
125	3 07.5	0.57	....	62.49	.02	124.96	2.27	83.35	41.68	124.98
150	3 45.0	0.82	....	74.99	.02	149.94	3.27	100.02	50.02	149.97
200	5 00.0	1.45	....	99.97	.03	199.85	5.81	133.39	66.72	199.93
250	6 15.0	2.27	....	124.95	.04	249.70	9.08	166.77	83.43	249.87
300	7 30.0	3.27	....	149.91	.05	299.49	13.07	200.18	100.16	299.77
350	8 45.0	4.45	....	174.86	.06	349.18	17.79	233.62	116.93	349.64
400	10 00.0	5.81	.01	199.80	.06	398.78	23.22	267.09	133.72	399.46
450	11 15.0	7.36	.01	224.71	.07	448.27	29.37	300.61	150.55	449.23
500	12 30.0	9.08	.01	249.60	.08	497.62	36.24	334.17	167.43	498.94
550	13 45.0	10.98	.01	274.47	.09	546.84	43.82	367.78	184.36	548.59
600	15 00.0	13.06	.01	299.32	.09	595.90	52.10	401.45	201.32	598.17
D = 5° 30′										
100	2°45.0′	0.40	....	50.00	.02	99.98	1.60	66.67	33.34	99.99
125	3 26.3	0.62	....	62.49	.02	124.96	2.50	83.35	41.68	124.98
150	4 07.5	0.90	....	74.99	.03	149.92	3.60	100.03	50.02	149.97
200	5 30.0	1.60	....	99.97	.04	199.82	6.40	133.40	66.73	199.92
250	6 52.5	2.50	....	124.94	.05	249.64	9.99	166.79	83.45	249.84
300	8 15.0	3.60	....	149.90	.06	299.38	14.38	200.22	100.20	299.72
350	9 37.5	4.90	.01	174.84	.07	349.01	19.56	233.68	116.98	349.56
400	11 00.0	6.39	.01	199.76	.08	398.53	25.53	267.18	133.80	399.34
450	12 22.5	8.09	.01	224.65	.09	447.90	32.29	300.74	150.67	449.07
500	13 45.0	9.98	.01	249.52	.10	497.13	39.83	334.35	167.59	498.72
550	15 07.5	12.07	.01	274.36	.10	546.16	48.15	368.02	184.56	548.30
600	16 30.0	14.36	.02	299.17	.11	595.04	57.26	401.75	201.59	597.79
D = 6°										
100	3°00.0′	0.44	....	50.00	.02	99.97	1.74	66.68	33.34	99.99
125	3 45.0	0.68	....	62.49	.03	124.95	2.73	83.35	41.68	124.98
150	4 30.0	0.98	....	74.98	.03	149.91	3.93	100.03	50.03	149.96
200	6 00.0	1.74	....	99.96	.05	199.78	6.98	133.41	66.74	199.90
250	7 30.0	2.73	....	124.93	.06	249.57	10.90	166.82	83.47	249.81
300	9 00.0	3.92	.01	149.88	.07	299.26	15.68	200.26	100.24	299.67
350	10 30.0	5.34	.01	174.80	.08	348.83	21.33	233.75	117.04	349.48
400	12 00.0	6.97	.01	199.71	.09	398.25	27.84	267.28	133.89	399.22
450	13 30.0	8.82	.01	224.59	.10	447.51	35.20	300.88	150.80	448.89
500	15 00.0	10.88	.01	249.43	.11	496.58	43.42	334.54	167.76	498.48
D = 6° 30′										
100	3°15.0′	0.47	....	49.99	.03	99.97	1.89	66.68	33.34	99.99
125	4 03.8	0.74	....	62.49	.03	124.94	2.95	83.36	41.69	124.97
150	4 52.5	1.06	....	74.98	.04	149.89	4.25	100.04	50.04	149.95
200	6 30.0	1.89	....	99.96	.05	199.74	7.56	133.42	66.75	199.89
250	8 07.5	2.95	....	124.91	.07	249.50	11.80	166.84	83.49	249.78
300	9 45.0	4.25	.01	149.85	.08	299.13	16.98	200.30	100.28	299.61
350	11 22.5	5.78	.01	174.77	.09	348.62	23.10	233.82	117.11	349.39
400	13 00.0	7.55	.01	199.66	.11	397.94	30.14	267.39	133.99	399.08
450	14 37.5	9.55	.02	224.51	.12	447.08	38.11	301.03	150.94	448.70
500	16 15.0	11.78	.02	249.33	.13	495.99	47.00	334.75	167.96	498.22

* Subtract from tabulated value for D_a when chord definition of D is used.

TABLE XI.—SELECTED SPIRALS

L_s	Δ	o		X_0		X	Y	L.T.	S.T.	L.C.
		D_a	*	D_a	*					
					D = 7°					
100	3°30.0′	0.51	....	49.99	.03	99.96	2.04	66.68	33.35	99.98
125	4 22.5	0.79	....	62.49	.04	124.93	3.18	83.36	41.69	124.97
150	5 15.0	1.15	....	74.98	.05	149.87	4.58	100.04	50.04	149.94
200	7 00.0	2.04	....	99.95	.06	199.70	8.14	133.44	66.76	199.87
250	8 45.0	3.18	.01	124.90	.08	249.42	12.70	166.87	83.52	249.74
300	10 30.0	4.58	.01	149.83	.09	298.99	18.28	200.35	100.32	299.55
350	12 15.0	6.23	.01	174.73	.11	348.40	24.86	233.89	117.18	349.29
400	14 00.0	8.13	.02	199.60	.12	397.62	32.44	267.51	134.10	398.94
450	15 45.0	10.28	.02	224.43	.14	446.61	41.01	301.20	151.09	448.49
500	17 30.0	12.68	.02	249.22	.15	495.36	50.56	334.98	168.16	497.93
					D = 8°					
100	4°00.0′	0.58	....	49.99	.04	99.95	2.33	66.68	33.35	99.98
125	5 00.0	0.91	....	62.48	.05	124.90	3.63	83.37	41.70	124.96
150	6 00.0	1.31	....	74.97	.06	149.83	5.23	100.06	50.05	149.93
200	8 00.0	2.33	.01	99.93	.08	199.61	9.30	133.47	66.79	199.83
250	10 00.0	3.63	.01	124.87	.10	249.24	14.51	166.93	83.58	249.66
300	12 00.0	5.23	.01	149.78	.12	298.69	20.88	200.46	100.42	299.42
350	14 00.0	7.11	.02	174.65	.14	347.92	28.38	234.07	117.33	349.07
400	16 00.0	9.28	.02	199.48	.16	396.89	37.03	267.76	134.33	398.62
450	18 00.0	11.74	.03	224.26	.18	445.58	46.79	301.57	151.42	448.03
500	20 00.0	14.48	.04	248.99	.20	493.94	57.68	335.49	168.63	497.30
					D = 9°					
100	4°30.0′	0.65	....	49.99	.05	99.94	2.62	66.69	33.35	99.97
125	5 37.5	1.02	....	62.48	.06	124.88	4.09	83.38	41.71	124.95
150	6 45.0	1.47	....	74.96	.08	149.79	5.88	100.07	50.07	149.91
200	9 00.0	2.62	.01	99.92	.10	199.51	10.45	133.51	66.82	199.78
250	11 15.0	4.09	.01	124.84	.13	249.04	16.32	167.00	83.64	249.57
300	13 30.0	5.88	.02	149.72	.15	298.34	23.47	200.58	100.53	299.26
350	15 45.0	7.99	.02	174.56	.18	347.36	31.90	234.26	117.51	348.83
400	18 00.0	10.43	.03	199.34	.20	396.07	41.59	268.06	134.60	398.25
450	20 15.0	13.19	.04	224.07	.23	444.41	52.54	301.99	151.81	447.51
500	22 30.0	16.27	.05	248.72	.25	492.34	64.73	336.07	169.15	496.58
					D = 10°					
100	5°00.0′	0.73	....	49.99	.06	99.92	2.91	66.69	33.36	99.97
125	6 15.0	1.14	....	62.48	.08	124.85	4.54	83.39	41.71	124.93
150	7 30.0	1.64	.01	74.96	.10	149.74	6.54	100.09	50.08	149.89
200	10 00.0	2.91	.01	99.90	.13	199.39	11.61	133.55	66.86	199.73
250	12 30.0	4.54	.02	124.80	.16	248.81	18.12	167.08	83.71	249.47
300	15 00.0	6.53	.02	149.66	.19	297.95	26.05	200.72	100.66	299.09
350	17 30.0	8.88	.03	174.46	.22	346.75	35.40	234.48	117.71	348.55
400	20 00.0	11.58	.04	199.19	.25	395.15	46.14	268.39	134.90	397.84
450	22 30.0	14.64	.06	223.85	.28	443.11	58.26	302.46	152.24	446.92
500	25 00.0	18.06	.07	248.42	.31	490.56	71.74	336.72	169.75	495.78

* Subtract from tabulated value for D_a when chord definition of D is used.

TABLE XI.—SELECTED SPIRALS

L_s	Δ	o		X_0		X	Y	L.T.	S.T.	L.C.
		D_a	*	D_a	*					
D = 11°										
80	4°24.0′	0.51	....	39.99	.06	79.95	2.05	53.35	26.68	79.98
100	5 30.0	0.80	....	49.98	.08	99.91	3.20	66.70	33.36	99.96
125	6 52.5	1.25	.01	62.47	.10	124.82	5.00	83.40	41.72	124.92
150	8 15.0	1.80	.01	74.95	.12	149.69	7.19	100.11	50.10	149.86
200	11 00.0	3.20	.01	99.88	.15	199.26	12.77	133.59	66.90	199.67
250	13 45.0	4.99	.02	124.76	.19	248.56	19.92	167.17	83.79	249.36
300	16 30.0	7.18	.03	149.59	.23	297.52	28.63	200.88	100.80	298.90
350	19 15.0	9.76	.04	174.34	.26	346.07	38.88	234.73	117.94	348.25
400	22 00.0	12.73	.06	199.02	.30	394.14	50.66	268.76	135.23	397.39
500	27 30.0	19.84	.09	248.09	.37	488.60	78.69	337.45	170.41	494.90
D = 12°										
80	4°48.0′	0.56	....	39.99	.07	79.94	2.23	53.35	26.68	79.98
100	6 00.0	0.87	....	49.98	.09	99.89	3.49	66.70	33.37	99.95
125	7 30.0	1.36	.01	62.46	.11	124.79	5.45	83.41	41.74	124.90
150	9 00.0	1.96	.01	74.94	.14	149.63	7.84	100.13	50.12	149.84
200	12 00.0	3.49	.02	99.85	.18	199.12	13.92	133.64	66.95	199.61
250	15 00.0	5.44	.03	124.72	.23	248.29	21.71	167.27	83.88	249.24
300	18 00.0	7.82	.04	149.51	.27	297.05	31.19	201.04	100.95	298.69
350	21 00.0	10.64	.06	174.22	.31	345.33	42.35	235.00	118.18	347.91
400	24 00.0	13.88	.08	198.84	.36	393.04	55.16	269.16	135.60	396.89
500	30 00.0	21.60	.12	247.73	.44	486.46	85.57	338.25	171.14	493.93
D = 13°										
80	5°12.0′	0.60	....	39.99	.09	79.93	2.42	53.36	26.69	79.97
100	6 30.0	0.94	.01	49.98	.11	99.87	3.78	66.71	33.37	99.94
125	8 07.5	1.48	.01	62.45	.13	124.75	5.90	83.42	41.75	124.89
150	9 45.0	2.13	.01	74.93	.16	149.57	8.49	100.15	50.14	149.81
200	13 00.0	3.77	.02	99.83	.21	198.97	15.07	133.69	67.00	199.54
250	16 15.0	5.89	.04	124.67	.27	248.00	23.50	167.37	83.98	249.11
300	19 30.0	8.47	.05	149.42	.32	296.54	33.75	201.23	101.12	298.46
350	22 45.0	11.52	.07	174.09	.37	344.52	45.80	235.29	118.45	347.55
400	26 00.0	15.01	.10	198.63	.42	391.84	59.62	269.60	136.00	396.35
500	32 30.0	23.36	.15	247.34	.51	484.15	92.39	339.13	171.95	492.89
D = 14°										
80	5°36.0′	0.64	....	39.99	.10	79.93	2.56	53.36	26.69	79.97
100	7 00.0	1.02	.01	49.98	.12	99.85	4.07	66.72	33.38	99.93
125	8 45.0	1.59	.01	62.45	.16	124.71	6.35	83.44	41.76	124.87
150	10 30.0	2.29	.02	74.92	.19	149.50	9.14	100.18	50.16	149.78
200	14 00.0	4.06	.03	99.80	.25	198.81	16.22	133.75	67.05	199.47
250	17 30.0	6.34	.05	124.61	.31	247.68	25.28	167.49	84.08	248.96
300	21 00.0	9.12	.07	149.33	.37	296.00	36.30	201.43	101.30	298.21
350	24 30.0	12.39	.09	173.94	.42	343.65	49.24	235.61	118.74	347.16
400	28 00.0	16.15	.12	198.42	.48	390.55	64.06	270.08	136.44	395.77
500	35 00.0	25.12	.18	246.92	.59	481.66	99.13	340.09	172.83	491.76

* Subtract from tabulated value for D_a when chord definition of D is used.

TABLE XI.—SELECTED SPIRALS

L_s	Δ	o D_a	o *	X_0 D_a	X_0 *	X	Y	L.T.	S.T.	L.C.
					D = 15°					
80	6°00.0′	0.70	.01	39.99	.11	79.91	2.79	53.36	26.69	79.96
100	7 30.0	1.09	.01	49.97	.14	99.83	4.36	66.73	33.39	99.92
125	9 22.5	1.70	.01	62.44	.18	124.67	6.80	83.45	41.77	124.85
150	11 15.0	2.45	.02	74.90	.21	149.42	9.79	100.20	50.18	149.74
200	15 00.0	4.35	.04	99.77	.28	198.63	17.37	133.82	67.11	199.39
250	18 45.0	6.79	.06	124.56	.35	247.34	27.06	167.61	84.19	248.81
300	22 30.0	9.76	.08	149.23	.42	295.41	38.84	201.64	101.49	297.95
350	26 15.0	13.26	.11	173.78	.48	342.72	52.65	235.95	119.05	346.75
400	30 00.0	17.28	.15	198.18	.55	389.17	68.46	270.60	136.92	395.15
500	37 30.0	26.86	.22	246.48	.67	479.00	105.79	341.14	173.78	490.55
					D = 16°					
80	6°24.0′	0.74	.01	39.98	.13	79.90	2.98	53.37	26.70	79.96
100	8 00.0	1.16	.01	49.97	.16	99.80	4.65	66.74	33.40	99.91
125	10 00.0	1.82	.02	62.44	.20	124.62	7.26	83.47	41.79	124.83
150	12 00.0	2.61	.03	74.89	.24	149.34	10.44	100.23	50.21	149.71
200	16 00.0	4.64	.05	99.74	.32	198.45	18.51	133.88	67.17	199.31
250	20 00.0	7.24	.07	124.50	.40	246.97	28.84	167.74	84.31	248.65
300	24 00.0	10.41	.10	149.13	.48	294.78	41.37	201.87	101.70	297.67
350	28 00.0	14.13	.14	173.62	.55	341.73	56.05	236.32	119.39	346.30
400	32 00.0	18.41	.18	197.94	.62	387.70	72.82	271.16	137.43	394.48
500	40 00.0	28.59	.27	246.00	.75	476.18	112.36	342.26	174.81	489.25
					D = 18°					
80	7°12.0′	0.84	.01	39.98	.15	79.87	3.48	53.38	26.71	79.94
100	9 00.0	1.31	.02	49.96	.21	99.75	5.23	66.75	33.41	99.89
125	11 15.0	2.04	.03	62.42	.26	124.52	8.16	83.50	41.82	124.79
150	13 30.0	2.94	.04	74.86	.31	149.17	11.73	100.29	50.27	149.63
200	18 00.0	5.22	.06	99.67	.41	198.04	20.80	134.03	67.30	199.12
250	22 30.0	8.14	.10	124.36	.50	246.17	32.36	168.03	84.58	248.29
300	27 00.0	11.69	.14	148.90	.60	293.41	46.38	202.38	102.16	297.05
350	31 30.0	15.86	.19	173.25	.69	339.57	62.77	237.14	120.13	345.32
400	36 00.0	20.65	.25	197.40	.77	384.50	81.44	272.40	138.56	393.03
500	45 00.0	32.02	.38	244.95	.93	470.02	125.24	344.78	177.12	486.43
					D = 20°					
60	6°00.0′	0.52	.01	29.99	.15	59.93	2.09	40.02	20.02	59.97
80	8 00.0	0.93	.01	39.97	.20	79.84	3.72	53.39	26.72	79.93
100	10 00.0	1.45	.02	49.95	.25	99.70	5.80	66.77	33.43	99.86
125	12 30.0	2.27	.03	62.40	.31	124.41	9.06	83.54	41.86	124.74
150	15 00.0	3.26	.05	74.83	.38	148.98	13.03	100.36	50.33	149.54
200	20 00.0	5.79	.09	99.60	.50	197.58	23.07	134.19	67.45	198.92
250	25 00.0	9.03	.14	124.21	.62	245.28	35.87	168.36	84.87	247.89
300	30 00.0	12.96	.20	148.64	.73	291.88	51.34	202.95	102.69	296.36
350	35 00.0	17.58	.27	172.85	.84	337.16	69.39	238.06	120.98	344.23
400	40 00.0	22.87	.34	196.80	.94	380.94	89.89	273.81	139.85	391.40

* Subtract from tabulated value for D_a when chord definition of D is used.

TABLE XII.—SPIRAL FUNCTIONS FOR $L_s=1$

Δ°	o		X_0		X	Y	L.T.	S.T.	L.C.
	D_a	*	D_a	*					
0.0	.00000	.0000	.50000	.0000	1.00000	.00000	.66667	.33333	1.00000
.1	015	00	000	01	1.00000	058	667	333	1.00000
.2	029	00	000	03	1.00000	116	667	334	1.00000
.3	044	00	000	04	.99999	175	667	334	1.00000
.4	058	00	000	05	.99999	233	667	334	1.00000
0.5	.00073	.0000	.50000	.0006	.99999	.00291	.66667	.33334	1.00000
.6	088	00	000	08	999	349	667	334	1.00000
.7	102	00	000	09	998	407	668	334	.99999
.8	117	00	000	10	998	465	668	334	999
.9	131	00	000	11	997	524	668	334	999
1.0	.00146	.0000	.49999	.0013	.99997	.00582	.66668	.33334	.99999
.1	161	00	999	14	996	640	668	335	998
.2	175	00	999	15	995	698	668	335	998
.3	190	00	999	17	995	756	669	335	998
.4	204	00	999	18	994	814	669	335	997
1.5	.00219	.0000	.49999	.0019	.99993	.00873	.66669	.33336	.99997
.6	233	00	999	20	992	931	669	336	997
.7	248	00	998	22	991	989	670	336	996
.8	262	00	998	23	990	.01047	670	337	996
.9	277	00	998	24	989	105	671	337	995
2.0	.00291	.0000	.49998	.0025	.99988	.01163	.66671	.33337	.99995
.1	305	00	998	27	987	222	671	338	994
.2	320	00	997	28	985	280	672	338	993
.3	334	01	997	29	984	338	672	339	993
.4	349	01	997	31	982	396	673	339	992
2.5	.00363	.0001	.49997	.0032	.99981	.01454	.66673	.33339	.99992
.6	377	01	996	33	979	512	674	340	991
.7	392	01	996	34	978	571	675	340	990
.8	406	01	996	36	976	629	675	341	990
.9	421	01	996	37	975	687	676	341	989
3.0	.00435	.0001	.49995	.0038	.99973	.01745	.66676	.33342	.99988
.1	450	01	995	39	971	803	677	343	987
.2	464	01	994	41	969	861	678	343	986
.3	479	01	994	42	967	919	678	344	985
.4	493	01	994	43	965	978	679	345	984
3.5	.00508	.0001	.49994	.0045	.99963	.02036	.66680	.33345	.99983
.6	523	01	993	46	961	094	681	346	982
.7	537	01	993	47	958	152	681	347	981
.8	552	02	993	48	956	210	682	347	980
.9	566	02	992	50	953	268	683	348	979
4.0	.00581	.0002	.49992	.0051	.99951	.02326	.66684	.33349	.99978
.1	596	02	991	52	948	384	685	350	977
.2	610	02	991	53	946	443	686	350	976
.3	625	02	991	55	943	501	686	351	975
.4	639	02	990	56	941	559	687	352	974
4.5	.00654	.0002	.49990	.0057	.99938	.02617	.66688	.33353	.99973
.6	669	02	989	59	935	675	689	354	971
.7	683	02	989	60	932	733	690	355	970
.8	698	03	988	61	930	791	691	356	969
.9	712	03	988	62	927	849	692	357	967
5.0	.00727	.0003	.49987	.0064	.99924	.02907	.66693	.33358	.99966

* Multiply functions (except * values) by the given value of L_S. When chord definition of degree of curve (D_c) is used, correct the calculated values of o or X_0 by subtracting the product of D_c and the * values. See page 354 for example.

TABLE XII.—SPIRAL FUNCTIONS FOR $L_s = 1$

Δ°	o		X_o		X	Y	L.T.	S.T.	L.C.
	D_a	*	D_a	*					
5.0	.00727	.0003	.49987	.0064	.99924	.02907	.66693	.33358	.99966
.1	742	03	987	65	921	965	694	359	965
.2	756	03	986	66	918	.03023	696	360	963
.3	771	03	986	67	914	082	697	361	962
.4	785	03	985	69	911	140	698	362	961
5.5	.00800	.0003	.49985	.0070	.99908	.03198	.66699	.33363	.99959
.6	814	03	984	71	904	256	700	364	958
.7	829	04	984	73	901	314	701	365	956
.8	843	04	983	74	897	372	703	366	954
.9	858	04	983	75	894	430	704	367	953
6.0	.00872	.0004	.49982	.0076	.99890	.03488	.66705	.33368	.99951
.1	887	04	981	78	886	546	706	369	950
.2	901	04	981	79	882	604	708	371	948
.3	916	04	980	80	879	662	709	372	946
.4	930	05	979	81	875	720	710	373	944
6.5	.00945	.0005	.49979	.0083	.99871	.03778	.66712	.33374	.99943
.6	960	05	978	84	867	836	713	376	941
.7	974	05	977	85	863	894	715	377	939
.8	989	05	976	86	859	952	716	378	937
.9	.01003	05	976	88	855	.04010	717	380	936
7.0	.01018	.0005	.49975	.0089	.99851	.04068	.66719	.33381	.99934
.1	033	06	974	90	846	126	720	382	932
.2	047	06	973	91	842	184	722	384	930
.3	062	06	973	93	838	242	724	385	928
.4	076	06	972	94	833	300	725	386	926
7.5	.01091	.0006	.49971	.0095	.99829	.04358	.66727	.33388	.99924
.6	105	06	970	97	824	416	728	389	922
.7	120	07	969	98	819	474	730	391	920
.8	134	07	969	99	815	532	732	392	918
.9	149	07	968	.0100	810	590	733	394	916
8.0	.01163	.0007	.49967	.0102	.99805	.04648	.66735	.33395	.99913
.1	178	07	966	03	800	706	737	397	911
.2	192	07	965	04	795	764	738	399	909
.3	207	08	965	05	790	822	740	400	907
.4	221	08	964	07	785	879	742	402	904
8.5	.01236	.0008	.49963	.0108	.99780	.04937	.66744	.33403	.99902
.6	250	08	962	09	775	995	745	405	900
.7	265	08	961	10	770	.05053	747	407	897
.8	279	09	961	12	764	111	749	409	895
.9	294	09	960	13	759	169	751	410	893
9.0	.01308	.0009	.49959	.0114	.99754	.05227	.66753	.33412	.99890
.1	323	09	958	16	748	285	755	414	888
.2	337	09	957	17	742	342	757	416	885
.3	352	10	956	18	737	400	759	417	883
.4	366	10	955	19	731	458	761	419	880
9.5	.01381	.0010	.49954	.0120	.99725	.05516	.66763	.33421	.99878
.6	395	10	953	22	719	574	765	423	875
.7	410	10	952	23	713	632	767	425	873
.8	424	11	951	24	708	690	769	427	870
.9	439	11	950	26	702	747	771	428	867
10.0	.01453	.0011	.49949	.0127	.99696	.05805	.66773	.33430	.99865

* Multiply functions (except * values) by the given value of L_S. When chord definition of degree of curve (D_c) is used, correct the calculated values of o or X_0 by subtracting the product of D_c and the * values. See page 354 for example.

TABLE XII.—SPIRAL FUNCTIONS FOR $L_s=1$

Δ°	o		X_0		X	Y	L.T.	S.T.	L.C.
	D_a	*	D_a	*					
10.0	.01453	.0011	.49949	.0127	.99696	.05805	.66773	.33430	.99865
.1	468	11	948	28	690	863	776	432	862
.2	482	12	947	29	684	921	778	434	859
.3	497	12	946	31	677	978	780	436	856
.4	511	12	945	32	671	.06036	782	438	854
10.5	.01526	.0012	.49944	.0133	.99665	.06094	.66784	.33440	.99851
.6	540	12	943	34	658	152	787	442	848
.7	555	13	942	36	652	210	789	444	845
.8	569	13	941	37	645	267	791	447	842
.9	584	13	940	38	639	325	794	449	839
11.0	.01598	.0013	.49939	.0139	.99632	.06383	.66796	.33451	.99836
.1	613	14	938	41	625	440	798	453	833
.2	627	14	937	42	618	498	801	455	830
.3	642	14	935	43	612	556	803	457	827
.4	656	14	934	44	605	614	806	460	824
11.5	.01671	.0015	.49933	.0146	.99598	.06671	.66808	.33462	.99821
.6	685	15	932	47	591	729	811	464	818
.7	700	15	931	48	584	787	813	466	815
.8	714	15	929	49	576	844	816	469	812
.9	729	16	928	51	569	902	818	471	808
12.0	.01743	.0016	.49927	.0152	.99562	.06959	.66821	.33473	.99805
.1	757	16	926	53	555	.07017	823	476	802
.2	772	16	924	54	547	075	826	478	799
.3	786	17	923	56	540	132	828	480	795
.4	801	17	922	57	532	190	831	483	792
12.5	.01815	.0017	.49921	.0158	.99525	.07248	.66834	.33485	.99789
.6	829	18	919	59	517	305	836	488	785
.7	844	18	918	60	509	363	839	490	782
.8	858	18	917	62	502	420	842	493	778
.9	873	18	915	63	494	478	845	495	775
13.0	.01887	.0019	.49914	.0164	.99486	.07535	.66847	.33498	.99771
.1	902	19	913	65	478	593	850	500	768
.2	916	19	911	67	470	650	853	503	764
.3	931	20	910	68	462	708	856	505	761
.4	945	20	909	69	454	765	859	508	757
13.5	.01960	.0020	.49908	.0170	.99446	.07823	.66862	.33511	.99753
.6	974	20	906	72	438	880	865	513	750
.7	989	21	905	73	430	938	868	516	746
.8	.02003	21	904	74	421	995	871	519	742
.9	018	21	902	75	413	.08053	874	521	739
14.0	.02032	.0022	.49901	.0177	.99405	.08110	.66877	.33524	.99735
.1	046	22	900	78	396	168	880	527	731
.2	061	22	898	79	387	225	883	530	727
.3	075	23	897	80	379	282	886	532	723
.4	090	23	895	82	370	340	889	535	720
14.5	.02104	.0023	.49894	.0183	.99362	.08397	.66892	.33538	.99716
.6	118	24	892	84	353	455	895	541	712
.7	133	24	891	85	344	512	898	544	708
.8	147	24	889	86	335	569	901	547	704
.9	162	25	888	88	326	627	904	550	700
15.0	.02176	.0025	.49886	.0189	.99317	.08684	.66908	.33553	.99696

* Multiply functions (except * values) by the given value of L_S. When chord definition of degree of curve (D_c) is used, correct the calculated values of *o* or X_0 by subtracting the product of D_c and the * values. See page 354 for example.

TABLE XII.—SPIRAL FUNCTIONS FOR $L_s=1$

Δ°	o		X_o		X	Y	L.T.	S.T.	L.C.
	D_a	*	D_a	*					
15.0	.02176	.0025	.49886	.0189	.99317	.08684	.66908	.33553	.99696
.1	190	25	884	90	308	741	911	556	692
.2	205	26	883	91	299	799	914	559	688
.3	219	26	881	93	289	856	918	561	683
.4	234	26	880	94	280	913	921	564	679
15.5	.02248	.0026	.49878	.0195	.99271	.08970	.66924	.33567	.99675
.6	262	27	876	96	261	.09028	928	571	671
.7	277	27	875	98	252	085	931	574	667
.8	291	28	873	99	242	142	934	577	662
.9	306	28	872	.0200	233	200	938	580	658
16.0	.02320	.0028	.49870	.0201	.99223	.09257	.66941	.33583	.99654
.1	335	29	868	02	213	314	945	586	649
.2	349	29	867	04	203	371	948	589	645
.3	364	29	865	05	194	428	952	593	641
.4	378	30	864	06	184	485	955	596	636
16.5	.02393	.0030	.49862	.0207	.99174	.09543	.66959	.33599	.99632
.6	407	30	860	09	164	600	962	602	627
.7	422	31	859	10	154	657	966	606	623
.8	436	31	857	1	143	714	970	609	618
.9	451	32	856	12	133	771	973	612	614
17.0	.02465	.0032	.49854	.0213	.99123	.09828	.66977	.33615	.99609
.1	479	32	852	15	113	885	981	619	605
.2	494	33	850	16	102	942	984	622	600
.3	508	33	849	17	092	999	988	626	595
.4	522	33	847	18	081	.10056	992	629	591
17.5	.02537	.0034	.49845	.0220	.99071	.10113	.66995	.33632	.99586
.6	551	34	843	21	060	170	999	636	581
.7	565	35	841	22	050	227	.67003	639	576
.8	579	35	840	23	039	284	007	643	572
.9	594	35	838	24	029	341	011	646	567
18.0	.02608	.0036	.49836	.0226	.99018	.10398	.67015	.33650	.99562
.1	622	36	834	27	007	455	019	654	557
.2	637	37	832	28	.98996	512	023	657	552
.3	651	37	830	29	985	569	027	661	547
.4	666	37	828	30	974	626	031	664	542
18.5	.02680	.0038	.49827	.0232	.98962	.10683	.67035	.33668	.99537
.6	694	38	825	33	951	740	039	672	532
.7	709	39	823	34	940	797	043	675	527
.8	723	39	821	35	929	854	047	679	522
.9	738	39	819	36	917	910	051	683	517
19.0	.02752	.0040	.49817	.0238	.98906	.10967	.67055	.33687	.99512
.1	766	40	815	39	894	.11024	059	690	507
.2	781	41	813	40	883	081	063	694	502
.3	795	41	811	41	871	138	067	698	497
.4	810	41	809	42	860	194	072	702	491
19.5	.02824	.0042	.49808	.0244	.98848	.11251	.67076	.33706	.99486
.6	838	42	806	45	836	308	080	709	481
.7	853	43	804	46	824	364	084	713	476
.8	867	43	802	47	812	421	089	717	470
.9	882	44	800	48	800	478	093	721	465
20.0	.02896	.0044	.49798	.0250	.98788	.11535	.67097	.33725	.99460

* Multiply functions (except * values) by the given value of L_S. When chord definition of degree of curve (D_c) is used, correct the calculated values of o or X_0 by subtracting the product of D_c and the * values. See page 354 for example.

TABLE XII.—SPIRAL FUNCTIONS FOR $L_s = 1$

Δ°	o		X_o		X	Y	L.T.	S.T.	L.C.
	D_a	*	D_a	*					
20.0	.02896	.0044	.49798	.0250	.98788	.11535	.67097	.33725	.99460
.1	910	44	796	51	776	591	102	729	454
.2	925	45	794	52	764	648	106	733	449
.3	939	45	791	53	752	705	111	737	443
.4	954	46	789	54	740	761	115	741	438
20.5	.02968	.0046	.49787	.0256	.98728	.11818	.67119	.33745	.99432
.6	982	47	785	57	715	875	124	749	427
.7	997	47	783	58	703	931	128	753	421
.8	.03011	48	781	59	690	988	133	758	415
.9	026	48	779	60	678	.12044	137	762	410
21.0	.03040	.0048	.49777	.0262	.98665	.12101	.67142	.33766	.99404
.1	054	49	775	63	652	157	147	770	399
.2	068	49	773	64	639	214	151	774	393
.3	083	50	770	65	627	270	156	779	387
.4	097	50	768	66	614	327	161	783	381
21.5	.03111	.0051	.49766	.0267	.98601	.12383	.67165	.33787	.99376
.6	125	51	764	68	588	439	170	791	370
.7	140	52	762	70	575	496	175	796	364
.8	154	52	759	71	562	552	180	800	358
.9	169	53	757	72	549	609	184	804	352
22.0	.03183	.0053	.49755	.0273	.98536	.12665	.67189	.33809	.99346
.1	197	54	753	75	523	721	194	813	340
.2	211	54	751	76	509	777	199	818	334
.3	226	55	749	77	496	834	204	822	328
.4	240	55	747	78	482	890	208	826	322
22.5	.03254	.0056	.49745	.0279	.98469	.12946	.67213	.33831	.99316
.6	268	56	743	81	455	.13002	218	835	310
.7	283	57	740	82	442	059	223	840	304
.8	297	57	738	83	428	115	228	844	298
.9	312	58	735	84	415	172	233	849	292
23.0	.03326	.0058	.49733	.0285	.98401	.13228	.67238	.33854	.99286
.1	340	59	731	86	387	284	243	858	279
.2	354	59	728	88	373	340	248	863	273
.3	369	60	726	89	359	396	254	868	267
.4	383	60	723	90	345	452	259	872	261
23.5	.03397	.0061	.49721	.0291	.98331	.13508	.67264	.33877	.99254
.6	411	61	719	92	316	564	269	882	248
.7	426	62	716	93	302	621	274	886	242
.8	440	62	714	95	288	677	280	891	235
.9	455	63	711	96	274	733	285	896	229
24.0	.03469	.0063	.49709	.0297	.98260	.13789	.67290	.33901	.99222
.1	483	64	707	98	245	845	295	906	216
.2	497	64	704	99	231	901	301	910	209
.3	512	65	702	.0300	216	957	306	915	203
.4	526	65	699	02	202	.14012	311	920	196
24.5	.03540	.0066	.49697	.0303	.98187	.14068	.67317	.33925	.99190
.6	554	66	694	04	172	124	322	930	183
.7	568	67	692	05	157	180	328	935	177
.8	583	67	689	06	143	236	333	940	170
.9	597	68	687	07	128	292	339	945	163
25.0	.03611	.0068	.49684	.0309	.98113	.14348	.67344	.33950	.99157

* Multiply functions (except * values) by the given value of L_S. When chord definition of degree of curve (D_c) is used, correct the calculated values of o or X_0 by subtracting the product of D_c and the * values. See page 354 for example.

TABLE XII.—SPIRAL FUNCTIONS FOR $L_S=1$

$\Delta°$	o		X_o		X	Y	L.T.	S.T.	L.C.
	D_a	*	D_a	*					
25.0	.03611	.0068	.49684	.0309	.98113	.14348	.67344	.33950	.99157
.1	625	69	681	10	098	404	350	955	150
.2	640	69	679	11	083	459	355	960	143
.3	654	70	676	12	068	515	361	965	136
.4	669	71	674	13	053	571	366	970	129
25.5	.03683	.0071	.49671	.0314	.98037	.14627	.67372	.33975	.99123
.6	697	72	668	15	022	682	378	981	116
.7	711	72	666	17	007	738	383	986	109
.8	725	73	663	18	.97991	794	389	991	102
.9	739	73	661	19	976	849	395	996	095
26.0	.03753	.0074	.49658	.0320	.97960	.14905	.67400	.34001	.99088
.1	767	74	656	21	945	961	406	007	081
.2	782	75	653	22	929	.15016	412	012	074
.3	796	76	651	23	913	072	418	017	067
.4	811	76	648	25	898	128	424	023	060
26.5	.03825	.0077	.49646	.0326	.97882	.15183	.67430	.34028	.99053
.6	839	77	643	27	866	239	435	033	046
.7	853	78	640	28	850	294	441	039	038
.8	868	78	638	29	834	350	447	044	031
.9	882	79	635	30	818	405	453	049	024
27.0	.03896	.0080	.49632	.0331	.97802	.15461	.67459	.34055	.99017
.1	910	80	629	33	786	516	465	060	009
.2	924	81	626	34	770	571	471	066	002
.3	939	81	624	35	753	627	477	071	.98995
.4	953	82	621	36	737	682	483	077	987
27.5	.03967	.0082	.49618	.0337	.97721	.15738	.67490	.34083	.98980
.6	981	83	615	38	704	793	496	088	973
.7	995	84	613	39	688	848	502	094	965
.8	.04009	84	610	40	671	903	508	099	958
.9	023	85	608	42	655	959	514	105	950
28.0	.04037	.0085	.49605	.0343	.97638	.16014	.67520	.34111	.98943
.1	051	86	602	44	621	069	527	116	935
.2	065	87	599	45	604	124	533	122	928
.3	080	87	596	46	588	180	539	128	920
.4	094	88	593	47	571	235	546	134	913
28.5	.04108	.0088	.49590	.0348	.97554	.16290	.67552	.34139	.98905
.6	122	89	587	49	537	345	558	145	897
.7	136	90	584	51	520	400	565	151	890
.8	151	90	582	52	503	455	571	157	882
.9	165	91	579	53	486	510	578	163	874
29.0	.04179	.0092	.49576	.0354	.97469	.16565	.67584	.34169	.98866
.1	193	92	573	55	452	620	591	165	859
.2	207	93	570	56	434	675	597	171	851
.3	222	93	567	57	417	730	604	177	843
.4	236	94	564	58	399	785	610	193	835
29.5	.04250	.0095	.49561	.0359	.97382	.16840	.67617	.34199	.98827
.6	264	95	558	61	364	895	623	205	819
.7	278	96	555	62	346	950	630	211	811
.8	293	97	552	63	329	.17005	637	217	803
.9	307	97	549	64	311	060	643	223	795
30.0	.04321	.0098	.49546	.0365	.97293	.17114	.67650	.34229	.98787

* Multiply functions (except * values) by the given value of L_S. When chord definition of degree of curve (D_c) is used, correct the calculated values of *o* or X_0 by subtracting the product of D_c and the * values. See page 354 for example.

TABLE XII.—SPIRAL FUNCTIONS FOR $L_s = 1$

$\Delta°$	o		X_0		X	Y	L.T.	S.T.	L.C.
	D_a	*	D_a	*					
30.0	.04321	.0098	.49546	.0365	.97293	.17114	.67650	.34229	.98787
.1	335	98	543	66	275	169	657	235	779
.2	349	99	540	67	257	224	664	241	771
.3	363	.0100	537	68	239	279	670	248	763
.4	377	00	534	69	221	333	677	254	755
30.5	.04391	.0101	.49531	.0371	.97203	.17388	.67684	.34260	.98746
.6	405	02	528	72	185	443	691	266	738
.7	419	02	525	73	167	498	698	273	730
.8	434	03	522	74	148	552	705	279	722
.9	448	04	519	75	130	607	712	285	713
31.0	.04462	.0104	.49516	.0376	.97112	.17661	.67719	.34292	.98705
.1	476	05	513	77	094	716	726	298	697
.2	490	06	510	78	075	770	733	304	688
.3	504	06	506	79	057	825	740	311	680
.4	518	07	503	80	038	879	747	317	672
31.5	.04532	.0108	.49500	.0381	.97020	.17934	.67754	.34324	.98663
.6	546	08	497	83	001	988	761	330	655
.7	560	09	494	84	.96982	.18043	768	337	646
.8	574	10	490	85	963	097	775	343	638
.9	588	10	487	86	944	152	783	350	629
32.0	.04602	.0111	.49484	.0387	.96926	.18206	.67790	.34356	.98621
.1	616	12	481	88	907	260	797	363	612
.2	630	12	478	89	887	315	804	370	603
.3	645	13	475	90	868	369	812	376	595
.4	659	14	472	91	849	424	819	383	586
32.5	.04673	.0114	.49469	.0392	.96830	.18478	.67826	.34390	.98577
.6	687	15	466	93	811	532	834	397	569
.7	701	16	462	94	791	586	841	403	560
.8	715	16	459	95	772	640	849	410	551
.9	729	17	455	97	752	694	856	417	542
33.0	.04743	.0118	.49452	.0398	.96733	.18748	.67863	.34424	.98534
.1	757	18	449	99	713	803	871	431	525
.2	771	19	445	.0400	694	857	878	438	516
.3	785	20	442	01	674	911	886	444	507
.4	799	21	438	02	655	965	894	451	498
33.5	.04813	.0121	.49435	.0403	.96635	.19019	.67901	.34458	.98489
.6	827	22	432	04	615	073	909	465	480
.7	841	23	429	05	595	127	916	472	471
.8	855	23	425	06	576	181	924	479	462
.9	869	24	422	07	556	234	932	486	453
34.0	.04883	.0125	.49419	.0408	.96536	.19288	.67939	.34493	.98444
.1	897	26	415	09	516	342	947	500	435
.2	911	26	412	10	496	396	955	508	425
.3	925	27	408	11	475	450	963	515	416
.4	939	28	405	12	455	504	971	522	407
34.5	.04953	.0128	.49401	.0413	.96435	.19557	.67979	.34529	.98398
.6	967	29	398	15	414	611	987	536	389
.7	981	30	395	16	394	665	994	544	379
.8	995	31	391	17	373	718	.68002	551	370
.9	.05009	31	388	18	353	772	010	558	361
35.0	.05023	.0132	.49385	.0419	.96332	.19826	.68018	.34565	.98351

* Multiply functions (except * values) by the given value of L_S. When chord definition of degree of curve (D_c) is used, correct the calculated values of *o* or X_0 by subtracting the product of D_c and the * values. See page 354 for example.

TABLE XII.—SPIRAL FUNCTIONS FOR $L_s=1$

Δ°	o		X_o		X	Y	L.T.	S.T.	L.C.
	D_a	*	D_a	*					
35.0	.05023	.0132	.49385	.0419	.96332	.19826	.68018	.34565	.98351
.1	037	33	381	20	311	879	026	573	342
.2	051	33	378	21	291	933	034	580	333
.3	065	34	374	22	270	987	042	587	323
.4	079	35	371	23	250	.20040	051	595	314
35.5	.05093	.0136	.49367	.0424	.96229	.20094	.68059	.34602	.98304
.6	107	36	363	25	208	147	067	610	295
.7	121	37	360	26	187	201	075	617	285
.8	135	38	356	27	166	254	083	625	276
.9	149	39	353	28	145	307	092	632	266
36.0	.05163	.0139	.49349	.0429	.96124	.20361	.68100	.34640	.98257
.1	177	40	345	30	103	414	108	647	247
.2	191	41	342	31	081	467	116	655	237
.3	204	42	338	32	060	521	125	663	227
.4	218	42	335	33	038	574	133	670	218
36.5	.05232	.0143	.49331	.0434	.96017	.20627	.68141	.34678	.98208
.6	246	44	327	35	.95996	680	150	686	198
.7	260	45	324	36	974	734	158	693	188
.8	273	45	320	37	953	787	167	701	179
.9	287	46	317	38	931	840	175	709	169
37.0	.05301	.0147	.49313	.0439	.95910	.20893	.68184	.34717	.98159
.1	315	48	309	40	888	946	192	725	149
.2	329	49	306	41	866	999	201	732	139
.3	343	49	302	42	844	.21052	210	740	129
.4	357	50	299	43	822	105	218	748	119
37.5	.05371	.0151	.49295	.0444	.95800	.21158	.68227	.34756	.98109
.6	385	52	291	45	778	211	236	764	099
.7	399	52	287	46	756	264	244	772	089
.8	413	53	284	47	734	317	253	780	079
.9	427	54	280	48	712	370	262	788	069
38.0	.05441	.0155	.49276	.0449	.95690	.21423	.68271	.34796	.98059
.1	455	56	272	50	668	475	279	804	049
.2	469	56	268	51	645	528	288	812	038
.3	482	57	264	52	623	581	297	820	028
.4	496	58	260	53	601	634	306	829	018
38.5	.05510	.0159	.49256	.0454	.95578	.21686	.68315	.34837	.98008
.6	524	59	252	55	556	739	324	845	.97997
.7	538	60	249	56	533	792	333	853	987
.8	551	61	245	57	511	844	342	861	977
.9	565	62	242	58	488	897	351	870	967
39.0	.05579	.0163	.49238	.0459	.95466	.21949	.68360	.34878	.97956
.1	593	63	234	60	443	.22002	369	886	946
.2	607	64	230	61	420	054	379	895	935
.3	620	65	226	62	397	107	388	903	925
.4	634	66	222	63	374	159	397	911	914
39.5	.05648	.0167	.49218	.0464	.95351	.22212	.68406	.34920	.97904
.6	662	68	214	65	328	264	415	928	893
.7	676	68	210	66	305	316	424	937	883
.8	690	69	207	67	281	369	434	945	872
.9	704	70	203	68	258	421	443	954	861
40.0	.05718	.0171	.49199	.0469	.95235	.22473	.68452	.34962	.97851

* Multiply functions (except * values) by the given value of L_S. When chord definition of degree of curve (D_c) is used, correct the calculated values of o or X_0 by subtracting the product of D_c and the * values. See page 354 for example.

TABLE XII.—SPIRAL FUNCTIONS FOR $L_s=1$

$\Delta°$	o		X_0		X	Y	L.T.	S.T.	L.C.
	D_a	*	D_a	*					
40.0	.05718	.0171	.49199	.0469	.95235	.22473	.68452	.34962	.97851
.1	732	72	195	70	212	526	462	971	840
.2	745	72	191	71	188	578	471	980	829
.3	759	73	187	72	165	630	481	988	819
.4	772	74	183	73	141	682	490	997	808
40.5	.05786	.0175	.49179	.0474	.95118	.22734	.68500	.35006	.97797
.6	800	76	175	75	094	786	509	014	786
.7	814	77	171	76	071	838	519	023	775
.8	827	77	167	77	047	890	528	032	765
.9	841	78	163	78	023	942	538	041	754
41.0	.05855	.0179	.49159	.0479	.95000	.22994	.68547	.35049	.97743
.1	869	80	155	80	.94976	.23046	557	058	732
.2	883	81	151	81	952	098	567	067	721
.3	896	82	146	82	928	150	577	076	710
.4	910	82	142	83	904	202	586	085	699
41.5	.05924	.0183	.49138	.0484	.94880	.23254	.68596	.35094	.97688
.6	938	84	134	85	856	306	606	103	677
.7	952	85	130	86	832	358	616	112	666
.8	965	86	126	87	807	409	626	121	655
.9	979	87	122	88	783	461	635	130	643
42.0	.05993	.0187	.49118	.0488	.94759	.23513	.68645	.35139	.97632
.1	.06007	88	114	89	734	564	655	148	621
.2	020	89	110	90	710	616	665	158	610
.3	034	90	105	91	685	667	675	167	599
.4	047	91	101	92	661	719	685	176	587
42.5	.06061	.0192	.49097	.0493	.94636	.23771	.68695	.35185	.97576
.6	075	93	093	94	612	822	706	194	565
.7	089	93	088	95	587	874	716	204	553
.8	102	94	084	96	562	925	726	213	542
.9	116	95	079	97	538	976	736	222	531
43.0	.06130	.0196	.49075	.0498	.94513	.24028	.68746	.35232	.97519
.1	144	97	071	99	488	079	756	241	508
.2	157	98	067	.0500	463	130	767	250	496
.3	171	99	062	01	438	182	777	260	485
.4	184	.0200	058	02	413	233	787	269	473
43.5	.06198	.0200	.49054	.0503	.94388	.24284	.68798	.35279	.97462
.6	212	01	050	03	363	335	808	288	450
.7	226	02	045	04	337	387	818	298	438
.8	239	03	041	05	312	438	829	307	427
.9	253	04	036	06	287	489	839	317	415
44.0	.06267	.0205	.49032	.0507	.94262	.24540	.68850	.35327	.97404
.1	281	06	028	08	236	591	860	336	392
.2	294	07	024	09	211	642	871	346	380
.3	308	08	019	10	185	693	882	356	368
.4	321	08	015	11	160	744	892	365	357
44.5	.06335	.0209	.49011	.0512	.94134	.24795	.68903	.35375	.97345
.6	349	10	007	13	108	846	914	385	333
.7	362	11	003	14	082	896	924	395	321
.8	376	12	.48998	14	057	947	935	405	309
.9	389	13	994	15	031	998	946	415	297
45.0	.06403	.0214	.48990	.0516	.94005	.25049	.68957	.35424	.97285

* Multiply functions (except * values) by the given value of L_S. When chord definition of degree of curve (D_c) is used, correct the calculated values of o or X_0 by subtracting the product of D_c and the * values. See page 354 for example.

TABLE XII.—SPIRAL FUNCTIONS FOR $L_s=1$

Δ°	o	X_o	X	Y	Δ°	o	X_o	X	Y
45.0	.06403	.48989	.94005	.25049	47.5	.06741	.48876	.93342	.26307
45.1	.06416	.48985	.93979	.25100	47.6	.06755	.48871	.93315	.26357
45.2	.06430	.48980	.93953	.25150	47.7	.06768	.48867	.93288	.26407
45.3	.06443	.48976	.93927	.25201	47.8	.06782	.48862	.93261	.26457
45.4	.06457	.48972	.93901	.25252	47.9	.06795	.48857	.93233	.26506
45.5	.06471	.48967	.93875	.25302	48.0	.06809	.48853	.93206	.26556
45.6	.06484	.48963	.93849	.25353	48.1	.06822	.48848	.93178	.26606
45.7	.06498	.48958	.93823	.25403	48.2	.06836	.48843	.93151	.26656
45.8	.06511	.48954	.93796	.25454	48.3	.06849	.48839	.93124	.26705
45.9	.06525	.48949	.93770	.25504	48.4	.06863	.48834	.93096	.26755
46.0	.06538	.48945	.93744	.25554	48.5	.06876	.48829	.93068	.26804
46.1	.06552	.48940	.93717	.25605	48.6	.06890	.48824	.93041	.26854
46.2	.06566	.48936	.93691	.25655	48.7	.06903	.48820	.93013	.26904
46.3	.06579	.48931	.93664	.25706	48.8	.06916	.48815	.92985	.26953
46.4	.06593	.48927	.93638	.25756	48.9	.06930	.48810	.92958	.27002
46.5	.06606	.48922	.93611	.25806	49.0	.06943	.48805	.92930	.27052
46.6	.06620	.48918	.93584	.25856	49.1	.06957	.48801	.92902	.27101
46.7	.06633	.48913	.93558	.25907	49.2	.06970	.48796	.92874	.27151
46.8	.06647	.48908	.93531	.25957	49.3	.06984	.48791	.92846	.27200
46.9	.06660	.48904	.93504	.26007	49.4	.06997	.48786	.92818	.27249
47.0	.06674	.48899	.93477	.26057	49.5	.07010	.48781	.92790	.27299
47.1	.06687	.48895	.93450	.26107	49.6	.07024	.48777	.92761	.27348
47.2	.06701	.48890	.93424	.26157	49.7	.07037	.48772	.92733	.27397
47.3	.06714	.48885	.93396	.26207	49.8	.07051	.48767	.92705	.27446
47.4	.06728	.48881	.93369	.26257	49.9	.07064	.48762	.92677	.27495
47.5	.06741	.48876	.93342	.26307	50.0	.07078	.48757	.92648	.27544

For values of Δ exceeding 45°, Table XII does not contain the * corrections for conversion to the A.R.E.A. spiral nor the functions for obtaining L.T., S.T., and L.C. In the event these missing values are needed, they may be obtained as follows:

Find L.T., S.T., and L.C. by solving triangle *ABC* (Fig. 5–4) knowing *X*, *Y*, and Δ.

Find * value for *o* from *=0.0730 vers Δ.

Find * value for X_0 from *=0.0730 sin Δ.

EXAMPLE. Given $L_s=500$, $D_c=30°$, $\Delta=75°$.

$o=0.10264\times 500-30\times 0.0730\times 0.74118=49.70$

$X_0=0.47276\times 500-30\times 0.0730\times 0.96593=234.26$

TABLE XII.—SPIRAL FUNCTIONS FOR $L_s=1$

Δ°	o	X_o	X	Y	Δ°	o	X_o	X	Y
50.0	.07078	.48757	.92648	.27544	55.0	.07741	.48503	.91170	.29952
50.1	.07091	.48752	.92620	.27593	55.1	.07754	.48498	.91139	.30000
50.2	.07104	.48748	.92592	.27642	55.2	.07768	.48492	.91109	.30047
50.3	.07118	.48743	.92563	.27691	55.3	.07781	.48487	.91078	.30094
50.4	.07131	.48738	.92534	.27740	55.4	.07794	.48482	.91047	.30141
50.5	.07144	.48733	.92506	.27789	55.5	.07807	.48476	.91016	.30188
50.6	.07158	.48728	.92477	.27838	55.6	.07820	.48471	.90985	.30235
50.7	.07171	.48723	.92449	.27887	55.7	.07833	.48466	.90954	.30282
50.8	.07184	.48718	.92420	.27936	55.8	.07846	.48460	.90923	.30329
50.9	.07198	.48713	.92391	.27984	55.9	.07859	.48455	.90892	.30376
51.0	.07211	.48708	.92362	.28033	56.0	.07872	.48449	.90860	.30423
51.1	.07224	.48703	.92333	.28082	56.1	.07886	.48444	.90829	.30470
51.2	.07238	.48698	.92304	.28130	56.2	.07899	.48439	.90798	.30516
51.3	.07251	.48693	.92276	.28179	56.3	.07912	.48433	.90767	.30563
51.4	.07264	.48688	.92246	.28228	56.4	.07925	.48428	.90735	.30610
51.5	.07278	.48683	.92217	.28276	56.5	.07938	.48422	.90704	.30657
51.6	.07291	.48678	.92188	.28325	56.6	.07951	.48417	.90672	.30703
51.7	.07304	.48673	.92159	.28373	56.7	.07964	.48412	.90641	.30750
51.8	.07318	.48668	.92130	.28422	56.8	.07977	.48406	.90609	.30796
51.9	.07331	.48663	.92101	.28470	56.9	.07990	.48400	.90578	.30843
52.0	.07344	.48658	.92071	.28518	57.0	.08003	.48395	.90546	.30890
52.1	.07358	.48653	.92042	.28567	57.1	.08016	.48390	.90514	.30936
52.2	.07371	.48648	.92012	.28615	57.2	.08029	.48384	.90483	.30982
52.3	.07384	.48643	.91983	.28663	57.3	.08042	.48379	.90451	.31029
52.4	.07398	.48638	.91954	.28712	57.4	.08056	.48373	.90419	.31075
52.5	.07411	.48633	.91924	.28760	57.5	.08068	.48368	.90387	.31121
52.6	.07424	.48628	.91894	.28808	57.6	.08082	.48362	.90356	.31168
52.7	.07437	.48623	.91865	.28856	57.7	.08095	.48356	.90324	.31214
52.8	.07451	.48618	.91835	.28904	57.8	.08108	.48351	.90292	.31260
52.9	.07464	.48612	.91805	.28952	57.9	.08121	.48345	.90260	.31306
53.0	.07477	.48607	.91776	.29000	58.0	.08134	.48340	.90227	.31352
53.1	.07490	.48602	.91746	.29048	58.1	.08147	.48334	.90195	.31398
53.2	.07504	.48597	.91716	.29096	58.2	.08160	.48329	.90163	.31444
53.3	.07517	.48592	.91686	.29144	58.3	.08173	.48323	.90131	.31490
53.4	.07530	.48587	.91656	.29192	58.4	.08186	.48318	.90099	.31536
53.5	.07543	.48582	.91626	.29240	58.5	.08199	.48312	.90066	.31582
53.6	.07557	.48576	.91596	.29287	58.6	.08212	.48306	.90034	.31628
53.7	.07570	.48571	.91566	.29335	58.7	.08224	.48301	.90002	.31674
53.8	.07583	.48566	.91536	.29383	58.8	.08238	.48295	.89969	.31720
53.9	.07596	.48561	.91506	.29430	58.9	.08250	.48290	.89937	.31765
54.0	.07609	.48556	.91475	.29478	59.0	.08263	.48284	.89904	.31811
54.1	.07623	.48550	.91445	.29526	59.1	.08276	.48278	.89872	.31857
54.2	.07636	.48545	.91415	.29573	59.2	.08289	.48272	.89839	.31902
54.3	.07649	.48540	.91384	.29621	59.3	.08302	.48267	.89806	.31948
54.4	.07662	.48535	.91354	.29668	59.4	.08315	.48261	.89774	.31994
54.5	.07675	.48529	.91323	.29716	59.5	.08328	.48256	.89741	.32039
54.6	.07688	.48524	.91293	.29763	59.6	.08341	.48250	.89708	.32084
54.7	.07702	.48519	.91262	.29810	59.7	.08354	.48244	.89675	.32130
54.8	.07715	.48514	.91232	.29858	59.8	.08367	.48238	.89642	.32175
54.9	.07728	.48508	.91201	.29905	59.9	.08380	.48233	.89610	.32221
55.0	.07741	.48503	.91170	.29952	60.0	.08393	.48227	.89577	.32266

TABLE XII.—SPIRAL FUNCTIONS FOR $L_s=1$

Δ°	o	X_o	X	Y	Δ°	o	X_o	X	Y
60.0	.08393	.48227	.89577	.32266	65.0	.09031	.47930	.87874	.34478
60.1	.08406	.48221	.89544	.32311	65.1	.09044	.47924	.87839	.34521
60.2	.08418	.48215	.89511	.32356	65.2	.09056	.47918	.87804	.34565
60.3	.08431	.48210	.89478	.32402	65.3	.09069	.47912	.87769	.34608
60.4	.08444	.48204	.89444	.32447	65.4	.09081	.47905	.87734	.34651
60.5	.08457	.48198	.89411	.32492	65.5	.09094	.47899	.87698	.34694
60.6	.08470	.48192	.89378	.32537	65.6	.09107	.47893	.87663	.34737
60.7	.08483	.48187	.89345	.32582	65.7	.09119	.47887	.87628	.34780
60.8	.08496	.48181	.89311	.32627	65.8	.09132	.47881	.87592	.34822
60.9	.08508	.48175	.89278	.32672	65.9	.09144	.47874	.87557	.34865
61.0	.08521	.48169	.89245	.32717	66.0	.09157	.47868	.87521	.34908
61.1	.08534	.48163	.89211	.32762	66.1	.09170	.47862	.87486	.34951
61.2	.08547	.48158	.89178	.32806	66.2	.09182	.47856	.87450	.34994
61.3	.08560	.48152	.89144	.32851	66.3	.09195	.47850	.87415	.35036
61.4	.08573	.48146	.89111	.32896	66.4	.09207	.47843	.87379	.35079
61.5	.08586	.48140	.89077	.32940	66.5	.09220	.47837	.87344	.35121
61.6	.08598	.48134	.89043	.32985	66.6	.09232	.47831	.87308	.35164
61.7	.08611	.48128	.89010	.33030	66.7	.09245	.47824	.87272	.35206
61.8	.08624	.48122	.88976	.33074	66.8	.09257	.47818	.87236	.35249
61.9	.08637	.48116	.88942	.33119	66.9	.09270	.47812	.87200	.35291
62.0	.08650	.48111	.88908	.33163	67.0	.09282	.47806	.87164	.35334
62.1	.08662	.48105	.88874	.33208	67.1	.09295	.47799	.87129	.35376
62.2	.08675	.48099	.88841	.33252	67.2	.09307	.47793	.87093	.35418
62.3	.08688	.48093	.88807	.33296	67.3	.09320	.47787	.87057	.35460
62.4	.08701	.48087	.88773	.33341	67.4	.09332	.47780	.87021	.35502
62.5	.08714	.48081	.88739	.33385	67.5	.09345	.47774	.86985	.35545
62.6	.08726	.48075	.88705	.33429	67.6	.09357	.47768	.86948	.35587
62.7	.08739	.48069	.88670	.33474	67.7	.09370	.47761	.86912	.35629
62.8	.08752	.48063	.88636	.33518	67.8	.09382	.47755	.86876	.35671
62.9	.08764	.48057	.88602	.33562	67.9	.09395	.47748	.86840	.35713
63.0	.08777	.48051	.88568	.33606	68.0	.09407	.47742	.86804	.35755
63.1	.08790	.48045	.88534	.33650	68.1	.09420	.47736	.86767	.35796
63.2	.08803	.48039	.88499	.33694	68.2	.09432	.47729	.86731	.35838
63.3	.08816	.48033	.88465	.33738	68.3	.09445	.47723	.86695	.35880
63.4	.08828	.48027	.88430	.33782	68.4	.09457	.47716	.86658	.35922
63.5	.08841	.48021	.88396	.33826	68.5	.09469	.47710	.86622	.35963
63.6	.08854	.48015	.88362	.33869	68.6	.09482	.47704	.86585	.36005
63.7	.08866	.48009	.88327	.33913	68.7	.09494	.47697	.86549	.36047
63.8	.08879	.48003	.88292	.33957	68.8	.09507	.47691	.86512	.36088
63.9	.08892	.47997	.88258	.34001	68.9	.09519	.47684	.86476	.36130
64.0	.08904	.47991	.88223	.34044	69.0	.09532	.47678	.86439	.36171
64.1	.08917	.47985	.88188	.34088	69.1	.09544	.47671	.86402	.36213
64.2	.08930	.47979	.88154	.34131	69.2	.09556	.47665	.86366	.36254
64.3	.08942	.47973	.88119	.34175	69.3	.09569	.47658	.86329	.36295
64.4	.08955	.47967	.88084	.34218	69.4	.09581	.47652	.86292	.36337
64.5	.08968	.47961	.88049	.34262	69.5	.09593	.47646	.86255	.36378
64.6	.08980	.47954	.88014	.34305	69.6	.09606	.47639	.86218	.36419
64.7	.08993	.47948	.87979	.34348	69.7	.09618	.47632	.86181	.36460
64.8	.09006	.47942	.87944	.34392	69.8	.09630	.47626	.86144	.36501
64.9	.09018	.47936	.87909	.34435	69.9	.09643	.47619	.86107	.36542
65.0	.09031	.47930	.87874	.34478	70.0	.09655	.47613	.86070	.36583

TABLE XII.—SPIRAL FUNCTIONS FOR $L_s=1$

Δ°	o	X_o	X	Y	Δ°	o	X_o	X	Y
70.0	.09655	.47613	.86070	.36583	75.0	.10264	.47276	.84172	.38576
70.1	.09668	.47606	.86033	.36624	75.1	.10276	.47269	.84133	.38614
70.2	.09680	.47600	.85996	.36665	75.2	.10288	.47262	.84094	.38653
70.3	.09692	.47593	.85959	.36706	75.3	.10300	.47256	.84055	.38691
70.4	.09704	.47587	.85922	.36747	75.4	.10312	.47249	.84016	.38730
70.5	.09717	.47580	.85885	.36788	75.5	.10324	.47242	.83977	.38768
70.6	.09729	.47574	.85847	.36828	75.6	.10336	.47235	.83938	.38807
70.7	.09741	.47567	.85810	.36869	75.7	.10348	.47228	.83899	.38845
70.8	.09754	.47560	.85773	.36910	75.8	.10360	.47221	.83860	.38883
70.9	.09766	.47554	.85735	.36950	75.9	.10372	.47214	.83821	.38922
71.0	.09778	.47547	.85698	.36991	76.0	.10384	.47207	.83782	.38960
71.1	.09790	.47540	.85660	.37032	76.1	.10396	.47200	.83742	.38998
71.2	.09803	.47534	.85623	.37072	76.2	.10408	.47193	.83703	.39036
71.3	.09815	.47527	.85586	.37112	76.3	.10420	.47186	.83664	.39074
71.4	.09827	.47521	.85548	.37153	76.4	.10432	.47179	.83625	.39112
71.5	.09840	.47514	.85510	.37193	76.5	.10444	.47172	.83585	.39150
71.6	.09852	.47507	.85473	.37233	76.6	.10456	.47165	.83546	.39188
71.7	.09864	.47501	.85435	.37274	76.7	.10468	.47158	.83506	.39226
71.8	.09876	.47494	.85397	.37314	76.8	.10480	.47150	.83467	.39264
71.9	.09888	.47487	.85360	.37354	76.9	.10492	.47144	.83427	.39302
72.0	.09901	.47480	.85322	.37394	77.0	.10504	.47136	.83388	.39340
72.1	.09913	.47474	.85284	.37434	77.1	.10516	.47129	.83348	.39377
72.2	.09925	.47467	.85246	.37474	77.2	.10528	.47122	.83309	.39415
72.3	.09937	.47460	.85208	.37514	77.3	.10540	.47115	.83269	.39452
72.4	.09950	.47454	.85170	.37554	77.4	.10551	.47108	.83230	.39490
72.5	.09962	.47447	.85132	.37594	77.5	.10563	.47101	.83190	.39528
72.6	.09974	.47440	.85094	.37634	77.6	.10575	.47094	.83150	.39565
72.7	.09986	.47434	.85056	.37674	77.7	.10587	.47087	.83110	.39602
72.8	.09998	.47427	.85018	.37713	77.8	.10599	.47080	.83070	.39640
72.9	.10010	.47420	.84980	.37753	77.9	.10611	.47072	.83031	.39677
73.0	.10023	.47413	.84942	.37792	78.0	.10622	.47065	.82991	.39714
73.1	.10035	.47406	.84904	.37832	78.1	.10634	.47058	.82951	.39752
73.2	.10047	.47400	.84866	.37872	78.2	.10646	.47051	.82911	.39789
73.3	.10059	.47393	.84828	.37911	78.3	.10658	.47044	.82871	.39826
73.4	.10071	.47386	.84789	.37951	78.4	.10670	.47037	.82831	.39863
73.5	.10083	.47379	.84751	.37990	78.5	.10682	.47030	.82791	.39900
73.6	.10095	.47372	.84713	.38029	78.6	.10693	.47022	.82751	.39937
73.7	.10108	.47366	.84674	.38069	78.7	.10705	.47015	.82711	.39974
73.8	.10120	.47359	.84636	.38108	78.8	.10717	.47008	.82671	.40011
73.9	.10132	.47352	.84597	.38147	78.9	.10729	.47001	.82631	.40048
74.0	.10144	.47345	.84559	.38186	79.0	.10741	.46994	.82590	.40084
74.1	.10156	.47338	.84520	.38226	79.1	.10752	.46986	.82550	.40121
74.2	.10168	.47331	.84482	.38264	79.2	.10764	.46979	.82510	.40158
74.3	.10180	.47325	.84443	.38304	79.3	.10776	.46972	.82470	.40194
74.4	.10192	.47318	.84404	.38343	79.4	.10788	.46965	.82430	.40231
74.5	.10204	.47311	.84366	.38382	79.5	.10799	.46958	.82389	.40268
74.6	.10216	.47304	.84327	.38420	79.6	.10811	.46950	.82349	.40304
74.7	.10228	.47297	.84288	.38459	79.7	.10823	.46943	.82308	.40340
74.8	.10240	.47290	.84250	.38498	79.8	.10835	.46936	.82268	.40377
74.9	.10252	.47283	.84211	.38537	79.9	.10846	.46928	.82227	.40413
75.0	.10264	.47276	.84172	.38576	80.0	.10858	.46921	.82187	.40450

TABLE XII.—SPIRAL FUNCTIONS FOR $L_s=1$

$\Delta°$	o	X_o	X	Y	$\Delta°$	o	X_o	X	Y
80.0	.10858	.46921	.82187	.40450	85.0	.11435	.46548	.80123	.42201
80.1	.10870	.46914	.82146	.40486	85.1	.11446	.46540	.80081	.42235
80.2	.10881	.46906	.82106	.40522	85.2	.11458	.46533	.80039	.42268
80.3	.10893	.46899	.82065	.40558	85.3	.11469	.46525	.79997	.42302
80.4	.10905	.46892	.82025	.40594	85.4	.11480	.46518	.79955	.42336
80.5	.10916	.46885	.81984	.40630	85.5	.11492	.46510	.79913	.42369
80.6	.10928	.46877	.81943	.40666	85.6	.11503	.46502	.79871	.42403
80.7	.10940	.46870	.81903	.40702	85.7	.11514	.46495	.79829	.42436
80.8	.10952	.46863	.81862	.40738	85.8	.11526	.46487	.79786	.42470
80.9	.10963	.46855	.81821	.40774	85.9	.11537	.46479	.79744	.42503
81.0	.10975	.46848	.81780	.40810	86.0	.11548	.46472	.79702	.42536
81.1	.10986	.46841	.81739	.40846	86.1	.11560	.46464	.79660	.42570
81.2	.10998	.46833	.81698	.40881	86.2	.11571	.46456	.79617	.42603
81.3	.11010	.46826	.81658	.40917	86.3	.11582	.46448	.79575	.42636
81.4	.11021	.46818	.81617	.40952	86.4	.11594	.46441	.79532	.42669
81.5	.11033	.46811	.81576	.40988	86.5	.11605	.46433	.79490	.42702
81.6	.11044	.46804	.81535	.41024	86.6	.11616	.46425	.79448	.42735
81.7	.11056	.46796	.81494	.41059	86.7	.11627	.46417	.79405	.42768
81.8	.11068	.46789	.81453	.41094	86.8	.11639	.46410	.79363	.42801
81.9	.11079	.46781	.81412	.41130	86.9	.11650	.46402	.79320	.42834
82.0	.11091	.46774	.81370	.41165	87.0	.11661	.46394	.79278	.42866
82.1	.11102	.46767	.81329	.41200	87.1	.11672	.46386	.79235	.42899
82.2	.11114	.46759	.81288	.41236	87.2	.11684	.46379	.79192	.42932
82.3	.11126	.46752	.81247	.41271	87.3	.11695	.46371	.79150	.42964
82.4	.11137	.46744	.81206	.41306	87.4	.11706	.46363	.79107	.42997
82.5	.11149	.46737	.81164	.41341	87.5	.11717	.46355	.79065	.43030
82.6	.11160	.46729	.81123	.41376	87.6	.11728	.46348	.79022	.43062
82.7	.11172	.46722	.81082	.41411	87.7	.11740	.46340	.78979	.43094
82.8	.11183	.46714	.81040	.41446	87.8	.11751	.46332	.78936	.43127
82.9	.11195	.46707	.80999	.41481	87.9	.11762	.46324	.78894	.43159
83.0	.11206	.46699	.80958	.41515	88.0	.11773	.46316	.78851	.43191
83.1	.11218	.46692	.80916	.41550	88.1	.11784	.46308	.78808	.43224
83.2	.11229	.46684	.80875	.41585	88.2	.11795	.46300	.78765	.43256
83.3	.11241	.46677	.80833	.41620	88.3	.11806	.46293	.78722	.43288
83.4	.11252	.46669	.80792	.41654	88.4	.11818	.46285	.78679	.43320
83.5	.11264	.46662	.80750	.41689	88.5	.11829	.46277	.78636	.43352
83.6	.11275	.46654	.80709	.41723	88.6	.11840	.46269	.78593	.43384
83.7	.11287	.46647	.80667	.41758	88.7	.11851	.46261	.78550	.43416
83.8	.11298	.46639	.80625	.41792	88.8	.11862	.46253	.78507	.43448
83.9	.11310	.46632	.80584	.41826	88.9	.11873	.46246	.78464	.43479
84.0	.11321	.46624	.80542	.41861	89.0	.11884	.46238	.78421	.43511
84.1	.11332	.46617	.80500	.41895	89.1	.11895	.46230	.78378	.43543
84.2	.11344	.46609	.80458	.41929	89.2	.11906	.46222	.78335	.43574
84.3	.11355	.46601	.80417	.41963	89.3	.11918	.46214	.78292	.43606
84.4	.11367	.46594	.80375	.41997	89.4	.11929	.46206	.78249	.43638
84.5	.11378	.46586	.80333	.42032	89.5	.11940	.46198	.78206	.43669
84.6	.11390	.46579	.80291	.42066	89.6	.11951	.46190	.78162	.43701
84.7	.11401	.46571	.80249	.42099	89.7	.11962	.46182	.78119	.43732
84.8	.11412	.46563	.80207	.42133	89.8	.11973	.46174	.78076	.43763
84.9	.11424	.46556	.80165	.42167	89.9	.11984	.46166	.78033	.43795
85.0	.11435	.46548	.80123	.42201	90.0	.11995	.46158	.77989	.43826

TABLE XIII.—COEFFICIENTS FOR CURVE WITH EQUAL SPIRALS

Δ ° ′	N	P	M	S
1 00	0.034906	0.000203	1.000051	0.017454
20	0.046540	0.000361	1.000090	0.023271
40	0.058173	0.000564	1.000141	0.029088
2 00	0.069805	0.000812	1.000203	0.034905
20	0.081435	0.001106	1.000277	0.040722
40	0.093064	0.001444	1.000361	0.046539
3 00	0.104691	0.001827	1.000457	0.052355
20	0.116316	0.002256	1.000564	0.058171
40	0.127938	0.002729	1.000682	0.063986
4 00	0.139558	0.003248	1.000812	0.069802
20	0.151175	0.003812	1.000953	0.075616
40	0.162789	0.004421	1.001106	0.081430
5 00	0.174400	0.005074	1.001269	0.087244
20	0.186007	0.005773	1.001444	0.093057
40	0.197611	0.006517	1.001630	0.098870
6 00	0.209210	0.007305	1.001827	0.104682
20	0.220805	0.008139	1.002036	0.110492
40	0.232396	0.009017	1.002255	0.116306
7 00	0.243982	0.009940	1.002486	0.122113
20	0.255563	0.010908	1.002728	0.127921
40	0.267139	0.011921	1.002982	0.133729
8 00	0.278709	0.012979	1.003247	0.139536
20	0.290274	0.014081	1.003523	0.145342
40	0.301832	0.015229	1.003811	0.151146
9 00	0.313358	0.016420	1.004108	0.156951
20	0.324931	0.017657	1.004419	0.162753
40	0.336471	0.018938	1.004739	0.168555
10 00	0.348004	0.020264	1.005072	0.174356
20	0.359530	0.021634	1.005415	0.180155
40	0.371049	0.023049	1.005770	0.185954
11 00	0.382560	0.024508	1.006135	0.191751
20	0.394063	0.026011	1.006511	0.197546
40	0.405558	0.027559	1.006900	0.203340
12 00	0.417046	0.029152	1.007300	0.209134
20	0.428524	0.030788	1.007710	0.214925
40	0.439994	0.032469	1.008131	0.220715
13 00	0.451455	0.034194	1.008564	0.226504
20	0.462907	0.035963	1.009008	0.232291
40	0.474350	0.037777	1.009464	0.238077
14 00	0.485783	0.039634	1.009930	0.243861
20	0.497206	0.041535	1.010407	0.249643
40	0.508619	0.043480	1.010895	0.255424
15 00	0.520022	0.045469	1.011395	0.261203
20	0.531414	0.047502	1.011906	0.266980
40	0.542795	0.049579	1.012428	0.272755
16 00	0.554166	0.051699	1.012961	0.278529
20	0.565525	0.053863	1.013505	0.284300
40	0.576873	0.056071	1.014061	0.290070
17 00	0.588209	0.058322	1.014627	0.295837
20	0.599534	0.060616	1.015204	0.301604
40	0.610846	0.062954	1.015792	0.307367
18 00	0.622146	0.065335	1.016391	0.313129
20	0.633433	0.067759	1.017002	0.318888
40	0.644708	0.070227	1.017624	0.324646
19 00	0.655969	0.072737	1.018256	0.330401
20	0.667217	0.075291	1.018899	0.336154
40	0.678452	0.077888	1.019554	0.341904
20 00	0.689673	0.080527	1.020220	0.347653
20	0.700881	0.083209	1.020896	0.353400
40	0.712074	0.085934	1.021584	0.359143

TABLE XIII.—COEFFICIENTS FOR CURVE WITH EQUAL SPIRALS

Δ	N	P	M	S
° ′				
21 00	0.723252	0.088702	1.022282	0.364884
30	0.739993	0.092933	1.023351	0.373492
22 00	0.756700	0.097260	1.024444	0.382093
30	0.773373	0.101682	1.025562	0.390690
23 00	0.790011	0.106198	1.026703	0.399280
30	0.806613	0.110810	1.027870	0.407864
24 00	0.823178	0.115515	1.029061	0.416441
30	0.839706	0.120315	1.030276	0.425013
25 00	0.856196	0.125208	1.031516	0.433578
30	0.872648	0.130195	1.032780	0.442137
26 00	0.889060	0.135275	1.034069	0.450689
30	0.905432	0.140447	1.035381	0.459234
27 00	0.921763	0.145712	1.036718	0.467773
30	0.938052	0.151069	1.038080	0.476303
28 00	0.954299	0.156518	1.039466	0.484827
30	0.970504	0.162058	1.040875	0.493345
29 00	0.986664	0.167689	1.042309	0.501854
30	1.002780	0.173410	1.043766	0.510356
30 00	1.018851	0.179222	1.045247	0.518851
30	1.034876	0.185124	1.046753	0.527338
31 00	1.050854	0.191115	1.048282	0.535816
30	1.066785	0.197195	1.049835	0.544286
32 00	1.082668	0.203365	1.051413	0.552749
30	1.098503	0.209623	1.053014	0.561203
33 00	1.114288	0.215967	1.054638	0.569649
30	1.130023	0.222399	1.056285	0.578086
34 00	1.145707	0.228919	1.057957	0.586514
30	1.161340	0.235525	1.059651	0.594934
35 00	1.176922	0.242218	1.061370	0.603346
30	1.192451	0.248997	1.063113	0.611748
36 00	1.207926	0.255861	1.064878	0.620141
30	1.223347	0.262809	1.066666	0.628524
37 00	1.238714	0.269842	1.068478	0.636899
30	1.254026	0.276959	1.070312	0.645265
38 00	1.269281	0.284160	1.072171	0.653619
30	1.284479	0.291444	1.074052	0.661964
39 00	1.299621	0.298809	1.075955	0.670301
30	1.314705	0.306257	1.077882	0.678627
40 00	1.329730	0.313788	1.079832	0.686942
30	1.344696	0.321399	1.081805	0.695248
41 00	1.359603	0.329090	1.083800	0.703544
30	1.374449	0.336862	1.085818	0.711829
42 00	1.389234	0.344714	1.087859	0.720103
30	1.403958	0.352645	1.089922	0.728368
43 00	1.418619	0.360654	1.092008	0.736621
30	1.433217	0.368741	1.094115	0.744862
44 00	1.447752	0.376906	1.096246	0.753094
30	1.462223	0.385148	1.098398	0.761314
45 00	1.476630	0.393466	1.100573	0.769523
30	1.490971	0.401861	1.102770	0.777721
46 00	1.505248	0.410331	1.104989	0.785908
30	1.519457	0.418875	1.107230	0.794083
47 00	1.533600	0.427495	1.109493	0.802246
30	1.547674	0.436187	1.111777	0.810397
48 00	1.561682	0.444954	1.114085	0.818537
30	1.575620	0.453792	1.116412	0.826664
49 00	1.589490	0.462704	1.118763	0.834780
30	1.603290	0.471686	1.121134	0.842884
50 00	1.617020	0.480740	1.123528	0.850976

TABLE XIV.—TANGENTS AND EXTERNALS FOR UNIT DOUBLE-SPIRAL CURVE

I°	Ts	Es	I°	Ts	Es	I°	Ts	Es
4	1.00028	.01164	42.5	1.03394	.13135	57.0	1.06399	.18536
6	1.00064	.01747	43.0	1.03479	.13309	.2	1.06449	.18617
8	1.00114	.02332	43.5	1.03566	.13484	.4	1.06499	.18698
10	1.00178	.02918	44.0	1.03653	.13660	.6	1.06550	.18778
12	1.00257	.03507	44.5	1.03742	.13836	.8	1.06600	.18859
13	1.00302	.03802	45.0	1.03831	.14013	58.0	1.06651	.18940
14	1.00350	.04098	45.5	1.03922	.14191	.2	1.06703	.19021
15	1.00402	.04396	46.0	1.04015	.14370	.4	1.06754	.19103
16	1.00458	.04696	46.5	1.04109	.14550	.6	1.06806	.19184
17	1.00518	.04992	47.0	1.04204	.14730	.8	1.06857	.19266
18	1.00581	.05292	47.5	1.04301	.14912	59.0	1.06909	.19348
19	1.00648	.05593	48.0	1.04399	.15094	.2	1.06962	.19431
20	1.00719	.05895	48.5	1.04498	.15277	.4	1.07015	.19513
21	1.00794	.06198	49.0	1.04598	.15460	.6	1.07068	.19596
22	1.00873	.06502	49.5	1.04701	.15645	.8	1.07121	.19679
23	1.00955	.06808	50.0	1.04804	.15831	60.0	1.07174	.19762
24	1.01042	.07115	.2	1.04845	.15905	.2	1.07228	.19846
25	1.01132	.07424	.4	1.04887	.15980	.4	1.07282	.19929
26	1.01226	.07734	.6	1.04929	.16055	.6	1.07337	.20013
27	1.01324	.08045	.8	1.04971	.16130	.8	1.07391	.20097
27.5	1.01375	.08201	51.0	1.05014	.16206	61.0	1.07446	.20181
28.0	1.01427	.08358	.2	1.05057	.16281	.2	1.07501	.20266
28.5	1.01480	.08515	.4	1.05100	.16356	.4	1.07556	.20350
29.0	1.01533	.08674	.6	1.05143	.16432	.6	1.07612	.20435
29.5	1.01588	.08831	.8	1.05186	.16508	.8	1.07668	.20519
30.0	1.01644	.08990	52.0	1.05230	.16584	62.0	1.07724	.20604
30.5	1.01700	.09149	.2	1.05274	.16660	.2	1.07781	.20690
31.0	1.01758	.09309	.4	1.05318	.16736	.4	1.07838	.20775
31.5	1.01817	.09469	.6	1.05362	.16813	.6	1.07895	.20861
32.0	1.01877	.09630	.8	1.05407	.16889	.8	1.07953	.20947
32.5	1.01938	.09791	53.0	1.05452	.16966	63.0	1.08010	.21034
33.0	1.02000	.09952	.2	1.05497	.17043	.2	1.08068	.21120
33.5	1.02064	.10114	.4	1.05542	.17120	.4	1.08126	.21207
34.0	1.02128	.10277	.6	1.05588	.17197	.6	1.08184	.21294
34.5	1.02194	.10440	.8	1.05634	.17275	.8	1.08243	.21381
35.0	1.02260	.10604	54.0	1.05680	.17352	64.0	1.08302	.21468
35.5	1.02327	.10768	.2	1.05726	.17430	.2	1.08361	.21555
36.0	1.02396	.10933	.4	1.05772	.17508	.4	1.08421	.21643
36.5	1.02466	.11099	.6	1.05819	.17586	.6	1.08481	.21731
37.0	1.02537	.11265	.8	1.05866	.17664	.8	1.08541	.21820
37.5	1.02609	.11432	55.0	1.05913	.17742	65.0	1.08602	.21908
38.0	1.02682	.11599	.2	1.05961	.17820	.2	1.08663	.21997
38.5	1.02756	.11767	.4	1.06009	.17899	.4	1.08724	.22086
39.0	1.02832	.11936	.6	1.06057	.17979	.6	1.08785	.22175
39.5	1.02909	.12105	.8	1.06105	.18058	.8	1.08847	.22265
40.0	1.02987	.12275	56.0	1.06153	.18137	66.0	1.08909	.22355
40.5	1.03066	.12446	.2	1.06201	.18217	.2	1.08971	.22445
41.0	1.03146	.12617	.4	1.06250	.18296	.4	1.09034	.22535
41.5	1.03227	.12789	.6	1.06299	.18376	.6	1.09097	.22625
42.0	1.03310	.12962	.8	1.06349	.18456	.8	1.09160	.22716

TABLE XIV.—TANGENTS AND EXTERNALS FOR UNIT DOUBLE-SPIRAL CURVE

I°	T_s	E_s	I°	T_s	E_s	I°	T_s	E_s
67.0	1.09223	.22807	75.0	1.12036	.26669	83.0	1.15453	.31048
.2	1.09287	.22898	.2	1.12124	.26772	.2	1.15547	.31165
.4	1.09351	.22990	.4	1.12192	.26875	.4	1.15642	.31283
.6	1.09416	.23082	.6	1.12270	.26978	.6	1.15738	.31401
.8	1.09481	.23174	.8	1.12348	.27082	.8	1.15834	.31520
68.0	1.09546	.23266	76.0	1.12427	.27186	84.0	1.15930	.31639
.2	1.09612	.23358	.2	1.12507	.27290	.2	1.16027	.31759
.4	1.09678	.23451	.4	1.12587	.27394	.4	1.16124	.31879
.6	1.09744	.23544	.6	1.12667	.27499	.6	1.16222	.31999
.8	1.09810	.23637	.8	1.12747	.27604	.8	1.16320	.32120
69.0	1.09876	.23731	77.0	1.12828	.27710	85.0	1.16418	.32241
.2	1.09943	.23825	.2	1.12909	.27816	.2	1.16517	.32363
.4	1.10010	.23919	.4	1.12991	.27923	.4	1.16617	.32485
.6	1.10078	.24013	.6	1.13074	.28030	.6	1.16717	.32607
.8	1.10146	.24108	.8	1.13157	.28137	.8	1.16818	.32730
70.0	1.10214	.24203	78.0	1.13240	.28244	86.0	1.16919	.32854
.2	1.10283	.24298	.2	1.13323	.28352	.2	1.17021	.32978
.4	1.10352	.24393	.4	1.13407	.28460	.4	1.17123	.33102
.6	1.10421	.24489	.6	1.13491	.28568	.6	1.17226	.33227
.8	1.10491	.24585	.8	1.13576	.28677	.8	1.17329	.33352
71.0	1.10561	.24681	79.0	1.13661	.28786	87.0	1.17433	.33478
.2	1.10632	.24777	.2	1.13746	.28896	.2	1.17537	.33605
.4	1.10703	.24874	.4	1.13832	.29006	.4	1.17642	.33732
.6	1.10774	.24971	.6	1.13918	.29116	.6	1.17748	.33859
.8	1.10854	.25069	.8	1.14005	.29226	.8	1.17854	.33987
72.0	1.10917	.25167	80.0	1.14092	.29337	88.0	1.17960	.34115
.2	1.10989	.25265	.2	1.14179	.29449	.2	1.18067	.34244
.4	1.11062	.25363	.4	1.14267	.29561	.4	1.18174	.34373
.6	1.11135	.25462	.6	1.14356	.29673	.6	1.18282	.34503
.8	1.11208	.25561	.8	1.14445	.29785	.8	1.18391	.34633
73.0	1.11281	.25660	81.0	1.14535	.29898	89.0	1.18500	.34764
.2	1.11355	.25760	.2	1.14625	.30011	.2	1.18609	.34895
.4	1.11429	.25860	.4	1.14715	.30125	.4	1.18719	.35027
.6	1.11504	.25960	.6	1.14805	.30239	.6	1.18830	.35159
.8	1.11579	.26060	.8	1.14896	.30353	.8	1.18942	.35292
74.0	1.11654	.26161	82.0	1.14988	.30468	90.0	1.19054	.35425
.2	1.11730	.26262	.2	1.15080	.30583	.2	1.19167	.35559
.4	1.11806	.26363	.4	1.15173	.30699	.4	1.19280	.35693
.6	1.11882	.26465	.6	1.15266	.30815	.6	1.19394	.35828
.8	1.11959	.26567	.8	1.15359	.30931	.8	1.19508	.35963

NOTES: (1) For any given I and selected spiral length, T_s and E_s for the double-spiral, or all-transitional, curve may be found by multiplying the tabulated values by the length of spiral. (2) If the given I does not correspond to a tabulated value, T_s and E_s may be found by simple interpolation. (3) Deflection angles for staking the curve may be computed, as for any spiral, by using $\Delta = \frac{I}{2}$.

TABLE XV.—DEFLECTION ANGLES FOR 10-CHORD SPIRAL

Δ	1	2	3	4	5	6	7	8	9	10
°	° ′	° ′	° ′	° ′	° ′	° ′	° ′	° ′	° ′	° ′
0.0	0 00.00	0 00.0	0 00.0	0 00.0	0 00.0	0 00.0	0 00.0	0 00.0	0 00.0	0 00.0
.1	00.02	00.1	00.2	00.3	00.5	00.7	01.0	01.3	01.6	02.0
.2	00.04	00.2	00.4	00.6	01.0	01.4	02.0	02.6	03.2	04.0
.3	00.06	00.2	00.5	01.0	01.5	02.2	02.9	03.8	04.9	06.0
.4	00.08	00.3	00.7	01.3	02.0	02.9	03.9	05.1	06.5	08.0
0.5	0 00.10	0 00.4	0 00.9	0 01.6	0 02.5	0 03.6	0 04.9	0 06.4	0 08.1	0 10.0
.6	00.12	00.5	01.1	01.9	03.0	04.3	05.9	07.7	09.7	12.0
.7	00.14	00.6	01.3	02.2	03.5	05.0	06.9	09.0	11.3	14.0
.8	00.16	00.6	01.4	02.6	04.0	05.8	07.8	10.2	13.0	16.0
.9	00.18	00.7	01.6	02.9	04.5	06.5	08.8	11.5	14.6	18.0
1.0	0 00.20	0 00.8	0 01.8	0 03.2	0 05.0	0 07.2	0 09.8	0 12.8	0 16.2	0 20.0
.1	00.22	00.9	02.0	03.5	05.5	07.9	10.8	14.1	17.8	22.0
.2	00.24	01.0	02.2	03.8	06.0	08.6	11.8	15.4	19.4	24.0
.3	00.26	01.0	02.3	04.2	06.5	09.4	12.7	16.6	21.1	26.0
.4	00.28	01.1	02.5	04.5	07.0	10.1	13.7	17.9	22.7	28.0
1.5	0 00.30	0 01.2	0 02.7	0 04.8	0 07.5	0 10.8	0 14.7	0 19.2	0 24.3	0 30.0
.6	00.32	01.3	02.9	05.1	08.0	11.5	15.7	20.5	25.9	32.0
.7	00.34	01.4	03.1	05.4	08.5	12.2	16.7	21.8	27.5	34.0
.8	00.36	01.4	03.2	05.8	09.0	13.0	17.6	23.0	29.2	36.0
.9	00.38	01.5	03.4	06.1	09.5	13.7	18.6	24.3	30.8	38.0
2.0	0 00.40	0 01.6	0 03.6	0 06.4	0 10.0	0 14.4	0 19.6	0 25.6	0 32.4	0 40.0
.1	00.42	01.7	03.8	06.7	10.5	15.1	20.6	26.9	34.0	42.0
.2	00.44	01.8	04.0	07.0	11.0	15.8	21.6	28.2	35.6	44.0
.3	00.46	01.8	04.1	07.4	11.5	16.6	22.5	29.4	37.3	46.0
.4	00.48	01.9	04.3	07.7	12.0	17.3	23.5	30.7	38.9	48.0
2.5	0 00.50	0 02.0	0 04.5	0 08.0	0 12.5	0 18.0	0 24.5	0 32.0	0 40.5	0 50.0
.6	00.52	02.1	04.7	08.3	13.0	18.7	25.5	33.3	42.1	52.0
.7	00.54	02.2	04.9	08.6	13.5	19.4	26.5	34.6	43.7	54.0
.8	00.56	02.2	05.0	09.0	14.0	20.2	27.4	35.8	45.4	56.0
.9	00.58	02.3	05.2	09.3	14.5	20.9	28.4	37.1	47.0	58.0
3.0	0 00.60	0 02.4	0 05.4	0 09.6	0 15.0	0 21.6	0 29.4	0 38.4	0 48.6	1 00.0
.1	00.62	02.5	05.6	09.9	15.5	22.3	30.4	39.7	50.2	02.0
.2	00.64	02.6	05.8	10.2	16.0	23.0	31.4	41.0	51.8	04.0
.3	00.66	02.6	05.9	10.6	16.5	23.8	32.3	42.2	53.5	06.0
.4	00.68	02.7	06.1	10.9	17.0	24.5	33.3	43.5	55.1	08.0
3.5	0 00.70	0 02.8	0 06.3	0 11.2	0 17.5	0 25.2	0 34.3	0 44.8	0 56.7	1 10.0
.6	00.72	02.9	06.5	11.5	18.0	25.9	35.3	46.1	58.3	12.0
.7	00.74	03.0	06.7	11.8	18.5	26.6	36.3	47.4	59.9	14.0
.8	00.76	03.0	06.8	12.2	19.0	27.4	37.2	48.6	1 01.6	16.0
.9	00.78	03.1	07.0	12.5	19.5	28.1	38.2	49.9	03.2	18.0
4.0	0 00.80	0 03.2	0 07.2	0 12.8	0 20.0	0 28.8	0 39.2	0 51.2	1 04.8	1 20.0
.1	00.82	03.3	07.4	13.1	20.5	29.5	40.2	52.5	06.4	22.0
.2	00.84	03.4	07.6	13.4	21.0	30.2	41.2	53.8	08.0	24.0
.3	00.86	03.4	07.7	13.8	21.5	31.0	42.1	55.0	09.7	26.0
.4	00.88	03.5	07.9	14.1	22.0	31.7	43.1	56.3	11.3	28.0
4.5	0 00.90	0 03.6	0 08.1	0 14.4	0 22.5	0 32.4	0 44.1	0 57.6	1 12.9	1 30.0
.6	00.92	03.7	08.3	14.7	23.0	33.1	45.1	58.9	14.5	32.0
.7	00.94	03.8	08.5	15.0	23.5	33.8	46.1	1 00.2	16.1	34.0
.8	00.96	03.8	08.6	15.4	24.0	34.6	47.0	01.4	17.8	36.0
.9	00.98	03.9	08.8	15.7	24.5	35.3	48.0	02.7	19.4	38.0
5.0	0 01.00	0 04.0	0 09.0	0 16.0	0 25.0	0 36.0	0 49.0	1 04.0	1 21.0	1 40.0
.1	01.02	04.1	09.2	16.3	25.5	36.7	50.0	05.3	22.6	42.0
.2	01.04	04.2	09.4	16.6	26.0	37.4	51.0	06.6	24.2	44.0
.3	01.06	04.2	09.5	17.0	26.5	38.2	51.9	07.8	25.9	46.0
.4	01.08	04.3	09.7	17.3	27.0	38.9	52.9	09.1	27.5	48.0
5.5	0 01.10	0 04.4	0 09.9	0 17.6	0 27.5	0 39.6	0 53.9	1 10.4	1 29.1	1 50.0
.6	01.12	04.5	10.1	17.9	28.0	40.3	54.9	11.7	30.7	52.0
.7	01.14	04.6	10.3	18.2	28.5	41.0	55.9	13.0	32.3	54.0
.8	01.16	04.6	10.4	18.6	29.0	41.8	56.8	14.2	34.0	56.0
.9	01.18	04.7	10.6	18.9	29.5	42.5	57.8	15.5	35.6	58.0

TABLE XV.—DEFLECTION ANGLES FOR 10-CHORD SPIRAL

Δ	1	2	3	4	5	6	7	8	9	10
°	° ′	° ′	° ′	° ′	° ′	° ′	° ′	° ′	° ′	° ′
6.0	0 01.20	0 04.8	0 10.8	0 19.2	0 30.0	0 43.2	0 58.8	1 16.8	1 37.2	2 00.0
.1	01.22	04.9	11.0	19.5	30.5	43.9	59.8	18.1	38.8	02.0
.2	01.24	05.0	11.2	19.8	31.0	44.6	1 00.8	19.4	40.4	04.0
.3	01.26	05.0	11.3	20.2	31.5	45.4	01.7	20.6	42.1	06.0
.4	01.28	05.1	11.5	20.5	32.0	46.1	02.7	21.9	43.7	08.0
6.5	0 01.30	0 05.2	0 11.7	0 20.8	0 32.5	0 46.8	1 03.7	1 23.2	1 45.3	2 10.0
.6	01.32	05.3	11.9	21.1	33.0	47.5	04.7	24.5	46.9	12.0
.7	01.34	05.4	12.1	21.4	33.5	48.2	05.7	25.8	48.5	14.0
.8	01.36	05.4	12.2	21.8	34.0	49.0	06.6	27.0	50.2	16.0
.9	01.38	05.5	12.4	22.1	34.5	49.7	07.6	28.3	51.8	18.0
7.0	0 01.40	0 05.6	0 12.6	0 22.4	0 35.0	0 50.4	1 08.6	1 29.6	1 53.4	2 20.0
.1	01.42	05.7	12.8	22.7	35.5	51.1	09.6	30.9	55.0	22.0
.2	01.44	05.8	13.0	23.0	36.0	51.8	10.6	32.2	56.6	24.0
.3	01.46	05.8	13.1	23.4	36.5	52.6	11.5	33.4	58.3	26.0
.4	01.48	05.9	13.3	23.7	37.0	53.3	12.5	34.7	59.9	28.0
7.5	0 01.50	0 06.0	0 13.5	0 24.0	0 37.5	0 54.0	1 13.5	1 36.0	2 01.5	2 30.0
.6	01.52	06.1	13.7	24.3	38.0	54.7	14.5	37.3	03.1	32.0
.7	01.54	06.2	13.9	24.6	38.5	55.4	15.5	38.6	04.7	34.0
.8	01.56	06.2	14.0	25.0	39.0	56.2	16.4	39.8	06.4	36.0
.9	01.58	06.3	14.2	25.3	39.5	56.9	17.4	41.1	08.0	38.0
8.0	0 01.60	0 06.4	0 14.4	0 25.6	0 40.0	0 57.6	1 18.4	1 42.4	2 09.6	2 40.0
.1	01.62	06.5	14.6	25.9	40.5	58.3	19.4	43.7	11.2	42.0
.2	01.64	06.6	14.8	26.2	41.0	59.0	20.4	45.0	12.8	44.0
.3	01.66	06.6	14.9	26.6	41.5	59.8	21.3	46.2	14.5	46.0
.4	01.68	06.7	15.1	26.9	42.0	1 00.5	22.3	47.5	16.1	48.0
8.5	0 01.70	0 06.8	0 15.3	0 27.2	0 42.5	1 01.2	1 23.3	1 48.8	2 17.7	2 50.0
.6	01.72	06.9	15.5	27.5	43.0	01.9	24.3	50.1	19.3	52.0
.7	01.74	07.0	15.7	27.8	43.5	02.6	25.3	51.4	20.9	54.0
.8	01.76	07.0	15.8	28.2	44.0	03.4	26.2	52.6	22.6	56.0
.9	01.78	07.1	16.0	28.5	44.5	04.1	27.2	53.9	24.2	58.0
9.0	0 01.80	0 07.2	0 16.2	0 28.8	0 45.0	1 04.8	1 28.2	1 55.2	2 25.8	3 00.0
.1	01.82	07.3	16.4	29.1	45.5	05.5	29.2	56.5	27.4	02.0
.2	01.84	07.4	16.6	29.4	46.0	06.2	30.2	57.8	29.0	04.0
.3	01.86	07.4	16.7	29.8	46.5	07.0	31.1	59.0	30.7	06.0
.4	01.88	07.5	16.9	30.1	47.0	07.7	32.1	2 00.3	32.3	08.0
9.5	0 01.90	0 07.6	0 17.1	0 30.4	0 47.5	1 08.4	1 33.1	2 01.6	2 33.9	3 10.0
.6	01.92	07.7	17.3	30.7	48.0	09.1	34.1	02.9	35.5	12.0
.7	01.94	07.8	17.5	31.0	48.5	09.8	35.1	04.2	37.1	14.0
.8	01.96	07.8	17.6	31.4	49.0	10.6	36.0	05.4	38.8	16.0
.9	01.98	07.9	17.8	31.7	49.5	11.3	37.0	06.7	40.4	18.0
10.0	0 02.00	0 08.0	0 18.0	0 32.0	0 50.0	1 12.0	1 38.0	2 08.0	2 42.0	3 20.0
.1	02.02	08.1	18.2	32.3	50.5	12.7	39.0	09.3	43.6	21.9
.2	02.04	08.2	18.4	32.6	51.0	13.4	40.0	10.6	45.2	23.9
.3	02.06	08.2	18.5	33.0	51.5	14.2	40.9	11.8	46.8	25.9
.4	02.08	08.3	18.7	33.3	52.0	14.9	41.9	13.1	48.5	27.9
10.5	0 02.10	0 08.4	0 18.9	0 33.6	0 52.5	1 15.6	1 42.9	2 14.4	2 50.1	3 29.9
.6	02.12	08.5	19.1	33.9	53.0	16.3	43.9	15.7	51.7	31.9
.7	02.14	08.6	19.3	34.2	53.5	17.0	44.9	17.0	53.3	33.9
.8	02.16	08.6	19.4	34.6	54.0	17.8	45.8	18.2	54.9	35.9
.9	02.18	08.7	19.6	34.9	54.5	18.5	46.8	19.5	56.6	37.9
11.0	0 02.20	0 08.8	0 19.8	0 35.2	0 55.0	1 19.2	1 47.8	2 20.8	2 58.2	3 39.9
.1	02.22	08.9	20.0	35.5	55.5	19.9	48.8	22.1	59.8	41.9
.2	02.24	09.0	20.2	35.8	56.0	20.6	49.8	23.4	3 01.4	43.9
.3	02.26	09.0	20.3	36.2	56.5	21.4	50.7	24.6	03.0	45.9
.4	02.28	09.1	20.5	36.5	57.0	22.1	51.7	25.9	04.7	47.9
11.5	0 02.30	0 09.2	0 20.7	0 36.8	0 57.5	1 22.8	1 52.7	2 27.2	3 06.3	3 49.9
.6	02.32	09.3	20.9	37.1	58.0	23.5	53.7	28.5	07.9	51.9
.7	02.34	09.4	21.1	37.4	58.5	24.2	54.7	29.7	09.5	53.9
.8	02.36	09.4	21.2	37.8	59.0	25.0	55.6	31.0	11.1	55.9
.9	02.38	09.5	21.4	38.1	59.5	25.7	56.6	32.3	12.7	57.9

TABLE XV.—DEFLECTION ANGLES FOR 10-CHORD SPIRAL

Δ	1	2	3	4	5	6	7	8	9	10
°	° ′	° ′	° ′	° ′	° ′	° ′	° ′	° ′	° ′	° ′
12.0	0 02.40	0 09.6	0 21.6	0 38.4	1 00.0	1 26.4	1 57.6	2 33.6	3 14.4	3 59.9
.1	02.42	09.7	21.8	38.7	00.5	27.1	58.6	34.9	16.0	4 01.9
.2	02.44	09.8	22.0	39.0	01.0	27.8	59.5	36.1	17.6	03.9
.3	02.46	09.8	22.1	39.4	01.5	28.6	2 00.5	37.4	19.2	05.9
.4	02.48	09.9	22.3	39.7	02.0	29.3	01.5	38.7	20.8	07.9
12.5	0 02.50	0 10.0	0 22.5	0 40.0	1 02.5	1 30.0	2 02.5	2 40.0	3 22.5	4 09.9
.6	02.52	10.1	22.7	40.3	03.0	30.7	03.5	41.3	24.1	11.9
.7	02.54	10.2	22.9	40.6	03.5	31.4	04.4	42.5	25.7	13.9
.8	02.56	10.2	23.0	41.0	04.0	32.2	05.4	43.8	27.3	15.9
.9	02.58	10.3	23.2	41.3	04.5	32.9	06.4	45.1	28.9	17.9
13.0	0 02.60	0 10.4	0 23.4	0 41.6	1 05.0	1 33.6	2 07.4	2 46.4	3 30.5	4 19.9
.1	02.62	10.5	23.6	41.9	05.5	34.3	08.4	47.7	32.2	21.9
.2	02.64	10.6	23.8	42.2	06.0	35.0	09.3	48.9	33.8	23.9
.3	02.66	10.6	23.9	42.6	06.5	35.8	10.3	50.2	35.4	25.9
.4	02.68	10.7	24.1	42.9	07.0	36.5	11.3	51.5	37.0	27.9
13.5	0 02.70	0 10.8	0 24.3	0 43.2	1 07.5	1 37.2	2 12.3	2 52.8	3 38.6	4 29.9
.6	02.72	10.9	24.5	43.5	08.0	37.9	13.3	54.0	40.3	31.9
.7	02.74	11.0	24.7	43.8	08.5	38.6	14.2	55.3	41.9	33.9
.8	02.76	11.0	24.8	44.2	09.0	39.4	15.2	56.6	43.5	35.9
.9	02.78	11.1	25.0	44.5	09.5	40.1	16.2	57.9	45.1	37.9
14.0	0 02.80	0 11.2	0 25.2	0 44.8	1 10.0	1 40.8	2 17.2	2 59.2	3 46.7	4 39.9
.1	02.82	11.3	25.4	45.1	10.5	41.5	18.2	3 00.4	48.4	41.9
.2	02.84	11.4	25.6	45.4	11.0	42.2	19.1	01.7	50.0	43.9
.3	02.86	11.4	25.7	45.8	11.5	43.0	20.1	03.0	51.6	45.9
.4	02.88	11.5	25.9	46.1	12.0	43.7	21.1	04.3	53.2	47.9
14.5	0 02.90	0 11.6	0 26.1	0 46.4	1 12.5	1 44.4	2 22.1	3 05.6	3 54.8	4 49.9
.6	02.92	11.7	26.3	46.7	13.0	45.1	23.1	06.8	56.4	51.9
.7	02.94	11.8	26.5	47.0	13.5	45.8	24.0	08.1	58.1	53.9
.8	02.96	11.8	26.6	47.4	14.0	46.6	25.0	09.4	59.7	55.8
.9	02.98	11.9	26.8	47.7	14.5	47.3	26.0	10.7	4 01.3	57.8
15.0	0 03.00	0 12.0	0 27.0	0 48.0	1 15.0	1 48.0	2 27.0	3 12.0	4 02.9	4 59.8
.1	03.02	12.1	27.2	48.3	15.5	48.7	28.0	13.2	04.5	5 01.8
.2	03.04	12.2	27.4	48.6	16.0	49.4	28.9	14.5	06.1	03.8
.3	03.06	12.2	27.5	49.0	16.5	50.2	29.9	15.8	07.8	05.8
.4	03.08	12.3	27.7	49.3	17.0	50.9	30.9	17.1	09.4	07.8
15.5	0 03.10	0 12.4	0 27.9	0 49.6	1 17.5	1 51.6	2 31.9	3 18.4	4 11.0	5 09.8
.6	03.12	12.5	28.1	49.9	18.0	52.3	32.9	19.6	12.6	11.8
.7	03.14	12.6	28.3	50.2	18.5	53.0	33.8	20.9	14.2	13.8
.8	02.16	12.6	28.4	50.6	19.0	53.7	34.8	22.2	15.9	15.8
.9	03.18	12.7	28.6	50.9	19.5	54.5	35.8	23.5	17.5	17.8
16.0	0 03.20	0 12.8	0 28.8	0 51.2	1 20.0	1 55.2	2 36.8	3 24.8	4 19.1	5 19.8
.1	03.22	12.9	29.0	51.5	20.5	55.9	37.8	26.0	20.7	21.8
.2	03.24	13.0	29.2	51.8	21.0	56.6	38.7	27.3	22.3	23.8
.3	03.26	13.0	29.3	52.2	21.5	57.3	39.7	28.6	23.9	25.8
.4	03.28	13.1	29.5	52.5	22.0	58.1	40.7	29.9	25.6	27.8
16.5	0 03.30	0 13.2	0 29.7	0 52.8	1 22.5	1 58.8	2 41.7	3 31.1	4 27.2	5 29.8
.6	03.32	13.3	29.9	53.1	23.0	59.5	42.7	32.4	28.8	31.8
.7	03.34	13.4	30.1	53.4	23.5	2 00.2	43.6	33.7	30.4	33.8
.8	03.36	13.4	30.2	53.8	24.0	00.9	44.6	35.0	32.0	35.8
.9	03.38	13.5	30.4	54.1	24.5	01.7	45.6	36.3	33.6	37.8
17.0	0 03.40	0 13.6	0 30.6	0 54.4	1 25.0	2 02.4	2 46.6	3 37.5	4 35.3	5 39.8
.1	03.42	13.7	30.8	54.7	25.5	03.1	47.5	38.8	36.9	41.8
.2	03.44	13.8	31.0	55.0	26.0	03.8	48.5	40.1	38.5	43.7
.3	03.46	13.8	31.1	55.4	26.5	04.5	49.5	41.4	40.1	45.7
.4	03.48	13.9	31.3	55.7	27.0	05.3	50.5	42.7	41.7	47.7
17.5	0 03.50	0 14.0	0 31.5	0 56.0	1 27.5	2 06.0	2 51.5	3 43.9	4 43.4	5 49.7
.6	03.52	14.1	31.7	56.3	28.0	06.7	52.4	45.2	45.0	51.7
.7	03.54	14.2	31.9	56.6	28.5	07.4	53.4	46.5	46.6	53.7
.8	03.56	14.2	32.0	57.0	29.0	08.1	54.4	47.8	48.2	55.7
.9	03.58	14.3	32.2	57.3	29.5	08.9	55.4	49.0	49.8	57.7

TABLE XV.—DEFLECTION ANGLES FOR 10-CHORD SPIRAL

Δ	1	2	3	4	5	6	7	8	9	10
°	° ′	° ′	° ′	° ′	° ′	° ′	° ′	° ′	° ′	° ′
18.0	0 03.60	0 14.4	0 32.4	0 57.6	1 30.0	2 09.6	2 56.4	3 50.3	4 51.5	5 59.7
.1	03.62	14.5	32.6	57.9	30.5	10.3	57.3	51.6	53.1	6 01.7
.2	03.64	14.6	32.8	58.2	31.0	11.0	58.3	52.9	54.7	03.7
.3	03.66	14.6	32.9	58.6	31.5	11.7	59.3	54.2	56.3	05.7
.4	03.68	14.7	33.1	58.9	32.0	12.5	3 00.3	55.4	57.9	07.7
18.5	0 03.70	0 14.8	0 33.3	0 59.2	1 32.5	2 13.2	3 01.3	3 56.7	4 59.5	6 09.7
.6	03.72	14.9	33.5	59.5	33.0	13.9	02.2	58.0	5 01.2	11.7
.7	03.74	15.0	33.7	59.8	33.5	14.6	03.2	59.3	02.8	13.7
.8	03.76	15.0	33.8	1 00.2	34.0	15.3	04.2	4 00.6	04.4	15.7
.9	03.78	15.1	34.0	00.5	34.5	16.1	05.2	01.8	06.0	17.7
19.0	0 03.80	0 15.2	0 34.2	1 00.8	1 35.0	2 16.8	3 06.2	4 03.1	5 07.6	6 19.7
.1	03.82	15.3	34.4	01.1	35.5	17.5	07.1	04.4	09.2	21.7
.2	03.84	15.4	34.6	01.4	36.0	18.2	08.1	05.7	10.9	23.7
.3	03.86	15.4	34.7	01.8	36.5	18.9	09.1	06.9	12.5	25.6
.4	03.88	15.5	34.9	02.1	37.0	19.7	10.1	08.2	14.1	27.6
19.5	0 03.90	0 15.6	0 35.1	1 02.4	1 37.5	2 20.4	3 11.1	4 09.5	5 15.7	6 29.6
.6	03.92	15.7	35.3	02.7	38.0	21.1	12.0	10.8	17.3	31.6
.7	03.94	15.8	35.5	03.0	38.5	21.8	13.0	12.1	18.9	33.6
.8	03.96	15.8	35.6	03.4	39.0	22.5	14.0	13.3	20.6	35.6
.9	03.98	15.9	35.8	03.7	39.5	23.3	15.0	14.6	22.2	37.6
20.0	0 04.00	0 16.0	0 36.0	1 04.0	1 40.0	2 24.0	3 16.0	4 15.9	5 23.8	6 39.6
.1	04.02	16.1	36.2	04.3	40.5	24.7	16.9	17.2	25.4	41.6
.2	04.04	16.2	36.4	04.6	41.0	25.4	17.9	18.5	27.0	43.6
.3	04.06	16.2	36.5	05.0	41.5	26.1	18.9	19.7	28.6	45.6
.4	04.08	16.3	36.7	05.3	42.0	26.9	19.9	21.0	30.3	47.6
20.5	0 04.10	0 16.4	0 36.9	1 05.6	1 42.5	2 27.6	3 20.9	4 22.3	5 31.9	6 49.6
.6	04.12	16.5	37.1	05.9	43.0	28.3	21.8	23.6	33.5	51.6
.7	04.14	16.6	37.3	06.2	43.5	29.0	22.8	24.8	35.1	53.6
.8	04.16	16.6	37.4	06.6	44.0	29.7	23.8	26.1	36.7	55.6
.9	04.18	16.7	37.6	06.9	44.5	30.5	24.8	27.4	38.3	57.5
21.0	0 04.20	0 16.8	0 37.8	1 07.2	1 45.0	2 31.2	3 25.8	4 28.7	5 40.0	6 59.5
.1	04.22	16.9	38.0	07.5	45.5	31.9	26.7	30.0	41.6	7 01.5
.2	04.24	17.0	38.2	07.8	46.0	32.6	27.7	31.2	43.2	03.5
.3	04.26	17.0	38.3	08.2	46.5	33.4	28.7	32.5	44.8	05.5
.4	04.28	17.1	38.5	08.5	47.0	34.1	29.7	33.8	46.4	07.5
21.5	0 04.30	0 17.2	0 38.7	1 08.8	1 47.5	2 34.8	3 30.7	4 35.1	5 48.0	7 09.5
.6	04.32	17.3	38.9	09.1	48.0	35.5	31.6	36.4	49.7	11.5
.7	04.34	17.4	39.1	09.4	48.5	36.2	32.6	37.6	51.3	13.5
.8	04.36	17.4	39.2	09.8	49.0	37.0	33.6	38.9	52.9	15.5
.9	04.38	17.5	39.4	10.1	49.5	37.7	34.6	40.2	54.5	17.5
22.0	0 04.40	0 17.6	0 39.6	1 10.4	1 50.0	2 38.4	3 35.5	4 41.5	5 56.1	7 19.5
.1	04.42	17.7	39.8	10.7	50.5	39.1	36.5	42.7	58.7	21.5
.2	04.44	17.8	40.0	11.0	51.0	39.8	37.5	44.0	59.3	23.5
.3	04.46	17.8	40.1	11.4	51.5	40.6	38.5	45.3	6 01.0	25.5
.4	04.48	17.9	40.3	11.7	52.0	41.3	39.5	46.6	02.6	27.5
22.5	0 04.50	0 18.0	0 40.5	1 12.0	1 52.5	2 42.0	3 40.5	4 47.9	6 04.2	7 29.4
.6	04.52	18.1	40.7	12.3	53.0	42.7	41.4	49.1	05.8	31.4
.7	04.54	18.2	40.9	12.6	53.5	43.4	42.4	50.4	07.4	33.4
.8	04.56	18.2	41.0	13.0	54.0	44.2	43.4	51.7	09.0	35.4
.9	04.58	18.3	41.2	13.3	54.5	44.9	44.4	53.0	10.7	37.4
23.0	0 04.60	0 18.4	0 41.4	1 13.6	1 55.0	2 45.6	3 45.3	4 54.2	6 12.3	7 39.4
.1	04.62	18.5	41.6	13.9	55.5	46.3	46.3	55.5	13.9	41.4
.2	04.64	18.6	41.8	14.2	56.0	47.0	47.3	56.8	15.5	43.4
.3	04.66	18.6	41.9	14.6	56.5	47.8	48.3	58.1	17.1	45.4
.4	04.68	18.7	42.1	14.9	57.0	48.5	49.2	59.4	18.7	47.4
23.5	0 04.70	0 18.8	0 42.3	1 15.2	1 57.5	2 49.2	3 50.2	5 00.6	6 20.4	7 49.4
.6	04.72	18.9	42.5	15.5	58.0	49.9	51.2	01.9	22.0	51.4
.7	04.74	19.0	42.7	15.8	58.5	50.6	52.2	03.2	23.6	53.3
.8	04.76	19.0	42.8	16.2	59.0	51.4	53.2	04.5	25.2	55.3
.9	04.78	19.1	43.0	16.5	59.5	52.1	54.2	05.7	26.8	57.3

TABLE XV.—DEFLECTION ANGLES FOR 10-CHORD SPIRAL

Δ	1	2	3	4	5	6	7	8	9	10
°	° ′	° ′	° ′	° ′	° ′	° ′	° ′	° ′	° ′	° ′
24.0	0 04.80	0 19.2	0 43.2	1 16.8	2 00.0	2 52.8	3 55.1	5 07.0	6 28.4	7 59.3
.1	04.82	19.3	43.4	17.1	00.5	53.5	56.1	08.3	30.1	8 01.3
.2	04.84	19.4	43.6	17.4	01.0	54.2	57.1	09.6	31.7	03.3
.3	04.86	19.4	43.7	17.8	01.5	55.0	58.1	10.9	33.3	05.3
.4	04.88	19.5	43.9	18.1	02.0	55.7	59.0	12.1	34.9	07.3
24.5	0 04.90	0 19.6	0 44.1	1 18.4	2 02.5	2 56.4	4 00.0	5 13.4	6 36.5	8 09.3
.6	04.92	19.7	44.3	18.7	03.3	57.1	01.0	14.7	38.1	11.3
.7	04.94	19.8	44.5	19.0	03.5	57.8	02.0	16.0	39.7	13.3
.8	04.96	19.8	44.6	19.4	04.0	58.6	03.0	17.2	41.3	15.3
.9	04.98	19.9	44.8	19.7	04.5	59.3	03.9	18.5	43.0	17.2
25.0	0 05.00	0 20.0	0 45.0	1 20.0	2 05.0	3 00.0	4 04.9	5 19.8	6 44.6	8 19.2
.1	05.02	20.1	45.2	20.3	05.5	00.7	05.9	21.1	46.2	21.2
.2	05.04	20.2	45.4	20.6	06.0	01.4	06.9	22.4	47.8	23.2
.3	05.06	20.2	45.5	21.0	06.5	02.1	07.9	23.6	49.4	25.2
.4	05.08	20.3	45.7	21.3	07.0	02.9	08.8	24.9	51.0	27.2
25.5	0 05.10	0 20.4	0 45.9	1 21.6	2 07.5	3 03.6	4 09.8	5 26.2	6 52.7	8 29.2
.6	05.12	20.5	46.1	21.9	08.0	04.3	10.8	27.5	54.3	31.2
.7	05.14	20.6	46.3	22.2	08.5	05.0	11.8	28.7	55.9	33.2
.8	05.16	20.6	46.4	22.6	09.0	05.8	12.8	30.0	57.5	35.2
.9	05.18	20.7	46.6	22.9	09.5	06.5	13.7	31.3	59.1	37.1
26.0	0 05.20	0 20.8	0 46.8	1 23.2	2 10.0	3 07.2	4 14.7	5 32.6	7 00.7	8 39.1
.1	05.22	20.9	47.0	23.5	10.5	07.9	15.7	33.9	02.3	41.1
.2	05.24	21.0	47.2	23.8	11.0	08.6	16.7	35.1	04.0	43.1
.3	05.26	21.0	47.3	24.2	11.5	09.3	17.6	36.4	05.6	45.1
.4	05.28	21.1	47.5	24.5	12.0	10.1	18.6	37.7	07.2	47.1
26.5	0 05.30	0 21.2	0 47.7	1 24.8	2 12.5	3 10.8	4 19.6	5 39.0	7 08.8	8 49.1
.6	05.32	21.3	47.9	25.1	13.0	11.5	20.6	40.2	10.4	51.1
.7	05.34	21.3	48.1	25.4	13.5	12.2	21.6	41.5	12.0	53.1
.8	05.36	21.4	48.2	25.8	14.0	13.0	22.5	42.8	13.7	55.1
.9	05.38	21.5	48.4	26.1	14.5	13.7	23.5	44.1	15.3	57.0
27.0	0 05.40	0 21.6	0 48.6	1 26.4	2 15.0	3 14.4	4 24.5	5 45.4	7 16.9	8 59.0
.2	05.44	21.8	49.0	27.0	16.0	15.8	26.5	47.9	20.1	9 03.0
.4	05.48	21.9	49.3	27.7	17.0	17.3	28.4	50.5	23.3	06.9
.6	05.52	22.1	49.7	28.3	18.0	18.7	30.4	53.0	26.6	10.9
.8	05.56	22.2	50.0	29.0	19.0	20.1	32.3	55.6	29.8	14.9
28.0	0 05.60	0 22.4	0 50.4	1 29.6	2 20.0	3 21.5	4 34.3	5 58.1	7 33.0	9 18.9
.2	05.64	22.6	50.8	30.2	21.0	23.0	36.2	6 00.7	36.3	22.9
.4	05.68	22.7	51.1	30.9	22.0	24.4	38.2	03.2	39.5	26.8
.6	05.72	22.9	51.5	31.5	23.0	25.9	40.2	05.8	42.7	30.8
.8	05.76	23.0	51.8	32.2	24.0	27.3	42.1	08.3	45.8	34.7
29.0	0 05.80	0 23.2	0 52.2	1 32.8	2 25.0	3 28.8	4 44.1	6 10.8	7 49.1	9 38.7
.2	05.84	23.4	52.6	33.4	26.0	30.2	46.0	13.4	52.3	42.7
.4	05.88	23.5	52.9	34.1	27.0	31.6	48.0	16.0	55.6	46.7
.6	05.92	23.7	53.3	34.7	28.0	33.1	49.9	18.5	58.8	50.7
.8	05.96	23.8	53.6	35.4	29.0	34.5	51.9	21.0	8 02.0	54.6
30.0	0 06.00	0 24.0	0 54.0	1 36.0	2 30.0	3 35.9	4 53.8	6 23.6	8 05.3	9 58.6
.2	06.04	24.2	54.4	36.6	31.0	37.4	55.8	26.2	08.5	10 02.6
.4	06.08	24.3	54.7	37.3	32.0	38.8	57.8	28.8	11.7	06.5
.6	06.12	24.5	55.1	37.9	33.0	40.3	59.7	31.3	15.0	10.5
.8	06.16	24.6	55.4	38.6	34.0	41.7	5 01.7	33.9	18.2	14.5
31.0	0 06.20	0 24.8	0 55.8	1 39.2	2 35.0	3 43.1	5 03.6	6 36.4	8 21.4	10 18.5
.2	06.24	25.0	56.2	39.8	36.0	44.6	05.6	39.0	24.6	22.5
.4	06.28	25.1	56.5	40.5	37.0	46.0	07.5	41.5	27.8	26.4
.6	06.32	25.3	56.9	41.1	38.0	47.5	09.5	44.0	31.0	30.4
.8	06.36	25.4	57.2	41.8	39.0	48.9	11.5	46.6	34.3	34.3
32.0	0 06.40	0 25.6	0 57.6	1 42.4	2 40.0	3 50.3	5 13.4	6 49.2	8 37.5	10 38.3
.2	06.44	25.8	58.0	43.0	41.0	51.8	15.4	51.7	40.7	42.3
.4	06.48	25.9	58.3	43.7	42.0	53.2	17.3	54.2	43.9	46.2
.6	06.52	26.1	58.7	44.3	43.0	54.7	19.3	56.8	47.1	50.2
.8	06.56	26.2	59.0	45.0	44.0	56.1	21.2	59.3	50.4	54.2

TABLE XV.—DEFLECTION ANGLES FOR 10-CHORD SPIRAL

Δ	1	2	3	4	5	6	7	8	9	10
°	° ′	° ′	° ′	° ′	° ′	° ′	° ′	° ′	° ′	° ′
33.0	0 06.60	0 26.4	0 59.4	1 45.6	2 45.0	3 57.5	5 23.2	7 01.9	8 53.6	10 58.1
.2	06.64	26.6	59.8	46.2	46.0	59.0	25.1	04.5	56.8	11 02.1
.4	06.68	26.7	1 00.1	46.9	47.0	4 00.4	27.1	07.0	9 00.0	06.1
.6	06.72	26.9	00.5	47.5	48.0	01.8	29.1	09.5	03.2	10.0
.8	06.76	27.0	00.8	48.2	49.0	03.3	31.0	12.1	06.5	14.0
34.0	0 06.80	0 27.2	1 01.2	1 48.8	2 50.0	4 04.7	5 33.0	7 14.7	9 09.7	11 18.0
.2	06.84	27.4	01.6	49.4	51.0	06.2	34.9	17.2	12.9	21.9
.4	06.88	27.5	01.9	50.1	52.0	07.6	36.9	19.7	16.1	25.9
.6	06.92	27.7	02.3	50.7	53.0	09.0	38.8	22.4	19.4	29.9
.8	06.96	27.8	02.6	51.4	54.0	10.5	40.8	24.9	22.6	33.8
35.0	0 07.00	0 28.0	1 03.0	1 52.0	2 55.0	4 11.9	5 42.8	7 27.4	9 25.8	11 37.8
.2	07.04	28.2	03.4	52.6	56.0	13.3	44.7	29.9	29.0	41.7
.4	07.08	28.3	03.7	53.3	57.0	14.8	46.7	32.6	32.3	45.7
.6	07.12	28.5	04.1	53.9	58.0	16.2	48.6	35.1	35.5	49.7
.8	07.16	28.6	04.4	54.6	59.0	17.7	50.6	37.6	38.7	53.6
36.0	0 07.20	0 28.8	1 04.8	1 55.2	3 00.0	4 19.1	5 52.5	7 40.2	9 41.9	11 57.6
.2	07.24	29.0	05.2	55.8	01.0	20.5	54.5	42.7	45.1	12 01.6
.4	07.28	29.1	05.5	56.5	02.0	22.0	56.5	45.2	48.3	05.5
.6	07.32	29.3	05.9	57.1	03.0	23.4	58.4	47.8	51.6	09.5
.8	07.36	29.4	06.2	57.8	04.0	24.8	6 00.4	50.3	54.8	13.4
37.0	0 07.40	0 29.6	1 06.6	1 58.4	3 05.0	4 26.3	6 02.3	7 52.9	9 58.0	12 17.4
.2	07.44	29.8	07.0	59.0	06.0	27.7	04.3	55.5	10 01.2	21.3
.4	07.48	29.9	07.3	59.7	07.0	29.2	06.2	58.0	04.4	25.3
.6	07.52	30.1	07.7	2 00.3	08.0	30.6	08.2	8 00.6	07.6	29.3
.8	07.56	30.2	08.0	01.0	09.0	32.0	10.1	03.1	10.9	33.2
38.0	0 07.60	0 30.4	1 08.4	2 01.6	3 10.0	4 33.5	6 12.1	8 05.6	10 14.1	12 37.1
.2	07.64	30.6	08.8	02.2	11.0	34.9	14.0	08.2	17.3	41.1
.4	07.68	30.7	09.1	02.9	12.0	36.3	16.0	10.7	20.5	45.0
.6	07.72	30.9	09.5	03.5	13.0	37.8	18.0	13.3	23.7	49.0
.8	07.76	31.0	09.8	04.2	14.0	39.2	19.9	15.8	26.9	52.9
39.0	0 07.80	0 31.2	1 10.2	2 04.8	3 15.0	4 40.7	6 21.9	8 18.4	10 30.1	12 56.9
.2	07.84	31.4	10.6	05.4	16.0	42.1	23.8	20.9	33.4	13 00.9
.4	07.88	31.5	10.9	06.1	17.0	43.5	25.8	23.5	36.6	04.8
.6	07.92	31.7	11.3	06.7	18.0	45.0	27.7	26.0	39.8	08.8
.8	07.96	31.8	11.6	07.4	19.0	46.4	29.7	28.6	43.0	12.7
40.0	0 08.00	0 32.0	1 12.0	2 08.0	3 20.0	4 47.9	6 31.6	8 31.1	10 46.2	13 16.7
.2	08.04	32.2	12.4	08.6	20.9	49.3	33.6	33.7	49.4	20.6
.4	08.08	32.3	12.7	09.3	21.9	50.7	35.5	36.2	52.7	24.6
.6	08.12	32.5	13.1	09.9	22.9	52.2	37.5	38.8	55.9	28.5
.8	08.16	32.6	13.4	10.6	23.9	53.6	39.5	41.3	59.1	32.4
41.0	0 08.20	0 32.8	1 13.8	2 11.2	3 24.9	4 55.0	6 41.4	8 43.9	11 02.3	13 36.4
.2	08.24	33.0	14.2	11.8	25.9	56.5	43.4	46.4	05.5	40.4
.4	08.28	33.1	14.5	12.5	26.9	57.9	45.3	49.0	08.7	44.3
.6	08.32	33.3	14.9	13.1	27.9	59.4	47.3	51.5	11.9	48.3
.8	08.36	33.4	15.2	13.8	28.9	5 00.8	49.2	54.0	15.1	52.2
42.0	0 08.40	0 33.6	1 15.6	2 14.4	3 29.9	5 02.2	6 51.1	8 56.5	11 18.3	13 56.1
.2	08.44	33.8	16.0	15.0	30.9	03.7	53.1	59.1	21.6	14 00.1
.4	08.48	33.9	16.3	15.7	31.9	05.1	55.1	9 01.7	24.8	04.0
.6	08.52	34.1	16.7	16.3	32.9	06.5	57.0	04.2	28.0	08.0
.8	08.56	34.2	17.0	17.0	33.9	08.0	59.0	06.8	31.2	11.9
43.0	0 08.60	0 34.4	1 17.4	2 17.6	3 34.9	5 09.4	7 00.9	9 09.3	11 34.4	14 15.8
.2	08.64	34.6	17.8	18.2	35.9	10.9	02.9	11.9	37.6	18.8
.4	08.68	34.7	18.1	18.9	36.9	12.3	04.8	14.4	40.8	23.7
.6	08.72	34.9	18.5	19.5	37.9	13.7	06.7	16.9	44.0	27.7
.8	08.76	35.0	18.8	20.2	38.9	15.2	08.7	19.5	47.2	31.6
44.0	0 08.80	0 35.2	1 19.2	2 20.8	3 39.9	5 16.6	7 10.7	9 22.0	11 50.4	14 35.5
.2	08.84	35.4	19.6	21.4	40.9	18.0	12.6	24.6	53.6	39.5
.4	08.88	35.5	19.9	22.1	41.9	19.5	14.6	27.1	56.8	43.4
.6	08.92	35.7	20.3	22.7	42.9	20.9	16.5	29.7	12 00.0	47.4
.8	08.96	35.8	20.6	23.4	43.9	22.4	18.5	32.2	03.3	51.3
45.0	0 09.00	0 36.0	1 21.0	2 24.0	3 44.9	5 23.8	7 20.4	9 34.8	12 06.5	14 55.2

TABLE XVI.—COEFFICIENTS OF a_1 FOR DEFLECTIONS TO ANY CHORD POINT ON SPIRAL*

DEFLECTION TO CHORD POINT NO.	TRANSIT AT CHORD POINT NUMBER																					DEFLECTION TO CHORD POINT NO.
	0	1	2	3	4	5	6	7	8	9	10	11	12	13	14	15	16	17	18	19	20	
T.S. = 0	0	2	8	18	32	50	72	98	128	162	200	242	288	338	392	450	512	578	648	722	800	T.S. = 0
1	1	0	5	14	27	44	65	90	119	152	189	230	275	324	377	434	495	560	629	702	779	1
2	4	4	0	8	20	36	56	80	108	140	176	216	260	308	360	416	476	540	608	680	756	2
3	9	10	7	0	11	26	45	68	95	126	161	200	243	290	341	396	455	518	585	656	731	3
4	16	18	16	10	0	14	32	54	80	110	144	182	224	270	320	374	432	494	560	630	704	4
5	25	28	27	22	13	0	17	38	63	92	125	162	203	248	297	350	407	468	533	602	675	5
6	36	40	40	36	28	16	0	20	44	72	104	140	180	224	272	324	380	440	504	572	644	6
7	49	54	55	52	45	34	19	0	23	50	81	116	155	198	245	296	351	410	473	540	611	7
8	64	70	72	70	64	54	40	22	0	26	56	90	128	170	216	266	320	378	440	506	576	8
9	81	88	91	90	85	76	63	46	25	0	29	62	99	140	185	234	287	344	405	470	539	9
10	100	108	112	112	108	100	88	72	52	28	0	32	68	108	152	200	252	308	368	432	500	10
11	121	130	135	136	133	126	115	100	81	58	31	0	35	74	117	164	215	270	329	392	459	11
12	144	154	160	162	160	154	144	130	112	90	64	34	0	38	80	126	176	230	288	350	416	12
13	169	180	187	190	189	184	175	162	145	124	99	70	37	0	41	86	135	188	245	306	371	13
14	196	208	216	220	220	216	208	196	180	160	136	108	76	40	0	44	92	144	200	260	324	14
15	225	238	247	252	253	250	243	232	217	198	175	148	117	82	43	0	47	98	153	212	275	15
16	256	270	280	286	288	286	280	270	256	238	216	190	160	126	88	46	0	50	104	162	224	16
17	289	304	315	322	325	324	319	310	297	280	259	234	205	172	135	94	49	0	53	110	171	17
18	324	340	352	360	364	364	360	352	340	324	304	280	252	220	184	144	100	52	0	56	116	18
19	361	378	391	400	405	406	403	396	385	370	351	328	301	270	235	196	153	106	55	0	59	19
20	400	418	432	442	448	450	448	442	432	418	400	378	352	322	288	250	208	162	112	58	0	20

* See section 5-7 for theory and example.

TABLE XVI-A.—CORRECTIONS C_s TO DEFLECTIONS FROM T.S.*

Δ	R=Ratio I to L_s					
	R=0.5	R=0.6	R=0.7	R=0.8	R=0.9	R=1.0
	C_s	C_s	C_s	C_s	C_s	C_s
Deg.	Sec.	Sec.	Sec.	Sec.	Sec.	Sec.
10	0.0	0.1	0.3	0.8	1.7	3.1
15	0.2	0.5	1.2	2.8	5.5	10.5
20	0.4	1.2	2.9	6.5	13.2	24.8
25	0.8	2.3	5.7	12.7	25.8	48.6
30	1.3	3.9	9.8	22.0	44.6	84.1
32	1.6	4.8	12.0	26.7	54.2	102.2
34	1.9	5.7	14.4	32.0	65.0	122.7
36	2.3	6.8	17.1	38.0	77.2	145.8
38	2.7	7.9	20.1	44.8	90.9	171.6
40	3.1	9.2	23.4	52.2	106.1	200.4
41	3.3	9.9	25.2	56.2	114.3	215.9
42	3.6	10.7	27.1	60.4	122.9	232.2
43	3.9	11.5	29.1	64.9	132.0	249.4
44	4.1	12.3	31.2	69.5	141.5	267.4
45	4.4	13.2	33.3	74.4	151.4	286.2
46	4.7	14.1	35.6	79.5	161.8	305.9
47	5.0	15.0	38.0	84.8	172.6	326.5
48	5.4	16.0	40.5	90.4	184.0	348.0
49	5.7	17.0	43.1	96.2	195.8	370.5
50	6.0	18.1	45.8	102.2	208.2	394.0
51	6.4	19.2	48.5	108.5	221.0	418.4
52	6.8	20.3	51.5	115.1	234.3	443.8
53	7.2	21.5	54.5	121.8	248.2	470.3
54	7.6	22.7	57.7	129.0	262.7	497.8
55	8.1	24.0	60.9	136.3	277.7	526.4
56	8.5	25.4	64.4	143.9	293.3	556.0
57	9.0	26.8	67.8	151.8	309.4	586.8
58	9.5	28.3	71.5	159.9	326.1	618.8
59	10.0	29.8	75.2	168.4	343.5	651.9
60	10.5	31.4	79.2	177.2	361.6	686.2

* See section 5-17 and Appendixes B and C for theory and examples.

TABLE XVI-B.—CORRECTIONS C_ϕ TO DEFLECTIONS FROM S.C.*

Δ	R = Ratio l to L_s							
	R=0.3	R=0.4	R=0.5	R=0.6	R=0.7	R=0.8	R=0.9	R=1.0
	C_ϕ	C_ϕ	C_ϕ	C_ϕ	C_ϕ	C_ϕ	C_ϕ	C_ϕ
Deg.	Sec.	Sec.	Sec.	Sec.	Sec.	Sec.	Sec.	Sec.
10	0.0	0.0	0.4	0.8	1.3	1.8	2.4	3.1
15	0.0	0.7	1.6	2.8	4.5	6.4	8.4	10.5
20	0.6	1.6	3.4	6.3	10.0	14.5	19.6	24.8
25	1.1	3.1	6.4	12.4	19.5	28.4	38.8	48.6
30	1.9	5.4	11.3	21.0	33.2	49.5	66.2	84.1
32	2.4	6.6	14.1	25.2	40.8	59.7	80.6	102.2
34	2.9	8.0	17.0	30.4	49.1	71.6	95.0	122.7
36	3.4	9.5	20.1	36.5	58.5	85.2	113.2	145.8
38	4.0	11.1	23.8	43.2	69.0	100.3	134.6	171.6
40	4.6	12.9	27.6	50.2	80.2	117.0	158.2	200.4
41	5.0	13.8	29.7	53.7	86.3	126.2	170.3	215.9
42	5.4	14.8	31.9	57.7	92.8	135.8	183.2	232.2
43	5.7	15.8	34.3	62.2	99.7	145.9	196.8	249.4
44	6.1	17.0	36.8	66.8	107.0	156.3	211.1	267.4
45	6.5	18.2	39.4	71.5	114.6	167.1	226.0	286.2
46	7.0	19.5	42.2	76.4	122.5	178.7	241.6	305.9
47	7.5	20.8	45.0	81.5	130.7	190.8	257.9	326.5
48	8.0	22.1	47.9	86.8	139.2	203.4	274.9	348.0
49	8.5	23.6	51.0	92.4	148.1	216.4	292.6	370.5
50	9.0	25.2	54.2	98.2	157.4	230.0	311.0	394.0
51	9.5	26.7	57.5	104.3	167.1	244.2	330.1	418.4
52	10.1	28.2	61.0	110.6	177.2	259.0	350.1	443.8
53	10.7	29.8	64.6	117.2	187.8	274.4	371.0	470.3
54	11.3	31.5	68.4	124.0	198.9	290.5	392.8	497.8
55	11.9	33.3	72.3	131.0	210.4	307.1	415.6	526.4
56	12.5	35.2	76.3	138.3	222.3	324.3	439.0	556.0
57	13.1	37.1	80.5	145.8	234.6	342.1	463.3	586.8
58	13.8	39.1	85.0	153.8	247.3	360.6	488.5	618.8
59	14.6	41.2	89.9	162.0	260.4	379.9	514.6	651.9
60	15.4	43.3	95.3	170.5	273.9	400.1	541.5	686.2

*See Appendix C for notation and examples.

TABLE XVI-C.—CORRECTIONS FOR COMBINING SPIRALS*

Δ_S	Δ_N	C_b	C_p	C_a	Δ_S	Δ_N	C_b	C_p	C_a
Deg.	Deg.	Sec.	Ft.	Sec.	Deg.	Deg.	Sec.	Ft.	Sec.
10	0	0	0	0	18	3	29.84	0.018	9.95
10	1	3.30	0.002	1.10	18	4	37.11	0.022	12.38
10	2	5.86	0.004	1.95	18	5	43.05	0.026	14.36
10	3	7.69	0.005	2.56	18	6	47.66	0.029	15.90
10	4	8.79	0.005	2.93	18	7	50.93	0.031	17.00
10	5	9.15	0.006	3.05	18	8	52.88	0.032	17.65
10	6	8.78	0.005	2.93	18	9	53.51	0.032	17.87
10	7	7.68	0.005	2.56	18	10	52.81	0.032	17.64
10	8	5.85	0.004	1.95	18	11	50.80	0.031	16.98
10	9	3.29	0.002	1.10	18	12	47.46	0.029	15.87
10	10	0	0	0	18	13	42.82	0.026	14.33
12	0	0	0	0	18	14	36.86	0.022	12.34
12	1	4.84	0.003	1.62	18	15	29.60	0.018	9.92
12	2	8.80	0.005	2.94	18	16	21.03	0.013	7.05
12	3	11.88	0.007	3.96	18	17	11.16	0.007	3.75
12	4	14.07	0.009	4.69	18	18	0	0	0
12	5	15.39	0.009	5.13	20	0	0	0	0
12	6	15.82	0.010	5.28	20	1	14.04	0.008	4.68
12	7	15.37	0.009	5.13	20	2	26.59	0.016	8.86
12	8	14.05	0.009	4.69	20	3	37.65	0.023	12.55
12	9	11.85	0.007	3.96	20	4	47.22	0.028	15.74
12	10	8.77	0.005	2.93	20	5	55.30	0.033	18.44
12	11	4.82	0.003	1.61	20	6	61.89	0.037	20.65
12	12	0	0	0	20	7	67.00	0.040	22.36
14	0	0	0	0	20	8	70.64	0.043	23.58
14	1	6.69	0.004	2.23	20	9	72.79	0.044	24.31
14	2	12.34	0.007	4.11	20	10	73.47	0.044	24.55
14	3	16.96	0.010	5.66	20	11	72.68	0.044	24.30
14	4	20.55	0.012	6.85	20	12	70.42	0.043	23.55
14	5	23.11	0.014	7.71	20	13	66.70	0.040	22.32
14	6	24.64	0.015	8.22	20	14	61.52	0.037	20.60
14	7	25.14	0.015	8.39	20	15	54.89	0.033	18.39
14	8	24.61	0.015	8.22	20	16	46.80	0.028	15.69
14	9	23.06	0.014	7.70	20	17	37.26	0.023	12.50
14	10	20.49	0.012	6.84	20	18	26.28	0.016	8.82
14	11	16.89	0.010	5.65	20	19	13.86	0.008	4.66
14	12	12.28	0.007	4.11	20	20	0	0	0
14	13	6.65	0.004	2.22	22	0	0	0	0
14	14	0	0	0	22	1	17.11	0.010	5.70
16	0	0	0	0	22	2	32.58	0.019	10.86
16	1	8.83	0.005	2.94	22	3	46.39	0.028	15.47
16	2	16.48	0.010	5.50	22	4	58.56	0.035	19.53
16	3	22.95	0.014	7.65	22	5	69.08	0.041	23.04
16	4	28.23	0.017	9.41	22	6	77.97	0.047	26.01
16	5	32.32	0.019	10.78	22	7	85.21	0.051	28.44
16	6	35.24	0.021	11.76	22	8	90.82	0.055	30.32
16	7	36.98	0.022	12.34	22	9	94.81	0.057	31.66
16	8	37.55	0.023	12.54	22	10	97.16	0.058	32.46
16	9	36.94	0.022	12.34	22	11	97.89	0.059	32.72
16	10	35.16	0.021	11.75	22	12	97.00	0.058	32.44
16	11	32.21	0.019	10.77	22	13	94.50	0.057	31.62
16	12	28.09	0.017	9.40	22	14	90.38	0.055	30.26
16	13	22.81	0.014	7.63	22	15	84.66	0.051	28.37
16	14	16.36	0.010	5.48	22	16	77.33	0.047	25.93
16	15	8.76	0.005	2.94	22	17	68.41	0.041	22.96
16	16	0	0	0	22	18	57.89	0.035	19.44
18	0	0	0	0	22	19	45.79	0.028	15.39
18	1	11.28	0.007	3.76	22	20	32.10	0.019	10.80
18	2	21.23	0.013	7.08	22	21	16.84	0.010	5.67
18	3	29.84	0.018	9.95	22	22	0	0	0

* See section 5-17 and Appendixes B and C for theory and examples.

TABLE XVI-C.—CORRECTIONS FOR COMBINING SPIRALS

Δ_S	Δ_N	C_b	C_p	C_a	Δ_S	Δ_N	C_b	C_p	C_a
Deg.	Deg.	Sec.	Ft.	Sec.	Deg.	Deg.	Sec.	Ft.	Sec.
22	22	0	0	0	28	6	137.48	0.082	45.86
24	0	0	0	0	28	7	152.96	0.091	51.04
24	1	20.50	0.012	6.83	28	8	166.33	0.099	55.53
24	2	39.19	0.023	13.06	28	9	177.60	0.106	59.31
24	3	56.07	0.033	18.70	28	10	186.77	0.112	62.40
24	4	71.15	0.043	23.73	28	11	193.84	0.116	64.79
24	5	84.43	0.051	28.16	28	12	198.83	0.119	66.50
24	6	95.91	0.057	31.99	28	13	201.73	0.121	67.50
24	7	105.59	0.063	35.24	28	14	202.56	0.122	67.82
24	8	113.48	0.068	37.88	28	15	201.31	0.121	67.45
24	9	119.59	0.072	39.94	28	16	198.00	0.119	66.39
24	10	123.92	0.074	41.40	28	17	192.64	0.116	64.64
24	11	126.46	0.076	42.27	28	18	185.22	0.112	62.20
24	12	127.23	0.077	42.55	28	19	175.77	0.106	59.08
24	13	126.24	0.076	42.24	28	20	164.27	0.099	55.26
24	14	123.47	0.074	41.34	28	21	150.75	0.091	50.76
24	15	118.95	0.072	39.86	28	22	135.21	0.082	45.58
24	16	112.68	0.068	37.78	28	23	117.66	0.071	39.70
24	17	104.65	0.063	35.12	28	24	98.10	0.060	33.14
24	18	94.88	0.057	31.87	28	25	76.54	0.047	25.89
24	19	83.38	0.051	28.03	28	26	53.00	0.032	17.95
24	20	70.14	0.043	23.60	28	27	27.49	0.017	9.32
24	21	55.18	0.033	18.58	28	28	0	0	0
24	22	38.50	0.023	12.98	30	0	0	0	0
24	23	20.10	0.012	6.78	30	1	32.60	0.019	10.87
24	24	0	0	0	30	2	62.90	0.037	20.97
26	0	0	0	0	30	3	90.91	0.054	30.31
26	1	24.21	0.014	8.07	30	4	116.61	0.069	38.88
26	2	46.44	0.028	15.48	30	5	140.03	0.083	46.70
26	3	66.71	0.040	22.24	30	6	161.16	0.096	53.76
26	4	85.01	0.051	28.35	30	7	180.02	0.107	60.07
26	5	101.35	0.060	33.80	30	8	196.60	0.117	65.63
26	6	115.74	0.069	38.61	30	9	210.91	0.126	70.43
26	7	128.16	0.077	42.77	30	10	222.96	0.133	74.49
26	8	138.64	0.083	46.28	30	11	232.75	0.139	77.80
26	9	147.18	0.088	49.15	30	12	240.29	0.144	80.36
26	10	153.77	0.092	51.38	30	13	245.59	0.147	82.18
26	11	158.43	0.095	52.96	30	14	248.66	0.149	83.26
26	12	161.16	0.097	53.90	30	15	249.49	0.150	83.59
26	13	161.96	0.097	54.20	30	16	248.10	0.149	83.19
26	14	160.85	0.097	53.86	30	17	244.50	0.147	82.04
26	15	157.82	0.095	52.88	30	18	238.70	0.144	80.16
26	16	152.88	0.092	51.26	30	19	230.69	0.139	77.54
26	17	146.04	0.088	49.01	30	20	220.49	0.133	74.18
26	18	137.31	0.083	46.11	30	21	208.11	0.126	70.08
26	19	126.68	0.077	42.58	30	22	193.56	0.117	65.24
26	20	114.18	0.069	38.41	30	23	176.84	0.107	59.67
26	21	99.80	0.060	33.61	30	24	157.97	0.096	53.36
26	22	83.55	0.051	28.16	30	25	136.95	0.083	46.31
26	23	65.43	0.040	22.08	30	26	113.79	0.069	38.53
26	24	45.47	0.028	15.36	30	27	88.51	0.054	30.00
26	25	23.65	0.014	8.00	30	28	61.11	0.037	20.74
26	26	0	0	0	30	29	31.60	0.019	10.74
28	0	0	0	0	30	30	0	0	0
28	1	28.24	0.017	9.41	31	0	0	0	0
28	2	54.34	0.032	18.12	31	1	34.91	0.021	11.64
28	3	78.32	0.047	26.11	31	2	67.44	0.040	22.48
28	4	100.16	0.060	33.40	31	3	97.58	0.058	32.53
28	5	119.88	0.071	39.98	31	4	125.34	0.074	41.80
28	6	137.48	0.082	45.86	31	5	150.73	0.089	50.27

TABLE XVI-C.—CORRECTIONS FOR COMBINING SPIRALS

Δ_S	Δ_N	C_b	C_p	C_a	Δ_S	Δ_N	C_b	C_p	C_a
Deg.	Deg.	Sec.	Ft.	Sec.	Deg.	Deg.	Sec.	Ft.	Sec.
31	5	150.73	0.089	50.27	32	32	0	0	0
31	6	173.75	0.103	57.96	33	0	0	0	0
31	7	194.40	0.115	64.87	33	1	39.78	0.023	13.26
31	8	212.70	0.126	71.00	33	2	77.01	0.045	25.67
31	9	228.65	0.136	76.36	33	3	111.68	0.066	37.23
31	10	242.26	0.144	80.94	33	4	143.80	0.085	47.95
31	11	253.53	0.151	84.74	33	5	173.37	0.102	57.82
31	12	262.46	0.157	87.78	33	6	200.41	0.118	66.86
31	13	269.08	0.161	90.04	33	7	224.92	0.133	75.06
31	14	273.37	0.163	91.53	33	8	246.90	0.146	82.42
31	15	275.36	0.165	92.26	33	9	266.37	0.158	88.95
31	16	275.04	0.165	92.22	33	10	283.32	0.168	94.66
31	17	272.43	0.163	91.41	33	11	297.77	0.177	99.53
31	18	267.54	0.161	89.84	33	12	309.72	0.184	103.58
31	19	260.36	0.157	87.51	33	13	319.19	0.190	106.81
31	20	250.92	0.151	84.41	33	14	326.18	0.194	109.21
31	21	239.22	0.144	80.55	33	15	330.69	0.197	110.80
31	22	225.27	0.136	75.93	33	16	332.74	0.199	111.56
31	23	209.07	0.126	70.54	33	17	332.33	0.199	111.51
31	24	190.64	0.115	64.40	33	18	329.48	0.197	110.64
31	25	170.00	0.103	57.49	33	19	324.19	0.194	108.96
31	26	147.13	0.089	49.82	33	20	316.48	0.190	106.46
31	27	122.07	0.074	41.38	33	21	306.34	0.184	103.51
31	28	94.81	0.058	32.18	33	22	293.80	0.177	99.03
31	29	65.38	0.040	22.22	33	23	278.86	0.168	94.09
31	30	33.77	0.021	11.49	33	24	261.53	0.158	88.34
31	31	0	0	0	33	25	241.83	0.146	81.78
32	0	0	0	0	33	26	219.76	0.133	74.40
32	1	37.30	0.022	12.44	33	27	195.34	0.118	66.21
32	2	72.14	0.043	24.05	33	28	168.57	0.102	57.21
32	3	104.50	0.062	34.84	33	29	139.47	0.085	47.40
32	4	134.40	0.079	44.82	33	30	108.05	0.066	36.77
32	5	161.84	0.096	53.98	33	31	74.32	0.045	25.33
32	6	186.83	0.111	62.33	33	32	38.30	0.023	13.07
32	7	209.37	0.124	69.87	33	33	0	0	0
32	8	229.47	0.136	76.60	34	0	0	0	0
32	9	247.14	0.147	82.53	34	1	42.35	0.025	14.12
32	10	262.38	0.156	87.66	34	2	82.05	0.048	27.35
32	11	275.20	0.164	91.99	34	3	119.11	0.070	39.71
32	12	285.61	0.170	95.52	34	4	153.53	0.090	51.20
32	13	293.61	0.175	98.25	34	5	185.33	0.109	61.81
32	14	299.21	0.179	100.18	34	6	214.50	0.127	71.56
32	15	302.42	0.181	101.33	34	7	241.06	0.142	80.44
32	16	303.25	0.181	101.68	34	8	265.00	0.157	88.46
32	17	301.71	0.181	101.24	34	9	286.34	0.169	95.62
32	18	297.80	0.179	100.00	34	10	305.09	0.181	101.93
32	19	291.53	0.175	97.98	34	11	321.25	0.191	107.38
32	20	282.91	0.170	95.17	34	12	334.83	0.199	111.98
32	21	271.96	0.164	91.57	34	13	345.84	0.206	115.72
32	22	258.68	0.156	87.19	34	14	354.28	0.211	118.62
32	23	243.07	0.147	82.02	34	15	360.17	0.215	120.67
32	24	225.16	0.136	76.06	34	16	363.52	0.217	121.88
32	25	204.95	0.124	69.31	34	17	364.32	0.218	122.24
32	26	182.45	0.111	61.77	34	18	362.60	0.217	121.77
32	27	157.67	0.096	53.45	34	19	358.37	0.215	120.44
32	28	130.63	0.079	44.34	34	20	351.62	0.211	118.28
32	29	101.33	0.062	34.44	34	21	342.38	0.206	115.28
32	30	69.78	0.043	23.75	34	22	330.65	0.199	111.44
32	31	36.00	0.022	12.27	34	23	316.44	0.191	106.77
32	32	0	0	0	34	24	299.77	0.181	101.25

TABLE XVI-C.—CORRECTIONS FOR COMBINING SPIRALS

Δ_S	Δ_N	C_b	C_p	C_a	Δ_S	Δ_N	C_b	C_p	C_a
Deg.	Deg.	Sec.	Ft.	Sec.	Deg.	Deg.	Sec.	Ft.	Sec.
34	24	299.77	0.181	101.25	36	12	388.03	0.230	129.77
34	25	280.64	0.169	94.90	36	13	402.36	0.238	134.63
34	26	259.07	0.157	87.71	36	14	413.95	0.246	138.60
34	27	235.07	0.142	79.68	36	15	422.83	0.251	141.66
34	28	208.65	0.127	70.82	36	16	428.99	0.255	143.83
34	29	179.82	0.109	61.11	36	17	432.45	0.258	145.10
34	30	148.59	0.090	50.57	36	18	433.21	0.258	145.47
34	31	114.99	0.070	39.19	36	19	431.30	0.258	144.95
34	32	79.10	0.048	26.97	36	20	426.72	0.255	143.54
34	33	40.68	0.025	13.94	36	21	419.47	0.251	141.24
34	34	0	0	0	36	22	409.58	0.246	138.04
35	0	0	0	0	36	23	397.05	0.238	133.96
35	1	45.00	0.026	15.00	36	24	381.90	0.230	128.99
35	2	87.27	0.051	29.09	36	25	364.13	0.219	123.13
35	3	126.80	0.074	42.28	36	26	343.76	0.207	116.38
35	4	163.62	0.096	54.56	36	27	320.81	0.194	108.74
35	5	197.71	0.116	65.94	36	28	295.28	0.179	100.21
35	6	229.10	0.135	76.43	36	29	267.19	0.162	90.80
35	7	257.79	0.152	86.02	36	30	236.55	0.144	80.50
35	8	283.78	0.167	94.73	36	31	203.38	0.124	69.31
35	9	307.08	0.181	102.55	36	32	167.68	0.102	57.23
35	10	327.70	0.194	109.48	36	33	129.49	0.079	44.26
35	11	345.64	0.205	115.53	36	34	88.80	0.054	30.40
35	12	360.93	0.214	120.70	36	35	45.63	0.028	15.65
35	13	373.56	0.222	125.00	36	36	0	0	0
35	14	383.54	0.228	128.42	37	0	0	0	0
35	15	390.88	0.233	130.96	37	1	50.57	0.030	16.86
35	16	395.60	0.236	132.64	37	2	98.23	0.057	32.75
35	17	397.69	0.237	133.44	37	3	142.98	0.084	47.67
35	18	397.18	0.237	133.37	37	4	184.83	0.108	61.63
35	19	394.07	0.236	132.44	37	5	223.79	0.131	74.64
35	20	388.36	0.233	130.64	37	6	259.86	0.152	86.69
35	21	380.08	0.228	127.98	37	7	293.05	0.172	97.79
35	22	369.24	0.222	124.45	37	8	323.38	0.190	107.95
35	23	355.84	0.214	120.06	37	9	350.84	0.207	117.16
35	24	339.89	0.205	114.80	37	10	375.45	0.221	125.43
35	25	321.40	0.194	108.68	37	11	397.22	0.234	132.77
35	26	300.40	0.181	101.70	37	12	416.14	0.246	139.17
35	27	276.89	0.167	93.86	37	13	432.25	0.256	144.63
35	28	250.88	0.152	85.15	37	14	445.54	0.264	149.17
35	29	222.39	0.135	75.58	37	15	456.02	0.270	152.78
35	30	191.42	0.116	65.14	37	16	463.70	0.275	155.47
35	31	158.00	0.096	53.85	37	17	468.60	0.279	157.23
35	32	122.13	0.074	41.68	37	18	470.72	0.280	158.07
35	33	83.83	0.051	28.66	37	19	470.08	0.280	157.98
35	34	43.12	0.026	14.76	37	20	466.68	0.279	156.98
35	35	0	0	0	37	21	460.55	0.275	155.07
36	0	0	0	0	37	22	451.68	0.270	152.23
36	1	47.74	0.028	15.91	37	23	440.10	0.264	148.48
36	2	92.66	0.054	30.89	37	24	425.81	0.256	143.82
36	3	134.76	0.079	44.93	37	25	408.83	0.246	138.24
36	4	174.05	0.102	58.04	37	26	389.17	0.234	131.75
36	5	210.53	0.124	70.22	37	27	366.84	0.221	124.34
36	6	244.22	0.144	81.47	37	28	341.86	0.207	116.02
36	7	275.12	0.162	91.81	37	29	314.24	0.190	106.79
36	8	303.23	0.179	101.22	37	30	284.00	0.172	96.64
36	9	328.57	0.194	109.72	37	31	251.14	0.152	85.58
36	10	351.15	0.207	117.31	37	32	215.69	0.131	73.61
36	11	370.96	0.219	123.99	37	33	177.65	0.108	60.72
36	12	388.03	0.230	129.77	37	34	137.05	0.084	46.92

TABLE XVI-C.—CORRECTIONS FOR COMBINING SPIRALS

Δ_S	Δ_N	C_b	C_p	C_a	Δ_S	Δ_N	C_b	C_p	C_a
Deg.	Deg.	Sec.	Ft.	Sec.	Deg.	Deg.	Sec.	Ft.	Sec.
37	34	137.05	0.084	46.92	39	16	537.14	0.318	180.09
37	35	93.90	0.057	32.20	39	17	545.15	0.323	182.91
37	36	48.21	0.030	16.56	39	18	550.21	0.327	184.75
37	37	0	0	0	39	19	552.34	0.328	185.62
38	0	0	0	0	39	20	551.54	0.328	185.52
38	1	53.49	0.031	17.83	39	21	547.84	0.327	184.45
38	2	103.98	0.061	34.66	39	22	541.25	0.323	182.41
38	3	151.47	0.088	50.50	39	23	531.77	0.318	179.40
38	4	195.97	0.114	65.35	39	24	519.43	0.311	175.43
38	5	237.48	0.139	79.21	39	25	504.23	0.302	170.49
38	6	276.03	0.162	92.08	39	26	486.19	0.292	164.58
38	7	311.60	0.183	103.98	39	27	465.32	0.280	157.71
38	8	344.22	0.202	114.90	39	28	441.64	0.266	149.88
38	9	373.89	0.220	124.86	39	29	415.16	0.251	141.08
38	10	400.61	0.236	133.84	39	30	385.90	0.233	131.31
38	11	424.41	0.250	141.86	39	31	353.88	0.214	120.59
38	12	445.28	0.263	148.91	39	32	319.10	0.194	108.90
38	13	463.24	0.274	155.00	39	33	281.59	0.171	96.24
38	14	478.30	0.283	160.14	39	34	241.35	0.147	82.62
38	15	490.46	0.290	164.32	39	35	198.42	0.121	68.03
38	16	499.75	0.296	167.55	39	36	152.80	0.093	52.48
38	17	506.16	0.301	169.83	39	37	104.51	0.064	35.96
38	18	509.72	0.303	171.16	39	38	53.57	0.033	18.46
38	19	510.42	0.304	171.54	39	39	0	0	0
38	20	508.29	0.303	170.98	40	0	0	0	0
38	21	503.34	0.301	169.47	40	1	59.60	0.035	19.87
38	22	495.57	0.296	167.02	40	2	116.02	0.067	38.68
38	23	485.00	0.290	163.63	40	3	169.26	0.098	56.43
38	24	471.65	0.283	159.30	40	4	219.32	0.128	73.13
38	25	455.52	0.274	154.02	40	5	266.22	0.155	88.79
38	26	436.64	0.263	147.81	40	6	309.96	0.181	103.40
38	27	415.01	0.250	140.66	40	7	350.55	0.205	116.98
38	28	390.64	0.236	132.57	40	8	388.01	0.227	129.52
38	29	363.56	0.220	123.55	40	9	422.34	0.247	141.04
38	30	333.78	0.202	113.58	40	10	453.55	0.266	151.52
38	31	301.30	0.183	102.67	40	11	481.65	0.283	160.98
38	32	266.15	0.162	90.83	40	12	506.65	0.298	169.43
38	33	228.35	0.139	78.05	40	13	528.57	0.311	176.86
38	34	187.90	0.114	64.32	40	14	547.40	0.323	183.28
38	35	144.82	0.088	49.66	40	15	563.18	0.332	188.68
38	36	99.14	0.061	34.05	40	16	575.89	0.340	193.08
38	37	50.86	0.031	17.50	40	17	585.57	0.346	196.47
38	38	0	0	0	40	18	592.21	0.351	198.86
39	0	0	0	0	40	19	595.84	0.354	200.24
39	1	56.50	0.033	18.83	40	20	596.46	0.354	200.63
39	2	109.91	0.064	36.64	40	21	594.09	0.354	200.02
39	3	160.22	0.093	53.42	40	22	588.74	0.351	198.42
39	4	207.46	0.121	69.18	40	23	580.42	0.346	195.82
39	5	251.62	0.147	83.92	40	24	569.16	0.340	192.22
39	6	292.72	0.171	97.65	40	25	554.96	0.332	187.64
39	7	330.77	0.194	110.38	40	26	537.83	0.323	182.06
39	8	365.76	0.214	122.10	40	27	517.80	0.311	175.49
39	9	397.72	0.233	132.81	40	28	494.87	0.298	167.94
39	10	426.64	0.251	142.53	40	29	469.06	0.283	159.39
39	11	452.55	0.266	151.26	40	30	440.40	0.266	149.85
39	12	475.44	0.280	159.00	40	31	408.89	0.247	139.33
39	13	495.34	0.292	165.74	40	32	374.55	0.227	127.81
39	14	512.25	0.302	171.51	40	33	337.39	0.205	115.31
39	15	526.18	0.311	176.29	40	34	297.44	0.181	101.81
39	16	537.14	0.318	180.09	40	35	254.71	0.155	87.33

TABLE XVI-C.—CORRECTIONS FOR COMBINING SPIRALS

Δ_S	Δ_N	C_b	C_p	C_a	Δ_S	Δ_N	C_b	C_p	C_a
Deg.	Deg.	Sec.	Ft.	Sec.	Deg.	Deg.	Sec.	Ft.	Sec.
40	35	254.71	0.155	87.33	42	11	542.77	0.317	181.41
40	36	209.22	0.128	71.85	42	12	572.23	0.335	191.36
40	37	160.98	0.098	55.38	42	13	598.42	0.351	200.23
40	38	110.02	0.067	37.92	42	14	621.36	0.365	208.03
40	39	56.36	0.035	19.46	42	15	641.06	0.377	214.77
40	40	0	0	0	42	16	657.52	0.387	220.44
41	0	0	0	0	42	17	670.77	0.395	225.05
41	1	62.80	0.036	20.93	42	18	680.81	0.402	228.60
41	2	122.32	0.071	40.78	42	19	687.67	0.407	231.10
41	3	178.56	0.104	59.53	42	20	691.34	0.409	232.54
41	4	231.54	0.134	77.21	42	21	691.85	0.410	232.93
41	5	281.26	0.163	93.81	42	22	689.22	0.409	232.27
41	6	327.74	0.191	109.33	42	23	683.44	0.407	230.56
41	7	370.97	0.216	123.79	42	24	674.55	0.402	227.81
41	8	410.98	0.240	137.19	42	25	662.56	0.395	224.01
41	9	447.76	0.262	149.52	42	26	647.47	0.387	219.16
41	10	481.34	0.282	160.81	42	27	629.31	0.377	213.28
41	11	511.72	0.300	171.04	42	28	608.10	0.365	206.35
41	12	538.91	0.316	180.22	42	29	583.85	0.351	198.38
41	13	562.92	0.331	188.35	42	30	556.57	0.335	189.37
41	14	583.77	0.343	195.45	42	31	526.29	0.317	179.32
41	15	601.46	0.354	201.50	42	32	493.02	0.298	168.23
41	16	616.02	0.363	206.53	42	33	456.77	0.276	156.10
41	17	627.44	0.371	210.51	42	34	417.58	0.253	142.93
41	18	635.74	0.376	213.47	42	35	375.45	0.228	128.72
41	19	640.94	0.380	215.40	42	36	330.41	0.201	113.46
41	20	643.05	0.381	216.30	42	37	282.47	0.172	97.16
41	21	642.09	0.381	216.18	42	38	231.65	0.141	79.82
41	22	638.06	0.380	215.03	42	39	177.97	0.109	61.44
41	23	630.98	0.376	212.86	42	40	121.46	0.074	42.01
41	24	620.86	0.371	209.68	42	41	62.13	0.038	21.53
41	25	607.73	0.363	205.47	42	42	0	0	0
41	26	591.58	0.354	200.25	43	0	0	0	0
41	27	572.45	0.343	194.01	43	1	69.48	0.040	23.16
41	28	550.35	0.331	186.76	43	2	135.48	0.078	45.16
41	29	525.29	0.316	178.49	43	3	198.02	0.114	66.02
41	30	497.28	0.300	169.20	43	4	257.10	0.149	85.73
41	31	466.35	0.282	158.90	43	5	312.74	0.181	104.30
41	32	432.51	0.262	147.59	43	6	364.94	0.211	121.75
41	33	395.78	0.240	135.26	43	7	413.72	0.240	138.06
41	34	356.18	0.216	121.92	43	8	459.09	0.267	153.25
41	35	313.72	0.191	107.55	43	9	501.05	0.291	167.32
41	36	268.42	0.163	92.18	43	10	539.62	0.314	180.27
41	37	220.30	0.134	75.78	43	11	574.80	0.335	192.12
41	38	169.38	0.104	58.37	43	12	606.62	0.354	202.86
41	39	115.67	0.071	39.93	43	13	635.08	0.371	212.49
41	40	59.21	0.036	20.48	43	14	660.19	0.387	221.03
41	41	0	0	0	43	15	681.97	0.400	228.47
42	0	0	0	0	43	16	700.43	0.412	234.82
42	1	66.09	0.038	22.03	43	17	715.58	0.421	240.08
42	2	128.80	0.074	42.94	43	18	727.44	0.429	244.25
42	3	188.15	0.109	62.73	43	19	736.02	0.434	247.34
42	4	244.14	0.141	81.41	43	20	741.34	0.438	249.35
42	5	296.77	0.172	98.98	43	21	743.40	0.440	250.28
42	6	346.06	0.201	115.45	43	22	742.23	0.440	250.13
42	7	392.03	0.228	130.82	43	23	737.84	0.438	248.91
42	8	434.67	0.253	145.10	43	24	730.25	0.434	246.61
42	9	474.00	0.276	158.28	43	25	719.46	0.429	243.24
42	10	510.03	0.298	170.39	43	26	705.51	0.421	238.80
42	11	542.77	0.317	181.41	43	27	688.39	0.412	233.29

TABLE XVI-C.—CORRECTIONS FOR COMBINING SPIRALS

Δ_S	Δ_N	C_b	C_p	C_a	Δ_S	Δ_N	C_b	C_p	C_a
Deg.	Deg.	Sec.	Ft.	Sec.	Deg.	Deg.	Sec.	Ft.	Sec.
43	27	688.39	0.412	233.29	44	42	133.43	0.082	46.32
43	28	668.14	0.400	226.72	44	43	68.17	0.042	23.71
43	29	644.76	0.387	219.07	44	44	0	0	0
43	30	618.28	0.371	210.36	45	0	0	0	0
43	31	588.72	0.354	200.59	45	1	76.53	0.044	25.51
43	32	556.08	0.335	189.74	45	2	149.40	0.086	49.80
43	33	520.39	0.314	177.84	45	3	218.61	0.126	72.88
43	34	481.67	0.291	164.86	45	4	284.17	0.163	94.76
43	35	439.94	0.267	150.82	45	5	346.10	0.199	115.43
43	36	395.22	0.240	135.71	45	6	404.40	0.233	134.91
43	37	347.52	0.211	119.53	45	7	459.08	0.265	153.19
43	38	296.86	0.181	102.29	45	8	510.16	0.295	170.29
43	39	243.27	0.149	83.98	45	9	557.64	0.323	186.21
43	40	186.77	0.114	64.59	45	10	601.55	0.349	200.96
43	41	127.38	0.078	44.14	45	11	641.88	0.373	214.54
43	42	65.11	0.040	22.60	45	12	678.66	0.395	226.95
43	43	0	0	0	45	13	711.90	0.415	238.20
44	0	0	0	0	45	14	741.61	0.433	248.29
44	1	72.96	0.042	24.32	45	15	767.80	0.449	257.22
44	2	142.34	0.082	47.45	45	16	790.49	0.463	265.01
44	3	208.17	0.120	69.40	45	17	809.70	0.475	271.65
44	4	270.45	0.156	90.18	45	18	825.42	0.484	277.15
44	5	329.18	0.190	109.79	45	19	837.70	0.492	281.50
44	6	384.39	0.222	128.23	45	20	846.53	0.498	284.72
44	7	436.07	0.252	145.52	45	21	851.93	0.502	286.80
44	8	484.25	0.281	161.65	45	22	853.92	0.504	287.76
44	9	528.93	0.307	176.63	45	23	852.52	0.504	287.58
44	10	570.12	0.331	190.46	45	24	847.74	0.502	286.27
44	11	607.84	0.354	203.16	45	25	839.59	0.498	283.84
44	12	642.09	0.374	214.72	45	26	828.10	0.492	280.28
44	13	672.90	0.393	225.15	45	27	813.29	0.484	275.61
44	14	700.27	0.409	234.45	45	28	795.17	0.475	269.81
44	15	724.21	0.424	242.62	45	29	773.76	0.463	262.89
44	16	744.75	0.437	249.68	45	30	749.08	0.449	254.85
44	17	761.88	0.447	255.61	45	31	721.14	0.433	245.69
44	18	775.64	0.456	260.43	45	32	689.98	0.415	235.41
44	19	786.03	0.463	264.14	45	33	655.60	0.395	224.02
44	20	793.06	0.468	266.74	45	34	618.02	0.373	211.51
44	21	796.76	0.471	268.24	45	35	577.28	0.349	197.88
44	22	797.13	0.472	268.62	45	36	533.38	0.323	183.14
44	23	794.20	0.471	267.91	45	37	486.35	0.295	167.27
44	24	787.97	0.468	266.10	45	38	436.20	0.265	150.29
44	25	778.47	0.463	263.18	45	39	382.98	0.233	132.19
44	26	765.71	0.456	259.17	45	40	326.68	0.199	112.97
44	27	749.71	0.447	254.07	45	41	267.34	0.163	92.62
44	28	730.49	0.437	247.87	45	42	204.98	0.126	71.16
44	29	708.06	0.424	240.57	45	43	139.62	0.086	48.56
44	30	682.45	0.409	232.19	45	44	71.29	0.044	24.85
44	31	653.66	0.393	222.71	45	45	0	0	0
44	32	621.73	0.374	212.14	46	0	0	0	0
44	33	586.66	0.354	200.47	46	1	80.21	0.046	26.74
44	34	548.48	0.331	187.72	46	2	156.66	0.090	52.22
44	35	507.21	0.307	173.87	46	3	229.34	0.131	76.46
44	36	462.86	0.281	158.93	46	4	298.28	0.171	99.46
44	37	415.47	0.252	142.90	46	5	363.49	0.209	121.23
44	38	365.04	0.222	125.78	46	6	424.98	0.245	141.77
44	39	311.60	0.190	107.56	46	7	482.75	0.278	161.09
44	40	255.17	0.156	88.24	46	8	536.82	0.310	179.19
44	41	195.77	0.120	67.83	46	9	587.20	0.339	196.08
44	42	133.43	0.082	46.32	46	10	633.90	0.367	211.77

TABLE XVI-C.—CORRECTIONS FOR COMBINING SPIRALS

Δ_S	Δ_N	C_b	C_p	C_a	Δ_S	Δ_N	C_b	C_p	C_a
Deg.	Deg.	Sec.	Ft.	Sec.	Deg.	Deg.	Sec.	Ft.	Sec.
46	10	633.90	0.367	211.77	47	22	973.26	0.573	327.96
46	11	676.95	0.392	226.26	47	23	975.15	0.575	328.93
46	12	716.34	0.416	239.54	47	24	973.48	0.575	328.72
46	13	752.10	0.437	251.64	47	25	968.27	0.573	327.33
46	14	784.23	0.457	262.55	47	26	959.54	0.568	324.75
46	15	812.75	0.474	272.28	47	27	947.32	0.562	321.01
46	16	837.68	0.489	280.82	47	28	931.61	0.554	316.08
46	17	859.03	0.502	288.20	47	29	912.45	0.543	309.98
46	18	876.81	0.514	294.40	47	30	889.84	0.531	302.71
46	19	891.05	0.523	299.43	47	31	863.81	0.516	294.27
46	20	901.75	0.530	303.29	47	32	834.38	0.500	284.66
46	21	908.94	0.535	305.99	47	33	801.57	0.481	273.88
46	22	912.62	0.538	307.53	47	34	765.41	0.460	261.93
46	23	912.83	0.539	307.92	47	35	725.91	0.437	248.81
46	24	909.56	0.538	307.14	47	36	683.09	0.412	234.52
46	25	902.85	0.535	305.22	47	37	636.98	0.385	219.06
46	26	892.71	0.530	302.14	47	38	587.60	0.356	202.43
46	27	879.15	0.523	297.92	47	39	534.98	0.325	184.63
46	28	862.20	0.514	292.54	47	40	479.13	0.292	165.66
46	29	841.88	0.502	286.02	47	41	420.08	0.256	145.52
46	30	818.20	0.489	278.35	47	42	357.86	0.219	124.21
46	31	791.19	0.474	269.54	47	43	292.49	0.179	101.72
46	32	760.85	0.457	259.59	47	44	223.99	0.137	78.06
46	33	727.23	0.437	248.49	47	45	152.40	0.094	53.22
46	34	690.32	0.416	236.24	47	46	77.72	0.048	27.20
46	35	650.17	0.392	222.86	47	47	0	0	0
46	36	606.78	0.367	208.33	48	0	0	0	0
46	37	560.17	0.339	192.66	48	1	87.86	0.050	29.29
46	38	510.38	0.310	175.84	48	2	171.76	0.098	57.26
46	39	457.43	0.278	157.88	48	3	251.70	0.144	83.92
46	40	401.32	0.245	138.77	48	4	327.69	0.187	109.27
46	41	342.10	0.209	118.52	48	5	399.75	0.229	133.32
46	42	279.78	0.171	97.12	48	6	467.88	0.268	156.08
46	43	214.39	0.131	74.56	48	7	532.11	0.305	177.56
46	44	145.94	0.090	50.86	48	8	592.44	0.340	197.76
46	45	74.47	0.046	26.01	48	9	648.89	0.373	216.68
46	46	0	0	0	48	10	701.46	0.404	234.34
47	0	0	0	0	48	11	750.18	0.433	250.73
47	1	83.99	0.048	28.00	48	12	795.06	0.459	265.86
47	2	164.11	0.094	54.71	48	13	836.11	0.484	279.75
47	3	240.37	0.137	80.14	48	14	873.35	0.506	292.38
47	4	312.79	0.179	104.30	48	15	906.78	0.526	303.77
47	5	381.37	0.219	127.20	48	16	936.44	0.544	313.93
47	6	446.14	0.256	148.83	48	17	962.33	0.560	322.84
47	7	507.09	0.292	169.21	48	18	984.47	0.574	330.53
47	8	564.24	0.325	188.35	48	19	1002.87	0.586	336.99
47	9	617.61	0.356	206.24	48	20	1017.56	0.595	342.23
47	10	667.21	0.385	222.89	48	21	1028.55	0.603	346.24
47	11	713.04	0.412	238.32	48	22	1035.86	0.608	349.04
47	12	755.14	0.437	252.52	48	23	1039.50	0.611	350.63
47	13	793.49	0.460	265.49	48	24	1039.50	0.612	351.00
47	14	828.13	0.481	277.25	48	25	1035.87	0.611	350.17
47	15	859.07	0.500	287.79	48	26	1028.64	0.608	348.13
47	16	886.32	0.516	297.13	48	27	1017.81	0.603	344.88
47	17	909.90	0.531	305.26	48	28	1003.42	0.595	340.43
47	18	929.82	0.543	312.19	48	29	985.48	0.586	334.78
47	19	946.10	0.554	317.92	48	30	964.02	0.574	327.94
47	20	958.76	0.562	322.46	48	31	939.04	0.560	319.89
47	21	967.81	0.568	325.80	48	32	910.58	0.544	310.64
47	22	973.26	0.573	327.96	48	33	878.66	0.526	300.20

TABLE XVI-C.—CORRECTIONS FOR COMBINING SPIRALS

Δ_S	Δ_N	C_b	C_p	C_a	Δ_S	Δ_N	C_b	C_p	C_a
Deg.	Deg.	Sec.	Ft.	Sec.	Deg.	Deg.	Sec.	Ft.	Sec.
48	33	878.66	0.526	300.20	49	43	458.83	0.280	159.55
48	34	843.30	0.506	288.57	49	44	390.40	0.239	136.04
48	35	804.52	0.484	275.74	49	45	318.72	0.195	111.29
48	36	762.34	0.459	261.71	49	46	243.81	0.150	85.32
48	37	716.79	0.433	246.49	49	47	165.70	0.102	58.11
48	38	667.88	0.404	230.08	49	48	84.42	0.052	29.67
48	39	615.66	0.373	212.46	49	49	0	0	0
48	40	560.12	0.340	193.66	50	0	0	0	0
48	41	501.32	0.305	173.65	50	1	95.92	0.054	31.98
48	42	439.25	0.268	152.45	50	2	187.67	0.106	62.56
48	43	373.96	0.229	130.05	50	3	275.26	0.156	91.77
48	44	305.47	0.187	106.45	50	4	358.69	0.204	119.61
48	45	233.80	0.144	81.65	50	5	437.99	0.249	146.08
48	46	158.98	0.098	55.64	50	6	513.16	0.292	171.19
48	47	81.04	0.050	28.42	50	7	584.22	0.333	194.95
48	48	0	0	0	50	8	651.19	0.372	217.37
49	0	0	0	0	50	9	714.07	0.409	238.45
49	1	91.84	0.052	30.61	50	10	772.88	0.443	258.19
49	2	179.61	0.102	59.88	50	11	827.64	0.475	276.61
49	3	263.32	0.150	87.79	50	12	878.36	0.505	293.71
49	4	342.99	0.195	114.37	50	13	925.05	0.533	309.50
49	5	418.62	0.239	139.62	50	14	967.74	0.558	323.98
49	6	490.22	0.280	163.54	50	15	1006.43	0.582	337.15
49	7	557.82	0.319	186.14	50	16	1041.15	0.603	349.02
49	8	621.42	0.356	207.43	50	17	1071.92	0.621	359.60
49	9	681.04	0.391	227.42	50	18	1098.74	0.638	368.88
49	10	736.68	0.423	246.10	50	19	1121.64	0.652	376.89
49	11	788.38	0.454	263.49	50	20	1140.63	0.665	383.60
49	12	836.13	0.482	279.60	50	21	1155.74	0.675	389.04
49	13	879.96	0.508	294.42	50	22	1166.98	0.682	393.21
49	14	919.88	0.532	307.96	50	23	1174.38	0.688	396.10
49	15	955.90	0.554	320.22	50	24	1177.94	0.691	397.73
49	16	988.04	0.573	331.22	50	25	1177.70	0.692	398.09
49	17	1016.33	0.590	340.95	50	26	1173.67	0.691	397.19
49	18	1040.77	0.606	349.43	50	27	1165.87	0.688	395.02
49	19	1061.38	0.619	356.64	50	28	1154.33	0.682	391.60
49	20	1078.18	0.630	362.61	50	29	1139.06	0.675	386.93
49	21	1091.19	0.638	367.32	50	30	1120.09	0.665	381.00
49	22	1100.42	0.645	370.79	50	31	1097.43	0.652	373.82
49	23	1105.90	0.649	373.02	50	32	1071.12	0.638	365.38
49	24	1107.65	0.651	374.00	50	33	1041.18	0.621	355.70
49	25	1105.67	0.651	373.75	50	34	1007.62	0.603	344.77
49	26	1100.00	0.649	372.27	50	35	970.48	0.582	332.59
49	27	1090.65	0.645	369.55	50	36	929.77	0.558	319.16
49	28	1077.65	0.638	365.60	50	37	885.52	0.533	304.48
49	29	1061.01	0.630	360.43	50	38	837.76	0.505	288.56
49	30	1040.76	0.619	354.03	50	39	786.50	0.475	271.40
49	31	1016.91	0.606	346.40	50	40	731.79	0.443	252.98
49	32	989.49	0.590	337.55	50	41	673.64	0.409	233.32
49	33	958.52	0.573	327.48	50	42	612.07	0.372	212.40
49	34	924.03	0.554	316.18	50	43	547.12	0.333	190.24
49	35	886.03	0.532	303.66	50	44	478.81	0.292	166.83
49	36	844.56	0.508	289.92	50	45	407.17	0.249	142.17
49	37	799.62	0.482	274.96	50	46	332.23	0.204	116.25
49	38	751.26	0.454	258.78	50	47	254.01	0.156	89.07
49	39	699.49	0.423	241.38	50	48	172.55	0.106	60.64
49	40	644.33	0.391	222.76	50	49	87.87	0.054	30.95
49	41	585.82	0.356	202.91	50	50	0	0	0
49	42	523.98	0.319	181.85	51	0	0	0	0
49	43	458.83	0.280	159.55	51	1	100.11	0.056	33.37

TABLE XVI-C.—CORRECTIONS FOR COMBINING SPIRALS

Δ_S	Δ_N	C_b	C_p	C_a	Δ_S	Δ_N	C_b	C_p	C_a
Deg.	Deg.	Sec.	Ft.	Sec.	Deg.	Deg.	Sec.	Ft.	Sec.
51	1	100.11	0.056	33.37	52	8	713.14	0.405	238.04
51	2	195.94	0.111	65.32	52	9	782.83	0.446	261.40
51	3	287.50	0.163	95.85	52	10	848.25	0.484	283.37
51	4	374.81	0.212	124.98	52	11	909.41	0.520	303.94
51	5	457.87	0.260	152.71	52	12	966.34	0.553	323.13
51	6	536.71	0.305	179.04	52	13	1019.03	0.584	340.94
51	7	611.33	0.348	204.00	52	14	1067.52	0.613	357.38
51	8	681.76	0.389	227.57	52	15	1111.83	0.639	372.45
51	9	748.00	0.427	249.77	52	16	1151.96	0.664	386.15
51	10	810.07	0.463	270.61	52	17	1187.93	0.685	398.51
51	11	867.98	0.497	290.10	52	18	1219.78	0.705	409.51
51	12	921.75	0.529	308.22	52	19	1247.50	0.722	419.16
51	13	971.41	0.558	325.01	52	20	1271.13	0.737	427.48
51	14	1016.95	0.585	340.45	52	21	1290.68	0.750	434.45
51	15	1058.40	0.610	354.55	52	22	1306.18	0.760	440.09
51	16	1095.79	0.633	367.33	52	23	1317.63	0.768	444.40
51	17	1129.11	0.653	378.78	52	24	1325.08	0.774	447.38
51	18	1158.40	0.671	388.91	52	25	1328.52	0.778	449.04
51	19	1183.67	0.687	397.72	52	26	1328.00	0.779	449.39
51	20	1204.94	0.700	405.22	52	27	1323.52	0.778	448.41
51	21	1222.23	0.712	411.42	52	28	1315.12	0.774	446.12
51	22	1235.56	0.721	416.31	52	29	1302.81	0.768	442.52
51	23	1244.95	0.728	419.90	52	30	1286.62	0.760	437.61
51	24	1250.41	0.732	422.19	52	31	1266.57	0.750	431.39
51	25	1251.98	0.734	423.18	52	32	1242.68	0.737	423.86
51	26	1249.66	0.734	422.89	52	33	1214.99	0.722	415.04
51	27	1243.48	0.732	421.31	52	34	1183.51	0.705	404.91
51	28	1233.47	0.728	418.44	52	35	1148.27	0.685	393.47
51	29	1219.65	0.721	414.29	52	36	1109.29	0.664	380.74
51	30	1202.03	0.712	408.85	52	37	1066.60	0.639	366.71
51	31	1180.64	0.700	402.14	52	38	1020.23	0.613	351.38
51	32	1155.51	0.687	394.15	52	39	970.21	0.584	334.74
51	33	1126.66	0.671	384.88	52	40	916.56	0.553	316.81
51	34	1094.10	0.653	374.34	52	41	859.30	0.520	297.58
51	35	1057.88	0.633	362.52	52	42	798.47	0.484	277.05
51	36	1018.00	0.610	349.43	52	43	734.09	0.446	255.22
51	37	974.50	0.585	335.06	52	44	666.20	0.405	232.09
51	38	927.40	0.558	319.42	52	45	594.82	0.363	207.65
51	39	876.74	0.529	302.51	52	46	519.98	0.318	181.92
51	40	822.52	0.497	284.33	52	47	441.71	0.271	154.87
51	41	764.78	0.463	264.87	52	48	360.04	0.221	126.52
51	42	703.56	0.427	244.14	52	49	275.01	0.169	96.86
51	43	638.86	0.389	222.13	52	50	186.63	0.115	65.89
51	44	570.73	0.348	198.84	52	51	94.95	0.059	33.60
51	45	499.19	0.305	174.28	52	52	0	0	0
51	46	424.27	0.260	148.44	53	0	0	0	0
51	47	346.00	0.212	121.32	53	1	108.80	0.061	36.27
51	48	264.41	0.163	92.92	53	2	213.10	0.120	71.04
51	49	179.53	0.111	63.24	53	3	312.92	0.176	104.33
51	50	91.38	0.056	32.26	53	4	408.28	0.230	136.14
51	51	0	0	0	53	5	499.19	0.282	166.48
52	0	0	0	0	53	6	585.65	0.331	195.37
52	1	104.40	0.059	34.80	53	7	667.70	0.378	222.80
52	2	204.41	0.115	68.14	53	8	745.33	0.423	248.79
52	3	300.05	0.169	100.04	53	9	818.58	0.465	273.34
52	4	391.33	0.221	130.49	53	10	887.44	0.505	296.46
52	5	478.27	0.271	159.51	53	11	951.95	0.542	318.15
52	6	560.87	0.318	187.10	53	12	1012.11	0.578	338.43
52	7	639.16	0.363	213.28	53	13	1067.95	0.611	357.30
52	8	713.14	0.405	238.04	53	14	1119.48	0.641	374.76

TABLE XVI-C.—CORRECTIONS FOR COMBINING SPIRALS

Δ_S	Δ_N	C_b	C_p	C_a	Δ_S	Δ_N	C_b	C_p	C_a
Deg.	Deg.	Sec.	Ft.	Sec.	Deg.	Deg.	Sec.	Ft.	Sec.
53	14	1119.48	0.641	374.76	54	19	1380.62	0.795	463.87
53	15	1166.72	0.669	390.83	54	20	1409.22	0.813	473.89
53	16	1209.69	0.695	405.50	54	21	1433.54	0.829	482.51
53	17	1248.40	0.719	418.78	54	22	1453.61	0.842	489.74
53	18	1282.88	0.740	430.69	54	23	1469.46	0.853	495.58
53	19	1313.14	0.758	441.21	54	24	1481.09	0.861	500.03
53	20	1339.21	0.775	450.36	54	25	1488.54	0.867	503.10
53	21	1361.11	0.789	458.14	54	26	1491.83	0.871	504.79
53	22	1378.85	0.801	464.56	54	27	1490.98	0.872	505.11
53	23	1392.46	0.810	469.62	54	28	1486.01	0.871	504.05
53	24	1401.96	0.817	473.33	54	29	1476.96	0.867	501.63
53	25	1407.37	0.822	475.68	54	30	1463.84	0.861	497.84
53	26	1408.71	0.824	476.68	54	31	1446.67	0.853	492.69
53	27	1406.01	0.824	476.34	54	32	1425.49	0.842	486.17
53	28	1399.29	0.822	474.66	54	33	1400.32	0.829	478.30
53	29	1388.57	0.817	471.63	54	34	1371.19	0.813	469.07
53	30	1373.88	0.810	467.27	54	35	1338.12	0.795	458.48
53	31	1355.24	0.801	461.57	54	36	1301.14	0.775	446.54
53	32	1332.67	0.789	454.54	54	37	1260.28	0.752	433.24
53	33	1306.20	0.775	446.17	54	38	1215.56	0.727	418.60
53	34	1275.86	0.758	436.48	54	39	1167.02	0.700	402.59
53	35	1241.67	0.740	425.46	54	40	1114.67	0.670	385.24
53	36	1203.66	0.719	413.11	54	41	1058.56	0.638	366.54
53	37	1161.85	0.695	399.43	54	42	998.71	0.603	346.48
53	38	1116.28	0.669	384.43	54	43	935.15	0.566	325.07
53	39	1066.96	0.641	368.10	54	44	867.90	0.526	302.31
53	40	1013.93	0.611	350.45	54	45	797.01	0.484	278.20
53	41	957.21	0.578	331.47	54	46	722.50	0.440	252.72
53	42	896.84	0.542	311.16	54	47	644.40	0.394	225.90
53	43	832.84	0.505	289.53	54	48	562.75	0.345	197.72
53	44	765.25	0.465	266.57	54	49	477.57	0.293	168.17
53	45	694.08	0.423	242.29	54	50	388.91	0.239	137.27
53	46	619.38	0.378	216.67	54	51	296.78	0.183	105.00
53	47	541.17	0.331	189.72	54	52	201.23	0.124	71.37
53	48	459.48	0.282	161.45	54	53	102.30	0.063	36.37
53	49	374.34	0.230	131.84	54	54	0	0	0
53	50	285.80	0.176	100.89	55	0	0	0	0
53	51	193.87	0.120	68.60	55	1	117.92	0.066	39.31
53	52	98.59	0.061	34.97	55	2	231.12	0.129	77.05
53	53	0	0	0	55	3	339.63	0.190	113.23
54	0	0	0	0	55	4	443.46	0.249	147.87
54	1	113.30	0.063	37.77	55	5	542.61	0.305	180.97
54	2	222.00	0.124	74.01	55	6	637.11	0.358	212.54
54	3	326.12	0.183	108.73	55	7	726.98	0.409	242.58
54	4	425.65	0.239	141.93	55	8	812.22	0.458	271.11
54	5	520.63	0.293	173.64	55	9	892.86	0.504	298.14
54	6	611.06	0.345	203.85	55	10	968.91	0.548	323.67
54	7	696.97	0.394	232.57	55	11	1040.39	0.590	347.71
54	8	778.36	0.440	259.81	55	12	1107.32	0.629	370.26
54	9	855.25	0.484	285.58	55	13	1169.71	0.665	391.34
54	10	927.66	0.526	309.89	55	14	1227.60	0.699	410.95
54	11	995.60	0.566	332.74	55	15	1280.98	0.731	429.09
54	12	1059.10	0.603	354.14	55	16	1329.89	0.760	445.78
54	13	1118.17	0.638	374.10	55	17	1374.34	0.787	461.02
54	14	1172.83	0.670	392.62	55	18	1414.36	0.811	474.81
54	15	1223.10	0.700	409.71	55	19	1449.96	0.833	487.16
54	16	1268.99	0.727	425.37	55	20	1481.16	0.853	498.08
54	17	1310.53	0.752	439.62	55	21	1508.00	0.870	507.56
54	18	1347.73	0.775	452.45	55	22	1530.48	0.885	515.63
54	19	1380.62	0.795	463.87	55	23	1548.64	0.897	522.27

TABLE XVI-C.—CORRECTIONS FOR COMBINING SPIRALS

Δ_S	Δ_N	C_b	C_p	C_a	Δ_S	Δ_N	C_b	C_p	C_a
Deg.	Deg.	Sec.	Ft.	Sec.	Deg.	Deg.	Sec.	Ft.	Sec.
55	23	1548.64	0.897	522.27	56	26	1665.37	0.968	563.47
55	24	1562.49	0.906	527.50	56	27	1668.45	0.971	565.19
55	25	1572.06	0.914	531.31	56	28	1667.23	0.973	565.48
55	26	1577.37	0.919	533.72	56	29	1661.72	0.971	564.34
55	27	1578.45	0.921	534.72	56	30	1651.97	0.968	561.77
55	28	1575.32	0.921	534.32	56	31	1637.98	0.961	557.79
55	29	1568.00	0.919	532.53	56	32	1619.80	0.953	552.39
55	30	1556.52	0.914	529.34	56	33	1597.43	0.942	545.57
55	31	1540.91	0.906	524.76	56	34	1570.93	0.928	537.34
55	32	1521.19	0.897	518.79	56	35	1540.30	0.912	527.69
55	33	1497.39	0.885	511.43	56	36	1505.59	0.893	516.64
55	34	1469.54	0.870	502.68	56	37	1466.81	0.872	504.18
55	35	1437.66	0.853	492.56	56	38	1424.00	0.849	490.31
55	36	1401.78	0.833	481.05	56	39	1377.20	0.822	475.04
55	37	1361.92	0.811	468.16	56	40	1326.42	0.794	458.36
55	38	1318.13	0.787	453.88	56	41	1271.70	0.763	440.27
55	39	1270.42	0.760	438.24	56	42	1213.06	0.729	420.78
55	40	1218.82	0.731	421.21	56	43	1150.55	0.693	399.89
55	41	1163.38	0.699	402.80	56	44	1084.20	0.655	377.59
55	42	1104.10	0.665	383.02	56	45	1014.02	0.614	353.88
55	43	1041.04	0.629	361.85	56	46	940.07	0.571	328.77
55	44	974.21	0.590	339.31	56	47	862.37	0.525	302.25
55	45	903.65	0.548	315.39	56	48	780.95	0.476	274.33
55	46	829.39	0.504	290.09	56	49	695.85	0.426	244.99
55	47	751.46	0.458	263.40	56	50	607.10	0.372	214.24
55	48	669.90	0.409	235.34	56	51	514.74	0.316	182.08
55	49	584.73	0.358	205.89	56	52	418.81	0.258	148.51
55	50	496.00	0.305	175.05	56	53	319.33	0.197	113.52
55	51	403.73	0.249	142.83	56	54	216.34	0.134	77.10
55	52	307.96	0.190	109.21	56	55	109.89	0.068	39.26
55	53	208.72	0.129	74.21	56	56	0	0	0
55	54	106.06	0.066	37.80	57	0	0	0	0
55	55	0	0	0	57	1	127.48	0.071	42.50
56	0	0	0	0	57	2	250.03	0.139	83.35
56	1	122.65	0.068	40.88	57	3	367.66	0.205	122.58
56	2	240.47	0.134	80.16	57	4	480.37	0.268	160.18
56	3	353.48	0.197	117.85	57	5	588.20	0.328	196.17
56	4	461.70	0.258	153.95	57	6	691.15	0.386	230.56
56	5	565.13	0.316	188.48	57	7	789.25	0.442	263.36
56	6	663.81	0.372	221.44	57	8	882.50	0.495	294.57
56	7	757.73	0.426	252.84	57	9	970.94	0.545	324.21
56	8	846.93	0.476	282.70	57	10	1054.57	0.593	352.28
56	9	931.42	0.525	311.02	57	11	1133.41	0.639	378.79
56	10	1011.21	0.571	337.80	57	12	1207.49	0.682	403.75
56	11	1086.32	0.614	363.06	57	13	1276.82	0.722	427.16
56	12	1156.78	0.655	386.80	57	14	1341.43	0.760	449.04
56	13	1222.59	0.693	409.03	57	15	1401.33	0.795	469.39
56	14	1283.79	0.729	429.75	57	16	1456.54	0.828	488.22
56	15	1340.38	0.763	448.99	57	17	1507.09	0.859	505.53
56	16	1392.40	0.794	466.73	57	18	1553.00	0.886	521.33
56	17	1439.85	0.822	482.98	57	19	1594.28	0.912	535.63
56	18	1482.77	0.849	497.77	57	20	1630.97	0.934	548.43
56	19	1521.17	0.872	511.08	57	21	1663.08	0.955	559.74
56	20	1555.07	0.893	522.92	57	22	1690.64	0.972	569.56
56	21	1584.50	0.912	533.30	57	23	1713.68	0.987	577.89
56	22	1609.48	0.928	542.23	57	24	1732.21	1.000	584.76
56	23	1630.04	0.942	549.70	57	25	1746.26	1.010	590.14
56	24	1646.19	0.953	555.74	57	26	1755.85	1.018	594.07
56	25	1657.96	0.961	560.32	57	27	1761.02	1.023	596.52
56	26	1665.37	0.968	563.47	57	28	1761.78	1.025	597.52

TABLE XVI-C.—CORRECTIONS FOR COMBINING SPIRALS

Δ_S	Δ_N	C_b	C_p	C_a	Δ_S	Δ_N	C_b	C_p	C_a
Deg.	Deg.	Sec.	Ft.	Sec.	Deg.	Deg.	Sec.	Ft.	Sec.
57	28	1761.78	1.025	597.52	58	29	1857.35	1.081	630.72
57	29	1758.17	1.025	597.06	58	30	1851.26	1.079	629.49
57	30	1750.20	1.023	595.15	58	31	1840.75	1.075	626.78
57	31	1737.92	1.018	591.79	58	32	1825.84	1.069	622.59
57	32	1721.33	1.010	586.98	58	33	1806.57	1.060	616.93
57	33	1700.48	1.000	580.73	58	34	1782.97	1.048	609.80
57	34	1675.40	0.987	573.04	58	35	1755.07	1.034	601.20
57	35	1646.10	0.972	563.90	58	36	1722.89	1.018	591.13
57	36	1612.62	0.955	553.33	58	37	1686.47	0.998	579.60
57	37	1574.98	0.934	541.32	58	38	1645.84	0.976	566.61
57	38	1533.23	0.912	527.88	58	39	1601.03	0.952	552.16
57	39	1487.39	0.886	513.01	58	40	1552.07	0.925	536.25
57	40	1437.48	0.859	496.70	58	41	1498.99	0.896	518.88
57	41	1383.55	0.828	478.96	58	42	1441.82	0.863	500.05
57	42	1325.62	0.795	459.79	58	43	1380.61	0.829	479.76
57	43	1263.73	0.760	439.19	58	44	1315.38	0.791	458.02
57	44	1197.91	0.722	417.15	58	45	1246.16	0.752	434.82
57	45	1128.18	0.682	393.69	58	46	1172.99	0.709	410.16
57	46	1054.59	0.639	368.79	58	47	1095.91	0.664	384.04
57	47	977.17	0.593	342.46	58	48	1014.95	0.617	356.46
57	48	895.96	0.545	314.70	58	49	930.14	0.567	327.42
57	49	810.98	0.495	285.50	58	50	841.53	0.514	296.91
57	50	722.27	0.442	254.86	58	51	749.15	0.459	264.95
57	51	629.87	0.386	222.79	58	52	653.03	0.401	231.51
57	52	533.82	0.328	189.27	58	53	553.21	0.340	196.61
57	53	434.14	0.268	154.31	58	54	449.73	0.278	160.24
57	54	330.89	0.205	117.91	58	55	342.63	0.212	122.40
57	55	224.09	0.139	80.06	58	56	231.95	0.144	83.08
57	56	113.78	0.071	40.76	58	57	117.73	0.073	42.28
57	57	0	0	0	58	58	0	0	0
58	0	0	0	0	59	0	0	0	0
58	1	132.44	0.073	44.15	59	1	137.50	0.076	45.84
58	2	259.82	0.144	86.62	59	2	269.84	0.149	89.96
58	3	382.17	0.212	127.42	59	3	397.03	0.220	132.37
58	4	499.50	0.278	166.56	59	4	519.08	0.288	173.08
58	5	611.82	0.340	204.05	59	5	636.01	0.353	212.12
58	6	719.16	0.401	239.90	59	6	747.84	0.416	249.47
58	7	821.53	0.459	274.13	59	7	854.58	0.476	285.16
58	8	918.94	0.514	306.74	59	8	956.27	0.533	319.19
58	9	1011.43	0.567	337.73	59	9	1052.90	0.588	351.58
58	10	1099.00	0.617	367.12	59	10	1144.52	0.640	382.32
58	11	1181.67	0.664	394.92	59	11	1231.12	0.690	411.44
58	12	1259.47	0.709	421.13	59	12	1312.75	0.737	438.94
58	13	1332.42	0.752	445.76	59	13	1389.40	0.782	464.82
58	14	1400.53	0.791	468.82	59	14	1461.12	0.823	489.10
58	15	1463.83	0.829	490.32	59	15	1527.91	0.863	511.78
58	16	1522.33	0.863	510.27	59	16	1589.80	0.899	532.87
58	17	1576.07	0.896	528.66	59	17	1646.81	0.933	552.38
58	18	1625.06	0.925	545.51	59	18	1698.97	0.965	570.31
58	19	1669.32	0.952	560.83	59	19	1746.30	0.993	586.68
58	20	1708.88	0.976	574.61	59	20	1788.82	1.019	601.48
58	21	1743.76	0.998	586.87	59	21	1826.56	1.043	614.72
58	22	1773.99	1.018	597.62	59	22	1859.54	1.064	626.42
58	23	1799.58	1.034	606.85	59	23	1887.78	1.082	636.57
58	24	1820.58	1.048	614.57	59	24	1911.32	1.098	645.18
58	25	1836.99	1.060	620.79	59	25	1930.18	1.111	652.26
58	26	1848.84	1.069	625.51	59	26	1944.38	1.121	657.80
58	27	1856.17	1.075	628.73	59	27	1953.95	1.129	661.83
58	28	1859.00	1.079	630.47	59	28	1958.92	1.134	664.33
58	29	1857.35	1.081	630.72	59	29	1959.32	1.137	665.31

TABLE XVI-C.—CORRECTIONS FOR COMBINING SPIRALS

Δ_S	Δ_N	C_b	C_p	C_a	Δ_S	Δ_N	C_b	C_p	C_a
Deg.	Deg.	Sec.	Ft.	Sec.	Deg.	Deg.	Sec.	Ft.	Sec.
59	29	1959.32	1.137	665.31	60	15	1593.58	0.897	533.77
59	30	1955.17	1.137	664.79	60	16	1658.96	0.936	556.04
59	31	1946.50	1.134	662.75	60	17	1719.34	0.972	576.70
59	32	1933.35	1.129	659.21	60	18	1774.77	1.005	595.75
59	33	1915.74	1.121	654.17	60	19	1825.25	1.035	613.19
59	34	1893.70	1.111	647.63	60	20	1870.82	1.063	629.04
59	35	1867.26	1.098	639.59	60	21	1911.51	1.089	643.30
59	36	1836.46	1.082	630.06	60	22	1947.32	1.111	655.97
59	37	1801.32	1.064	619.03	60	23	1978.31	1.131	667.07
59	38	1761.87	1.043	606.52	60	24	2004.47	1.148	676.60
59	39	1718.16	1.019	592.51	60	25	2025.86	1.163	684.56
59	40	1670.20	0.993	577.02	60	26	2042.48	1.175	690.97
59	41	1618.04	0.965	560.04	60	27	2054.38	1.184	695.82
59	42	1561.70	0.933	541.58	60	28	2061.57	1.191	699.11
59	43	1501.22	0.899	521.63	60	29	2064.09	1.195	700.86
59	44	1436.64	0.863	500.20	60	30	2061.97	1.196	701.07
59	45	1367.99	0.823	477.28	60	31	2055.23	1.195	699.74
59	46	1295.31	0.782	452.88	60	32	2043.90	1.191	696.87
59	47	1218.62	0.737	427.00	60	33	2028.02	1.184	692.47
59	48	1137.97	0.690	399.62	60	34	2007.61	1.175	686.54
59	49	1053.40	0.640	370.76	60	35	1982.71	1.163	679.09
59	50	964.93	0.588	340.41	60	36	1953.35	1.148	670.12
59	51	872.61	0.533	308.58	60	37	1919.56	1.131	659.62
59	52	776.48	0.476	275.25	60	38	1881.36	1.111	647.60
59	53	676.57	0.416	240.43	60	39	1838.81	1.089	634.07
59	54	572.92	0.353	204.11	60	40	1791.92	1.063	619.03
59	55	465.58	0.288	166.30	60	41	1740.74	1.035	602.47
59	56	354.57	0.220	126.98	60	42	1685.30	1.005	584.39
59	57	239.94	0.149	86.16	60	43	1625.62	0.972	564.81
59	58	121.74	0.076	43.84	60	44	1561.75	0.936	543.71
59	59	0	0	0	60	45	1493.73	0.897	521.10
60	0	0	0	0	60	46	1421.58	0.856	496.98
60	1	142.69	0.078	47.56	60	47	1345.35	0.812	471.35
60	2	280.09	0.154	93.37	60	48	1265.07	0.766	444.21
60	3	412.23	0.227	137.44	60	49	1180.78	0.716	415.55
60	4	539.11	0.298	179.76	60	50	1092.51	0.665	385.38
60	5	660.76	0.366	220.37	60	51	1000.32	0.610	353.70
60	6	777.19	0.431	259.26	60	52	904.22	0.553	320.49
60	7	888.42	0.493	296.45	60	53	804.27	0.493	285.77
60	8	994.48	0.553	331.95	60	54	700.50	0.431	249.53
60	9	1095.38	0.610	365.76	60	55	592.95	0.366	211.77
60	10	1191.14	0.665	397.89	60	56	481.67	0.298	172.48
60	11	1281.78	0.716	428.37	60	57	366.69	0.227	131.66
60	12	1367.32	0.766	457.18	60	58	248.06	0.154	89.31
60	13	1447.79	0.812	484.35	60	59	125.82	0.078	45.42
60	14	1523.20	0.856	509.88	60	60	0	0	0
60	15	1593.58	0.897	533.77					

TABLE XVI-D.—CORRECTIONS C_s TO DEFLECTIONS FOR SIMPLE SPIRAL. TRANSIT AT T.S.; SIGHT AT S.C.

$\Delta=\Delta_N$	C_s	$\Delta=\Delta_N$	C_s	$\Delta=\Delta_N$	C_s
Deg.	Sec.	Deg.	Sec.	Deg.	Sec.
1	0.00	21	28.75	41	215.93
2	0.02	22	33.07	42	232.25
3	0.08	23	37.80	43	249.40
4	0.20	24	42.96	44	267.37
5	0.39	25	48.58	45	286.21
6	0.67	26	54.66	46	305.92
7	1.06	27	61.24	47	326.52
8	1.59	28	68.32	48	348.05
9	2.26	29	75.94	49	370.52
10	3.10	30	84.11	50	393.96
11	4.12	31	92.84	51	418.38
12	5.35	32	102.17	52	443.80
13	6.81	33	112.10	53	470.26
14	8.50	34	122.67	54	497.77
15	10.46	35	133.88	55	526.35
16	12.70	36	145.76	56	556.03
17	15.24	37	158.33	57	586.83
18	18.09	38	171.61	58	618.78
19	21.28	39	185.63	59	651.89
20	24.83	40	200.39	60	686.19

TABLE XVII.—LEVEL SECTIONS $\frac{1}{4}:1$

HEIGHT	CUBIC YARDS PER 100-FT STATION							
	BASE 14	BASE 16	BASE 18	BASE 20	BASE 24	BASE 30	BASE 40	PER FT OF ADDED BASE
0.5	26	30	34	37	45	56	74	1.85
1.0	53	60	68	75	90	112	149	3.70
1.5	80	91	102	113	135	169	224	5.56
2.0	107	122	137	152	181	226	300	7.41
2.5	135	154	172	191	228	284	376	9.26
3.0	163	186	208	231	275	342	453	11.11
3.5	193	219	245	271	322	400	530	12.96
4.0	222	252	281	311	370	459	607	14.81
4.5	252	285	319	352	419	519	685	16.67
5.0	282	319	356	394	468	579	764	18.52
5.5	313	354	395	435	517	639	843	20.37
6.0	344	389	433	478	567	700	922	22.22
6.5	376	424	472	521	617	761	1002	24.07
7.0	408	460	512	564	668	823	1083	25.93
7.5	441	497	552	608	719	885	1163	27.78
8.0	474	533	593	652	770	948	1245	29.63
8.5	508	571	634	696	822	1011	1326	31.48
9.0	542	608	675	742	875	1075	1408	33.33
9.5	576	647	717	787	928	1139	1491	35.18
10.0	611	685	759	833	981	1204	1574	37.04
10.5	647	724	802	880	1035	1269	1658	38.89
11.0	682	764	845	927	1090	1334	1742	40.74
11.5	719	804	889	974	1144	1400	1826	42.59
12.0	756	844	933	1022	1200	1467	1911	44.44
12.5	793	885	978	1071	1256	1534	1996	46.30
13.0	831	926	1023	1119	1312	1601	2082	48.15
13.5	869	969	1069	1169	1369	1669	2169	50.00
14.0	907	1010	1115	1219	1426	1737	2256	51.85
14.5	947	1054	1161	1269	1484	1806	2343	53.70
15.0	986	1096	1208	1319	1542	1875	2431	55.56
15.5	1026	1141	1256	1371	1600	1945	2519	57.41
16.0	1067	1184	1304	1422	1659	2015	2607	59.26
16.5	1108	1230	1352	1474	1719	2085	2696	61.11
17.0	1149	1274	1401	1527	1779	2157	2786	62.96
17.5	1191	1321	1450	1580	1839	2228	2876	64.82
18.0	1233	1366	1500	1633	1900	2300	2967	66.67
18.5	1276	1413	1550	1687	1961	2373	3058	68.52
19.0	1319	1460	1601	1742	2023	2445	3149	70.37
19.5	1363	1508	1652	1797	2085	2519	3241	72.22
20	1407	1556	1704	1852	2148	2593	3333	74.07
21	1497	1653	1808	1964	2275	2742	3520	77.78
22	1589	1752	1915	2078	2404	2893	3707	81.48
23	1682	1853	2023	2194	2534	3045	3897	85.19
24	1778	1956	2133	2311	2667	3200	4089	88.89
25	1875	2060	2245	2431	2801	3357	4282	92.59
26	1974	2166	2359	2552	2937	3514	4478	96.30
27	2075	2274	2475	2675	3075	3675	4675	100.00
28	2178	2384	2593	2800	3215	3837	4874	103.70
29	2282	2496	2712	2927	3356	4001	5075	107.41
30	2389	2611	2833	3056	3500	4167	5278	111.11
31	2497	2726	2956	3186	3645	4334	5482	114.81
32	2607	2844	3081	3319	3793	4504	5689	118.52
33	2719	2964	3208	3453	3942	4675	5897	122.22
34	2833	3085	3337	3589	4093	4848	6108	125.93
35	2949	3208	3468	3727	4245	5023	6320	129.63
36	3067	3333	3600	3867	4400	5200	6533	133.33
37	3186	3460	3734	4008	4556	5379	6749	137.04
38	3307	3589	3870	4152	4715	5559	6967	140.74
39	3431	3719	4008	4297	4875	5742	7186	144.44
40	3556	3852	4148	4444	5037	5926	7407	148.15

TABLE XVII.—LEVEL SECTIONS ½:1

HEIGHT	CUBIC YARDS PER 100-FT STATION							
	BASE 14	BASE 16	BASE 18	BASE 20	BASE 24	BASE 30	BASE 40	PER FT OF ADDED BASE
0.5	26	30	34	38	45	56	75	1.85
1.0	54	61	69	76	91	113	150	3.70
1.5	82	93	104	115	138	171	226	5.56
2.0	111	126	141	156	185	230	304	7.41
2.5	141	160	178	197	234	289	382	9.26
3.0	172	194	217	239	283	350	461	11.11
3.5	204	230	256	282	334	412	541	12.96
4.0	237	267	296	326	385	474	622	14.81
4.5	271	304	338	371	438	538	704	16.67
5.0	306	343	380	417	491	602	787	18.52
5.5	341	382	423	463	545	667	871	20.37
6.0	378	422	467	511	600	733	956	22.22
6.5	415	463	512	560	656	800	1041	24.07
7.0	454	506	557	609	713	869	1128	25.93
7.5	493	549	604	660	771	938	1215	27.78
8.0	533	593	652	711	830	1007	1304	29.63
8.5	575	638	700	763	889	1078	1393	31.48
9.0	617	683	750	817	950	1150	1483	33.33
9.5	660	730	800	871	1012	1223	1575	35.18
10.0	704	778	852	926	1074	1296	1667	37.04
10.5	749	826	904	982	1138	1371	1760	38.89
11.0	794	876	957	1039	1202	1446	1854	40.74
11.5	841	926	1012	1097	1267	1523	1949	42.59
12.0	889	978	1067	1156	1333	1600	2044	44.44
12.5	937	1030	1123	1215	1400	1678	2141	46.30
13.0	987	1083	1180	1276	1469	1757	2239	48.15
13.5	1037	1138	1238	1338	1538	1838	2338	50.00
14.0	1089	1193	1296	1400	1607	1919	2437	51.85
14.5	1141	1248	1356	1463	1678	2000	2538	53.70
15.0	1194	1306	1417	1528	1750	2083	2639	55.56
15.5	1249	1363	1478	1593	1823	2167	2741	57.41
16.0	1304	1422	1541	1659	1896	2252	2844	59.26
16.5	1360	1482	1604	1726	1971	2338	2949	61.11
17.0	1417	1543	1669	1794	2046	2424	3054	62.96
17.5	1475	1604	1734	1863	2123	2512	3160	64.82
18.0	1533	1667	1800	1933	2200	2600	3267	66.67
18.5	1593	1730	1867	2004	2278	2689	3375	68.52
19.0	1654	1794	1935	2076	2357	2780	3483	70.37
19.5	1715	1860	2004	2149	2438	2871	3593	72.22
20	1778	1926	2074	2222	2519	2963	3704	74.07
21	1906	2061	2217	2372	2683	3115	3928	77.78
22	2037	2200	2363	2526	2852	3341	4156	81.48
23	2172	2343	2513	2683	3024	3535	4387	85.19
24	2311	2489	2667	2844	3200	3733	4622	88.89
25	2454	2639	2824	3009	3380	3935	4861	92.59
26	2600	2793	2985	3178	3563	4141	5104	96.30
27	2750	2950	3150	3350	3750	4350	5350	100.00
28	2904	3111	3319	3526	3941	4563	5600	103.70
29	3061	3276	3491	3706	4135	4780	5854	107.41
30	3222	3444	3667	3889	4333	5000	6111	111.11
31	3387	3617	3846	4076	4535	5224	6372	114.81
32	3556	3793	4030	4267	4741	5452	6637	118.52
33	3728	3972	4217	4461	4950	5683	6906	122.22
34	3904	4156	4407	4659	5163	5919	7178	125.93
35	4083	4343	4602	4861	5380	6157	7454	129.63
36	4267	4533	4800	5067	5600	6400	7733	133.33
37	4454	4728	5002	5276	5824	6646	8017	137.04
38	4644	4926	5207	5489	6052	6896	8304	140.74
39	4839	5128	5417	5706	6283	7150	8594	144.44
40	5037	5333	5630	5926	6519	7407	8889	148.15

TABLE XVII.—LEVEL SECTIONS **1:1**

HEIGHT	CUBIC YARDS PER 100-FT STATION							
	BASE 14	BASE 16	BASE 18	BASE 20	BASE 24	BASE 30	BASE 40	PER FT OF ADDED BASE
0.5	27	31	34	38	45	56	75	1.85
1.0	56	63	70	78	93	115	152	3.70
1.5	81	97	108	119	142	175	231	5.56
2.0	119	133	148	163	193	237	311	7.41
2.5	153	171	190	208	245	301	394	9.26
3.0	189	211	233	256	300	367	478	11.11
3.5	227	253	279	305	356	434	564	12.96
4.0	267	296	326	356	415	504	652	14.81
4.5	308	342	375	408	475	575	742	16.67
5.0	352	389	426	463	537	648	833	18.52
5.5	397	438	479	519	601	723	927	20.37
6.0	444	489	533	578	667	800	1022	22.22
6.5	494	542	590	638	734	879	1119	24.07
7.0	544	596	648	700	804	959	1219	25.93
7.5	597	653	708	764	875	1042	1319	27.78
8.0	652	711	770	830	948	1126	1422	29.63
8.5	708	771	834	897	1023	1212	1527	31.48
9.0	767	833	900	967	1100	1300	1633	33.33
9.5	827	897	968	1038	1179	1390	1742	35.18
10.0	889	963	1037	1111	1259	1481	1852	37.04
10.5	953	1031	1108	1186	1342	1575	1964	38.89
11.0	1019	1100	1181	1263	1426	1670	2078	40.74
11.5	1086	1171	1256	1342	1512	1768	2194	42.59
12.0	1156	1244	1333	1422	1600	1867	2311	44.44
12.5	1227	1319	1412	1505	1690	1968	2431	46.30
13.0	1300	1396	1493	1589	1781	2070	2552	48.15
13.5	1375	1475	1575	1675	1875	2175	2675	50.00
14.0	1452	1556	1659	1763	1970	2281	2800	51.85
14.5	1531	1638	1745	1853	2068	2390	2927	53.70
15.0	1611	1722	1833	1944	2167	2500	3056	55.56
15.5	1694	1808	1923	2038	2268	2612	3186	57.41
16.0	1778	1896	2015	2133	2370	2726	3319	59.26
16.5	1864	1986	2108	2231	2475	2842	3453	61.11
17.0	1952	2078	2204	2330	2581	2959	3589	62.96
17.5	2042	2171	2301	2431	2690	3079	3727	64.82
18.0	2133	2267	2400	2533	2800	3200	3867	66.67
18.5	2227	2364	2501	2638	2912	3323	4008	68.52
19.0	2322	2463	2604	2744	3026	3448	4152	70.37
19.5	2419	2564	2708	2853	3142	3575	4297	72.22
20	2519	2667	2815	2963	3259	3704	4444	74.07
21	2722	2878	3033	3189	3500	3967	4744	77.78
22	2933	3096	3259	3422	3748	4237	5052	81.48
23	3152	3322	3493	3663	4004	4515	5367	85.19
24	3378	3556	3733	3911	4267	4800	5689	88.89
25	3611	3796	3981	4167	4537	5093	6019	92.59
26	3852	4044	4237	4430	4815	5393	6356	96.30
27	4100	4300	4500	4700	5100	5700	6700	100.00
28	4356	4563	4770	4978	5393	6015	7052	103.70
29	4619	4833	5048	5263	5693	6337	7411	107.41
30	4889	5111	5333	5556	6000	6667	7778	111.11
31	5167	5396	5626	5856	6315	7004	8152	114.81
32	5452	5689	5926	6163	6637	7348	8533	118.52
33	5744	5989	6233	6478	6967	7700	8922	122.22
34	6044	6296	6548	6800	7304	8059	9319	125.93
35	6352	6611	6870	7130	7648	8426	9722	129.63
36	6667	6933	7200	7467	8000	8800	10133	133.33
37	6989	7263	7537	7811	8359	9181	10552	137.04
38	7319	7600	7881	8163	8726	9570	10978	140.74
39	7656	7944	8233	8522	9100	9967	11411	144.44
40	8000	8296	8593	8889	9481	10370	11852	148.15

TABLE XVII.—LEVEL SECTIONS $1\frac{1}{2}:1$

HEIGHT	CUBIC YARDS PER 100-FT STATION							
	BASE 14	BASE 16	BASE 18	BASE 20	BASE 24	BASE 30	BASE 40	PER FT OF ADDED BASE
0.5	27	31	35	38	46	57	75	1.85
1.0	57	65	72	80	94	117	154	3.70
1.5	90	101	113	124	146	179	235	5.56
2.0	126	141	156	170	200	244	319	7.41
2.5	164	183	201	220	257	313	405	9.26
3.0	206	228	250	272	317	383	494	11.11
3.5	250	275	301	327	379	457	587	12.96
4.0	296	326	356	385	444	533	681	14.81
4.5	346	379	413	446	513	613	779	16.67
5.0	398	435	472	509	583	694	880	18.52
5.5	453	494	535	575	657	779	983	20.37
6.0	511	556	600	644	733	867	1089	22.22
6.5	572	620	668	716	813	957	1198	24.07
7.0	635	687	739	791	894	1050	1309	25.93
7.5	701	757	813	868	979	1146	1424	27.78
8.0	770	830	889	948	1067	1244	1541	29.63
8.5	841	905	968	1031	1157	1346	1661	31.48
9.0	917	983	1050	1117	1250	1450	1783	33.33
9.5	994	1064	1135	1205	1346	1557	1909	35.18
10.0	1074	1148	1222	1296	1444	1667	2037	37.04
10.5	1157	1235	1313	1390	1546	1779	2168	38.89
11.0	1243	1324	1406	1487	1650	1894	2302	40.74
11.5	1331	1416	1501	1587	1757	2013	2438	42.59
12.0	1422	1511	1600	1689	1867	2133	2578	44.44
12.5	1516	1609	1701	1794	1979	2257	2720	46.30
13.0	1613	1709	1806	1902	2094	2383	2865	48.15
13.5	1713	1813	1913	2013	2213	2513	3013	50.00
14.0	1815	1919	2022	2126	2333	2644	3163	51.85
14.5	1920	2027	2135	2242	2457	2779	3316	53.70
15.0	2028	2139	2250	2361	2583	2917	3472	55.56
15.5	2138	2253	2368	2483	2713	3057	3631	57.41
16.0	2252	2370	2489	2607	2844	3200	3793	59.26
16.5	2368	2490	2613	2735	2979	3346	3957	61.11
17.0	2487	2613	2739	2865	3117	3494	4124	62.96
17.5	2609	2738	2868	2998	3257	3646	4294	64.82
18.0	2733	2867	3000	3133	3400	3800	4467	66.67
18.5	2861	2998	3135	3272	3546	3957	4642	68.52
19.0	2991	3131	3272	3413	3694	4117	4820	70.37
19.5	3124	3268	3413	3557	3846	4279	5001	72.22
20	3259	3407	3556	3704	4000	4444	5185	74.07
21	3539	3694	3850	4006	4317	4783	5561	77.78
22	3830	3993	4156	4319	4644	5133	5948	81.48
23	4131	4302	4472	4643	4983	5494	6346	85.19
24	4444	4622	4800	4978	5333	5867	6756	88.89
25	4769	4954	5139	5324	5694	6250	7176	92.59
26	5104	5296	5489	5681	6067	6644	7607	96.30
27	5450	5650	5850	6050	6450	7050	8050	100.00
28	5807	6015	6222	6430	6844	7467	8504	103.70
29	6176	6391	6606	6820	7250	7894	8969	107.41
30	6556	6778	7000	7222	7667	8333	9444	111.11
31	6946	7176	7406	7635	8094	8783	9931	114.81
32	7348	7585	7822	8059	8533	9244	10430	118.52
33	7761	8006	8250	8494	8983	9717	10939	122.22
34	8185	8437	8689	8941	9444	10200	11459	125.93
35	8620	8880	9139	9398	9917	10694	11991	129.63
36	9067	9333	9600	9867	10400	11200	12533	133.33
37	9524	9798	10072	10346	10894	11717	13087	137.04
38	9993	10274	10556	10837	11400	12244	13652	140.74
39	10472	10761	11050	11339	11917	12783	14228	144.44
40	10963	11259	11556	11852	12444	13333	14815	148.15

TABLE XVII.—LEVEL SECTIONS

2:1

HEIGHT	CUBIC YARDS PER 100-FT STATION							
	BASE 14	BASE 16	BASE 18	BASE 20	BASE 24	BASE 30	BASE 40	PER FT OF ADDED BASE
0.5	28	31	35	39	46	57	76	1.85
1.0	59	67	74	81	96	119	156	3.70
1.5	94	106	117	128	150	183	239	5.56
2.0	133	148	163	178	207	252	326	7.41
2.5	176	194	213	231	269	324	417	9.26
3.0	222	244	267	289	333	400	511	11.11
3.5	272	298	324	350	402	480	609	12.96
4.0	326	356	385	415	474	563	711	14.81
4.5	383	417	450	483	550	650	817	16.67
5.0	444	481	519	556	630	741	926	18.52
5.5	509	550	591	631	714	835	1039	20.37
6.0	578	622	667	711	800	933	1156	22.22
6.5	650	698	746	794	891	1035	1276	24.07
7.0	726	778	830	881	985	1141	1400	25.93
7.5	806	861	917	972	1083	1250	1528	27.78
8.0	889	948	1007	1067	1185	1363	1659	29.63
8.5	976	1039	1102	1165	1291	1480	1794	31.48
9.0	1067	1133	1200	1267	1400	1600	1933	33.33
9.5	1161	1231	1302	1372	1513	1724	2076	35.18
10.0	1259	1333	1407	1481	1630	1852	2222	37.04
10.5	1361	1439	1517	1594	1750	1983	2372	38.89
11.0	1467	1548	1630	1711	1874	2119	2526	40.74
11.5	1576	1661	1746	1831	2002	2257	2683	42.59
12.0	1689	1778	1867	1956	2133	2400	2844	44.44
12.5	1806	1898	1991	2083	2269	2546	3009	46.30
13.0	1926	2022	2119	2215	2407	2696	3178	48.15
13.5	2050	2150	2250	2350	2550	2850	3350	50.00
14.0	2178	2281	2385	2489	2696	3007	3526	51.85
14.5	2309	2417	2524	2631	2846	3169	3706	53.70
15.0	2444	2556	2667	2778	3000	3333	3889	55.56
15.5	2583	2698	2813	2928	3157	3502	4076	57.41
16.0	2726	2844	2963	3081	3319	3674	4267	59.26
16.5	2872	2994	3117	3239	3483	3850	4461	61.11
17.0	3022	3148	3274	3400	3652	4030	4659	62.96
17.5	3176	3306	3435	3565	3824	4213	4861	64.82
18.0	3333	3467	3600	3733	4000	4400	5067	66.67
18.5	3494	3631	3769	3906	4180	4591	5276	68.52
19.0	3659	3800	3941	4081	4363	4785	5489	70.37
19.5	3828	3972	4117	4261	4550	4983	5706	72.22
20	4000	4148	4296	4444	4741	5185	5926	74.07
21	4356	4511	4667	4822	5133	5600	6378	77.78
22	4730	4889	5052	5215	5541	6030	6844	81.48
23	5111	5281	5452	5622	5963	6474	7326	85.19
24	5511	5689	5867	6044	6400	6933	7822	88.89
25	5926	6111	6296	6481	6852	7407	8333	92.59
26	6356	6548	6741	6933	7319	7896	8859	96.30
27	6800	7000	7200	7400	7800	8400	9400	100.00
28	7259	7467	7674	7881	8296	8919	9956	103.70
29	7733	7948	8163	8378	8807	9452	10526	107.41
30	8222	8444	8667	8889	9333	10000	11111	111.11
31	8726	8956	9185	9415	9874	10563	11711	114.81
32	9244	9481	9719	9956	10430	11141	12326	118.52
33	9778	10022	10267	10511	11000	11733	12956	122.22
34	10326	10578	10830	11081	11585	12341	13600	125.93
35	10889	11148	11407	11667	12185	12963	14259	129.63
36	11467	11733	12000	12267	12800	13600	14933	133.33
37	12059	12333	12607	12881	13430	14252	15622	137.04
38	12667	12948	13230	13511	14074	14919	16326	140.74
39	13289	13578	13867	14156	14733	15600	17044	144.44
40	13926	14222	14519	14815	15407	16296	17778	148.15

TABLE XVII.—LEVEL SECTIONS 3:1

HEIGHT	CUBIC YARDS PER 100-FT STATION							
	BASE 14	BASE 16	BASE 18	BASE 20	BASE 24	BASE 30	BASE 40	PER FT OF ADDED BASE
0.5	29	32	36	40	47	58	77	1.85
1.0	63	70	78	85	100	122	159	3.70
1.5	103	114	125	136	158	192	247	5.56
2.0	148	163	178	193	222	267	341	7.41
2.5	199	218	236	255	292	347	440	9.26
3.0	256	278	300	322	367	433	544	11.11
3.5	318	344	369	395	447	525	655	12.96
4.0	385	415	444	474	533	622	770	14.81
4.5	458	492	525	558	625	725	892	16.67
5.0	537	574	611	648	722	833	1019	18.52
5.5	621	662	703	744	825	947	1151	20.37
6.0	711	756	800	844	933	1067	1289	22.22
6.5	806	855	903	951	1047	1192	1432	24.07
7.0	907	959	1011	1063	1167	1322	1581	25.93
7.5	1014	1069	1125	1181	1292	1458	1736	27.78
8.0	1126	1185	1244	1304	1422	1600	1896	29.63
8.5	1244	1306	1369	1432	1558	1747	2062	31.48
9.0	1367	1433	1500	1567	1700	1900	2233	33.33
9.5	1495	1566	1636	1706	1847	2058	2410	35.18
10.0	1630	1704	1778	1852	2000	2222	2593	37.04
10.5	1769	1847	1925	2003	2158	2392	2781	38.89
11.0	1915	1996	2078	2159	2322	2567	2974	40.74
11.5	2066	2150	2236	2321	2492	2747	3173	42.59
12.0	2222	2311	2400	2489	2667	2933	3378	44.44
12.5	2384	2477	2569	2662	2847	3125	3588	46.30
13.0	2552	2648	2744	2841	3033	3322	3804	48.15
13.5	2725	2825	2925	3025	3225	3525	4025	50.00
14.0	2904	3007	3111	3215	3422	3733	4252	51.85
14.5	3088	3195	3303	3410	3625	3947	4484	53.70
15.0	3278	3389	3500	3611	3833	4167	4722	55.56
15.5	3473	3588	3703	3818	4047	4392	4966	57.41
16.0	3674	3793	3911	4030	4267	4622	5215	59.26
16.5	3881	4003	4125	4247	4492	4858	5469	61.11
17.0	4093	4219	4344	4470	4722	5100	5730	62.96
17.5	4310	4440	4569	4699	4958	5347	5995	64.82
18.0	4533	4667	4800	4933	5200	5600	6267	66.67
18.5	4762	4899	5036	5173	5447	5858	6544	68.52
19.0	4996	5137	5278	5419	5700	6122	6826	70.37
19.5	5236	5381	5525	5669	5958	6392	7114	72.22
20	5481	5630	5778	5926	6222	6667	7407	74.07
21	5989	6144	6300	6456	6767	7233	8011	77.78
22	6519	6681	6844	7007	7333	7822	8637	81.48
23	7070	7241	7411	7581	7922	8433	9285	85.19
24	7644	7822	8000	8178	8533	9067	9956	88.89
25	8241	8426	8611	8796	9167	9722	10648	92.59
26	8859	9052	9244	9437	9822	10400	11363	96.30
27	9500	9700	9900	10100	10500	11100	12100	100.00
28	10163	10370	10578	10785	11200	11822	12859	103.70
29	10848	11063	11278	11493	11922	12567	13641	107.41
30	11556	11778	12000	12222	12667	13333	14444	111.11
31	12285	12515	12744	12974	13433	14122	15270	114.81
32	13037	13274	13511	13748	14222	14933	16119	118.52
33	13811	14056	14300	14544	15033	15767	16989	122.22
34	14607	14859	15111	15363	15867	16622	17881	125.93
35	15426	15685	15944	16204	16722	17500	18796	129.63
36	16267	16533	16800	17067	17600	18400	19733	133.33
37	17130	17404	17678	17952	18500	19322	20693	137.04
38	18015	18296	18578	18859	19422	20267	21674	140.74
39	18922	19211	19500	19789	20367	21233	22678	144.44
40	19852	20148	20444	20741	21333	22222	23704	148.15

The coefficients in the tables on the opposite page provide a rapid method of correcting the level-section quantities of Table XVII for transverse ground slopes.

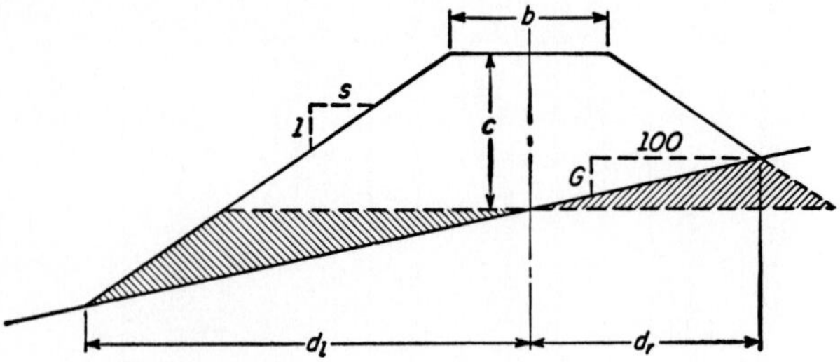

From the figure,

$$d_l = \frac{b}{2} + sc + \frac{sd_lG}{100}$$

or

$$d_l = \frac{(b/2) + sc}{1 - (sG/100)} \tag{1}$$

Similarly,

$$d_r = \frac{(b/2) + sc}{1 + (sG/100)} \tag{2}$$

From equation 6-3, the area of the level section is $A_L = c(b + cs)$. On a transverse slope, the level-section area is increased by one triangular area and decreased by another, as shown by the cross-hatched areas in the figure. The resulting net area A is

$$A = A_L + \frac{1}{2}\left(\frac{b}{2} + sc\right)\left(\frac{d_lG}{100}\right) - \frac{1}{2}\left(\frac{b}{2} + sc\right)\left(\frac{d_rG}{100}\right)$$

which reduces to

$$A = A_L + \frac{G}{200}\left(\frac{b}{2} + sc\right)(d_l - d_r) \tag{3}$$

When equations 1 and 2 are substituted in equation 3, the net area A reduces to

$$A = A_L + \left(c + \frac{b}{2s}\right)^2\left[\frac{s^3}{(100/G)^2 - s^2}\right] \tag{4}$$

Therefore,

$$\text{Cr yd}^3 \text{ per 100 ft} = \text{level-section yardage} + C\left(c + \frac{b}{2s}\right)^2$$

TABLE XVII —A. VALUES OF b/2s

BASE b	SIDE SLOPE RATIO, s ¼:1	½:1	1:1	1½:1	2:1	3:1
14	28	14	7	4.67	3.50	2.33
16	32	16	8	5.33	4	2.67
18	36	18	9	6	4.50	3
20	40	20	10	6.67	5	3.33
24	48	24	12	8	6	4
30	60	30	15	10	7.5	5
40	80	40	20	13.3	10	6.67
50	100	50	25	16.7	12.5	8.33
60	120	60	30	20	15	10

TABLE XVII.—B. VALUES OF C

GROUND SLOPE G	SIDE SLOPE RATIO, s ¼:1	½:1	1:1	1½:1	2:1	3:1
5%	0.0001	0.0012	0.0093	0.0313	0.0748	0.2558
10%	0.0006	0.0046	0.0374	0.1278	0.3086	1.099
15%	0.0013	0.0105	0.0852	0.2962	0.7325	3.821
20%	0.0023	0.0187	0.1543	0.5494	1.411	6.250
25%	0.0036	0.0294	0.2469	0.9090	2.469	14.29
30%	0.0052	0.0426	0.3663	1.411	4.167	47.37
35%	0.0071	0.0585	0.5170	2.113	7.117	
40%	0.0093	0.0771	0.7054	3.125	13.17	
45%	0.0119	0.0988	0.9405	4.650	31.58	
50%	0.0147	0.1234	1.235	7.143		
60%	0.0213	0.1831	2.083	23.68		
70%	0.0292	0.2585	3.557			
80%	0.0386	0.3526	6.584			
90%	0.0494	0.4702	15.79			
100%	0.0617	0.6172				

where

$$C = \left(\frac{100}{27}\right)\left[\frac{s^3}{(100/G)^2 - s^2}\right]$$

Example

Given: a cross section on a uniform transverse ground slope G of 20%. Other data are: base $b = 30$ ft; center cut or fill $c = 15$ ft; side slope ralio $s = 1\frac{1}{2}:1$.

From Table XVII,

Level-section quantity = 2917 yd³ per sta.

From Tables XVII.-A and XVII.-B,

Correction for slope

$(15 + 10)^2(0.5494) =$ 343 yd³ per sta.

Yardage corrected for slope = 3260 yd³ per sta.

TABLE XVIII.—TRIANGULAR PRISMS. CUBIC YARDS PER 50 FEET

HEIGHT OR WIDTH	WIDTH OR HEIGHT								
	1	2	3	4	5	6	7	8	9
0.1	0.09	0.19	0.28	0.37	0.46	0.56	0.65	0.74	0.83
.2	0.19	0.37	0.56	0.74	0.93	1.11	1.30	1.48	1.67
.3	0.28	0.56	0.83	1.11	1.39	1.67	1.94	2.22	2.50
.4	0.37	0.74	1.11	1.48	1.85	2.22	2.59	2.96	3.33
0.5	0.46	0.93	1.39	1.85	2.31	2.78	3.24	3.70	4.17
.6	0.56	1.11	1.67	2.22	2.78	3.33	3.89	4.44	5.00
.7	0.65	1.30	1.94	2.59	3.24	3.89	4.54	5.19	5.83
.8	0.74	1.48	2.22	2.96	3.70	4.44	5.19	5.93	6.67
.9	0.83	1.67	2.50	3.33	4.17	5.00	5.83	6.67	7.50
1.0	0.93	1.85	2.78	3.70	4.63	5.56	6.48	7.41	8.33
.1	1.02	2.04	3.06	4.07	5.09	6.11	7.13	8.15	9.17
.2	1.11	2.22	3.33	4.44	5.56	6.67	7.78	8.89	10.00
.3	1.20	2.41	3.61	4.81	6.02	7.22	8.43	9.63	10.83
.4	1.30	2.59	3.89	5.19	6.48	7.78	9.07	10.37	11.67
1.5	1.39	2.78	4.17	5.56	6.94	8.33	9.72	11.11	12.50
.6	1.48	2.96	4.44	5.93	7.41	8.89	10.37	11.85	13.33
.7	1.57	3.15	4.72	6.30	7.87	9.44	11.02	12.59	14.17
.8	1.67	3.33	5.00	6.67	8.33	10.00	11.67	13.33	15.00
.9	1.76	3.52	5.28	7.04	8.80	10.56	12.31	14.07	15.83
2.0	1.85	3.70	5.56	7.41	9.26	11.11	12.96	14.81	16.67
.1	1.94	3.89	5.83	7.78	9.72	11.67	13.61	15.56	17.50
.2	2.04	4.07	6.11	8.15	10.19	12.22	14.26	16.30	18.33
.3	2.13	4.26	6.39	8.52	10.65	12.78	14.91	17.04	19.17
.4	2.22	4.44	6.67	8.89	11.11	13.33	15.56	17.78	20.00
2.5	2.31	4.63	6.94	9.26	11.57	13.89	16.20	18.52	20.83
.6	2.41	4.81	7.22	9.63	12.04	14.44	16.85	19.26	21.67
.7	2.50	5.00	7.50	10.00	12.50	15.00	17.50	20.00	22.50
.8	2.59	5.19	7.78	10.37	12.96	15.56	18.15	20.74	23.33
.9	2.69	5.37	8.06	10.74	13.43	16.11	18.80	21.48	24.17
3.0	2.78	5.56	8.33	11.11	13.89	16.67	19.44	22.22	25.00
.1	2.87	5.74	8.61	11.48	14.35	17.22	20.09	22.96	25.83
.2	2.96	5.93	8.89	11.85	14.81	17.78	20.74	23.70	26.67
.3	3.06	6.11	9.17	12.22	15.28	18.33	21.39	24.44	27.50
.4	3.15	6.30	9.44	12.59	15.74	18.89	22.04	25.19	28.33
3.5	3.24	6.48	9.72	12.96	16.20	19.44	22.69	25.93	29.17
.6	3.33	6.67	10.00	13.33	16.67	20.00	23.33	26.67	30.00
.7	3.43	6.85	10.28	13.70	17.13	20.56	23.98	27.41	30.83
.8	3.52	7.04	10.56	14.07	17.59	21.11	24.63	28.15	31.67
.9	3.61	7.22	10.83	14.44	18.06	21.67	25.28	28.89	32.50
4.0	3.70	7.41	11.11	14.81	18.52	22.22	25.93	29.63	33.33
.1	3.80	7.59	11.39	15.19	19.98	22.78	26.57	30.37	34.17
.2	3.89	7.78	11.67	15.56	19.44	23.33	27.22	31.11	35.00
.3	3.98	7.96	11.94	15.93	19.91	23.89	27.87	31.85	35.83
.4	4.07	8.15	12.22	16.30	20.37	24.44	28.52	32.59	36.67
4.5	4.17	8.33	12.50	16.67	20.83	25.00	29.17	33.33	37.50
.6	4.26	8.52	12.78	17.04	21.30	25.56	29.81	34.07	38.33
.7	4.35	8.70	13.06	17.41	21.76	26.11	30.46	34.81	39.17
.8	4.44	8.89	13.33	17.78	22.22	26.67	31.11	35.56	40.00
.9	4.54	9.07	13.61	18.15	22.69	27.22	31.76	36.30	40.83
5.0	4.63	9.26	13.89	18.52	23.15	27.78	32.41	37.04	41.67
.1	4.72	9.44	14.17	18.89	23.61	28.33	33.06	37.78	42.50
.2	4.81	9.63	14.44	19.26	24.07	28.89	33.70	38.52	43.33
.3	4.91	9.81	14.72	19.63	24.54	29.44	34.35	39.26	44.17
.4	5.00	10.00	15.00	20.00	25.00	30.00	35.00	40.00	45.00
5.5	5.09	10.19	15.28	20.37	25.46	30.56	35.65	40.74	45.83
.6	5.19	10.37	15.56	20.74	25.93	31.11	36.30	41.48	46.67
.7	5.28	10.56	15.83	21.11	26.39	31.67	36.94	42.22	47.50
.8	5.37	10.74	16.11	21.48	26.85	32.22	37.59	42.96	48.33
.9	5.46	10.93	16.39	21.85	27.31	32.78	38.24	43.70	49.17

TABLE XVIII.—TRIANGULAR PRISMS. CUBIC YARDS PER 50 FEET

HEIGHT OR WIDTH	WIDTH OR HEIGHT								
	1	2	3	4	5	6	7	8	9
6.0	5.56	11.11	16.67	22.22	27.78	33.33	38.89	44.44	50.00
.1	5.65	11.30	16.94	22.59	28.24	33.89	39.54	45.19	50.83
.2	5.74	11.48	17.22	22.96	28.70	34.44	40.19	45.93	51.67
.3	5.83	11.67	17.50	23.33	29.17	35.00	40.83	46.67	52.50
.4	5.93	11.85	17.78	23.70	29.63	35.56	41.48	47.41	53.33
6.5	6.02	12.04	18.06	24.07	30.09	36.11	42.13	48.15	54.17
.6	6.11	12.22	18.33	24.44	30.56	36.67	42.78	48.89	55.00
.7	6.20	12.41	18.61	24.81	31.02	37.22	43.43	49.63	55.83
.8	6.30	12.59	18.89	25.19	31.48	37.78	44.07	50.37	56.67
.9	6.39	12.78	19.17	25.56	31.94	38.33	44.72	51.11	57.50
7.0	6.48	12.96	19.44	25.93	32.41	38.89	45.37	51.85	58.33
.1	6.57	13.15	19.72	26.30	32.87	39.44	46.02	52.59	59.17
.2	6.67	13.33	20.00	26.67	33.33	40.00	46.67	53.33	60.00
.3	6.76	13.52	20.28	27.04	33.80	40.56	47.31	54.07	60.83
.4	6.85	13.70	20.56	27.41	34.26	41.11	47.96	54.81	61.67
7.5	6.94	13.89	20.83	27.78	34.72	41.67	48.61	55.56	62.50
.6	7.04	14.07	21.11	28.15	35.19	42.22	49.26	56.30	63.33
.7	7.13	14.26	21.39	28.52	35.65	42.78	49.91	57.04	64.17
.8	7.22	14.44	21.67	28.89	36.11	43.33	50.56	57.78	65.00
.9	7.31	14.63	21.94	29.26	36.57	43.89	51.20	58.52	65.83
8.0	7.41	14.81	22.22	29.63	37.04	44.44	51.85	59.26	66.67
.1	7.50	15.00	22.50	30.00	37.50	45.00	52.50	60.00	67.50
.2	7.59	15.19	22.78	30.37	37.96	45.56	53.15	60.74	68.33
.3	7.69	15.37	23.06	30.74	38.43	46.11	53.80	61.48	69.17
.4	7.78	15.56	23.33	31.11	38.89	46.67	54.44	62.22	70.00
8.5	7.87	15.74	23.61	31.48	39.35	47.22	55.09	62.96	70.83
.6	7.96	15.93	23.89	31.85	39.81	47.78	55.74	63.70	71.67
.7	8.06	16.11	24.17	32.22	40.28	48.33	56.39	64.44	72.50
.8	8.15	16.30	24.44	32.59	40.74	48.89	57.04	65.19	73.33
.9	8.24	16.48	24.72	32.96	41.20	49.44	57.69	65.93	74.17
9.0	8.33	16.67	25.00	33.33	41.67	50.00	58.33	66.67	75.00
.1	8.43	16.85	25.28	33.70	42.13	50.56	58.98	67.41	75.83
.2	8.52	17.04	25.56	34.07	42.59	51.11	59.63	68.15	76.67
.3	8.61	17.22	25.83	34.44	43.06	51.67	60.28	68.89	77.50
.4	8.70	17.41	26.11	34.81	43.52	52.22	60.93	69.63	78.33
9.5	8.80	17.59	26.39	35.19	43.98	52.78	61.57	70.37	79.17
.6	8.89	17.78	26.67	35.56	44.44	53.33	62.22	71.11	80.00
.7	8.98	17.96	26.94	35.93	44.91	53.89	62.87	71.85	80.83
.8	9.07	18.15	27.22	36.30	45.37	54.44	63.52	72.59	81.67
.9	9.17	18.33	27.50	36.67	45.83	55.00	64.17	73.33	82.50
10.0	9.26	18.52	27.78	37.04	46.30	55.56	64.81	74.07	83.33
.1	9.35	18.70	28.06	37.41	46.76	56.11	65.46	74.81	84.17
.2	9.44	18.89	28.33	37.78	47.22	56.67	66.11	75.56	85.00
.3	9.54	19.07	28.61	38.15	47.69	57.22	66.76	76.30	85.83
.4	9.63	19.26	28.89	38.52	48.15	57.78	67.41	77.04	86.67
10.5	9.72	19.44	29.17	38.89	48.61	58.33	68.06	77.78	87.50
.6	9.81	19.63	29.44	39.26	49.07	58.89	68.70	78.52	88.33
.7	9.91	19.81	29.72	39.63	49.54	59.44	69.35	79.26	89.17
.8	10.00	20.00	30.00	40.00	50.00	60.00	70.00	80.00	90.00
.9	10.09	20.19	30.28	40.37	50.46	60.56	70.65	80.74	90.83
11.0	10.19	20.37	30.56	40.74	50.93	61.11	71.30	81.48	91.67
.1	10.28	20.56	30.83	41.11	51.39	61.67	71.94	82.22	92.50
.2	10.37	20.74	31.11	41.48	51.85	62.22	72.59	82.96	93.33
.3	10.46	20.93	31.39	41.85	52.31	62.78	73.24	83.70	94.17
.4	10.56	21.11	31.67	42.22	52.78	63.33	73.89	84.44	95.00
11.5	10.65	21.30	31.94	42.59	53.24	63.89	74.54	85.19	95.83
.6	10.74	21.48	32.22	42.96	53.70	64.44	75.19	85.93	96.67
.7	10.83	21.67	32.50	43.33	54.17	65.00	75.83	86.67	97.50
.8	10.93	21.85	32.78	43.70	54.63	65.56	76.48	87.41	98.33
.9	11.02	22.04	33.06	44.07	55.09	66.11	77.13	88.15	99.17

TABLE XVIII.—TRIANGULAR PRISMS. CUBIC YARDS PER 50 FEET

HEIGHT OR WIDTH	WIDTH OR HEIGHT								
	1	2	3	4	5	6	7	8	9
12.0	11.11	22.22	33.33	44.44	55.56	66.67	77.78	88.89	100.00
.1	11.20	22.41	33.61	44.81	56.02	67.22	78.43	89.63	100.83
.2	11.30	22.59	33.89	45.19	56.48	67.78	79.07	90.37	101.67
.3	11.39	22.78	34.17	45.56	56.94	68.33	79.72	91.11	102.50
.4	11.48	22.96	34.44	45.93	57.41	68.89	80.37	91.85	103.33
12.5	11.57	23.15	34.72	46.30	57.87	69.44	81.02	92.59	104.17
.6	11.67	23.33	35.00	46.67	58.33	70.00	81.67	93.33	105.00
.7	11.76	23.52	35.28	47.04	58.80	70.56	82.31	94.07	105.83
.8	11.85	23.70	35.56	47.41	59.26	71.11	82.96	94.81	106.67
.9	11.94	23.89	35.83	47.78	59.72	71.67	83.61	95.56	107.50
13.0	12.04	24.07	36.11	48.15	60.19	72.22	84.26	96.30	108.33
.1	12.13	24.26	36.39	48.52	60.65	72.78	84.91	97.04	109.17
.2	12.22	24.44	36.67	48.89	61.11	73.33	85.56	97.78	110.00
.3	12.31	24.63	36.94	49.26	61.57	73.89	86.20	98.52	110.83
.4	12.41	24.81	37.22	49.63	62.04	74.44	86.85	99.26	111.67
13.5	12.50	25.00	37.50	50.00	62.50	75.00	87.50	100.00	112.50
.6	12.59	25.19	37.78	50.37	62.96	75.56	88.15	100.74	113.33
.7	12.69	25.37	38.06	50.74	63.43	76.11	88.80	101.48	114.17
.8	12.78	25.56	38.33	51.11	63.89	76.67	89.44	102.22	115.00
.9	12.87	25.74	38.61	51.48	64.35	77.22	90.09	102.96	115.83
14.0	12.96	25.93	38.89	51.85	64.81	77.78	90.74	103.70	116.67
.1	13.06	26.11	39.17	52.22	65.28	78.33	91.39	104.44	117.50
.2	13.15	26.30	39.44	52.59	65.74	78.89	92.04	105.19	118.33
.3	13.24	26.48	39.72	52.96	66.20	79.44	92.69	105.93	119.17
.4	13.33	26.67	40.00	53.33	66.67	80.00	93.33	106.67	120.00
14.5	13.43	26.85	40.28	53.70	67.13	80.56	93.98	107.41	120.83
.6	13.52	27.04	40.56	54.07	67.59	81.11	94.63	108.15	121.67
.7	13.61	27.22	40.83	54.44	68.06	81.67	95.28	108.89	122.50
.8	13.70	27.41	41.11	54.81	68.52	82.22	95.93	109.63	123.33
.9	13.80	27.59	41.39	55.19	68.98	82.78	96.57	110.37	124.17
15.0	13.89	27.78	41.67	55.56	69.44	83.33	97.22	111.11	125.00
.1	13.98	27.96	41.94	55.93	69.91	83.89	97.87	111.85	125.83
.2	14.07	28.15	42.22	56.30	70.37	84.44	98.52	112.59	126.67
.3	14.17	28.33	42.50	56.67	70.83	85.00	99.17	113.33	127.50
.4	14.26	28.52	42.78	57.04	71.30	85.56	99.81	114.07	128.33
15.5	14.35	28.70	43.06	57.41	71.76	86.11	100.46	114.81	129.17
.6	14.44	28.89	43.33	57.78	72.22	86.67	101.11	115.56	130.00
.7	14.54	29.07	43.61	58.15	72.69	87.22	101.76	116.30	130.83
.8	14.63	29.26	43.89	58.52	73.15	87.78	102.41	117.04	131.67
.9	14.72	29.44	44.17	58.89	73.61	88.33	103.06	117.78	132.50
16.0	14.81	29.63	44.44	59.26	74.07	88.89	103.70	118.52	133.33
.1	14.91	29.81	44.72	59.63	74.54	89.44	104.35	119.26	134.17
.2	15.00	30.00	45.00	60.00	75.00	90.00	105.00	120.00	135.00
.3	15.09	30.19	45.28	60.37	75.46	90.56	105.65	120.74	135.83
.4	15.19	30.37	45.56	60.74	75.93	91.11	106.30	121.48	136.67
16.5	15.28	30.56	45.83	61.11	76.39	91.67	106.94	122.22	137.50
.6	15.37	30.74	46.11	61.48	76.85	92.22	107.59	122.96	138.33
.7	15.46	30.93	46.39	61.85	77.31	92.78	108.24	123.70	139.17
.8	15.56	31.11	46.67	62.22	77.78	93.33	108.89	124.44	140.00
.9	15.65	31.30	46.94	62.59	78.24	93.89	109.54	125.19	140.83
17.0	15.74	31.48	47.22	62.96	78.70	94.44	110.19	125.93	141.67
.1	15.83	31.67	47.50	63.33	79.17	95.00	110.83	126.67	142.50
.2	15.93	31.85	47.78	63.70	79.63	95.56	111.48	127.41	143.33
.3	16.02	32.04	48.06	64.07	80.09	96.11	112.13	128.15	144.17
.4	16.11	32.22	48.33	64.44	80.56	96.67	112.78	128.89	145.00
17.5	16.20	32.41	48.61	64.81	81.02	97.22	113.43	129.63	145.83
.6	16.30	32.59	48.89	65.19	81.48	97.78	114.07	130.37	146.67
.7	16.39	32.78	49.17	65.56	81.94	98.33	114.72	131.11	147.50
.8	16.48	32.96	49.44	65.93	82.41	98.89	115.37	131.85	148.33
.9	16.57	33.15	49.72	66.30	82.87	99.44	116.02	132.59	149.17

TABLE XVIII.—TRIANGULAR PRISMS. CUBIC YARDS PER 50 FEET

HEIGHT OR WIDTH	WIDTH OR HEIGHT								
	1	2	3	4	5	6	7	8	9
18.0	16.67	33.33	50.00	66.67	83.33	100.00	116.67	133.33	150.00
.1	16.76	33.52	50.28	67.04	83.80	100.56	117.31	134.07	150.83
.2	16.85	33.70	50.56	67.41	84.26	101.11	117.96	134.81	151.67
.3	16.94	33.89	50.83	67.78	84.72	101.67	118.61	135.56	152.50
.4	17.04	34.07	51.11	68.15	85.19	102.22	119.26	136.30	153.33
18.5	17.13	34.26	51.39	68.52	85.65	102.78	119.91	137.04	154.17
.6	17.22	34.44	51.67	68.89	86.11	103.33	120.56	137.78	155.00
.7	17.31	34.63	51.94	69.26	86.57	103.89	121.20	138.52	155.83
.8	17.41	34.81	52.22	69.63	87.04	104.44	121.85	139.26	156.67
.9	17.50	35.00	52.50	70.00	87.50	105.00	122.50	140.00	157.50
19.0	17.59	35.19	52.78	70.37	87.96	105.56	123.15	140.74	158.33
.1	17.69	35.37	53.06	70.74	88.43	106.11	123.80	141.48	159.17
.2	17.78	35.56	53.33	71.11	88.89	106.67	124.44	142.22	160.00
.3	17.87	35.74	53.61	71.48	89.35	107.22	125.09	142.96	160.83
.4	17.96	35.93	53.89	71.85	89.81	107.78	125.74	143.70	161.67
19.5	18.06	36.11	54.17	72.22	90.28	108.33	126.39	144.44	162.50
.6	18.15	36.30	54.44	72.59	90.74	108.89	127.04	145.19	163.33
.7	18.24	36.48	54.72	72.96	91.20	109.44	127.69	145.93	164.17
.8	18.33	36.67	55.00	73.33	91.67	110.00	128.33	146.67	165.00
.9	18.43	36.85	55.28	73.70	92.13	110.56	128.98	147.41	165.83
20.0	18.52	37.04	55.56	74.07	92.59	111.11	129.63	148.15	166.67
.1	18.61	37.22	55.83	74.44	93.06	111.67	130.28	148.89	167.50
.2	18.70	37.41	56.11	74.81	93.52	112.22	130.93	149.63	168.33
.3	18.80	37.59	56.39	75.19	93.98	112.78	131.57	150.37	169.17
.4	18.89	37.78	56.67	75.56	94.44	113.33	132.22	151.11	170.00
20.5	18.98	37.96	56.94	75.93	94.91	113.89	132.87	151.85	170.83
.6	19.07	38.15	57.22	76.30	95.37	114.44	133.52	152.59	171.67
.7	19.17	38.33	57.50	76.67	95.83	115.00	134.17	153.33	172.50
.8	19.26	38.52	57.78	77.04	96.30	115.56	134.81	154.07	173.33
.9	19.35	38.70	58.06	77.41	96.76	116.11	135.46	154.81	174.17
21.0	19.44	38.89	58.33	77.78	97.22	116.67	136.11	155.56	175.00
.1	19.54	39.07	58.61	78.15	97.69	117.22	136.76	156.30	175.83
.2	19.63	39.26	58.89	78.52	98.15	117.78	137.41	157.04	176.67
.3	19.72	39.44	59.17	78.89	98.61	118.33	138.06	157.78	177.50
.4	19.81	39.63	59.44	79.26	99.07	118.89	138.70	158.52	178.33
21.5	19.91	39.81	59.72	79.63	99.54	119.44	139.35	159.26	179.17
.6	20.00	40.00	60.00	80.00	100.00	120.00	140.00	160.00	180.00
.7	20.09	40.19	60.28	80.37	100.46	120.56	140.65	160.74	180.83
.8	20.19	40.37	60.56	80.74	100.93	121.11	141.30	161.48	181.67
.9	20.28	40.56	60.83	81.11	101.39	121.67	141.94	162.22	182.50
22.0	20.37	40.74	61.11	81.48	101.85	122.22	142.59	162.96	183.33
.1	20.46	40.93	61.39	81.85	102.31	122.78	143.24	163.70	184.17
.2	20.56	41.11	61.67	82.22	102.78	123.33	143.89	164.44	185.00
.3	20.65	41.30	61.94	82.59	103.24	123.89	144.54	165.19	185.83
.4	20.74	41.48	62.22	82.96	103.70	124.44	145.19	165.93	186.67
22.5	20.83	41.67	62.50	83.33	104.17	125.00	145.83	166.67	187.50
.6	20.93	41.85	62.78	83.70	104.63	125.56	146.48	167.41	188.33
.7	21.02	42.04	63.06	84.07	105.09	126.11	147.13	168.15	189.17
.8	21.11	42.22	63.33	84.44	105.56	126.67	147.78	168.89	190.00
.9	21.20	42.41	63.61	84.81	106.02	127.22	148.43	169.63	190.83
23.0	21.30	42.59	63.89	85.19	106.48	127.78	149.07	170.37	191.67
.1	21.39	42.78	64.17	85.56	106.94	128.33	149.72	171.11	192.50
.2	21.48	42.96	64.44	85.93	107.41	128.89	150.37	171.85	193.33
.3	21.57	43.15	64.72	86.30	107.87	129.44	151.02	172.59	194.17
.4	21.67	43.33	65.00	86.67	108.33	130.00	151.67	173.33	195.00
23.5	21.76	43.52	65.28	87.04	108.80	130.56	152.31	174.07	195.83
.6	21.85	43.70	65.56	87.41	109.26	131.11	152.96	174.81	196.67
.7	21.94	43.89	65.83	87.78	109.72	131.67	153.61	175.56	197.50
.8	22.04	44.07	66.11	88.15	110.19	132.22	154.26	176.30	198.33
.9	22.13	44.26	66.39	88.52	110.65	132.78	154.91	177.04	199.17

TABLE XVIII.—TRIANGULAR PRISMS. CUBIC YARDS PER 50 FEET

HEIGHT OR WIDTH	WIDTH OR HEIGHT								
	1	2	3	4	5	6	7	8	9
24.0	22.22	44.44	66.67	88.89	111.11	133.33	155.56	177.78	200.00
.1	22.31	44.63	66.94	89.26	111.57	133.89	156.20	178.52	200.83
.2	22.41	44.81	67.22	89.63	112.04	134.44	156.85	179.26	201.67
.3	22.50	45.00	67.50	90.00	112.50	135.00	157.50	180.00	202.50
.4	22.59	45.19	67.78	90.37	112.96	135.56	158.15	180.74	203.33
24.5	22.69	45.37	68.06	90.74	113.43	136.11	158.80	181.48	204.17
.6	22.78	45.56	68.33	91.11	113.89	136.67	159.44	182.22	205.00
.7	22.87	45.74	68.61	91.48	114.35	137.22	160.09	182.96	205.83
.8	22.96	45.93	68.89	91.85	114.81	137.78	160.74	183.70	206.67
.9	23.06	46.11	69.17	92.22	115.28	138.33	161.39	184.44	207.50
25.0	23.15	46.30	69.44	92.59	115.74	138.89	162.04	185.19	208.33
.1	23.24	46.48	69.72	92.96	116.20	139.44	162.69	185.93	209.17
.2	23.33	46.67	70.00	93.33	116.67	140.00	163.33	186.67	210.00
.3	23.43	46.85	70.28	93.70	117.13	140.56	163.98	187.41	210.83
.4	23.52	47.04	70.56	94.07	117.59	141.11	164.63	188.15	211.67
25.5	23.61	47.22	70.83	94.44	118.06	141.67	165.28	188.89	212.50
.6	23.70	47.41	71.11	94.81	118.52	142.22	165.93	189.63	213.33
.7	23.80	47.59	71.39	95.19	118.98	142.78	166.57	190.37	214.17
.8	23.89	47.78	71.67	95.56	119.44	143.33	167.22	191.11	215.00
.9	23.98	47.96	71.94	95.93	119.91	143.89	167.87	191.85	215.83
26.0	24.07	48.15	72.22	96.30	120.37	144.44	168.52	192.59	216.67
.1	24.17	48.33	72.50	96.67	120.83	145.00	169.17	193.33	217.50
.2	24.26	48.52	72.78	97.04	121.30	145.56	169.81	194.07	218.33
.3	24.35	48.70	73.06	97.41	121.76	146.11	170.46	194.81	219.17
.4	24.44	48.89	73.33	97.78	122.22	146.67	171.11	195.56	220.00
26.5	24.54	49.07	73.61	98.15	122.69	147.22	171.76	196.30	220.83
.6	24.63	49.26	73.89	98.52	123.15	147.78	172.41	197.04	221.67
.7	24.72	49.44	74.17	98.89	123.61	148.33	173.06	197.78	222.50
.8	24.81	49.63	74.44	99.26	124.07	148.89	173.70	198.52	223.33
.9	24.91	49.81	74.72	99.63	124.54	149.44	174.35	199.26	224.17
27.0	25.00	50.00	75.00	100.00	125.00	150.00	175.00	200.00	225.00
.1	25.09	50.19	75.28	100.37	125.46	150.56	175.65	200.74	225.83
.2	25.19	50.37	75.56	100.74	125.93	151.11	176.30	201.48	226.67
.3	25.28	50.56	75.83	101.11	126.39	151.67	176.94	202.22	227.50
.4	25.37	50.74	76.11	101.48	126.85	152.22	177.59	202.96	228.33
27.5	25.46	50.93	76.39	101.85	127.31	152.78	178.24	203.70	229.17
.6	25.56	51.11	76.67	102.22	127.78	153.33	178.89	204.44	230.00
.7	25.65	51.30	76.94	102.59	128.24	153.89	179.54	205.19	230.83
.8	25.74	51.48	77.22	102.96	128.70	154.44	180.19	205.93	231.67
.9	25.83	51.67	77.50	103.33	129.17	155.00	180.83	206.67	232.50
28.0	25.93	51.85	77.78	103.70	129.63	155.56	181.48	207.41	233.33
.1	26.02	52.04	78.06	104.07	130.09	156.11	182.13	208.15	234.17
.2	26.11	52.22	78.33	104.44	130.56	156.67	182.78	208.89	235.00
.3	26.20	52.41	78.61	104.81	131.02	157.22	183.43	209.63	235.83
.4	26.30	52.59	78.89	105.19	131.48	157.78	184.07	210.37	236.67
28.5	26.39	52.78	79.17	105.56	131.94	158.33	184.72	211.11	237.50
.6	26.48	52.96	79.44	105.93	132.41	158.89	185.37	211.85	238.33
.7	26.57	53.15	79.72	106.30	132.87	159.44	186.02	212.59	239.17
.8	26.67	53.33	80.00	106.67	133.33	160.00	186.67	213.33	240.00
.9	26.76	53.52	80.28	107.04	133.80	160.56	187.31	214.07	240.83
29.0	26.85	53.70	80.56	107.41	134.26	161.11	187.96	214.81	241.67
.1	26.94	53.89	80.83	107.78	134.72	161.67	188.61	215.56	242.50
.2	27.04	54.07	81.11	108.15	135.19	162.22	189.26	216.30	243.33
.3	27.13	54.26	81.39	108.52	135.65	162.78	189.91	217.04	244.17
.4	27.22	54.44	81.67	108.89	136.11	163.33	190.56	217.78	245.00
29.5	27.31	54.63	81.94	109.26	136.57	163.89	191.20	218.52	245.83
.6	27.41	54.81	82.22	109.63	137.04	164.44	191.85	219.26	246.67
.7	27.50	55.00	82.50	110.00	137.50	165.00	192.50	220.00	247.50
.8	27.59	55.19	82.78	110.37	137.96	165.56	193.15	220.74	248.33
.9	27.69	55.37	83.06	110.74	138.43	166.11	193.80	221.48	249.17

TABLE XVIII.—TRIANGULAR PRISMS. CUBIC YARDS PER 50 FEET

HEIGHT OR WIDTH	WIDTH OR HEIGHT								
	1	2	3	4	5	6	7	8	9
30.0	27.78	55.56	83.33	111.11	138.89	166.67	194.44	222.22	250.00
.1	27.87	55.74	83.61	111.48	139.35	167.22	195.09	222.96	250.83
.2	27.96	55.93	83.89	111.85	139.81	167.78	195.74	223.70	251.67
.3	28.06	56.11	84.17	112.22	140.28	168.33	196.39	224.44	252.50
.4	28.15	56.30	84.44	112.59	140.74	168.89	197.04	225.19	253.33
30.5	28.24	56.48	84.72	112.96	141.20	169.44	197.69	225.93	254.17
.6	28.33	56.67	85.00	113.33	141.67	170.00	198.33	226.67	255.00
.7	28.43	56.85	85.28	113.70	142.13	170.56	198.98	227.41	255.83
.8	28.52	57.04	85.56	114.07	142.59	171.11	199.63	228.15	256.67
.9	28.61	57.22	85.83	114.44	143.06	171.67	200.28	228.89	257.50
31.0	28.70	57.41	86.11	114.81	143.52	172.22	200.93	229.63	258.33
.1	28.80	57.59	86.39	115.19	143.98	172.78	201.57	230.37	259.17
.2	28.89	57.78	86.67	115.56	144.44	173.33	202.22	231.11	260.00
.3	28.98	57.96	86.94	115.93	144.91	173.89	202.87	231.85	260.83
.4	29.07	58.15	87.22	116.30	145.37	174.44	203.52	232.59	261.67
31.5	29.17	58.33	87.50	116.67	145.83	175.00	204.17	233.33	262.50
.6	29.26	58.52	87.78	117.04	146.30	175.56	204.81	234.07	263.33
.7	29.35	58.70	88.06	117.41	146.76	176.11	205.46	234.81	264.17
.8	29.44	58.89	88.33	117.78	147.22	176.67	206.11	235.56	265.00
.9	29.54	59.07	88.61	118.15	147.69	177.22	206.76	236.30	265.83
32.0	29.63	59.26	88.89	118.52	148.15	177.78	207.41	237.04	266.67
.1	29.72	59.44	89.17	118.89	148.61	178.33	208.06	237.78	267.50
.2	29.81	59.63	89.44	119.26	149.07	178.89	208.70	238.52	268.33
.3	29.91	59.81	89.72	119.63	149.54	179.44	209.35	239.26	269.17
.4	30.00	60.00	90.00	120.00	150.00	180.00	210.00	240.00	270.00
32.5	30.09	60.19	90.28	120.37	150.46	180.56	210.65	240.74	270.83
.6	30.19	60.37	90.56	120.74	150.93	181.11	211.30	241.48	271.67
.7	30.28	60.56	90.83	121.11	151.39	181.67	211.94	242.22	272.50
.8	30.37	60.74	91.11	121.48	151.85	182.22	212.59	242.96	273.33
.9	30.46	60.93	91.39	121.85	152.31	182.78	213.24	243.70	274.17
33.0	30.56	61.11	91.67	122.22	152.78	183.33	213.89	244.44	275.00
.1	30.65	61.30	91.94	122.59	153.24	183.89	214.54	245.19	275.83
.2	30.74	61.48	92.22	122.96	153.70	184.44	215.19	245.93	276.67
.3	30.83	61.67	92.50	123.33	154.17	185.00	215.83	246.67	277.50
.4	30.93	61.85	92.78	123.70	154.63	185.56	216.48	247.41	278.33
33.5	31.02	62.04	93.06	124.07	155.09	186.11	217.13	248.15	279.17
.6	31.11	62.22	93.33	124.44	155.56	186.67	217.78	248.89	280.00
.7	31.20	62.41	93.61	124.81	156.02	187.22	218.43	249.63	280.83
.8	31.30	62.59	93.89	125.19	156.48	187.78	219.07	250.37	281.67
.9	31.39	62.78	94.17	125.56	156.94	188.33	219.72	251.11	282.50
34.0	31.48	62.96	94.44	125.93	157.41	188.89	220.37	251.85	283.33
.1	31.57	63.15	94.72	126.30	157.87	189.44	221.02	252.59	284.17
.2	31.67	63.33	95.00	126.67	158.33	190.00	221.67	253.33	285.00
.3	31.76	63.52	95.28	127.04	158.80	190.56	222.31	254.07	285.83
.4	31.85	63.70	95.56	127.41	159.26	191.11	222.96	254.81	286.67
34.5	31.94	63.89	95.83	127.78	159.72	191.67	223.61	255.56	287.50
.6	32.04	64.07	96.11	128.15	160.19	192.22	224.26	256.30	288.33
.7	32.13	64.26	96.39	128.52	160.65	192.78	224.91	257.04	289.17
.8	32.22	64.44	96.67	128.89	161.11	193.33	225.56	257.78	290.00
.9	32.31	64.63	96.94	129.26	161.57	193.89	226.20	258.52	290.83
35.0	32.41	64.81	97.22	129.63	162.04	194.44	226.85	259.26	291.67
.1	32.50	65.00	97.50	130.00	162.50	195.00	227.50	260.00	292.50
.2	32.59	65.19	97.78	130.37	162.96	195.56	228.15	260.74	293.33
.3	32.69	65.37	98.06	130.74	163.43	196.11	228.80	261.48	294.17
.4	32.78	65.56	98.33	131.11	163.89	196.67	229.44	262.22	295.00
35.5	32.87	65.74	98.61	131.48	164.35	197.22	230.09	262.96	295.83
.6	32.96	65.93	98.89	131.85	164.81	197.78	230.74	263.70	296.67
.7	33.06	66.11	99.17	132.22	165.28	198.33	231.39	264.44	297.50
.8	33.15	66.30	99.44	132.59	165.74	198.89	232.04	265.19	298.33
.9	33.24	66.48	99.72	132.96	166.20	199.44	232.69	265.93	299.17

TABLE XVIII.—TRIANGULAR PRISMS. CUBIC YARDS PER 50 FEET

HEIGHT OR WIDTH	WIDTH OR HEIGHT								
	1	2	3	4	5	6	7	8	9
36.0	33.33	66.67	100.00	133.33	166.67	200.00	233.33	266.67	300.00
.1	33.43	66.85	100.28	133.70	167.13	200.56	233.98	267.41	300.83
.2	33.52	67.04	100.56	134.07	167.59	201.11	234.63	268.15	301.67
.3	33.61	67.22	100.83	134.44	168.06	201.67	235.28	268.89	302.50
.4	33.70	67.41	101.11	134.81	168.52	202.22	235.93	269.63	303.33
36.5	33.80	67.59	101.39	135.19	168.98	202.78	236.57	270.37	304.17
.6	33.89	67.78	101.67	135.56	169.44	203.33	237.22	271.11	305.00
.7	33.98	67.96	101.94	135.93	169.91	203.89	237.87	271.85	305.83
.8	34.07	68.15	102.22	136.30	170.37	204.44	238.52	272.59	306.67
.9	34.17	68.33	102.50	136.67	170.83	205.00	239.17	273.33	307.50
37.0	34.26	68.52	102.78	137.04	171.30	205.56	239.81	274.07	308.33
.1	34.35	68.70	103.06	137.41	171.76	206.11	240.46	274.81	309.17
.2	34.44	68.89	103.33	137.78	172.22	206.67	241.11	275.56	310.00
.3	34.54	69.07	103.61	138.15	172.69	207.22	241.76	276.30	310.83
.4	34.63	69.26	103.89	138.52	173.15	207.78	242.41	277.04	311.67
37.5	34.72	69.44	104.17	138.89	173.61	208.33	243.06	277.78	312.50
.6	34.81	69.63	104.44	139.26	174.07	208.89	243.70	278.52	313.33
.7	34.91	69.81	104.72	139.63	174.54	209.44	244.35	279.26	314.17
.8	35.00	70.00	105.00	140.00	175.00	210.00	245.00	280.00	315.00
.9	35.09	70.19	105.28	140.37	175.46	210.56	245.65	280.74	315.83
38.0	35.19	70.37	105.56	140.74	175.93	211.11	246.30	281.48	316.67
.1	35.28	70.56	105.83	141.11	176.39	211.67	246.94	282.22	317.50
.2	35.37	70.74	106.11	141.48	176.85	212.22	247.59	282.96	318.33
.3	35.46	70.93	106.39	141.85	177.31	212.78	248.24	283.70	319.17
.4	35.56	71.11	106.67	142.22	177.78	213.33	248.89	284.44	320.00
38.5	35.65	71.30	106.94	142.59	178.24	213.89	249.54	285.19	320.83
.6	35.74	71.48	107.22	142.96	178.70	214.44	250.19	285.93	321.67
.7	35.83	71.67	107.50	143.33	179.17	215.00	250.83	286.67	322.50
.8	35.93	71.85	107.78	143.70	179.63	215.56	251.48	287.41	323.33
.9	36.02	72.04	108.06	144.07	180.09	216.11	252.13	288.15	324.17
39.0	36.11	72.22	108.33	144.44	180.56	216.67	252.78	288.89	325.00
.1	36.20	72.41	108.61	144.81	181.02	217.22	253.43	289.63	325.83
.2	36.30	72.59	108.89	145.19	181.48	217.78	254.07	290.37	326.67
.3	36.39	72.78	109.17	145.56	181.94	218.33	254.72	291.11	327.50
.4	36.48	72.96	109.44	145.93	182.41	218.89	255.37	291.85	328.33
39.5	36.57	73.15	109.72	146.30	182.87	219.44	256.02	292.59	329.17
.6	36.67	73.33	110.00	146.67	183.33	220.00	256.67	293.33	330.00
.7	36.76	73.52	110.28	147.04	183.80	220.56	257.31	294.07	330.83
.8	36.85	73.70	110.56	147.41	184.26	221.11	257.96	294.81	331.67
.9	36.94	73.89	110.83	147.78	184.72	221.67	258.61	295.56	332.50
40.0	37.04	74.07	111.11	148.15	185.19	222.22	259.26	296.30	333.33
.1	37.13	74.26	111.39	148.52	185.65	222.78	259.91	297.04	334.17
.2	37.22	74.44	111.67	148.89	186.11	223.33	260.56	297.78	335.00
.3	37.31	74.63	111.94	149.26	186.57	223.89	261.20	298.52	335.83
.4	37.41	74.81	112.22	149.63	187.04	224.44	261.85	299.26	336.67
40.5	37.50	75.00	112.50	150.00	187.50	225.00	262.50	300.00	337.50
.6	37.59	75.19	112.78	150.37	187.96	225.56	263.15	300.74	338.33
.7	37.69	75.37	113.06	150.74	188.43	226.11	263.80	301.48	339.17
.8	37.78	75.56	113.33	151.11	188.89	226.67	264.44	302.22	340.00
.9	37.87	75.74	113.61	151.48	189.35	227.22	265.09	302.96	340.83
41.0	37.96	75.93	113.89	151.85	189.81	227.78	265.74	303.70	341.67
.1	38.06	76.11	114.17	152.22	190.28	228.33	266.39	304.44	342.50
.2	38.15	76.30	114.44	152.59	190.74	228.89	267.04	305.19	343.33
.3	38.24	76.48	114.72	152.96	191.20	229.44	267.69	305.93	344.17
.4	38.33	76.67	115.00	153.33	191.67	230.00	268.33	306.67	345.00
41.5	38.43	76.85	115.28	153.70	192.13	230.56	268.98	307.41	345.83
.6	38.52	77.04	115.56	154.07	192.59	231.11	269.63	308.15	346.67
.7	38.61	77.22	115.83	154.44	193.06	231.67	270.28	308.89	347.50
.8	38.70	77.41	116.11	154.81	193.52	232.22	270.93	309.63	348.33
.9	38.80	77.59	116.39	155.19	193.98	232.78	271.57	310.37	349.17

TABLE XVIII.—TRIANGULAR PRISMS. CUBIC YARDS PER 50 FEET

HEIGHT OR WIDTH	WIDTH OR HEIGHT								
	1	2	3	4	5	6	7	8	9
42.0	38.89	77.78	116.67	155.56	194.44	233.33	272.22	311.11	350.00
.1	38.98	77.96	116.94	155.93	194.91	233.89	272.87	311.85	350.83
.2	39.07	78.15	117.22	156.30	195.37	234.44	273.52	312.59	351.67
.3	39.17	78.33	117.50	156.67	195.83	235.00	274.17	313.33	352.50
.4	39.26	78.52	117.78	157.04	196.30	235.56	274.81	314.07	353.33
42.5	39.35	78.70	118.06	157.41	196.76	236.11	275.46	314.81	354.17
.6	39.44	78.89	118.33	157.78	197.22	236.67	276.11	315.56	355.00
.7	39.54	79.07	118.61	158.15	197.69	237.22	276.76	316.30	355.83
.8	39.63	79.26	118.89	158.52	198.15	237.78	277.41	317.04	356.67
.9	39.72	79.44	119.17	158.89	198.61	238.33	278.06	317.78	357.50
43.0	39.81	79.63	119.44	159.26	199.07	238.89	278.70	318.52	358.33
.1	39.91	79.81	119.72	159.63	199.54	239.44	279.35	319.26	359.17
.2	40.00	80.00	120.00	160.00	200.00	240.00	280.00	320.00	360.00
.3	40.09	80.19	120.28	160.37	200.46	240.56	280.65	320.74	360.83
.4	40.19	80.37	120.56	160.74	200.93	241.11	281.30	321.48	361.67
43.5	40.28	80.56	120.83	161.11	201.39	241.67	281.94	322.22	362.50
.6	40.37	80.74	121.11	161.48	201.85	242.22	282.59	322.96	363.33
.7	40.46	80.93	121.39	161.85	202.31	242.78	283.24	323.70	364.17
.8	40.56	81.11	121.67	162.22	202.78	243.33	283.89	324.44	365.00
.9	40.65	81.30	121.94	162.59	203.24	243.89	284.54	325.19	365.83
44.0	40.74	81.48	122.22	162.96	203.70	244.44	285.19	325.93	366.67
.1	40.83	81.67	122.50	163.33	204.17	245.00	285.83	326.67	367.50
.2	40.93	81.85	122.78	163.70	204.63	245.56	286.48	327.41	368.33
.3	41.02	82.04	123.06	164.07	205.09	246.11	287.13	328.15	369.17
.4	41.11	82.22	123.33	164.44	205.56	246.67	287.78	328.89	370.00
44.5	41.20	82.41	123.61	164.81	206.02	247.22	288.43	329.63	370.83
.6	41.30	82.59	123.89	165.19	206.48	247.78	289.07	330.37	371.67
.7	41.39	82.78	124.17	165.56	206.94	248.33	289.72	331.11	372.50
.8	41.48	82.96	124.44	165.93	207.41	248.89	290.37	331.85	373.33
.9	41.57	83.15	124.72	166.30	207.87	249.44	291.02	332.59	374.17
45.0	41.67	83.33	125.00	166.67	208.33	250.00	291.67	333.33	375.00
.1	41.76	83.52	125.28	167.04	208.80	250.56	292.31	334.07	375.83
.2	41.85	83.70	125.56	167.41	209.26	251.11	292.96	334.81	376.67
.3	41.94	83.89	125.83	167.78	209.72	251.67	293.61	335.56	377.50
.4	42.04	84.07	126.11	168.15	210.19	252.22	294.26	336.30	378.33
45.5	42.13	84.26	126.39	168.52	210.65	252.78	294.91	337.04	379.17
.6	42.22	84.44	126.67	168.89	211.11	253.33	295.56	337.78	380.00
.7	42.31	84.63	126.94	169.26	211.57	253.89	296.20	338.52	380.83
.8	42.41	84.81	127.22	169.63	212.04	254.44	296.85	339.26	381.67
.9	42.50	85.00	127.50	170.00	212.50	255.00	297.50	340.00	382.50
46.0	42.59	85.19	127.78	170.37	212.96	255.56	298.15	340.74	383.33
.1	42.69	85.37	128.06	170.74	213.43	256.11	298.80	341.48	384.17
.2	42.78	85.56	128.33	171.11	213.89	256.67	299.44	342.22	385.00
.3	42.87	85.74	128.61	171.48	214.35	257.22	300.09	342.96	385.83
.4	42.96	85.93	128.89	171.85	214.81	257.78	300.74	343.70	386.67
46.5	43.06	86.11	129.17	172.22	215.28	258.33	301.39	344.44	387.50
.6	43.15	86.30	129.44	172.59	215.74	258.89	302.04	345.19	388.33
.7	43.24	86.48	129.72	172.96	216.20	259.44	302.69	345.93	389.17
.8	43.33	86.67	130.00	173.33	216.67	260.00	303.33	346.67	390.00
.9	43.43	86.85	130.28	173.70	217.13	260.56	303.98	347.41	390.83
47.0	43.52	87.04	130.56	174.07	217.59	261.11	304.63	348.15	391.67
.1	43.61	87.22	130.83	174.44	218.06	261.67	305.28	348.89	392.50
.2	43.70	87.41	131.11	174.81	218.52	262.22	305.93	349.63	393.33
.3	43.80	87.59	131.39	175.19	218.98	262.78	306.57	350.37	394.17
.4	43.89	87.78	131.67	175.56	219.44	263.33	307.22	351.11	395.00
47.5	43.98	87.96	131.94	175.93	219.91	263.89	307.87	351.85	395.83
.6	44.07	88.15	132.22	176.30	220.37	264.44	308.52	352.59	396.67
.7	44.17	88.33	132.50	176.67	220.83	265.00	309.17	353.33	397.50
.8	44.26	88.52	132.78	177.04	221.30	265.56	309.81	354.07	398.33
.9	44.35	88.70	133.06	177.41	221.76	266.11	310.46	354.81	399.17

TABLE XVIII.—TRIANGULAR PRISMS. CUBIC YARDS PER 50 FEET

HEIGHT OR WIDTH	WIDTH OR HEIGHT								
	1	2	3	4	5	6	7	8	9
48.0	44.44	88.89	133.33	177.78	222.22	266.67	311.11	355.56	400.00
.1	44.54	89.07	133.61	178.15	222.69	267.22	311.76	356.30	400.83
.2	44.63	89.26	133.89	178.52	223.15	267.78	312.41	357.04	401.67
.3	44.72	89.44	134.17	178.89	223.61	268.33	313.06	357.78	402.50
.4	44.81	89.63	134.44	179.26	224.07	268.89	313.70	358.52	403.33
48.5	44.91	89.81	134.72	179.63	224.54	269.44	314.35	359.26	404.17
.6	45.00	90.00	135.00	180.00	225.00	270.00	315.00	360.00	405.00
.7	45.09	90.19	135.28	180.37	225.46	270.56	315.65	360.74	405.83
.8	45.19	90.37	135.56	180.74	225.93	271.11	316.30	361.48	406.67
.9	45.28	90.56	135.83	181.11	226.39	271.67	316.94	352.22	407.50
43.0	45.37	90.74	136.11	181.48	226.85	272.22	317.59	362.96	408.33
.1	45.46	90.93	136.39	181.85	227.31	272.78	318.24	363.70	409.17
.2	45.56	91.11	136.67	182.22	227.78	273.33	318.89	364.44	410.00
.3	45.65	91.30	136.94	182.59	228.24	273.89	319.54	365.19	410.83
.4	45.74	91.48	137.22	182.96	228.70	274.44	320.19	365.93	411.67
49.5	45.83	91.67	137.50	183.33	229.17	275.00	320.83	366.67	412.50
.6	45.93	91.85	137.78	183.70	229.63	275.56	321.48	367.41	413.33
.7	46.02	92.04	138.06	184.07	230.09	276.11	322.13	368.15	414.17
.8	46.11	92.22	138.33	184.44	230.56	276.67	322.78	368.89	415.00
.9	46.20	92.41	138.61	184.81	231.02	277.22	323.43	369.63	415.83
50.0	46.30	92.59	138.89	185.19	231.48	277.78	324.07	370.37	416.67
.1	46.39	92.78	139.17	185.56	231.94	278.33	324.72	371.11	417.50
.2	46.48	92.96	139.44	185.93	232.41	278.89	325.37	371.85	418.33
.3	46.57	93.15	139.72	186.30	232.87	279.44	326.02	372.59	419.17
.4	46.67	93.33	140.00	186.67	233.33	280.00	326.67	373.33	420.00
50.5	46.76	93.52	140.28	187.04	233.80	280.56	327.31	374.07	420.83
.6	46.85	93.70	140.56	187.41	234.26	281.11	327.96	374.81	421.67
.7	46.94	93.89	140.83	187.78	234.72	281.67	328.61	375.56	422.50
.8	47.04	94.07	141.11	188.15	235.19	282.22	329.26	376.30	423.33
.9	47.13	94.26	141.39	188.52	235.65	282.78	329.91	377.04	424.17
51.0	47.22	94.44	141.67	188.89	236.11	283.33	330.56	377.78	425.00
.1	47.31	94.63	141.94	189.26	236.57	283.89	331.20	378.52	425.83
.2	47.41	94.81	142.22	189.63	237.04	284.44	331.85	379.26	426.67
.3	47.50	95.0C	142.50	190.00	237.50	285.00	332.50	380.00	427.50
.4	47.59	95.19	142.78	190.37	237.96	285.56	333.15	380.74	428.33
51.5	47.69	95.37	143.06	190.74	238.43	286.11	333.80	381.48	429.17
.6	47.78	95.56	143.33	191.11	238.89	286.67	334.44	382.22	430.00
.7	47.87	95.74	143.61	191.48	239.35	287.22	335.09	382.96	430.83
.8	47.96	95.93	143.89	191.85	239.81	287.78	335.74	383.70	431.67
.9	48.06	96.11	144.17	192.22	240.28	288.33	336.39	384.44	432.50
52.0	48.15	96.30	144.44	192.59	240.74	288.89	337.04	385.19	433.33
.1	48.24	96.48	144.72	192.96	241.20	289.44	337.69	385.93	434.17
.2	48.33	96.67	145.00	193.33	241.67	290.00	338.33	386.67	435.00
.3	48.43	96.85	145.28	193.70	242.13	290.56	338.98	387.41	435.83
.4	48.52	97.04	145.56	194.07	242.59	291.11	339.63	388.15	436.67
52.5	48.61	97.22	145.83	194.44	243.06	291.67	340.28	388.89	437.50
.6	48.70	97.41	146.11	194.81	243.52	292.22	340.93	389.63	438.33
.7	48.80	97.59	146.39	195.19	243.98	292.78	341.57	390.37	439.17
.8	48.89	97.78	146.67	195.56	244.44	293.33	342.22	391.11	440.00
.9	48.98	97.96	146.94	195.93	244.91	293.89	342.87	391.85	440.83
53.0	49.07	98.15	147.22	196.30	245.37	294.44	343.52	392.59	441.67
.1	49.17	98.33	147.50	196.67	245.83	295.00	344.17	393.33	442.50
.2	49.26	98.52	147.78	197.04	246.30	295.56	344.81	394.07	443.33
.3	49.35	98.70	148.06	197.41	246.76	296.11	345.46	394.81	444.17
.4	49.44	98.89	148.33	197.78	247.22	296.67	346.11	395.56	445.00
53.5	49.54	99.07	148.61	198.15	247.69	297.22	346.76	396.30	445.83
.6	49.63	99.26	148.89	198.52	248.15	297.78	347.41	397.04	446.67
.7	49.72	99.44	149.17	198.89	248.61	298.33	348.06	397.78	447.50
.8	49.81	99.63	149.44	199.26	249.07	298.89	348.70	398.52	448.33
.9	49.91	99.81	149.72	199.63	249.54	299.44	349.35	399.26	449.17

TABLE XVIII.—TRIANGULAR PRISMS. CUBIC YARDS PER 50 FEET

HEIGHT OR WIDTH	WIDTH OR HEIGHT								
	1	2	3	4	5	6	7	8	9
54.0	50.00	100.00	150.00	200.00	250.00	300.00	350.00	400.00	450.00
.1	50.09	100.19	150.28	200.37	250.46	300.56	350.65	400.74	450.83
.2	50.19	100.37	150.56	200.74	250.93	301.11	351.30	401.48	451.67
.3	50.28	100.56	150.83	201.11	251.39	301.67	351.94	402.22	452.50
.4	50.37	100.74	151.11	201.48	251.85	302.22	352.59	402.96	453.33
54.5	50.46	100.93	151.39	201.85	252.31	302.78	353.24	403.70	454.17
.6	50.56	101.11	151.67	202.22	252.78	303.33	353.89	404.44	455.00
.7	50.65	101.30	151.94	202.59	253.24	303.89	354.54	405.19	455.83
.8	50.74	101.48	152.22	202.96	253.70	304.44	355.19	405.93	456.67
.9	50.83	101.67	152.50	203.33	254.17	305.00	355.83	406.67	457.50
55.0	50.93	101.85	152.78	203.70	254.63	305.56	356.48	407.41	458.33
.1	51.02	102.04	153.06	204.07	255.09	306.11	357.13	408.15	459.17
.2	51.11	102.22	153.33	204.44	255.56	306.67	357.78	408.89	460.00
.3	51.20	102.41	153.61	204.81	256.02	307.22	358.43	409.63	460.83
.4	51.30	102.59	153.89	205.19	256.48	307.78	359.07	410.37	461.67
55.5	51.39	102.78	154.17	205.56	256.94	308.33	359.72	411.11	462.50
.6	51.48	102.96	154.44	205.93	257.41	308.89	360.37	411.85	463.33
.7	51.57	103.15	154.72	206.30	257.87	309.44	361.02	412.59	464.17
.8	51.67	103.33	155.00	206.67	258.33	310.00	361.67	413.33	465.00
.9	51.76	103.52	155.28	207.04	258.80	310.56	362.31	414.07	465.83
56.0	51.85	103.70	155.56	207.41	259.26	311.11	362.96	414.81	466.67
.1	51.94	103.89	155.83	207.78	259.72	311.67	363.61	415.56	467.50
.2	52.04	104.07	156.11	208.15	260.19	312.22	364.26	416.30	468.33
.3	52.13	104.26	156.39	208.52	260.65	312.78	364.91	417.04	469.17
.4	52.22	104.44	156.67	208.89	261.11	313.33	365.56	417.78	470.00
56.5	52.31	104.63	156.94	209.26	261.57	313.89	366.20	418.52	470.83
.6	52.41	104.81	157.22	209.63	262.04	314.44	366.85	419.26	471.67
.7	52.50	105.00	157.50	210.00	262.50	315.00	367.50	420.00	472.50
.8	52.59	105.19	157.78	210.37	262.96	315.56	368.15	420.74	473.33
.9	52.69	105.37	158.06	210.74	263.43	316.11	368.80	421.48	474.17
57.0	52.78	105.56	158.33	211.11	263.89	316.67	369.44	422.22	475.00
.1	52.87	105.74	158.61	211.48	264.35	317.22	370.09	422.96	475.83
.2	52.96	105.93	158.89	211.85	264.81	317.78	370.74	423.70	476.67
.3	53.06	106.11	159.17	212.22	265.28	318.33	371.39	424.44	477.50
.4	53.15	106.30	159.44	212.59	265.74	318.89	372.04	425.19	478.33
57.5	53.24	106.48	159.72	212.96	266.20	319.44	372.69	425.93	479.17
.6	53.33	106.67	160.00	213.33	266.67	320.00	373.33	426.67	480.00
.7	53.43	106.85	160.28	213.70	267.13	320.56	373.98	427.41	480.83
.8	53.52	107.04	160.56	214.07	267.59	321.11	374.63	428.15	481.67
.9	53.61	107.22	160.83	214.44	268.06	321.67	375.28	428.89	482.50
58.0	53.70	107.41	161.11	214.81	268.52	322.22	375.93	429.63	483.33
.1	53.80	107.59	161.39	215.19	268.98	322.78	376.57	430.37	484.17
.2	53.89	107.78	161.67	215.56	269.44	323.33	377.22	431.11	485.00
.3	53.98	107.96	161.94	215.93	269.91	323.89	377.87	431.85	485.83
.4	54.07	108.15	162.22	216.30	270.37	324.44	378.52	432.59	486.67
58.5	54.17	108.33	162.50	216.67	270.83	325.00	379.17	433.33	487.50
.6	54.26	108.52	162.78	217.04	271.30	325.56	379.81	434.07	488.33
.7	54.35	108.70	163.06	217.41	271.76	326.11	380.46	434.81	489.17
.8	54.44	108.89	163.33	217.78	272.22	326.67	381.11	435.56	490.00
.9	54.54	109.07	163.61	218.15	272.69	327.22	381.76	436.30	490.83
59.0	54.63	109.26	163.89	218.52	273.15	327.78	382.41	437.04	491.67
.1	54.72	109.44	164.17	218.89	273.61	328.33	383.06	437.78	492.50
.2	54.81	109.63	164.44	219.26	274.07	328.89	383.70	438.52	493.33
.3	54.91	109.81	164.72	219.63	274.54	329.44	384.35	439.26	494.17
.4	55.00	110.00	165.00	220.00	275.00	330.00	385.00	440.00	495.00
59.5	55.09	110.19	165.28	220.37	275.46	330.56	385.65	440.74	495.83
.6	55.19	110.37	165.56	220.74	275.93	331.11	386.30	441.48	496.67
.7	55.28	110.56	165.83	221.11	276.39	331.67	386.94	442.22	497.50
.8	55.37	110.74	166.11	221.48	276.85	332.22	387.59	442.96	498.33
.9	55.46	110.93	166.39	221.85	277.31	332.78	388.24	443.70	499.17

TABLE XIX.—CUBIC YARDS PER 100-FOOT STATION

CU YD ↓ / →	0	100	200	300	400	500	600	700	800	900	1000	1100
	SUM OF END AREAS IN SQ FT											
0.0	0	54	108	162	216	270	324	378	432	486	540	594
1.9	1	55	109	163	217	271	325	379	433	487	541	595
3.7	2	56	110	164	218	272	326	380	434	488	542	596
5.6	3	57	111	165	219	273	327	381	435	489	543	597
7.4	4	58	112	166	220	274	328	382	436	490	544	598
9.3	5	59	113	167	221	275	329	383	437	491	545	599
11.1	6	60	114	168	222	276	330	384	438	492	546	600
13.0	7	61	115	169	223	277	331	385	439	493	547	601
14.8	8	62	116	170	224	278	332	386	440	494	548	602
16.7	9	63	117	171	225	279	333	387	441	495	549	603
18.5	10	64	118	172	226	280	334	388	442	496	550	604
20.4	11	65	119	173	227	281	335	389	443	497	551	605
22.2	12	66	120	174	228	282	336	390	444	498	552	606
24.1	13	67	121	175	229	283	337	391	445	499	553	607
25.9	14	68	122	176	230	284	338	392	446	500	554	608
27.8	15	69	123	177	231	285	339	393	447	501	555	609
29.6	16	70	124	178	232	286	340	394	448	502	556	610
31.5	17	71	125	179	233	287	341	395	449	503	557	611
33.3	18	72	126	180	234	288	342	396	450	504	558	612
35.2	19	73	127	181	235	289	343	397	451	505	559	613
37.0	20	74	128	182	236	290	344	398	452	506	560	614
38.9	21	75	129	183	237	291	345	399	453	507	561	615
40.7	22	76	130	184	238	292	346	400	454	508	562	616
42.6	23	77	131	185	239	293	347	401	455	509	563	617
44.4	24	78	132	186	240	294	348	402	456	510	564	618
46.3	25	79	133	187	241	295	349	403	457	511	565	619
48.1	26	80	134	188	242	296	350	404	458	512	566	620
50.0	27	81	135	189	243	297	351	405	459	513	567	621
51.9	28	82	136	190	244	298	352	406	460	514	568	622
53.7	29	83	137	191	245	299	353	407	461	515	569	623
55.6	30	84	138	192	246	300	354	408	462	516	570	624
57.4	31	85	139	193	247	301	355	409	463	517	571	625
59.3	32	86	140	194	248	302	356	410	464	518	572	626
61.1	33	87	141	195	249	303	357	411	465	519	573	627
63.0	34	88	142	196	250	304	358	412	466	520	574	628
64.8	35	89	143	197	251	305	359	413	467	521	575	629
66.7	36	90	144	198	252	306	360	414	468	522	576	630
68.5	37	91	145	199	253	307	361	415	469	523	577	631
70.4	38	92	146	200	254	308	362	416	470	524	578	632
72.2	39	93	147	201	255	309	363	417	471	525	579	633
74.1	40	94	148	202	256	310	364	418	472	526	580	634
75.9	41	95	149	203	257	311	365	419	473	527	581	635
77.8	42	96	150	204	258	312	366	420	474	528	582	636
79.6	43	97	151	205	259	313	367	421	475	529	583	637
81.5	44	98	152	206	260	314	368	422	476	530	584	638
83.3	45	99	153	207	261	315	369	423	477	531	585	639
85.2	46	100	154	208	262	316	370	424	478	532	586	640
87.0	47	101	155	209	263	317	371	425	479	533	587	641
88.9	48	102	156	210	264	318	372	426	480	534	588	642
90.7	49	103	157	211	265	319	373	427	481	535	589	643
92.6	50	104	158	212	266	320	374	428	482	536	590	644
94.4	51	105	159	213	267	321	375	429	483	537	591	645
96.3	52	106	160	214	268	322	376	430	484	538	592	646
98.1	53	107	161	215	269	323	377	431	485	539	593	647
	0	100	200	300	400	500	600	700	800	900	1000	1100

FROM SUM OF END AREAS

1200	1300	1400	1500	1600	1700	1800	1900	2000	2100	← CU YD ↓
SUM OF END AREAS IN SQ FT										
648	702	756	810	864	918	972	1026	1080	1134	0.0
649	703	757	811	865	919	973	1027	1081	1135	1.9
650	704	758	812	866	920	974	1028	1082	1136	3.7
651	705	759	813	867	921	975	1029	1083	1137	5.6
652	706	760	814	868	922	976	1030	1084	1138	7.4
653	707	761	815	869	923	977	1031	1085	1139	9.3
654	708	762	816	870	924	978	1032	1086	1140	11.1
655	709	763	817	871	925	979	1033	1087	1141	13.0
656	710	764	818	872	926	980	1034	1088	1142	14.8
657	711	765	819	873	927	981	1035	1089	1143	16.7
658	712	766	820	874	928	982	1036	1090	1144	18.5
659	713	767	821	875	929	983	1037	1091	1145	20.4
660	714	768	822	876	930	984	1038	1092	1146	22.2
661	715	769	823	877	931	985	1039	1093	1147	24.1
662	716	770	824	878	932	986	1040	1094	1148	25.9
663	717	771	825	879	933	987	1041	1095	1149	27.8
664	718	772	826	880	934	988	1042	1096	1150	29.6
665	719	773	827	881	935	989	1043	1097	1151	31.5
666	720	774	828	882	936	990	1044	1098	1152	33.3
667	721	775	829	883	937	991	1045	1099	1153	35.2
668	722	776	830	884	938	992	1046	1100	1154	37.0
669	723	777	831	885	939	993	1047	1101	1155	38.9
670	724	778	832	886	940	994	1048	1102	1156	40.7
671	725	779	833	887	941	995	1049	1103	1157	42.6
672	726	780	834	888	942	996	1050	1104	1158	44.4
673	727	781	835	889	943	997	1051	1105	1159	46.3
674	728	782	836	890	944	998	1052	1106	1160	48.1
675	729	783	837	891	945	999	1053	1107	1161	50.0
676	730	784	838	892	946	1000	1054	1108	1162	51.9
677	731	785	839	893	947	1001	1055	1109	1163	53.7
678	732	786	840	894	948	1002	1056	1110	1164	55.6
679	733	787	841	895	949	1003	1057	1111	1165	57.4
680	734	788	842	896	950	1004	1058	1112	1166	59.3
681	735	789	843	897	951	1005	1059	1113	1167	61.1
682	736	790	844	898	952	1006	1060	1114	1168	63.0
683	737	791	845	899	953	1007	1061	1115	1169	64.8
684	738	792	846	900	954	1008	1062	1116	1170	66.7
685	739	793	847	901	955	1009	1063	1117	1171	68.5
686	740	794	848	902	956	1010	1064	1118	1172	70.4
687	741	795	849	903	957	1011	1065	1119	1173	72.2
688	742	796	850	904	958	1012	1066	1120	1174	74.1
689	743	797	851	905	959	1013	1067	1121	1175	75.9
690	744	798	852	906	960	1014	1068	1122	1176	77.8
691	745	799	853	907	961	1015	1069	1123	1177	79.6
692	746	800	854	908	962	1016	1070	1124	1178	81.5
693	747	801	855	909	963	1017	1071	1125	1179	83.3
694	748	802	856	910	964	1018	1072	1126	1180	85.2
695	749	803	857	911	965	1019	1073	1127	1181	87.0
696	750	804	858	912	966	1020	1074	1128	1182	88.9
697	751	805	859	913	967	1021	1075	1129	1183	90.7
698	752	806	860	914	968	1022	1076	1130	1184	92.6
699	753	807	861	915	969	1023	1077	1131	1185	94.4
700	754	808	862	916	970	1024	1078	1132	1186	96.3
701	755	809	863	917	971	1025	1079	1133	1187	98.1
1200	1300	1400	1500	1600	1700	1800	1900	2000	2100	

Table XX. Natural Sines, Cosines

EXPLANATION

This table is adapted to solution of route-surveying problems without a trig calculator. Its special advantages are: (1) six significant figures; (2) most-frequently used functions in one table; (3) "streamline" interpolation for multiples of 10″ and 15″; (4) chances for mistakes reduced by elimination of column headings at bottom of page and by use of only one degree per page.

Precise angle and distance measurements in route surveying justify one more figure than in the usual 5-place table, but hardly justify 7-place tables.

In a large number of curve formulas the sine and versine of the same angle appear. Other combinations are sin-tan-cos and sin-cos-tan-vers. The appearance of these functions in the same table should save time. The omission of cotangents is of little consequence since they are rarely used in route surveying: if needed, they are equivalent to the tangent of the complement.

The most serious defect of most tables of natural functions is the absence of a convenient aid in interpolating for seconds. This defect is remedied in Table XX, largely through the use of sets of corrections for multiples of 10″ and 15″. These multiples are the ones most frequently needed when single or double angles are turned with a 20″, 30″, or 1′ instrument.

Example

Find the tangent of 28°18′45″, Answer: 0.538444 + 281 = 0.538725.

The foregoing is the quickest method and is sufficiently precise for most purposes. However, because the sets of corrections are exact at the middle of the indicated range, slightly greater precision would be obtained in this example by adding a correction of 282, since the 45″ correction at 28°30′ is 283. This method gives 0.538726. Many computers would prefer to subtract the 15″ correction of 94 from the tangent of 28°19′ (an excellent method), giving also 0.538726. Use of these more precise methods is recommended for finding exsecants of angles between 45° and 60°, and for tangents and exsecants of angles above 70°.

Obviously, corrections for any number of seconds could be obtained speedily, if the field work justifies, by combining the tabulated corrections and shifting the decimal point.

VERSINES, EXSECANTS, AND TANGENTS
0°

'	SINE	CORR. FOR SEC. +	COSINE	CORR. FOR SEC. − +	VERSINE	EXSEC	CORR. FOR SEC. +	CORR. FOR SEC. +	TANGENT	'
0	Zero		One		Zero	Zero			Zero	0
1	.000291		One		Zero	Zero			.000291	1
2	0582		One		Zero	Zero			0582	2
3	0873		One		Zero	Zero			0873	3
4	1164		.999999		.000001	.000001			1164	4
5	.001454		.999999		.000001	.000001			.001454	5
6	1745		9998		0002	0001			1745	6
7	2036		9998		0002	0002			2036	7
8	2327		9997	By	0003	0003	By		2327	8
9	2618		9997	Inspec-	0003	0003	Inspec-		2618	9
10	.002909		.999996	tion	.000004	.000004	tion		.002909	10
11	3200	" Corr.	9995		0005	0005		" Corr.	3200	11
12	3491	10 48	9994		0006	0006		10 48	3491	12
13	3782	15 73	9993		0007	0007		15 73	3782	13
14	4072	20 97	9992		0008	0008		20 97	4072	14
15	.004363	30 145	.999990		.000010	.000010		30 145	.004363	15
16	4654	40 194	9989		0011	0011		40 194	4654	16
17	4945	45 218	9988		0012	0012		45 218	4945	17
18	5236	50 242	9986		0014	0014		50 242	5236	18
19	5527		9985		0015	0015			5527	19
20	.005818		.999983		.000017	.000017			.005818	20
21	6109		9981		0019	0019			6109	21
22	6400		9980		0020	0020			6400	22
23	6690		9978		0022	0022			6690	23
24	6981		9976		0024	0024			6981	24
25	.007272		.999974		.000026	.000026			.007272	25
26	7563		9971		0029	0029			7563	26
27	7854		9969		0031	0031			7854	27
28	8145		9967		0033	0033			8145	28
29	8436		9964		0036	0036			8436	29
30	.008726		.999962		.000038	.000038			.008727	30
31	9017		9959		0041	0041			9018	31
32	9308		9957		0043	0043			9309	32
33	9599		9954		0046	0046			9600	33
34	9890		9951		0049	0049			9890	34
35	.010181		.999948		.000052	.000052			.010181	35
36	0472		9945		0055	0055			0472	36
37	0763		9942		0058	0058			0763	37
38	1054		9939		0061	0061			1054	38
39	1344		9936		0064	0064			1345	39
40	.011635		.999932		.000068	.000068			.011636	40
41	1926	" Corr.	9929		0071	0071		" Corr.	1927	41
42	2217	10 48	9925		0075	0075		10 48	2218	42
43	2508	15 73	9922		0078	0078		15 73	2509	43
44	2799	20 97	9918		0082	0082		20 97	2800	44
45	.013090	30 145	.999914		.000086	.000086		30 145	.013091	45
46	3380	40 194	9910		0090	0090		40 194	3382	46
47	3671	45 218	9906		0094	0094		45 218	3673	47
48	3962	50 242	9902		0098	0098		50 242	3964	48
49	4253		9898		0102	0102			4254	49
50	.014544		.999894		.000106	.000106			.014545	50
51	4835		9890		0110	0110			4836	51
52	5126		9886		0114	0114			5127	52
53	5416		9881		0119	0119			5418	53
54	5707		9877		0123	0123			5709	54
55	.015998		.999872		.000128	.000128			.016000	55
56	6289		9867		0133	0133			6291	56
57	6580		9862		0138	0138			6582	57
58	6871		9858		0142	0142			6873	58
59	7162		9853		0147	0147			7164	59
60	.017452		.999848		.000152	.000152			.017455	60

TABLE XX.—NATURAL SINES, COSINES,

1°

'	SINE	CORR. FOR SEC.	COSINE	CORR. FOR SEC.	VERSINE	EXSEC	CORR. FOR SEC.	CORR. FOR SEC.	TANGENT	'
0	.017452	+	.999848	− +	.000152	.000152	+	+	.017455	0
1	7743		9843		0157	0157			7746	1
2	8034		9837		0163	0163			8037	2
3	8325		9832		0168	0168			8328	3
4	8616		9827		0173	0173			8619	4
5	.018907		.999821		.000179	.000179			.018910	5
6	9197		9816		0184	0184			9201	6
7	9488		9810		0190	0190			9492	7
8	9779		9804		0196	0196			9783	8
9	.020070		9799		0201	0201			.020074	9
10	.020361		.999793		.000207	.000207			.020365	10
11	0652	" Corr.	9787	" Corr.	0213	0213	" Corr.	" Corr.	0656	11
12	0942	10 48	9781	10 1	0219	0219	10 1	10 49	0947	12
13	1233	15 73	9774	15 2	0226	0226	15 2	15 73	1238	13
14	1524	20 97	9768	20 2	0232	0232	20 2	20 97	1529	14
15	.021815	30 145	.999762	30 3	.000238	.000238	30 3	30 146	.021820	15
16	2106	40 194	9756	40 4	0244	0244	40 4	40 194	2111	16
17	2396	45 218	9749	45 5	0251	0251	45 5	45 218	2402	17
18	2687	50 242	9743	50 5	0257	0257	50 5	50 243	2693	18
19	2978		9736		0264	0264			2984	19
20	.023269		.999729		.000271	.000271			.023275	20
21	3560		9722		0278	0278			3566	21
22	3851		9716		0284	0284			3857	22
23	4141		9709		0291	0291			4148	23
24	4432		9702		0298	0299			4439	24
25	.024723		.999694		.000306	.000306			.024730	25
26	5014		9687		0313	0313			5022	26
27	5305		9680		0320	0320			5313	27
28	5595		9672		0328	0328			5604	28
29	5886		9665		0335	0335			5895	29
30	.026177		.999657		.000343	.000343			.026186	30
31	6468		9650		0350	0350			6477	31
32	6758		9642		0358	0358			6768	32
33	7049		9634		0366	0366			7059	33
34	7340		9626		0374	0374			7350	34
35	.027631		.999618		.000382	.000382			.027641	35
36	7922		9610		0390	0390			7932	36
37	8212		9602		0398	0398			8224	37
38	8503		9594		0406	0406			8515	38
39	8794		9585		0415	0415			8806	39
40	.029085		.999577		.000423	.000423			.029097	40
41	9376	" Corr.	9568	" Corr.	0432	0432	" Corr.	" Corr.	9388	41
42	9666	10 48	9560	10 1	0440	0440	10 1	10 49	9679	42
43	9957	15 73	9551	15 2	0449	0449	15 2	15 73	9970	43
44	.030248	20 97	9542	20 3	0458	0458	20 3	20 97	.030262	44
45	.030538	30 145	.999534	30 4	.000466	.000467	30 4	30 146	.030553	45
46	0829	40 194	9525	40 6	0475	0476	40 6	40 194	0844	46
47	1120	45 218	9516	45 7	0484	0485	45 7	45 218	1135	47
48	1411	50 242	9507	50 7	0493	0494	50 7	50 243	1426	48
49	1701		9497		0503	0503			1717	49
50	.031992		.999488		.000512	.000512			.032009	50
51	2283		9479		0521	0521			2300	51
52	2574		9469		0531	0531			2591	52
53	2864		9460		0540	0540			2882	53
54	3155		9450		0550	0550			3173	54
55	.033446		.999440		.000559	.000560			.033465	55
56	3737		9431		0569	0570			3756	56
57	4027		9421		0579	0579			4047	57
58	4318		9411		0589	0589			4338	58
59	4609		9401		0599	0599			4629	59
60	.034899		.999391		.000609	.000609			.034921	60

VERSINES, EXSECANTS, AND TANGENTS

2°

'	SINE	CORR. FOR SEC.	COSINE	CORR. FOR SEC.	VERSINE	EXSEC	CORR. FOR SEC.	CORR. FOR SEC.	TANGENT	'
0	.034900	+	.999391	− +	.000609	.000610	+	+	.034921	0
1	5190		9381		0619	0620			5212	1
2	5481		9370		0630	0630			5503	2
3	5772		9360		0640	0640			5794	3
4	6062		9350		0650	0651			6086	4
5	.036353		.999339		.000661	.000661			.036377	5
6	6644		9328		0672	0672			6668	6
7	6934		9318		0682	0683			6960	7
8	7225		9307		0693	0694			7251	8
9	7516		9296		0704	0704			7542	9
10	.037806		.999285		.000715	.000715			.037834	10
11	8097	" Corr.	9274	" Corr.	0726	0726	" Corr.	" Corr.	8125	11
12	8388	10 48	9263	10 2	0737	0738	10 2	10 49	8416	12
13	8678	15 73	9252	15 3	0748	0749	15 3	15 73	8707	13
14	8969	20 97	9240	20 4	0760	0760	20 4	20 97	8999	14
15	.039260	30 145	.999229	30 6	.000771	.000772	30 6	30 146	.039290	15
16	9550	40 194	9218	40 8	0782	0783	40 8	40 194	9581	16
17	9841	45 218	9206	45 9	0794	0795	45 9	45 219	9873	17
18	.040132	50 242	9194	50 10	0806	0806	50 10	50 243	.040164	18
19	0422		9183		0817	0818			0456	19
20	.040713		.999171		.000829	.000830			.040747	20
21	1004		9159		0841	0842			1038	21
22	1294		9147		0853	0854			1330	22
23	1585		9135		0865	0866			1621	23
24	1876		9123		0877	0878			1912	24
25	.042166		.999111		.000889	.000890			.042204	25
26	2457		9098		0902	0902			2495	26
27	2748		9086		0914	0915			2787	27
28	3038		9073		0927	0927			3078	28
29	3329		9061		0939	0940			3370	29
30	.043619		.999048		.000952	.000953			.043661	30
31	3910		9036		0964	0965			3952	31
32	4201		9023		0977	0978			4244	32
33	4491		9010		0990	0991			4535	33
34	4782		8997		1003	1004			4827	34
35	.045072		.998984		.001016	.001017			.045118	35
36	5363		8971		1029	1030			5410	36
37	5654		8957		1043	1044			5701	37
38	5944		8944		1056	1057			5993	38
39	6235		8931		1069	1070			6284	39
40	.046525		.998917		.001083	.001084			.046576	40
41	6816	" Corr.	8904	" Corr.	1096	1098	" Corr.	" Corr.	6867	41
42	7106	10 48	8890	10 2	1110	1111	10 2	10 49	7159	42
43	7397	15 73	8876	15 3	1124	1125	15 3	15 73	7450	43
44	7688	20 97	8862	20 5	1138	1139	20 5	20 97	7742	44
45	.047978	30 145	.998848	30 7	.001152	.001153	30 7	30 146	.048033	45
46	8269	40 194	8834	40 9	1166	1167	40 9	40 194	8325	46
47	8559	45 218	8820	45 10	1180	1181	45 10	45 219	8617	47
48	8850	50 242	8806	50 12	1194	1195	50 12	50 243	8908	48
49	9140		8792		1208	1210			9200	49
50	.049431		.998778		.001222	.001224			.049491	50
51	9721		8763		1237	1238			9783	51
52	.050012		8749		1251	1253			.050075	52
53	0302		8734		1266	1268			0366	53
54	0593		8719		1281	1282			0658	54
55	.050884		.998705		.001295	.001297			.050950	55
56	1174		8690		1310	1312			1241	56
57	1464		8675		1325	1327			1533	57
58	1755		8660		1340	1342			1824	58
59	2046		8645		1355	1357			2116	59
60	.052336		.998630		.001370	.001372			.052408	60

TABLE XX.—NATURAL SINES, COSINES,

3°

'	SINE	CORR. FOR SEC.	COSINE	CORR. FOR SEC.	VERSINE	EXSEC	CORR. FOR SEC.	CORR. FOR SEC.	TANGENT	'
0	.052336	+	.998630	− +	.001370	.001372	+	+	.052408	0
1	2626		8614		1386	1388			2700	1
2	2917		8599		1401	1403			2991	2
3	3207		8584		1416	1418			3283	3
4	3498		8568		1432	1434			3575	4
5	.053788		.998552		.001448	.001450			.053866	5
6	4079		8537		1463	1466			4158	6
7	4369		8521		1479	1481			4450	7
8	4660		8505		1495	1497			4742	8
9	4950		8489		1511	1513			5033	9
10	.055241		.998473		.001527	.001529			.055325	10
11	5531	" Corr.	8457	" Corr.	1543	1545	" Corr.	" Corr.	5617	11
12	5822	10 48	8441	10 3	1559	1562	10 3	10 49	5909	12
13	6112	15 73	8424	15 4	1576	1578	15 4	15 73	6200	13
14	6402	20 97	8408	20 5	1592	1594	20 5	20 97	6492	14
15	.056693	30 145	.998392	30 8	.001608	.001611	30 8	30 146	.056784	15
16	6983	40 194	8375	40 11	1625	1628	40 11	40 195	7076	16
17	7274	45 218	8358	45 12	1642	1644	45 12	45 219	7368	17
18	7564	50 242	8342	50 14	1658	1661	50 14	50 243	7660	18
19	7854		8325		1675	1678			7952	19
20	.058145		.998308		.001692	.001695			.058243	20
21	8435		8291		1709	1712			8535	21
22	8726		8274		1726	1729			8827	22
23	9016		8257		1743	1746			9119	23
24	9306		8240		1760	1763			9411	24
25	.059597		.998222		.001778	.001781			.059703	25
26	9887		8205		1795	1798			9995	26
27	.060178		8188		1812	1816			.060287	27
28	0468		8170		1830	1833			0579	28
29	0758		8152		1848	1851			0871	29
30	.061048		.998135		.001865	.001869			.061163	30
31	1339		8117		1883	1887			1455	31
32	1629		8099		1901	1904			1747	32
33	1920		8081		1919	1922			2039	33
34	2210		8063		1937	1941			2331	34
35	.062500		.998045		.001955	.001959			.062623	35
36	2790		8027		1973	1977			2915	36
37	3081		8008		1992	1996			3207	37
38	3371		7990		2010	2014			3499	38
39	3661		7972		2028	2033			3791	39
40	.063952		.997953		.002047	.002051			.064083	40
41	4242	" Corr.	7934	" Corr.	2066	2070	" Corr.	" Corr.	4375	41
42	4532	10 48	7916	10 3	2084	2089	10 3	10 49	4667	42
43	4823	15 73	7897	15 5	2103	2108	15 5	15 73	4959	43
44	5113	20 97	7878	20 6	2122	2127	20 6	20 97	5251	44
45	.065403	30 145	.997859	30 9	.002141	.002146	30 9	30 146	.065544	45
46	5693	40 194	7840	40 13	2160	2165	40 13	40 195	5836	46
47	5984	45 218	7821	45 14	2179	2184	45 14	45 219	6128	47
48	6274	50 242	7802	50 16	2198	2203	50 16	50 243	6420	48
49	6564		7782		2218	2223			6712	49
50	.066854		.997763		.002237	.002242			.067004	50
51	7145		7743		2257	2262			7297	51
52	7435		7724		2276	2282			7589	52
53	7725		7704		2296	2301			7881	53
54	8015		7684		2316	2321			8173	54
55	.068306		.997664		.002336	.002341			.068465	55
56	8596		7644		2356	2361			8758	56
57	8886		7624		2376	2381			9050	57
58	9176		7604		2396	2401			9342	58
59	9466		7584		2416	2422			9634	59
60	.069756		.997564		.002436	.002442			.069927	60

VERSINES, EXSECANTS, AND TANGENTS

4°

'	SINE	CORR. FOR SEC. +	COSINE	CORR. FOR SEC. − +	VERSINE	EXSEC	CORR. FOR SEC. +	CORR. FOR SEC. +	TANGENT	'
0	.069756		.997564		.002436	.002442			.069927	0
1	.070047		7544		2456	2462			.070219	1
2	0337		7523		2477	2483			0512	2
3	0627		7503		2497	2504			0804	3
4	0917		7482		2518	2524			1096	4
5	.071207		.997462		.002538	.002545			.071388	5
6	1497		7441		2559	2566			1681	6
7	1788		7420		2580	2587			1973	7
8	2078		7399		2601	2608			2266	8
9	2368		7378		2622	2629			2558	9
10	.072658		.997357		.002643	.002650			.072850	10
11	2948	" Corr.	7336	" Corr.	2664	2671	" Corr.	" Corr.	3143	11
12	3238	10 48	7314	10 4	2686	2693	10 4	10 49	3435	12
13	3528	15 73	7293	15 5	2707	2714	15 5	15 73	3728	13
14	3818	20 97	7272	20 7	2728	2736	20 7	20 97	4020	14
15	.074108	30 145	.997250	30 11	.002750	.002757	30 11	30 146	.074313	15
16	4399	40 193	7229	40 14	2771	2779	40 14	40 195	4605	16
17	4689	45 218	7207	45 16	2793	2801	45 16	45 219	4898	17
18	4979	50 242	7185	50 18	2815	2823	50 18	50 244	5190	18
19	5269		7163		2837	2845			5483	19
20	.075559		.997141		.002859	.002867			.075776	20
21	5849		7119		2881	2889			6068	21
22	6139		7097		2903	2911			6361	22
23	6429		7075		2925	2934			6653	23
24	6719		7053		2947	2956			6946	24
25	.077009		.997030		.002970	.002979			.077238	25
26	7299		7008		2992	3001			7531	26
27	7589		6985		3015	3024			7824	27
28	7879		6963		3037	3046			8116	28
29	8169		6940		3060	3069			8409	29
30	.078459		.996917		.003083	.003092			.078702	30
31	8749		6894		3106	3115			8994	31
32	9039		6872		3128	3138			9287	32
33	9329		6848		3152	3162			9580	33
34	9619		6825		3175	3185			9873	34
35	.079909		.996802		.003198	.003208			.080165	35
36	.080199		6779		3221	3232			0458	36
37	0489		6756		3244	3255			0751	37
38	0779		6732		3268	3279			1044	38
39	1069		6708		3292	3302			1336	39
40	.081359		.996685		.003315	.003326			.081629	40
41	1649	" Corr.	6661	" Corr.	3339	3350	" Corr.	" Corr.	1922	41
42	1939	10 48	6637	10 4	3363	3374	10 4	10 49	2215	42
43	2228	15 72	6614	15 6	3386	3398	15 6	15 73	2508	43
44	2518	20 97	6590	20 8	3410	3422	20 8	20 98	2801	44
45	.082808	30 145	.996566	30 12	.003434	.003446	30 12	30 146	.083094	45
46	3098	40 193	6541	40 16	3459	3471	40 16	40 195	3387	46
47	3388	45 217	6517	45 18	3483	3495	45 18	45 220	3679	47
48	3678	50 242	6493	50 20	3507	3520	50 20	50 244	3972	48
49	3968		6468		3532	3544			4265	49
50	.084258		.996444		.003556	.003569			.084558	50
51	4547		6420		3580	3593			4851	51
52	4837		6395		3605	3618			5144	52
53	5127		6370		3630	3643			5437	53
54	5417		6345		3655	3668			5730	54
55	.085707		.996320		.003680	.003693			.086023	55
56	5997		6295		3705	3718			6316	56
57	6286		6270		3730	3744			6609	57
58	6576		6245		3755	3769			6902	58
59	6866		6220		3780	3794			7196	59
60	.087156		.996195		.003805	.003820			.087489	60

TABLE XX.—NATURAL SINES, COSINES,

5°

'	SINE	CORR. FOR SEC.	COSINE	CORR. FOR SEC.	VERSINE	EXSEC	CORR. FOR SEC.	CORR. FOR SEC.	TANGENT	'
0	.087156	+	.996195	− +	.003805	.003820	+	+	.087489	0
1	7446		6169		3831	3845			7782	1
2	7735		6144		3856	3871			8075	2
3	8025		6118		3882	3897			8368	3
4	8315		6093		3907	3923			8661	4
5	.088605		.996067		.003933	.003949			.088954	5
6	8894		6041		3959	3975			9248	6
7	9184		6015		3985	4001			9541	7
8	9474		5989		4011	4027			9834	8
9	9764		5963		4037	4053			.090127	9
10	.090053		5937		4063	4080			0421	10
11	0343	" Corr.	5911	" Corr.	4089	4106	" Corr.	" Corr.	0714	11
12	0633	10 48	5884	10 4	4116	4133	10 4	10 49	1007	12
13	0922	15 72	5858	15 7	4142	4159	15 7	15 73	1300	13
14	1212	20 97	5832	20 9	4168	4186	20 9	20 98	1594	14
15	.091502	30 145	.995805	30 13	.004195	.004213	30 13	30 147	.091887	15
16	1791	40 193	5778	40 18	4222	4240	40 18	40 196	2180	16
17	2081	45 217	5752	45 20	4248	4267	45 20	45 220	2474	17
18	2371	50 241	5725	50 22	4275	4294	50 22	50 244	2767	18
19	2660		5698		4302	4321			3061	19
20	.092950		.995671		.004329	.004348			.093354	20
21	3240		5644		4356	4375			3647	21
22	3529		5616		4384	4403			3941	22
23	3819		5589		4411	4430			4234	23
24	4108		5562		4438	4458			4528	24
25	.094398		.995534		.004466	.004486			.094821	25
26	4688		5507		4493	4513			5115	26
27	4977		5480		4520	4541			5408	27
28	5267		5452		4548	4569			5702	28
29	5556		5424		4576	4597			5996	29
30	.095846		.995396		.004604	.004625			.096289	30
31	6135		5368		4632	4653			6583	31
32	6425		5340		4660	4682			6876	32
33	6714		5312		4688	4710			7170	33
34	7004		5284		4716	4738			7464	34
35	.097293		.995256		.004744	.004767			.097757	35
36	7583		5227		4773	4796			8051	36
37	7872		5199		4801	4824			8345	37
38	8162		5170		4830	4853			8638	38
39	8451		5142		4858	4882			8932	39
40	.098741		.995113		.004887	.004911			.099226	40
41	9030	" Corr.	5084	" Corr.	4916	4940	" Corr.	" Corr.	9520	41
42	9320	10 48	5056	10 5	4944	4969	10 5	10 49	9813	42
43	9609	15 72	5027	15 7	4973	4998	15 7	15 73	.100107	43
44	9899	20 96	4998	20 10	5002	5028	20 10	20 98	0401	44
45	.100188	30 145	.994968	30 15	.005032	.005057	30 15	30 147	.100695	45
46	0478	40 193	4939	40 20	5061	5086	40 20	40 196	0989	46
47	0767	45 217	4910	45 22	5090	5116	45 22	45 220	1282	47
48	1056	50 241	4881	50 24	5119	5146	50 24	50 245	1576	48
49	1346		4851		5149	5175			1870	49
50	.101635		.994822		.005178	.005205			.102164	50
51	1924		4792		5208	5235			2458	51
52	2214		4762		5238	5265			2752	52
53	2503		4733		5267	5295			3046	53
54	2792		4703		5297	5325			3340	54
55	.103082		.994673		.005327	.005356			.103634	55
56	3371		4643		5357	5386			3928	56
57	3660		4613		5387	5416			4222	57
58	3950		4582		5418	5447			4516	58
59	4239		4552		5448	5478			4810	59
60	.104528		.994522		.005478	.005508			.105104	60

VERSINES, EXSECANTS, AND TANGENTS

6°

'	SINE	CORR. FOR SEC. +	COSINE	CORR. FOR SEC. − +	VERSINE	EXSEC	CORR. FOR SEC. +	CORR. FOR SEC. +	TANGENT	'
0	.104528		.994522		.005478	.005508			.105104	0
1	4818		4491		5509	5539			5398	1
2	5107		4461		5539	5570			5692	2
3	5396		4430		5570	5601			5987	3
4	5686		4400		5600	5632			6281	4
5	.105975		.994369		.005631	.005663			.106575	5
6	6264		4338		5662	5694			6869	6
7	6553		4307		5693	5726			7163	7
8	6842		4276		5724	5757			7458	8
9	7132		4245		5755	5788			7752	9
10	.107421		.994214		.005786	.005820			.108046	10
11	7710	" Corr.	4182	" Corr.	5818	5852	" Corr.	" Corr.	8340	11
12	7999	10 48	4151	10 5	5849	5883	10 5	10 49	8635	12
13	8288	15 72	4120	15 8	5880	5915	15 8	15 74	8929	13
14	8578	20 96	4088	20 11	5912	5947	20 11	20 98	9223	14
15	.108867	30 145	.994056	30 16	.005944	.005979	30 16	30 147	.109518	15
16	9156	40 193	4025	40 21	5975	6011	40 21	40 196	9812	16
17	9445	45 217	3993	45 24	6007	6044	45 24	45 221	.110107	17
18	9734	50 241	3961	50 26	6039	6076	50 27	50 245	0401	18
19	.110023		3929		6071	6108			0696	19
20	.110313		.993897		.006103	.006140			.110990	20
21	0602		3865		6135	6173			1284	21
22	0891		3833		6167	6206			1579	22
23	1180		3800		6200	6238			1873	23
24	1469		3768		6232	6271			2168	24
25	.111758		.993736		.006264	.006304			.112462	25
26	2047		3703		6297	6337			2757	26
27	2336		3670		6330	6370			3052	27
28	2625		3638		6362	6403			3346	28
29	2914		3605		6395	6436			3641	29
30	.113203		.993572		.006428	.006470			.113936	30
31	3492		3539		6461	6503			4230	31
32	3781		3506		6494	6537			4525	32
33	4070		3473		6527	6570			4820	33
34	4359		3440		6560	6604			5114	34
35	.114648		.993406		.006594	.006638			.115409	35
36	4937		3373		6627	6671			5704	36
37	5226		3339		6661	6705			5999	37
38	5515		3306		6694	6739			6294	38
39	5804		3272		6728	6774			6588	39
40	.116093		.993238		.006762	.006808			.116883	40
41	6382	" Corr.	3204	" Corr.	6796	6842	" Corr.	" Corr.	7178	41
42	6671	10 48	3171	10 6	6829	6876	10 6	10 49	7473	42
43	6960	15 72	3137	15 9	6863	6911	15 9	15 74	7768	43
44	7248	20 96	3103	20 11	6897	6945	20 12	20 98	8063	44
45	.117537	30 144	.993069	30 17	.006931	.006980	30 17	30 147	.118358	45
46	7826	40 193	3034	40 23	6966	7015	40 23	40 197	8653	46
47	8115	45 217	3000	45 26	7000	7049	45 26	45 221	8948	47
48	8404	50 241	2966	50 29	7034	7084	50 29	50 246	9243	48
49	8693		2931		7069	7119			9538	49
50	.118982		.992896		.007104	.007154			.119833	50
51	9270		2862		7138	7190			.120128	51
52	9559		2827		7173	7225			0423	52
53	9848		2792		7208	7260			0718	53
54	.120137		2757		7243	7296			1013	54
55	.120426		.992722		.007278	.007331			.121308	55
56	0714		2687		7313	7367			1604	56
57	1003		2652		7348	7402			1899	57
58	1292		2617		7383	7438			2194	58
59	1581		2582		7418	7474			2489	59
60	.121869		.992546		.007454	.007510			.122785	60

TABLE XX.—NATURAL SINES, COSINES,

7°

'	SINE	CORR. FOR SEC.	COSINE	CORR. FOR SEC.	VERSINE	EXSEC	CORR. FOR SEC.	CORR. FOR SEC.	TANGENT	'
		+		− +			+	+		
0	.121869		.992546		.007454	.007510			.122785	0
1	2158		2511		7489	7546			3080	1
2	2447		2475		7525	7582			3375	2
3	2736		2439		7561	7618			3670	3
4	3024		2404		7596	7654			3966	4
5	.123313		.992368		.007632	.007691			.124261	5
6	3602		2332		7668	7727			4557	6
7	3890		2296		7704	7764			4852	7
8	4179		2260		7740	7800			5147	8
9	4467		2224		7776	7837			5443	9
10	.124756		.992187		.007813	.007874			.125738	10
11	5045	" Corr.	2151	" Corr.	7849	7911	" Corr.	" Corr.	6034	11
12	5333	10 48	2115	10 6	7885	7948	10 6	10 49	6329	12
13	5622	15 72	2078	15 9	7922	7985	15 9	15 74	6625	13
14	5910	20 96	2042	20 12	7958	8022	20 12	20 99	6920	14
15	.126199	30 144	.992005	30 18	.007995	.008060	30 19	30 148	.127216	15
16	6488	40 192	1968	40 25	8032	8097	40 25	40 197	7512	16
17	6776	45 216	1931	45 28	8069	8134	45 28	45 222	7807	17
18	7065	50 241	1894	50 31	8106	8172	50 31	50 246	8103	18
19	7353		1857		8143	8209			8399	19
20	.127642		.991820		.008180	.008247			.128694	20
21	7930		1783		8217	8285			8990	21
22	8219		1746		8254	8323			9286	22
23	8507		1709		8291	8361			9582	23
24	8796		1671		8329	8399			9877	24
25	.129084		.991634		.008366	.008437			.130173	25
26	9372		1596		8404	8475			0469	26
27	9661		1558		8442	8514			0765	27
28	9949		1521		8479	8552			1061	28
29	.130238		1483		8517	8590			1357	29
30	.130526		.991445		.008555	.008629			.131652	30
31	0815		1407		8593	8668			1948	31
32	1103		1369		8631	8706			2244	32
33	1391		1331		8669	8745			2540	33
34	1680		1292		8708	8784			2836	34
35	.131968		.991254		.008746	.008823			.133132	35
36	2256		1216		8784	8862			3428	36
37	2545		1177		8823	8902			3725	37
38	2833		1138		8862	8941			4021	38
39	3121		1100		8900	8980			4317	39
40	.133410		.991061		.008939	.009020			.134613	40
41	3698	" Corr.	1022	" Corr.	8978	9059	" Corr.	" Corr.	4909	41
42	3986	10 48	0983	10 7	9017	9099	10 7	10 49	5205	42
43	4274	15 72	0944	15 10	9056	9139	15 10	15 74	5502	43
44	4563	20 96	0905	20 13	9095	9178	20 13	20 99	5798	44
45	.134851	30 144	.990866	30 20	.009134	.009218	30 20	30 148	.136094	45
46	5139	40 192	0827	40 26	9173	9258	40 27	40 198	6390	46
47	5427	45 216	0787	45 29	9213	9298	45 30	45 222	6687	47
48	5716	50 240	0748	50 33	9252	9339	50 33	50 247	6983	48
49	6004		0708		9292	9379			7279	49
50	.136292		.990669		.009331	.009419			.137576	50
51	6580		0629		9371	9460			7872	51
52	6868		0589		9411	9500			8168	52
53	7156		0549		9451	9541			8465	53
54	7444		0510		9490	9582			8762	54
55	.137733		.990469		.009531	.009622			.139058	55
56	8021		0429		9571	9663			9354	56
57	8309		0389		9611	9704			9651	57
58	8597		0349		9651	9745			9948	58
59	8885		0308		9692	9786			.140244	59
60	.139173		.990268		.009732	.009828			.140541	60

VERSINES, EXSECANTS, AND TANGENTS
8°

'	SINE	CORR. FOR SEC.	COSINE	CORR. FOR SEC.	VERSINE	EXSEC	CORR. FOR SEC.	CORR. FOR SEC.	TANGENT	'
		+		− +			+	+		
0	.139173		.990268		.009732	.009828			.140541	0
1	9461		0228		9772	9869			0838	1
2	9749		0187		9813	9910			1134	2
3	.140037		0146		9854	9952			1431	3
4	0325		0106		9894	9993			1728	4
5	.140613		.990065		.009935	.010035			.142024	5
6	0901		0024		9976	0077			2321	6
7	1189		.989983		.010017	0119			2618	7
8	1477		9942		0058	0161			2915	8
9	1765		9900		0100	0203			3212	9
10	.142053		.989859		.010141	.010245			.143508	10
11	2341	" Corr.	9818	" Corr.	0182	0287	" Corr.	" Corr.	3805	11
12	2629	10 48	9776	10 7	0224	0329	10 7	10 50	4102	12
13	2917	15 72	9735	15 10	0265	0372	15 11	15 74	4399	13
14	3205	20 96	9693	20 14	0307	0414	20 14	20 99	4696	14
15	.143493	30 144	.989651	30 21	.010349	.010457	30 21	30 149	.144993	15
16	3780	40 192	9610	40 28	0390	0500	40 28	40 198	5290	16
17	4068	45 216	9568	45 31	0432	0542	45 32	45 223	5587	17
18	4356	50 240	9526	50 35	0474	0585	50 36	50 248	5884	18
19	4644		9484		0516	0628			6181	19
20	.144932		.989442		.010558	.010671			.146478	20
21	5220		9399		0601	0714			6776	21
22	5508		9357		0643	0757			7073	22
23	5795		9315		0685	0801			7370	23
24	6083		9272		0728	0844			7667	24
25	.146371		.989230		.010770	.010888			.147964	25
26	6658		9187		0813	0931			8262	26
27	6946		9144		0856	0975			8559	27
28	7234		9102		0898	1018			8856	28
29	7522		9059		0941	1062			9154	29
30	.147809		.989016		.010984	.011106			.149451	30
31	8097		8973		1027	1150			9748	31
32	8385		8930		1070	1194			.150046	32
33	8672		8886		1114	1238			0343	33
34	8960		8843		1157	1283			0641	34
35	.149248		.988800		.011200	.011327			.150938	35
36	9535		8756		1244	1372			1236	36
37	9823		8713		1287	1416			1533	37
38	.150111		8669		1331	1461			1831	38
39	0398		8626		1374	1505			2128	39
40	.150686		.988582		.011418	.011550			.152426	40
41	0973	" Corr.	8538	" Corr.	1462	1595	" Corr.	" Corr.	2724	41
42	1261	10 48	8494	10 7	1506	1640	10 8	10 50	3022	42
43	1548	15 72	8450	15 11	1550	1685	15 11	15 74	3319	43
44	1836	20 96	8406	20 15	1594	1730	20 15	20 99	3617	44
45	.152123	30 144	.988362	30 22	.011638	.011776	30 23	30 149	.153915	45
46	2411	40 192	8317	40 29	1683	1821	40 30	40 199	4212	46
47	2698	45 216	8273	45 33	1727	1866	45 34	45 223	4510	47
48	2986	50 240	8228	50 37	1772	1912	50 38	50 248	4808	48
49	3273		8184		1816	1958			5106	49
50	.153561		.988139		.011861	.012003			.155404	50
51	3848		8094		1906	2049			5702	51
52	4136		8050		1950	2095			6000	52
53	4423		8005		1995	2141			6298	53
54	4710		7960		2040	2187			6596	54
55	.154998		.987915		.012085	.012233			.156894	55
56	5285		7870		2130	2279			7192	56
57	5572		7824		2176	2326			7490	57
58	5860		7779		2221	2372			7788	58
59	6147		7734		2266	2418			8086	59
60	.156434		.987688		.012312	.012465			.158384	60

TABLE XX.—NATURAL SINES, COSINES,

9°

'	SINE	CORR. FOR SEC.	COSINE	CORR. FOR SEC.	VERSINE	EXSEC	CORR. FOR SEC.	CORR. FOR SEC.	TANGENT	'
		+		− +			+	+		
0	.156434		.987688		.012312	.012465			.158384	0
1	6722		7643		2357	2512			8683	1
2	7009		7597		2403	2559			8981	2
3	7296		7551		2449	2606			9279	3
4	7584		7506		2494	2652			9577	4
5	.157871		.987460		.012540	.012700			.159876	5
6	8158		7414		2586	2747			.160174	6
7	8445		7368		2632	2794			0472	7
8	8732		7322		2678	2841			0771	8
9	9020		7275		2725	2889			1069	9
10	.159307		.987229		.012771	.012936			.161368	10
11	9594	" Corr.	7183	" Corr.	2817	2984	" Corr.	" Corr.	1666	11
12	9881	10 48	7136	10 8	2864	3031	10 8	10 50	1965	12
13	.160168	15 72	7090	15 12	2910	3079	15 12	15 75	2263	13
14	0456	20 96	7043	20 16	2957	3127	20 16	20 100	2562	14
15	.160743	30 144	.986996	30 23	.013004	.013175	30 24	30 149	.162860	15
16	1030	40 191	6950	40 31	3050	3223	40 32	40 199	3159	16
17	1317	45 215	6903	45 35	3097	3271	45 36	45 224	3458	17
18	1604	50 239	6856	50 39	3144	3319	50 40	50 249	3756	18
19	1891		6809		3191	3368			4055	19
20	.162178		.986762		.013238	.013416			.164354	20
21	2465		6714		3286	3465			4652	21
22	2752		6667		3333	3513			4951	22
23	3039		6620		3380	3562			5250	23
24	3326		6572		3428	3611			5549	24
25	.163613		.986525		.013475	.013660			.165848	25
26	3900		6477		3523	3708			6147	26
27	4187		6429		3571	3757			6446	27
28	4474		6382		3618	3807			6745	28
29	4761		6334		3666	3856			7044	29
30	.165048		.986286		.013714	.013905			.167343	30
31	5334		6238		3762	3954			7642	31
32	5621		6189		3811	4004			7941	32
33	5908		6141		3859	4054			8240	33
34	6195		6093		3907	4103			8539	34
35	.166482		.986044		.013956	.014153			.168838	35
36	6769		5996		4004	4203			9137	36
37	7056		5948		4052	4253			9437	37
38	7342		5899		4101	4303			9736	38
39	7629		5850		4150	4353			.170035	39
40	.167916		.985801		.014199	.014403			.170334	40
41	8203	" Corr.	5752	" Corr.	4248	4454	" Corr.	" Corr.	0634	41
42	8489	10 48	5704	10 8	4296	4504	10 8	10 50	0933	42
43	8776	15 72	5654	15 12	4346	4554	15 13	15 75	1232	43
44	9063	20 96	5605	20 16	4395	4605	20 17	20 100	1532	44
45	.169350	30 143	.985556	30 25	.014444	.014656	30 25	30 150	.171831	45
46	9636	40 191	5507	40 33	4493	4706	40 34	40 200	2131	46
47	9923	45 215	5457	45 37	4543	4757	45 38	45 225	2430	47
48	.170210	50 239	5408	50 41	4592	4808	50 42	50 250	2730	48
49	0496		5358		4642	4859			3030	49
50	.170783		.985309		.014691	.014910			.173329	50
51	1069		5259		4741	4962			3629	51
52	1356		5209		4791	5013			3928	52
53	1642		5159		4841	5064			4228	53
54	1929		5109		4891	5116			4528	54
55	.172216		.985059		.014941	.015167			.174828	55
56	2502		5009		4991	5219			5128	56
57	2789		4959		5041	5271			5427	57
58	3075		4909		5091	5323			5727	58
59	3362		4858		5142	5375			6027	59
60	.173648		.984808		.015192	.015427			.176327	60

* Multiply functions (except * values) by the given value of L_S. When chord definition of degree of curve (D_c) is used, correct the calculated values of o or X_0 by subtracting the product of D_c and the * values. See page 354 for example.

VERSINES, EXSECANTS, AND TANGENTS

10°

'	SINE	CORR. FOR SEC.	COSINE	CORR. FOR SEC.	VERSINE	EXSEC	CORR. FOR SEC.	CORR. FOR SEC.	TANGENT	'
		+		− +			+	+		
0	.173648		.984808		.015192	.015427			.176327	0
1	3935		4757		5243	5479			6630	1
2	4221		4707		5293	5531			6927	2
3	4508		4656		5344	5583			7227	3
4	4794		4605		5395	5636			7527	4
5	.175080		.984554		.015446	.015688			.177827	5
6	5367		4503		5497	5741			8127	6
7	5653		4452		5548	5793			8427	7
8	5940		4401		5599	5846			8727	8
9	6226		4350		5650	5899			9028	9
10	.176512		.984298		.015702	.015952			.179328	10
11	6798	" Corr.	4247	" Corr.	5753	6005	" Corr.	" Corr.	9628	11
12	7085	10 48	4196	10 9	5804	6058	10 9	10 50	9928	12
13	7371	15 72	4144	15 13	5856	6111	15 13	15 75	.180229	13
14	7657	20 95	4092	20 17	5908	6165	20 18	20 100	0529	14
15	.177944	30 143	.984041	30 26	.015959	.016218	30 27	30 150	.180830	15
16	8230	40 191	3989	40 35	6011	6272	40 36	40 200	1130	16
17	8516	45 215	3937	45 39	6063	6325	45 40	45 225	1430	17
18	8802	50 239	3885	50 43	6115	6379	50 45	50 250	1731	18
19	9088		3833		6167	6433			2031	19
20	.179375		.983781		.016219	.016486			.182332	20
21	9661		3729		6271	6540			2632	21
22	9947		3676		6324	6595			2933	22
23	.180233		3624		6376	6649			3234	23
24	0519		3572		6428	6703			3534	24
25	.180805		.983519		.016481	.016757			.183835	25
26	1091		3466		6534	6812			4136	26
27	1377		3414		6586	6866			4436	27
28	1664		3361		6639	6921			4737	28
29	1950		3308		6692	6976			5038	29
30	.182236		.983255		.016745	.017030			.185339	30
31	2522		3202		6798	7085			5640	31
32	2808		3149		6851	7140			5941	32
33	3094		3096		6904	7195			6242	33
34	3380		3042		6958	7250			6543	34
35	.183665		.982989		.017011	.017306			.186844	35
36	3951		2935		7065	7361			7145	36
37	4237		2882		7118	7416			7446	37
38	4523		2828		7172	7472			7747	38
39	4809		2774		7226	7528			8048	39
40	.185095		.982721		.017279	.017583			.188350	40
41	5381	" Corr.	2667	" Corr.	7333	7639	" Corr.	" Corr.	8651	41
42	5667	10 48	2613	10 9	7387	7695	10 9	10 50	8952	42
43	5952	15 71	2559	15 14	7441	7751	15 14	15 75	9253	43
44	6238	20 95	2505	20 18	7495	7807	20 19	20 100	9555	44
45	.186524	30 143	.982450	30 27	.017550	.017863	30 28	30 151	.189856	45
46	6810	40 190	2396	40 36	7604	7919	40 37	40 201	.190157	46
47	7096	45 214	2342	45 41	7658	7976	45 42	45 226	0459	47
48	7381	50 238	2287	50 45	7713	8032	50 47	50 251	0760	48
49	7667		2233		7767	8089			1062	49
50	.187953		.982178		.017822	.018145			.191363	50
51	8238		2123		7877	8202			1665	51
52	8524		2069		7931	8259			1966	52
53	8810		2014		7986	8316			2268	53
54	9095		1959		8041	8373			2570	54
55	.189381		.981904		.018096	.018430			.192871	55
56	9667		1848		8152	8487			3173	56
57	9952		1793		8207	8544			3475	57
58	.190238		1738		8262	8602			3777	58
59	0523		1683		8317	8659			4078	59
60	.190809		.981627		.018373	.018717			.194380	60

TABLE XX.—NATURAL SINES, COSINES,

11°

'	SINE	CORR. FOR SEC. +	COSINE	CORR. FOR SEC. − +	VERSINE	EXSEC	CORR. FOR SEC. +	CORR. FOR SEC. +	TANGENT	'
0	.190809		.981627		.018373	.018717			.194380	0
1	1094		1572		8428	8774			4682	1
2	1380		1516		8484	8832			4984	2
3	1666		1460		8540	8890			5286	3
4	1951		1404		8596	8948			5588	4
5	.192236		.981349		.018651	.019006			.195890	5
6	2522		1293		8707	9064			6192	6
7	2807		1237		8763	9122			6494	7
8	3093		1180		8820	9180			6796	8
9	3378		1124		8876	9239			7099	9
10	.193664		.981068		.018932	.019297			.197401	10
11	3949	" Corr.	1012	" Corr.	8988	9356	" Corr.	" Corr.	7703	11
12	4234	10 48	0955	10 9	9045	9415	10 10	10 50	8005	12
13	4520	15 71	0899	15 14	9101	9473	15 15	15 76	8308	13
14	4805	20 95	0842	20 19	9158	9532	20 20	20 101	8610	14
15	.195090	30 143	.980785	30 28	.019215	.019591	30 29	30 151	.198912	15
16	5376	40 190	0728	40 38	9272	9650	40 39	40 202	9215	16
17	5661	45 214	0672	45 43	9328	9709	45 44	45 227	9517	17
18	5946	50 238	0615	50 47	9385	9769	50 49	50 252	9820	18
19	6231		0558		9442	9828			.200122	19
20	.196517		.980500		.019500	.019887			.200425	20
21	6802		0443		9557	9947			0727	21
22	7087		0386		9614	.020006			1030	22
23	7372		0329		9671	0066			1333	23
24	7657		0271		9729	0126			1635	24
25	.197942		.980214		.019786	.020186			.201938	25
26	8228		0156		9844	0246			2241	26
27	8513		0098		9902	0306			2544	27
28	8798		0040		9960	0366			2846	28
29	9083		.979983		.020017	0426			3149	29
30	.199368		.979925		.020075	.020487			.203452	30
31	9653		9867		0133	0547			3755	31
32	9938		9809		0191	0608			4058	32
33	.200223		9750		0250	0668			4361	33
34	0508		9692		0308	0729			4664	34
35	.200793		.979634		.020366	.020790			.204967	35
36	1078		9575		0425	0851			5270	36
37	1363		9517		0483	0912			5574	37
38	1648		9458		0542	0973			5877	38
39	1933		9399		0601	1034			6180	39
40	.202218		.979341		.020659	.021095			.206483	40
41	2502	" Corr.	9282	" Corr.	0718	1157	" Corr.	" Corr.	6787	41
42	2787	10 47	9223	10 10	0777	1218	10 10	10 51	7090	42
43	3072	15 71	9164	15 15	0836	1280	15 15	15 76	7393	43
44	3357	20 95	9105	20 20	0895	1341	20 21	20 101	7697	44
45	.203642	30 142	.979046	30 30	.020954	.021403	30 31	30 152	.208000	45
46	3926	40 190	8986	40 40	1014	1465	40 41	40 202	8304	46
47	4211	45 214	8927	45 44	1073	1527	45 46	45 228	8607	47
48	4496	50 237	8867	50 49	1133	1589	50 51	50 253	8911	48
49	4781		8808		1192	1651			9214	49
50	.205066		.978748		.021252	.021713			.209518	50
51	5350		8689		1311	1776			9822	51
52	5635		8629		1371	1838			.210126	52
53	5920		8569		1431	1900			0429	53
54	6204		8509		1491	1963			0733	54
55	.206489		.978449		.021551	.022026			.211037	55
56	6773		8389		1611	2088			1341	56
57	7058		8329		1671	2151			1645	57
58	7343		8268		1732	2214			1949	58
59	7627		8208		1792	2277			2252	59
60	.207912		.978148		.021852	.022341			.212557	60

VERSINES, EXSECANTS, AND TANGENTS

12°

'	SINE	CORR. FOR SEC.	COSINE	CORR. FOR SEC.	VERSINE	EXSEC	CORR. FOR SEC.	CORR. FOR SEC.	TANGENT	'
		+		− +			+	+		
0	.207912		.978148		.021852	.022341			.212557	0
1	8196		8087		1913	2404			2861	1
2	8481		8026		1974	2467			3165	2
3	8765		7966		2034	2531			3469	3
4	9050		7905		2095	2594			3773	4
5	.209334		.977844		.022156	.022658			.214077	5
6	9619		7783		2217	2722			4381	6
7	9903		7722		2278	2785			4686	7
8	.210187		7661		2339	2849			4990	8
9	0472		7600		2400	2913			5294	9
10	.210756		.977539		.022461	.022977			.215599	10
11	1040	" Corr.	7477	" Corr.	2523	3042	" Corr.	" Corr.	5903	11
12	1325	10 47	7416	10 10	2584	3106	10 11	10 51	6208	12
13	1609	15 71	7354	15 15	2646	3170	15 16	15 76	6512	13
14	1893	20 95	7293	20 21	2707	3235	20 21	20 102	6817	14
15	.212178	30 142	.977231	30 31	.022769	.023299	30 32	30 152	.217121	15
16	2462	40 189	7169	40 41	2831	3364	40 43	40 203	7426	16
17	2746	45 213	7108	45 46	2892	3429	45 49	45 228	7731	17
18	3030	50 237	7046	50 51	2954	3494	50 54	50 254	8035	18
19	3315		6984		3016	3559			8340	19
20	.213599		.976922		.023078	.023624			.218645	20
21	3883		6859		3141	3689			8950	21
22	4167		6797		3203	3754			9254	22
23	4451		6735		3265	3820			9559	23
24	4735		6672		3328	3885			9864	24
25	.215019		.976610		.023390	.023950			.220169	25
26	5304		6547		3453	4016			0474	26
27	5588		6484		3516	4082			0779	27
28	5872		6422		3578	4148			1084	28
29	6156		6359		3641	4214			1390	29
30	.216440		.976296		.023704	.024280			.221695	30
31	6724		6233		3767	4346			2000	31
32	7008		6170		3830	4412			2305	32
33	7292		6107		3893	4478			2610	33
34	7575		6044		3956	4544			2916	34
35	.217859		.975980		.024020	.024611			.223221	35
36	8143		5917		4083	4678			3526	36
37	8427		5853		4147	4744			3832	37
38	8711		5790		4210	4811			4137	38
39	8995		5726		4274	4878			4443	39
40	.219279		.975662		.024338	.024945			.224748	40
41	9562	" Corr.	5598	" Corr.	4402	5012	" Corr.	" Corr.	5054	41
42	9846	10 47	5534	10 11	4466	5079	10 11	10 51	5360	42
43	.220130	15 71	5471	15 16	4529	5146	15 17	15 76	5665	43
44	0414	20 95	5406	20 21	4594	5214	20 22	20 102	5971	44
45	.220697	30 142	.975342	30 32	.024658	.025281	30 34	30 153	.226277	45
46	0981	40 189	5278	40 43	4722	5349	40 45	40 204	6583	46
47	1265	45 213	5214	45 48	4786	5416	45 51	45 229	6888	47
48	1548	50 236	5149	50 53	4851	5484	50 56	50 255	7194	48
49	1832		5085		4915	5552			7500	49
50	.222116		.975020		.024980	.025620			.227806	50
51	2399		4956		5044	5688			8112	51
52	2683		4891		5109	5756			8418	52
53	2967		4826		5174	5824			8724	53
54	3250		4761		5239	5892			9031	54
55	.223534		.974696		.025304	.025961			.229337	55
56	3817		4631		5369	6029			9643	56
57	4101		4566		5434	6098			9949	57
58	4384		4501		5499	6166			.230256	58
59	4668		4436		5564	6235			0562	59
60	.224951		.974370		.025630	.026304			.230868	60

TABLE XX.—NATURAL SINES, COSINES,

13°

'	SINE	CORR. FOR SEC.	COSINE	CORR. FOR SEC.	VERSINE	EXSEC	CORR. FOR SEC.	CORR. FOR SEC.	TANGENT	'
0	.224951	+	.974370	− +	.025630	.026304	+	+	.230868	0
1	5234		4305		5695	6373			1175	1
2	5518		4239		5761	6442			1481	2
3	5801		4173		5827	6511			1788	3
4	6085		4108		5892	6581			2094	4
5	.226368		.974042		.025958	.026650			.232401	5
6	6651		3976		6024	6719			2707	6
7	6935		3910		6090	6789			3014	7
8	7218		3844		6156	6859			3321	8
9	7501		3778		6222	6928			3627	9
10	.227784		.973712		.026288	.026998			.233934	10
11	8068	" Corr.	3645	" Corr.	6355	7068	" Corr.	" Corr.	4241	11
12	8351	10 47	3579	10 11	6421	7138	10 12	10 51	4548	12
13	8634	15 71	3512	15 17	6488	7208	15 18	15 77	4855	13
14	8917	20 94	3446	20 22	6554	7278	20 23	20 102	5162	14
15	.229200	30 142	.973379	30 33	.026621	.027349	30 35	30 153	.235469	15
16	9484	40 189	3312	40 44	6688	7419	40 47	40 205	5776	16
17	9767	45 212	3246	45 50	6754	7490	45 53	45 230	6083	17
18	.230050	50 236	3179	50 56	6821	7560	50 59	50 256	6390	18
19	0333		3112		6888	7631			6697	19
20	.230616		.973045		.026955	.027702			.237004	20
21	0899		2978		7022	7773			7312	21
22	1182		2910		7090	7844			7619	22
23	1465		2843		7157	7915			7926	23
24	1748		2776		7224	7986			8234	24
25	.232031		.972708		.027292	.028057			.238541	25
26	2314		2641		7359	8129			8848	26
27	2597		2573		7427	8200			9156	27
28	2880		2506		7494	8272			9464	28
29	3162		2438		7562	8343			9771	29
30	.233445		.972370		.027630	.028415			.240079	30
31	3728		2302		7698	8487			0386	31
32	4011		2234		7766	8559			0694	32
33	4294		2166		7834	8631			1002	33
34	4577		2098		7902	8703			1310	34
35	.234859		.972029		.027971	.028776			.241618	35
36	5142		1961		8039	8848			1926	36
37	5425		1893		8107	8920			2233	37
38	5708		1824		8176	8993			2541	38
39	5990		1755		8245	9066			2849	39
40	.236273		.971687		.028313	.029138			.243158	40
41	6556	" Corr.	1618	" Corr.	8382	9211	" Corr.	" Corr.	3466	41
42	6838	10 47	1549	10 12	8451	9284	10 12	10 51	3774	42
43	7121	15 71	1480	15 17	8520	9357	15 18	15 77	4082	43
44	7403	20 94	1411	20 23	8589	9430	20 24	20 103	4390	44
45	.237686	30 141	.971342	30 35	.028658	.029503	30 37	30 154	.244698	45
46	7968	40 188	1273	40 46	8727	9577	40 49	40 206	5007	46
47	8251	45 212	1204	45 52	8796	9650	45 55	45 231	5315	47
48	8534	50 235	1134	50 58	8866	9724	50 61	50 257	5624	48
49	8816		1065		8935	9797			5932	49
50	.239098		.970995		.029005	.029871			.246240	50
51	9381		0926		9074	9945			6549	51
52	9663		0856		9144	.030019			6858	52
53	9946		0786		9214	0093			7166	53
54	.240228		0716		9284	0167			7475	54
55	.240510		.970647		.029353	.030241			.247784	55
56	0793		0577		9423	0315			8092	56
57	1075		0506		9494	0390			8401	57
58	1357		0436		9564	0464			8710	58
59	1640		0366		9634	0539			9019	59
60	.241922		.970296		.029704	.030614			.249328	60

VERSINES, EXSECANTS, AND TANGENTS

14°

'	SINE	CORR. FOR SEC.	COSINE	CORR. FOR SEC.	VERSINE	EXSEC	CORR. FOR SEC.	CORR. FOR SEC.	TANGENT	'
0	.241922	+	.970296	− +	.029704	.030614	+	+	.249328	0
1	2204		0225		9775	0688			9637	1
2	2486		0155		9845	0763			9946	2
3	2768		0084		9916	0838			.250255	3
4	3051		0014		9986	0913			0564	4
5	.243333		.969943		.030057	.030989			.250873	5
6	3615		9872		0128	1064			1183	6
7	3897		9801		0199	1139			1492	7
8	4179		9730		0270	1215			1801	8
9	4461		9659		0341	1290			2111	9
10	.244743		.969588		.030412	.031366			.252420	10
11	5025	" Corr.	9517	" Corr.	0483	1442	" Corr.	" Corr.	2729	11
12	5307	10 47	9445	10 12	0555	1518	10 13	10 52	3039	12
13	5589	15 70	9374	15 18	0626	1594	15 19	15 77	3348	13
14	5871	20 94	9302	20 24	0698	1670	20 25	20 103	3658	14
15	.246153	30 141	.969231	30 36	.030769	.031746	30 38	30 155	.253968	15
16	6435	40 188	9159	40 48	0841	1822	40 51	40 206	4277	16
17	6717	45 211	9088	45 54	0912	1898	45 57	45 232	4587	17
18	6999	50 235	9016	50 60	0984	1975	50 64	50 258	4897	18
19	7281		8944		1056	2052			5207	19
20	.247563		.968872		.031128	.032128			.255516	20
21	7844		8800		1200	2205			5826	21
22	8126		8728		1272	2282			6136	22
23	8408		8656		1344	2359			6446	23
24	8690		8583		1417	2436			6756	24
25	.248972		.968511		.031489	.032513			.257066	25
26	9253		8438		1562	2590			7377	26
27	9535		8366		1634	2668			7687	27
28	9817		8293		1707	2745			7997	28
29	.250098		8220		1780	2823			8307	29
30	.250380		.968148		.031852	.032900			.258618	30
31	0662		8075		1925	2978			8928	31
32	0943		8002		1998	3056			9238	32
33	1225		7929		2071	3134			9549	33
34	1506		7856		2144	3212			9859	34
35	.251788		.967782		.032218	.033290			.260170	35
36	2069		7709		2291	3368			0480	36
37	2351		7636		2364	3447			0791	37
38	2632		7562		2438	3525			1102	38
39	2914		7489		2511	3604			1413	39
40	.253195		.967415		.032585	.033682			.261723	40
41	3477	" Corr.	7342	" Corr.	2658	3761	" Corr.	" Corr.	2034	41
42	3758	10 47	7268	10 12	2732	3840	10 13	10 52	2345	42
43	4039	15 70	7194	15 19	2806	3919	15 20	15 78	2656	43
44	4321	20 94	7120	20 25	2880	3998	20 26	20 104	2967	44
45	.254602	30 141	.967046	30 37	.032954	.034077	30 40	30 156	.263278	45
46	4883	40 188	6972	40 49	3028	4156	40 53	40 207	3589	46
47	5164	45 211	6898	45 56	3102	4236	45 59	45 233	3900	47
48	5446	50 234	6823	50 62	3177	4315	50 66	50 259	4211	48
49	5727		6749		3251	4395			4523	49
50	.256008		.966675		.033325	.034474			.264834	50
51	6289		6600		3400	4554			5145	51
52	6570		6526		3474	4634			5457	52
53	6852		6451		3549	4714			5768	53
54	7133		6376		3624	4794			6079	54
55	.257414		.966301		.033699	.034874			.266391	55
56	7695		6226		3774	4954			6702	56
57	7976		6151		3849	5035			7014	57
58	8257		6076		3924	5115			7326	58
59	8538		6001		3999	5196			7637	59
60	.258819		.965926		.034074	.035276			.267949	60

TABLE XX.—NATURAL SINES, COSINES,

15°

′	SINE	CORR. FOR SEC. +	COSINE	CORR. FOR SEC. − +	VERSINE	EXSEC	CORR. FOR SEC. +	CORR. FOR SEC. +	TANGENT	′
0	.258819		.965926		.034074	.035276			.267949	0
1	9100		5850		4150	5357			8261	1
2	9381		5775		4225	5438			8573	2
3	9662		5700		4300	5519			8885	3
4	9943		5624		4376	5600			9197	4
5	.260224		.965548		.034452	.035681			.269509	5
6	0504		5473		4527	5762			9821	6
7	0785		5397		4603	5844			.270133	7
8	1066		5321		4679	5925			0445	8
9	1347		5245		4755	6006			0757	9
10	.261628		.965169		.034831	.036088			.271069	10
11	1908	" Corr.	5093	" Corr.	4907	6170	" Corr.	" Corr.	1382	11
12	2189	10 47	5016	10 13	4984	6252	10 14	10 52	1694	12
13	2470	15 70	4940	15 19	5060	6334	15 21	15 78	2006	13
14	2751	20 94	4864	20 25	5136	6416	20 27	20 104	2319	14
15	.263031	30 140	.964787	30 38	.035213	.036498	30 41	30 156	.272631	15
16	3312	40 187	4711	40 51	5289	6580	40 55	40 208	2944	16
17	3592	45 210	4634	45 57	5366	6662	45 62	45 234	3256	17
18	3873	50 234	4557	50 64	5443	6745	50 68	50 260	3569	18
19	4154		4481		5519	6828			3882	19
20	.264434		.964404		.035596	.036910			.274194	20
21	4715		4327		5673	6993			4507	21
22	4995		4250		5750	7076			4820	22
23	5276		4173		5827	7159			5133	23
24	5556		4095		5905	7242			5446	24
25	.265837		.964018		.035982	.037325			.275759	25
26	6117		3941		6059	7408			6072	26
27	6397		3863		6137	7492			6385	27
28	6678		3786		6214	7575			6698	28
29	6958		3708		6292	7658			7011	29
30	.267238		.963630		.036370	.037742			.277324	30
31	7519		3553		6447	7826			7638	31
32	7799		3475		6525	7910			7951	32
33	8079		3397		6603	7994			8265	33
34	8359		3319		6681	8078			8578	34
35	.268640		.963241		.036759	.038162			.278892	35
36	8920		3163		6837	8246			9205	36
37	9200		3084		6916	8331			9519	37
38	9480		3006		6994	8415			9832	38
39	9760		2928		7072	8500			.280146	39
40	.270040		.962849		.037151	.038584			.280460	40
41	0320	" Corr.	2770	" Corr.	7230	8669	" Corr.	" Corr.	0774	41
42	0600	10 47	2692	10 13	7308	8754	10 14	10 52	1087	42
43	0880	15 70	2613	15 20	7387	8839	15 21	15 79	1401	43
44	1160	20 93	2534	20 26	7466	8924	20 28	20 105	1715	44
45	.271440	30 140	.962455	30 40	.037545	.039009	30 43	30 157	.282029	45
46	1720	40 187	2376	40 53	7624	9095	40 57	40 209	2343	46
47	2000	45 210	2297	45 59	7703	9180	45 64	45 235	2657	47
48	2280	50 233	2218	50 66	7782	9266	50 71	50 262	2972	48
49	2560		2139		7861	9351			3286	49
50	.272840		.962059		.037941	.039437			.283600	50
51	3120		1980		8020	9523			3914	51
52	3400		1900		8100	9608			4229	52
53	3679		1821		8179	9694			4543	53
54	3959		1741		8259	9781			4858	54
55	.274239		.961662		.038338	.039867			.285172	55
56	4519		1582		8418	9953			5487	56
57	4798		1502		8498	.040040			5801	57
58	5078		1422		8578	0126			6116	58
59	5358		1342		8658	0213			6431	59
60	.275637		.961262		.038738	.040299			.286745	60

VERSINES, EXSECANTS, AND TANGENTS

16°

'	SINE	CORR. FOR SEC. +	COSINE	CORR. FOR SEC. − +	VERSINE	EXSEC	CORR. FOR SEC. +	CORR. FOR SEC. +	TANGENT	'
0	.275637		.961262		.038738	.040299			.286745	0
1	5917		1182		8818	0386			7060	1
2	6197		1101		8899	0473			7375	2
3	6476		1021		8979	0560			7690	3
4	6756		0940		9060	0647			8005	4
5	.277035		.960860		.039140	.040735			.288320	5
6	7315		0779		9221	0822			8635	6
7	7594		0698		9302	0909			8950	7
8	7874		0618		9382	0997			9266	8
9	8153		0537		9463	1084			9581	9
10	.278432		.960456		.039544	.041172			.289896	10
11	8712	" Corr.	0375	" Corr.	9625	1260	" Corr.	" Corr.	.290211	11
12	8991	10 47	0294	10 14	9706	1348	10 15	10 53	0527	12
13	9270	15 70	0212	15 20	9788	1436	15 22	15 79	0842	13
14	9550	20 93	0131	20 27	9869	1524	20 29	20 105	1158	14
15	.279829	30 140	.960050	30 41	.039950	.041613	30 44	30 158	.291473	15
16	.280108	40 186	.959968	40 54	.040032	1701	40 59	40 210	1789	16
17	0388	45 209	9887	45 61	0113	1789	45 66	45 237	2105	17
18	0667	50 233	9805	50 68	0195	1878	50 74	50 263	2420	18
19	0946		9724		0276	1967			2736	19
20	.281225		.959642		.040358	.042055			.293052	20
21	1504		9560		0440	2144			3368	21
22	1783		9478		0522	2233			3684	22
23	2062		9396		0604	2322			4000	23
24	2342		9314		0686	2412			4316	24
25	.282620		.959232		.040768	.042501			.294632	25
26	2900		9150		0850	2590			4948	26
27	3178		9067		0933	2680			5264	27
28	3458		8985		1015	2769			5581	28
29	3736		8902		1098	2859			5897	29
30	.284015		.958820		.041180	.042949			.296214	30
31	4294		8737		1263	3039			6530	31
32	4573		8654		1346	3129			6846	32
33	4852		8572		1428	3219			7163	33
34	5131		8489		1511	3309			7480	34
35	.285410		.958406		.041594	.043400			.297796	35
36	5688		8323		1677	3490			8113	36
37	5967		8239		1761	3580			8430	37
38	6246		8156		1844	3671			8746	38
39	6525		8073		1927	3762			9063	39
40	.286803		.957990		.042010	.043853			.299380	40
41	7082	" Corr.	7906	" Corr.	2094	3944	" Corr.	" Corr.	9697	41
42	7360	10 46	7822	10 14	2178	4035	10 15	10 53	.300014	42
43	7639	15 70	7739	15 21	2261	4126	15 23	15 79	0332	43
44	7918	20 93	7655	20 28	2345	4217	20 30	20 106	0649	44
45	.288196	30 139	.957571	30 42	.042429	.044309	30 46	30 159	.300966	45
46	8475	40 186	7488	40 56	2512	4400	40 61	40 212	1283	46
47	8753	45 209	7404	45 63	2596	4492	45 69	45 238	1600	47
48	9032	50 232	7320	50 70	2680	4583	50 76	50 264	1918	48
49	9310		7235		2765	4675			2235	49
50	.289589		.957151		.042849	.044767			.302553	50
51	9867		7067		2933	4859			2870	51
52	.290146		6982		3018	4951			3188	52
53	0424		6898		3102	5043			3506	53
54	0702		6814		3186	5136			3823	54
55	.290980		.956729		.043271	.045228			.304141	55
56	1259		6644		3356	5321			4459	56
57	1537		6560		3440	5413			4777	57
58	1815		6475		3525	5506			5095	58
59	2094		6390		3610	5599			5413	59
60	.292372		.956305		.043695	.045692			.305731	60

TABLE XX.—NATURAL SINES, COSINES, 17°

′	SINE	CORR. FOR SEC. +	COSINE	CORR. FOR SEC. − +	VERSINE	EXSEC	CORR. FOR SEC. +	CORR. FOR SEC. +	TANGENT	′
0	.292372		.956305		.043695	.045692			.305731	0
1	2650		6220		3780	5785			6049	1
2	2928		6134		3866	5878			6367	2
3	3206		6049		3951	5971			6685	3
4	3484		5964		4036	6065			7003	4
5	.293762		.955878		.044122	.046158			.307322	5
6	4040		5793		4207	6252			7640	6
7	4318		5707		4293	6345			7959	7
8	4596		5622		4378	6439			8277	8
9	4874		5536		4464	6533			8596	9
10	.295152		.955450		.044550	.046627			.308914	10
11	5430	" Corr.	5364	" Corr.	4636	6721	" Corr.	" Corr.	9233	11
12	5708	10 46	5278	10 14	4722	6815	10 16	10 53	9552	12
13	5986	15 69	5192	15 22	4808	6910	15 24	15 80	9870	13
14	6264	20 93	5106	20 29	4894	7004	20 32	20 106	.310189	14
15	.296542	30 139	.955020	30 43	.044980	.047099	30 47	30 159	.310508	15
16	6819	40 185	4934	40 58	5066	7193	40 63	40 213	0827	16
17	7097	45 208	4847	45 65	5153	7288	45 71	45 239	1146	17
18	7375	50 231	4761	50 72	5239	7383	50 79	50 266	1465	18
19	7653		4674		5326	7478			1784	19
20	.297930		.954588		.045412	.047573			.312104	20
21	8208		4501		5499	7668			2423	21
22	8486		4414		5586	7763			2742	22
23	8763		4327		5673	7859			3062	23
24	9041		4240		5760	7954			3381	24
25	.299318		.954153		.045847	.048050			.313700	25
26	9596		4066		5934	8145			4020	26
27	9873		3979		6021	8241			4340	27
28	.300151		3892		6108	8337			4659	28
29	0428		3804		6196	8433			4979	29
30	.300706		.953717		.046283	.048529			.315299	30
31	0983		3629		6371	8625			5619	31
32	1261		3542		6458	8722			5938	32
33	1538		3454		6546	8818			6258	33
34	1815		3366		6634	8915			6578	34
35	.302093		.953279		.046721	.049011			.316899	35
36	2370		3191		6809	9108			7219	36
37	2647		3103		6897	9205			7539	37
38	2924		3015		6985	9302			7859	38
39	3202		2926		7074	9399			8179	39
40	.303479		.952838		.047162	.049496			.318500	40
41	3756	" Corr.	2750	" Corr.	7250	9593	" Corr.	" Corr.	8820	41
42	4033	10 46	2662	10 15	7338	9691	10 16	10 53	9141	42
43	4310	15 69	2573	15 22	7427	9788	15 24	15 80	9461	43
44	4587	20 92	2484	20 30	7516	9886	20 33	20 107	9782	44
45	.304864	30 138	.952396	30 44	.047604	.049984	30 49	30 160	.320102	45
46	5141	40 185	2307	40 59	7693	.050082	40 65	40 214	0423	46
47	5418	45 208	2218	45 67	7782	0179	45 73	45 241	0744	47
48	5695	50 231	2129	50 74	7871	0277	50 82	50 267	1065	48
49	5972		2040		7960	0376			1386	49
50	.306249		.951951		.048049	.050474			.321707	50
51	6526		1862		8138	0572			2028	51
52	6803		1773		8227	0671			2349	52
53	7080		1684		8316	0769			2670	53
54	7357		1594		8406	0868			2991	54
55	.307633		.951505		.048495	.050967			.323312	55
56	7910		1415		8585	1066			3634	56
57	8187		1326		8674	1165			3955	57
58	8464		1236		8764	1264			4277	58
59	8740		1146		8854	1363			4598	59
60	.309017		.951056		.048944	.051462			.324920	60

VERSINES, EXSECANTS, AND TANGENTS
18°

'	SINE	CORR. FOR SEC.	COSINE	CORR. FOR SEC.	VERSINE	EXSEC	CORR. FOR SEC.	CORR. FOR SEC.	TANGENT	'
		+		− +			+	+		
0	.309017		.951056		.048944	.051462			.324920	0
1	9294		0967		9033	1562			5241	1
2	9570		0877		9123	1661			5563	2
3	9847		0786		9214	1761			5885	3
4	.310123		0696		9304	1861			6207	4
5	.310400		.950606		.049394	.051960			.326528	5
6	0676		0516		9484	2060			6850	6
7	0953		0425		9575	2160			7172	7
8	1229		0335		9665	2261			7494	8
9	1506		0244		9756	2361			7816	9
10	.311782		.950154		.049846	.052461			.328139	10
11	2059	" Corr.	0063	" Corr.	9937	2562	" Corr.	" Corr.	8461	11
12	2335	10 46	.949972	10 15	.050028	2662	10 17	10 54	8783	12
13	2611	15 69	9881	15 23	0119	2763	15 25	15 81	9106	13
14	2888	20 92	9790	20 30	0210	2864	20 34	20 107	9428	14
15	.313164	30 138	.949699	30 46	.050301	.052965	30 51	30 161	.329750	15
16	3440	40 184	9608	40 61	0392	3066	40 67	40 215	.330073	16
17	3716	45 207	9517	45 68	0483	3167	45 76	45 242	0396	17
18	3992	50 230	9426	50 76	0574	3269	50 84	50 269	0718	18
19	4269		9334		0666	3370			1041	19
20	.314545		.949243		.050757	.053471			.331364	20
21	4821		9151		0849	3573			1687	21
22	5097		9060		0940	3675			2010	22
23	5373		8968		1032	3776			2333	23
24	5649		8876		1124	3878			2656	24
25	.315925		.948784		.051216	.053980			.332979	25
26	6201		8692		1308	4083			3302	26
27	6477		8600		1400	4185			3625	27
28	6753		8508		1492	4287			3948	28
29	7029		8416		1584	4390			4272	29
30	.317305		.948324		.051676	.054492			.334595	30
31	7580		8231		1769	4595			4919	31
32	7856		8139		1861	4698			5242	32
33	8132		8046		1954	4801			5566	33
34	8408		7954		2046	4904			5890	34
35	.318684		.947861		.052139	.055007			.336213	35
36	8959		7768		2232	5110			6537	36
37	9235		7676		2324	5213			6861	37
38	9511		7583		2417	5317			7185	38
39	9786		7490		2510	5420			7509	39
40	.320062		.947397		.052603	.055524			.337833	40
41	0337	" Corr.	7304	" Corr.	2696	5628	" Corr.	" Corr.	8157	41
42	0613	10 46	7210	10 16	2790	5732	10 17	10 54	8481	42
43	0888	15 69	7117	15 23	2883	5836	15 26	15 81	8806	43
44	1164	20 92	7024	20 31	2976	5940	20 35	20 108	9130	44
45	.321440	30 138	.946930	30 47	.053070	.056044	30 52	30 162	.339454	45
46	1715	40 184	6837	40 62	3163	6148	40 70	40 216	9779	46
47	1990	45 207	6743	45 70	3257	6253	45 78	45 243	.340103	47
48	2266	50 229	6649	50 78	3351	6358	50 87	50 270	0428	48
49	2541		6556		3444	6462			0752	49
50	.322816		.946462		.053538	.056567			.341077	50
51	3092		6368		3632	6672			1402	51
52	3367		6274		3726	6777			1727	52
53	3642		6180		3820	6882			2052	53
54	3917		6085		3915	6987			2376	54
55	.324193		.945991		.054009	.057092			.342702	55
56	4468		5897		4103	7198			3027	56
57	4743		5802		4198	7303			3352	57
58	5018		5708		4292	7409			3677	58
59	5293		5613		4387	7515			4002	59
60	.325568		.945519		.054481	.057621			.344328	60

TABLE XX.—NATURAL SINES, COSINES,

19°

'	SINE	CORR. FOR SEC.	COSINE	CORR. FOR SEC.	VERSINE	EXSEC	CORR. FOR SEC.	CORR. FOR SEC.	TANGENT	'
		+		− +			+	+		
0	.325568		.945519		.054481	.057621			.344328	0
1	5843		5424		4576	7727			4653	1
2	6118		5329		4671	7833			4978	2
3	6393		5234		4766	7939			5304	3
4	6668		5139		4861	8045			5630	4
5	.326943		.945044		.054956	.058152			.345955	5
6	7218		4949		5051	8258			6281	6
7	7493		4654		5146	8365			6607	7
8	7768		4758		5242	8472			6933	8
9	8042		4663		5337	8579			7259	9
10	.328317		.944568		.055432	.058686			.347585	10
11	8592	" Corr.	4472	" Corr.	5528	8793	" Corr.	" Corr.	7911	11
12	8867	10 46	4376	10 16	5624	8900	10 18	10 54	8237	12
13	9141	15 69	4281	15 24	5719	9007	15 27	15 82	8563	13
14	9416	20 92	4185	20 32	5815	9115	20 36	20 109	8889	14
15	.329691	30 137	.944089	30 48	.055911	.059222	30 54	30 163	.349216	15
16	9965	40 183	3993	40 64	6007	9330	40 72	40 218	9542	16
17	.330240	45 206	3897	45 72	6103	9438	45 81	45 245	9868	17
18	0514	50 229	3801	50 80	6199	9545	50 90	50 272	.350195	18
19	0789		3705		6295	9653			0522	19
20	.331063		.943608		.056392	.059762			.350848	20
21	1338		3512		6488	9870			1175	21
22	1612		3416		6584	9978			1502	22
23	1887		3319		6681	.060086			1829	23
24	2161		3223		6777	0195			2156	24
25	.332436		.943126		.056874	.060304			.352483	25
26	2710		3029		6971	0412			2810	26
27	2984		2932		7068	0521			3137	27
28	3258		2836		7164	0630			3464	28
29	3533		2739		7261	0740			3791	29
30	.333807		.942642		.057358	.060849			.354119	30
31	4081		2544		7456	0958			4446	31
32	4355		2447		7553	1068			4773	32
33	4629		2350		7650	1177			5101	33
34	4903		2252		7748	1287			5429	34
35	.335178		.942155		.057845	.061397			.355756	35
36	5452		2058		7942	1506			6084	36
37	5726		1960		8040	1616			6412	37
38	6000		1862		8138	1726			6740	38
39	6274		1764		8236	1837			7068	39
40	.336548		.941666		.058334	.061947			.357396	40
41	6821	" Corr.	1569	" Corr.	8431	2058	" Corr.	" Corr.	7724	41
42	7095	10 46	1470	10 16	8530	2168	10 19	10 55	8052	42
43	7369	15 68	1372	15 25	8628	2279	15 28	15 82	8380	43
44	7643	20 91	1274	20 33	8726	2390	20 37	20 109	8708	44
45	.337917	30 137	.941176	30 49	.058824	.062500	30 56	30 164	.359037	45
46	8190	40 183	1078	40 66	8922	2612	40 74	40 219	9365	46
47	8464	45 205	0979	45 74	9021	2723	45 83	45 246	9694	47
48	8738	50 228	0881	50 82	9119	2834	50 93	50 274	.360022	48
49	9012		0782		9218	2945			0351	49
50	.339285		.940684		.059316	.063057			.360680	50
51	9559		0585		9415	3168			1008	51
52	9832		0486		9514	3280			1337	52
53	.340106		0387		9613	3392			1666	53
54	0380		0288		9712	3504			1995	54
55	.340653		.940189		.059811	.063616			.362324	55
56	0926		0090		9910	3728			2653	56
57	1200		.939991		.060009	3840			2982	57
58	1473		9891		0109	3953			3312	58
59	1747		9792		0208	4065			3641	59
60	.342020		.939693		.060307	.064178			.363970	60

VERSINES, EXSECANTS, AND TANGENTS

20°

'	SINE	CORR. FOR SEC.	COSINE	CORR. FOR SEC.	VERSINE	EXSEC	CORR. FOR SEC.	CORR. FOR SEC.	TANGENT	'
0	.342020	+	.939693	− +	.060307	.064178	+	+	.363970	0
1	2294		9593		0407	4290			4300	1
2	2567		9494		0506	4403			4629	2
3	2840		9394		0606	4516			4959	3
4	3113		9294		0706	4629			5288	4
5	.343386		.939194		.060806	.064742			.365618	5
6	3660		9094		0906	4856			5948	6
7	3933		8994		1006	4969			6278	7
8	4206		8894		1106	5083			6608	8
9	4479		8794		1206	5196			6938	9
10	.344752		.938694		.061306	.065310			.367268	10
11	5025	" Corr.	8593	" Corr.	1407	5424	" Corr.	" Corr.	7598	11
12	5298	10 45	8493	10 17	1507	5538	10 19	10 55	7928	12
13	5571	15 68	8392	15 25	1608	5652	15 29	15 83	8259	13
14	5844	20 91	8292	20 34	1708	5766	20 38	20 110	8589	14
15	.346117	30 136	.938191	30 50	.061809	.065881	30 57	30 165	.368920	15
16	6390	40 182	8091	40 67	1909	5995	40 76	40 220	9250	16
17	6663	45 205	7990	45 76	2010	6110	45 86	45 248	9581	17
18	6936	50 227	7889	50 84	2111	6224	50 95	50 275	9911	18
19	7208		7788		2212	6339			.370242	19
20	.347481		.937687		.062313	.066454			.370573	20
21	7754		7586		2414	6569			0904	21
22	8027		7485		2515	6684			1235	22
23	8299		7383		2617	6799			1566	23
24	8572		7282		2718	6915			1897	24
25	.348845		.937181		.062819	.067030			.372228	25
26	9117		7079		2921	7146			2559	26
27	9390		6977		3023	7262			2890	27
28	9662		6876		3124	7377			3222	28
29	9935		6774		3226	7493			3553	29
30	.350207		.936672		.063328	.067609			.373885	30
31	0480		6570		3430	7726			4216	31
32	0752		6468		3532	7842			4548	32
33	1025		6366		3634	7958			4880	33
34	1297		6264		3736	8075			5212	34
35	.351569		.936162		.063838	.068191			.375543	35
36	1842		6060		3940	8308			5875	36
37	2114		5957		4043	8425			6207	37
38	2386		5855		4145	8542			6539	38
39	2658		5752		4248	8659			6872	39
40	.352931		.935650		.064350	.068776			.377204	40
41	3203	" Corr.	5547	" Corr.	4453	8894	" Corr.	" Corr.	7536	41
42	3475	10 45	5444	10 17	4556	9011	10 20	10 55	7868	42
43	3747	15 68	5341	15 26	4659	9129	15 29	15 83	8201	43
44	4019	20 91	5238	20 34	4762	9246	20 39	20 111	8534	44
45	.354291	30 136	.935135	30 52	.064865	.069364	30 59	30 166	.378866	45
46	4563	40 181	5032	40 69	4968	9482	40 79	40 222	9199	46
47	4835	45 204	4929	45 77	5071	9600	45 88	45 250	9532	47
48	5107	50 227	4826	50 86	5174	9718	50 98	50 277	9864	48
49	5379		4722		5278	9836			.380197	49
50	.355651		.934619		.065381	.069955			.380530	50
51	5923		4515		5485	.070073			0863	51
52	6194		4412		5588	0192			1196	52
53	6466		4308		5692	0311			1530	53
54	6738		4204		5796	0430			1863	54
55	.357010		.934101		.065899	.070548			.382196	55
56	7281		3997		6003	0668			2530	56
57	7553		3893		6107	0787			2863	57
58	7825		3789		6211	0906			3197	58
59	8096		3685		6315	1025			3530	59
60	.358368		.933580		.066420	.071145			.383864	60

TABLE XX.—NATURAL SINES, COSINES,
21°

'	SINE	CORR. FOR SEC.	COSINE	CORR. FOR SEC.	VERSINE	EXSEC	CORR. FOR SEC.	CORR. FOR SEC.	TANGENT	'
		+		− +			+	+		
0	.358368		.933580		.066420	.071145			.383864	0
1	8640		3476		6524	1265			4198	1
2	8911		3372		6628	1384			4532	2
3	9182		3267		6733	1504			4866	3
4	9454		3163		6837	1624			5200	4
5	.359725		.933058		.066942	.071744			.385534	5
6	9997		2954		7046	1865	" Corr.	" Corr.	5868	6
7	.360268		2849		7151	1985	10 20	10 56	6202	7
8	0540		2744		7256	2106	15 30	15 84	6536	8
9	0811		2639		7361	2226	20 40	20 112	6871	9
10	.361082		.932534		.067466	.072347	30 60	30 167	.387205	10
11	1353	" Corr.	2429	" Corr.	7571	2468	40 81	40 223	7540	11
12	1625	10 45	2324	10 18	7676	2589	45 91	45 251	7874	12
13	1896	15 68	2219	15 26	7781	2710	50 101	50 279	8209	13
14	2167	20 90	2113	20 35	7887	2831			8544	14
15	.362438	30 136	.932008	30 53	.067992	.072952			.388879	15
16	2709	40 181	1902	40 70	8098	3074			9214	16
17	2980	45 203	1797	45 79	8203	3195			9549	17
18	3251	50 226	1691	50 88	8309	3317			9884	18
19	3522		1586		8414	3439			.390219	19
20	.363793		.931480		.068520	.073561			.390554	20
21	4064		1374		8626	3683			0889	21
22	4335		1268		8732	3805			1225	22
23	4606		1162		8838	3927			1560	23
24	4877		1056		8944	4050			1896	24
25	.365148		.930950		.069050	.074172			.392231	25
26	5418		0843		9157	4295	" Corr.	" Corr.	2567	26
27	5689		0737		9263	4417	10 21	10 56	2903	27
28	5960		0631		9369	4540	15 31	15 84	3239	28
29	6231		0524		9476	4663	20 41	20 112	3574	29
30	.366501		.930418		.069582	.074786	30 62	30 168	.393910	30
31	6772		0311		9689	4910	40 82	40 224	4246	31
32	7042		0204		9796	5033	45 92	45 252	4583	32
33	7313		0097		9903	5156	50 103	50 280	4919	33
34	7584		.929990		.070010	5280			5255	34
35	.367854		.929884		.070116	.075404			.395592	35
36	8125		9776		0224	5527			5928	36
37	8395		9669		0331	5651			6264	37
38	8665		9562		0438	5775			6601	38
39	8936		9455		0545	5900			6938	39
40	.369206		.929348		.070652	.076024			.397275	40
41	9476	" Corr.	9240	" Corr.	0760	6148			7611	41
42	9747	10 45	9133	10 18	0867	6273			7948	42
43	.370017	15 68	9025	15 27	0975	6397			8285	43
44	0287	20 90	8917	20 36	1083	6522			8622	44
45	.370557	30 135	.928810	30 54	.071190	.076647			.398960	45
46	0828	40 180	8702	40 72	1298	6772	" Corr.	" Corr.	9297	46
47	1098	45 203	8594	45 81	1406	6897	10 21	10 56	9634	47
48	1368	50 225	8486	50 90	1514	7022	15 31	15 84	9972	48
49	1638		8378		1622	7148	20 42	20 113	.400309	49
50	.371908		.928270		.071730	.077273	30 63	30 169	.400646	50
51	2178		8161		1839	7399	40 84	40 225	0984	51
52	2448		8053		1947	7525	45 94	45 253	1322	52
53	2718		7945		2055	7650	50 105	50 281	1660	53
54	2988		7836		2164	7776			1997	54
55	.373258		.927728		.072272	.077902			.402335	55
56	3528		7619		2381	8029			2673	56
57	3797		7510		2490	8155			3012	57
58	4067		7402		2598	8282			3350	58
59	4337		7293		2707	8408			3688	59
60	.374607		.927184		.072816	.078535			.404026	60

VERSINES, EXSECANTS, AND TANGENTS
22°

'	SINE	CORR. FOR SEC. +	COSINE	CORR. FOR SEC. − +	VERSINE	EXSEC	CORR. FOR SEC. +	CORR. FOR SEC. +	TANGENT	'
0	.374607		.927184		.072816	.078535			.404026	0
1	4876		7075		2925	8662			4365	1
2	5146		6966		3034	8788			4703	2
3	5416		6857		3143	8916			5042	3
4	5685		6747		3253	9043			5380	4
5	.375955		.926638		.073362	.079170			.405719	5
6	6224		6529		3471	9298	" Corr.	" Corr.	6058	6
7	6494		6419		3581	9425	10 21	10 57	6397	7
8	6763		6310		3690	9553	15 32	15 85	6736	8
9	7033		6200		3800	9680	20 43	20 113	7075	9
10	.377302		.926090		.073910	.079808	30 64	30 170	.407414	10
11	7571	" Corr.	5980	" Corr.	4020	9936	40 85	40 226	7753	11
12	7841	10 45	5871	10 18	4129	.080065	45 96	45 254	8092	12
13	8110	15 67	5761	15 28	4239	0193	50 107	50 283	8432	13
14	8379	20 90	5651	20 37	4349	0321			8771	14
15	.378649	30 135	.925540	30 55	.074460	.080450			.409111	15
16	8918	40 180	5430	40 73	4570	0578			9450	16
17	9187	45 202	5320	45 83	4680	0707			9790	17
18	9456	50 224	5210	50 92	4790	0836			.410130	18
19	9725		5099		4901	0965			0470	19
20	.379994		.924989		.075011	.081094			.410810	20
21	.380263		4878		5122	1223			1150	21
22	0532		4768		5232	1353			1490	22
23	0801		4657		5343	1482			1830	23
24	1070		4546		5454	1612			2170	24
25	.381339		.924435		.075565	.081742			.412511	25
26	1608		4324		5676	1872	" Corr.	" Corr.	2851	26
27	1877		4213		5787	2002	10 22	10 57	3192	27
28	2146		4102		5898	2132	15 33	15 85	3532	28
29	2415		3991		6009	2262	20 43	20 114	3873	29
30	.382683		.923880		.076120	.082392	30 65	30 170	.414214	30
31	2952		3768		6232	2523	40 87	40 227	4554	31
32	3221		3657		6343	2653	45 98	45 256	4895	32
33	3490		3545		6455	2784	50 109	50 284	5236	33
34	3758		3434		6566	2915			5577	34
35	.384027		.923322		.076678	.083046			.415919	35
36	4295		3210		6790	3177			6260	36
37	4564		3098		6902	3308			6601	37
38	4832		2986		7014	3440			6943	38
39	5101		2874		7126	3571			7284	39
40	.385369		.922762		.077238	.083702			.417626	40
41	5638	" Corr.	2650	" Corr.	7350	3834			7967	41
42	5906	10 45	2538	10 19	7462	3966			8309	42
43	6174	15 67	2426	15 28	7574	4098			8651	43
44	6443	20 89	2313	20 37	7687	4230			8993	44
45	.386711	30 134	.922201	30 56	.077799	.084362			.419335	45
46	6979	40 179	2088	40 75	7912	4495	" Corr.	" Corr.	9677	46
47	7247	45 201	1976	45 84	8024	4627	10 22	10 57	.420019	47
48	7516	50 224	1863	50 94	8137	4760	15 33	15 86	0361	48
49	7784		1750		8250	4892	20 44	20 114	0704	49
50	.388052		.921638		.078362	.085025	30 66	30 171	.421046	50
51	8320		1525		8475	5158	40 89	40 228	1388	51
52	8588		1412		8588	5291	45 100	45 257	1731	52
53	8856		1299		8701	5424	50 111	50 285	2074	53
54	9124		1185		8815	5558			2416	54
55	.389392		.921072		.078928	.085691			.422759	55
56	9660		0959		9041	5825			3102	56
57	9928		0846		9154	5958			3445	57
58	.390196		0732		9268	6092			3788	58
59	0463		0618		9382	6226			4132	59
60	.390731		.920505		.079495	.086360			.424475	60

TABLE XX.—NATURAL SINES, COSINES,

23°

'	SINE	CORR. FOR SEC. +	COSINE	CORR. FOR SEC. − +	VERSINE	EXSEC	CORR. FOR SEC. +	CORR. FOR SEC. +	TANGENT	'
0	.390731		.920505		.079495	.086360			.424475	0
1	0999		0391		9609	6495			4818	1
2	1267		0277		9723	6629			5162	2
3	1534		0164		9836	6763			5505	3
4	1802		0050		9950	6898			5849	4
5	.392070		.919936		.080064	.087033			.426192	5
6	2337		9822		0178	7168	" Corr.	" Corr.	6536	6
7	2605		9707		0293	7302	10 23	10 57	6880	7
8	2872		9593		0407	7438	15 34	15 86	7224	8
9	3140		9479		0521	7573	20 45	20 115	7568	9
10	.393407		.919364		.080636	.087708	30 68	30 172	.427912	10
11	3674	" Corr.	9250	" Corr.	0750	7844	40 90	40 230	8256	11
12	3942	10 45	9135	10 19	0865	7979	45 102	45 258	8600	12
13	4209	15 67	9021	15 29	0979	8115	50 113	50 287	8945	13
14	4477	20 89	8906	20 38	1094	8251			9289	14
15	.394744	30 134	.918791	30 57	.081209	.088387			.429634	15
16	5011	40 178	8676	40 77	1324	8523			9978	16
17	5278	45 200	8561	45 86	1439	8659			.430323	17
18	5546	50 223	8446	50 96	1554	8795			0668	18
19	5813		8331		1669	8932			1013	19
20	.396080		.918216		.081784	.089068			.431358	20
21	6347		8101		1899	9205			1703	21
22	6614		7986		2014	9342			2048	22
23	6881		7870		2130	9479			2393	23
24	7148		7755		2245	9616			2739	24
25	.397415		.917639		.082361	.089753			.433084	25
26	7682		7523		2477	9890	" Corr.	" Corr.	3430	26
27	7949		7408		2592	.090028	10 23	10 58	3775	27
28	8216		7292		2708	0166	15 34	15 86	4121	28
29	8482		7176		2824	0303	20 46	20 115	4466	29
30	.398749		.917060		.082940	.090441	30 69	30 173	.434812	30
31	9016		6944		3056	0579	40 92	40 231	5158	31
32	9282		6828		3172	0717	45 104	45 259	5504	32
33	9549		6712		3288	0855	50 115	50 288	5850	33
34	9816		6596		3404	0994			6197	34
35	.400082		.916479		.083521	.091132			.436543	35
36	0349		6363		3637	1271			6889	36
37	0616		6246		3754	1410			7236	37
38	0882		6130		3870	1548			7582	38
39	1149		6013		3987	1688			7929	39
40	.401415		.915896		.084104	.091827			.438276	40
41	1681	" Corr.	5780	" Corr.	4220	1966			8622	41
42	1948	10 44	5663	10 20	4337	2105			8969	42
43	2214	15 67	5546	15 29	4454	2245			9316	43
44	2480	20 89	5429	20 39	4571	2384			9663	44
45	.402747	30 133	.915312	30 59	.084688	.092524			.440010	45
46	3013	40 178	5194	40 78	4806	2664	" Corr.	" Corr.	0358	46
47	3279	45 200	5077	45 88	4923	2804	10 23	10 56	0705	47
48	3545	50 222	4960	50 98	5040	2944	15 35	15 87	1053	48
49	3811		4842		5158	3085	20 47	20 116	1400	49
50	.404078		.914725		.085275	.093225	30 70	30 174	.441748	50
51	4344		4607		5393	3366	40 94	40 232	2095	51
52	4610		4490		5510	3506	45 105	45 261	2443	52
53	4876		4372		5628	3647	50 117	50 290	2791	53
54	5142		4254		5746	3788			3139	54
55	.405408		.914136		.085864	.093929			.443487	55
56	5673		4018		5982	4070			3835	56
57	5939		3900		6100	4212			4183	57
58	6205		3782		6218	4353			4532	58
59	6471		3664		6336	4495			4880	59
60	.406737		.913546		.086454	.094636			.445229	60

VERSINES, EXSECANTS, AND TANGENTS

24°

′	SINE	CORR. FOR SEC. +	COSINE	CORR. FOR SEC. − +	VERSINE	EXSEC	CORR. FOR SEC. +	CORR. FOR SEC. +	TANGENT	′
0	.406737		.913546		.086454	.094636			.445229	0
1	7002		3427		6573	4778			5577	1
2	7268		3309		6691	4920			5926	2
3	7534		3190		6810	5062			6275	3
4	7799		3072		6928	5204			6624	4
5	.408065		.912953		.087047	.095347			.446973	5
6	8330		2834		7166	5489	″ Corr.	″ Corr.	7322	6
7	8596		2715		7285	5632	10 24	10 58	7671	7
8	8862		2596		7404	5775	15 36	15 87	8020	8
9	9127		2478		7522	5917	20 48	20 116	8369	9
10	.409392		.912358		.087642	.096060	30 72	30 175	.448719	10
11	9658	″ Corr.	2239	″ Corr.	7761	6204	40 95	40 233	9068	11
12	9923	10 44	2120	10 20	7880	6347	45 107	45 262	9418	12
13	.410188	15 66	2001	15 30	7999	6490	50 119	50 291	9768	13
14	0454	20 88	1882	20 40	8118	6634			.450117	14
15	.410719	30 133	.911762	30 60	.088238	.096777			.450467	15
16	0984	40 177	1642	40 80	8358	6921			0817	16
17	1249	45 199	1523	45 90	8477	7065			1167	17
18	1514	50 221	1403	50 100	8597	7209			1517	18
19	1780		1284		8716	7353			1868	19
20	.412044		.911164		.088836	.097498			.452218	20
21	2310		1044		8956	7642			2568	21
22	2574		0924		9076	7787			2919	22
23	2840		0804		9196	7931			3269	23
24	3104		0684		9316	8076			3620	24
25	.413369		.910564		.089436	.098221			.453971	25
26	3634		0443		9557	8366	″ Corr.	″ Corr.	4322	26
27	3899		0323		9677	8511	10 24	10 59	4673	27
28	4164		0202		9798	8657	15 36	15 88	5024	28
29	4428		0082		9918	8802	20 49	20 117	5375	29
30	.414693		.909961		.090039	.098948	30 73	30 176	.455726	30
31	4958		9841		0159	9094	40 97	40 234	6078	31
32	5223		9720		0280	9240	45 109	45 263	6429	32
33	5487		9599		0401	9386	50 121	50 293	6781	33
34	5752		9478		0522	9532			7132	34
35	.416016		.909357		.090643	.099678			.457484	35
36	6281		9236		0764	9824			7836	36
37	6545		9115		0885	9971			8188	37
38	6810		8994		1006	.100118			8540	38
39	7074		8872		1128	0264			8892	39
40	.417338		.908751		.091249	.100411			.459244	40
41	7603	″ Corr.	8630	″ Corr.	1370	0558			9596	41
42	7867	10 44	8508	10 20	1492	0706			9949	42
43	8131	15 66	8387	15 30	1613	0853			.460301	43
44	8396	20 88	8265	20 41	1735	1000			0654	44
45	.418660	30 132	.908143	30 61	.091857	.101148			.461006	45
46	8924	40 176	8021	40 81	1979	1296	″ Corr.	″ Corr.	1359	46
47	9188	45 198	7900	45 91	2100	1444	10 25	10 59	1712	47
48	9452	50 220	7778	50 101	2222	1592	15 37	15 88	2065	48
49	9716		7655		2345	1740	20 49	20 118	2418	49
50	.419980		.907533		.092467	.101888	30 74	30 177	.462771	50
51	.420244		7411		2589	2036	40 99	40 235	3124	51
52	0508		7289		2711	2185	45 111	45 265	3478	52
53	0772		7166		2834	2334	50 123	50 294	3831	53
54	1036		7044		2956	2482			4184	54
55	.421300		.906922		.093078	.102631			.464538	55
56	1563		6799		3201	2780			4892	56
57	1827		6676		3324	2930			5246	57
58	2091		6554		3446	3079			5600	58
59	2355		6431		3569	3228			5954	59
60	.422618		.906308		.093692	.103378			.466308	60

TABLE XX.—NATURAL SINES, COSINES,

25°

'	SINE	CORR. FOR SEC. +	COSINE	CORR. FOR SEC. − +	VERSINE	EXSEC	CORR. FOR SEC. +	CORR. FOR SEC. +	TANGENT	'
0	.422618		.906308		.093692	.103378			.466308	0
1	2882		6185		3815	3528			6662	1
2	3146		6062		3938	3678			7016	2
3	3409		5939		4061	3828			7370	3
4	3672		5815		4185	3978			7725	4
5	.423936		.905692		.094308	.104128			.468080	5
6	4199		5569		4431	4278	" Corr.	" Corr.	8434	6
7	4463		5445		4555	4429	10 25	10 59	8789	7
8	4726		5322		4678	4580	15 38	15 89	9144	8
9	4990		5198		4802	4730	20 50	20 118	9499	9
10	.425253		.905075		.094925	.104881	30 76	30 178	.469854	10
11	5516	" Corr.	4951	" Corr.	5049	5032	40 101	40 236	.470209	11
12	5779	10 44	4827	10 21	5173	5184	45 113	45 266	0564	12
13	6042	15 66	4703	15 31	5297	5335	50 126	50 296	0920	13
14	6306	20 88	4579	20 41	5421	5486			1275	14
15	.426569	30 132	.904455	30 62	.095545	.105638			.471631	15
16	6832	40 175	4331	40 83	5669	5790			1986	16
17	7095	45 197	4207	45 93	5793	5942			2342	17
18	7358	50 219	4082	50 103	5918	6094			2698	18
19	7621		3958		6042	6246			3054	19
20	.427884		.903834		.096166	.106398			.473410	20
21	8147		3709		6291	6551			3766	21
22	8410		3585		6415	6703			4122	22
23	8672		3460		6540	6856			4478	23
24	8935		3335		6665	7009			4835	24
25	.429198		.903210		.096790	.107162			.475191	25
26	9461		3086		6914	7315	" Corr.	" Corr.	5548	26
27	9723		2961		7039	7468	10 26	10 60	5905	27
28	9986		2836		7164	7621	15 38	15 89	6262	28
29	.430248		2710		7290	7775	20 51	20 119	6618	29
30	.430511		.902585		.097415	.107928	30 77	30 179	.476976	30
31	0774		2460		7540	8082	40 103	40 238	7333	31
32	1036		2335		7665	8236	45 115	45 268	7690	32
33	1299		2209		7791	8390	50 128	50 298	8047	33
34	1561		2084		7916	8544			8405	34
35	.431823		.901958		.098042	.108699			.478762	35
36	2086		1832		8168	8853			9120	36
37	2348		1707		8293	9008			9477	37
38	2610		1581		8419	9163			9835	38
39	2873		1455		8545	9318			.480193	39
40	.433135		.901329		.098671	.109473			.480551	40
41	3397	" Corr.	1203	" Corr.	8797	9628			0909	41
42	3659	10 44	1077	10 21	8923	9783			1268	42
43	3921	15 65	0951	15 32	9049	9938			1626	43
44	4183	20 87	0825	20 42	9175	.110094			1984	44
45	.434445	30 131	.900698	30 63	.099302	.110250			.482343	45
46	4707	40 175	0572	40 84	9428	0406	" Corr.	" Corr.	2701	46
47	4969	45 196	0445	45 95	9555	0562	10 26	10 60	3060	47
48	5231	50 218	0319	50 105	9681	0718	15 39	15 90	3419	48
49	5493		0192		9808	0874	20 52	20 120	3778	49
50	.435755		.900065		.099935	.111030	30 78	30 180	.484137	50
51	6017		.899939		.100061	1187	40 104	40 239	4496	51
52	6278		9812		0188	1344	45 117	45 269	4855	52
53	6540		9685		0315	1500	50 130	50 299	5214	53
54	6802		9558		0442	1657			5574	54
55	.437063		.899431		.100569	.111814			.485933	55
56	7325		9304		0696	1972			6293	56
57	7587		9176		0824	2129			6653	57
58	7848		9049		0951	2286			7013	58
59	8110		8922		1078	2444			7373	59
60	.438371		.898794		.101206	.112602			.487733	60

VERSINES, EXSECANTS, AND TANGENTS

26°

′	SINE	CORR. FOR SEC. +	COSINE	CORR. FOR SEC. − +	VERSINE	EXSEC	CORR. FOR SEC. +	CORR. FOR SEC. +	TANGENT	′
0	.438371		.898794		.101206	.112602			.487733	0
1	8633		8666		1334	2760			8093	1
2	8894		8539		1461	2918			8453	2
3	9155		8411		1589	3076			8813	3
4	9417		8283		1717	3234			9174	4
5	.439678		.898156		.101844	.113393			.489534	5
6	9939		8028		1972	3552	″ Corr.	″ Corr.	9895	6
7	.440200		7900		2100	3710	10 27	10 60	.490256	7
8	0462		7772		2228	3869	15 40	15 90	0617	8
9	0723		7643		2357	4028	20 53	20 120	0978	9
10	.440984		.897515		.102485	.114187	30 80	30 181	.491339	10
11	1245	″ Corr.	7387	″ Corr.	2613	4347	40 106	40 241	1700	11
12	1506	10 43	7258	10 21	2742	4506	45 119	45 271	2061	12
13	1767	15 65	7130	15 32	2870	4666	50 133	50 301	2422	13
14	2028	20 87	7001	20 43	2999	4826			2784	14
15	.442289	30 130	.896873	30 64	.103127	.114985			.493145	15
16	2550	40 174	6744	40 86	3256	5145			3507	16
17	2810	45 196	6615	45 97	3385	5306			3869	17
18	3071	50 217	6486	50 107	3514	5466			4231	18
19	3332		6358		3642	5626			4593	19
20	.443593		.896228		.103772	.115787			.494955	20
21	3853		6099		3901	5948			5317	21
22	4114		5970		4030	6108			5679	22
23	4375		5841		4159	6269			6042	23
24	4635		5712		4288	6431			6404	24
25	.444896		.895582		.104418	.116592			.496767	25
26	5156		5453		4547	6753	″ Corr.	″ Corr.	7130	26
27	5417		5323		4677	6915	10 27	10 61	7492	27
28	5677		5194		4806	7077	15 41	15 91	7855	28
29	5938		5064		4936	7238	20 54	20 121	8218	29
30	.446198		.894934		.105066	.117400	30 81	30 182	.498582	30
31	6458		4804		5196	7562	40 108	40 242	8945	31
32	6718		4675		5325	7725	45 122	45 272	9308	32
33	6979		4545		5455	7887	50 135	50 303	9672	33
34	7239		4415		5585	8050			.500035	34
35	.447499		.894284		.105716	.118212			.500399	35
36	7759		4154		5846	8375			0763	36
37	8019		4024		5976	8538			1127	37
38	8279		3894		6106	8701			1491	38
39	8539		3763		6237	8865			1855	39
40	.448799		.893633		.106367	.119028			.502219	40
41	9059	″ Corr.	3502	″ Corr.	6498	9192			2583	41
42	9319	10 43	3371	10 22	6629	9355			2948	42
43	9579	15 65	3241	15 33	6759	9519			3312	43
44	9839	20 87	3110	20 44	6890	9683			3677	44
45	.450098	30 130	.892979	30 65	.107021	.119847			.504042	45
46	0358	40 173	2848	40 87	7152	.120012	″ Corr.	″ Corr.	4406	46
47	0618	45 195	2717	45 98	7283	0176	10 27	10 61	4771	47
48	0878	50 216	2586	50 109	7414	0340	15 41	15 91	5136	48
49	1137		2455		7545	0505	20 55	20 122	5502	49
50	.451397		.892323		.107677	.120670	30 82	30 183	.505867	50
51	1656		2192		7808	0835	40 110	40 244	6232	51
52	1916		2061		7939	1000	45 124	45 274	6598	52
53	2175		1929		8071	1165	50 137	50 304	6963	53
54	2435		1798		8202	1331			7329	54
55	.452694		.891666		.108334	.121496			.507695	55
56	2954		1534		8466	1662			8061	56
57	3213		1402		8598	1828			8427	57
58	3472		1270		8730	1994			8793	58
59	3731		1138		8862	2160			9159	59
60	.453990		.891006		.108994	.122326			.509525	60

TABLE XX.—NATURAL SINES, COSINES,

27°

'	SINE	CORR. FOR SEC.	COSINE	CORR. FOR SEC.	VERSINE	EXSEC	CORR. FOR SEC.	CORR. FOR SEC.	TANGENT	'
0	.453990	+	.891006	− +	.108994	.122326	+	+	.509525	0
1	4250		0874		9126	2493			9892	1
2	4509		0742		9258	2659			.510258	2
3	4768		0610		9390	2826			0625	3
4	5027		0478		9522	2993			0992	4
5	.455286		.890345		.109655	.123160			.511359	5
6	5545		0213		9787	3327	" Corr.	" Corr.	1726	6
7	5804		0080		9920	3494	10 28	10 61	2093	7
8	6063		.889948		.110052	3662	15 42	15 92	2460	8
9	6322		9815		0185	3829	20 56	20 122	2828	9
10	.456580		.889682		.110318	.123997	30 84	30 184	.513195	10
11	6839	" Corr.	9549	" Corr.	0451	4165	40 112	40 245	3562	11
12	7098	10 43	9416	10 22	0584	4333	45 126	45 276	3930	12
13	7357	15 65	9283	15 33	0717	4501	50 140	50 306	4298	13
14	7615	20 86	9150	20 44	0850	4669			4666	14
15	.457874	30 129	.889017	30 67	.110983	.124838			.515034	15
16	8132	40 172	8884	40 89	1116	5006			5402	16
17	8391	45 194	8751	45 100	1249	5175			5770	17
18	8650	50 215	8617	50 111	1383	5344			6138	18
19	8908		8484		1516	5513			6507	19
20	.459166		.888350		.111650	.125682			.516876	20
21	9425		8217		1783	5851			7244	21
22	9683		8083		1917	6021			7613	22
23	9942		7949		2051	6190			7982	23
24	.460200		7815		2185	6360			8351	24
25	.460458		.887682		.112318	.126530			.518720	25
26	0716		7548		2452	6700	" Corr.	" Corr.	9089	26
27	0974		7413		2587	6870	10 28	10 62	9458	27
28	1232		7279		2721	7041	15 43	15 92	9828	28
29	1491		7145		2855	7211	20 57	20 123	.520197	29
30	.461749		.887011		.112989	.127382	30 85	30 185	.520567	30
31	2007		6876		3124	7553	40 114	40 246	0937	31
32	2265		6742		3258	7724	45 128	45 277	1307	32
33	2522		6608		3392	7895	50 142	50 308	1677	33
34	2780		6473		3527	8066			2047	34
35	.463038		.886338		.113662	.128237			.522417	35
36	3296		6204		3796	8409			2787	36
37	3554		6069		3931	8581			3158	37
38	3812		5934		4066	8752			3528	38
39	4069		5799		4201	8924			3899	39
40	.464327		.885664		.114336	.129096			.524270	40
41	4584	" Corr.	5529	" Corr.	4471	9269			4641	41
42	4842	10 43	5394	10 23	4606	9441			5012	42
43	5100	15 64	5258	15 34	4742	9614			5383	43
44	5357	20 86	5123	20 45	4877	9786			5754	44
45	.465614	30 129	.884988	30 68	.115012	.129959			.526126	45
46	5872	40 172	4852	40 90	5148	.130132	" Corr.	" Corr.	6497	46
47	6129	45 193	4717	45 102	5283	0306	10 29	10 62	6868	47
48	6387	50 214	4581	50 113	5419	0479	15 43	15 93	7240	48
49	6644		4445		5555	0652	20 58	20 124	7612	49
50	.466901		.884310		.115690	.130826	30 87	30 186	.527984	50
51	7158		4174		5826	1000	40 116	40 248	8356	51
52	7416		4038		5962	1174	45 130	45 279	8728	52
53	7673		3902		6098	1348	50 145	50 310	9100	53
54	7930		3766		6234	1522			9473	54
55	.468187		.883630		.116370	.131696			.529845	55
56	8444		3493		6507	1871			.530218	56
57	8701		3357		6643	2045			0591	57
58	8958		3221		6779	2220			0963	58
59	9215		3084		6916	2395			1336	59
60	.469472		.882948		.117052	.132570			.531709	60

VERSINES, EXSECANTS, AND TANGENTS

28°

'	SINE	CORR. FOR SEC.	COSINE	CORR. FOR SEC.	VERSINE	EXSEC	CORR. FOR SEC.	CORR. FOR SEC.	TANGENT	'
		+		− +			+	+		
0	.469472		.882948		.117052	.132570			.531709	0
1	9728		2811		7189	2745			2083	1
2	9985		2674		7326	2921			2456	2
3	.470242		2538		7462	3096			2829	3
4	0499		2401		7599	3272			3203	4
5	.470755		.882264		.117736	.133448			.533576	5
6	1012		2127		7873	3624	" Corr.	" Corr.	3950	6
7	1268		1990		8010	3800	10 29	10 62	4324	7
8	1525		1853		8147	3976	15 44	15 94	4698	8
9	1782		1716		8284	4153	20 59	20 125	5072	9
10	.472038		.881578		.118422	.134329	30 88	30 187	.535446	10
11	2294	" Corr.	1441	" Corr.	8559	4506	40 118	40 249	5821	11
12	2551	10 43	1304	10 23	8696	4683	45 133	45 281	6195	12
13	2807	15 64	1166	15 34	8834	4860	50 147	50 312	6570	13
14	3063	20 85	1028	20 46	8972	5037			6945	14
15	.473320	30 128	.880891	30 69	.119109	.135215			.537319	15
16	3576	40 171	0753	40 92	9247	5392			7694	16
17	3832	45 192	0615	45 103	9385	5570			8069	17
18	4088	50 213	0477	50 115	9523	5748			8444	18
19	4344		0339		9661	5926			8820	19
20	.474600		.880201		.119799	.136104			.539195	20
21	4856		0063		9937	6282			9571	21
22	5112		.879925		.120075	6460			9946	22
23	5368		9787		0213	6639			.540322	23
24	5624		9649		0351	6818			0698	24
25	.475880		.879510		.120490	.136996			.541074	25
26	6136		9372		0628	7176	" Corr.	" Corr.	1450	26
27	6392		9233		0767	7355	10 30	10 63	1826	27
28	6647		9095		0905	7534	15 45	15 94	2203	28
29	6903		8956		1044	7714	20 60	20 126	2579	29
30	.477159		.878817		.121183	.137893	30 90	30 188	.542956	30
31	7414		8678		1322	8073	40 120	40 251	3332	31
32	7670		8539		1461	8253	45 135	45 283	3709	32
33	7926		8400		1600	8433	50 150	50 314	4086	33
34	8181		8261		1739	8613			4463	34
35	.478436		.878122		.121878	.138794			.544840	35
36	8692		7983		2017	8974			5218	36
37	8947		7844		2156	9155			5595	37
38	9203		7704		2296	9336			5973	38
39	9458		7565		2435	9517			6350	39
40	.479713		.877425		.122575	.139698			.546728	40
41	9968	" Corr.	7286	" Corr.	2714	9879			7106	41
42	.480224	10 42	7146	10 23	2854	.140061			7484	42
43	0479	15 64	7006	15 35	2994	0242			7862	43
44	0734	20 85	6867	20 47	3133	0424			8240	44
45	.480989	30 127	.876727	30 70	.123273	.140606			.548619	45
46	1244	40 170	6587	40 93	3413	0788	" Corr.	" Corr.	8997	46
47	1499	45 191	6447	45 105	3553	0971	10 30	10 63	9376	47
48	1754	50 212	6307	50 117	3693	1153	15 46	15 95	9755	48
49	2009		6166		3834	1336	20 61	20 126	.550134	49
50	.482263		.876026		.123974	.141518	30 91	30 190	.550512	50
51	2518		5886		4114	1701	40 122	40 253	0892	51
52	2773		5746		4254	1884	45 137	45 284	1271	52
53	3028		5605		4395	2067	50 152	50 316	1650	53
54	3282		5464		4536	2251			2030	54
55	.483537		.875324		.124676	.142434			.552409	55
56	3792		5183		4817	2618			2789	56
57	4046		5042		4958	2802			3169	57
58	4301		4902		5098	2986			3549	58
59	4555		4761		5239	3170			3929	59
60	.484810		.874620		.125380	.143354			.554309	60

TABLE XX.—NATURAL SINES, COSINES,
29°

'	SINE	CORR. FOR SEC.	COSINE	CORR. FOR SEC.	VERSINE	EXSEC	CORR. FOR SEC.	CORR. FOR SEC.	TANGENT	'
0	.484810	+	.874620	− +	.125380	.143354	+	+	.554309	0
1	5064		4479		5521	3538			4689	1
2	5318		4338		5662	3723			5070	2
3	5573		4196		5804	3908			5450	3
4	5827		4055		5945	4093			5831	4
5	.486081		.873914		.126086	.144278			.556212	5
6	6335		3772		6228	4463	" Corr.	" Corr.	6593	6
7	6590		3631		6369	4648	10 31	10 64	6974	7
8	6844		3489		6511	4834	15 46	15 95	7355	8
9	7098		3348		6652	5020	20 62	20 127	7736	9
10	.487352		.873206		.126794	.145206	30 93	30 191	.558118	10
11	7606	" Corr.	3064	" Corr.	6936	5392	40 124	40 254	8499	11
12	7860	10 42	2922	10 24	7078	5578	45 140	45 286	8881	12
13	8114	15 63	2780	15 36	7220	5764	50 155	50 318	9263	13
14	8367	20 85	2638	20 47	7362	5950			9645	14
15	.488621	30 127	.872496	30 71	.127504	.146137			.560027	15
16	8875	40 169	2354	40 95	7646	6324			0409	16
17	9129	45 190	2212	45 107	7788	6511			0791	17
18	9382	50 211	2069	50 118	7931	6698			1174	18
19	9636		1927		8073	6885			1556	19
20	.489890		.871784		.128216	.147073			.561939	20
21	.490143		1642		8358	7260			2322	21
22	0397		1499		8501	7448			2705	22
23	0650		1357		8643	7636			3088	23
24	0904		1214		8786	7824			3471	24
25	.491157		.871071		.128929	.148012			.563854	25
26	1410		0928		9072	8200	" Corr.	" Corr.	4238	26
27	1664		0785		9215	8389	10 32	10 64	4621	27
28	1917		0642		9358	8578	15 47	15 96	5005	28
29	2170		0499		9501	8766	20 63	20 128	5389	29
30	.492424		.870356		.129644	.148956	30 95	30 192	.565773	30
31	2677		0212		9788	9145	40 126	40 256	6157	31
32	2930		0069		9931	9334	45 143	45 288	6541	32
33	3183		.869926		.130074	9524	50 158	50 320	6925	33
34	3436		9782		0218	9713			7310	34
35	.493689		.869639		.130361	.149903			.567694	35
36	3942		9495		0505	.150093			8079	36
37	4195		9351		0649	0283			8464	37
38	4448		9207		0793	0473			8849	38
39	4700		9064		0936	0664			9234	39
40	.494953		.868920		.131080	.150854			.569619	40
41	5206	" Corr.	8776	" Corr.	1224	1045			.570004	41
42	5459	10 42	8632	10 24	1368	1236			0390	42
43	5711	15 63	8487	15 36	1513	1427			0776	43
44	5964	20 84	8343	20 48	1657	1618			1161	44
45	.496216	30 126	.868199	30 72	.131801	.151810			.571547	45
46	6469	40 168	8054	40 96	1946	2002	" Corr.	" Corr.	1933	46
47	6722	45 189	7910	45 108	2090	2193	10 32	10 64	2319	47
48	6974	50 210	7766	50 120	2234	2385	15 48	15 97	2705	48
49	7226		7621		2379	2577	20 64	20 129	3092	49
50	.497479		.867476		.132524	.152769	30 96	30 193	.573478	50
51	7731		7331		2669	2962	40 128	40 258	3865	51
52	7983		7187		2813	3154	45 144	45 290	4252	52
53	8236		7042		2958	3347	50 160	50 322	4638	53
54	8488		6897		3103	3540			5026	54
55	.498740		.866752		.133248	.153733			.575413	55
56	8992		6607		3393	3926			5800	56
57	9244		6461		3539	4120			6187	57
58	9496		6316		3684	4313			6575	58
59	9748		6171		3829	4507			6962	59
60	.500000		.866025		.133975	.154700			.577350	60

VERSINES, EXSECANTS, AND TANGENTS
30°

'	SINE	CORR. FOR SEC. +	COSINE	CORR. FOR SEC. − +	VERSINE	EXSEC	CORR. FOR SEC. +	CORR. FOR SEC. +	TANGENT	'
0	.500000		.866025		.133975	.154700			.577350	0
1	0252		5880		4120	4894			7738	1
2	0504		5734		4266	5089			8126	2
3	0756		5589		4411	5283			8514	3
4	1007		5443		4557	5478			8903	4
5	.501259		.865297		.134703	.155672			.579291	5
6	1511		5151		4849	5867	" Corr.	" Corr.	9680	6
7	1762		5006		4994	6062	10 33	10 65	.580068	7
8	2014		4860		5140	6257	15 49	15 97	0457	8
9	2266		4713		5287	6452	20 65	20 130	0846	9
10	.502517		.864567		.135433	.156648	30 98	30 195	.581235	10
11	2768	" Corr.	4421	" Corr.	5579	6844	40 130	40 259	1624	11
12	3020	10 42	4275	10 24	5725	7039	45 147	45 292	2014	12
13	3271	15 63	4128	15 37	5872	7235	50 163	50 324	2403	13
14	3523	20 84	3982	20 49	6018	7432			2793	14
15	.503774	30 126	.863836	30 73	.136164	.157628			.583183	15
16	4025	40 167	3689	40 98	6311	7824			3573	16
17	4276	45 188	3542	45 110	6458	8021			3963	17
18	4528	50 209	3396	50 122	6604	8218			4353	18
19	4779		3249		6751	8415			4743	19
20	.505030		.863102		.136898	.158612			.585134	20
21	5281		2955		7045	8809			5524	21
22	5532		2808		7192	9006			5915	22
23	5783		2661		7339	9204			6306	23
24	6034		2514		7486	9402			6696	24
25	.506285		.862366		.137634	.159600			.587088	25
26	6536		2219		7781	9798	" Corr.	" Corr.	7479	26
27	6786		2072		7928	9996	10 33	10 65	7870	27
28	7037		1924		8076	.160195	15 50	15 98	8262	28
29	7288		1777		8223	0393	20 66	20 131	8653	29
30	.507538		.861629		.138371	.160592	30 100	30 196	.589045	30
31	7789		1482		8518	0791	40 133	40 261	9437	31
32	8040		1334		8666	0990	45 149	45 294	9829	32
33	8290		1186		8814	1189	50 166	50 327	.590221	33
34	8541		1038		8962	1389			0613	34
35	.508791		.860890		.139110	.161588			.591006	35
36	9041		0742		9258	1788			1398	36
37	9292		0594		9406	1988			1791	37
38	9542		0446		9554	2188			2184	38
39	9792		0298		9702	2389			2577	39
40	.510043		.860149		.139851	.162589			.592970	40
41	0293	" Corr.	0001	" Corr.	9999	2790			3363	41
42	0543	10 42	.859852	10 25	.140148	2990			3756	42
43	0793	15 62	9704	15 37	0296	3191			4150	43
44	1043	20 83	9555	20 50	0445	3392			4544	44
45	.511293	30 125	.859406	30 74	.140594	.163594			.594938	45
46	1543	40 167	9258	40 99	0742	3795	" Corr.	" Corr.	5331	46
47	1793	45 187	9109	45 112	0891	3997	10 34	10 66	5726	47
48	2043	50 208	8960	50 124	1040	4199	15 51	15 99	6120	48
49	2293		8811		1189	4401	20 67	20 132	6514	49
50	.512542		.858662		.141338	.164603	30 101	30 197	.596908	50
51	2792		8513		1487	4805	40 135	40 263	7303	51
52	3042		8364		1636	5008	45 152	45 296	7698	52
53	3292		8214		1786	5210	50 169	50 329	8093	53
54	3541		8065		1935	5413			8488	54
55	.513791		.857916		.142084	.165616			.598883	55
56	4040		7766		2234	5819			9278	56
57	4290		7616		2384	6022			9674	57
58	4539		7467		2533	6226			.600069	58
59	4789		7317		2683	6430			0465	59
60	.515038		.857167		.142833	.166633			.600861	60

TABLE XX.—NATURAL SINES, COSINES,

31°

′	SINE	CORR. FOR SEC. +	COSINE	CORR. FOR SEC. − +	VERSINE	EXSEC	CORR. FOR SEC. +	CORR. FOR SEC. +	TANGENT	′
0	.515038		.857167		.142833	.166633			.600861	0
1	5287		7017		2983	6837			1257	1
2	5537		6868		3132	7042			1653	2
3	5786		6718		3282	7246			2049	3
4	6035		6567		3433	7450			2445	4
5	.516284		.856417		.143583	.167655			.602842	5
6	6533		6267		3733	7860	" Corr.	" Corr.	3239	6
7	6782		6117		3883	8065	10 34	10 66	3635	7
8	7031		5966		4034	8270	15 51	15 99	4032	8
9	7280		5816		4184	8476	20 69	20 132	4429	9
10	.517529		.855666		.144334	.168681	30 103	30 199	.604827	10
11	7778	" Corr.	5515	" Corr.	4485	8887	40 137	40 265	5224	11
12	8027	10 41	5364	10 25	4636	9093	45 154	45 298	5622	12
13	8276	15 62	5214	15 38	4786	9299	50 171	50 331	6019	13
14	8525	20 83	5063	20 50	4937	9505			6417	14
15	.518773	30 124	.854912	30 76	.145088	.169711			.606815	15
16	9022	40 166	4761	40 101	5239	9918			7213	16
17	9270	45 186	4610	45 113	5390	.170124			7611	17
18	9519	50 207	4459	50 126	5541	0331			8010	18
19	9768		4308		5692	0538			8408	19
20	.520016		.854156		.145844	.170746			.608807	20
21	0265		4005		5995	0953			9205	21
22	0513		3854		6146	1161			9604	22
23	0761		3702		6298	1368			.610003	23
24	1010		3551		6449	1576			0403	24
25	.521258		.853399		.146601	.171784			.610802	25
26	1506		3248		6752	1993	" Corr.	" Corr.	1201	26
27	1754		3096		6904	2201	10 35	10 67	1601	27
28	2002		2944		7056	2410	15 52	15 100	2001	28
29	2250		2792		7208	2619	20 70	20 133	2401	29
30	.522499		.852640		.147360	.172828	30 105	30 200	.612801	30
31	2747		2488		7512	3037	40 139	40 267	3201	31
32	2994		2336		7664	3246	45 157	45 300	3601	32
33	3242		2184		7816	3456	50 174	50 333	4002	33
34	3490		2032		7968	3665			4402	34
35	.523738		.851879		.148121	.173875			.614803	35
36	3986		1727		8273	4085			5204	36
37	4234		1574		8426	4295			5605	37
38	4481		1422		8578	4506			6006	38
39	4729		1269		8731	4716			6408	39
40	.524977		.851117		.148883	.174927			.616809	40
41	5224	" Corr.	0964	" Corr.	9036	5138			7211	41
42	5472	10 41	0811	10 26	9189	5349			7613	42
43	5719	15 62	0658	15 38	9342	5560			8014	43
44	5966	20 82	0505	20 51	9495	5772			8417	44
45	.526214	30 124	.850352	30 77	.149648	.175983			.618819	45
46	6461	40 165	0199	40 102	9801	6195	" Corr.	" Corr.	9221	46
47	6708	45 186	0046	45 115	9954	6407	10 35	10 67	9624	47
48	6956	50 206	.849893	50 128	.150107	6619	15 53	15 101	.620026	48
49	7203		9739		0261	6831	20 71	20 134	0429	49
50	.527450		.849586		.150414	.177044	30 106	30 202	.620832	50
51	7697		9432		0568	7257	40 142	40 269	1235	51
52	7944		9279		0721	7469	45 160	45 302	1638	52
53	8191		9125		0875	7682	50 177	50 336	2042	53
54	8438		8972		1028	7896			2445	54
55	.528685		.848818		.151182	.178109			.622849	55
56	8932		8664		1336	8322			3253	56
57	9179		8510		1490	8536			3657	57
58	9426		8356		1644	8750			4061	58
59	9673		8202		1798	8964			4465	59
60	.529919		.848048		.151952	.179178			.624869	60

VERSINES, EXSECANTS, AND TANGENTS
32°

'	SINE	CORR. FOR SEC.	COSINE	CORR. FOR SEC.	VERSINE	EXSEC	CORR. FOR SEC.	CORR. FOR SEC.	TANGENT	'
0	.529919	+	.848048	− +	.151952	.179178	+	+	.624869	0
1	.530166		7894		2106	9393	" Corr.	" Corr.	5274	1
2	0412		7740		2260	9607	10 36	10 68	5679	2
3	0659		7585		2415	9822	15 54	15 101	6083	3
4	0906		7431		2569	.180037	20 72	20 135	6488	4
5	.531152		.847276		.152724	.180252	30 108	30 203	.626894	5
6	1399		7122		2878	0468	40 144	40 270	7299	6
7	1645		6967		3033	0683	45 162	45 304	7704	7
8	1891		6813		3187	0899	50 180	50 338	8110	8
9	2138		6658		3342	1115			8516	9
10	.532384		.846503		.153497	.181331			.628922	10
11	2630	" Corr.	6348	" Corr.	3652	1547			9327	11
12	2876	10 41	6193	10 26	3807	1763			9734	12
13	3122	15 61	6038	15 39	3962	1980			.630140	13
14	3368	20 82	5883	20 52	4117	2197			0546	14
15	.533614	30 123	.845728	30 78	.154272	.182414			.630953	15
16	3860	40 164	5573	40 103	4427	2631	" Corr.	" Corr.	1360	16
17	4106	45 185	5417	45 116	4583	2848	10 36	10 68	1767	17
18	4352	50 205	5262	50 129	4738	3065	15 55	15 102	2174	18
19	4598		5106		4894	3283	20 73	20 136	2581	19
20	.534844		.844951		.155049	.183501	30 109	30 204	.632988	20
21	5090		4795		5205	3719	40 146	40 272	3396	21
22	5336		4640		5360	3937	45 164	45 306	3804	22
23	5581		4484		5516	4155	50 182	50 340	4211	23
24	5827		4328		5672	4374			4619	24
25	.536072		.844172		.155828	.184593			.635027	25
26	6318		4016		5984	4812			5436	26
27	6563		3860		6140	5031			5844	27
28	6809		3704		6296	5250			6253	28
29	7054		3548		6452	5469			6661	29
30	.537300		.843391		.156609	.185689			.637070	30
31	7545		3235		6765	5909	" Corr.	" Corr.	7479	31
32	7790		3079		6921	6129	10 37	10 68	7888	32
33	8035		2922		7078	6349	15 55	15 103	8298	33
34	8281		2766		7234	6569	20 74	20 137	8707	34
35	.538526		.842609		.157391	.186790	30 111	30 205	.639117	35
36	8771		2452		7548	7011	40 147	40 273	9527	36
37	9016		2296		7704	7232	45 166	45 308	9937	37
38	9261		2139		7861	7453	50 184	50 342	.640347	38
39	9506		1982		8018	7674			0757	39
40	.539751		.841825		.158175	.187895			.641167	40
41	9996	" Corr.	1668	" Corr.	8332	8117			1578	41
42	.540240	10 41	1511	10 26	8489	8339			1989	42
43	0485	15 61	1354	15 39	8646	8561			2399	43
44	0730	20 82	1196	20 52	8804	8783			2810	44
45	.540974	30 122	.841039	30 79	.158961	.189006			.643222	45
46	1219	40 163	0882	40 105	9118	9228	" Corr.	" Corr.	3633	46
47	1464	45 183	0724	45 118	9276	9451	10 37	10 69	4044	47
48	1708	50 204	0567	50 131	9433	9674	15 56	15 103	4456	48
49	1953		0409		9591	9897	20 75	20 137	4868	49
50	.542197		.840251		.159749	.190120	30 112	30 206	.645280	50
51	2442		0094		9906	0344	40 149	40 275	5692	51
52	2686		.839936		.160064	0567	45 168	45 309	6104	52
53	2930		9778		0222	0791	50 186	50 344	6516	53
54	3174		9620		0380	1015			6929	54
55	.543419		.839462		.160538	.191239			.647342	55
56	3663		9304		0696	1464			7755	56
57	3907		9146		0854	1688			8168	57
58	4151		8987		1013	1913			8581	58
59	4395		8829		1171	2138			8994	59
60	.544639		.838671		.161329	.192363			.649408	60

TABLE XX.—NATURAL SINES, COSINES,

33°

'	SINE	CORR. FOR SEC.	COSINE	CORR. FOR SEC.	VERSINE	EXSEC	CORR. FOR SEC.	CORR. FOR SEC.	TANGENT	'
0	.544639	+	.838671	− +	.161329	.192363	+	+	.649408	0
1	4883		8512		1488	2589	" Corr.	" Corr.	9821	1
2	5127		8354		1646	2814	10 38	10 69	.650235	2
3	5371		8195		1805	3040	15 57	15 104	0649	3
4	5614		8036		1964	3266	20 76	20 138	1063	4
5	.545858		.837878		.162122	.193492	30 113	30 207	.651477	5
6	6102		7719		2281	3718	40 151	40 276	1892	6
7	6346		7560		2440	3945	45 170	45 311	2306	7
8	6589		7401		2599	4171	50 189	50 346	2721	8
9	6833		7242		2758	4398			3136	9
10	.547076		.837083		.162917	.194625			.653551	10
11	7320	" Corr.	6924	" Corr.	3076	4852			3966	11
12	7563	10 41	6764	10 27	3236	5080			4382	12
13	7807	15 61	6605	15 40	3395	5307			4797	13
14	8050	20 81	6446	20 53	3554	5535			5213	14
15	.548293	30 122	.836286	30 80	.163714	.195763			.655629	15
16	8536	40 162	6127	40 106	3873	5991	" Corr.	" Corr.	6045	16
17	8780	45 182	5967	45 120	4033	6219	10 38	10 69	6461	17
18	9023	50 203	5807	50 133	4193	6448	15 57	15 104	6877	18
19	9266		5648		4352	6677	20 77	20 139	7294	19
20	.549509		.835488		.164512	.196906	30 115	30 209	.657710	20
21	9752		5328		4672	7135	40 153	40 278	8127	21
22	9995		5168		4832	7364	45 172	45 313	8544	22
23	.550238		5008		4992	7593	50 191	50 348	8961	23
24	0481		4848		5152	7823			9378	24
25	.550724		.834688		.165312	.198053			.659796	25
26	0966		4528		5472	8283			.660214	26
27	1209		4367		5633	8513			0631	27
28	1452		4207		5793	8744			1049	28
29	1694		4046		5954	8974			1467	29
30	.551937		.833886		.166114	.199205			.661886	30
31	2180		3725		6275	9436	" Corr.	" Corr.	2304	31
32	2422		3565		6435	9667	10 39	10 70	2722	32
33	2664		3404		6596	9898	15 58	15 105	3141	33
34	2907		3243		6757	.200130	20 77	20 140	3560	34
35	.553149		.833082		.166918	.200362	30 116	30 210	.663979	35
36	3392		2921		7079	0594	40 155	40 280	4398	36
37	3634		2760		7240	0826	45 174	45 315	4818	37
38	3876		2599		7401	1058	50 194	50 350	5237	38
39	4118		2438		7562	1291			5657	39
40	.554360		.832277		.167723	.201523			.666077	40
41	4602	" Corr.	2116	" Corr.	7884	1756			6497	41
42	4844	10 40	1954	10 27	8046	1989			6917	42
43	5086	15 60	1793	15 40	8207	2223			7337	43
44	5328	20 81	1631	20 54	8369	2456			7758	44
45	.555570	30 121	.831470	30 81	.168530	.202690			.668179	45
46	5812	40 161	1308	40 108	8692	2924	" Corr.	" Corr.	8600	46
47	6054	45 181	1146	45 121	8854	3158	10 39	10 70	9020	47
48	6296	50 202	0984	50 135	9016	3392	15 59	15 106	9442	48
49	6537		0823		9177	3626	20 78	20 141	9863	49
50	.556779		.830661		.169339	.203861	30 118	30 211	.670284	50
51	7021		0499		9501	4096	40 157	40 281	0706	51
52	7262		0337		9663	4331	45 176	45 316	1128	52
53	7504		0174		9826	4566	50 196	50 352	1550	53
54	7745		0012		9988	4801			1972	54
55	.557986		.829850		.170150	.205037			.672394	55
56	8228		9688		0312	5273			2817	56
57	8469		9525		0475	5509			3240	57
58	8710		9363		0637	5745			3662	58
59	8952		9200		0800	5981			4085	59
60	.559193		.829038		.170962	.206218			.674508	60

VERSINES, EXSECANTS, AND TANGENTS

34°

'	SINE	CORR. FOR SEC.	COSINE	CORR. FOR SEC.	VERSINE	EXSEC	CORR. FOR SEC.	CORR. FOR SEC.	TANGENT	'
0	.559193	+	.829038	− +	.170962	.206218	+	+	.674508	0
1	9434		8875		1125	6455	" Corr.	" Corr.	4932	1
2	9675		8712		1288	6692	10 40	10 71	5355	2
3	9916		8549		1451	6929	15 60	15 106	5779	3
4	.560157		8386		1614	7166	20 79	20 141	6203	4
5	.560398		.828223		.171777	.207404	30 119	30 212	.676627	5
6	0639		8060		1940	7642	40 159	40 283	7051	6
7	0880		7897		2103	7879	45 179	45 318	7475	7
8	1121		7734		2266	8118	50 198	50 354	7900	8
9	1361		7571		2429	8356			8324	9
10	.561602		.827407		.172593	.208594			.678749	10
11	1843	" Corr.	7244	" Corr.	2756	8833			9174	11
12	2083	10 40	7081	10 27	2919	9072			9599	12
13	2324	15 60	6917	15 41	3083	9311			.680025	13
14	2564	20 80	6753	20 55	3247	9550			0450	14
15	.562805	30 120	.826590	30 82	.173410	.209790			.680876	15
16	3045	40 159	6426	40 109	3574	.210030	" Corr.	" Corr.	1302	16
17	3286	45 180	6262	45 123	3738	0270	10 40	10 71	1728	17
18	3526	50 200	6098	50 136	3902	0510	15 60	15 107	2154	18
19	3766		5934		4066	0750	20 80	20 142	2580	19
20	.564007		.825770		.174230	.210990	30 121	30 213	.683007	20
21	4247		5606		4394	1231	40 161	40 285	3433	21
22	4487		5442		4558	1472	45 181	45 320	3860	22
23	4727		5278		4722	1713	50 201	50 356	4287	23
24	4967		5114		4886	1954			4714	24
25	.565207		.824949		.175051	.212196			.685142	25
26	5447		4785		5215	2438			5569	26
27	5687		4620		5380	2680			5997	27
28	5927		4456		5544	2922			6425	28
29	6166		4291		5709	3164			6853	29
30	.566406		.824126		.175874	.213406			.687281	30
31	6646		3961		6039	3649	" Corr.	" Corr.	7709	31
32	6886		3796		6204	3892	10 41	10 72	8138	32
33	7125		3632		6368	4135	15 61	15 107	8567	33
34	7365		3467		6533	4378	20 81	20 143	8996	34
35	.567604		.823302		.176698	.214622	30 122	30 215	.689425	35
36	7844		3136		6864	4866	40 163	40 286	9854	36
37	8083		2971		7029	5109	45 183	45 322	.690283	37
38	8322		2806		7194	5354	50 203	50 358	0713	38
39	8562		2640		7360	5598			1142	39
40	.568801		.822475		.177525	.215842			.691572	40
41	9040	" Corr.	2310	" Corr.	7690	6087			2003	41
42	9280	10 39	2144	10 28	7856	6332			2433	42
43	9519	15 60	1978	15 41	8022	6577			2863	43
44	9758	20 80	1813	20 55	8187	6822			3294	44
45	.569997	30 119	.821647	30 83	.178353	.217068			.693725	45
46	.570236	40 159	1481	40 110	8519	7314	" Corr.	" Corr.	4156	46
47	0475	45 179	1315	45 124	8685	7559	10 41	10 72	4587	47
48	0714	50 199	1149	50 138	8851	7806	15 62	15 108	5018	48
49	0952		0983		9017	8052	20 81	20 144	5450	49
50	.571191		.820817		.179183	.218298	30 124	30 216	.695681	50
51	1430		0651		9349	8545	40 165	40 288	6313	51
52	1669		0485		9515	8792	45 185	45 324	6745	52
53	1907		0318		9682	9039	50 206	50 360	7177	53
54	2146		0152		9848	9286			7610	54
55	.572384		.819985		.180015	.219534			.698042	55
56	2623		9819		0181	9782			8475	56
57	2861		9652		0348	.220030			8908	57
58	3100		9486		0514	0278			9341	58
59	3338		9319		0681	0526			9774	59
60	.573576		.819152		.180848	.220775			.700208	60

TABLE XX.—NATURAL SINES, COSINES,

35°

'	SINE	CORR. FOR SEC.	COSINE	CORR. FOR SEC.	VERSINE	EXSEC	CORR. FOR SEC.	CORR. FOR SEC.	TANGENT	'
0	.573576	+	.819152	− +	.180848	.220775	+	+	.700208	0
1	3815		8985		1015	1023	" Corr.	" Corr.	0641	1
2	4053		8818		1182	1272	10 42	10 72	1075	2
3	4291		8651		1349	1522	15 63	15 109	1509	3
4	4529		8484		1516	1771	20 83	20 145	1943	4
5	.574767		.818317		.181683	.222020	30 125	30 217	.702377	5
6	5005		8150		1850	2270	40 167	40 290	2812	6
7	5243		7982		2018	2520	45 188	45 326	3246	7
8	5481		7815		2185	2770	50 208	50 362	3681	8
9	5719		7648		2352	.3021			4116	9
10	.575957		.817480		.182520	.223271			.704552	10
11	6195	" Corr.	7312	" Corr.	2688	3522			4987	11
12	6432	10 40	7145	10 28	2855	3773			5422	12
13	6670	15 59	6977	15 42	3023	4024			5858	13
14	6908	20 79	6809	20 56	3191	4276			6294	14
15	.577145	30 119	.816642	30 84	.183358	.224527			.706730	15
16	7383	40 158	6474	40 112	3526	4779	" Corr.	" Corr.	7166	16
17	7620	45 178	6306	45 126	3694	5031	10 42	10 73	7603	17
18	7858	50 198	6138	50 140	3862	5284	15 63	15 109	8040	18
19	8095		5970		4030	5536	20 84	20 146	8476	19
20	.578332		.815801		.184199	.225789	30 127	30 219	.708913	20
21	8570		5633		4367	6042	40 169	40 292	9350	21
22	8807		5465		4535	6295	45 190	45 328	9788	22
23	9044		5296		4704	6548	50 211	50 365	.710225	23
24	9281		5128		4872	6802			0663	24
25	.579518		.814959		.185041	.227055			.711101	25
26	9755		4791		5209	7309			1539	26
27	9992		4622		5378	7563			1977	27
28	.580229		4453		5547	7818			2416	28
29	0466		4284		5716	8072			2854	29
30	.580703		.814116		.185884	.228327			.713293	30
31	0940		3947		6053	8582	" Corr.	" Corr.	3732	31
32	1176		3778		6222	8837	10 43	10 73	4171	32
33	1413		3608		6392	9092	15 64	15 110	4611	33
34	1650		3439		6561	9348	20 85	20 147	5050	34
35	.581886		.813270		.186730	.229604	30 128	30 220	.715490	35
36	2123		3101		6899	9860	40 171	40 293	5930	36
37	2360		2931		7069	.230116	45 192	45 330	6370	37
38	2596		2762		7238	0372	50 214	50 367	6810	38
39	2832		2592		7408	0629			7250	39
40	.583069		.812423		.187577	.230886			.717691	40
41	3305	" Corr.	2253	" Corr.	7747	1143			8132	41
42	3541	10 39	2084	10 28	7916	1400			8573	42
43	3777	15 59	1914	15 42	8086	1658			9014	43
44	4014	20 79	1744	20 57	8256	1916			9455	44
45	.584250	30 118	.811574	30 85	.188426	.232174			.719897	45
46	4486	40 157	1404	40 113	8596	2432	" Corr.	" Corr.	.720339	46
47	4722	45 177	1234	45 127	8766	2690	10 43	10 74	0781	47
48	4958	50 197	1064	50 142	8936	2949	15 65	15 111	1223	48
49	5194		0894		9106	3207	20 87	20 148	1665	49
50	.585429		.810723		.189277	.233466	30 130	30 222	.722108	50
51	5665		0553		9447	3726	40 173	40 295	2550	51
52	5901		0383		9617	3985	45 195	45 332	2993	52
53	6137		0212		9788	4245	50 216	50 369	3436	53
54	6372		0042		9958	4504			3879	54
55	.586608		.809871		.190129	.234764			.724323	55
56	6844		9700		0300	5025			4766	56
57	7079		9530		0470	5285			5210	57
58	7314		9359		0641	5546			5654	58
59	7550		9188		0812	5807			6098	59
60	.587785		.809017		.190983	.236068			.726542	60

VERSINES, EXSECANTS, AND TANGENTS

36°

'	SINE	CORR. FOR SEC.	COSINE	CORR. FOR SEC.	VERSINE	EXSEC	CORR. FOR SEC.	CORR. FOR SEC.	TANGENT	'
0	.587785	+	.809017	− +	.190983	.236068	+	+	.726542	0
1	8021		8846		1154	6329	" Corr.	" Corr.	6987	1
2	8256		8675		1325	6591	10 44	10 74	7432	2
3	8491		8504		1496	6853	15 66	15 111	7877	3
4	8726		8332		1668	7115	20 88	20 149	8322	4
5	.588961		.808161		.191839	.237377	30 131	30 223	.728767	5
6	9196		7990		2010	7639	40 175	40 297	9212	6
7	9431		7818		2182	7902	45 197	45 334	9658	7
8	9666		7647		2353	8165	50 219	50 372	.730104	8
9	9901		7475		2525	8428			0550	9
10	.590136		.807304		.192696	.238691			.730996	10
11	0371	" Corr.	7132	" Corr.	2868	8955			1443	11
12	0606	10 39	6960	10 29	3040	9218			1889	12
13	0840	15 59	6788	15 43	3212	9482			2336	13
14	1075	20 76	6617	20 57	3383	9746			2783	14
15	.591310	30 117	.806445	30 86	.193555	.240011			.733230	15
16	1544	40 156	6273	40 115	3727	0275	" Corr.	" Corr.	3678	16
17	1779	45 176	6100	45 129	3900	0540	10 44	10 75	4125	17
18	2013	50 195	5928	50 143	4072	0805	15 67	15 112	4573	18
19	2248		5756		4244	1070	20 89	20 150	5021	19
20	.592482		.805584		.194416	.241336	30 133	30 224	.735469	20
21	2716		5411		4589	1602	40 177	40 299	5917	21
22	2950		5239		4761	1868	45 200	45 337	6366	22
23	3185		5066		4934	2134	50 222	50 374	6815	23
24	3419		4894		5106	2400			7264	24
25	.593653		.804721		.195279	.242666			.737713	25
26	3887		4548		5452	2933			8162	26
27	4121		4376		5624	3200			8612	27
28	4355		4203		5797	3468			9061	28
29	4589		4030		5970	3735			9511	29
30	.594823		.803857		.196143	.244003			.739961	30
31	5057		3684		6316	4270	" Corr.	" Corr.	.740411	31
32	5290		3511		6489	4538	10 45	10 75	0862	32
33	5524		3338		6662	4807	15 67	15 113	1312	33
34	5758		3164		6836	5075	20 90	20 151	1763	34
35	.595991		.802991		.197009	.245344	30 135	30 226	.742214	35
36	6225		2818		7182	5613	40 180	40 301	2666	36
37	6458		2644		7356	5882	45 202	45 339	3117	37
38	6692		2470		7530	6152	50 225	50 376	3569	38
39	6925		2297		7703	6421			4020	39
40	.597159		.802123		.197877	.246691			.744472	40
41	7392	" Corr.	1950	" Corr.	8050	6961			4925	41
42	7625	10 39	1776	10 29	8224	7232			5377	42
43	7858	15 58	1602	15 44	8398	7502			5830	43
44	8092	20 78	1428	20 58	8572	7773			6282	44
45	.598325	30 117	.801254	30 87	.198746	.248044			.746735	45
46	8558	40 155	1080	40 116	8920	8315	" Corr.	" Corr.	7189	46
47	8791	45 175	0906	45 131	9094	8587	10 45	10 76	7642	47
48	9024	50 194	0731	50 145	9269	8858	15 68	15 114	8096	48
49	9256		0557		9443	9130	20 91	20 152	8549	49
50	.599489		.800383		.199617	.249402	30 136	30 227	.749003	50
51	9722		0208		9792	9675	40 182	40 303	9458	51
52	9955		0034		9966	9947	45 205	45 341	9912	52
53	.600188		.799859		.200141	.250220	50 227	50 379	.750366	53
54	0420		9685		0315	0493			0821	54
55	.600653		.799510		.200490	.250766			.751276	55
56	0885		9335		0665	1040			1731	56
57	1118		9160		0840	1313			2187	57
58	1350		8986		1014	1587			2642	58
59	1583		8810		1190	1861			3098	59
60	.601815		.798636		.201364	.252136			.753554	60

TABLE XX.—NATURAL SINES, COSINES,

37°

′	SINE	CORR. FOR SEC. +	COSINE	CORR. FOR SEC. − +	VERSINE	EXSEC	CORR. FOR SEC. +	CORR. FOR SEC. +	TANGENT	′
0	.601815		.798636		.201364	.252136			.753554	0
1	2047		8460		1540	2410	" Corr.	" Corr.	4010	1
2	2280		8285		1715	2685	10 46	10 76	4467	2
3	2512		8110		1890	2960	15 69	15 114	4923	3
4	2744		7935		2065	3235	20 92	20 153	5380	4
5	.602976		.797759		.202241	.253511	30 138	30 229	.755837	5
6	3208		7584		2416	3786	40 184	40 305	6294	6
7	3440		7408		2592	4062	45 207	45 343	6751	7
8	3672		7233		2767	4339	50 230	50 381	7209	8
9	3904		7057		2943	4615			7667	9
10	.604136		.796882		.203118	.254892			.758125	10
11	4367	" Corr.	6706	" Corr.	3294	5168			8583	11
12	4599	10 39	6530	10 29	3470	5446			9041	12
13	4831	15 58	6354	15 44	3646	5723			9500	13
14	5062	20 77	6178	20 59	3822	6000			9959	14
15	.605294	30 116	.796002	30 88	.203998	.256278			.760418	15
16	5526	40 154	5826	40 117	4174	6556	" Corr.	" Corr.	0877	16
17	5757	45 174	5650	45 132	4350	6835	10 47	10 77	1336	17
18	5988	50 193	5474	50 147	4526	7113	15 70	15 115	1796	18
19	6220		5297		4703	7392	20 93	20 154	2256	19
20	.606451		.795121		.204879	.257670	30 140	30 230	.762716	20
21	6682		4944		5056	7950	40 186	40 307	3176	21
22	6914		4768		5232	8229	45 210	45 345	3636	22
23	7145		4591		5409	8509	50 233	50 384	4097	23
24	7376		4415		5585	8788			4558	24
25	.607607		.794238		.205762	.259069			.765019	25
26	7838		4061		5939	9349			5480	26
27	8069		3884		6116	9629			5941	27
28	8300		3707		6293	9910			6403	28
29	8531		3530		6470	.260191			6865	29
30	.608761		.793353		.206647	.260472			.767327	30
31	8992		3176		6824	0754	" Corr.	" Corr.	7789	31
32	9223		2999		7001	1036	10 47	10 77	8252	32
33	9454		2822		7178	1318	15 71	15 116	8714	33
34	9684		2644		7356	1600	20 94	20 155	9177	34
35	.609915		.792467		.207533	.261882	30 142	30 232	.769640	35
36	.610145		2290		7710	2165	40 189	40 309	.770104	36
37	0376		2112		7888	2448	45 212	45 348	0567	37
38	0606		1934		8066	2731	50 236	50 386	1031	38
39	0836		1757		8243	3014			1495	39
40	.611067		.791579		.208421	.263298			.771959	40
41	1297	" Corr.	1401	" Corr.	8599	3581			2423	41
42	1527	10 38	1224	10 30	8776	3865			2888	42
43	1757	15 57	1046	15 45	8954	4150			3353	43
44	1987	20 77	0868	20 59	9132	4434			3818	44
45	.612217	30 115	.790690	30 89	.209310	.264719			.774283	45
46	2447	40 153	0512	40 119	9488	5004	" Corr.	" Corr.	4748	46
47	2677	45 173	0333	45 134	9667	5289	10 48	10 78	5214	47
48	2907	50 192	0155	50 148	9845	5574	15 72	15 117	5680	48
49	3137		.789977		.210023	5860	20 95	20 156	6146	49
50	.613367		.789798		.210202	.266146	30 143	30 233	.776612	50
51	3596		9620		0380	6432	40 191	40 311	7078	51
52	3826		9441		0559	6719	45 215	45 350	7545	52
53	4056		9263		0737	7005	50 239	50 389	8012	53
54	4285		9084		0916	7292			8479	54
55	.614515		.788905		.211095	.267579			.778946	55
56	4744		8727		1273	7866			9414	56
57	4974		8548		1452	8154			9881	57
58	5203		8369		1631	8442			.780349	58
59	5432		8190		1810	8730			0817	59
60	.615662		.788011		.211989	.269018			.781286	60

VERSINES, EXSECANTS, AND TANGENTS

38°

'	SINE	CORR. FOR SEC.	COSINE	CORR. FOR SEC.	VERSINE	EXSEC	CORR. FOR SEC.	CORR. FOR SEC.	TANGENT	'
0	.615662	+	.788011	− +	.211989	.269018	+	+	.781286	0
1	5891		7832		2168	9307	" Corr.	" Corr.	1754	1
2	6120		7652		2348	9596	10 48	10 78	2223	2
3	6349		7473		2527	9884	15 73	15 117	2692	3
4	6578		7294		2706	.270174	20 97	20 157	3161	4
5	.616807		.787114		.212886	.270463	30 145	30 235	.783630	5
6	7036		6935		3065	0753	40 193	40 313	4100	6
7	7265		6756		3244	1043	45 218	45 352	4570	7
8	7494		6576		3424	1333	50 242	50 392	5040	8
9	7722		6396		3604	1624			5510	9
10	.617951		.786216		.213784	.271914			.785981	10
11	8180	" Corr.	6037	" Corr.	3963	2205			6452	11
12	8408	10 38	5857	10 30	4143	2496			6922	12
13	8637	15 57	5677	15 45	4323	2788			7394	13
14	8866	20 76	5497	20 60	4503	3079			7865	14
15	.619094	30 114	.785317	30 90	.214683	.273371			.788336	15
16	9322	40 152	5137	40 120	4863	3663	" Corr.	" Corr.	8808	16
17	9551	45 171	4957	45 135	5043	3956	10 49	10 79	9280	17
18	9779	50 190	4776	50 150	5224	4248	15 73	15 118	9752	18
19	.620007		4596		5404	4541	20 98	20 158	.790225	19
20	.620236		.784416		.215584	.274834	30 147	30 237	.790698	20
21	0464		4235		5765	5128	40 196	40 316	1170	21
22	0692		4055		5945	5421	45 220	45 355	1643	22
23	0920		3874		6126	5715	50 245	50 394	2117	23
24	1148		·3694		6306	6009			2590	24
25	.621376		.783513		.216487	.276303			.793064	25
26	1604		3332		6668	6598			3538	26
27	1831		3151		6849	6893			4012	27
28	2059		2970		7030	7188			4486	28
29	2287		2789		7211	7483			4961	29
30	.622515		.782608		.217392	.277779			.795436	30
31	2742		2427		7573	8074	" Corr.	" Corr.	5911	31
32	2970		2246		7754	8370	10 50	10 79	6386	32
33	3197		2065		7935	8667	15 74	15 119	6862	33
34	3425		1883		8117	8963	20 99	20 159	7337	34
35	.623652		.781702		.218298	.279260	30 149	30 238	.797813	35
36	3880		1520		8480	9557	40 198	40 318	8290	36
37	4107		1339		8661	9854	45 223	45 357	8766	37
38	4334		1157		8843	.280152	50 248	50 397	9242	38
39	4561		0976		9024	0450			9719	39
40	.624788		.780794		.219206	.280748			.800196	40
41	5016	" Corr.	0612	" Corr.	9388	1046			0674	41
42	5243	10 38	0430	10 30	9570	1344			1151	42
43	5470	15 57	0248	15 46	9752	1643			1629	43
44	5697	20 76	0066	20 61	9934	1942			2107	44
45	.625924	30 113	.779884	30 91	.220116	.282241			.802585	45
46	6150	40 151	9702	40 121	0298	2541	" Corr.	" Corr.	3063	46
47	6377	45 170	9520	45 137	0480	2840	10 50	10 80	3542	47
48	6604	50 189	9338	50 152	0662	3140	15 75	15 120	4021	48
49	6830		9156		0844	3441	20 100	20 160	4500	49
50	.627057		.778973		.221027	.283741	30 151	30 240	.804979	50
51	7284		8791		1209	4042	40 201	40 320	5458	51
52	7510		8608		1392	4343	45 226	45 360	5938	52
53	7737		8426		1574	4644	50 251	50 400	6418	53
54	7963		8243		1757	4944			6898	54
55	.628189		.778060		.221940	.285247			.807379	55
56	8416		7878		2122	5549			7859	56
57	8642		7695		2305	5851			8340	57
58	8868		7512		2488	6154			8821	58
59	9094		7329		2671	6457			9302	59
60	.629320		.777146		.222854	.286760			.809784	60

TABLE XX.—NATURAL SINES, COSINES,

39°

'	SINE	CORR. FOR SEC.	COSINE	CORR. FOR SEC.	VERSINE	EXSEC	CORR. FOR SEC.	CORR. FOR SEC.	TANGENT	'
0	.629320	+	.777146	− +	.222854	.286760	+	+	.809784	0
1	9546		6963		3037	7063	" Corr.	" Corr.	.810266	1
2	9772		6780		3220	7366	10 51	10 80	0748	2
3	9998		6596		3404	7670	15 76	15 121	1230	3
4	.630224		6413		3587	7974	20 102	20 161	1712	4
5	.630450		.776230		.223770	.288278	30 152	30 242	.812195	5
6	0676		6046		3954	8583	40 203	40 322	2678	6
7	0902		5863		4137	8888	45 229	45 362	3161	7
8	1127		5679		4321	9192	50 254	50 403	3644	8
9	1353		5496		4504	9498			4128	9
10	.631578		.775312		.224688	.269803			.814612	10
11	1804	" Corr.	5128	" Corr.	4872	.290109			5096	11
12	2029	10 38	4944	10 31	5056	0415			5580	12
13	2255	15 56	4761	15 46	5239	0721			6065	13
14	2480	20 75	4577	20 61	5423	1028			6549	14
15	.632705	30 113	.774393	30 92	.225607	.291335			.817034	15
16	2931	40 150	4209	40 123	5791	1642	" Corr.	" Corr.	7520	16
17	3156	45 169	4024	45 138	5976	1949	10 51	10 81	8005	17
18	3381	50 188	3840	50 153	6160	2256	15 77	15 121	8490	18
19	3606		3656		6344	2564	20 103	20 162	8976	19
20	.633831		.773472		.226528	.292872	30 154	30 243	.819462	20
21	4056		3287		6713	3181	40 205	40 324	9949	21
22	4281		3103		6897	3489	45 231	45 364	.820435	22
23	4506		2918		7082	3798	50 256	50 405	0922	23
24	4730		2734		7266	4107			1409	24
25	.634955		.772549		.227451	.294416			.821896	25
26	5180		2364		7636	4726	" Corr.	" Corr.	2384	26
27	5405		2179		7821	5036	10 52	10 81	2872	27
28	5629		1994		8006	5346	15 78	15 122	3360	28
29	5854		1810		8190	5656	20 104	20 163	3848	29
30	.636078		.771625		.228375	.295967	30 155	30 244	.824336	30
31	6303		1440		8560	6278	40 207	40 326	4825	31
32	6527		1254		8746	6589	45 233	45 367	5314	32
33	6751		1069		8931	6900	50 259	50 407	5803	33
34	6976		0884		9116	7212			6292	34
35	.637200		.770699		.229301	.297524			.826782	35
36	7424		0513		9487	7836			7272	36
37	7648		0328		9672	8149			7762	37
38	7872		0142		9858	8461			8252	38
39	8096		.769957		.230043	8774			8743	39
40	.638320		.769771		.230229	.299088			.829234	40
41	8544	" Corr.	9585	" Corr.	0415	9401	" Corr.	" Corr.	9725	41
42	8768	10 37	9400	10 31	0600	9715	10 52	10 82	.830216	42
43	8992	15 56	9214	15 46	0786	.300029	15 78	15 123	0708	43
44	9215	20 75	9028	20 62	0972	0343	20 105	20 164	1199	44
45	.639439	30 112	.768842	30 93	.231158	.300658	30 157	30 246	.831691	45
46	9663	40 149	8656	40 124	1344	0972	40 209	40 328	2183	46
47	9886	45 168	8470	45 139	1530	1288	45 235	45 369	2676	47
48	.640110	50 186	8284	50 155	1716	1603	50 262	50 410	3169	48
49	0333		8097		1903	1918			3662	49
50	.640557		.767911		.232089	.302234			.834155	50
51	0780		7725		2275	2550	" Corr.	" Corr.	4648	51
52	1003		7538		2462	2867	10 53	10 82	5142	52
53	1226		7352		2648	3183	15 79	15 124	5636	53
54	1450		7165		2835	3500	20 106	20 165	6130	54
55	.641673		.766978		.233022	.303818	30 159	30 247	.836624	55
56	1896		6792		3208	4135	40 211	40 330	7119	56
57	2119		6605		3395	4453	45 238	45 371	7614	57
58	2342		6418		3582	4771	50 264	50 412	8109	58
59	2565		6231		3769	5089			8604	59
60	.642788		.766044		.233956	.305407			.839100	60

VERSINES, EXSECANTS, AND TANGENTS

40°

'	SINE	CORR. FOR SEC.	COSINE	CORR. FOR SEC.	VERSINE	EXSEC	CORR. FOR SEC.	CORR. FOR SEC.	TANGENT	'
0	.642788	+	.766044	− +	.233956	.305407	+	+	.839100	0
1	3010		5857		4143	5726	" Corr.	" Corr.	9596	1
2	3233		5670		4330	6045	10 53	10 83	.840092	2
3	3456		5483		4517	6364	15 80	15 124	0588	3
4	3678		5296		4704	6684	20 107	20 166	1084	4
5	.643901		.765109		.234891	.307004	30 160	30 249	.841581	5
6	4124		4921		5079	7324	40 214	40 332	2078	6
7	4346		4734		5266	7644	45 240	45 373	2576	7
8	4568		4546		5454	7965	50 267	50 414	3073	8
9	4791		4359		5641	8286			3571	9
10	.645013		.764171		.235829	.308607			.844069	10
11	5236	" Corr.	3984	" Corr.	6016	8928			4567	11
12	5458	10 37	3796	10 31	6204	9250			5066	12
13	5680	15 55	3608	15 47	6392	9572			5564	13
14	5902	20 74	3420	20 63	6580	9894			6063	14
15	.646124	30 111	.763232	30 94	.236768	.310217			.846562	15
16	6346	40 148	3044	40 125	6956	0540	" Corr.	" Corr.	7062	16
17	6568	45 166	2856	45 141	7144	0863	10 54	10 83	7562	17
18	6790	50 185	2668	50 157	7332	1186	15 81	15 125	8062	18
19	7012		2480		7520	1510	20 108	20 167	8562	19
20	.647233		.762292		.237708	.311833	30 162	30 250	.849062	20
21	7455		2104		7896	2158	40 216	40 333	9563	21
22	7677		1915		8085	2482	45 243	45 375	.850064	22
23	7898		1727		8273	2807	50 270	50 417	0565	23
24	8120		1538		8462	3132			1067	24
25	.648341		.761350		.238650	.313457			.851568	25
26	8563		1161		8839	3782	" Corr.	" Corr.	2070	26
27	8784		0972		9028	4108	10 54	10 84	2573	27
28	9006		0784		9216	4434	15 82	15 126	3075	28
29	9227		0595		9405	4760	20 109	20 168	3578	29
30	.649448		.760406		.239594	.315087	30 163	30 262	.854081	30
31	9669		0217		9783	5414	40 218	40 335	4584	31
32	9890		0028		9972	5741	45 245	45 377	5087	32
33	.650111		.759839		.240161	6068	50 272	50 419	5591	33
34	0332		9650		0350	6396			6095	34
35	.650553		.759461		.240539	.316724			.856599	35
36	0774		9271		0729	7052			7104	36
37	0995		9082		0918	7381			7608	37
38	1216		8893		1107	7710			8113	38
39	1437		8703		1297	8039			8618	39
40	.651657		.758514		.241486	.318368			.859124	40
41	1878	" Corr.	8324	" Corr.	1676	8698	" Corr.	" Corr.	9630	41
42	2098	10 37	8134	10 32	1866	9027	10 55	10 84	.860136	42
43	2319	15 55	7945	15 47	2055	9358	15 83	15 127	0642	43
44	2539	20 73	7755	20 63	2245	9688	20 110	20 169	1148	44
45	.652760	30 110	.757565	30 95	.242435	.320019	30 165	30 253	.861655	45
46	2980	40 147	7375	40 127	2625	0350	40 220	40 337	2162	46
47	3200	45 165	7185	45 142	2815	0681	45 248	45 380	2669	47
48	3421	50 184	6995	50 158	3005	1013	50 275	50 422	3177	48
49	3641		6805		3195	1344			3685	49
50	.653861		.756615		.243385	.321676			.864193	50
51	4081		6425		3575	2009	" Corr.	" Corr.	4701	51
52	4301		6234		3766	2342	10 56	10 85	5209	52
53	4521		6044		3956	2674	15 83	15 127	5718	53
54	4741		5854		4146	3008	20 111	20 170	6227	54
55	.654961		.755663		.244337	.323341	30 167	30 255	.866736	55
56	5180		5472		4528	3675	40 222	40 339	7246	56
57	5400		5282		4718	4009	45 250	45 382	7756	57
58	5620		5091		4909	4344	50 278	50 424	8266	58
59	5840		4900		5100	4678			8776	59
60	.656059		.754710		.245290	.325013			.869287	60

TABLE XX.—NATURAL SINES, COSINES,

41°

'	SINE	CORR. FOR SEC.	COSINE	CORR. FOR SEC.	VERSINE	EXSEC	CORR. FOR SEC.	CORR. FOR SEC.	TANGENT	'
0	.656059	+	.754710	− +	.245290	.325013	+	+	.869287	0
1	6278		4519		5481	5348	" Corr.	" Corr.	9798	1
2	6498		4328		5672	5684	10 56	10 85	.870309	2
3	6717		4137		5863	6019	15 84	15 128	0820	3
4	6937		3946		6054	6355	20 112	20 171	1332	4
5	.657156		.753755		.246245	.326692	30 168	30 256	.871844	5
6	7375		3563		6437	7028	40 225	40 342	2356	6
7	7594		3372		6628	7365	45 253	45 384	2868	7
8	7814		3181		6819	7702	50 281	50 427	3381	8
9	8033		2989		7011	8040			3894	9
10	.658252		.752798		.247202	.328378			.874407	10
11	8471	" Corr.	2606	" Corr.	7394	8716			4920	11
12	8690	10 36	2415	10 32	7585	9054			5434	12
13	8908	15 55	2223	15 48	7777	9392			5948	13
14	9127	20 73	2032	20 64	7968	9731			6462	14
15	.659346	30 109	.751840	30 96	.248160	.330071			.876976	15
16	9564	40 146	1648	40 128	8352	0410	" Corr.	" Corr.	7491	16
17	9783	45 164	1456	45 144	8544	0750	10 57	10 86	8006	17
18	.660002	50 182	1264	50 160	8736	1090	15 85	15 129	8522	18
19	0220		1072		8928	1430	20 113	20 172	9037	19
20	.660439		.750880		.249120	.331771	30 170	30 258	.879553	20
21	0657		0688		9312	2112	40 227	40 343	.880069	21
22	0875		0496		9504	2453	45 255	45 387	0585	22
23	1094		0303		9697	2794	50 284	50 429	1102	23
24	1312		0111		9889	3136			1619	24
25	.661530		.749919		.250081	.333478			.882136	25
26	1748		9726		0274	3820	" Corr.	" Corr.	2653	26
27	1966		9534		0466	4163	10 57	10 86	3171	27
28	2184		9341		0659	4506	15 86	15 130	3689	28
29	2402		9148		0852	4849	20 115	20 173	4207	29
30	.662620		.748956		.251044	.335192	30 172	30 259	.884725	30
31	2838		8763		1237	5536	40 229	40 346	5244	31
32	3056		8570		1430	5880	45 258	45 389	5763	32
33	3273		8377		1623	6225	50 286	50 432	6282	33
34	3491		8184		1816	6569			6802	34
35	.663709		.747991		.252009	.336914			.887322	35
36	3926		7798		2202	7259			7842	36
37	4144		7605		2395	7605			8362	37
38	4361		7412		2588	7951			8882	38
39	4578		7218		2782	8297			9403	39
40	.664796		.747025		.252975	.338643			.889924	40
41	5013	" Corr.	6832	" Corr.	3168	8990	" Corr.	" Corr.	.890446	41
42	5230	10 36	6638	10 32	3362	9337	10 58	10 87	0968	42
43	5448	15 54	6445	15 48	3555	9684	15 87	15 130	1489	43
44	5665	20 72	6251	20 65	3749	.340032	20 116	20 174	2012	44
45	.665882	30 108	.746057	30 97	.253943	.340380	30 174	30 261	.892534	45
46	6099	40 145	5864	40 129	4136	0728	40 232	40 348	3057	46
47	6316	45 163	5670	45 145	4330	1076	45 260	45 391	3580	47
48	6532	50 181	5476	50 161	4524	1425	50 289	50 435	4103	48
49	6749		5282		4718	1774			4627	49
50	.666966		.745088		.254912	.342123			.895151	50
51	7183		4894		5106	2473	" Corr.	" Corr.	5675	51
52	7399		4700		5300	2823	10 58	10 87	6199	52
53	7616		4506		5494	3173	15 88	15 131	6724	53
54	7833		4312		5688	3523	20 117	20 175	7249	54
55	.668049		.744117		.255883	.343874	30 175	30 263	.897774	55
56	8266		3923		6077	4225	40 234	40 350	8299	56
57	8482		3728		6272	4577	45 263	45 394	8825	57
58	8698		3534		6466	4928	50 292	50 438	9351	58
59	8914		3339		6661	5280			9878	59
60	.669131		.743145		.256855	.345633			.900404	60

VERSINES, EXSECANTS, AND TANGENTS
42°

'	SINE	CORR. FOR SEC.	COSINE	CORR. FOR SEC.	VERSINE	EXSEC	CORR. FOR SEC.	CORR. FOR SEC.	TANGENT	'
0	.669131	+	.743145	− +	.256855	.345633	+	+	.900404	0
1	9347		2950		7050	5985	" Corr.	" Corr.	0931	1
2	9563		2755		7245	6338	10 59	10 88	1458	2
3	9779		2561		7439	6691	15 89	15 132	1985	3
4	9995		2366		7634	7045	20 118	20 176	2513	4
5	.670211		.742171		.257829	.347399	30 177	30 264	.903041	5
6	0427		1976		8024	7753	40 236	40 352	3569	6
7	0642		1781		8219	8107	45 266	45 396	4098	7
8	0858		1586		8414	8462	50 295	50 440	4627	8
9	1074		1390		8610	8817			5156	9
10	.671290		.741195		.258805	.349172			.905685	10
11	1505	" Corr.	1000	" Corr.	9000	9528	" Corr.	" Corr.	6215	11
12	1721	10 36	0805	10 33	9195	9884	10 59	10 88	6745	12
13	1936	15 54	0609	15 49	9391	.350240	15 89	15 133	7275	13
14	2152	20 72	0414	20 65	9586	0596	20 119	20 177	7805	14
15	.672367	30 108	.740218	30 98	.259782	.350953	30 179	30 266	.908336	15
16	2582	40 144	0022	40 130	9978	1310	40 238	40 354	8867	16
17	2797	45 161	.739827	45 147	.260173	1668	45 268	45 398	9398	17
18	3012	50 179	9631	50 163	0369	2025	50 298	50 443	9930	18
19	3228		9435		0565	2383			.910462	19
20	.673443		.739239		.260761	.352742			.910994	20
21	3658		9044		0956	3100	" Corr.	" Corr.	1526	21
22	3873		8848		1152	3459	10 60	10 89	2059	22
23	4088		8652		1348	3818	15 90	15 133	2592	23
24	4302		8455		1545	4178	20 120	20 178	3126	24
25	.674517		.738259		.261741	.354538	30 180	30 267	.913659	25
26	4732		8063		1937	4898	40 240	40 356	4193	26
27	4947		7867		2133	5256	45 270	45 400	4727	27
28	5161		7670		2330	5619	50 300	50 445	5262	28
29	5376		7474		2526	5980			5796	29
30	.675590		.737277		.262723	.356342			.916331	30
31	5805		7081		2919	6703	" Corr.	" Corr.	6866	31
32	6019		6884		3116	7065	10 61	10 89	7402	32
33	6233		6688		3312	7428	15 91	15 134	7938	33
34	6448		6491		3509	7790	20 121	20 179	8474	34
35	.676662		.736294		.263706	.358153	30 182	30 268	.919010	35
36	6876		6097		3903	8516	40 242	40 358	9547	36
37	7090		5900		4100	8880	45 272	45 403	.920084	37
38	7304		5703		4297	9244	50 303	50 447	0621	38
39	7518		5506		4494	9608			1159	39
40	.677732		.735309		.264691	.359972			.921697	40
41	7946	" Corr.	5112	" Corr.	4888	.360337	" Corr.	" Corr.	2235	41
42	8160	10 36	4915	10 33	5085	0702	10 61	10 90	2773	42
43	8373	15 53	4717	15 49	5283	1068	15 92	15 135	3312	43
44	8587	20 71	4520	20 66	5480	1433	20 122	20 180	3851	44
45	.678801	30 107	.734322	30 99	.265678	.361800	30 183	30 270	.924390	45
46	9014	40 142	4125	40 132	5875	2166	40 244	40 360	4930	46
47	9228	45 160	3928	45 148	6072	2532	45 275	45 405	5470	47
48	9441	50 178	3730	50 165	6270	2899	50 305	50 450	6010	48
49	9655		3532		6468	3267			6551	49
50	.679868		.733334		.266666	.363634			.927091	50
51	.680081		3137		6863	4002	" Corr.	" Corr.	7632	51
52	0295		2939		7061	4370	10 62	10 90	8174	52
53	0508		2741		7259	4739	15 92	15 136	8715	53
54	0721		2543		7457	5108	20 123	20 181	9257	54
55	.680934		.732345		.267655	.365477	30 185	30 271	.929800	55
56	1147		2147		7853	5846	40 246	40 362	.930342	56
57	1360		1949		8051	6216	45 277	45 407	0885	57
58	1573		1750		8250	6586	50 308	50 452	1428	58
59	1786		1552		8448	6957			1971	59
60	.681998		.731354		.268646	.367328			.932515	60

TABLE XX.—NATURAL SINES, COSINES,

43°

'	SINE	CORR. FOR SEC. +	COSINE	CORR. FOR SEC. − +	VERSINE	EXSEC	CORR. FOR SEC. +	CORR. FOR SEC. +	TANGENT	'
0	.681998		.731354		.268646	.367328			.932515	0
1	2211		1155		8845	7698	" Corr.	" Corr.	3059	1
2	2424		0957		9043	8070	10 62	10 91	3603	2
3	2636		0758		9242	8442	15 93	15 136	4148	3
4	2849		0560		9440	8814	20 124	20 182	4693	4
5	.683061		.730361		.269639	.369186	30 186	30 273	.935238	5
6	3274		0162		9838	9559	40 248	40 364	5783	6
7	3486		.729964		.270036	9932	45 280	45 409	6329	7
8	3698		9765		0235	.370305	50 311	50 454	6875	8
9	3911		9566		0434	0678			7422	9
10	.684123		.729367		.270633	.371052			.937968	10
11	4335	" Corr.	9168	" Corr.	0832	1427	" Corr.	" Corr.	8515	11
12	4547	10 35	8969	10 33	1031	1801	10 63	10 91	9062	12
13	4759	15 53	8770	15 50	1230	2176	15 94	15 137	9610	13
14	4971	20 71	8570	20 66	1430	2551	20 125	20 183	.940158	14
15	.685183	30 106	.728371	30 100	.271629	.372927	30 188	30 274	.940706	15
16	5395	40 141	8172	40 133	1828	3303	40 250	40 366	1254	16
17	5607	45 159	7972	45 150	2028	3679	45 282	45 411	1803	17
18	5818	50 176	7773	50 166	2227	4055	50 313	50 457	2352	18
19	6030		7573		2427	4432			2902	19
20	.686242		.727374		.272626	.374809			.943451	20
21	6453		7174		2826	5187	" Corr.	" Corr.	4001	21
22	6665		6974		3026	5564	10 63	10 92	4552	22
23	6876		6774		3226	5943	15 95	15 138	5102	23
24	7088		6575		3425	6321	20 126	20 184	5653	24
25	.687299		.726375		.273625	.376700	30 190	30 276	.946204	25
26	7510		6175		3825	7079	40 253	40 368	6756	26
27	7721		5975		4025	7458	45 284	45 414	7307	27
28	7932		5775		4225	7838	50 316	50 459	7860	28
29	8144		5575		4425	8218			8412	29
30	.688355		.725374		.274626	.378598			.948965	30
31	8566		5174		4826	8979	" Corr.	" Corr.	9518	31
32	8776		4974		5026	9360	10 64	10 92	.950071	32
33	8987		4773		5227	9742	15 96	15 139	0624	33
34	9198		4573		5427	.380123	20 127	20 185	1178	34
35	.689409		.724372		.275628	.380505	30 191	30 277	.951733	35
36	9620		4172		5828	0888	40 255	40 370	2287	36
37	9830		3971		6029	1270	45 287	45 416	2842	37
38	.690041		3770		6230	1653	50 319	50 462	3397	38
39	0251		3570		6430	2037			3953	39
40	.690462		.723369		.276631	.382420			.954508	40
41	0672	" Corr.	3168	" Corr.	6832	2804	" Corr.	" Corr.	5064	41
42	0882	10 35	2967	10 34	7033	3189	10 64	10 93	5621	42
43	1093	15 53	2766	15 50	7234	3573	15 96	15 139	6177	43
44	1303	20 70	2565	20 67	7435	3958	20 129	20 186	6734	44
45	.691513	30 105	.722364	30 101	.277636	.384344	30 193	30 279	.957292	45
46	1723	40 140	2163	40 134	7837	4729	40 257	40 372	7849	46
47	1933	45 158	1962	45 151	8038	5115	45 289	45 418	8407	47
48	2143	50 175	1760	50 168	8240	5502	50 321	50 465	8966	48
49	2353		1559		8441	5888			9524	49
50	.692563		.721357		.278643	.386275			.960083	50
51	2773		1156		8844	6663	" Corr.	" Corr.	0642	51
52	2982		0954		9046	7050	10 65	10 93	1202	52
53	3192		0753		9247	7438	15 97	15 140	1761	53
54	3402		0551		9449	7827	20 130	20 187	2322	54
55	.693611		.720349		.279651	.388215	30 194	30 280	.962882	55
56	3821		0148		9852	8604	40 259	40 374	3443	56
57	4030		.719946		.280054	8994	45 292	45 421	4004	57
58	4240		9744		0256	9383	50 324	50 467	4565	58
59	4449		9542		0458	9773			5127	59
60	.694658		.719340		.280660	.390164			.965689	60

VERSINES, EXSECANTS, AND TANGENTS
44°

'	SINE	CORR. FOR SEC.	COSINE	CORR. FOR SEC.	VERSINE	EXSEC	CORR. FOR SEC.	CORR. FOR SEC.	TANGENT	'
0	.694658	+	.719340	− +	.280660	.390164	+	+	.965689	0
1	4868		9138		0862	0554	" Corr.	" Corr.	6251	1
2	5077		8936		1064	0945	10 65	10 94	6814	2
3	5286		8733		1267	1337	15 98	15 141	7377	3
4	5495		8531		1469	1728	20 131	20 188	7940	4
5	.695704		.718329		.281671	.392120	30 196	30 282	.968504	5
6	5913		8126		1874	2513	40 262	40 376	9067	6
7	6122		7924		2076	2905	45 294	45 423	9632	7
8	6330		7721		2279	3298	50 327	50 470	.970196	8
9	6539		7519		2481	3692			0761	9
10	.696748		.717316		.282684	.394086			.971326	10
11	6956	" Corr.	7113	" Corr.	2887	4480	" Corr.	" Corr.	1892	11
12	7165	10 35	6911	10 34	3089	4874	10 66	10 94	2458	12
13	7374	15 52	6708	15 51	3292	5269	15 99	15 142	3024	13
14	7582	20 69	6505	20 68	3495	5664	20 132	20 189	3590	14
15	.697790	30 104	.716302	30 101	.283698	.396059	30 198	30 284	.974157	15
16	7999	40 139	6099	40 135	3901	6455	40 264	40 378	4724	16
17	8207	45 156	5896	45 152	4104	6851	45 297	45 425	5291	17
18	8415	50 174	5693	50 169	4307	7248	50 329	50 473	5859	18
19	8623		5490		4510	7644			6427	19
20	.698832		.715286		.284714	.398042			.976996	20
21	9040		5083		4917	8439	" Corr.	" Corr.	7564	21
22	9248		4880		5120	8837	10 67	10 95	8133	22
23	9456		4676		5324	9235	15 100	15 143	8703	23
24	9663		4473		5527	9634	20 133	20 190	9272	24
25	.699871		.714269		.285731	.400032	30 200	30 285	.979842	25
26	.700079		4066		5934	0432	40 266	40 380	.980413	26
27	0287		3862		6138	0831	45 299	45 428	0983	27
28	0494		3658		6342	1231	50 333	50 475	1554	28
29	0702		3454		6546	1632			2126	29
30	.700909		.713250		.286750	.402032			.982697	30
31	1117		3046		6954	2433	" Corr.	" Corr.	3269	31
32	1324		2843		7157	2834	10 67	10 96	3842	32
33	1531		2638		7362	3236	15 101	15 143	4414	33
34	1739		2434		7566	3638	20 134	20 191	4987	34
35	.701946		.712230		.287770	.404040	30 201	30 287	.985560	35
36	2153		2026		7974	4443	40 268	40 382	6134	36
37	2360		1822		8178	4846	45 302	45 430	6708	37
38	2567		1617		8383	5249	50 336	50 478	7282	38
39	2774		1413		8587	5653			7857	39
40	.702981		.711209		.288791	.406057			.988432	40
41	3188	" Corr.	1004	" Corr.	8996	6462	" Corr.	" Corr.	9007	41
42	3395	10 34	0800	10 34	9200	6866	10 68	10 96	9582	42
43	3601	15 52	0595	15 51	9405	7272	15 102	15 144	.990158	43
44	3808	20 69	0390	20 68	9610	7677	20 135	20 192	0735	44
45	.704015	30 103	.710185	30 102	.289815	.408083	30 203	30 288	.991311	45
46	4221	40 138	.709981	40 137	.290019	8489	40 271	40 385	1888	46
47	4428	45 155	9776	45 154	0224	8896	45 305	45 433	2465	47
48	4634	50 172	9571	50 171	0429	9303	50 338	50 481	3043	48
49	4841		9366		0634	9710			3621	49
50	.705047		.709161		.290839	.410118			.994199	50
51	5253		8956		1044	0526	" Corr.	" Corr.	4778	51
52	5459		8750		1250	0934	10 68	10 97	5357	52
53	5666		8545		1455	1343	15 102	15 145	5936	53
54	5872		8340		1660	1752	20 137	20 193	6515	54
55	.706078		.708134		.291866	.412161	30 205	30 290	.997095	55
56	6284		7929		2071	2571	40 273	40 387	7676	56
57	6489		7724		2276	2981	45 307	45 435	8256	57
58	6695		7518		2482	3392	50 341	50 484	8837	58
59	6901		7312		2688	3802			9418	59
60	.707107		.707107		.292893	.414214			1.00000	60

TABLE XX.—NATURAL SINES, COSINES,

45°

'	SINE	CORR. FOR SEC.	COSINE	CORR. FOR SEC.	VERSINE	EXSEC	CORR. FOR SEC.	CORR. FOR SEC.	TANGENT	'
		+		− +			+	+		
0	.707107		.707107		.292893	.414214			1.00000	0
1	7312		6901		3099	4625	" Corr.		.00058	1
2	7518		6695		3305	5037	10 69		.00116	2
3	7724		6489		3511	5449	15 103		.00175	3
4	7929		6284		3716	5862	20 138		.00233	4
5	.708134		.706078		.293922	.416275	30 207		1.00291	5
6	8340		5872		4128	6688	40 276		.00350	6
7	8545		5666		4334	7102	45 310		.00408	7
8	8750		5459		4541	7516	50 344		.00467	8
9	8956		5253		4747	7931			.00525	9
10	.709161		.705047		.294953	.418345			1.00583	10
11	9366	" Corr.	4841	" Corr.	5159	8760	" Corr.	" Corr.	.00642	11
12	9571	10 34	4634	10 34	5366	9176	10 69	10 10	.00701	12
13	9776	15 51	4428	15 52	5572	9592	15 104	15 15	.00759	13
14	9981	20 68	4221	20 69	5779	.420008	20 139	20 20	.00818	14
15	.710185	30 102	.704015	30 103	.295985	.420425	30 208	30 29	1.00876	15
16	0390	40 137	3808	40 138	6192	0842	40 278	40 39	.00935	16
17	0595	45 154	3601	45 155	6399	1259	45 313	45 44	.00994	17
18	0800	50 171	3395	50 172	6605	1677	50 347	50 49	.01053	18
19	1004		3188		6812	2095			.01112	19
20	.711209		.702981		.297019	.422513			1.01170	20
21	1413		2774		7226	2932	" Corr.		.01229	21
22	1617		2567		7433	3351	10 70		.01288	22
23	1822		2360		7640	3771	15 105		.01347	23
24	2026		2153		7847	4191	20 140		.01406	24
25	.712230		.701946		.298054	.424611	30 210		1.01465	25
26	2434		1739		8261	5032	40 280		.01524	26
27	2638		1531		8469	5453	45 316		.01583	27
28	2843		1324		8676	5874	50 351		.01642	28
29	3046		1117		8883	6296			.01702	29
30	.713250		.700909		.299091	.426718			1.01761	30
31	3454		0702		9298	7141	" Corr.		.01820	31
32	3658		0494		9506	7564	10 71		.01879	32
33	3862		0287		9713	7987	15 106		.01939	33
34	4066		0079		9921	8410	20 141		.01998	34
35	.714269		.699871		.300129	.428834	30 212		1.02057	35
36	4473		9663		0337	9259	40 283		.02117	36
37	4676		9456		0544	9684	45 318		.02176	37
38	4880		9248		0752	.430109	50 354		.02236	38
39	5083		9040		0960	0534			.02295	39
40	.715286		.698832		.301168	.430960			1.02355	40
41	5490	" Corr.	8623	" Corr.	1377	1386	" Corr.	" Corr.	.02414	41
42	5693	10 34	8415	10 35	1585	1813	10 71	10 10	.02474	42
43	5896	15 51	8207	15 52	1793	2240	15 107	15 15	.02533	43
44	6099	20 68	7999	20 69	2001	2667	20 143	20 20	.02593	44
45	.716302	30 101	.697790	30 104	.302210	.433095	30 214	30 30	1.02653	45
46	6505	40 135	7582	40 139	2418	3523	40 285	40 40	.02713	46
47	6708	45 152	7374	45 156	2626	3952	45 321	45 45	.02772	47
48	6911	50 169	7165	50 174	2835	4380	50 357	50 50	.02832	48
49	7113		6956		3044	4810			.02892	49
50	.717316		.696748		.303252	.435239			1.02952	50
51	7519		6539		3461	5669	" Corr.		.03012	51
52	7721		6330		3670	6100	10 72		.03072	52
53	7924		6122		3878	6530	15 108		.03132	53
54	8126		5913		4087	6962	20 144		.03192	54
55	.718329		.695704		.304296	.437393	30 216		1.03252	55
56	8531		5495		4505	7825	40 288		.03312	56
57	8733		5286		4714	8257	45 324		.03372	57
58	8936		5077		4923	8690	50 360		.03433	58
59	9138		4868		5132	9123			.03493	59
60	.719340		.694658		.305342	.439556			1.03553	60

VERSINES, EXSECANTS, AND TANGENTS
46°

′	SINE	CORR. FOR SEC.	COSINE	CORR. FOR SEC.	VERSINE	EXSEC	CORR. FOR SEC.	CORR. FOR SEC.	TANGENT	′
0	.719340	+	.694658	− +	.305342	.439556	+	+	1.03553	0
1	9542		4449		5551	9990	″ Corr.		.03613	1
2	9744		4240		5760	.440425	10 73		.03674	2
3	9946		4030		5970	0859	15 109		.03734	3
4	.720148		3821		6179	1294	20 145		.03794	4
5	.720349		.693611		.306389	.441730	30 218		1.03855	5
6	0551		3402		6598	2165	40 290		.03915	6
7	0753		3192		6808	2601	45 327		.03976	7
8	0954		2982		7018	3038	50 363		.04036	8
9	1156		2773		7227	3475			.04097	9
10	.721357		.692563		.307437	.443912			1.04158	10
11	1559	″ Corr.	2353	″ Corr.	7647	4350	″ Corr.	″ Corr.	.04218	11
12	1760	10 34	2143	10 35	7857	4788	10 73	10 10	.04279	12
13	1962	15 50	1933	15 53	8067	5226	15 110	15 15	.04340	13
14	2163	20 67	1723	20 70	8277	5665	20 147	20 20	.04401	14
15	.722364	30 101	.691513	30 105	.308487	.446104	30 220	30 30	1.04461	15
16	2565	40 134	1303	40 140	8697	6544	40 293	40 41	.04522	16
17	2766	45 151	1093	45 158	8907	6984	45 330	45 46	.04583	17
18	2967	50 168	0882	50 175	9118	7424	50 366	50 51	.04644	18
19	3168		0672		9328	7865			.04705	19
20	.723369		.690462		.309538	.448306			1.04766	20
21	3570		0251		9749	8748	″ Corr.		.04827	21
22	3770		0041		9959	9190	10 74		.04888	22
23	3971		.689830		.310170	9632	15 111		.04949	23
24	4172		9620		0380	.450075	20 148		.05010	24
25	.724372		.689409		.310591	.450518	30 222		1.05072	25
26	4573		9198		0802	0962	40 296		.05133	26
27	4773		8987		1013	1406	45 333		.05194	27
28	4974		8776		1224	1850	50 370		.05255	28
29	5174		8566		1434	2295			.05317	29
30	.725374		.688355		.311645	.452740			1.05378	30
31	5575		8144		1856	3185	″ Corr.		.05439	31
32	5775		7932		2068	3631	10 75		.05501	32
33	5975		7721		2279	4077	15 112		.05562	33
34	6175		7510		2490	4524	20 149		.05624	34
35	.726375		.687299		.312701	.454971	30 224		1.05685	35
36	6575		7088		2912	5419	40 298		.05747	36
37	6774		6876		3124	5867	45 336		.05809	37
38	6974		6665		3335	6315	50 373		.05870	38
39	7174		6453		3547	6764			.05932	39
40	.727374		.686242		.313758	.457213			1.05994	40
41	7573	″ Corr.	6030	″ Corr.	3970	7662	″ Corr.	″ Corr.	.06056	41
42	7773	10 33	5818	10 35	4182	8112	10 75	10 10	.06117	42
43	7972	15 50	5607	15 53	4393	8562	15 113	15 15	.06179	43
44	8172	20 66	5395	20 71	4605	9013	20 150	20 21	.06241	44
45	.728371	30 100	.685183	30 106	.314817	.459464	30 226	30 31	1.06303	45
46	8570	40 133	4971	40 141	5029	9916	40 301	40 41	.06365	46
47	8770	45 150	4759	45 159	5241	.460368	45 339	45 46	.06427	47
48	8969	50 166	4547	50 176	5453	0820	50 376	50 52	.06489	48
49	9168		4335		5665	1273			.06551	49
50	.729367		.684123		.315677	.461726			1.06613	50
51	9566		3911		6089	2179	″ Corr.		.06676	51
52	9765		3698		6302	2633	10 76		.06738	52
53	9964		3486		6514	3088	15 114		.06800	53
54	.730162		3274		6726	3542	20 152		.06862	54
55	.730361		.683061		.316939	.463997	30 228		1.06925	55
56	0560		2849		7151	4453	40 304		.06987	56
57	0758		2636		7364	4909	45 342		.07049	57
58	0957		2424		7576	5365	50 380		.07112	58
59	1155		2211		7789	5822			.07174	59
60	.731354		.681998		.318002	.466279			1.07237	60

TABLE XX.—NATURAL SINES, COSINES,

47°

'	SINE	CORR. FOR SEC.	COSINE	CORR. FOR SEC.	VERSINE	EXSEC	CORR. FOR SEC.	CORR. FOR SEC.	TANGENT	'
0	.731354	+	.681998	− +	.318002	.466279	+	+	1.07237	0
1	1552		1786		8214	6737	" Corr.		.07299	1
2	1750		1573		8427	7195	10 77		.07362	2
3	1949		1360		8640	7653	15 115		.07425	3
4	2147		1147		8853	8112	20 153		.07487	4
5	.732345		.680934		.319066	.468571	30 230		1.07550	5
6	2543		0721		9279	9031	40 306		.07613	6
7	2741		0508		9492	9491	45 345		.07676	7
8	2939		0295		9705	9951	50 383		.07738	8
9	3137		0081		9919	.470412			.07801	9
10	.733334		.679868		.320132	.470874			1.07864	10
11	3532	" Corr.	9655	" Corr.	0345	1335	" Corr.	" Corr.	.07927	11
12	3730	10 33	9441	10 36	0559	1798	10 77	10 11	.07990	12
13	3928	15 49	9228	15 53	0772	2260	15 116	15 16	.08053	13
14	4125	20 66	9014	20 71	0986	2723	20 155	20 21	.08116	14
15	.734322	30 99	.678801	30 107	.321199	.473186	30 232	30 32	1.08179	15
16	4520	40 132	8587	40 142	1413	3650	40 309	40 42	.08243	16
17	4717	45 148	8373	45 160	1627	4114	45 348	45 47	.08306	17
18	4915	50 165	8160	50 178	1840	4579	50 386	50 53	.08369	18
19	5112		7946		2054	5044			.08432	19
20	.735309		.677732		.322268	.475510			1.08496	20
21	5506		7518		2482	5975	" Corr.		.08559	21
22	5703		7304		2696	6442	10 78		.08622	22
23	5900		7090		2910	6908	15 117		.08686	23
24	6097		6876		3124	7376	20 156		.08749	24
25	.736294		.676662		.323338	.477843	30 234		1.08813	25
26	6491		6448		3552	8311	40 312		.08876	26
27	6688		6233		3767	8780	45 351		.08940	27
28	6884		6019		3981	9248	50 390		.09003	28
29	7081		5805		4195	9718			.09067	29
30	.737277		.675590		.324410	.480187			1.09131	30
31	7474		5376		4624	0657	" Corr.		.09195	31
32	7670		5161		4839	1128	10 79		.09258	32
33	7867		4947		5053	1599	15 118		.09322	33
34	8063		4732		5268	2070	20 157		.09386	34
35	.738259		.674517		.325483	.482542	30 236		1.09450	35
36	8455		4302		5698	3014	40 315		.09514	36
37	8652		4088		5912	3487	45 354		.09578	37
38	8848		3873		6127	3960	50 393		.09642	38
39	,9044		3658		6342	4433			.09706	39
40	.739239		.673443		.326557	.484907			1.09770	40
41	9435	" Corr.	3228	" Corr.	6772	5382	" Corr.	" Corr.	.09834	41
42	9631	10 33	3012	10 36	6988	5856	10 79	10 11	.09899	42
43	9827	15 49	2797	15 54	7203	6332	15 119	15 16	.09963	43
44	.740022	20 65	2582	20 72	7418	6807	20 159	20 21	.10027	44
45	.740218	30 98	.672367	30 108	.327633	.487283	30 238	30 32	1.10091	45
46	0414	40 130	2152	40 144	7848	7760	40 318	40 43	.10156	46
47	0609	45 147	1936	45 161	8064	8237	45 357	45 48	.10220	47
48	0805	50 163	1721	50 179	8279	8714	50 397	50 54	.10285	48
49	1000		1505		8495	9192			.10349	49
50	.741195		.671290		.328710	.489670			1.10414	50
51	1390		1074		8926	.490149	" Corr.		.10478	51
52	1586		0858		9142	0628	10 80		.10543	52
53	1781		0642		9358	1108	15 120		.10608	53
54	1976		0427		9573	1588	20 160		.10672	54
55	.742171		.670211		.329789	.492068	30 240		1.10737	55
56	2366		.669995		.330005	2549	40 320		.10802	56
57	2561		9779		0221	3030	45 361		.10867	57
58	2755		9563		0437	3512	50 401		.10931	58
59	2950		9347		0653	3994			.10996	59
60	.743145		.669131		.330869	.494476			1.11061	60

VERSINES, EXSECANTS, AND TANGENTS

48°

'	SINE	CORR. FOR SEC. +	COSINE	CORR. FOR SEC. − +	VERSINE	EXSEC	CORR. FOR SEC. +	CORR. FOR SEC. +	TANGENT	'
0	.743145		.669131		.330869	.494476			1.11061	0
1	3339		8914		1086	4960	" Corr.		.11126	1
2	3534		8698		1302	5443	10 81		.11191	2
3	3728		8482		1518	5927	15 121		.11256	3
4	3923		8266		1734	6411	20 162		.11321	4
5	.744117		.668049		.331951	.496896	30 243		1.11387	5
6	4312		7833		2167	7381	40 323		.11452	6
7	4506		7616		2384	7867	45 364		.11517	7
8	4700		7399		2601	8353	50 404		.11582	8
9	4894		7183		2817	8840			.11648	9
10	.745088		.666966		.333034	.499327			1.11713	10
11	5282	" Corr.	6749	" Corr.	3251	9814	" Corr.	" Corr.	.11778	11
12	5476	10 32	6532	10 36	3468	.500302	10 82	10 11	.11844	12
13	5670	15 48	6316	15 54	3684	0790	15 122	15 16	.11909	13
14	5864	20 65	6099	20 72	3901	1279	20 163	20 22	.11975	14
15	.746057	30 97	.665882	30 108	.334118	.501768	30 245	30 33	1.12041	15
16	6251	40 129	5665	40 145	4335	2258	40 326	40 44	.12106	16
17	6445	45 145	5448	45 163	4552	2748	45 367	45 49	.12172	17
18	6638	50 161	5230	50 181	4770	3239	50 408	50 55	.12238	18
19	6832		5013		4987	3730			.12303	19
20	.747025		.664796		.335204	.504221			1.12369	20
21	7218		4578		5422	4713	" Corr.		.12435	21
22	7412		4361		5639	5205	10 82		.12501	22
23	7605		4144		5856	5698	15 124		.12567	23
24	7798		3926		6074	6192	20 165		.12633	24
25	.747991		.663709		.336291	.506685	30 247		1.12699	25
26	8184		3491		6509	7179	40 329		.12765	26
27	8377		3273		6727	7674	45 371		.12831	27
28	8570		3056		6944	8169	50 412		.12897	28
29	8763		2838		7162	8664			.12963	29
30	.748956		.662620		.337380	.509160			1.13029	30
31	9148		2402		7598	9657	" Corr.		.13096	31
32	9341		2184		7816	.510154	10 83		.13162	32
33	9534		1966		8034	0651	15 125		.13228	33
34	9726		1748		8252	1149	20 166		.13295	34
35	.749919		.661530		.338470	.511647	30 249		1.13361	35
36	.750111		1312		8688	2146	40 332		.13428	36
37	0303		1094		8906	2645	45 374		.13494	37
38	0496		0875		9125	3145	50 416		.13561	38
39	0688		0657		9343	3645			.13627	39
40	.750880		.660439		.339561	.514145			1.13694	40
41	1072	" Corr.	0220	" Corr.	9780	4646	" Corr.	" Corr.	.13761	41
42	1264	10 32	0002	10 36	9998	5148	10 84	10 11	.13828	42
43	1456	15 48	.659783	15 55	.340217	5650	15 126	15 17	.13894	43
44	1648	20 64	9564	20 73	0436	6152	20 168	20 22	.13961	44
45	.751840	30 96	.659346	30 109	.340654	.516655	30 252	30 33	1.14028	45
46	2032	40 128	9127	40 146	0873	7158	40 336	40 45	.14095	46
47	2223	45 144	8908	45 164	1092	7662	45 378	45 50	.14162	47
48	2415	50 160	8690	50 182	1310	8166	50 419	50 56	.14229	48
49	2606		8471		1529	8671			.14296	49
50	.752798		.658252		.341748	.519176			1.14363	50
51	2989		8033		1967	9682	" Corr.		.14430	51
52	3181		7814		2186	.520188	10 85		.14498	52
53	3372		7594		2406	0694	15 127		.14565	53
54	3563		7375		2625	1201	20 169		.14632	54
55	.753755		.657156		.342844	.521709	30 254		1.14699	55
56	3946		6937		3063	2217	40 339		.14767	56
57	4137		6717		3283	2725	45 381		.14834	57
58	4328		6498		3502	3234	50 423		.14902	58
59	4519		6278		3722	3743			.14969	59
60	.754710		.656059		.343941	.524253			1.15037	60

TABLE XX.—NATURAL SINES, COSINES,

49°

′	SINE	CORR. FOR SEC. +	COSINE	CORR. FOR SEC. − +	VERSINE	EXSEC	CORR. FOR SEC. +	CORR. FOR SEC. +	TANGENT	′
0	.754710		.656059		.343941	.524253			1.15037	0
1	4900		5840		4160	4763	" Corr.		.15104	1
2	5091		5620		4380	5274	10 85		.15172	2
3	5282		5400		4600	5785	15 128		.15240	3
4	5472		5180		4820	6297	20 171		.15308	4
5	.755663		.654961		.345039	.526809	30 256		1.15375	5
6	5854		4741		5259	7322	40 342		.15443	6
7	6044		4521		5479	7835	45 384		.15511	7
8	6234		4301		5699	8349	50 427		.15579	8
9	6425		4081		5919	8863			.15647	9
10	.756615		.653861		.346139	.529377			1.15715	10
11	6805	" Corr.	3641	" Corr.	6359	9892	" Corr.	" Corr.	.15783	11
12	6995	10 32	3421	10 37	6579	.530408	10 86	10 11	.15851	12
13	7185	15 47	3200	15 55	6800	0924	15 129	15 17	.15919	13
14	7375	20 63	2980	20 73	7020	1440	20 172	20 23	.15987	14
15	.757565	30 95	.652760	30 110	.347240	.531957	30 259	30 34	1.16056	15
16	7755	40 127	2539	40 147	7461	2475	40 345	40 46	.16124	16
17	7945	45 142	2319	45 165	7681	2992	45 388	45 51	.16192	17
18	8134	50 158	2098	50 184	7902	3511	50 431	50 57	.16261	18
19	8324		1878		8122	4030			.16329	19
20	.758514		.651657		.348343	.534549			1.16398	20
21	8703		1437		8563	5069	" Corr.		.16466	21
22	8893		1216		8784	5589	10 87		.16535	22
23	9082		0995		9005	6110	15 131		.16603	23
24	9271		0774		9226	6631	20 174		.16672	24
25	.759461		.650553		.349447	.537153	30 261		1.16741	25
26	9650		0332		9668	7675	40 348		.16809	26
27	9839		0111		9889	8198	45 392		.16878	27
28	.760028		.649890		.350110	8721	50 435		.16947	28
29	0217		9669		0331	9245			.17016	29
30	.760406		.649448		.350552	.539769			1.17085	30
31	0595		9227		0773	.540294	" Corr.		.17154	31
32	0784		.9006		0994	0819	10 88		.17223	32
33	0972		8784		1216	1344	15 132		.17292	33
34	1161		8563		1437	1871	20 176		.17361	34
35	.761350		.648341		.351659	.542397	30 264		1.17430	35
36	1538		8120		1880	2924	40 351		.17500	36
37	1727		7898		2102	3452	45 395		.17569	37
38	1915		7677		2323	3980	50 439		.17638	38
39	2104		7455		2545	4509			.17708	39
40	.762292		.647233		.352767	.545038			1.17777	40
41	2480	" Corr.	7012	" Corr.	2988	5567	" Corr.	" Corr.	.17846	41
42	2668	10 31	6790	10 37	3210	6097	10 89	10 12	.17916	42
43	2856	15 47	6568	15 55	3432	6628	15 133	15 17	.17986	43
44	3044	20 63	6346	20 74	3654	7159	20 177	20 23	.18055	44
45	.763232	30 94	.646124	30 111	.353876	.547691	30 266	30 35	1.18125	45
46	3420	40 125	5902	40 148	4098	8223	40 355	40 46	.18194	46
47	3608	45 141	5680	45 166	4320	8755	45 399	45 52	.18264	47
48	3796	50 157	5458	50 185	4542	9288	50 443	50 58	.18334	48
49	3984		5236		4764	9822			.18404	49
50	.764171		.645013		.354987	.550356			1.18474	50
51	4359		4791		5209	0890	" Corr.		.18544	51
52	4546		4568		5432	1425	10 89		.18614	52
53	4734		4346		5654	1961	15 134		.18684	53
54	4921		4124		5876	2497	20 179		.18754	54
55	.765109		.643901		.356099	.553034	30 269		1.18824	55
56	5296		3678		6322	3571	40 358		.18894	56
57	5483		3456		6544	4108	45 403		.18964	57
58	5670		3233		6767	4646	50 448		.19035	58
59	5857		3010		6990	5185			.19105	59
60	.766044		642788		.357212	.555724			1.19175	60

VERSINES, EXSECANTS, AND TANGENTS
50°

'	SINE	CORR. FOR SEC.	COSINE	CORR. FOR SEC.	VERSINE	EXSEC	CORR. FOR SEC.	CORR. FOR SEC.	TANGENT	'
0	.766044	+	.642788	− +	.357212	.555724	+	+	1.19175	0
1	6231		2565		7435	6263	" Corr.		.19246	1
2	6418		2342		7658	6804	10 90		.19316	2
3	6605		2119		7881	7344	15 136		.19387	3
4	6792		1896		8104	7885	20 181		.19457	4
5	.766978		.641673		.358327	.558427	30 271		1.19528	5
6	7165		1450		8550	8969	40 361		.19599	6
7	7352		1226		8774	9512	45 407		.19669	7
8	7538		1003		8997	.560055	50 452		.19740	8
9	7725		0780		9220	0598			.19811	9
10	.767911		.640557		.359443	.561142			1.19882	10
11	8097	" Corr.	0333	" Corr.	9667	1687	" Corr.	" Corr.	.19953	11
12	8284	10 31	0110	10 37	9890	2232	10 91	10 12	.20024	12
13	8470	15 46	.639886	15 56	.360114	2778	15 137	15 18	.20095	13
14	8656	20 62	9663	20 75	0337	3324	20 182	20 24	.20166	14
15	.768842	30 93	.639439	30 112	.360561	.563871	30 274	30 36	1.20237	15
16	9028	40 124	9215	40 149	0785	4418	40 365	40 47	.20308	16
17	9214	45 139	8992	45 168	1008	4966	45 410	45 53	.20379	17
18	9400	50 155	8768	50 186	1232	5514	50 456	50 59	.20451	18
19	9585		8544		1456	6063			.20522	19
20	.769771		.638320		.361680	.566612			1.20593	20
21	9957		8096		1904	7162	" Corr.		.20665	21
22	.770142		7872		2128	7712	10 92		.20736	22
23	0328		7648		2352	8263	15 138		.20808	23
24	0513		7424		2576	8814	20 184		.20879	24
25	.770699		.637200		.362800	.569366	30 276		1.20951	25
26	0884		6976		3024	9919	40 368		.21023	26
27	1069		6751		3249	.570472	45 414		.21094	27
28	1254		6527		3473	1025	50 460		.21166	28
29	1440		6303		3697	1579			.21238	29
30	.771625		.636078		.363922	.572134			1.21310	30
31	1810		5854		4146	2689	" Corr.		.21382	31
32	1994		5629		4371	3244	10 93		.21454	32
33	2179		5405		4595	3800	15 139		.21526	33
34	2364		5180		4820	4357	20 186		.21598	34
35	.772549		.634955		.365045	.574914	30 279		1.21670	35
36	2734		4730		5270	5472	40 372		.21742	36
37	2918		4506		5494	6030	45 418		.21814	37
38	3103		4281		5719	6589	50 465		.21886	38
39	3287		4056		5944	7148			.21959	39
40	.773472		.633831		.366169	.577708			1.22031	40
41	3656	" Corr.	3606	" Corr.	6394	8268	" Corr.	" Corr.	.22104	41
42	3840	10 31	3381	10 38	6619	8829	10 94	10 12	.22176	42
43	4024	15 46	3156	15 56	6844	9390	15 141	15 18	.22249	43
44	4209	20 61	2931	20 75	7069	9952	20 188	20 24	.22321	44
45	.774393	30 92	.632705	30 113	.367295	.580515	30 281	30 36	1.22394	45
46	4577	40 123	2480	40 150	7520	1078	40 375	40 48	.22467	46
47	4761	45 138	2255	45 169	7745	1641	45 422	45 55	.22539	47
48	4944	50 153	2029	50 188	7971	2205	50 469	50 61	.22612	48
49	5128		1804		8196	2770			.22685	49
50	.775312		.631578		.368422	.583335			1.22758	50
51	5496		1353		8647	3900	" Corr.		.22831	51
52	5679		1127		8873	4467	10 95		.22904	52
53	5863		0902		9098	5033	15 142		.22977	53
54	6046		0676		9324	5601	20 189		.23050	54
55	.776230		.630450		.369550	.586168	30 284		1.23123	55
56	6413		0224		9776	6737	40 379		.23196	56
57	6596		.629998		.370002	7306	45 426		.23270	57
58	6780		9772		0228	7875	50 474		.23343	58
59	6963		9546		0454	8445			.23416	59
60	.777146		.629320		.370680	.589016			1.23490	60

TABLE XX.—NATURAL SINES, COSINES,

51°

'	SINE	CORR. FOR SEC. +	COSINE	CORR. FOR SEC. − +	VERSINE	EXSEC	CORR. FOR SEC. +	CORR. FOR SEC. +	TANGENT	'
0	.777146		.629320		.370680	.589016			1.23490	0
1	7329		9094		0906	9587	" Corr.		.23563	1
2	7512		8868		1132	.590158	10 96		.23637	2
3	7695		8642		1358	0731	15 143		.23710	3
4	7878		8416		1584	1303	20 191		.23784	4
5	.778060		.628189		.371811	.591877	30 287		1.23858	5
6	8243		7963		2037	2450	40 382		.23931	6
7	8426		7737		2263	3025	45 430		.24005	7
8	8608		7510		2490	3600	50 478		.24079	8
9	8791		7284		2716	4175			.24153	9
10	.778973		.627057		.372943	.594751			1.24227	10
11	9156	" Corr.	6830	" Corr.	3170	5328	" Corr.	" Corr.	.24301	11
12	9338	10 30	6604	10 38	3396	5905	10 97	10 12	.24375	12
13	9520	15 46	6377	15 57	3623	6482	15 145	15 19	.24449	13
14	9702	20 61	6150	20 76	3850	7061	20 193	20 25	.24523	14
15	.779884	30 91	.625924	30 113	.374076	.597639	30 290	30 37	1.24597	15
16	.780066	40 121	5697	40 151	4303	8219	40 386	40 50	.24672	16
17	0248	45 137	5470	45 170	4530	8799	45 434	45 56	.24746	17
18	0430	50 152	5243	50 189	4757	9379	50 483	50 62	.24820	18
19	0612		5016		4984	9960			.24895	19
20	.780794		.624788		.375212	.600542			1.24969	20
21	0976		4561		5439	1124	" Corr.		.25044	21
22	1157		4334		5666	1706	10 97		.25118	22
23	1339		4107		5893	2290	15 146		.25193	23
24	1520		3880		6120	2873	20 195		.25268	24
25	.781702		.623652		.376348	.603458	30 292		1.25343	25
26	1883		3425		6575	4043	40 390		.25417	26
27	2065		3197		6803	4628	45 439		.25492	27
28	2246		2970		7030	5214	50 487		.25567	28
29	2427		2742		7258	5801			.25642	29
30	.782608		.622515		.377485	.606388			1.25717	30
31	2789		2287		7713	6976	" Corr.		.25792	31
32	2970		2059		7941	7564	10 98		.25867	32
33	3151		1831		8169	8153	15 148		.25943	33
34	3332		1604		8396	8742	20 197		.26018	34
35	.783513		.621376		.378624	.609332	30 295		1.26093	35
36	3694		1148		8852	9923	40 394		.26169	36
37	3874		0920		9080	.610514	45 443		.26244	37
38	4055		0692		9308	1106	50 492		.26320	38
39	4235		0464		9536	1698			.26395	39
40	.784416		.620236		.379764	.612291			1.26471	40
41	4596	" Corr.	0007	" Corr.	9993	2884	" Corr.	" Corr.	.26546	41
42	4776	10 30	.619779	10 38	.380221	3478	10 99	10 13	.26622	42
43	4957	15 45	9551	15 57	0449	4073	15 149	15 19	.26698	43
44	5137	20 60	9322	20 76	0678	4668	20 199	20 25	.26774	44
45	.785317	30 90	.619094	30 114	.380906	.615264	30 298	30 38	1.26849	45
46	5497	40 120	8866	40 152	1134	5860	40 397	40 51	.26925	46
47	5677	45 135	8637	45 171	1363	6457	45 447	45 57	.27001	47
48	5857	50 150	8408	50 190	1592	7054	50 497	50 63	.27077	48
49	6037		8180		1820	7652			.27153	49
50	.786216		.617951		.382049	.618251			1.27230	50
51	6396		7722		2278	8850	" Corr.		.27306	51
52	6576		7494		2506	9450	10 100		.27382	52
53	6756		7265		2735	.620050	15 151		.27458	53
54	6935		7036		2964	0651	20 201		.27535	54
55	.787114		.616807		.383193	.621253	30 301		1.27611	55
56	7294		6578		3422	1855	40 401		.27688	56
57	7473		6349		3651	2458	45 452		.27764	57
58	7652		6120		3880	3061	50 502		.27841	58
59	7832		5891		4109	3665			.27917	59
60	.788011		.615662		.384338	.624269			1.27994	60

VERSINES, EXSECANTS, AND TANGENTS

52°

'	SINE	CORR. FOR SEC.	COSINE	CORR. FOR SEC.	VERSINE	EXSEC	CORR. FOR SEC.	CORR. FOR SEC.	TANGENT	'
0	.788011	+	.615662	− +	.384338	.624269	+	+	1.27994	0
1	8190		5432		4568	4874	" Corr.		.28071	1
2	8369		5203		4797	5480	10 101		.28148	2
3	8548		4974		5026	6086	15 152		.28225	3
4	8727		4744		5256	6693	20 203		.28302	4
5	.788905		.614515		.385485	.627300	30 304		1.28379	5
6	9084		4285		5715	7908	40 405		.28456	6
7	9263		4056		5944	8517	45 456		.28533	7
8	9441		3826		6174	9126	50 507		.28610	8
9	9620		3596		6404	9736			.28687	9
10	.789798		.613367		.386633	.630346			1.28764	10
11	9977	" Corr.	3137	" Corr.	6863	0957	" Corr.	" Corr.	.28842	11
12	.790155	10 30	2907	10 38	7093	1569	10 102	10 13	.28919	12
13	0333	15 45	2677	15 57	7323	2181	15 153	15 19	.28997	13
14	0512	20 59	2447	20 77	7553	2794	20 205	20 26	.29074	14
15	.790690	30 89	.612217	30 115	.387783	.633407	30 307	30 39	1.29152	15
16	0868	40 119	1987	40 153	8013	4021	40 409	40 52	.29229	16
17	1046	45 134	1757	45 173	8243	4636	45 460	45 58	.29307	17
18	1224	50 148	1527	50 192	8473	5251	50 512	50 65	.29385	18
19	1401		1297		8703	5866			.29463	19
20	.791579		.611067		.388933	.636483			1.29541	20
21	1757		0836		9164	7100	" Corr.		.29618	21
22	1934		0606		9394	7717	10 103		.29696	22
23	2112		0376		9624	8336	15 155		.29775	23
24	2290		0145		9855	8954	20 207		.29853	24
25	.792467		.609915		.390085	.639574	30 310		1.29931	25
26	2644		9684		0316	.640194	40 413		.30009	26
27	2822		9454		0546	0814	45 465		.30087	27
28	2999		9223		0777	1435	50 517		.30166	28
29	3176		8992		1008	2057			.30244	29
30	.793353		.608761		.391239	.642680			1.30323	30
31	3530		8531		1469	3303	" Corr.		.30401	31
32	3707		8300		1700	3926	10 104		.30480	32
33	3884		8069		1931	4551	15 157		.30558	33
34	4061		7838		2162	5175	20 209		.30637	34
35	.794238		.607607		.392393	.645801	30 313		1.30716	35
36	4415		7376		2624	6427	40 417		.30795	36
37	4591		7145		2855	7054	45 470		.30873	37
38	4768		6914		3086	7681	50 522		.30952	38
39	4944		6682		3318	8309			.31031	39
40	.795121		.606451		.393549	.648938			1.31110	40
41	5297	" Corr.	6220	" Corr.	3780	9567	" Corr.	" Corr.	.31190	41
42	5474	10 29	5988	10 39	4012	.650197	10 105	10 13	.31269	42
43	5650	15 44	5757	15 58	4243	0827	15 158	15 20	.31348	43
44	5826	20 59	5526	20 77	4474	1458	20 211	20 26	.31427	44
45	.796002	30 88	.605294	30 116	.394706	.652090	30 316	30 40	1.31507	45
46	6178	40 117	5062	40 154	4938	2722	40 421	40 53	.31586	46
47	6354	45 132	4831	45 174	5169	3355	45 474	45 60	.31666	47
48	6530	50 147	4599	50 193	5401	3988	50 527	50 66	.31745	48
49	6706		4367		5633	4623			.31825	49
50	.796882		.604136		.395864	.655258			1.31904	50
51	7057		3904		6096	5893	" Corr.		.31984	51
52	7233		3672		6328	6529	10 106		.32064	52
53	7408		3440		6560	7166	15 160		.32144	53
54	7584		3208		6792	7803	20 213		.32224	54
55	.797759		.602976		.397024	.658441	30 319		1.32304	55
56	7935		2744		7256	9080	40 426		.32384	56
57	8110		2512		7488	9719	45 479		.32464	57
58	8285		2280		7720	.660359	50 532		.32544	58
59	8460		2047		7953	0999			.32624	59
60	.798636		.601815		.398185	.661640			1.32704	60

TABLE XX.—NATURAL SINES, COSINES,

53°

'	SINE	CORR. FOR SEC. +	COSINE	CORR. FOR SEC. − +	VERSINE	EXSEC	CORR. FOR SEC. +	CORR. FOR SEC. +	TANGENT	'
0	.798636		.601815		.398185	.661640			1.32704	0
1	8810		1583		8417	2282	" Corr.		.32785	1
2	8986		1350		8650	2924	10 107		.32865	2
3	9160		1118		8882	3567	15 161		.32946	3
4	9335		0885		9115	4211	20 215		.33026	4
5	.799510		.600653		.399347	.664855	30 322		1.33107	5
6	9685		0420		9580	5500	40 430		.33188	6
7	9859		0188		9812	6146	45 484		.33268	7
8	.800031		.599955		.400045	6792	50 537		.33349	8
9	0208		9722		0278	7439			.33430	9
10	.800383		.599489		.400511	.668086			1.33511	10
11	0557	" Corr.	9256	" Corr.	0744	8734	" Corr.	" Corr.	.33592	11
12	0731	10 29	9024	10 39	0976	9383	10 108	10 14	.33673	12
13	0906	15 44	8791	15 58	1209	.670033	15 163	15 20	.33754	13
14	1080	20 58	8558	20 78	1442	0683	20 217	20 27	.33835	14
15	.801254	30 87	.598325	30 117	.401675	.671334	30 326	30 41	1.33916	15
16	1428	40 116	8092	40 155	1908	1985	40 434	40 54	.33998	16
17	1602	45 131	7858	45 175	2142	2637	45 489	45 61	.34079	17
18	1776	50 145	7625	50 194	2375	3290	50 543	50 68	.34160	18
19	1950		7392		2608	3943			.34242	19
20	.802123		.597159		.402841	.674597			1.34323	20
21	2297		6925		3075	5252	" Corr.		.34405	21
22	2470		6692		3308	5907	10 110		.34487	22
23	2644		6458		3542	6563	15 164		.34568	23
24	2818		6225		3775	7220	20 219		.34650	24
25	.802991		.595991		.404009	.677877	30 329		1.34732	25
26	3164		5758		4242	8535	40 438		.34814	26
27	3338		5524		4476	9193	45 493		.34896	27
28	3511		5290		4710	9852	50 548		.34978	28
29	3684		5057		4943	.680512			.35060	29
30	.803857		.594823		.405177	.681173			1.35142	30
31	4030		4589		5411	1834	" Corr.		.35224	31
32	4203		4355		5645	2496	10 111		.35307	32
33	4376		4121		5879	3159	15 166		.35389	33
34	4548		3887		6113	3822	20 222		.35472	34
35	.804721		.593653		.406347	.684486	30 332		1.35554	35
36	4894		3419		6581	5150	40 443		.35637	36
37	5066		3185		6815	5816	45 498		.35719	37
38	5239		2950		7050	6481	50 554		.35802	38
39	5411		2716		7284	7148			.35885	39
40	.805584		.592482		.407518	.687815			1.35968	40
41	5756	" Corr.	2248	" Corr.	7752	8483	" Corr.	" Corr.	.36051	41
42	5928	10 29	2013	10 39	7987	9152	10 112	10 14	.36134	42
43	6100	15 43	1779	15 59	8221	9821	15 168	15 21	.36217	43
44	6273	20 57	1544	20 78	8456	.690491	20 224	20 28	.36300	44
45	.806445	30 86	.591310	30 117	.408690	.691161	30 336	30 42	1.36383	45
46	6617	40 115	1075	40 156	8925	1833	40 447	40 55	.36466	46
47	6788	45 129	0840	45 176	9160	2504	45 503	45 62	.36549	47
48	6960	50 143	0606	50 195	9394	3177	50 559	50 69	.36633	48
49	7132		0371		9629	3850			.36716	49
50	.807304		.590136		.409864	.694524			1.36800	50
51	7475		.589901		.410099	5199	" Corr.		.36883	51
52	7647		9666		0334	5874	10 113		.36967	52
53	7818		9431		0569	6550	15 170		.37050	53
54	7990		9196		0804	7227	20 226		.37134	54
55	.808161		.588961		.411039	.697904	30 339		1.37218	55
56	8332		8726		1274	8582	40 452		.37302	56
57	8504		8491		1509	9261	45 509		.37386	57
58	8675		8256		1744	9941	50 565		.37470	58
59	8846		8021		1979	.700621			.37554	59
60	.809017		.587785		.412215	.701302			1.37638	60

VERSINES, EXSECANTS, AND TANGENTS
54°

'	SINE	CORR. FOR SEC.	COSINE	CORR. FOR SEC.	VERSINE	EXSEC	CORR. FOR SEC.	CORR. FOR SEC.	TANGENT	'
		+		− +			+	+		
0	.809017		.587785		.412215	.701302			1.37638	0
1	9188		7550		2450	1983	" Corr.		.37722	1
2	9359		7314		2686	2665	10 114		.37807	2
3	9530		7079		2921	3348	15 171		.37891	3
4	9700		6844		3156	4032	20 228		.37976	4
5	.809871		.586608		.413392	.704716	30 342		1.38060	5
6	.810042		6372		3628	5401	40 457		.38145	6
7	0212		6137		3863	6087	45 514		.38229	7
8	0383		5901		4099	6773	50 571		.38314	8
9	0553		5665		4335	7460			.38399	9
10	.810723		.585429		.414571	.708148			1.38484	10
11	0894	" Corr.	5194	" Corr.	4806	8836	" Corr.	" Corr.	.38568	11
12	1064	10 28	4958	10 39	5042	9525	10 115	10 14	.38653	12
13	1234	15 42	4722	15 59	5278	.710215	15 173	15 21	.38738	13
14	1404	20 57	4486	20 79	5514	0906	20 231	20 28	.38824	14
15	.811574	30 85	.584250	30 118	.415750	.711597	30 346	30 43	1.38909	15
16	1744	40 113	4014	40 157	5986	2289	40 461	40 57	.38994	16
17	1914	45 127	3777	45 177	6223	2982	45 519	45 64	.39079	17
18	2084	50 142	3541	50 197	6459	3675	50 577	50 71	.39165	18
19	2253		3305		6695	4369			.39250	19
20	.812423		.583069		.416931	.715064			1.39336	20
21	2592		2832		7168	5759	" Corr.		.39421	21
22	2762		2596		7404	6456	10 116		.39507	22
23	2931		2360		7640	7152	15 175		.39593	23
24	3101		2123		7877	7850	20 233		.39679	24
25	.813270		.581886		.418114	.718548	30 350		1.39764	25
26	3439		1650		8350	9248	40 466		.39850	26
27	3608		1413		8587	9947	45 524		.39936	27
28	3778		1176		8824	.720648	50 583		.40022	28
29	3947		0940		9060	1349			.40109	29
30	.814116		.580703		.419297	.722051			1.40195	30
31	4284		0466		9534	2753	" Corr.		.40281	31
32	4453		0229		9771	3457	10 118		.40367	32
33	4622		.579992		.420008	4161	15 177		.40454	33
34	4791		9755		0245	4866	20 235		.40540	34
35	.814959		.579518		.420482	.725571	30 353		1.40627	35
36	5128		9281		0719	6277	40 471		.40714	36
37	5296		9044		0956	6984	45 530		.40800	37
38	5465		8807		1193	7692	50 588		.40887	38
39	5633		8570		1430	8400			.40974	39
40	.815801		.578332		.421668	.729110			1.41061	40
41	5970	" Corr.	8095	" Corr.	1905	9820	" Corr.	" Corr.	.41148	41
42	6138	10 28	7858	10 40	2142	.730530	10 119	10 15	.41235	42
43	6306	15 42	7620	15 59	2380	1241	15 178	15 22	.41322	43
44	6474	20 56	7383	20 79	2617	1954	20 238	20 29	.41409	44
45	.816642	30 84	.577145	30 119	.422855	.732666	30 357	30 44	1.41497	45
46	6809	40 112	6908	40 158	3092	3380	40 476	40 58	.41584	46
47	6977	45 126	6670	45 178	3330	4094	45 535	45 66	.41672	47
48	7145	50 140	6432	50 198	3568	4809	50 595	50 73	.41759	48
49	7312		6195		3805	5525			.41847	49
50	.817480		.575957		.424043	.736241			1.41934	50
51	7648		5719		4281	6958	" Corr.		.42022	51
52	7815		5481		4519	7676	10 120		.42110	52
53	7982		5243		4757	8395	15 180		.42198	53
54	8150		5005		4995	9114	20 240		.42286	54
55	.818317		.574767		.425233	.739835	30 360		1.42374	55
56	8484		4529		5471	.740556	40 480		.42462	56
57	8651		4291		5709	1277	45 541		.42550	57
58	8818		4053		5947	2000	50 601		.42638	58
59	8985		3815		6185	2723			.42726	59
60	.819152		.573576		.426424	.743447			1.42815	60

TABLE XX.—NATURAL SINES, COSINES,

55°

'	SINE	CORR. FOR SEC. +	COSINE	CORR. FOR SEC. − +	VERSINE	EXSEC	CORR. FOR SEC. +	CORR. FOR SEC. +	TANGENT	'
0	.819152		.573576		.426424	.743447			1.42815	0
1	9319		3338		6662	4172	" Corr.		.42903	1
2	9486		3100		6900	4897	10 121		.42992	2
3	9652		2861		7139	5623	15 182		.43080	3
4	9819		2623		7377	6350	20 243		.43169	4
5	.819985		.572384		.427616	.747078	30 364		1.43258	5
6	.820152		2146		7854	7806	40 486		.43347	6
7	0318		1907		8093	8535	45 546		.43436	7
8	0485		1669		8331	9265	50 607		.43525	8
9	0651		1430		8570	9996			.43614	9
10	.820817		.571191		.428809	.750727			1.43703	10
11	0983	" Corr.	0952	" Corr.	9048	1460	" Corr.	" Corr.	.43792	11
12	1149	10 28	0714	10 39	9286	2192	10 123	10 15	.43881	12
13	1315	15 41	0475	15 60	9525	2926	15 184	15 22	.43970	13
14	1481	20 55	0236	20 80	9764	3661	20 245	20 30	.44060	14
15	.821647	30 83	.569997	30 119	.430003	.754396	30 368	30 45	1.44149	15
16	1813	40 110	9758	40 159	0242	5132	40 491	40 60	.44239	16
17	1978	45 124	9519	45 179	0481	5869	45 552	45 67	.44329	17
18	2144	50 138	9280	50 199	0720	6606	50 613	50 75	.44418	18
19	2310		9040		0960	7345			.44508	19
20	.822475		.568801		.431199	.758084			1.44598	20
21	2640		8562		1438	8824	" Corr.		.44688	21
22	2806		8322		1678	9564	10 124		.44778	22
23	2971		8083		1917	.760306	15 186		.44868	23
24	3136		7844		2156	1048	20 248		.44958	24
25	.823302		.567604		.432396	.761791	30 372		1.45048	25
26	3467		7365		2635	2534	40 496		.45139	26
27	3632		7125		2875	3279	45 558		.45229	27
28	3796		6886		3114	4024	50 620		.45320	28
29	3961		6646		3354	4770			.45410	29
30	.824126		.566406		.433594	.765517			1.45501	30
31	4291		6166		3834	6265	" Corr.		.45592	31
32	4456		5927		4073	7013	10 125		.45682	32
33	4620		5687		4313	7762	15 188		.45773	33
34	4785		5447		4553	8512	20 251		.45864	34
35	.824949		.565207		.434793	.769263	30 376		1.45955	35
36	5114		4967		5033	.770015	40 501		.46046	36
37	5278		4727		5273	0767	45 564		.46137	37
38	5442		4487		5513	1520	50 626		.46229	38
39	5606		4247		5753	2274			.46320	39
40	.825770		.564007		.435993	.773029			1.46411	40
41	5934	" Corr.	3766	" Corr.	6234	3784	" Corr.	" Corr.	.46503	41
42	6098	10 27	3526	10 40	6474	4541	10 127	10 15	.46595	42
43	6262	15 41	3286	15 60	6714	5298	15 190	15 23	.46686	43
44	6426	20 55	3045	20 80	6955	6056	20 253	20 31	.46778	44
45	.826590	30 82	.562805	30 120	.437195	.776815	30 380	30 46	1.46870	45
46	6753	40 109	2564	40 159	7436	7574	40 506	40 61	.46962	46
47	6917	45 123	2324	45 180	7676	8334	45 570	45 69	.47054	47
48	7081	50 136	2083	50 200	7917	9096	50 633	50 77	.47146	48
49	7244		1843		8157	9857			.47238	49
50	.827407		.561602		.438398	.780620			1.47330	50
51	7571		1361		8639	1384	" Corr.		.47422	51
52	7734		1121		8879	2148	10 128		.47514	52
53	7897		0880		9120	2913	15 192		.47607	53
54	8060		0639		9361	3679	20 256		.47699	54
55	.828223		.560398		.439602	.784446	30 384		1.47792	55
56	8386		0157		9843	5213	40 512		.47885	56
57	8549		.559916		.440084	5982	45 576		.47977	57
58	8712		9675		0325	6751	50 640		.48070	58
59	8875		9434		0566	7521			.48163	59
60	.829038		.559193		.440807	.788292			1.48256	60

VERSINES, EXSECANTS, AND TANGENTS
56°

'	SINE	CORR. FOR SEC. +	COSINE	CORR. FOR SEC. −	VERSINE +	EXSEC	CORR. FOR SEC. +	CORR. FOR SEC. +	TANGENT	'
0	.829038		.559193		.440807	.788292			1.48256	0
1	9200		8952		1048	9063	" Corr.		.48349	1
2	9363		8710		1290	9836	10 129		.48442	2
3	9525		8469		1531	.790609	15 194		.48536	3
4	9688		8228		1772	1383	20 259		.48629	4
5	.829850		.557986		.442014	.792158	30 388		1.48722	5
6	.830012		7745		2255	2934	40 517		.48816	6
7	0174		7504		2496	3710	45 582		.48909	7
8	0337		7262		2738	4488	50 646		.49003	8
9	0499		7021		2979	5266			.49097	9
10	.830661		.556779		.443221	.796045			1.49190	10
11	0823	" Corr.	6537	" Corr.	3463	6825	" Corr.	" Corr.	.49284	11
12	0984	10 27	6296	10 40	3704	7605	10 131	10 16	.49378	12
13	1146	15 40	6054	15 60	3946	8387	15 196	15 24	.49472	13
14	1308	20 54	5812	20 81	4188	9169	20 261	20 31	.49566	14
15	.831470	30 81	.555570	30 121	.444430	.799952	30 392	30 47	1.49661	15
16	1631	40 108	5328	40 161	4672	.800736	40 523	40 63	.49755	16
17	1793	45 121	5086	45 181	4914	1521	45 588	45 71	.49849	17
18	1954	50 135	4844	50 202	5156	2307	50 653	50 79	.49944	18
19	2116		4602		5398	3094			.50038	19
20	.832277		.554360		.445640	.803881			1.50133	20
21	2438		4118		5882	4669	" Corr.		.50228	21
22	2599		3876		6124	5458	10 132		.50322	22
23	2760		3634		6366	6248	15 198		.50417	23
24	2921		3392		6608	7039	20 264		.50512	24
25	.833082		.553149		.446851	.807830	30 396		1.50607	25
26	3243		2907		7093	8623	40 528		.50702	26
27	3404		2664		7336	9416	45 594		.50797	27
28	3565		2422		7578	.810210	50 660		.50893	28
29	3725		2180		7820	1005			.50988	29
30	.833886		.551937		.448063	.811801			1.51084	30
31	4046		1694		8306	2598	" Corr.		.51179	31
32	4207		1452		8548	3395	10 133		.51275	32
33	4367		1209		8791	4194	15 200		.51370	33
34	4528		0966		9034	4993	20 267		.51466	34
35	.834688		.550724		.449276	.815793	30 400		1.51562	35
36	4848		0481		9519	6594	40 534		.51658	36
37	5008		0238		9762	7396	45 601		.51754	37
38	5168		.549995		.450005	8198	50 667		.51850	38
39	5328		9752		0248	9002			.51946	39
40	.835488		.549509		.450491	.819806			1.52043	40
41	5648	" Corr.	9266	" Corr.	0734	.820612	" Corr.	" Corr.	.52139	41
42	5807	10 27	9023	10 41	0977	1418	10 135	10 16	.52235	42
43	5967	15 40	8780	15 61	1220	2225	15 202	15 24	.52332	43
44	6127	20 53	8536	20 81	1464	3033	20 270	20 32	.52429	44
45	.836286	30 80	.548293	30 122	.451707	.823842	30 405	30 48	1.52525	45
46	6446	40 106	8050	40 162	1950	4651	40 540	40 66	.52622	46
47	6605	45 120	7807	45 182	2193	5462	45 607	45 73	.52719	47
48	6764	50 133	7563	50 203	2437	6273	50 675	50 81	.52816	48
49	6924		7320		2680	7085			.52913	49
50	.837083		.547076		.452924	.827898	" Corr.		1.53010	50
51	7242		6833		3167	8712	10 136		.53107	51
52	7401		6589		3411	9527	15 205		.53205	52
53	7560		6346		3654	.830343	20 273		.53302	53
54	7719		6102		3898	1160			.53400	54
55	.837878		.545858		.454142	.831977	30 409		1.53497	55
56	8036		5614		4386	2796	40 546		.53595	56
57	8195		5371		4629	3615	45 614		.53693	57
58	8354		5127		4873	4435	50 682		.53791	58
59	8512		4883		5117	5256			.53888	59
60	.838671		.544639		.455361	.836078			1.53986	60

TABLE XX.—NATURAL SINES, COSINES,

57°

'	SINE	CORR. FOR SEC. +	COSINE	CORR. FOR SEC. − +	VERSINE	EXSEC	CORR. FOR SEC. +	CORR. FOR SEC. +	TANGENT	'
0	.838671		.544639		.455361	.836078			1.53986	0
1	8829		4395		5605	6901	" Corr.		.54085	1
2	8987		4151		5849	7725	10 138		.54183	2
3	9146		3907		6093	8550	15 207		.54281	3
4	9304		3663		6337	9375	20 276		.54379	4
5	.839462		.543419		.456581	.840202	30 414		1.54478	5
6	9620		3174		6826	1029	40 552		.54576	6
7	9778		2930		7070	1857	45 621		.54675	7
8	9936		2686		7314	2687	50 689		.54774	8
9	.840094		2442		7558	3517			.54873	9
10	.840251		.542197		.457803	.844348			1.54972	10
11	0409	" Corr.	1953	" Corr.	8047	5180	" Corr.	" Corr.	.55071	11
12	0567	10 26	1708	10 41	8292	6012	10 139	10 17	.55170	12
13	0724	15 39	1464	15 61	8536	6846	15 209	15 25	.55269	13
14	0882	20 52	1219	20 82	8781	7681	20 279	20 33	.55368	14
15	.841039	30 79	.540974	30 122	.459026	.848516	30 418	30 50	1.55467	15
16	1196	40 105	0730	40 163	9270	9352	40 558	40 66	.55567	16
17	1354	45 118	0485	45 183	9515	.850190	45 627	45 75	.55666	17
18	1511	50 131	0240	50 204	9760	1028	50 697	50 83	.55766	18
19	1668		.539996		.460004	1867			.55866	19
20	.841825		.539751		.460249	.852707			1.55966	20
21	1982		9506		0494	3548	" Corr.		.56065	21
22	2139		9261		0739	4390	10 141		.56165	22
23	2296		9016		0984	5233	15 211		.56265	23
24	2452		8771		1229	6077	20 282		.56366	24
25	.842609		.538526		.461474	.856922	30 423		1.56466	25
26	2766		8281		1719	7767	40 564		.56566	26
27	2922		8035		1965	8614	45 634		.56667	27
28	3079		7790		2210	9461	50 705		.56767	28
29	3235		7545		2455	.860310			.56868	29
30	.843391		.537300		.462700	.861159			1.56969	30
31	3548		7054		2946	2009	" Corr.		.57069	31
32	3704		6809		3191	2860	10 142		.57170	32
33	3860		6563		3437	3713	15 214		.57271	33
34	4016		6318		3682	4566	20 285		.57372	34
35	.844172		.536072		.463928	.865420	30 427		1.57474	35
36	4328		5827		4173	6275	40 570		.57575	36
37	4484		5581		4419	7131	45 641		.57676	37
38	4640		5336		4664	7988	50 712		.57778	38
39	4795		5090		4910	8845			.57879	39
40	.844951		.534844		.465156	.869704			1.57981	40
41	5106	" Corr.	4598	" Corr.	5402	.870564	" Corr.	" Corr.	.58083	41
42	5262	10 26	4352	10 41	5648	1424	10 144	10 17	.58184	42
43	5417	15 39	4106	15 61	5894	2286	15 216	15 26	.58286	43
44	5573	20 52	3860	20 82	6140	3148	20 288	20 34	.58388	44
45	.845728	30 78	.533614	30 123	.466386	.874012	30 432	30 51	1.58490	45
46	5883	40 103	3368	40 164	6632	4876	40 576	40 68	.58593	46
47	6038	45 116	3122	45 185	6878	5742	45 648	45 77	.58695	47
48	6193	50 129	2876	50 205	7124	6608	50 720	50 85	.58797	48
49	6348		2630		7370	7476			.58900	49
50	.846503		.532384		.467616	.878344			1.59002	50
51	6658		2138		7862	9213	" Corr.		.59105	51
52	6813		1891		8109	.880083	10 145		.59208	52
53	6967		1645		8355	0954	15 218		.59311	53
54	7122		1399		8601	1827	20 291		.59414	54
55	.847276		.531152		.468848	.882700	30 437		1.59517	55
56	7431		0906		9094	3574	40 583		.59620	56
57	7585		0659		9341	4449	45 655		.59723	57
58	7740		0412		9588	5325	50 728		.59826	58
59	7894		0166		9834	6202			.59930	59
60	.848048		.529919		.470081	.887080			1.60033	60

VERSINES, EXSECANTS, AND TANGENTS

58°

'	SINE	CORR. FOR SEC.	COSINE	CORR. FOR SEC.	VERSINE	EXSEC	CORR. FOR SEC.	CORR. FOR SEC.	TANGENT	'
		+		− +			+	+		
0	.848048		.529919		.470081	.887080			1.60033	0
1	8202		9673		0327	7959	" Corr.		.60137	1
2	8356		9426		0574	8839	10 147		.60241	2
3	8510		9179		0821	9720	15 221		.60345	3
4	8664		8932		1068	.890602	20 295		.60449	4
5	.848818		.528685		.471315	.891484	30 442		1.60553	5
6	8972		8438		1562	2368	40 589		.60657	6
7	9125		8191		1809	3253	45 663		.60761	7
8	9279		7944		2056	4139	50 737		.60865	8
9	9432		7697		2303	5026			.60970	9
10	.849586		.527450		.472550	.895914			1.61074	10
11	9739	" Corr.	7203	" Corr.	2797	6803	" Corr.	" Corr.	.61179	11
12	9893	10 26	6956	10 41	3044	7692	10 149	10 18	.61283	12
13	.850046	15 38	6708	15 62	3292	8583	15 223	15 26	.61388	13
14	0199	20 51	6461	20 82	3539	9475	20 298	20 35	.61493	14
15	.850352	30 77	.526214	30 124	.473786	.900368	30 447	30 53	1.61598	15
16	0505	40 102	5966	40 165	4034	1262	40 596	40 70	.61703	16
17	0658	45 115	5719	45 186	4281	2156	45 670	45 79	.61808	17
18	0811	50 128	5472	50 206	4528	3052	50 745	50 88	.61914	18
19	0964		5224		4776	3949			.62019	19
20	.851117		.524977		.475023	.904847			1.62125	20
21	1269		4729		5271	5746	" Corr.		.62230	21
22	1422		4481		5519	6646	10 151		.62336	22
23	1574		4234		5766	7546	15 226		.62442	23
24	1727		3986		6014	8448	20 301		.62548	24
25	.851879		.523738		.476262	.909351	30 452		1.62654	25
26	2032		3490		6510	.910255	40 603		.62760	26
27	2184		3242		6758	1160	45 678		.62866	27
28	2336		2994		7006	2066	50 753		.62972	28
29	2488		2747		7253	2973			.63079	29
30	.852640		.522499		.477501	.913881			1.63185	30
31	2792		2250		7750	4790	" Corr.		.63292	31
32	2944		2002		7996	5700	10 152		.63398	32
33	3096		1754		8246	6611	15 229		.63505	33
34	3248		1506		8494	7523	20 305		.63612	34
35	.853399		.521258		.478742	.918436	30 457		1.63719	35
36	3551		1010		8990	9350	40 609		.63826	36
37	3702		0761		9239	.920266	45 686		.63934	37
38	3854		0513		9487	1182	50 762		.64041	38
39	4005		0265		9735	2099			.64148	39
40	.854156		.520016		.479984	.923017			1.64256	40
41	4308	" Corr.	.519768	" Corr.	.480232	3937	" Corr.	" Corr.	.64363	41
42	4459	10 25	9519	10 41	0481	4857	10 154	10 18	.64471	42
43	4610	15 38	9270	15 62	0730	5778	15 231	15 27	.64579	43
44	4761	20 50	9022	20 83	0978	6701	20 308	20 36	.64687	44
45	.854912	30 76	.518773	30 124	.481227	.927624	30 462	30 54	1.64795	45
46	5063	40 101	8525	40 166	1475	8549	40 616	40 72	.64903	46
47	5214	45 113	8276	45 186	1724	9475	45 693	45 81	.65011	47
48	5364	50 126	8027	50 207	1973	.930401	50 770	50 90	.65120	48
49	5515		7778		2222	1329			.65228	49
50	.855666		.517529		.482471	.932258			1.65337	50
51	5816		7280		2720	3188	" Corr.		.65445	51
52	5966		7031		2969	4118	10 156		.65554	52
53	6117		6782		3218	5050	15 234		.65663	53
54	6267		6533		3467	5984	20 312		.65772	54
55	.856417		.516284		.483716	.936918	30 468		1.65881	55
56	6567		6035		3965	7853	40 623		.65990	56
57	6718		5786		4214	8789	45 701		.66099	57
58	6868		5537		4463	9726	50 779		.66209	58
59	7017		5287		4713	.940665			.66318	59
60	.857167		.515038		.484962	.941604			1.66428	60

TABLE XX.—NATURAL SINES, COSINES,

59°

'	SINE	CORR. FOR SEC. +	COSINE	CORR. FOR SEC. − +	VERSINE	EXSEC	CORR. FOR SEC. +	CORR. FOR SEC. +	TANGENT	'
0	.857167		.515038		.484962	.941604			1.66428	0
1	7317		4789		5211	2544	" Corr.		.66538	1
2	7467		4539		5461	3486	10 158		.66647	2
3	7616		4290		5710	4429	15 236		.66757	3
4	7766		4040		5960	5372	20 315		.66867	4
5	.857916		.513791		.486209	.946317	30 473		1.66978	5
6	8065		3541		6459	7263	40 630		.67088	6
7	8214		3292		6708	8210	45 709		.67198	7
8	8364		3042		6958	9158	50 788		.67309	8
9	8513		2792		7208	.950108			.67419	9
10	.858662		.512542		.487458	.951058			1.67530	10
11	8811	" Corr.	2293	" Corr.	7707	2009	" Corr.	" Corr.	.67641	11
12	8960	10 25	2043	10 42	7957	2962	10 159	10 19	.67752	12
13	9109	15 37	1793	15 62	8207	3915	15 239	15 28	.67863	13
14	9258	20 50	1543	20 83	8457	4870	20 319	20 37	.67974	14
15	.859406	30 74	.511293	30 125	.488707	.955825	30 478	30 56	1.68085	15
16	9555	40 99	1043	40 167	8957	6782	40 638	40 74	.68196	16
17	9704	45 112	0793	45 187	9207	7740	45 718	45 83	.68308	17
18	9852	50 124	0543	50 208	9457	8699	50 797	50 93	.68419	18
19	.860001		0293		9707	9659			.68531	19
20	.860149		.510043		.489957	.960621			1.68643	20
21	0298		.509792		.490208	1583	" Corr.		.68754	21
22	0446		9542		0458	2546	10 161		.68866	22
23	0594		9292		0708	3511	15 242		.68979	23
24	0742		9041		0959	4477	20 323		.69091	24
25	.860890		.508791		.491209	.965444	30 484		1.69203	25
26	1038		8541		1459	6411	40 645		.69316	26
27	1186		8290		1710	7380	45 726		.69428	27
28	1334		8040		1960	8351	50 807		.69541	28
29	1482		7789		2211	9322			.69653	29
30	.861629		.507538		.492462	.970294			1.69766	30
31	1777		7288		2712	1268	" Corr.		.69879	31
32	1924		7037		2963	2243	10 163		.69992	32
33	2072		6786		3214	3218	15 245		.70106	33
34	2219		6536		3464	4195	20 326		.70219	34
35	.862366		.506285		.493715	.975174	30 490		1.70332	35
36	2514		6034		3966	6153	40 653		.70446	36
37	2661		5783		4217	7133	45 734		.70560	37
38	2808		5532		4468	8115	50 816		.70673	38
39	2955		5281		4719	9097			.70787	39
40	.863102		.505030		.494970	.980081			1.70901	40
41	3249	" Corr.	4779	" Corr.	5221	1066	" Corr.	" Corr.	.71015	41
42	3396	10 24	4528	10 42	5472	2052	10 165	10 19	.71129	42
43	3542	15 37	4276	15 63	5724	3039	15 248	15 29	.71244	43
44	3689	20 49	4025	20 84	5975	4028	20 330	20 38	.71358	44
45	.863836	30 73	.503774	30 126	.496226	.985017	30 495	30 57	1.71473	45
46	3982	40 98	3523	40 167	6477	6008	40 660	40 76	.71588	46
47	4128	45 110	3271	45 188	6729	7000	45 743	45 86	.71702	47
48	4275	50 122	3020	50 209	6980	7993	50 826	50 96	.71817	48
49	4421		2768		7232	8987			.71932	49
50	.864567		.502517		.497483	.989982			1.72047	50
51	4713		2266		7734	.990979	" Corr.		.72163	51
52	4860		2014		7986	1976	10 167		.72278	52
53	5006		1762		8238	2975	15 251		.72393	53
54	5151		1511		8489	3975	20 334		.72509	54
55	.865297		.501259		.498741	.994976	30 501		1.72625	55
56	5443		1007		8993	5979	40 668		.72741	56
57	5589		0756		9244	6982	45 752		.72857	57
58	5734		0504		9496	7987	50 835		.72973	58
59	5880		0252		9748	8993			.73089	59
60	.866025		.500000		.500000	1.00000			1.73205	60

VERSINES, EXSECANTS, AND TANGENTS

60°

'	SINE	CORR. FOR SEC.	COSINE	CORR. FOR SEC.	VERSINE	EXSEC	CORR. FOR SEC.	CORR. FOR SEC.	TANGENT	'
		+		− +			+	+		
0	.866025		.500000		.500000	1.00000			1.73205	0
1	6171		9748		0252	.00101			.73321	1
2	6316		9496		0504	.00202			.73438	2
3	6461		9244		0756	.00303			.73555	3
4	6607		8992		1008	.00404			.73671	4
5	.866752		.498740		.501260	1.00505			1.73788	5
6	6897		8488		1512	.00607	" Corr.	" Corr.	.73905	6
7	7042		8236		1764	.00708	10 17	10 20	.74022	7
8	7187		7983		2017	.00810	15 25	15 29	.74140	8
9	7331		7731		2269	.00912	20 34	20 39	.74257	9
10	.867476		.497479		.502521	1.01014	30 51	30 59	1.74375	10
11	7621	" Corr.	7226	" Corr.	2774	.01116	40 68	40 78	.74492	11
12	7766	10 24	6974	10 42	3026	.01218	45 76	45 88	.74610	12
13	7910	15 36	6722	15 63	3278	.01320	50 85	50 98	.74728	13
14	8054	20 48	6469	20 84	3531	.01422			.74846	14
15	.868199	30 72	.496216	30 126	.503784	1.01525			1.74964	15
16	8343	40 96	5964	40 168	4036	.01628			.75082	16
17	8487	45 108	5711	45 189	4289	.01730			.75200	17
18	8632	50 120	5459	50 210	4541	.01833			.75319	18
19	8776		5206		4794	.01936			.75437	19
20	.868920		.494953		.505047	1.02039			1.75556	20
21	9064		4700		5300	.02143			.75675	21
22	9207		4448		5552	.02246			.75794	22
23	9351		4195		5805	.02349			.75913	23
24	9495		3942		6058	.02453			.76032	24
25	.869639		.493689		.506311	1.02557			1.76151	25
26	9782		3436		6564	.02661	" Corr.	" Corr.	.76271	26
27	9926		3183		6817	.02765	10 17	10 20	.76390	27
28	.870069		2930		7070	.02869	15 26	15 30	.76510	28
29	0212		2677		7323	.02973	20 35	20 40	.76630	29
30	.870356		.492424		.507576	1.03077	30 52	30 60	1.76749	30
31	0499		2170		7830	.03182	40 70	40 80	.76869	31
32	0642		1917		8083	.03286	45 78	45 90	.76990	32
33	0785		1664		8336	.03391	50 87	50 100	.77110	33
34	0928		1410		8590	.03496			.77230	34
35	.871071		.491157		.508843	1.03601			1.77351	35
36	1214		0904		9096	.03706			.77471	36
37	1357		0650		9350	.03811			.77592	37
38	1499		0397		9603	.03916			.77713	38
39	1642		0143		9857	.04022			.77834	39
40	.871784		.489890		.510110	1.04128			1.77955	40
41	1927	" Corr.	9636	" Corr.	0364	.04233			.78077	41
42	2069	10 24	9382	10 42	0618	.04339			.78198	42
43	2212	15 36	9129	15 63	0871	.04445			.78319	43
44	2354	20 47	8875	20 85	1125	.04551			.78441	44
45	.872496	30 71	.488621	30 127	.511379	1.04658			1.78563	45
46	2638	40 95	8367	40 169	1633	.04764	" Corr.	" Corr.	.78685	46
47	2780	45 107	8114	45 190	1886	.04870	10 18	10 20	.78807	47
48	2922	50 118	7860	50 211	2140	.04977	15 27	15 31	.78929	48
49	3064		7606		2394	.05084	20 36	20 41	.79051	49
50	.873206		.487352		.512648	1.05191	30 53	30 61	1.79174	50
51	3348		7098		2902	.05298	40 71	40 82	.79296	51
52	3489		6844		3156	.05405	45 80	45 92	.79419	52
53	3631		6590		3410	.05512	50 89	50 102	.79542	53
54	3772		6335		3665	.05619			.79665	54
55	.873914		.486081		.513919	1.05727			1.79788	55
56	4055		5827		4173	.05835			.79911	56
57	4196		5573		4427	.05942			.80034	57
58	4338		5318		4682	.06050			.80158	58
59	4479		5064		4936	.06158			.80281	59
60	.874620		.484810		.515190	1.06267			1.80405	60

TABLE XX.—NATURAL SINES, COSINES,

61°

'	SINE	CORR. FOR SEC. +	COSINE	CORR. FOR SEC. − +	VERSINE	EXSEC	CORR. FOR SEC. +	CORR. FOR SEC. +	TANGENT	'
0	.874620		.484810		.515190	1.06267			1.80405	0
1	4761		4555		5445	.06375			.80529	1
2	4902		4301		5699	.06483			.80653	2
3	5042		4046		5954	.06592			.80777	3
4	5183		3792		6208	.06701			.80901	4
5	.875324		.483537		.516463	1.06809			1.81025	5
6	5464		3282		6718	.06918	" Corr.	" Corr.	.81150	6
7	5605		3028		6972	.07027	10 18	10 21	.81274	7
8	5746		2773		7227	.07137	15 27	15 31	.81399	8
9	5886		2518		7482	.07246	20 37	20 42	.81524	9
10	.876026		.482263		.517737	1.07356	30 55	30 63	1.81649	10
11	6166	" Corr.	2009	" Corr.	7991	.07465	40 73	40 83	.81774	11
12	6307	10 23	1754	10 42	8246	.07575	45 82	45 94	.81899	12
13	6447	15 35	1499	15 64	8501	.07685	50 91	50 104	.82025	13
14	6587	20 47	1244	20 85	8756	.07795			.82150	14
15	.876727	30 70	.480989	30 127	.519011	1.07905			1.82276	15
16	6867	40 93	0734	40 170	9266	.08015			.82402	16
17	7006	45 105	0479	45 191	9521	.08126			.82528	17
18	7146	50 117	0224	50 212	9776	.08236			.82654	18
19	7286		.479968		.520032	.08347			.82780	19
20	.877425		.479713		.520287	1.08458			1.82906	20
21	7565		9458		0542	.08569			.83033	21
22	7704		9203		0797	.08680			.83159	22
23	7844		8947		1053	.08791			.83286	23
24	7983		8692		1308	.08903			.83413	24
25	.878122		.478436		.521564	1.09014			1.83540	25
26	8261		8181		1819	.09126	" Corr.	" Corr.	.83667	26
27	8400		7926		2074	.09238	10 19	10 21	.83794	27
28	8539		7670		2330	.09350	15 28	15 32	.83922	28
29	8678		7414		2586	.09462	20 37	20 43	.84049	29
30	.878817		.477159		.522841	1.09574	30 56	30 64	1.84177	30
31	8956		6903		3097	.09686	40 75	40 85	.84305	31
32	9095		6647		3353	.09799	45 84	45 96	.84433	32
33	9233		6392		3608	.09911	50 94	50 106	.84561	33
34	9372		6136		3864	.10024			.84689	34
35	.879510		.475880		.524120	1.10137			1.84818	35
36	9649		5624		4376	.10250			.84946	36
37	9787		5368		4632	.10363			.85075	37
38	9925		5112		4888	.10477			.85204	38
39	.880063		4856		5144	.10590			.85333	39
40	.880201		.474600		.525400	1.10704			1.85462	40
41	0339	" Corr.	4344	" Corr.	5656	.10817			.85591	41
42	0477	10 23	4088	10 43	5912	.10931			.85720	42
43	0615	15 34	3832	15 64	6168	.11045			.85850	43
44	0753	20 46	3576	20 85	6424	.11159			.85979	44
45	.880891	30 69	.473320	30 128	.526680	1.11274			1.86109	45
46	1028	40 92	3063	40 171	6937	.11388	" Corr.	" Corr.	.86239	46
47	1166	45 103	2807	45 192	7193	.11503	10 19	10 22	.86369	47
48	1304	50 115	2551	50 213	7449	.11617	15 29	15 33	.86499	48
49	1441		2294		7706	.11732	20 38	20 44	.86630	49
50	.881578		.472038		.527962	1.11847	30 58	30 65	1.86760	50
51	1716		1782		8218	.11963	40 77	40 87	.86891	51
52	1853		1525		8475	.12078	45 86	45 98	.87021	52
53	1990		1268		8732	.12193	50 96	50 109	.87152	53
54	2127		1012		8988	.12309			.87283	54
55	.882264		.470755		.529245	1.12425			1.87415	55
56	2401		0499		9501	.12540			.87546	56
57	2538		0242		9758	.12657			.87677	57
58	2674		.469985		.530015	.12773			.87809	58
59	2811		9728		0272	.12889			.87941	59
60	.882948		.469472		.530528	1.13005			1.88073	60

VERSINES, EXSECANTS, AND TANGENTS
62°

'	SINE	CORR. FOR SEC. +	COSINE	CORR. FOR SEC. − +	VERSINE	EXSEC	CORR. FOR SEC. +	CORR. FOR SEC. +	TANGENT	'
0	.882948		.469472		.530528	1.13005			1.88073	0
1	3084		9215		0785	.13122			.88205	1
2	3221		8958		1042	.13239			.88337	2
3	3357		8701		1299	.13356			.88469	3
4	3493		8444		1556	.13473			.88602	4
5	.883630		.468187		.531813	1.13590			1.88734	5
6	3766		7930		2070	.13707	" Corr.	" Corr.	.88867	6
7	3902		7673		2327	.13825	10 20	10 22	.89000	7
8	4038		7416		2584	.13942	15 30	15 33	.89133	8
9	4174		7158		2842	.14060	20 39	20 44	.89266	9
10	.884310		.466901		.533099	1.14178	30 59	30 67	1.89400	10
11	4445	" Corr.	6644	" Corr.	3356	.14296	40 79	40 89	.89533	11
12	4581	10 23	6387	10 43	3613	.14414	45 89	45 100	.89667	12
13	4717	15 34	6129	15 64	3871	.14533	50 98	50 111	.89801	13
14	4852	20 45	5872	20 86	4128	.14651			.89935	14
15	.884988	30 68	.465614	30 129	.534386	1.14770			1.90069	15
16	5123	40 90	5357	40 172	4643	.14889			.90203	16
17	5258	45 102	5100	45 193	4900	.15008			.90337	17
18	5394	50 113	4842	50 214	5158	.15127			.90472	18
19	5529		4584		5416	.15246			.90607	19
20	.885664		.464327		.535673	1.15366			1.90741	20
21	5799		4069		5931	.15485			.90876	21
22	5934		3812		6188	.15605			.91012	22
23	6069		3554		6446	.15725			.91147	23
24	6204		3296		6704	.15845			.91282	24
25	.886338		.463038		.536962	1.15965			1.91418	25
26	6473		2780		7220	.16085	" Corr.	" Corr.	.91554	26
27	6608		2522		7478	.16206	10 20	10 23	.91690	27
28	6742		2265		7735	.16326	15 30	15 34	.91826	28
29	6876		2007		7993	.16447	20 40	20 45	.91962	29
30	.887011		.461749		.538251	1.16568	30 61	30 68	1.92098	30
31	7145		1491		8509	.16689	40 81	40 91	.92235	31
32	7279		1232		8768	.16810	45 91	45 102	.92371	32
33	7413		0974		9026	.16932	50 101	50 114	.92508	33
34	7548		0716		9284	.17053			.92645	34
35	.887682		.460458		.539542	1.17175			1.92782	35
36	7815		0200		9800	.17297			.92920	36
37	7949		.459942		.540058	.17419			.93057	37
38	8083		9683		0317	.17541			.93195	38
39	8217		9425		0575	.17663			.93332	39
40	.888350		.459166		.540834	1.17786			1.93470	40
41	8484	" Corr.	8908	" Corr.	1092	.17909			.93608	41
42	8617	10 22	8650	10 43	1350	.18031			.93746	42
43	8751	15 33	8391	15 65	1609	.18154			.93885	43
44	8884	20 44	8132	20 86	1868	.18277			.94023	44
45	.889017	30 67	.457874	30 129	.542126	1.18401			1.94162	45
46	9150	40 89	7615	40 172	2385	.18524	" Corr.	" Corr.	.94301	46
47	9283	45 100	7357	45 194	2643	.18648	10 21	10 23	.94440	47
48	9416	50 111	7098	50 215	2902	.18772	15 31	15 35	.94579	48
49	9549		6839		3161	.18895	20 41	20 47	.94718	49
50	.889682		.456580		.543420	1.19019	30 62	30 70	1.94858	50
51	9815		6322		3678	.19144	40 83	40 93	.94997	51
52	9948		6063		3937	.19268	45 93	45 105	.95137	52
53	.890080		5804		4196	.19393	50 104	50 116	.95277	53
54	0213		5545		4455	.19517			.95417	54
55	.890345		.455286		.544714	1.19642			1.95557	55
56	0478		5027		4973	.19767			.95698	56
57	0610		4768		5232	.19892			.95838	57
58	0742		4509		5491	.20018			.95979	58
59	0874		4250		5750	.20143			.96120	59
60	.891006		.453990		.546010	1.20269			1.96261	60

TABLE XX.—NATURAL SINES, COSINES,

63°

′	SINE	CORR. FOR SEC. +	COSINE	CORR. FOR SEC. − +	VERSINE	EXSEC	CORR. FOR SEC. +	CORR. FOR SEC. +	TANGENT	′
0	.891006		.453990		.546010	1.20269			1.96261	0
1	1138		3731		6269	.20395			.96402	1
2	1270		3472		6528	.20521			.96544	2
3	1402		3213		6787	.20647			.96685	3
4	1534		2954		7046	.20773			.96827	4
5	.891666		.452694		.547306	1.20900			1.96969	5
6	1798		2435		7565	.21026	" Corr.	" Corr.	.97111	6
7	1929		2175		7825	.21153	10 21	10 24	.97253	7
8	2061		1916		8084	.21280	15 32	15 36	.97395	8
9	2192		1656		8344	.21407	20 42	20 48	.97538	9
10	.892323		.451397		.548603	1.21535	30 64	30 71	1.97680	10
11	2455	" Corr.	1137	" Corr.	8863	.21662	40 85	40 95	.97823	11
12	2586	10 22	0878	10 43	9122	.21790	45 96	45 107	.97966	12
13	2717	15 33	0618	15 65	9382	.21918	50 106	50 119	.98110	13
14	2848	20 44	0358	20 87	9642	.22045			.98253	14
15	.892979	30 65	.450098	30 130	.549902	1.22174			1.98396	15
16	3110	40 87	.449839	40 173	.550161	.22302			.98540	16
17	3241	45 98	9579	45 195	0421	.22430			.98684	17
18	3371	50 109	9319	50 216	0681	.22559			.98828	18
19	3502		9059		0941	.22688			.98972	19
20	.893633		.448799		.551201	1.22817			1.99116	20
21	3763		8539		1461	.22946			.99261	21
22	3894		8279		1721	.23075			.99406	22
23	4024		8019		1981	.23205			.99550	23
24	4154		7759		2241	.23334			.99695	24
25	.894284		.447499		.552501	1.23464			1.99841	25
26	4415		7239		2761	.23594	" Corr.	" Corr.	.99986	26
27	4545		6979		3021	.23724	10 22	10 24	2.00131	27
28	4675		6718		3282	.23855	15 33	15 37	.00277	28
29	4804		6458		3542	.23985	20 44	20 49	.00423	29
30	.894934		.446198		.553802	1.24116	30 65	30 73	2.00569	30
31	5064		5938		4062	.24247	40 87	40 97	.00715	31
32	5194		5677		4323	.24378	45 98	45 110	.00862	32
33	5323		5417		4583	.24509	50 109	50 122	.01008	33
34	5453		5156		4844	.24640			.01155	34
35	.895582		.444896		.555104	1.24772			2.01302	35
36	5712		4635		5365	.24903			.01449	36
37	5841		4375		5625	.25035			.01596	37
38	5970		4114		5886	.25167			.01743	38
39	6099		3853		6147	.25300			.01891	39
40	.896228		.443593		.556407	1.25432			2.02039	40
41	6358	" Corr.	3332	" Corr.	6668	.25565			.02187	41
42	6486	10 21	3071	10 43	6929	.25697			.02335	42
43	6615	15 32	2810	15 65	7190	.25830			.02483	43
44	6744	20 43	2550	20 87	7450	.25963			.02631	44
45	.896873	30 64	.442289	30 130	.557711	1.26097			2.02780	45
46	7001	40 86	2028	40 174	7972	.26230	" Corr.	" Corr.	.02929	46
47	7130	45 97	1767	45 196	8233	.26364	10 22	10 25	.03078	47
48	7258	50 107	1506	50 217	8494	.26498	15 34	15 37	.03227	48
49	7387		1245		8755	.26632	20 45	20 50	.03376	49
50	.897515		.440984		.559016	1.26766	30 67	30 75	2.03526	50
51	7643		0723		9277	.26900	40 90	40 100	.03675	51
52	7772		0462		9538	.27035	45 101	45 112	.03825	52
53	7900		0200		9800	.27169	50 112	50 125	.03975	53
54	8028		.439939		.560061	.27304			.04125	54
55	.898156		.439678		.560322	1.27439			2.04276	55
56	8283		9417		0583	.27574			.04426	56
57	8411		9155		0845	.27710			.04577	57
58	8539		8894		1106	.27845			.04728	58
59	8666		8633		1367	.27981			.04879	59
60	.898794		.438371		.561629	1.28117			2.05030	60

VERSINES, EXSECANTS, AND TANGENTS

64°

'	SINE	CORR. FOR SEC.	COSINE	CORR. FOR SEC.	VERSINE	EXSEC	CORR. FOR SEC.	CORR. FOR SEC.	TANGENT	'
		+		− +			+	+		
0	.898794		.438371		.561629	1.28117			2.05030	0
1	8922		8110		1890	.28253	" Corr.	" Corr.	.05182	1
2	9049		7848		2152	.28390	10 23	10 25	.05333	2
3	9176		7587		2413	.28526	15 34	15 38	.05485	3
4	9304		7325		2675	.28663	20 46	20 51	.05637	4
5	.899431		.437063		.562937	1.28800	30 69	30 76	2.05790	5
6	9558		6802		3198	.28937	40 92	40 102	.05942	6
7	9685		6540		3460	.29074	45 103	45 114	.06094	7
8	9812		6278		3722	.29211	50 114	50 127	.06247	8
9	9939		6017		3983	.29349			.06400	9
10	.900065		.435755		.564245	1.29487			2.06553	10
11	0192	" Corr.	5493	" Corr.	4507	.29625			.06706	11
12	0319	10 21	5231	10 44	4769	.29763			.06860	12
13	0445	15 32	4969	15 65	5031	.29901			.07014	13
14	0572	20 42	4707	20 87	5293	.30040			.07167	14
15	.900698	30 63	.434445	30 131	.565555	1.30179			2.07321	15
16	0825	40 84	4183	40 175	5817	.30318	" Corr.	" Corr.	.07476	16
17	0951	45 95	3921	45 196	6079	.30457	10 23	10 26	.07630	17
18	1077	50 105	3659	50 218	6341	.30596	15 35	15 39	.07785	18
19	1203		3397		6603	.30735	20 47	20 52	.07939	19
20	.901329		.433135		.566865	1.30875	30 70	30 78	2.08094	20
21	1455		2873		7127	.31015	40 93	40 104	.08250	21
22	1581		2610		7390	.31155	45 105	45 117	.08405	22
23	1707		2348		7652	.31295	50 117	50 130	.08560	23
24	1832		2086		7914	.31436			.08716	24
25	.901958		.431823		.568177	1.31576			2.08872	25
26	2084		1561		8439	.31717			.09028	26
27	2209		1299		8701	.31858			.09184	27
28	2335		1036		8964	.31999			.09341	28
29	2460		0774		9226	.32140			.09498	29
30	.902585		.430511		.569489	1.32282			2.09654	30
31	2710		0248		9752	.32424	" Corr.	" Corr.	.09811	31
32	2836		.429986		.570014	.32566	10 24	10 26	.09969	32
33	2961		9723		0277	.32708	15 36	15 40	.10126	33
34	3086		9461		0539	.32850	20 48	20 53	.10284	34
35	.903210		.429198		.570802	1.32993	30 72	30 79	2.10442	35
36	3335		8935		1065	.33135	40 95	40 106	.10600	36
37	3460		8672		1328	.33278	45 107	45 119	.10758	37
38	3585		8410		1590	.33422	50 119	50 132	.10916	38
39	3709		8147		1853	.33565			.11075	39
40	.903834		.427884		.572116	1.33708			2.11233	40
41	3958	" Corr.	7621	" Corr.	2379	.33852			.11392	41
42	4082	10 21	7358	10 44	2642	.33996			.11552	42
43	4207	15 31	7095	15 66	2905	.34140			.11711	43
44	4331	20 41	6832	20 88	3168	.34284			.11871	44
45	.904455	30 62	.426569	30 132	.573431	1.34429			2.12030	45
46	4579	40 83	6306	40 175	3694	.34573	" Corr.	" Corr.	.12190	46
47	4703	45 93	6042	45 197	3958	.34718	10 24	10 27	.12350	47
48	4827	50 103	5779	50 219	4221	.34863	15 37	15 40	.12511	48
49	4951		5516		4484	.35009	20 49	20 54	.12671	49
50	.905075		.425253		.574747	1.35154	30 73	30 81	2.12832	50
51	5198		4990		5010	.35300	40 97	40 108	.12993	51
52	5322		4726		5274	.35446	45 110	45 121	.13154	52
53	5445		4463		5537	.35592	50 122	50 134	.13316	53
54	5569		4199		5801	.35738			.13477	54
55	.905692		.423936		.576064	1.35885			2.13639	55
56	5815		3672		6328	.36031			.13801	56
57	5939		3409		6591	.36178			.13963	57
58	6062		3146		6854	.36325			.14125	58
59	6185		2882		7118	.36473			.14288	59
60	.906308		.422618		.577382	1.36620			2.14451	60

TABLE XX.—NATURAL SINES, COSINES,
65°

′	SINE	CORR. FOR SEC.	COSINE	CORR. FOR SEC.	VERSINE	EXSEC	CORR. FOR SEC.	CORR. FOR SEC.	TANGENT	′
		+		− +			+	+		
0	.906308		.422618		.577382	1.36620			2.14451	0
1	6431		2355		7645	.36768	″ Corr.	″ Corr.	.14614	1
2	6554		2091		7909	.36916	10 25	10 27	.14777	2
3	6676		1827		8173	.37064	15 37	15 41	.14940	3
4	6799		1563		8437	.37212	20 50	20 55	.15104	4
5	.906922		.421300		.578700	1.37361	30 75	30 82	2.15268	5
6	7044		1036		8964	.37509	40 99	40 110	.15432	6
7	7166		0772		9228	.37658	45 112	45 123	.15596	7
8	7289		0508		9492	.37808	50 124	50 137	.15760	8
9	7411		0244		9756	.37957			.15925	9
10	.907533		.419980		.580020	1.38106			2.16090	10
11	7655	″ Corr.	9716	″ Corr.	0284	.38256			.16255	11
12	7778	10 20	9452	10 44	0548	.38406			.16420	12
13	7900	15 30	9188	15 66	0812	.38556			.16585	13
14	8021	20 41	8924	20 88	1076	.38707			.16751	14
15	.908143	30 61	.418660	30 132	.581340	1.38857			2.16917	15
16	8265	40 81	8396	40 176	1604	.39008	″ Corr.	″ Corr.	.17083	16
17	8387	45 91	8131	45 198	1869	.39159	10 25	10 28	.17249	17
18	8508	50 101	7867	50 220	2133	.39311	15 38	15 42	.17416	18
19	8630		7603		2397	.39462	20 51	20 56	.17582	19
20	.908751		.417338		.582662	1.39614	30 76	30 84	2.17749	20
21	8872		7074		2926	.39766	40 102	40 112	.17916	21
22	8994		6810		3190	.39918	45 114	45 126	.18084	22
23	9115		6545		3455	.40070	50 127	50 140	.18251	23
24	9236		6281		3719	.40222			.18419	24
25	.909357		.416016		.583984	1.40375			2.18587	25
26	9478		5752		4248	.40528			.18755	26
27	9599		5487		4513	.40681			.18923	27
28	9720		5223		4777	.40835			.19092	28
29	9841		4958		5042	.40988			.19261	29
30	.909961		.414693		.585307	1.41142			2.19430	30
31	.910082		4428		5572	.41296	″ Corr.	″ Corr.	.19599	31
32	0202		4164		5836	.41450	10 26	10 28	.19769	32
33	0323		3899		6101	.41605	15 39	15 43	.19938	33
34	0443		3634		6366	.41760	20 52	20 57	.20108	34
35	.910564		.413369		.586631	1.41914	30 78	30 85	2.20278	35
36	0684		3104		6896	.42070	40 104	40 114	.20449	36
37	0804		2840		7160	.42225	45 117	45 128	.20619	37
38	0924		2574		7426	.42380	50 130	50 142	.20790	38
39	1044		2310		7690	.42536			.20961	39
40	.911164		.412044		.587956	1.42692			2.21132	40
41	1284	″ Corr.	1780	″ Corr.	8220	.42848			.21304	41
42	1403	10 20	1514	10 44	8486	.43005			.21475	42
43	1523	15 30	1249	15 66	8751	.43162			.21647	43
44	1642	20 40	0984	20 88	9016	.43318			.21819	44
45	.911762	30 60	.410719	30 133	.589281	1.43476			2.21992	45
46	1882	40 80	0454	40 177	9546	.43633	″ Corr.	″ Corr.	.22164	46
47	2001	45 90	0188	45 199	9812	.43790	10 26	10 29	.22337	47
48	2120	50 100	.409923	50 221	.590077	.43948	15 40	15 44	.22510	48
49	2239		9658		0342	.44106	20 53	20 58	.22683	49
50	.912358		.409392		.590608	1.44264	30 79	30 87	2.22857	50
51	2478		9127		0873	.44423	40 106	40 116	.23030	51
52	2596		8862		1138	.44582	45 119	45 131	.23204	52
53	2715		8596		1404	.44741	50 132	50 145	.23378	53
54	2834		8330		1670	.44900			.23553	54
55	.912953		.408065		.591935	1.45059			2.23727	55
56	3072		7799		2201	.45219			.23902	56
57	3190		7534		2466	.45378			.24077	57
58	3309		7268		2732	.45539			.24252	58
59	3427		7002		2998	.45699			.24428	59
60	.913546		.406737		.593263	1.45859			2.24604	60

VERSINES, EXSECANTS, AND TANGENTS
66°

'	SINE	CORR. FOR SEC.	COSINE	CORR. FOR SEC.	VERSINE	EXSEC	CORR. FOR SEC.	CORR. FOR SEC.	TANGENT	'
		+		− +			+	+		
0	.913546		.406737		.593263	1.45859			2.24604	0
1	3664		6471		3529	.46020	" Corr.	" Corr.	.24780	1
2	3782		6205		3795	.46181	10 27	10 30	.24956	2
3	3900		5939		4061	.46342	15 41	15 44	.25132	3
4	4018		5673		4327	.46504	20 54	20 59	.25309	4
5	.914136		.405408		.594592	1.46665	30 81	30 89	2.25486	5
6	4254		5142		4858	.46827	40 108	40 118	.25663	6
7	4372		4876		5124	.46989	45 122	45 133	.25840	7
8	4490		4610		5390	.47152	50 135	50 148	.26018	8
9	4607		4344		5656	.47314			.26196	9
10	.914725		.404078		.595922	1.47477			2.26374	10
11	4842	" Corr.	3811	" Corr.	6189	.47640			.26552	11
12	4960	10 20	3545	10 44	6455	.47804			.26730	12
13	5077	15 29	3279	15 67	6721	.47967			.26909	13
14	5194	20 39	3013	20 89	6987	.48131			.27088	14
15	.915312	30 59	.402747	30 133	.597253	1.48295			2.27267	15
16	5429	40 78	2480	40 178	7520	.48459	" Corr.	" Corr.	.27447	16
17	5546	45 88	2214	45 200	7786	.48624	10 27	10 30	.27626	17
18	5663	50 98	1948	50 222	8052	.48789	15 41	15 45	.27806	18
19	5780		1681		8319	.48954	20 55	20 60	.27987	19
20	.915896		.401415		.598585	1.49119	30 82	30 90	2.28167	20
21	6013		1149		8851	.49284	40 110	40 120	.28348	21
22	6130		0882		9118	.49450	45 124	45 135	.28528	22
23	6246		0616		9384	.49616	50 137	50 150	.28710	23
24	6363		0349		9651	.49782			.28891	24
25	.916479		.400082		.599918	1.49948			2.29073	25
26	6596		.399816		.600184	.50115	" Corr.	" Corr.	.29254	26
27	6712		9549		0451	.50282	10 28	10 31	.29437	27
28	6828		9282		0718	.50449	15 42	15 46	.29619	28
29	6944		9016		0984	.50617	20 56	20 61	.29801	29
30	.917060		.398749		.601251	1.50784	30 84	30 92	2.29984	30
31	7176		8482		1518	.50952	40 112	40 122	.30167	31
32	7292		8216		1784	.51120	45 126	45 137	.30351	32
33	7408		7949		2051	.51289	50 140	50 153	.30534	33
34	7523		7682		2318	.51457			.30718	34
35	.917639		.397415		.602585	1.51626			2.30902	35
36	7755		7148		2852	.51795			.31086	36
37	7870		6881		3119	.51965			.31271	37
38	7986		6614		3386	.52134			.31456	38
39	8101		6347		3653	.52304			.31641	39
40	.918216		.396080		.603920	1.52474			2.31826	40
41	8331	" Corr.	5813	" Corr.	4187	.52645	" Corr.	" Corr.	.32012	41
42	8446	10 19	5546	10 45	4454	.52815	10 28	10 31	.32197	42
43	8561	15 29	5278	15 67	4722	.52986	15 43	15 46	.32383	43
44	8676	20 38	5011	20 89	4989	.53157	20 57	20 62	.32570	44
45	.918791	30 57	.394744	30 134	.605256	1.53329	30 85	30 93	2.32756	45
46	8906	40 77	4477	40 178	5523	.53500	40 114	40 124	.32943	46
47	9021	45 86	4209	45 200	5791	.53672	45 128	45 139	.33130	47
48	9135	50 96	3942	50 223	6058	.53845	50 142	50 155	.33317	48
49	9250		3674		6326	.54017			.33505	49
50	.919364		.393407		.606593	1.54190			2.33693	50
51	9479		3140		6860	.54363	" Corr.	" Corr.	.33881	51
52	9593		2872		7128	.54536	10 29	10 32	.34069	52
53	9707		2605		7395	.54709	15 43	15 47	.34258	53
54	9822		2337		7663	.54883	20 58	20 63	.34447	54
55	.919936		.392070		.607930	1.55057	30 87	30 95	2.34636	55
56	.920050		1802		8198	.55231	40 116	40 126	.34825	56
57	0164		1534		8466	.55405	45 130	45 142	.35015	57
58	0277		1267		8733	.55580	50 145	50 158	.35205	58
59	0391		0999		9001	.55755			.35395	59
60	.920505		.390731		.609269	1.55930			2.35585	60

TABLE XX.—NATURAL SINES, COSINES,

67°

'	SINE	CORR. FOR SEC.	COSINE	CORR. FOR SEC.	VERSINE	EXSEC	CORR. FOR SEC.	CORR. FOR SEC.	TANGENT	'
		+		− +			+	+		
0	.920505		.390731		.609269	1.55930			2.35585	0
1	0618		0463		9537	.56106	" Corr.	" Corr.	.35776	1
2	0732		0196		9804	.56282	10 30	10 32	.35967	2
3	0846		.389928		.610072	.56458	15 44	15 48	.36158	3
4	0959		9660		0340	.56634	20 59	20 64	.36349	4
5	.921072		.389392		.610608	1.56811	30 89	30 96	2.36541	5
6	1185		9124		0876	.56988	40 118	40 129	.36733	6
7	1299		8856		1144	.57165	45 133	45 145	.36925	7
8	1412		8588		1412	.57342	50 148	50 161	.37118	8
9	1525		8320		1680	.57520			.37311	9
10	.921638		.388052		.611948	1.57698			2.37504	10
11	1750	" Corr.	7784	" Corr.	2216	.57876			.37697	11
12	1863	10 19	7516	10 45	2484	.58054			.37891	12
13	1976	15 28	7247	15 67	2753	.58233			.38084	13
14	2088	20 37	6979	20 89	3021	.58412			.38279	14
15	.922201	30 56	.386711	30 134	.613289	1.58591			2.38473	15
16	2313	40 75	6443	40 179	3557	.58771	" Corr.	" Corr.	.38668	16
17	2426	45 84	6174	45 201	3826	.58950	10 30	10 33	.38862	17
18	2538	50 94	5906	50 224	4094	.59130	15 45	15 49	.39058	18
19	2650		5638		4362	.59311	20 60	20 65	.39253	19
20	.922762		.385369		.614631	1.59491	30 90	30 98	2.39449	20
21	2874		5101		4899	.59672	40 120	40 130	.39645	21
22	2986		4832		5168	.59853	45 135	45 147	.39841	22
23	3098		4564		5436	.60035	50 150	50 163	.40038	23
24	3210		4295		5705	.60217			.40235	24
25	.923322		.384027		.615973	1.60399			2.40432	25
26	3434		3758		6242	.60581	" Corr.	" Corr.	.40629	26
27	3545		3490		6510	.60763	10 31	10 33	.40827	27
28	3657		3221		6779	.60946	15 46	15 50	.41025	28
29	3768		2952		7048	.61129	20 61	20 66	.41223	29
30	.923880		.382683		.617317	1.61313	30 92	30 99	2.41421	30
31	3991		2415		7585	.61496	40 122	40 133	.41620	31
32	4102		2146		7854	.61680	45 138	45 149	.41819	32
33	4213		1877		8123	.61864	50 153	50 166	.42019	33
34	4324		1608		8392	.62049			.42218	34
35	.924435		.381339		.618661	1.62234			2.42418	35
36	4546		1070		8930	.62419			.42618	36
37	4657		0801		9199	.62604			.42819	37
38	4768		0532		9468	.62790			.43019	38
39	4878		0263		9737	.62976			.43220	39
40	.924989		.379994		.620006	1.63162			2.43422	40
41	5099	" Corr.	9725	" Corr.	0275	.63348	" Corr.	" Corr.	.43623	41
42	5210	10 18	9456	10 45	0544	.63535	10 31	10 34	.43825	42
43	5320	15 28	9187	15 67	0813	.63722	15 47	15 51	.44027	43
44	5430	20 37	8918	20 90	1082	.63909	20 62	20 67	.44230	44
45	.925540	30 55	.378649	30 135	.621351	1.64097	30 94	30 101	2.44433	45
46	5651	40 73	8379	40 180	1621	.64285	40 125	40 135	.44636	46
47	5761	45 83	8110	45 202	1890	.64473	45 140	45 152	.44839	47
48	5871	50 92	7841	50 224	2159	.64662	50 156	50 168	.45043	48
49	5980		7571		2429	.64851			.45246	49
50	.926090		.377302		.622698	1.65040			2.45451	50
51	6200		7033		2967	.65229	" Corr.	" Corr.	.45655	51
52	6310		6763		3237	.65419	10 32	10 34	.45860	52
53	6419		6494		3506	.65609	15 48	15 51	.46065	53
54	6529		6224		3776	.65799	20 64	20 69	.46270	54
55	.926638		.375955		.624045	1.65989	30 95	30 103	2.46476	55
56	6747		5685		4315	.66180	40 127	40 137	.46682	56
57	6857		5416		4584	.66371	45 143	45 154	.46888	57
58	6966		5146		4854	.66563	50 159	50 171	.47095	58
59	7075		4876		5124	.66755			.47302	59
60	.927184		.374607		.625393	1.66947			2.47509	60

VERSINES, EXSECANTS, AND TANGENTS
68°

'	SINE	CORR. FOR SEC.	COSINE	CORR. FOR SEC.	VERSINE	EXSEC	CORR. FOR SEC.	CORR. FOR SEC.	TANGENT	'
0	.927184	+	.374607	− +	.625393	1.66947	+	+	2.47509	0
1	7293		4337		5663	.67139	" Corr.	" Corr.	.47716	1
2	7402		4067		5933	.67332	10 32	10 35	.47924	2
3	7510		3797		6203	.67525	15 48	15 52	.48132	3
4	7619		3528		6472	.67718	20 65	20 70	.48340	4
5	.927728		.373258		.626742	1.67911	30 97	30 104	2.48549	5
6	7836		2988		7012	.68105	40 129	40 139	.48758	6
7	7945		2718		7282	.68299	45 145	45 157	.48967	7
8	8053		2448		7552	.68494	50 162	50 174	.49177	8
9	8161		2178		7822	.68689			.49386	9
10	.928270		.371908		.628092	1.68884			2.49597	10
11	8378	" Corr.	1638	" Corr.	8362	.69079	" Corr.	" Corr.	.49807	11
12	8486	10 18	1368	10 45	8632	.69275	10 33	10 35	.50018	12
13	8594	15 27	1098	15 68	8902	.69471	15 49	15 53	.50229	13
14	8702	20 36	0828	20 90	9172	.69667	20 66	20 71	.50440	14
15	.928810	30 54	.370557	30 135	.629443	1.69864	30 98	30 106	2.50652	15
16	8917	40 72	0287	40 180	9713	.70061	40 131	40 141	.50864	16
17	9025	45 81	0017	45 203	9983	.70258	45 148	45 159	.51076	17
18	9133	50 90	.369747	50 225	.630253	.70455	50 164	50 177	.51289	18
19	9240		9476		0524	.70653			.51502	19
20	.929348		.369206		.630794	1.70851			2.51715	20
21	9455		8936		1064	.71050	" Corr.	" Corr.	.51929	21
22	9562		8665		1335	.71249	10 33	10 36	.52142	22
23	9669		8395		1605	.71448	15 50	15 54	.52357	23
24	9776		8125		1875	.71647	20 67	20 72	.52571	24
25	.929884		.367854		.632146	1.71847	30 100	30 108	2.52786	25
26	9990		7584		2416	.72047	40 133	40 143	.53001	26
27	.930097		7313		2687	.72247	45 150	45 161	.53217	27
28	0204		7042		2958	.72448	50 167	50 179	.53432	28
29	0311		6772		3228	.72649			.53648	29
30	.930418		.366501		.633499	1.72850			2.53865	30
31	0524		6231		3769	.73052	" Corr.	" Corr.	.54082	31
32	0631		5960		4040	.73254	10 34	10 36	.54299	32
33	0737		5689		4311	.73456	15 51	15 55	.54516	33
34	0843		5418		4582	.73659	20 68	20 73	.54734	34
35	.930950		.365148		.634852	1.73862	30 102	30 109	2.54952	35
36	1056		4877		5123	.74065	40 136	40 146	.55170	36
37	1162		4606		5394	.74269	45 152	45 164	.55389	37
38	1268		4335		5665	.74473	50 169	50 182	.55608	38
39	1374		4064		5936	.74677			.55827	39
40	.931480		.363793		.636207	1.74881			2.56046	40
41	1586	" Corr.	3522	" Corr.	6478	.75086	" Corr.	" Corr.	.56266	41
42	1691	10 18	3251	10 45	6749	.75292	10 34	10 37	.56487	42
43	1797	15 26	2980	15 68	7020	.75497	15 52	15 55	.56707	43
44	1902	20 35	2709	20 90	7291	.75703	20 69	20 74	.56928	44
45	.932008	30 53	.362438	30 136	.637562	1.75909	30 103	30 111	2.57150	45
46	2113	40 70	2167	40 181	7833	.76116	40 138	40 148	.57371	46
47	2219	45 79	1896	45 203	8104	.76323	45 155	45 166	.57593	47
48	2324	50 88	1625	50 226	8375	.76530	50 172	50 185	.57815	48
49	2429		1353		8647	.76737			.58038	49
50	.932534		.361082		.638918	1.76945			2.58261	50
51	2639		0811		9189	.77154	" Corr.	" Corr.	.58484	51
52	2744		0540		9460	.77362	10 35	10 37	.58708	52
53	2849		0268		9732	.77571	15 52	15 56	.58932	53
54	2954		.359997		.640003	.77780	20 70	20 75	.59156	54
55	.933058		.359725		.640275	1.77990	30 105	30 112	2.59381	55
56	3163		9454		0546	.78200	40 140	40 150	.59606	56
57	3267		9182		0818	.78410	45 157	45 169	.59831	57
58	3372		8911		1089	.78621	50 175	50 187	.60057	58
59	3476		8640		1360	.78832			.60283	59
60	.933580		.358368		.641632	1.79043			2.60509	60

TABLE XX.—NATURAL SINES, COSINES,

69°

'	SINE	CORR. FOR SEC.	COSINE	CORR. FOR SEC.	VERSINE	EXSEC	CORR. FOR SEC.	CORR. FOR SEC.	TANGENT	'
		+		− +			+	+		
0	.933580		.358368		.641632	1.79043			2.60509	0
1	3685		8096		1904	.79254	" Corr.	" Corr.	.60736	1
2	3789		7825		2175	.79466	10 36	10 38	.60963	2
3	3893		7553		2447	.79679	15 53	15 57	.61190	3
4	3997		7281		2719	.79891	20 71	20 76	.61418	4
5	.934101		.357010		.642990	1.80104	30 107	30 114	2.61646	5
6	4204		6738		3262	.80318	40 142	40 152	.61874	6
7	4308		6466		3534	.80531	45 160	45 171	.62103	7
8	4412		6194		3806	.80746	50 178	50 190	.62332	8
9	4515		5923		4077	.80960			.62561	9
10	.934619		.355651		.644349	1.81175			2.62791	10
11	4722	" Corr.	5379	" Corr.	4621	.81390	" Corr.	" Corr.	.63021	11
12	4826	10 17	5107	10 45	4893	.81605	10 36	10 39	.63252	12
13	4929	15 26	4835	15 68	5165	.81821	15 54	15 58	.63483	13
14	5032	20 34	4563	20 91	5437	.82037	20 72	20 77	.63714	14
15	.935135	30 52	.354291	30 136	.645709	1.82254	30 108	30 116	2.63945	15
16	5238	40 69	4019	40 181	5981	.82471	40 145	40 156	.64177	16
17	5341	45 77	3747	45 204	6253	.82688	45 163	45 174	.64410	17
18	5444	50 86	3475	50 227	6525	.82906	50 181	50 193	.64642	18
19	5547		3203		6797	.83124			.64875	19
20	.935650		.352931		.647069	1.83342			2.65109	20
21	5752		2658		7342	.83561	" Corr.	" Corr.	.65342	21
22	5855		2386		7614	.83780	10 37	10 39	.65576	22
23	5957		2114		7886	.83999	15 55	15 59	.65811	23
24	6060		1842		8158	.84219	20 73	20 78	.66046	24
25	.936162		.351569		.648431	1.84439	30 110	30 118	2.66281	25
26	6264		1297		8703	.84659	40 147	40 157	.66516	26
27	6366		1025		8975	.84880	45 165	45 177	.66752	27
28	6468		0752		9248	.85102	50 184	50 196	.66989	28
29	6570		0480		9520	.85323			.67225	29
30	.936672		.350207		.649793	1.85545			2.67462	30
31	6774		.349935		.650065	.85767	" Corr.	" Corr.	.67700	31
32	6876		9662		0338	.85990	10 37	10 40	.67937	32
33	6977		9390		0610	.86213	15 56	15 60	.68175	33
34	7079		9117		0883	.86437	20 75	20 80	.68414	34
35	.937181		.348845		.651155	1.86661	30 112	30 120	2.68653	35
36	7282		8572		1428	.86885	40 149	40 160	.68892	36
37	7383		8299		1701	.87109	45 168	45 179	.69131	37
38	7485		8027		1973	.87334	50 187	50 199	.69371	38
39	7586		7754		2246	.87560			.69612	39
40	.937687		.347481		.652519	1.87785			2.69853	40
41	7788	" Corr.	7208	" Corr.	2792	.88011	" Corr.	" Corr.	.70094	41
42	7889	10 17	6936	10 45	3064	.88238	10 38	10 40	.70335	42
43	7990	15 25	6663	15 68	3337	.88465	15 57	15 61	.70577	43
44	8091	20 34	6390	20 91	3610	.88692	20 76	20 81	.70819	44
45	.938191	30 50	.346117	30 136	.653883	1.88920	30 114	30 121	2.71062	45
46	8292	40 67	5844	40 182	4156	.89148	40 152	40 162	.71305	46
47	8392	45 76	5571	45 205	4429	.89376	45 171	45 182	.71548	47
48	8493	50 84	5298	50 227	4702	.89605	50 190	50 202	.71792	48
49	8593		5025		4975	.89834			.72036	49
50	.938694		.344752		.655248	1.90063			2.72281	50
51	8794		4479		5521	.90293	" Corr.	" Corr.	.72526	51
52	8894		4206		5794	.90524	10 39	10 41	.72771	52
53	8994		3933		6067	.90754	15 58	15 62	.73017	53
54	9094		3660		6340	.90986	20 77	20 82	.73263	54
55	.939194		.343386		.656614	1.91217	30 116	30 123	2.73509	55
56	9294		3113		6887	.91449	40 156	40 165	.73756	56
57	9394		2840		7160	.91681	45 174	45 185	.74004	57
58	9494		2567		7433	.91914	50 193	50 206	.74251	58
59	9593		2294		7706	.92147			.74499	59
60	.939693		.342020		.657980	1.92380			2.74748	60

VERSINES, EXSECANTS, AND TANGENTS
70°

'	SINE	CORR. FOR SEC.	COSINE	CORR. FOR SEC.	VERSINE	EXSEC	CORR. FOR SEC.	CORR. FOR SEC.	TANGENT	'
0	.939693	+	.342020	− +	.657980	1.92380	+	+	2.74748	0
1	9792		1747		8253	.92614	" Corr.	" Corr.	.74997	1
2	9891		1473		8527	.92849	10 39	10 42	.75246	2
3	9991		1200		8800	.93083	15 59	15 63	.75496	3
4	.940090		0926		9074	.93318	20 78	20 84	.75746	4
5	.940189		.340653		.659347	1.93554	30 118	30 125	2.75996	5
6	0288		0380		9620	.93790	40 157	40 167	.76247	6
7	0387		0106		9894	.94026	45 177	45 188	.76498	7
8	0486		.339832		.660168	.94263	50 196	50 209	.76750	8
9	0585		9559		0441	.94500			.77002	9
10	.940684		.339285		.660715	1.94737			2.77254	10
11	0782	" Corr.	9012	" Corr.	0988	.94975	" Corr.	" Corr.	.77507	11
12	0881	10 16	8738	10 46	1262	.95213	10 40	10 42	.77761	12
13	0979	15 25	8464	15 68	1536	.95452	15 60	15 64	.78014	13
14	1078	20 33	8190	20 91	1810	.95691	20 80	20 85	.78269	14
15	.941176	30 49	.337917	30 137	.662083	1.95931	30 120	30 127	2.78523	15
16	1274	40 66	7643	40 183	2357	.96171	40 160	40 170	.78778	16
17	1372	45 74	7369	45 205	2631	.96411	45 180	45 191	.79033	17
18	1470	50 82	7095	50 228	2905	.96652	50 200	50 212	.79289	18
19	1569		6821		3179	.96893			.79545	19
20	.941666		.336548		.663452	1.97135			2.79802	20
21	1764		6274		3726	.97377	" Corr.	" Corr.	.80059	21
22	1862		6000		4000	.97619	10 41	10 43	.80316	22
23	1960		5726		4274	.97862	15 61	15 65	.80574	23
24	2058		5452		4548	.98106	20 81	20 86	.80833	24
25	.942155		.335178		.664822	1.98349	30 122	30 130	2.81091	25
26	2252		4903		5097	.98594	40 163	40 173	.81350	26
27	2350		4629		5371	.98838	45 183	45 194	.81610	27
28	2447		4355		5645	.99083	50 203	50 216	.81870	28
29	2544		4081		5919	.99329			.82130	29
30	.942642		.333807		.666193	1.99574			2.82391	30
31	2739		3533		6467	.99821	" Corr.	" Corr.	.82653	31
32	2836		3258		6742	2.00067	10 41	10 44	.82914	32
33	2932		2984		7016	.00315	15 62	15 66	.83176	33
34	3029		2710		7290	.00562	20 83	20 88	.83439	34
35	.943126		.332436		.667564	2.00810	30 124	30 132	2.83702	35
36	3223		2161		7839	.01059	40 166	40 176	.83965	36
37	3319		1887		8113	.01308	45 186	45 198	.84229	37
38	3416		1612		8388	.01557	50 207	50 220	.84494	38
39	3512		1338		8662	.01807			.84758	39
40	.943608		.331063		.668937	2.02057			2.85023	40
41	3705	" Corr.	0789	" Corr.	9211	.02308	" Corr.	" Corr.	.85289	41
42	3801	10 16	0514	10 46	9486	.02559	10 42	10 45	.85555	42
43	3897	15 24	0240	15 69	9760	.02810	15 63	15 67	.85822	43
44	3993	20 32	.329965	20 92	.670035	.03062	20 84	20 89	.86089	44
45	.944089	30 48	.329691	30 137	.670309	2.03315	30 126	30 134	2.86356	45
46	4185	40 64	9416	40 183	0584	.03568	40 169	40 179	.86624	46
47	4281	45 72	9141	45 206	0859	.03821	45 190	45 201	.86892	47
48	4376	50 80	8867	50 229	1133	.04075	50 211	50 223	.87161	48
49	4472		8592		1408	.04329			.87430	49
50	.944568		.328317		.671683	2.04584			2.87700	50
51	4663		8042		1958	.04839	" Corr.	" Corr.	.87970	51
52	4758		7768		2232	.05094	10 43	10 45	.88240	52
53	4854		7493		2507	.05350	15 64	15 68	.88511	53
54	4949		7218		2782	.05607	20 86	20 91	.88783	54
55	.945044		.326943		.673057	2.05864	30 129	30 136	2.89055	55
56	5139		6668		3332	.06121	40 172	40 182	.89327	56
57	5234		6393		3607	.06379	45 193	45 204	.89600	57
58	5329		6118		3882	.06637	50 214	50 227	.89873	58
59	5424		5843		4157	.06896			.90147	59
60	.945519		.325568		.674432	2.07155			2.90421	60

TABLE XX.—NATURAL SINES, COSINES,

71°

'	SINE	CORR. FOR SEC.	COSINE	CORR. FOR SEC.	VERSINE	EXSEC	CORR. FOR SEC.	CORR. FOR SEC.	TANGENT	'
		+		− +			+	+		
0	.945519		.325568		.674432	2.07155			2.90421	0
1	5613		5293		4707	.07415	" Corr.	" Corr.	.90696	1
2	5708		5018		4982	.07675	10 44	10 46	.90971	2
3	5802		4743		5257	.07936	15 66	15 69	.91246	3
4	5897		4468		5532	.08197	20 87	20 92	.91523	4
5	.945991		.324193		.675807	2.08459	30 131	30 139	2.91799	5
6	6085		3917		6083	.08721	40 175	40 185	.92076	6
7	6180		3642		6358	.08983	45 197	45 208	.92354	7
8	6274		3367		6633	.09246	50 218	50 231	.92632	8
9	6368		3092		6908	.09510			.92910	9
10	.946462		.322816		.677184	2.09774			2.93189	10
11	6556	" Corr.	2541	" Corr.	7459	.10038	" Corr.	" Corr.	.93468	11
12	6649	10 16	2266	10 46	7734	.10303	10 44	10 47	.93748	12
13	6743	15 23	1990	15 69	8010	.10568	15 67	15 70	.94028	13
14	6837	20 31	1715	20 92	8285	.10834	20 89	20 94	.94309	14
15	.946930	30 47	.321440	30 138	.678560	2.11101	30 133	30 141	2.94590	15
16	7024	40 62	1164	40 184	8836	.11367	40 178	40 188	.94872	16
17	7117	45 70	0888	45 207	9112	.11635	45 200	45 211	.95155	17
18	7210	50 78	0613	50 229	9387	.11903	50 222	50 235	.95437	18
19	7304		0337		9663	.12171			.95720	19
20	.947397		.320062		.679938	2.12440			2.96004	20
21	7490		.319786		.680214	.12709	" Corr.	" Corr.	.96288	21
22	7583		9511		0489	.12979	10 45	10 48	.96573	22
23	7676		9235		0765	.13249	15 68	15 72	.96858	23
24	7768		8959		1041	.13520	20 91	20 96	.97144	24
25	.947861		.318684		.681316	2.13791	30 136	30 143	2.97430	25
26	7954		8408		1592	.14063	40 181	40 191	.97717	26
27	8046		8132		1868	.14335	45 204	45 215	.98004	27
28	8139		7856		2144	.14608	50 226	50 239	.98292	28
29	8231		7580		2420	.14881			.98580	29
30	.948324		.317305		.682695	2.15155			2.98868	30
31	8416		7029		2971	.15429	" Corr.	" Corr.	.99158	31
32	8508		6753		3247	.15704	10 46	10 49	.99447	32
33	8600		6477		3523	.15979	15 69	15 73	.99738	33
34	8692		6201		3799	.16255	20 92	20 97	3.00028	34
35	.948784		.315925		.684075	2.16531	30 138	30 146	3.00319	35
36	8876		5649		4351	.16808	40 185	40 194	.00611	36
37	8968		5373		4627	.17085	45 208	45 219	.00903	37
38	9060		5097		4903	.17363	50 231	50 243	.01196	38
39	9151		4821		5179	.17641			.01489	39
40	.949243		.314545		.685455	2.17920			3.01783	40
41	9334	" Corr.	4269	" Corr.	5731	.18199	" Corr.	" Corr.	.02077	41
42	9426	10 15	3992	10 46	6008	.18479	10 47	10 49	.02372	42
43	9517	15 23	3716	15 69	6284	.18759	15 70	15 74	.02667	43
44	9608	20 30	3440	20 92	6560	.19040	20 94	20 99	.02963	44
45	.949699	30 46	.313164	30 138	.686836	2.19322	30 141	30 148	3.03260	45
46	9790	40 61	2888	40 184	7112	.19604	40 188	40 198	.03556	46
47	9881	45 68	2611	45 207	7389	.19886	45 211	45 223	.03854	47
48	9972	50 76	2335	50 230	7665	.20169	50 235	50 247	.04152	48
49	.950063		2059		7941	.20453			.04450	49
50	.950154		.311782		.688218	2.20737			3.04749	50
51	0244		1506		8494	.21021	" Corr.	" Corr.	.05049	51
52	0335		1229		8771	.21306	10 48	10 50	.05349	52
53	0425		0953		9047	.21592	15 72	15 76	.05649	53
54	0516		0676		9324	.21878	20 96	20 101	.05950	54
55	.950606		.310400		.689600	2.22165	30 144	30 151	3.06252	55
56	0696		0123		9877	.22452	40 192	40 201	.06554	56
57	0786		.309847		.690153	.22740	45 215	45 227	.06857	57
58	0877		9570		0430	.23028	50 239	50 252	.07160	58
59	0967		9294		0706	.23317			.07464	59
60	.951056		.309017		.690983	2.23607			3.07768	60

VERSINES, EXSECANTS, AND TANGENTS
72°

′	SINE	CORR. FOR SEC.	COSINE	CORR. FOR SEC.	VERSINE	EXSEC	CORR. FOR SEC.	CORR. FOR SEC.	TANGENT	′
0	.951056	+	.309017	− +	.690983	2.23607	+	+	3.07768	0
1	1146		8740		1260	.23897	″ Corr.	″ Corr.	.08073	1
2	1236		8464		1536	.24187	10 49	10 51	.08379	2
3	1326		8187		1813	.24478	15 73	15 77	.08685	3
4	1415		7910		2090	.24770	20 98	20 103	.08991	4
5	.951505		.307633		.692367	2.25062	30 146	30 154	3.09298	5
6	1594		7357		2643	.25355	40 195	40 205	.09606	6
7	1684		7080		2920	.25648	45 220	45 231	.09914	7
8	1773		6803		3197	.25942	50 244	50 256	.10223	8
9	1862		6526		3474	.26237			.10532	9
10	.951951		.306249		.693751	2.26531			3.10842	10
11	2040	″ Corr.	5972	″ Corr.	4028	.26827	″ Corr.	″ Corr.	.11153	11
12	2129	10 15	5695	10 46	4305	.27123	10 50	10 52	.11464	12
13	2218	15 22	5418	15 69	4582	.27420	15 75	15 78	.11775	13
14	2307	20 30	5141	20 92	4859	.27717	20 99	20 104	.12087	14
15	.952396	30 44	.304864	30 138	.695136	2.28015	30 149	30 157	3.12400	15
16	2484	40 59	4587	40 185	5413	.28313	40 199	40 209	.12713	16
17	2573	45 67	4310	45 208	5690	.28612	45 224	45 235	.13027	17
18	2662	50 74	4033	50 231	5967	.28912	50 249	50 261	.13341	18
19	2750		3756		6244	.29212			.13656	19
20	.952838		.303479		.696521	2.29512			3.13972	20
21	2926		3202		6798	.29814	″ Corr.	″ Corr.	.14288	21
22	3015		2924		7076	.30115	10 51	10 53	.14605	22
23	3103		2647		7353	.30418	15 76	15 80	.14922	23
24	3191		2370		7630	.30721	20 101	20 106	.15240	24
25	.953279		.302093		.697907	2.31024	30 152	30 159	3.15558	25
26	3366		1815		8185	.31328	40 203	40 213	.15877	26
27	3454		1538		8462	.31633	45 228	45 239	.16197	27
28	3542		1261		8739	.31939	50 253	50 266	.16517	28
29	3629		0983		9017	.32244			.16838	29
30	.953717		.300706		.699294	2.32551			3.17159	30
31	3804		0428		9572	.32858	″ Corr.	″ Corr.	.17481	31
32	3892		0151		9849	.33166	10 52	10 54	.17804	32
33	3979		.299873		.700127	.33474	15 78	15 81	.18127	33
34	4066		9596		0404	.33783	20 103	20 108	.18451	34
35	.954153		.299318		.700682	2.34092	30 155	30 162	3.18775	35
36	4240		9041		0959	.34403	40 207	40 217	.19100	36
37	4327		8763		1237	.34713	45 233	45 244	.19426	37
38	4414		8486		1514	.35025	50 258	50 271	.19752	38
39	4501		8208		1792	.35336			.20079	39
40	.954588		.297930		.702070	2.35649			3.20406	40
41	4674	″ Corr.	7653	″ Corr.	2347	.35962	″ Corr.	″ Corr.	.20734	41
42	4761	10 14	7375	10 46	2625	.36276	10 53	10 55	.21063	42
43	4847	15 22	7097	15 69	2903	.36590	15 79	15 83	.21392	43
44	4934	20 29	6819	20 93	3181	.36905	20 105	20 110	.21722	44
45	.955020	30 43	.296542	30 139	.703458	2.37221	30 158	30 166	3.22053	45
46	5106	40 58	6264	40 185	3736	.37537	40 211	40 221	.22384	46
47	5192	45 65	5986	45 208	4014	.37854	45 237	45 248	.22715	47
48	5278	50 72	5708	50 231	4292	.38171	50 263	50 276	.23048	48
49	5364		5430		4570	.38489			.23381	49
50	.955450		.295152		.704848	2.38808			3.23714	50
51	5536		4874		5126	.39128	″ Corr.	″ Corr.	.24049	51
52	5622		4596		5404	.39448	10 54	10 56	.24383	52
53	5707		4318		5682	.39768	15 81	15 84	.24719	53
54	5793		4040		5960	.40089	20 107	20 112	.25055	54
55	.955878		.293762		.706238	2.40411	30 161	30 169	3.25392	55
56	5964		3484		6516	.40734	40 215	40 225	.25729	56
57	6049		3206		6794	.41057	45 242	45 253	.26067	57
58	6134		2928		7072	.41381	50 269	50 281	.26406	58
59	6220		2650		7350	.41705			.26745	59
60	.956305		.292372		.707628	2.42030			3.27085	60

TABLE XX.—NATURAL SINES, COSINES,

73°

'	SINE	CORR. FOR SEC.	COSINE	CORR. FOR SEC.	VERSINE	EXSEC	CORR. FOR SEC.	CORR. FOR SEC.	TANGENT	'
		+		− +			+	+		
0	.956305		.292372		.707628	2.42030			3.27085	0
1	6390		2094		7906	.42356	" Corr.	" Corr.	.27426	1
2	6475		1815		8185	.42683	10 55	10 57	.27767	2
3	6560		1537		8463	.43010	15 82	15 86	.28109	3
4	6644		1259		8741	.43337	20 110	20 115	.28452	4
5	.956729		.290980		.709020	2.43666	30 164	30 172	3.28795	5
6	6814		0702		9298	.43995	40 219	40 229	.29139	6
7	6898		0424		9576	.44324	45 247	45 258	.29483	7
8	6982		0146		9854	.44655	50 274	50 287	.29829	8
9	7067		.289867		.710133	.44986			.30174	9
10	.957151		.289589		.710411	2.45317			3.30521	10
11	7235	" Corr.	9310	" Corr.	0690	.45650	" Corr.	" Corr.	.30868	11
12	7320	10 14	9032	10 46	0968	.45983	10 56	10 58	.31216	12
13	7404	15 21	8753	15 70	1247	.46316	15 84	15 88	.31565	13
14	7488	20 28	8475	20 93	1525	.46651	20 112	20 117	.31914	14
15	.957571	30 42	.288196	30 139	.711804	2.46986	30 168	30 175	3.32264	15
16	7655	40 56	7918	40 186	2082	.47321	40 224	40 234	.32614	16
17	7739	45 63	7639	45 209	2361	.47658	45 252	45 263	.32965	17
18	7822	50 70	7360	50 232	2640	.47995	50 280	50 292	.33317	18
19	7906		7082		2918	.48333			.33670	19
20	.957990		.286803		.713197	2.48671			3.34023	20
21	8073		6525		3475	.49010	" Corr.	" Corr.	.34377	21
22	8156		6246		3754	.49350	10 57	10 59	.34732	22
23	8239		5967		4033	.49691	15 86	15 89	.35087	23
24	8323		5688		4312	.50032	20 114	20 119	.35443	24
25	.958406		.285410		.714590	2.50374	30 171	30 179	3.35800	25
26	8489		5131		4869	.50716	40 228	40 236	.36158	26
27	8572		4852		5148	.51060	45 257	45 268	.36516	27
28	8654		4573		5427	.51404	50 285	50 298	.36875	28
29	8737		4294		5706	.51748			.37234	29
30	.958820		.284015		.715985	2.52094			3.37594	30
31	8902		3736		6264	.52440	" Corr.	" Corr.	.37955	31
32	8985		3458		6542	.52787	10 58	10 61	.38317	32
33	9067		3178		6822	.53134	15 87	15 91	.38679	33
34	9150		2900		7100	.53482	20 117	20 122	.39042	34
35	.959232		.282620		.717380	2.53831	30 175	30 182	3.39406	35
36	9314		2342		7658	.54181	40 233	40 243	.39771	36
37	9396		2062		7938	.54531	45 262	45 273	.40136	37
38	9478		1783		8217	.54883	50 291	50 304	.40502	38
39	9560		1504		8496	.55234			.40869	39
40	.959642		.281225		.718775	2.55587			3.41236	40
41	9724	" Corr.	0946	" Corr.	9054	.55940	" Corr.	" Corr.	.41604	41
42	9805	10 14	0667	10 47	9333	.56294	10 59	10 62	.41973	42
43	9887	15 20	0388	15 70	9612	.56649	15 89	15 93	.42343	43
44	9968	20 27	0108	20 93	9892	.57005	20 119	20 124	.42713	44
45	.960050	30 41	.279829	30 140	.720171	2.57361	30 178	30 186	3.43084	45
46	0131	40 54	9550	40 186	0450	.57718	40 238	40 248	.43456	46
47	0212	45 61	9270	45 209	0730	.58076	45 268	45 279	.43829	47
48	0294	50 68	8991	50 233	1009	.58434	50 297	50 310	.44202	48
49	0375		8712		1288	.58794			.44576	49
50	.960456		.278432		.721568	2.59154			3.44951	50
51	0537		8153		1847	.59514	" Corr.	" Corr.	.45327	51
52	0618		7874		2126	.59876	10 61	10 63	.45703	52
53	0698		7594		2406	.60238	15 91	15 95	.46080	53
54	0779		7315		2685	.60601	20 122	20 126	.46458	54
55	.960860		.277035		.722965	2.60965	30 182	30 190	3.46837	55
56	0940		6756		3244	.61330	40 243	40 253	.47216	56
57	1021		6476		3524	.61695	45 273	45 285	.47596	57
58	1101		6196		3804	.62061	50 304	50 316	.47977	58
59	1182		5917		4083	.62428			.48359	59
60	.961262		.275637		.724363	2.62796			3.48741	60

VERSINES, EXSECANTS, AND TANGENTS
74°

'	SINE	CORR. FOR SEC. +	COSINE	CORR. FOR SEC. − +	VERSINE	EXSEC	CORR. FOR SEC. +	CORR. FOR SEC. +	TANGENT	'
0	.961262		.275637		.724363	2.62796			3.48741	0
1	1342		5358		4642	.63164	" Corr.	" Corr.	.49125	1
2	1422		5078		4922	.63533	10 62	10 65	.49509	2
3	1502		4798		5202	.63903	15 93	15 97	.49894	3
4	1582		4519		5481	.64274	20 124	20 129	.50279	4
5	.961662		.274239		.725761	2.64645	30 186	30 194	3.50666	5
6	1741		3959		6041	.65018	40 248	40 258	.51053	6
7	1821		3679		6321	.65391	45 279	45 290	.51441	7
8	1900		3400		6600	.65765	50 310	50 323	.51829	8
9	1980		3120		6880	.66140			.52219	9
10	.962059		.272840		.727160	2.66515			3.52609	10
11	2139	" Corr.	2560	" Corr.	7440	.66892	" Corr.	" Corr.	.53001	11
12	2218	10 13	2280	10 47	7720	.67269	10 63	10 66	.53393	12
13	2297	15 20	2000	15 70	8000	.67647	15 95	15 99	.53785	13
14	2376	20 26	1720	20 93	8280	.68025	20 127	20 132	.54179	14
15	.962455	30 40	.271440	30 140	.728560	2.68405	30 190	30 198	3.54573	15
16	2534	40 53	1160	40 187	8840	.68785	40 254	40 263	.54968	16
17	2613	45 59	0880	45 210	9120	.69167	45 285	45 296	.55364	17
18	2692	50 66	0600	50 233	9400	.69549	50 317	50 329	.55761	18
19	2770		0320		9680	.69931			.56159	19
20	.962849		.270040		.729960	2.70315			3.56557	20
21	2928		.269760		.730240	.70700	" Corr.	" Corr.	.56957	21
22	3006		9480		0520	.71085	10 65	10 67	.57357	22
23	3084		9200		0800	.71471	15 97	15 101	.57758	23
24	3163		8920		1080	.71858	20 130	20 134	.58160	24
25	.963241		.268640		.731360	2.72246	30 194	30 202	3.58562	25
26	3319		8359		1641	.72635	40 259	40 269	.58966	26
27	3397		8079		1921	.73024	45 292	45 303	.59370	27
28	3475		7799		2201	.73414	50 324	50 336	.59775	28
29	3553		7519		2481	.73806			.60181	29
30	.963630		.267238		.732762	2.74198			3.60588	30
31	3708		6958		3042	.74591	" Corr.	" Corr.	.60996	31
32	3786		6678		3322	.74984	10 66	10 69	.61405	32
33	3863		6397		3603	.75379	15 99	15 103	.61814	33
34	3941		6117		3883	.75775	20 132	20 137	.62224	34
35	.964018		.265837		.734163	2.76171	30 199	30 206	3.62636	35
36	4095		5556		4444	.76568	40 265	40 275	.63048	36
37	4173		5276		4724	.76966	45 298	45 309	.63461	37
38	4250		4995		5005	.77365	50 331	50 343	.63874	38
39	4327		4715		5285	.77765			.64289	39
40	.964404		.264434		.735566	2.78166			3.64705	40
41	4481	" Corr.	4154	" Corr.	5846	.78568	" Corr.	" Corr.	.65121	41
42	4557	10 13	3873	10 47	6127	.78970	10 68	10 70	.65538	42
43	4634	15 19	3592	15 70	6408	.79374	15 102	15 105	.65957	43
44	4711	20 25	3312	20 94	6688	.79778	20 135	20 140	.66376	44
45	.964787	30 38	.263031	30 140	.736969	2.80183	30 203	30 210	3.66796	45
46	4864	40 51	2751	40 187	7249	.80589	40 271	40 281	.67217	46
47	4940	45 57	2470	45 210	7530	.80996	45 305	45 316	.67638	47
48	5016	50 64	2189	50 234	7811	.81404	50 338	50 351	.68061	48
49	5093		1908		8092	.81813			.68485	49
50	.965169		.261628		.738372	2.82223			3.68909	50
51	5245		1347		8653	.82633	" Corr.	" Corr.	.69335	51
52	5321		1066		8934	.83045	10 69	10 72	.69761	52
53	5397		0785		9215	.83457	15 104	15 107	.70188	53
54	5473		0504		9496	.83871	20 138	20 143	.70616	54
55	.965548		.260224		.739776	2.84285	30 208	30 215	3.71046	55
56	5624		.259943		.740057	.84700	40 277	40 287	.71476	56
57	5700		9662		0338	.85116	45 311	45 322	.71907	57
58	5775		9381		0619	.85533	50 346	50 358	.72338	58
59	5850		9100		0900	.85951			.72771	59
60	.965926		.258819		.741181	2.86370			3.73205	60

TABLE XX.—NATURAL SINES, COSINES,

75°

'	SINE	CORR. FOR SEC.	COSINE	CORR. FOR SEC.	VERSINE	EXSEC	CORR. FOR SEC.	CORR. FOR SEC.	TANGENT	'
		+		− +			+	+		
0	.965926		.258819		.741181	2.86370			3.73205	0
1	6001		8538		1462	.86790	" Corr.	" Corr.	.73640	1
2	6076		8257		1743	.87211	10 71	10 73	.74075	2
3	6151		7976		2024	.87633	15 106	15 110	.74512	3
4	6226		7695		2305	.88056	20 142	20 146	.74950	4
5	.966301		.257414		.742586	2.88479	30 212	30 220	3.75388	5
6	6376		7133		2867	.88904	40 283	40 293	.75828	6
7	6451		6852		3148	.89330	45 319	45 330	.76268	7
8	6526		6570		3430	.89756	50 354	50 366	.76709	8
9	6600		6289		3711	.90184			.77152	9
10	.966675		.256008		.743992	2.90612			3.77595	10
11	6749	" Corr.	5727	" Corr.	4273	.91042	" Corr.	" Corr.	.78040	11
12	6823	10 12	5446	10 47	4554	.91473	10 72	10 75	.78485	12
13	6898	15 19	5164	15 70	4836	.91904	15 109	15 112	.78931	13
14	6972	20 25	4883	20 94	5117	.92337	20 145	20 150	.79378	14
15	.967046	30 37	.254602	30 141	.745398	2.92770	30 217	30 225	3.79827	15
16	7120	40 49	4321	40 188	5679	.93204	40 290	40 299	.80276	16
17	7194	45 56	4039	45 211	5961	.93640	45 326	45 337	.80726	17
18	7268	50 62	3758	50 234	6242	.94076	50 362	50 374	.81177	18
19	7342		3477		6523	.94514			.81630	19
20	.967415		.253195		.746805	2.94952			3.82083	20
21	7489		2914		7086	.95392	" Corr.	" Corr.	.82537	21
22	7562		2632		7368	.95832	10 74	10 77	.82992	22
23	7636		2351		7649	.96274	15 111	15 115	.83449	23
24	7709		2069		7931	.96716	20 148	20 153	.83906	24
25	.967782		.251788		.748212	2.97160	30 222	30 230	3.84364	25
26	7856		1506		8494	.97604	40 296	40 306	.84824	26
27	7929		1225		8775	.98050	45 333	45 344	.85284	27
28	8002		0943		9057	.98497	50 370	50 383	.85745	28
29	8075		0662		9338	.98944			.86208	29
30	.968148		.250380		.749620	2.99393			3.86671	30
31	8220		0098		9902	.99843	" Corr.	" Corr.	.87136	31
32	8293		.249817		.750183	3.00293	10 76	10 78	.87601	32
33	8366		9535		0465	.00745	15 114	15 117	.88068	33
34	8438		9253		0747	.01198	20 152	20 157	.88536	34
35	.968511		.248972		.751028	3.01652	30 227	30 235	3.89004	35
36	8583		8690		1310	.02107	40 303	40 313	.89474	36
37	8656		8408		1592	.02563	45 341	45 352	.89945	37
38	8728		8126		1874	.03020	50 379	50 391	.90417	38
39	8800		7844		2156	.03479			.90890	39
40	.968872		.247563		.752437	3.03938			3.91364	40
41	8944	" Corr.	7281	" Corr.	2719	.04398	" Corr.	" Corr.	.91839	41
42	9016	10 12	6999	10 47	3001	.04860	10 78	10 80	.92316	42
43	9088	15 18	6717	15 70	3283	.05322	15 116	15 120	.92793	43
44	9159	20 24	6435	20 94	3565	.05786	20 155	20 160	.93271	44
45	.969231	30 36	.246153	30 141	.753847	3.06251	30 233	30 240	3.93751	45
46	9302	40 48	5871	40 188	4129	.06717	40 310	40 320	.94232	46
47	9374	45 54	5589	45 211	4411	.07184	45 349	45 360	.94713	47
48	9445	50 60	5307	50 235	4693	.07652	50 388	50 400	.95196	48
49	9517		5025		4975	.08121			.95680	49
50	.969588		.244743		.755257	3.08591			3.96165	50
51	9659		4461		5539	.09063	" Corr.	" Corr.	.96651	51
52	9730		4179		5821	.09535	10 79	10 82	.97139	52
53	9801		3897		6103	.10009	15 119	15 123	.97627	53
54	9872		3615		6385	.10484	20 159	20 164	.98117	54
55	.969943		.243333		.756667	3.10960	30 239	30 246	3.98607	55
56	.970014		3051		6949	.11437	40 318	40 328	.99099	56
57	0084		2768		7232	.11915	45 358	45 369	.99592	57
58	0155		2486		7514	.12394	50 398	50 410	4.00086	58
59	0225		2204		7796	.12875			.00582	59
60	.970296		.241922		.758078	3.13357			4.01078	60

VERSINES, EXSECANTS, AND TANGENTS
76°

'	SINE	CORR. FOR SEC. +	COSINE	CORR. FOR SEC. − +	VERSINE	EXSEC	CORR. FOR SEC. +	CORR. FOR SEC. +	TANGENT	'
0	.970296		.241922		.758078	3.13357			4.01078	0
1	0366		1640		8360	.13839	" Corr.	" Corr.	.01576	1
2	0436		1357		8643	.14323	10 81	10 84	.02074	2
3	0506		1075		8925	.14809	15 122	15 126	.02574	3
4	0577		0793		9207	.15295	20 163	20 168	.03076	4
5	.970647		.240510		.759490	3.15782	30 244	30 252	4.03578	5
6	0716		0228		9772	.16271	40 326	40 336	.04081	6
7	0786		.239946		.760054	.16761	45 367	45 378	.04586	7
8	0856		9663		0337	.17252	50 407	50 420	.05092	8
9	0926		9381		0619	.17744			.05599	9
10	.970995		.239098		.760902	3.18238			4.06107	10
11	1065	" Corr.	8816	" Corr.	1184	.18733	" Corr.	" Corr.	.06616	11
12	1134	10 12	8534	10 47	1466	.19228	10 83	10 86	.07127	12
13	1204	15 17	8251	15 71	1749	.19725	15 125	15 129	.07639	13
14	1273	20 23	7968	20 94	2032	.20224	20 167	20 172	.08152	14
15	.971342	30 35	.237686	30 141	.762314	3.20723	30 250	30 258	4.08666	15
16	1411	40 46	7403	40 188	2597	.21224	40 334	40 344	.09182	16
17	1480	45 52	7121	45 212	2879	.21726	45 376	45 387	.09699	17
18	1549	50 58	6838	50 235	3162	.22229	50 417	50 430	.10216	18
19	1618		6556		3444	.22734			.10736	19
20	.971687		.236273		.763727	3.23239			4.11256	20
21	1755		5990		4010	.23746	" Corr.	" Corr.	.11778	21
22	1824		5708		4292	.24255	10 85	10 88	.12301	22
23	1893		5425		4575	.24764	15 128	15 132	.12825	23
24	1961		5142		4858	.25275	20 171	20 176	.13350	24
25	.972029		.234859		.765141	3.25787	30 257	30 264	4.13877	25
26	2098		4577		5423	.26300	40 342	40 352	.14405	26
27	2166		4294		5706	.26814	45 385	45 396	.14934	27
28	2234		4011		5989	.27330	50 428	50 440	.15465	28
29	2302		3728		6272	.27847			.15997	29
30	.972370		.233445		.766555	3.28366			4.16530	30
31	2438		3162		6838	.28885	" Corr.	" Corr.	.17064	31
32	2506		2880		7120	.29406	10 88	10 90	.17600	32
33	2573		2597		7403	.29929	15 132	15 135	.18137	33
34	2641		2314		7686	.30452	20 175	20 180	.18675	34
35	.972708		.232031		.767969	3.30977	30 263	30 270	4.19215	35
36	2776		1748		8252	.31503	40 351	40 361	.19756	36
37	2843		1465		8535	.32031	45 395	45 406	.20298	37
38	2910		1182		8818	.32560	50 438	50 451	.20842	38
39	2978		0899		9101	.33090			.21387	39
40	.973045		.230616		.769384	3.33622			4.21933	40
41	3112	" Corr.	0333	" Corr.	9667	.34154	" Corr.	" Corr.	.22481	41
42	3179	10 11	0050	10 47	9950	.34689	10 90	10 92	.23030	42
43	3246	15 17	.229767	15 71	.770233	.35224	15 135	15 139	.23580	43
44	3312	20 22	9484	20 94	0516	.35761	20 180	20 185	.24132	44
45	.973379	30 33	.229200	30 142	.770800	3.36299	30 270	30 277	4.24685	45
46	3446	40 44	8917	40 189	1083	.36839	40 360	40 370	.25239	46
47	3512	45 50	8634	45 212	1366	.37380	45 405	45 416	.25795	47
48	3579	50 56	8351	50 236	1649	.37923	50 450	50 462	.26352	48
49	3645		8068		1932	.38466			.26911	49
50	.973712		.227784		.772216	3.39012			4.27471	50
51	3778		7501		2499	.39558	" Corr.	" Corr.	.28032	51
52	3844		7218		2782	.40106	10 92	10 95	.28595	52
53	3910		6935		3065	.40656	15 138	15 142	.29159	53
54	3976		6651		3349	.41206	20 185	20 189	.29724	54
55	.974042		.226368		.773632	3.41759	30 277	30 284	4.30291	55
56	4108		6085		3915	.42312	40 369	40 379	.30860	56
57	4173		5801		4199	.42867	45 415	45 426	.31430	57
58	4239		5518		4482	.43424	50 461	50 474	.32001	58
59	4305		5234		4766	.43982			.32573	59
60	.974370		.224951		.775049	3.44541			4.33148	60

TABLE XX.—NATURAL SINES, COSINES,

77°

'	SINE	CORR. FOR SEC. +	COSINE	CORR. FOR SEC. − +	VERSINE	EXSEC	DIFF. 10"		TANGENT	'
0	.974370		.224951		.775049	3.44541	93.5	95.8	4.33148	0
1	4436		4668		5332	.45102	93.7	96.2	.33723	1
2	4501		4384		5616	.45664	94.0	96.5	.34300	2
3	4566		4101		5899	.46228	94.2	96.7	.34879	3
4	4631		3817		6183	.46793	94.5	96.8	.35459	4
5	.974696		.223534		.776466	3.47360	94.7	97.2	4.36040	5
6	4761		3250		6750	.47928	95.0	97.3	.36623	6
7	4826		2967		7033	.48498	95.2	97.7	.37207	7
8	4891		2683		7317	.49069	95.5	98.0	.37793	8
9	4956		2399		7601	.49642	95.7	98.0	.38381	9
10	.975020		.222116		.777884	3.50216	95.8	98.5	4.38969	10
11	5085	" Corr.	1832	" Corr.	8168	.50781	96.2	98.7	.39560	11
12	5149	10 11	1548	10 47	8452	.51368	96.5	98.8	.40152	12
13	5214	15 16	1265	15 71	8735	.51947	96.7	99.2	.40745	13
14	5278	20 21	0981	20 95	9019	.52527	97.0	99.3	.41340	14
15	.975342	30 32	.220697	30 142	.779303	3.53109	97.2	99.7	4.41936	15
16	5406	40 43	0414	40 189	9586	.53692	97.5	100.0	.42534	16
17	5471	45 48	0130	45 213	9870	.54277	97.7	100.2	.43134	17
18	5534	50 53	.219846	50 236	.780154	.54863	98.0	100.5	.43735	18
19	5598		9562		0438	.55451	98.3	100.7	.44338	19
20	.975662		.219279		.780721	3.56041	98.5	101.0	4.44942	20
21	5726		8995		1005	.56632	98.7	101.2	.45548	21
22	5790		8711		1289	.57224	99.2	101.5	.46155	22
23	5853		8427		1573	.57819	99.2	101.7	.46764	23
24	5917		8143		1857	.58414	99.7	102.0	.47374	24
25	.975980		.217859		.782141	3.59012	99.8	102.3	4.47986	25
26	6044		7575		2425	.59611	100.0	102.5	.48600	26
27	6107		7292		2708	.60211	100.3	102.8	.49215	27
28	6170		7008		2992	.60813	100.7	103.2	.49832	28
29	6233		6724		3276	.61417	101.0	103.3	.50451	29
30	.976296		.216440		.783560	3.62023	101.2	103.7	4.51071	30
31	6359		6156		3844	.62630	101.3	103.8	.51693	31
32	6422		5872		4128	.63238	101.8	104.2	.52316	32
33	6484		5588		4412	.63849	102.0	104.5	.52941	33
34	6547		5304		4696	.64461	102.2	104.7	.53568	34
35	.976610		.215019		.784981	3.65074	102.7	105.0	4.54196	35
36	6672		4735		5265	.65690	102.8	105.3	.54826	36
37	6735		4451		5549	.66307	103.0	105.5	.55458	37
38	6797		4167		5833	.66925	103.3	105.8	.56091	38
39	6859		3883		6117	.67545	103.7	106.2	.56726	39
40	.976922		.213599		.786401	3.68167	104.0	106.3	4.57363	40
41	6984	" Corr.	3315	" Corr.	6685	.68791	104.3	106.7	.58001	41
42	7046	10 10	3030	10 47	6970	.69417	104.5	107.0	.58641	42
43	7108	15 15	2746	15 71	7254	.70044	104.8	107.3	.59283	43
44	7169	20 21	2462	20 95	7538	.70673	105.0	107.5	.59927	44
45	.977231	30 31	.212178	30 142	.787822	3.71303	105.3	107.8	4.60572	45
46	7293	40 41	1893	40 189	8107	.71935	105.7	108.2	.61219	46
47	7354	45 46	1609	45 213	8391	.72569	106.0	108.3	.61868	47
48	7416	50 51	1325	50 237	8675	.73205	106.3	108.8	.62518	48
49	7477		1040		8960	.73843	106.5	109.0	.63171	49
50	.977539		.210756		.789244	3.74482	106.8	109.2	4.63825	50
51	7600		0472		9528	.75123	107.2	109.7	.64480	51
52	7661		0187		9813	.75766	107.5	109.8	.65138	52
53	7722		.209903		.790097	.76411	107.7	110.2	.65797	53
54	7783		9619		0381	.77057	108.0	110.5	.66458	54
55	.977844		.209334		.790666	3.77705	108.3	110.8	4.67121	55
56	7905		9050		0950	.78355	108.7	111.0	.67786	56
57	7966		8765		1235	.79007	109.0	111.5	.68452	57
58	8026		8481		1519	.79661	109.2	111.7	.69121	58
59	8087		8196		1804	.80316	109.3	112.0	.69791	59
60	.978148		.207912		.792088	3.80973			4.70463	60

VERSINES, EXSECANTS, AND TANGENTS

78°

'	SINE	CORR. FOR SEC. +		COSINE	CORR. FOR SEC. − +		VERSINE	EXSEC	DIFF. 10"		TANGENT	'
0	.978148			.207912			.792088	3.80973	110.0	112.3	4.70463	0
1	8208			7627			2373	.81633	110.2	112.7	.71137	1
2	8268			7343			2657	.82294	110.3	112.8	.71813	2
3	8329			7058			2942	.82956	110.8	113.3	.72490	3
4	8389			6773			3227	.83621	111.2	113.5	.73170	4
5	.978449			.206489			.793511	3.84288	111.3	113.8	4.73851	5
6	8509			6204			3796	.84956	111.8	114.2	.74534	6
7	8569			5920			4080	.85627	112.0	114.5	.75219	7
8	8629			5635			4365	.86299	112.3	114.8	.75906	8
9	8689			5350			4650	.86973	112.7	115.2	.76595	9
10	.978748			.205066			.794934	3.87649	113.0	115.3	4.77286	10
11	8808	"	Corr.	4781	"	Corr.	5219	.88327	113.3	115.8	.77978	11
12	8867	10	10	4496	10	47	5504	.89007	113.7	116.2	.78673	12
13	8927	15	15	4211	15	71	5789	.89689	114.0	116.3	.79370	13
14	8986	20	20	3926	20	95	6074	.90373	114.2	116.8	.80068	14
15	.979046	30	30	.203642	30	142	.796358	3.91058	114.7	117.0	4.80769	15
16	9105	40	40	3357	40	190	6643	.91746	115.0	117.3	.81471	16
17	9164	45	44	3072	45	214	6928	.92436	115.3	117.8	.82175	17
18	9223	50	49	2787	50	237	7213	.93128	115.5	118.0	.82882	18
19	9282			2502			7498	.93821	116.0	118.3	.83590	19
20	.979341			.202218			.797782	3.94517	116.3	118.8	4.84300	20
21	9399			1933			8067	.95215	116.5	119.0	.85013	21
22	9458			1648			8352	.95914	117.0	119.5	.85727	22
23	9517			1363			8637	.96616	117.3	119.7	.86444	23
24	9575			1078			8922	.97320	117.5	120.0	.87162	24
25	.979634			.200793			.799207	3.98025	118.0	120.5	4.87882	25
26	9692			0508			9492	.98733	118.3	120.8	.88605	26
27	9750			0223			9777	.99443	118.7	121.0	.89330	27
28	9809			.199938			.800062	4.00155	119.0	121.5	.90056	28
29	9867			9653			0347	.00869	119.3	121.8	.90785	29
30	.979925			.199368			.800632	4.01585	119.7	122.2	4.91516	30
31	9983			9083			0917	.02303	120.2	122.5	.92249	31
32	.980040			8798			1202	.03024	120.3	122.8	.92984	32
33	0098			8513			1487	.03746	120.8	123.2	.93721	33
34	0156			8228			1772	.04471	121.0	123.5	.94460	34
35	.980214			.197942			.802058	4.05197	121.5	124.0	4.95201	35
36	0271			7657			2343	.05926	121.8	124.2	.95945	36
37	0329			7372			2628	.06657	122.2	124.7	.96690	37
38	0386			7087			2913	.07390	122.5	125.0	.97438	38
39	0443			6802			3198	.08125	123.0	125.3	.98188	39
40	.980500			.196517			.803483	4.08863	123.2	125.8	4.98940	40
41	0558	"	Corr.	6231	"	Corr.	3769	.09602	123.7	126.0	.99695	41
42	0615	10	9	5946	10	48	4054	.10344	124.0	126.5	5.00451	42
43	0672	15	14	5661	15	71	4339	.11088	124.5	126.8	.01210	43
44	0728	20	19	5376	20	95	4624	.11835	124.7	127.2	.01971	44
45	.980785	30	28	.195090	30	143	.804910	4.12583	125.2	127.5	5.02734	45
46	0842	40	38	4805	40	190	5195	.13334	125.5	128.0	.03499	46
47	0899	45	43	4520	45	214	5480	.14087	125.8	128.3	.04267	47
48	0955	50	47	4234	50	238	5766	.14842	126.2	128.7	.05037	48
49	1012			3949			6051	.15599	126.7	129.2	.05809	49
50	.981068			.193664			.806336	4.16359	127.0	129.3	5.06584	50
51	1124			3378			6622	.17121	127.5	129.8	.07360	51
52	1180			3093			6907	.17886	127.7	130.3	.08139	52
53	1237			2807			7193	.18652	128.2	130.5	.08921	53
54	1293			2522			7478	.19421	128.7	131.0	.09704	54
55	.981349			.192236			.807764	4.20193	128.8	131.5	5.10490	55
56	1404			1951			8049	.20966	129.3	131.7	.11279	56
57	1460			1666			8334	.21742	129.7	132.2	.12069	57
58	1516			1380			8620	.22520	130.2	132.7	.12862	58
59	1572			1094			8906	.23301	130.5	132.8	.13658	59
60	.981627			.190809			.809191	4.24084			5.14455	60

TABLE XX.—NATURAL SINES, COSINES,

79°

'	SINE	CORR. FOR SEC. +	COSINE	CORR. FOR SEC. − +	VERSINE	EXSEC	DIFF. 10"	DIFF. 10"	TANGENT	'
0	.981627		.190809		.809191	4.24084	131.0	133.5	5.14455	0
1	1683		0523		9477	.24870	131.3	133.7	.15256	1
2	1738		0238		9762	.25658	131.7	134.2	.16058	2
3	1793		.189952		.810048	.26448	132.2	134.7	.16863	3
4	1848		9667		0333	.27241	132.5	134.8	.17671	4
5	.981904		.189381		.810619	4.28036	132.8	135.5	5.18480	5
6	1959		9095		0905	.28833	133.5	135.7	.19293	6
7	2014		8810		1190	.29634	133.7	136.3	.20107	7
8	2069		8524		1476	.30436	134.2	136.5	.20925	8
9	2123		8238		1762	.31241	134.7	137.0	.21744	9
10	.982178		.187953		.812047	4.32049	135.0	137.5	5.22566	10
11	2233	" Corr.	7667	" Corr.	2333	.32859	135.3	137.8	.23391	11
12	2287	10 9	7381	10 48	2619	.33671	135.8	138.3	.24218	12
13	2342	15 14	7096	15 71	2904	.34486	136.3	138.7	.25048	13
14	2396	20 18	6810	20 95	3190	.35304	136.7	139.2	.25880	14
15	.982450	30 27	.186524	30 143	.813476	4.36124	137.2	139.7	5.26715	15
16	2505	40 36	6238	40 190	3762	.36947	137.5	140.0	.27553	16
17	2559	45 41	5952	45 214	4048	.37772	138.0	140.3	.28393	17
18	2613	50 45	5667	50 238	4333	.38600	138.3	140.8	.29235	18
19	2667		5381		4619	.39430	138.8	141.3	.30080	19
20	.982721		.185095		.814905	4.40263	139.3	141.7	5.30928	20
21	2774		4809		5191	.41099	139.7	142.2	.31778	21
22	2828		4523		5477	.41937	140.2	142.7	.32631	22
23	2882		4237		5763	.42778	140.7	143.0	.33487	23
24	2935		3951		6049	.43622	141.0	143.5	.34345	24
25	.982989		.183665		.816335	4.44468	141.5	144.0	5.35206	25
26	3042		3380		6620	.45317	142.0	144.3	.36070	26
27	3096		3094		6906	.46169	142.3	144.8	.36936	27
28	3149		2808		7192	.47023	143.0	145.3	.37805	28
29	3202		2522		7478	.47881	143.2	145.8	.38677	29
30	.983255		.182236		.817764	4.48740	143.8	146.2	5.39552	30
31	3308		1950		8050	.49603	144.2	146.7	.40429	31
32	3361		1664		8336	.50468	144.8	147.2	.41309	32
33	3414		1377		8623	.51337	145.2	147.7	.42192	33
34	3466		1091		8909	.52208	145.5	148.0	.43078	34
35	.983519		.180805		.819195	4.53081	146.2	148.5	5.43966	35
36	3572		0519		9481	.53958	146.5	149.0	.44857	36
37	3624		0233		9767	.54837	147.2	149.5	.45751	37
38	3676		.179947		.820053	.55720	147.5	150.0	.46648	38
39	3729		9661		0339	.56605	148.0	150.5	.47548	39
40	.983781		.179375		.820625	4.57493	148.3	150.8	5.48451	40
41	3833	" Corr.	9088	" Corr.	0912	.58383	149.0	151.3	.49356	41
42	3885	10 9	8802	10 48	1198	.59277	149.5	152.0	.50264	42
43	3937	15 13	8516	15 72	1484	.60174	149.8	152.3	.51176	43
44	3989	20 17	8230	20 95	1770	.61073	150.5	152.8	.52090	44
45	.984041	30 26	.177944	30 143	.822056	4.61976	150.8	153.3	5.53007	45
46	4092	40 35	7657	40 191	2343	.62881	151.5	154.0	.53927	46
47	4144	45 39	7371	45 215	2629	.63790	151.8	154.3	.54851	47
48	4196	50 43	7085	50 239	2915	.64701	152.5	154.8	.55777	48
49	4247		6798		3202	.65616	152.8	155.3	.56706	49
50	.984298		.176512		.823488	4.66533	153.5	155.8	5.57638	50
51	4350		6226		3774	.67454	153.8	156.3	.58573	51
52	4401		5940		4060	.68377	154.5	156.8	.59511	52
53	4452		5653		4347	.69304	155.0	157.5	.60452	53
54	4503		5367		4633	.70234	155.3	157.8	.61397	54
55	.984554		.175080		.824920	4.71166	156.0	158.5	5.62344	55
56	4605		4794		5206	.72102	156.5	158.8	.63295	56
57	4656		4508		5492	.73041	157.0	159.5	.64248	57
58	4707		4221		5779	.73983	157.7	160.0	.65205	58
59	4757		3935		6065	.74929	158.0	160.5	.66165	59
60	.984808		.173648		.826352	4.75877			5.67128	60

VERSINES, EXSECANTS, AND TANGENTS

80°

'	SINE	CORR. FOR SEC.	COSINE	CORR. FOR SEC.	VERSINE	EXSEC	DIFF. 10"	DIFF. 10"	TANGENT	'
0	.984808	+	.173648	− +	.826352	4.75877	158.7	161.0	5.67128	0
1	4858		3362		6638	.76829	159.2	161.7	.68094	1
2	4909		3075		6925	.77784	159.7	162.2	.69064	2
3	4959		2789		7211	.78742	160.2	162.7	.70037	3
4	5009		2502		7498	.79703	160.7	163.2	.71013	4
5	.985059		.172216		.827784	4.80667	161.3	163.7	5.71992	5
6	5109		1929		8071	.81635	161.8	164.3	.72974	6
7	5159		1642		8358	.82606	162.5	164.8	.73960	7
8	5209		1356		8644	.83581	162.8	165.3	.74949	8
9	5259		1069		8931	.84558	163.5	166.0	.75941	9
10	.985309		.170783		.829217	4.85539	164.2	166.5	5.76937	10
11	5358	" Corr.	0496	" Corr.	9504	.86524	164.5	167.0	.77936	11
12	5408	10 8	0210	10 48	9790	.87511	165.2	167.7	.78938	12
13	5457	15 12	.169923	15 72	.830077	.88502	165.8	168.2	.79944	13
14	5507	20 16	9636	20 96	0364	.89497	166.3	168.8	.80953	14
15	.985556	30 25	.169350	30 143	.830650	4.90495	166.8	169.3	5.81966	15
16	5605	40 33	9063	40 191	0937	.91496	167.5	169.8	.82982	16
17	5654	45 37	8776	45 215	1224	.92501	168.0	170.5	.84001	17
18	5704	50 41	8489	50 239	1511	.93509	168.7	171.2	.85024	18
19	5752		8203		1797	.94521	169.3	171.5	.86051	19
20	.985801		.167916		.832084	4.95536	169.8	172.3	5.87080	20
21	5850		7629		2371	.96555	170.3	172.8	.88114	21
22	5899		7342		2658	.97577	171.0	173.3	.89151	22
23	5948		7056		2944	.98603	171.7	174.2	.90191	23
24	5996		6769		3231	.99633	172.2	174.5	.91236	24
25	.986044		.166482		.833518	5.00666	172.7	175.3	5.92283	25
26	6093		6195		3805	.01702	173.5	175.8	.93335	26
27	6141		5908		4092	.02743	174.0	176.3	.94390	27
28	6189		5621		4379	.03787	174.5	177.0	.95448	28
29	6238		5334		4666	.04834	175.3	177.7	.96510	29
30	.986286		.165048		.834952	5.05886	175.8	178.3	5.97576	30
31	6334		4761		5239	.06941	176.5	179.0	.98646	31
32	6382		4474		5526	.08000	177.0	179.5	.99720	32
33	6429		4187		5813	.09062	177.7	180.2	6.00797	33
34	6477		3900		6100	.10128	178.5	180.7	.01878	34
35	.986525		.163613		.836387	5.11199	179.0	181.5	6.02962	35
36	6572		3326		6674	.12273	179.5	182.0	.04051	36
37	6620		3039		6961	.13350	180.3	182.8	.05143	37
38	6667		2752		7248	.14432	180.8	183.3	.06240	38
39	6714		2465		7535	.15517	181.7	184.0	.07340	39
40	.986762		.162178		.837822	5.16607	182.2	184.7	6.08444	40
41	6809	" Corr.	1891	" Corr.	8109	.17700	182.8	185.3	.09552	41
42	6856	10 8	1604	10 48	8396	.18797	183.7	185.8	.10664	42
43	6903	15 12	1317	15 72	8683	.19898	184.3	186.7	.11779	43
44	6950	20 16	1030	20 96	8970	.21004	184.8	187.3	.12899	44
45	.986996	30 23	.160743	30 144	.839257	5.22113	185.5	188.0	6.14023	45
46	7043	40 31	0456	40 191	9544	.23226	186.2	188.7	.15151	46
47	7090	45 35	0168	45 215	9832	.24343	186.8	189.3	.16283	47
48	7136	50 39	.159881	50 239	.840119	.25464	187.7	190.0	.17419	48
49	7183		9594		0406	.26590	188.2	190.7	.18559	49
50	.987229		.159307		.840693	5.27719	189.0	191.3	6.19703	50
51	7275		9020		0980	.28853	189.7	192.0	.20851	51
52	7322		8732		1268	.29991	190.3	192.8	.22003	52
53	7368		8445		1555	.31133	191.0	193.5	.23160	53
54	7414		8158		1842	.32279	191.7	194.2	.24321	54
55	.987460		.157871		.842129	5.33429	192.5	194.8	6.25486	55
56	7506		7584		2416	.34584	193.2	195.7	.26655	56
57	7551		7296		2704	.35743	193.8	196.3	.27829	57
58	7597		7009		2991	.36906	194.5	197.0	.29007	58
59	7643		6722		3278	.38073	195.3	197.7	.30189	59
60	.987688		.156434		.843566	5.39245			6.31375	60

TABLE XX.—NATURAL SINES, COSINES,

81°

′	SINE	CORR. FOR SEC. +	COSINE	CORR. FOR SEC. − +	VERSINE	EXSEC	DIFF. 10″		TANGENT	′
0	.987688		.156434		.843566	5.39245	196.2	198.5	6.31375	0
1	7734		6147		3853	.40422	196.7	199.2	.32566	1
2	7779		5860		4140	.41602	197.5	200.0	.33761	2
3	7824		5572		4428	.42787	198.3	200.7	.34961	3
4	7870		5285		4715	.43977	199.0	201.5	.36165	4
5	.987915		.154998		.845002	5.45171	199.7	202.2	6.37374	5
6	7960		4710		5290	.46369	200.5	202.8	.38587	6
7	8005		4423		5577	.47572	201.2	203.7	.39804	7
8	8050		4136		5864	.48779	202.0	204.5	.41026	8
9	8094		3848		6152	.49991	202.8	205.2	.42253	9
10	.988139		.153561		.846439	5.51208	203.5	206.0	6.43484	10
11	8184	″ Corr.	3273	″ Corr.	6727	.52429	204.3	206.8	.44720	11
12	8228	10 7	2986	10 48	7014	.53655	205.2	207.5	.45961	12
13	8273	15 11	2698	15 72	7302	.54886	205.8	208.4	.47206	13
14	8317	20 15	2411	20 96	7589	.56121	206.7	209.0	.48456	14
15	.988362	30 22	.152123	30 144	.847877	5.57361	207.5	210.0	6.49710	15
16	8406	40 29	1836	40 192	8164	.58606	208.2	210.7	.50970	16
17	8450	45 33	1548	45 216	8452	.59855	209.2	211.5	.52234	17
18	8494	50 37	1261	50 240	8739	.61110	209.8	212.3	.53503	18
19	8538		0973		9027	.62369	210.7	213.0	.54777	19
20	.988582		.150686		.849314	5.63633	211.5	214.0	6.56055	20
21	8626		0398		9602	.64902	212.3	214.7	.57339	21
22	8669		0111		9889	.66176	213.0	215.7	.58627	22
23	8713		.149823		.850177	.67454	214.0	216.3	.59921	23
24	8756		9535		0465	.68738	214.8	217.3	.61219	24
25	.988800		.149248		.850752	5.70027	215.7	218.0	6.62523	25
26	8843		8960		1040	.71321	216.9	218.8	.63831	26
27	8886		8672		1328	.72620	217.3	219.8	.65144	27
28	8930		8385		1615	.73924	218.2	220.7	.66463	28
29	8973		8097		1903	.75233	219.0	221.5	.67787	29
30	.989016		.147809		.852191	5.76547	219.8	222.3	6.69116	30
31	9059		7522		2478	.77866	220.8	223.2	.70450	31
32	9102		7234		2766	.79191	221.7	224.0	.71789	32
33	9144		6946		3054	.80521	222.5	225.0	.73133	33
34	9187		6658		3342	.81856	223.3	225.8	.74483	34
35	.989230		.146371		.853629	5.83196	224.3	226.8	6.75838	35
36	9272		6083		3917	.84542	225.2	227.5	.77199	36
37	9315		5795		4205	.85893	226.2	228.7	.78564	37
38	9357		5508		4492	.87250	227.0	229.3	.79936	38
39	9399		5220		4780	.88612	227.8	230.3	.81312	39
40	.989442		.144932		.855068	5.89979	228.8	231.3	6.82694	40
41	9484	″ Corr.	4644	″ Corr.	5356	.91352	229.8	232.2	.84082	41
42	9526	10 7	4356	10 48	5644	.92731	230.7	233.2	.85475	42
43	9568	15 10	4068	15 72	5932	.94115	231.7	234.0	.86874	43
44	9610	20 14	3780	20 96	6220	.95505	232.5	235.0	.88278	44
45	.989651	30 21	.143493	30 144	.856507	5.96900	233.5	236.0	6.89688	45
46	9693	40 28	3205	40 192	6795	.98301	234.5	236.8	.91104	46
47	9735	45 31	2917	45 216	7083	.99708	235.3	237.8	.92525	47
48	9776	50 35	2629	50 240	7371	6.01120	236.3	238.8	.93952	48
49	9818		2341		7659	.02538	237.3	239.7	.95385	49
50	.989859		.142053		.857947	6.03962	238.3	240.8	6.96823	50
51	9900		1765		8235	.05392	239.3	241.7	.98268	51
52	9942		1477		8523	.06828	240.2	242.7	.99718	52
53	9983		1189		8811	.08269	241.3	243.8	7.01174	53
54	.990024		0901		9099	.09717	242.3	244.7	.02637	54
55	.990065		.140613		.859387	6.11171	243.2	245.7	7.04105	55
56	0106		0325		9675	.12630	244.3	246.7	.05579	56
57	0146		0037		9963	.14096	245.3	247.8	.07059	57
58	0187		.139749		.860251	.15568	246.3	248.7	.08546	58
59	0228		9461		0539	.17046	247.3	249.8	.10038	59
60	.990268		.139173		.860827	6.18530			7.11537	60

VERSINES, EXSECANTS, AND TANGENTS

82°

'	SINE	CORR. FOR SEC.	COSINE	CORR. FOR SEC.	VERSINE	EXSEC	DIFF. 10"	DIFF. 10"	TANGENT	'
0	.990268	+	.139173	− +	.860827	6.18530	248.3	250.8	7.11537	0
1	0308		8885		1115	.20020	249.5	251.8	.13042	1
2	0349		8597		1403	.21517	250.3	253.0	.14553	2
3	0389		8309		1691	.23019	251.7	253.8	.16071	3
4	0429		8021		1979	.24529	252.5	255.2	.17594	4
5	.990469		.137733		.862267	6.26044	253.7	256.0	7.19125	5
6	0510		7444		2556	.27566	254.8	257.2	.20661	6
7	0549		7156		2844	.29095	255.8	258.3	.22204	7
8	0589		6868		3132	.30630	256.8	259.3	.23754	8
9	0629		6580		3420	.32171	258.0	260.5	.25310	9
10	.990669		.136292		.863708	6.33719	259.2	261.5	7.26873	10
11	0708	" Corr.	6004	" Corr.	3996	.35274	260.2	262.7	.28442	11
12	0748	10 7	5716	10 48	4284	.36835	261.3	263.7	.30018	12
13	0787	15 10	5427	15 72	4573	.38403	262.5	265.0	.31600	13
14	0827	20 13	5139	20 96	4861	.39978	263.7	266.0	.33190	14
15	.990866	30 20	.134851	30 144	.865149	6.41560	264.7	267.2	7.34786	15
16	0905	40 26	4563	40 192	5437	.43148	265.8	268.3	.36389	16
17	0944	45 29	4274	45 216	5726	.44743	267.2	269.5	.37999	17
18	0983	50 33	3986	50 240	6014	.46346	268.2	270.7	.39616	18
19	1022		3698		6302	.47955	269.3	271.8	.41240	19
20	.991061		.133410		.866590	6.49571	270.5	273.0	7.42871	20
21	1100		3121		6879	.51194	271.8	274.2	.44509	21
22	1138		2833		7167	.52825	272.8	275.3	.46154	22
23	1177		2545		7455	.54462	274.2	276.5	.47806	23
24	1216		2256		7744	.56107	275.3	277.8	.49465	24
25	.991254		.131968		.868032	6.57759	276.5	279.0	7.51132	25
26	1292		1680		8320	.59418	277.8	280.2	.52806	26
27	1331		1391		8609	.61085	279.0	281.5	.54487	27
28	1369		1103		8897	.62759	280.3	282.8	.56176	28
29	1407		0815		9185	.64441	281.5	283.8	.57872	29
30	.991445		.130526		.869474	6.66130	282.7	285.3	7.59575	30
31	1483		0238		9762	.67826	284.0	286.3	.61287	31
32	1521		.129949		.870051	.69530	285.3	287.8	.63005	32
33	1558		9661		0339	.71242	286.7	289.0	.64732	33
34	1596		9372		0628	.72962	287.8	290.3	.66466	34
35	.991634		.129084		.870916	6.74689	289.2	291.2	7.68208	35
36	1671		8796		1204	.76424	290.5	293.0	.69957	36
37	1709		8507		1493	.78167	291.8	294.2	.71715	37
38	1746		8219		1781	.79918	293.2	295.7	.73480	38
39	1783		7930		2070	.81677	294.3	296.8	.75254	39
40	.991820		.127642		.872358	6.83443	295.8	298.3	7.77035	40
41	1857	" Corr.	7353	" Corr.	2647	.85218	297.2	299.5	.78825	41
42	1894	10 6	7065	10 48	2935	.87001	298.5	301.0	.80622	42
43	1931	15 9	6776	15 72	3224	.88792	300.0	302.3	.82428	43
44	1968	20 12	6488	20 96	3512	.90592	301.3	303.7	.84242	44
45	.992005	30 18	.126199	30 144	.873801	6.92400	302.7	305.2	7.86064	45
46	2042	40 25	5910	40 192	4090	.94216	304.0	306.5	.87895	46
47	2078	45 28	5622	45 216	4378	.96040	305.5	308.0	.89734	47
48	2115	50 31	5333	50 241	4667	.97873	306.8	309.3	.91582	48
49	2151		5045		4955	.99714	308.3	310.7	.93438	49
50	.992187		.124756		.875244	7.01564	309.8	312.3	7.95302	50
51	2224		4467		5533	.03423	311.3	313.7	.97176	51
52	2260		4179		5821	.05291	312.7	315.0	.99058	52
53	2296		3890		6110	.07167	314.2	316.7	8.00948	53
54	2332		3602		6398	.09052	315.7	318.0	.02848	54
55	.992368		.123313		.876687	7.10946	317.2	319.7	8.04756	55
56	2404		3024		6976	.12849	318.5	321.0	.06674	56
57	2439		2736		7264	.14760	320.2	322.7	.08600	57
58	2475		2447		7553	.16681	321.8	324.2	.10536	58
59	2511		2158		7842	.18612	323.2	325.7	.12481	59
60	.992546		.121869		.878131	7.20551			8.14435	60

TABLE XX.—NATURAL SINES, COSINES,

83°

′	SINE	CORR. FOR SEC. +	COSINE	CORR. FOR SEC. − +	VERSINE	EXSEC	DIFF. 10″	DIFF. 10″	TANGENT	′
0	.992546		.121869		.878131	7.20551	324.8	327.2	8.14435	0
1	2582		1581		8419	.22500	326.2	328.7	.16398	1
2	2617		1292		8708	.24457	328.0	330.3	.18370	2
3	2652		1003		8997	.26425	329.5	332.0	.20352	3
4	2687		0714		9286	.28402	331.0	333.5	.22344	4
5	.992722		.120426		.879574	7.30388	332.7	335.0	8.24345	5
6	2757		0137		9863	.32384	334.3	336.8	.26355	6
7	2792		.119848		.880152	.34390	335.8	338.3	.28376	7
8	2827		9559		0441	.36405	337.7	340.0	.30406	8
9	2862		9270		0730	.38431	339.2	341.7	.32446	9
10	.992896		.118982		.881018	7.40466	340.8	343.2	8.34496	10
11	2931	″ Corr.	8693	″ Corr.	1307	.42511	342.5	345.0	.36555	11
12	2966	10 6	8404	10 48	1596	.44566	344.3	346.7	.38625	12
13	3000	15 9	8115	15 72	1885	.46632	345.8	348.3	.40705	13
14	3034	20 11	7826	20 96	2174	.48707	347.7	350.2	.42795	14
15	.993068	30 17	.117537	30 144	.882463	7.50793	349.3	351.8	8.44896	15
16	3103	40 23	7248	40 193	2752	.52889	351.2	353.5	.47007	16
17	3137	45 26	6960	45 217	3040	.54996	352.8	355.2	.49128	17
18	3171	50 29	6671	50 241	3329	.57113	354.7	357.2	.51259	18
19	3204		6382		3618	.59241	356.3	358.8	.53402	19
20	.993238		.116093		.883907	7.61379	358.2	360.5	8.55555	20
21	3272		5804		4196	.63528	360.0	362.5	.57718	21
22	3306		5515		4485	.65688	361.8	364.2	.59893	22
23	3339		5226		4774	.67859	363.7	366.2	.62078	23
24	3373		4937		5063	.70041	365.5	367.8	.64275	24
25	.993406		.114648		.885352	7.72234	367.3	369.8	8.66482	25
26	3440		4359		5641	.74438	369.2	371.7	.68701	26
27	3473		4070		5930	.76653	371.2	373.5	.70931	27
28	3506		3781		6219	.78880	373.0	375.5	.73172	28
29	3539		3492		6508	.81118	374.8	377.3	.75425	29
30	.993572		.113203		.886797	7.83367	376.8	379.2	8.77689	30
31	3605		2914		7086	.85628	378.8	381.3	.79964	31
32	3638		2625		7375	.87901	380.8	383.2	.82252	32
33	3670		2336		7664	.90186	382.7	385.2	.84551	33
34	3703		2047		7953	.92482	384.8	387.2	.86862	34
35	.993736		.111758		.888242	7.94791	386.7	389.2	8.89185	35
36	3768		1469		8531	.97111	388.8	391.2	.91520	36
37	3800		1180		8820	.99444	390.7	393.3	.93867	37
38	3833		0891		9109	8.01788	393.0	395.2	.96227	38
39	3865		0602		9398	.04146	394.8	397.5	.98598	39
40	.993897		.110313		.889687	8.06515	397.0	399.3	9.00983	40
41	3929	″ Corr.	0023	″ Corr.	9977	.08897	399.2	401.7	.03379	41
42	3961	10 5	.109734	10 48	.890266	.11292	401.2	403.7	.05789	42
43	3993	15 8	9445	15 72	0555	.13699	403.5	405.8	.08211	43
44	4025	20 11	9156	20 96	0844	.16120	405.5	407.8	.10646	44
45	.994056	30 16	.108867	30 145	.891133	8.18553	407.7	410.2	9.13093	45
46	4088	40 21	8578	40 193	1422	.20999	410.0	412.3	.15554	46
47	4120	45 24	8288	45 217	1712	.23459	412.0	414.7	.18028	47
48	4151	50 26	7999	50 241	2001	.25931	414.3	416.7	.20516	48
49	4182		7710		2290	.28417	416.7	419.0	.23016	49
50	.994214		.107421		.892579	8.30917	418.8	421.3	9.25530	50
51	4245		7132		2868	.33430	421.2	423.5	.28058	51
52	4276		6842		3158	.35957	423.3	425.8	.30599	52
53	4307		6553		3447	.38497	425.8	428.3	.33154	53
54	4338		6264		3736	.41052	428.0	430.5	.35724	54
55	.994369		.105975		.894025	8.43620	430.5	432.8	9.38307	55
56	4400		5686		4314	.46203	432.8	435.2	.40904	56
57	4430		5396		4604	.48800	435.2	437.7	.43515	57
58	4461		5107		4893	.51411	437.7	440.0	.46141	58
59	4491		4818		5182	.54037	440.0	442.5	.48781	59
60	994522		.104528		.895472	8.56677			9.51436	60

VERSINES, EXSECANTS, AND TANGENTS

84°

'	SINE	CORR. FOR SEC.	COSINE	CORR. FOR SEC.	VERSINE	EXSEC	DIFF. 10"	DIFF. 10"	TANGENT	'
		+		− +						
0	.994522		.104528		.895472	8.56677	442.5	445.0	9.51436	0
1	4552		4239		5761	.59332	445.0	447.5	.54106	1
2	4582		3950		6050	.62002	447.5	449.8	.56791	2
3	4613		3660		6340	.64687	450.0	452.5	.59490	3
4	4643		3371		6629	.67387	452.7	455.0	.62205	4
5	.994673		.103082		.896918	8.70103	455.0	457.5	9.64935	5
6	4703		2792		7208	.72833	457.7	460.2	.67680	6
7	4733		2503		7497	.75579	460.3	462.7	.70441	7
8	4762		2214		7786	.78341	463.1	465.3	.73217	8
9	4792		1924		8076	.81119	465.5	468.0	.76009	9
10	.994822		.101635		.898365	8.83912	468.3	470.7	9.78817	10
11	4851	" Corr.	1346	" Corr.	8654	.86722	470.8	473.5	.81641	11
12	4881	10 5	1056	10 48	8944	.89547	473.7	476.0	.84482	12
13	4910	15 7	0767	15 72	9233	.92389	476.5	478.8	.87338	13
14	4939	20 10	0478	20 96	9522	.95248	479.2	481.7	.90211	14
15	.994968	30 15	.100188	30 145	.899812	8.98123	482.0	484.3	9.93101	15
16	4998	40 20	.099899	40 193	.900101	9.01015	484.7	487.2	.96007	16
17	5027	45 22	9609	45 217	0391	.03923	487.7	- - -	.98930	17
18	5056	50 24	9320	50 241	0680	.06849	490.5	49.3	10.0187	18
19	5084		9030		0970	.09792	493.3	49.5	.0483	19
20	.995113		.098741		.901259	9.12752	496.3	50.0	10.0780	20
21	5142		8451		1549	.15730	499.2	50.2	.1080	21
22	5170		8162		1838	.18725	502.3	50.3	.1381	22
23	5199		7872		2128	.21739	505.2	50.8	.1683	23
24	5227		7583		2417	.24770	508.2	51.0	.1988	24
25	.995256		.097293		.902707	9.27819	511.3	51.3	10.2294	25
26	5284		7004		2996	.30887	514.2	51.8	.2602	26
27	5312		6714		3286	.33973	517.3	51.8	.2913	27
28	5340		6425		3575	.37077	520.7	52.3	.3224	28
29	5368		6135		3865	.40201	523.7	52.7	.3538	29
30	.995396		.095846		.904154	9.43343	527.0	53.0	10.3854	30
31	5424		5556		4444	.46505	530.0	53.2	.4172	31
32	5452		5267		4733	.49685	533.5	53.7	.4491	32
33	5480		4977		5023	.52886	536.7	53.8	.4813	33
34	5507		4688		5312	.56106	540.0	54.3	.5136	34
35	.995534		.094398		.905602	9.59346	543.2	54.5	10.5462	35
36	5562		4108		5892	.62605	546.5	54.8	.5789	36
37	5589		3819		6181	.65885	550.2	55.3	.6118	37
38	5616		3529		6471	.69186	553.5	55.5	.6450	38
39	5644		3240		6760	.72507	557.0	56.0	.6783	39
40	.995671		.092950		.907050	9.75849	560.5	56.3	10.7119	40
41	5698	" Corr.	2660	" Corr.	7340	.79212	564.0	56.7	.7457	41
42	5725	10 4	2371	10 48	7629	.82596	567.5	57.0	.7797	42
43	5752	15 7	2081	15 72	7919	.86001	571.2	57.3	.8139	43
44	5778	20 9	1791	20 97	8209	.89428	574.8	57.7	.8483	44
45	.995805	30 13	.091502	30 145	.908498	9.92877	578.5	58.2	10.8829	45
46	5832	40 18	1212	40 193	8788	.96348	582.2	58.3	.9178	46
47	5858	45 20	0922	45 217	9078	.99841	- - -	59.0	.9528	47
48	5884	50 22	0633	50 241	9367	10.0336	58.8	59.2	.9882	48
49	5911		0343		9657	.0689	59.3	59.5	11.0237	49
50	.995937		.090053		.909947	10.1045	59.8	60.0	11.0594	50
51	5963		.089764		.910236	.1404	60.2	60.3	.0954	51
52	5989		9474		0526	.1765	60.5	60.8	.1316	52
53	6015		9184		0816	.2128	60.8	61.2	.1681	53
54	6041		8894		1106	.2493	61.3	61.5	.2048	54
55	.996067		.088605		.911395	10.2861	61.7	62.0	11.2417	55
56	6093		8315		1685	.3231	62.2	62.3	.2789	56
57	6118		8025		1975	.3604	62.5	62.8	.3163	57
58	6144		7735		2265	.3979	63.0	63.2	.3540	58
59	6169		7446		2554	.4357	63.3	63.7	.3919	59
60	.996195		.087156		.912844	10.4737			11.4301	60

TABLE XX.—NATURAL SINES, COSINES,

85°

'	SINE	CORR. FOR SEC. +	COSINE	CORR. FOR SEC. − +	VERSINE	EXSEC	DIFF. 10"	TANGENT	'
0	.996195		.087156		.912844	10.4737	63.9	11.4301	0
1	6220		6866		3134	.5120	64.3	.4685	1
2	6245		6576		3424	.5505	64.8	.5072	2
3	6270		6286		3714	.5893	65.3	.5461	3
4	6295		5997		4003	.6284	65.7	.5853	4
5	.996320		.085707		.914293	10.6677	66.1	11.6248	5
6	6345		5417		4583	.7073	66.5	.6645	6
7	6370		5127		4873	.7471	67.0	.7045	7
8	6395		4837		5163	.7873	67.4	.7448	8
9	6420		4547		5453	.8277	67.9	.7853	9
10	.996444		.084258		.915742	10.8684	68.3	11.8262	10
11	6468	" Corr.	3968	" Corr.	6032	.9093	68.8	.8673	11
12	6493	10 4	3678	10 48	6322	.9506	69.3	.9087	12
13	6517	15 6	3388	15 72	6612	.9921	69.8	.9504	13
14	6541	20 8	3098	20 97	6902	11.0340	70.3	.9923	14
15	.996566	30 12	.082808	30 145	.917192	11.0761	70.8	12.0346	15
16	6590	40 16	2518	40 193	7482	.1185	71.3	.0772	16
17	6614	45 18	2228	45 217	7772	.1612	71.8	.1201	17
18	6637	50 20	1938	50 242	8062	.2043	72.3	.1632	18
19	6661		1649		8351	.2476	72.8	.2067	19
20	.996685		.081359		.918641	11.2913	73.3	12.2505	20
21	6708		1069		8931	.3352	73.9	.2946	21
22	6732		0779		9221	.3795	74.5	.3390	22
23	6756		0489		9511	.4241	75.0	.3838	23
24	6779		0199		9801	.4690	75.5	.4288	24
25	.996802		.079909		.920091	11.5142	76.1	12.4742	25
26	6825		9619		0381	.5598	76.6	.5199	26
27	6848		9329		0671	.6057	77.2	.5660	27
28	6872		9039		0961	.6520	77.8	.6124	28
29	6894		8749		1251	.6986	78.3	.6591	29
30	.996917		.078459		.921541	11.7455	78.9	12.7062	30
31	6940		8169		1831	.7928	79.5	.7536	31
32	6963		7879		2121	.8404	80.2	.8014	32
33	6985		7589		2411	.8884	80.8	.8496	33
34	7008		7299		2701	.9368	81.4	.8981	34
35	.997030		.077009		.922991	11.9855	82.0	12.9469	35
36	7053		6719		3281	12.0346	82.6	.9962	36
37	7075		6429		3571	.0840	83.3	13.0458	37
38	7097		6139		3861	.1339	83.9	.0958	38
39	7119		5849		4151	.1841	84.5	.1461	39
40	.997141		.075559		.924441	12.2347	85.1	13.1969	40
41	7163	" Corr.	5269	" Corr.	4731	.2857	85.8	.2480	41
42	7185	10 4	4979	10 48	5021	.3371	86.4	.2996	42
43	7207	15 5	4689	15 73	5311	.3889	87.1	.3515	43
44	7229	20 7	4399	20 97	5601	.4411	87.8	.4039	44
45	.997250	30 11	.074108	30 145	.925892	12.4937	88.5	13.4566	45
46	7272	40 14	3818	40 193	6182	.5468	89.2	.5098	46
47	7293	45 16	3528	45 218	6472	.6002	89.9	.5634	47
48	7314	50 18	3238	50 242	6762	.6541	90.6	.6174	48
49	7336		2948		7052	.7084	91.3	.6719	49
50	.997357		.072658		.927342	12.7631	92.1	13.7267	50
51	7378		2368		7632	.8183	92.8	.7821	51
52	7399		2078		7922	.8739	93.6	.8378	52
53	7420		1788		8212	.9300	94.4	.8940	53
54	7441		1497		8503	.9865	95.2	.9507	54
55	.997462		.071207		.928793	13.0435	95.9	14.0079	55
56	7482		0917		9083	.1010	96.6	.0655	56
57	7503		0627		9373	.1589	97.5	.1235	57
58	7523		0337		9663	.2173	98.3	.1821	58
59	7544		0047		9953	.2762	99.2	.2411	59
60	.997564		.069756		.930244	13.3356		14.3007	60

VERSINES, EXSECANTS, AND TANGENTS

86°

'	SINE	CORR. FOR SEC.	COSINE	CORR. FOR SEC.	VERSINE	EXSEC	DIFF. 10"	TANGENT	'
0	.997564	+	.069756	− +	.930244	13.3356	99.9	14.3007	0
1	7584		9466		0534	.3955	100.8	.3607	1
2	7604		9176		0824	.4559	101.7	.4212	2
3	7624		8886		1114	.5168	102.5	.4823	3
4	7644		8596		1404	.5782	103.4	.5438	4
5	.997664		.068306		.931694	13.6401	104.3	14.6059	5
6	7684		8015		1985	.7026	105.2	.6685	6
7	7704		7725		2275	.7656	106.0	.7317	7
8	7724		7435		2565	.8291	106.9	.7954	8
9	7743		7145		2855	.8932	107.9	.8596	9
10	.997763		.066854		.933146	13.9579	108.9	14.9244	10
11	7782	" Corr.	6564	" Corr.	3436	14.0231	109.8	.9898	11
12	7802	10 3	6274	10 48	3726	.0889	110.8	15.0557	12
13	7821	15 5	5984	15 73	4016	.1553	111.8	.1222	13
14	7840	20 6	5693	20 97	4307	.2222	112.8	.1893	14
15	.997859	30 9	.065403	30 145	.934597	14.2898	113.8	15.2571	15
16	7878	40 13	5113	40 194	4887	.3579	114.8	.3254	16
17	7897	45 14	4823	45 218	5177	.4267	115.8	.3943	17
18	7916	50 16	4532	50 242	5468	.4961	116.9	.4638	18
19	7934		4242		5758	.5661	117.9	.5340	19
20	.997953		.063952		.936048	14.6368	118.9	15.6048	20
21	7972		3661		6339	.7081	120.1	.6762	21
22	7990		3371		6629	.7801	121.2	.7483	22
23	8008		3081		6919	.8527	122.3	.8211	23
24	8027		2790		7210	.9260	123.4	.8945	24
25	.998045		.062500		.937500	14.9999	124.6	15.9687	25
26	8063		2210		7790	15.0746	125.8	16.0435	26
27	8081		1920		8080	.1500	126.9	.1190	27
28	8099		1629		8371	.2261	128.2	.1952	28
29	8117		1339		8661	.3029	129.4	.2722	29
30	.998135		.061048		.938952	15.3804	130.6	16.3499	30
31	8152		0758		9242	.4587	131.9	.4283	31
32	8170		0468		9532	.5377	133.1	.5075	32
33	8188		0178		9822	.6175	134.4	.5874	33
34	8205		.059887		.940113	.6981	135.7	.6681	34
35	.998222		.059597		.940403	15.7794	137.1	16.7496	35
36	8240		9306		0694	.8616	138.4	.8319	36
37	8257		9016		0984	.9446	139.8	.9150	37
38	8274		8726		1274	16.0283	141.2	.9990	38
39	8291		8435		1565	.1130	142.6	17.0837	39
40	.998308		.058145		.941855	16.1984	144.1	17.1693	40
41	8325	" Corr.	7854	" Corr.	2146	.2848	145.5	.2558	41
42	8342	10 3	7564	10 48	2436	.3720	146.9	.3432	42
43	8358	15 4	7274	15 73	2726	.4600	148.4	.4314	43
44	8375	20 5	6983	20 97	3017	.5490	150.0	.5205	44
45	.998392	30 8	.056693	30 145	.943307	16.6389	151.5	17.6106	45
46	8408	40 11	6402	40 194	3598	.7298	153.0	.7015	46
47	8424	45 12	6112	45 218	3888	.8215	154.7	.7934	47
48	8441	50 14	5822	50 242	4178	.9142	156.4	.8863	48
49	8457		5531		4469	17.0079	158.0	.9802	49
50	.998473		.055241		.944759	17.1026	159.6	18.0750	50
51	8489		4950		5050	.1983	161.3	.1708	51
52	8505		4660		5340	.2950	163.0	.2677	52
53	8521		4369		5631	.3927	164.8	.3655	53
54	8537		4079		5921	.4915	166.6	.4645	54
55	.998552		.053788		.946212	17.5914	168.4	18.5645	55
56	8568		3498		6502	.6923	170.2	.6656	56
57	8584		3207		6793	.7944	172.0	.7678	57
58	8599		2917		7083	.8975	174.0	.8711	58
59	8614		2626		7374	18.0019	175.9	.9755	59
60	.998630		.052336		.947664	18.1073		19.0811	60

TABLE XX.—NATURAL SINES, COSINES,

87°

'	SINE	CORR. FOR SEC.	COSINE	CORR. FOR SEC.	VERSINE	EXSEC	DIFF. 10"	TANGENT	'
0	.998630	+	.052336	− +	.947664	18.1073	177.9	19.0811	0
1	8645		2046		7954	.2140	179.9	.1879	1
2	8660		1755		8245	.3218	181.9	.2959	2
3	8675		1464		8536	.4309	184.0	.4051	3
4	8690		1174		8826	.5412	186.1	.5156	4
5	.998705		.050884		.949116	18.6528	188.2	19.6273	5
6	8719		0593		9407	.7656	190.4	.7403	6
7	8734		0302		9698	.8798	192.6	.8546	7
8	8749		0012		9988	.9952	194.9	.9702	8
9	8763		.049721		.950279	19.1121	197.2	20.0872	9
10	.998778		.049431		.950569	19.2303	199.4	20.2056	10
11	8792	" Corr.	9140	" Corr.	0860	.3499	201.8	.3253	11
12	8806	10 2	8850	10 48	1150	.4709	204.3	.4465	12
13	8820	15 3	8559	15 73	1441	.5934	206.8	.5691	13
14	8834	20 5	8269	20 97	1731	.7174	209.4	.6932	14
15	.998848	30 7	.047978	30 145	.952022	19.8428	211.9	20.8188	15
16	8862	40 9	7688	40 194	2312	.9698	214.4	.9460	16
17	8876	45 10	7397	45 218	2603	20.0984	216.9	21.0747	17
18	8890	50 12	7106	50 242	2894	.2285	219.7	.2049	18
19	8904		6816		3184	.3603	222.4	.3369	19
20	.998917		.046525		.953475	20.4937	225.3	21.4704	20
21	8931		6235		3765	.6288	228.2	.6056	21
22	8944		5944		4056	.7656	231.0	.7426	22
23	8957		5654		4346	.9041	233.9	.8813	23
24	8971		5363		4637	21.0444	237.0	22.0217	24
25	.998984		.045072		.954928	21.1865	240.1	22.1640	25
26	8997		4782		5218	.3305	243.2	.3081	26
27	9010		4491		5509	.4764	246.3	.4541	27
28	9023		4201		5799	.6241	249.7	.6020	28
29	9036		3910		6090	.7739	253.0	.7519	29
30	.999048		.043619		.956381	21.9256	256.4	22.9038	30
31	9061		3329		6671	22.0794	259.8	23.0577	31
32	9073		3038		6962	.2352	263.4	.2137	32
33	9086		2748		7252	.3932	267.0	.3718	33
34	9098		2457		7543	.5533	270.6	.5321	34
35	.999111		.042166		.957834	22.7156	274.5	23.6945	35
36	9123		1876		8124	.8802	278.3	.8593	36
37	9135		1585		8415	23.0471	282.3	24.0263	37
38	9147		1294		8706	.2164	286.2	.1957	38
39	9159		1004		8996	.3880	290.3	.3675	39
40	.999171		.040713		.959287	23.5621	294.3	24.5418	40
41	9183	" Corr.	0422	" Corr.	9578	.7387	298.7	.7185	41
42	9194	10 2	0132	10 48	9868	.9179	303.2	.8978	42
43	9206	15 3	.039841	15 73	.960160	24.0997	307.5	25.0798	43
44	9218	20 4	9550	20 97	0450	.2841	312.1	.2644	44
45	.999229	30 6	.039260	30 145	.960740	24.4713	316.8	25.4517	45
46	9240	40 8	8969	40 194	1031	.6613	321.6	.6418	46
47	9252	45 9	8678	45 218	1322	.8542	326.3	.8348	47
48	9263	50 10	8388	50 242	1612	25.0499	331.4	26.0307	48
49	9274		8097		1903	.2487	336.5	.2296	49
50	.999285		.037806		.962194	25.4505	341.7	26.4316	50
51	9296		7516		2484	.6555	347.0	.6367	51
52	9307		7225		2775	.8636	352.5	.8450	52
53	9318		6934		3066	26.0750	358.1	27.0566	53
54	9328		6644		3356	.2898	363.8	.2715	54
55	.999339		.036353		.963647	26.5080	369.7	27.4899	55
56	9350		6062		3938	.7298	375.7	.7117	56
57	9360		5772		4228	.9551	381.9	.9372	57
58	9370		5481		4519	27.1842	388.2	28.1664	58
59	9381		5190		4810	.4170	394.7	.3994	59
60	.999391		.034900		.965100	27.6537		28.6363	60

VERSINES, EXSECANTS, AND TANGENTS
88°

′	SINE	CORR. FOR SEC. +	COSINE	CORR. FOR SEC. − +	VERSINE	EXSEC	DIFF. 10″	TANGENT	′
0	.999391		.034900		.965100	27.6537	401	28.6363	0
1	9401		4609		5391	27.8944	408	28.8771	1
2	9411		4318		5682	28.1392	415	29.1220	2
3	9421		4027		5973	28.3881	422	29.3711	3
4	9431		3737		6263	28.6414	429	29.6245	4
5	.999440		.033446		.966554	28.8990	437	29.8823	5
6	9450		3155		6845	29.1612	445	30.1446	6
7	9460		2864		7136	29.4280	453	30.4116	7
8	9469		2574		7426	29.6996	461	30.6833	8
9	9479		2283		7717	29.9761	469	30.9599	9
10	.999488		.031992		.968008	30.2576	478	31.2416	10
11	9497	″ Corr.	1702	″ Corr.	8298	30.5442	487	31.5284	11
12	9507	10 1	1411	10 48	8589	30.8362	496	31.8205	12
13	9516	15 2	1120	15 73	8880	31.1337	505	32.1181	13
14	9525	20 3	0829	20 97	9171	31.4367	515	32.4213	14
15	.999534	30 4	.030538	30 145	.969462	31.7455	525	32.7303	15
16	9542	40 6	0248	40 194	9752	32.0603	535	33.0452	16
17	9551	45 7	.029957	45 218	.970043	32.3812	545	33.3662	17
18	9560	50 7	9666	50 242	0334	32.7083	556	33.6935	18
19	9568		9376		0624	33.0420	567	34.0273	19
20	.999577		.029085		.970915	33.3823	579	34.3678	20
21	9585		8794		1206	33.7295	591	34.7151	21
22	9594		8503		1497	34.0838	603	35.0695	22
23	9602		8212		1788	34.4454	615	35.4313	23
24	9610		7922		2078	34.8145	628	35.8006	24
25	.999618		.027631		.972369	35.1914	641	36.1776	25
26	9626		7340		2660	35.5763	655	36.5627	26
27	9634		7049		2951	35.9695	670	36.9560	27
28	9642		6758		3242	36.3713	685	37.3579	28
29	9650		6468		3532	36.7818	700	37.7686	29
30	.999657		.026177		.973823	37.2016	715	38.1885	30
31	9665		5886		4114	37.6307	731	38.6177	31
32	9672		5595		4405	38.0696	748	39.0568	32
33	9680		5305		4695	38.5185	766	39.5059	33
34	9687		5014		4986	38.9780	784	39.9655	34
35	.999694		.024723		.975277	39.4482	803	40.4358	35
36	9702		4432		5568	39.9296	822	40.9174	36
37	9709		4141		5859	40.4227	842	41.4106	37
38	9716		3851		6149	40.9277	863	41.9158	38
39	9722		3560		6440	41.4452	884	42.4335	39
40	.999729		.023269		.976731	41.9757	907	42.9641	40
41	9736	″ Corr.	2978	″ Corr.	7022	42.5196	930	43.5081	41
42	9743	10 1	2687	10 48	7313	43.0775	954	44.0661	42
43	9749	15 2	2396	15 73	7604	43.6498	979	44.6386	43
44	9756	20 2	2106	20 97	7894	44.2372	1005	45.2261	44
45	.999762	30 3	.021815	30 145	.978185	44.8403	1032	45.8294	45
46	9768	40 4	1524	40 194	8476	45.4596	1061	46.4489	46
47	9774	45 5	1233	45 218	8767	46.0960	1090	47.0853	47
48	9781	50 5	0942	50 242	9058	46.7500	1121	47.7395	48
49	9787		0652		9348	47.4224	1153	48.4121	49
50	.999793		.020361		.979639	48.1141	1186	49.1039	50
51	9799		0070		9930	48.8258	1221	49.8157	51
52	9804		.019779		.980221	49.5584	1258	50.5485	52
53	9810		9488		0512	50.3129	1296	51.3032	53
54	9816		9197		0803	51.0903	1336	52.0807	54
55	.999821		.018907		.981093	51.8916	1378	52.8821	55
56	9827		8616		1384	52.7179	1421	53.7086	56
57	9832		8325		1675	53.5705	1467	54.5613	57
58	9837		8034		1966	54.4505	1515	55.4415	58
59	9843		7743		2257	55.3595	1566	56.3506	59
60	.999848		.017452		.982548	56.2987		57.2900	60

TABLE XX.—NATURAL TRIGONOMETRIC FUNCTIONS

89°

'	SINE	CORR. FOR SEC. +	COSINE	CORR. FOR SEC. − +	VERSINE	EXSEC	DIFF. 10"	TANGENT	'
0	.999848		.017452		.982548	56.2987	1619	57.2900	0
1	9853		7162		2838	57.2698	1674	58.2612	1
2	9858		6871		3129	58.2743	1733	59.2659	2
3	9862		6580		3420	59.3141	1795	60.3058	3
4	9867		6289		3711	60.3910	1860	61.3829	4
5	.999872		.015998		.984002	61.5072	1929	62.4992	5
6	9877		5707		4293	62.6646	2002	63.6567	6
7	9881		5416		4584	63.8657	2079	64.8580	7
8	9886		5126		4874	65.1130	2161	66.1055	8
9	9890		4835		5165	66.4093	2247	67.4019	9
10	.999894		.014544		.985456	67.7574	2339	68.7501	10
11	9898	By	4253	" Corr.	5747	69.1605	2436	70.1533	11
12	9902	Inspec-	3962	10 48	6038	70.6221	2540	71.6151	12
13	9906	tion	3671	15 73	6329	72.1458	2650	73.1390	13
14	9910		3380	20 97	6620	73.7359	2768	74.7292	14
15	.999914		.013090	30 145	.986910	75.3966	2894	76.3900	15
16	9918		2799	40 194	7201	77.1327	3028	78.1263	16
17	9922		2508	45 218	7492	78.9497	3173	79.9434	17
18	9925		2217	50 242	7783	80.8532	3327	81.8470	18
19	9929		1926		8074	82.8495	3494	83.8435	19
20	.999932		.011635		.988365	84.9456	3673	85.9398	20
21	9936		1344		8656	87.1492	3866	88.1436	21
22	9939		1054		8946	89.4689	4075	90.4633	22
23	9942		0763		9237	91.9139	4302	92.9085	23
24	9945		0472		9528	94.4947	4547	95.4895	24
25	.999948		.010181		.989819	97.2230	482	98.2179	25
26	9951		.009890		.990110	100.112	511	101.107	26
27	9954		9599		0401	103.176	543	104.171	27
28	9957		9308		0692	106.431	578	107.426	28
29	9959		9017		0983	109.897	616	110.892	29
30	.999962		.008726		.991274	113.593	659	114.589	30
31	9964		8436		1564	117.544	706	118.540	31
32	9967		8145		1855	121.778	758	122.774	32
33	9969		7854		2146	126.325	816	127.321	33
34	9971		7563		2437	131.222	881	132.219	34
35	.999974		.007272		.992728	136.511	955	137.507	35
36	9976		6981		3019	142.241	1.038	143.237	36
37	9978		6690		3310	148.468	1.132	149.465	37
38	9980		6400		3600	155.262	1.240	156.259	38
39	9981		6109		3891	162.703	1.364	163.700	39
40	.999983		.005818		.994182	170.888	1.508	171.885	40
41	9985		5527	" Corr.	4473	179.935	1.675	180.932	41
42	9986		5236	10 48	4764	189.987	1.872	190.984	42
43	9988		4945	15 73	5055	201.221	2.107	202.219	43
44	9989		4654	20 97	5346	213.860	2.387	214.858	44
45	.999990		.004363	30 145	.995637	228.184	2.728	229.182	45
46	9992		4072	40 194	5928	244.554	3.148	245.552	46
47	9993		3782	45 218	6218	263.443	3.673	264.441	47
48	9994		3491	50 242	6509	285.479	4.341	286.478	48
49	9995		3200		6800	311.523	5.209	312.521	49
50	.999996		.002909		.997091	342.775	6.366	343.774	50
51	9997		2618		7382	380.972	7.958	381.971	51
52	9997		2327		7673	428.719	10.231	429.718	52
53	9998		2036		7964	490.107	13.642	491.106	53
54	9998		1745		8255	571.958	19.099	572.957	54
55	.999999		.001454		.998546	686.550	26.648	687.549	55
56	9999		1164		8836	858.437	47.75	859.436	56
57	One		0873		9127	1144.92	95.49	1145.92	57
58	One		0582		9418	1717.87	286.48	1718.87	58
59	One		0291		9709	3436.75		3437.75	59
60	One		Zero		One	Infin.		Infin.	60

Table XXI. TRIGONOMETRIC FORMULAS

Let A = angle BAC = arc BF, and let radius $AF = AB = AH = 1$. Then,

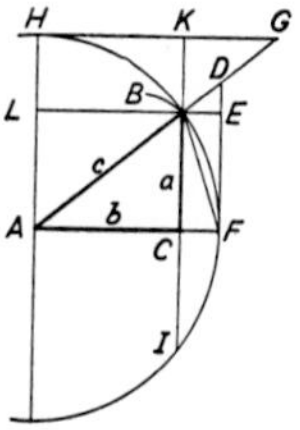

$\sin A = BC$	$\csc A = AG$
$\cos A = AC$	$\sec A = AD$
$\tan A = DF$	$\cot A = HG$
$\text{vers } A = CF = BE$	$\text{covers } A = BK = LH$
$\text{exsec } A = BD$	$\text{coexsec } A = BG$
$\text{chord } A = BF$	$\text{chord } 2A = BI = 2\,BC$

$$\tan \tfrac{1}{2}A = \frac{CF}{BC} = \frac{\text{vers } A}{\sin A}$$

In the right-angled triangle ABC, let $AB = c$, $BC = a$, $CA = b$. Then,

1. $\sin A = \dfrac{a}{c}$
2. $\cos A = \dfrac{b}{c}$
3. $\tan A = \dfrac{a}{b}$
4. $\cot A = \dfrac{b}{a}$
5. $\sec A = \dfrac{c}{b}$
6. $\csc A = \dfrac{c}{a}$
7. $\text{vers } A = 1 - \cos A = \dfrac{c-b}{c} = \text{covers } B$
8. $\text{exsec } A = \sec A - 1 = \dfrac{c-b}{b} = \text{coexsec } B$
9. $\text{covers } A = \dfrac{c-a}{c} = \text{vers } B$

12. $b = c \cos A = a \cot A$
13. $c = \dfrac{a}{\sin A} = \dfrac{b}{\cos A}$
14. $a = c \cos B = b \cot B$
15. $b = c \sin B = a \tan B$
16. $c = \dfrac{a}{\cos B} = \dfrac{b}{\sin B}$
17. $a = \sqrt{c^2 - b^2} = \sqrt{(c-b)(c+b)}$
18. $b = \sqrt{c^2 - a^2} = \sqrt{(c-a)(c+a)}$
19. $c = \sqrt{a^2 + b^2}$
20. $C = 90^\circ = A + B$

Table XX (*continued*)

10. $\text{coexsec } A = \dfrac{c-a}{a} = \text{exsec } B$

21. $\text{Area} = \frac{1}{2}ab$

11. $a = c \sin A = b \tan A$

GENERAL FORMULAS

22. $\sin A = 2 \sin \frac{1}{2}A \cos \frac{1}{2}A = \sqrt{1 - \cos^2 A} = \tan A \cos A$

23. $\cos A = 2 \cos^2 \frac{1}{2}A - 1 = 1 - 2 \sin^2 \frac{1}{2}A = \cos^2 \frac{1}{2}A - \sin^2 \frac{1}{2}A$

24. $\tan A = \dfrac{\sin A}{\cos A} = \dfrac{\sin 2A}{1 + \cos 2A} = \dfrac{\text{vers } 2A}{\sin 2A}$

25. $\cot A = \dfrac{\cos A}{\sin A} = \dfrac{\sin 2A}{1 - \cos 2A} = \dfrac{\sin 2A}{\text{vers } 2A}$

26. $\text{vers } A = 1 - \cos A = \sin A \tan \frac{1}{2}A = 2 \sin^2 \frac{1}{2}A$

27. $\text{exsec } A = \sec A - 1 = \tan A \tan \frac{1}{2}A = \dfrac{\text{vers } A}{\cos A}$

Table XXI. TRIGONOMETRIC FORMULAS

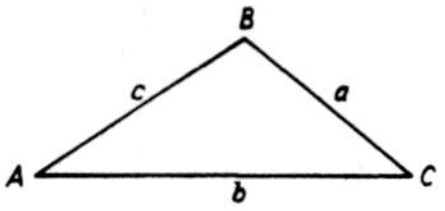

NO.	GIVEN	SOUGHT	FORMULA
28.	A, B, a	C, b, c	$C = 180° - (A + B)$
			$b = \dfrac{a}{\sin A} \times \sin B$
			$c = \dfrac{a}{\sin A} \times \sin (A + B) = \dfrac{a}{\sin A} \times \sin C$
		Area	$\text{Area} = \frac{1}{2}ab \sin C = \dfrac{a^2 \sin B \sin C}{2 \sin A}$
29.	A, a, b	B, C, c	$\sin B = \dfrac{\sin A}{a} \times b$
			$C = 180° - (A + B)$
			$c = \dfrac{a}{\sin A} \times \sin C$
		Area	$\text{Area} = \frac{1}{2}ab \sin C$
30.	C, a, b	c	$c = \sqrt{a^2 + b^2 - 2ab \cos C}$
31.		$\frac{1}{2}(A + B)$	$\frac{1}{2}(A + B) = 90° - \frac{1}{2}C$
32.		$\frac{1}{2}(A - B)$	$\tan \frac{1}{2}(A - B) = \dfrac{a - b}{a + b} \times \tan \frac{1}{2}(A + B)$
33.		A, B	$A = \frac{1}{2}(A + B) + \frac{1}{2}(A - B)$
			$B = \frac{1}{2}(A + B) - \frac{1}{2}(A - B)$
34.		c	$c = (a + b) \times \dfrac{\cos \frac{1}{2}(A + B)}{\cos \frac{1}{2}(A - B)}$
			$= (a - b) \times \dfrac{\sin \frac{1}{2}(A + B)}{\sin \frac{1}{2}(A - B)}$
35.		Area	$\text{Area} = \frac{1}{2}ab \sin C$

36.	a, b, c	A	Let $s = \dfrac{a+b+c}{2}$
37.			$\sin \frac{1}{2}A = \sqrt{\dfrac{(s-b)(s-c)}{bc}}$
			$\cos \frac{1}{2}A = \sqrt{\dfrac{s(s-a)}{bc}}$
			$\tan \frac{1}{2}A = \sqrt{\dfrac{(s-b)(s-c)}{s(s-a)}}$
38.			$\sin A = \dfrac{2\sqrt{s(s-a)(s-b)(s-c)}}{bc}$
			$\cos A = \dfrac{b^2+c^2-a^2}{2bc}$
39.		Area	Area $= \sqrt{s(s-a)(s-b)(s-c)}$

APPENDIXES

Appendix A Theory of the Simple Spiral

A-1. Derivation of Coordinates *X* and *Y*

From Fig. 5-3, $dx = dl \cos \delta$ and $dy = dl \sin \delta$. When the cosine and sine are expressed as infinite series and angle δ is in radians,

$$dx = dl\left[1 - \frac{\delta^2}{2!} + \frac{\delta^4}{4!} - \frac{\delta^6}{6!} + \cdots\right]$$

and

$$dy = dl\left[\delta - \frac{\delta^3}{3!} + \frac{\delta^5}{5!} - \frac{\delta^7}{7!} + \cdots\right]$$

But, from formula 5-7, $\delta = (l/L_s)^2\Delta$. Therefore,

$$dx = dl\left[1 - \frac{\Delta^2}{2!}\left(\frac{l}{L_s}\right)^4 + \frac{\Delta^4}{4!}\left(\frac{l}{L_s}\right)^8 - \frac{\Delta^6}{6!}\left(\frac{l}{L_s}\right)^{12} + \cdots\right]$$

The result obtained by integrating and substituting $(L_s/l)^2\delta$ for Δ is

$$x = l\left[1 - \frac{\delta^2}{5(2!)} + \frac{\delta^4}{9(4!)} - \frac{\delta^6}{13(6!)} + \cdots\right] \tag{A-1}$$

By following a similar procedure,

$$y = l\left[\frac{\delta}{3} - \frac{\delta^3}{7(3!)} + \frac{\delta^5}{11(5!)} - \frac{\delta^7}{15(7!)} + \cdots\right] \tag{A-2}$$

Formulas A-1 and A-2 yield the coordinates of any point, P, on a simple spiral a distance l ft from the T.S., as shown in Fig. 5-3. The coordinates X and Y at the end of a simple spiral (see Fig. 5-2) are found from these equations by substituting Δ for δ and L_s for l. Selected values of X and Y are given in Tables XI and XII, Part III.

A-2. Derivation of Coordinates x_c and o_c

From Fig. 5-3, $x_c = x - r \sin \delta$ and $o_c = y - r(1 - \cos \delta)$. Substituting $1/2\delta$ for r, expressing sine and cosine as infinite series, integrating, and reducing gives:

$$x_c = l\left[\frac{1}{2} - \frac{\delta^2}{10(3!)} + \frac{\delta^4}{18(5!)} - \frac{\delta^6}{26(7!)} + \cdots\right] \tag{A-3}$$

and

$$o_c = l\left[\frac{\delta}{6(2!)} - \frac{\delta^3}{14(4!)} + \frac{\delta^5}{22(6!)} - \frac{\delta^7}{30(8!)} + \cdots\right] \tag{A-4}$$

Formulas A-3 and A-4 yield the coordinates of the offset T.C. for any point, P, on a simple spiral a distance l ft from the T.S., as shown in Fig. 5-3. The coordinates X_0 and o of the offset T.C. from the end of a simple spiral (see Fig. 5-2) are found from these formulas by substituting Δ for δ and L_s for l. The coordinates X_0 and o may also be found from formulas 5-3a and 5-4a. Selected values of X_0 and o are given in Tables XI and XII, Part III.

A-3. Derivation of Correction C_s in $a = \frac{1}{3}\delta - C_s$

Dividing formula A-2 by formula A-1 gives:

$$\tan a = \frac{y}{x} = \frac{\delta}{3} + \frac{\delta^3}{105} + \frac{26\,\delta^5}{155{,}925} - \frac{17\,\delta^7}{3{,}378{,}375} - \cdots$$

But

$$a = \tan a - \tfrac{1}{3}\tan^3 a + \tfrac{1}{5}\tan^5 a - \cdots$$

By substituting the value of tan a, the angle a becomes

$$a = \frac{\delta}{3} - \frac{8\,\delta^3}{2835} - \frac{32\,\delta^5}{467{,}775} - \frac{128\,\delta^7}{83{,}284{,}288} - \cdots \qquad \text{(A-5)}$$

or

$$a = \tfrac{1}{3}\,\delta - \text{a small correction } C_s$$

Radians being converted to seconds, the value of the correction C_s is

$$C_s = 0.00309\delta^3 + 0.00228\delta^5(10)^{-5} + \cdots \qquad \text{(A-6)}$$

where C_s is in seconds and δ is in degrees.

Table XVI-D contains values of C_s for the end points of simple spirals. The corrections to intermediate points, with transit at the T.S., have been computed by considering each intermediate point to be the end point of a shorter spiral. These corrections are listed in Table XVI-A.

A-4. Source of the Corrections Marked * for Conversion to A.R.E.A. Spiral

In the case of a spiraled curve, changing the definition of the degree of the simple curve from D_a to D_c affects only the coordinates of the offset T.C.

In Table XI the * corrections to be subtracted from the coordinates of the offset T.C. based upon D_a, in order to obtain those based upon D_c, are:

$$(X - R_a \sin \Delta) - (X - R_c \sin \Delta) = (R_c - R_a) \sin \Delta \qquad \text{(A-7)}$$

and

$$(Y - R_a \operatorname{vers} \Delta) - (Y - R_c \operatorname{vers} \Delta) = (R_c - R_a) \operatorname{vers} \Delta \qquad \text{(A-8)}$$

In Table XII the * coefficients come from the following relations:

Total correction to $X_0 = (R_c - R_a) \sin \Delta = {*}$ times D

Total correction to $o = (R_c - R_a) \operatorname{vers} \Delta = {*}$ times D

Appendix B Theory of the Combining Spiral*

A *combining spiral* connects and is tangent to two circular arcs having different radii of curvature. At each point of tangency the radius of curvature of the spiral equals that of the circular arc to which it connects.

In sections 5-17 and 5-18 it was shown that the theory of the osculating circle uses approximations that may lead to significant errors in the case of sharp spirals. "The derivations [herein] replace the theory of the osculating circle with mathematical exactness which permits the precise layout of a highway spiral between the branches of a compound curve regardless of sharpness.

"The derivations themselves are cumbersome and extensive. They are given in outline form but in sufficient detail to permit checking." (The author

* Based on "The Highway Spiral for Combining Curves of Different Radii" by Paul Hartman, *Transactions, ASCE*, Vol. 122, 1957, Paper No. 2867, pp. 389–399; and adapted to this book by permission of the author. Quotations on this page taken from Paper No. 2867.

of this book has verified the derivations and has extended the analysis to include some formulas not given in Paper No. 2867.)

B-1. Derivation of Coordinates *X* and *Y*

This section yields the coordinates of the flatter end of a combining spiral with respect to an origin at the sharper end (see Fig. B-1).

The degree of curvature of a spiral varies uniformly along its length. In Fig. B-1 the degree of curvature at any point P, a distance l ft from point A' is

$$D_p = D_S - (D_S - D_L)\frac{l}{l_s} \tag{B-1}$$

where l_s is the length of the combining spiral $A'C'$.

Referring to Fig. B-2, which shows the first portion of Fig. B-1 enlarged,

$$dl = r\,d\delta = \frac{100\,d\delta}{D_p} \tag{B-2}$$

Substituting for D_p its value from B-1 and integrating,

$$\delta = \frac{D_S l}{100} - (D_S - D_L)\frac{l^2}{200 l_s} \tag{B-3}$$

But, from formula 5-34, $(D_S - D_L)(l_s/200)$ equals the nominal spiral angle Δ_N. Therefore,

$$\delta = \frac{D_S l}{100} - \Delta_N\left(\frac{l}{l_s}\right)^2 \tag{B-4}$$

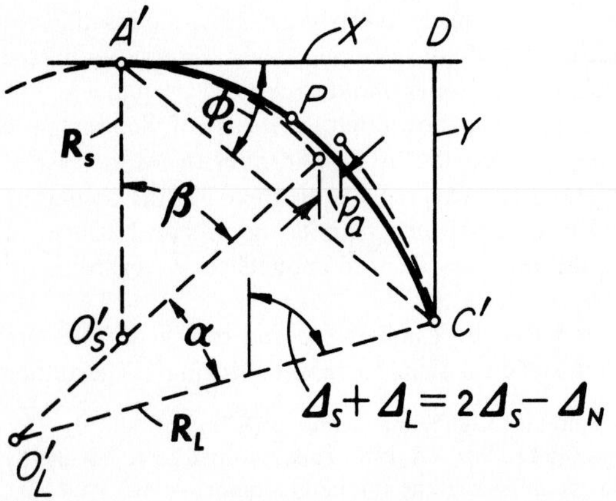

Figure B-1

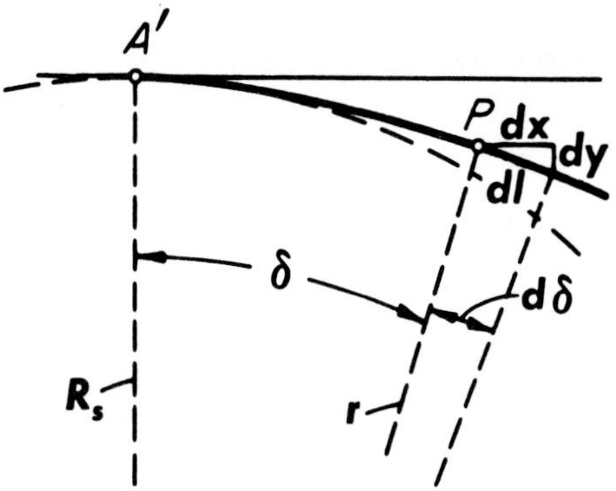

Figure B-2

Also,

$$dx = dl \cos \delta \tag{B-5}$$

and

$$dy = dl \sin \delta \tag{B-6}$$

The X and Y coordinates are obtained by carrying out the following mathematical steps:

1. Substitute the value of δ from formula B-4 in formulas B-5 and B-6.
2. Replace cosine and sine by their series expansions.
3. Integrate to obtain expressions for x and y.
4. Substitute l_s for l. Also factor out $l_s D_S/200$ and replace it by Δ_S.

The resulting formulas give the *coordinates of the flatter end* of a combining spiral with respect to an *origin at the sharper end:*

$$X = l_s\Big(1 - \frac{2}{3}\Delta_S{}^2 + \frac{1}{2}\Delta_S\Delta_N - \frac{1}{10}\Delta_N{}^2 + \frac{2}{15}\Delta_S{}^4 - \frac{2}{9}\Delta_S{}^3\Delta_N + \frac{1}{7}\Delta_S{}^2\Delta_N{}^2$$
$$- \frac{1}{24}\Delta_S\Delta_N{}^3 + \frac{1}{216}\Delta_N{}^4 - \frac{4}{315}\Delta_S{}^6 + \frac{1}{30}\Delta_S{}^5\Delta_N - \frac{1}{27}\Delta_S{}^4\Delta_N{}^2$$
$$+ \frac{1}{45}\Delta_S{}^3\Delta_N{}^3 - \frac{1}{132}\Delta_S{}^2\Delta_N{}^4 + \frac{1}{720}\Delta_S\Delta_N{}^5 - \frac{1}{9360}\Delta_N{}^6\Big) \tag{B-7}$$

$$Y = l_s\Big(\Delta_S - \frac{1}{3}\Delta_N - \frac{1}{3}\Delta_S{}^3 + \frac{2}{5}\Delta_S{}^2\Delta_N - \frac{1}{6}\Delta_S\Delta_N{}^2 + \frac{1}{42}\Delta_N{}^3 + \frac{2}{45}\Delta_S{}^5$$
$$- \frac{2}{21}\Delta_S{}^4\Delta_N + \frac{1}{12}\Delta_S{}^3\Delta_N{}^2 - \frac{1}{27}\Delta_S{}^2\Delta_N{}^3 + \frac{1}{120}\Delta_S\Delta_N{}^4 - \frac{1}{1320}\Delta_N{}^5\Big) \tag{B-8}$$

in which the angles are in radians.

By means of a similar analysis, the *coordinates of the sharper end* of a combining spiral with respect to an *origin at the flatter end* become

$$X = l_s\left(1 - \frac{2}{3}\Delta_L^2 - \frac{1}{2}\Delta_L\Delta_N - \frac{1}{10}\Delta_N^2 + \frac{2}{15}\Delta_L^4 + \frac{2}{9}\Delta_L^3\Delta_N + \frac{1}{7}\Delta_L^2\Delta_N^2\right.$$

$$+ \frac{1}{24}\Delta_L\Delta_N^3 + \frac{1}{216}\Delta_N^4 - \frac{4}{315}\Delta_L^6 - \frac{1}{30}\Delta_L^5\Delta_N - \frac{1}{27}\Delta_L^4\Delta_N^2$$

$$\left.- \frac{1}{45}\Delta_L^3\Delta_N^3 - \frac{1}{132}\Delta_L^2\Delta_N^4 - \frac{1}{720}\Delta_L\Delta_N^5 - \frac{1}{9360}\Delta_N^6\right) \quad \text{(B-9)}$$

$$Y = l_s\left(\Delta_L + \frac{1}{3}\Delta_N - \frac{1}{3}\Delta_L^3 - \frac{2}{5}\Delta_L^2\Delta_N - \frac{1}{6}\Delta_L\Delta_N^2 - \frac{1}{42}\Delta_N^3 + \frac{2}{45}\Delta_L^5\right.$$

$$\left.+ \frac{2}{21}\Delta_L^4\Delta_N + \frac{1}{12}\Delta_L^3\Delta_N^2 + \frac{1}{27}\Delta_L^2\Delta_N^3 + \frac{1}{120}\Delta_L\Delta_N^4 + \frac{1}{1320}\Delta_N^5\right) \quad \text{(B-10)}$$

in which the angles are in radians. These two sets of coordinates are interrelated in the same way as those for the simple spiral (see Fig. 5-2 along with formulas 5-3 and 5-4).

B-2. Derivation of Correction C_ϕ in $C_\phi = C_a + C_s$

The correction C_ϕ, needed to obtain the corrected deflection angle ϕ_c (Fig. B-1), is found by first dividing formula B-8 by formula B-7. This gives

$$\tan\phi_c = \Delta_S - \frac{1}{3}\Delta_N + \frac{1}{3}\Delta_S^3 - \frac{29}{90}\Delta_S^2\Delta_N$$

$$+ \frac{1}{10}\Delta_S\Delta_N^2 - \frac{1}{105}\Delta_N^3 + \frac{2}{15}\Delta_S^5 \ldots \quad \text{(B-11)}$$

Substituting formula B-11 in

$$\phi_c = \tan\phi_c - \tfrac{1}{3}\tan^3\phi_c + \tfrac{1}{5}\tan^5\phi_c \cdots$$

and reducing yields

$$\phi_c = \Delta_S - \frac{1}{3}\Delta_N + \frac{1}{90}\Delta_S^2\Delta_N - \frac{1}{90}\Delta_S\Delta_N^2 + \frac{8}{2835}\Delta_N^3$$

$$+ \frac{1}{945}\Delta_S^4\Delta_N - \frac{2}{945}\Delta_S^3\Delta_N^2 + \frac{181}{113400}\Delta_S^2\Delta_N^3$$

$$- \frac{61\frac{1}{2}}{113400}\Delta_S\Delta_N^4 \ldots \quad \text{(B-12)}$$

But $\Delta_S - \frac{1}{3}\Delta_N$ is the deflection angle to the flatter end of the spiral by the osculating-circle theory. Consequently the remaining terms constitute the correction C_ϕ. The final expression is found by converting C_ϕ to seconds and Δ_S and Δ_N to degrees, giving

$$\begin{aligned} C_\phi &= 0.012185\Delta_S\Delta_N(\Delta_S - \Delta_N) + 0.3535\Delta_S{}^3\Delta_N(\Delta_S - 2\Delta_N)10^{-6} \\ &\quad + 0.2946\Delta_S\Delta_N{}^3(181\Delta_S - 61\Delta_N)10^{-8} + 0.00309\Delta_N{}^3 \\ &\quad + 0.00228\Delta_N{}^5\, 10^{-5} \end{aligned} \tag{B-13}$$

This is the same formula denoted as 5-37 in Chapter 5, numerical values for which are listed in Tables XVI-C and XVI-D. Section 5-17 and Appendix C contain detailed explanations of the use of these tables in conjunction with Tables XVI, XVI-A, and XVI-B.

B-3. Derivation of β to Yield C_b

The expression for β is found from Fig. B-1 by means of the traverse method. In the closed traverse $O_S'A'DC'O_L'O_S'$, with 0° azimuth assumed in direction $O_S'A'$,

$$\begin{aligned} \Sigma \text{ latitudes} &= 0 \\ &= R_S - Y - R_L \cos(2\Delta_S - \Delta_N) + (R_L - R_S - p_a)\cos\beta \end{aligned}$$

$$\begin{aligned} \Sigma \text{ departures} &= 0 \\ &= X - R_L \sin(2\Delta_S - \Delta_N) + (R_L - R_S - p_a)\sin\beta \end{aligned}$$

Solving these equations for β,

$$\cos\beta = \frac{Y + R_L\cos(2\Delta_S - \Delta_N) - R_S}{R_L - R_S - p_a} \tag{B-14}$$

$$\sin\beta = \frac{R_L\sin(2\Delta_S - \Delta_N) - X}{R_L - R_S - p_a} \tag{B-15}$$

Dividing formula B-15 by formula B-14,

$$\tan\beta = \frac{R_L\sin(2\Delta_S - \Delta_N) - X}{Y + R_L\cos(2\Delta_S - \Delta_N) - R_S} \tag{B-16}$$

All terms in this fraction may be expressed in terms of Δ_S and Δ_N by making these substitutions: for sine and cosine, their series expansions of $(2\Delta_S - \Delta_N)$; for X and Y, the expressions in formulas B-7 and B-8; for R_L, its equivalent, $l_s/[2(\Delta_S - \Delta_N)]$; and for R_S, its equivalent, $l_s/2\Delta_S$. After collecting terms and dividing, the result is

$$\begin{aligned} \tan\beta &= \Delta_S + \frac{1}{3}\Delta_S{}^3 + \frac{2}{15}\Delta_S{}^5 + \frac{1}{30}\Delta_S{}^2\Delta_N - \frac{1}{30}\Delta_S\Delta_N{}^2 + \frac{23}{630}\Delta_S{}^4\Delta_N \\ &\quad - \frac{5}{126}\Delta_S{}^3\Delta_N{}^2 + \frac{17}{7560}\Delta_S{}^2\Delta_N{}^3 + \frac{1}{1080}\Delta_S\Delta_N{}^4 \cdots \end{aligned} \tag{B-17}$$

Substituting formula B-17 in $\beta = \tan\beta - \frac{1}{3}\tan^3\beta + \frac{1}{5}\tan^5\beta \cdots$ and reducing yields

$$\beta = \Delta_S + \frac{1}{30}\Delta_S{}^2\Delta_N - \frac{1}{30}\Delta_S\Delta_N{}^2 + \frac{1}{315}\Delta_S{}^4\Delta_N - \frac{2}{315}\Delta_S{}^3\Delta_N{}^2$$

$$+ \frac{17}{7560}\Delta_S{}^2\Delta_N{}^3 + \frac{1}{1080}\Delta_S\Delta_N{}^4 \cdots \qquad \text{(B-18)}$$

In formula B-18, the first term Δ_S is the central angle of the sharper arc, as assumed in the osculating-circle theory (Fig. 5-8). The remaining terms comprise the correction C_b, or the amount by which β exceeds Δ_S.

An expression for C_b in a form suitable for computation is

$$C_b = 0.03655\Delta_S\Delta_N(\Delta_S - \Delta_N)\{1 + 121\ [24\Delta_S(\Delta_S - \Delta_N)$$
$$- 7\Delta_N{}^2]\,10^{-8}\}\cdots \qquad \text{(B-19)}$$

in which Δ_S and Δ_N are in degrees and C_b is in seconds. This is the same formula denoted as 5-35 in Chapter 5, numerical values for which are listed in Table XVI-C. Examples are found in section 5-17.

B-4. Derivation of p_a to Yield C_p

Solving formula B-15 for p_a,

$$p_a = \frac{X - R_L \sin(2\Delta_S - \Delta_N) + (R_L - R_S)\sin\beta}{\sin\beta} \qquad \text{(B-20)}$$

All terms in this fraction may be expressed in terms of Δ_S and Δ_N. Sin β is replaced by the sine series expansion of β, where β comes from formula B-18. The equivalents of the remaining terms are the same as those used in developing formula B-17. After collecting terms and dividing, the result is

$$p_a = l_s\left(\frac{1}{12}\Delta_N - \frac{1}{240}\Delta_S{}^2\Delta_N + \frac{1}{240}\Delta_S\Delta_N{}^2 - \frac{1}{336}\Delta_N{}^3 + \frac{1}{10{,}080}\Delta_S{}^4\Delta_N\right.$$

$$\left. - \frac{1}{5040}\Delta_S{}^3\Delta_N{}^2 + \frac{1}{18{,}900}\Delta_S{}^2\Delta_N{}^3 + \frac{1}{21{,}600}\Delta_S\Delta_N{}^4 + \frac{1}{15{,}840}\Delta_N{}^5\cdots\right)$$

(B-21)

Now compare this combining spiral with an *equivalent* simple spiral. To be "equivalent," the terms l_s, Δ_N, and Δ_S in the combining spiral become L_s, Δ, and zero in the simple spiral, and p_a becomes the throw o. That is, formula B-21 reduces to

$$\text{Throw } o = L_s\left(\frac{1}{12}\Delta - \frac{1}{336}\Delta^3 + \frac{1}{15{,}840}\Delta^5\cdots\right) \qquad \text{(B-22)}$$

This expression becomes the same as formula A-4 when o_c, l, and δ in that formula are replaced by o, L_s, and Δ. It follows, therefore, that p_a is less than the throw o of an equivalent simple spiral by an amount C_p, expressed as

$$C_p = l_s\left(\frac{1}{240}\Delta_S\Delta_N{}^2 - \frac{1}{240}\Delta_S{}^2\Delta_N + \frac{1}{10{,}080}\Delta_S{}^4\Delta_N \cdots\right) \quad \text{(B-23)}$$

An expression for C_p in a form suitable for computation is

$$C_p = 0.2215\, l_s\Delta_S\Delta_N(\Delta_S - \Delta_N)10^{-7} \cdots \quad \text{(B-24)}$$

in which C_p and l_s are in feet and Δ_S and Δ_N are in degrees. (The fifth powers of products of $\Delta_S\Delta_N$ are omitted because they are negligible in highway practice, even for the very sharp spirals sometimes used on interchange ramps.) Numerical values of C_p are listed in Table XVI-C; examples are found in section 5-17.

Appendix C
Exact Deflection Angles for Simple and Combining Spirals

The method of determining the exact deflection angles along simple and combining spirals depends on the location of the transit. Three locations are possible: (1) at the T.S.; (2) at the S.C.; and (3) at any intermediate point on the spiral.

In Fig. C-1, AC is a simple spiral of length L_s which may be staked (completely or in part) from A, C, or one or more intermediate points on the spiral. If a portion of the simple spiral (as P_1P_2) is used as a transition between two arcs of a compound curve, that portion is a combining spiral of length l_s. A combining spiral cannot begin at A; if so, it would become a simple spiral. However, in computations the combining spiral P_1P_2 is treated as the latter portion of a simple spiral of length AP_2; that is, point P_2 becomes the S.C. for purposes of computation.

The following notation is used herein. Formula numbers in parentheses are those in the text which involve the term defined:

L_s = length of simple spiral, T.S. to S.C.; (5-2)

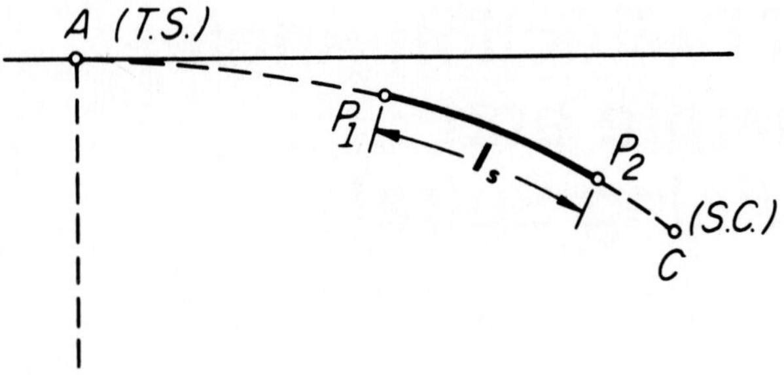

Figure C-1

D = degree of curve at the end of a simple spiral; (5-2)

Δ = central angle of spiral L_s; (5-2)

l = length of portion of spiral sighted over;

R = ratio of l to L_s;

l_s = length of a combining spiral;

D_L = degree of curve at flatter end of spiral l;

D_S = degree of curve at sharper end of spiral l;

D_t = degree of curve at the transit setup;

Δ_S = central angle for curve of degree D_S and length l;

Δ_N = nominal central angle of a combining spiral; (5-34)

C_s = angular correction listed in Tables XVI-A and XVI-D; (A-6)

C_ϕ = angular correction listed in Tables XVI-B and XVI-C; (5-37)

C_a = angular correction listed in Table XVI-C; (5-38)

ϕ_f = deflection angle, by osculating-circle theory, to a point having a greater degree of curve; (5-23)

ϕ_b = deflection angle, by osculating-circle theory, to a point having a lesser degree of curve; (5-22)

ϕ_c = the exact (or corrected) deflection angle for the specified condition.

The following relations are also needed in this analysis:

$$R = \frac{l}{L_s} \tag{C-1}$$

$$\Delta_S = \frac{lD_S}{200} \tag{C-2}$$

$$\Delta_N = R^2\Delta \tag{C-3}$$

Case I. Transit at T.S. of a Simple Spiral

Assume that D, L_s, and Δ are known and that the exact deflection angles, ϕ_c, to specified points on the spiral are required. (In this case, $D_L = D_t =$ zero.)

Prepare the tabular form of computation shown at the end of this appendix and follow these steps:

1. List values of l and compute values of R from formula C-1.
2. Compute values of D_S from $D_S = RD$.
3. Compute values of Δ_S from formula C-2.
4. Compute values of Δ_N from formula C-3.
5. Take values of C_s by interpolation from Table XVI-A, using Δ and values of R as arguments. As an alternate method (essential when $\Delta > 60°$) follow step 6.
6. Take values of C_s by interpolation from Table XVI-D, using values of Δ_N as arguments.
7. Compute values of $\Delta/3(l/L_s)^2$.
8. Obtain values of ϕ_c from $\phi_c = \Delta/3(l/L_s)^2 - C_s$.

Case II. Transit at S.C. of a Simple Spiral

Assume that D, L_s, and Δ are known and that the exact deflection angles, ϕ_c, to specified points on the spiral are required.

Prepare the tabular form of computation shown at the end of this appendix and follow these steps:

1. List values of l and compute values of R from formula C-1.
2. List the constant value of D_S (in this case $D_S = D_t = D$).
3. Compute values of Δ_S from formula C-2.
4. Compute values of Δ_N from formula C-3.
5. Take values of C_ϕ by interpolation from Table XVI-B, using Δ and values of R as arguments. As an alternate method (essential when $\Delta > 60°$) follow steps 6, 7, and 8.
6. Take values of C_a by interpolation from Table XVI-C, using values of Δ_S and Δ_N as arguments.
7. Take values of C_s by interpolation from Table XVI-D, using values of Δ_N as arguments.
8. Compute values of C_ϕ from $C_\phi = C_a + C_s$.
9. Compute values of ϕ_b.
10. Obtain values of ϕ_c from $\phi_c = \phi_b + C_\phi$.

Case III. Transit at Intermediate Point on Spiral

Assume that D, L_s, Δ, and D_t are known and that the exact deflection angles, ϕ_c, to specified points on the spiral are required.

Prepare the tabular form of computation shown at the end of this appendix and follow these steps:

1. List values of l and compute values of R from formula C-1.
2. Compute values of D_S from $D_S = D_L + RD$.
3. Compute values of Δ_S from formula C-2.
4. Compute values of Δ_N from formula C-3.
5. Take values of C_a by interpolation from Table XVI-C, using values of Δ_S and Δ_N as arguments.
6. Take values of C_s by interpolation from Table XVI-D, using values of Δ_N as arguments.
7. Compute values of C_ϕ from $C_\phi = C_a + C_s$.
8. Compute values of ϕ_f.
9. Obtain values of ϕ_c from $\phi_c = \phi_f - C_\phi$.

Example of Case 1. Simple Spiral. Transit at T.S.

Transit at T.S.
(Orient by sight along tangent with 0°00′ on vernier.)

SIGHT AT POINT	l (ft)	R	D_S (deg)	Δ_S (deg)	Δ_N (deg)	C_s (s)	$\frac{\Delta}{3}\left(\frac{l}{L_s}\right)^2$	ϕ_c
0 (T.S.)	0		0				0°00′	0°00′00″
1	60	0.067	$\frac{2}{3}$	0.2	0.2	0	0°04′	0°04′00″
2	120	0.133	$1\frac{1}{3}$	0.8	0.8	0	0°16′	0°16′00″
3	180	0.200	2	1.8	1.8	0	0°36′	0°36′00″
4	240	0.267	$2\frac{2}{3}$	3.2	3.2	0	1°04′	1°04′00″
5	300	0.333	$3\frac{1}{3}$	5.0	5.0	0	1°40′	1°40′00″
6	360	0.400	4	7.2	7.2	1	2°24′	2°23′59″
7	420	0.467	$4\frac{2}{3}$	9.8	9.8	3	3°16′	3°15′57″
8	480	0.533	$5\frac{1}{3}$	12.8	12.8	7	4°16′	4°15′53″
9	540	0.600	6	16.2	16.2	13	5°24′	5°23′47″
10	600	0.667	$6\frac{2}{3}$	20.0	20.0	25	6°40′	6°39′35″
11	660	0.733	$7\frac{1}{3}$	24.2	24.2	44	8°04′	8°03′16″
12	720	0.800	8	28.8	28.8	74	9°36′	9°34′46″
13	780	0.867	$8\frac{2}{3}$	33.8	33.8	121	11°16′	11°13′59″
14	840	0.933	$9\frac{1}{3}$	39.2	39.2	189	13°04′	13°00′51″
15 (S.C.)	900	1	10	45.0	45.0	286	15°00′	14°55′14″

Transit at S.C. (For staking curve beyond S.C., orient by backsight to T.S. with 30°04′46″ on vernier, plunge and turn to 0°00′.)

$D = 10°$; $L_s = 900$ ft; $\Delta = 45°$. Spiral to be staked using fifteen 60-ft chords. (See section 5-7 for another version of this spiral.)

Example of Case II. Simple Spiral. Transit at S.C.

Transit at S.C. (Orient with 0°00′ along local tangent at S.C. by sight to T.S. with 30°04′46″ on vernier.)

SIGHT AT POINT	l (ft)	R	D_S (deg)	Δ_S (deg)	Δ_N (deg)	C_a (s)	C_s (s)	C_ϕ (s)	ϕ_b	ϕ_c
15 (S.C.)	0		10						0°00′	0°00′00″
14	60	0.067	10	3	0.2	0	0	0	2°56′	2°56′00″
13	120	0.133	10	6	0.8	0	0	0	5°44′	5°44′00″
12	180	0.200	10	9	1.8	2	0	2	8°24′	8°24′02″
11	240	0.267	10	12	3.2	4	0	4	10°56′	10°56′04″
10	300	0.333	10	15	5.0	9+	0+	10	13°20′	13°20′10″
9	360	0.400	10	18	7.2	17	1	18	15°36′	15°36′18″
8	420	0.467	10	21	9.8	29	3	32	17°44′	17°44′32″
7	480	0.533	10	24	12.8	43	7	50	19°44′	19°44′50″
6	540	0.600	10	27	16.2	59	13	72	21°36′	21°37′12″
5	600	0.667	10	30	20.0	74	25	99	23°20′	23°21′39″
4	660	0.733	10	33	24.2	88	44	132	24°56′	24°58′12″
3	720	0.800	10	36	28.8	93	74	167	26°24′	26°26′47″
2	780	0.867	10	39	33.8	85	121	206	27°44′	27°47′26″
1	840	0.933	10	42	39.2	57	189	246	28°56′	29°00′06″
0 (T.S.)	900	1	10	45	45.0	0	286	286	30°00′	30°04′46″

Same spiral as a previous example of Case I.

Example of Case III. Transit at Intermediate Point on a Spiral

Required to set full and half-stations on a combining spiral between sta. 81 on an 8° curve and sta. 85 on a 24° curve. The spiral is to be staked completely from a setup at sta. 81.

From these data, l_s is the latter portion of a simple spiral for which $D = 24°$, $L_s = 600$ ft, and $\Delta = 72°$. With reference to Fig. C-1, this combining spiral runs between points P_1 and C; that is, points P_2 and C coincide.

If two intermediate setups on spiral AC were required (as at P_1 and P_2), the computations would require two applications of Case III. The first portion of the combining spiral would be treated as the latter portion of a simple spiral of length AP_2; and the second portion (P_2 to C), as the latter portion of a simple spiral of length AC. In practice such a case would rarely require the deflection angles to be computed by the exact theory, since the effect of multiple setups on a spiral is to shorten the sights and make the angular corrections negligible. The simpler osculating-circle theory would suffice, as explained in section 5-7.

Example of Case III. Transit at Intermediate Point on a Spiral.

Transit at sta. 81. (Orient with 0°00′ along local tangent.)

SIGHT AT POINT	l (ft)	R	D_S (deg)	Δ_S (deg)	Δ_N (deg)	C_a (s)	C_s (s)	C_ϕ (s)	ϕ_f	ϕ_c
81	0		8						0°00′	0°00′00″
+50	50	0.083	10	2.5	0.5	0	0	0	2°10′	2°10′00″
82	100	0.167	12	6.0	2.0	1	0	1	4°40′	4°39′59″
+50	150	0.250	14	10.5	4.5	4	0	4	7°30′	7°29′56″
83	200	0.333	16	16.0	8.0	12	2	14	10°40′	10°39′46″
+50	250	0.417	18	22.5	12.5	35	6	41	14°10′	14°09′19″
84	300	0.500	20	30.0	18.0	80	18	98	18°00′	17°58′22″
+50	350	0.583	22	38.5	24.5	165	46	211	22°10′	22°06′29″
85	400	0.667	24	48.0	32.0	311	102	413	26°40′	26°33′07″

Transit at sta. 85. (For staking curve beyond sta. 85, orient by backsight to sta. 81 with 37°26′53″ on vernier, plunge and turn to 0°00′.)[a]

(Spiral staked from setup at sta. 81. Data on previous page.)

[a] Found from $\phi_c = \phi_b + C_\phi$ (Case II), where $\phi_b = 37°20'$ and $C_\phi = 413''$.

Index